Computational Methods in Applied Sciences

Volume 53

Series Editor

Eugenio Oñate, Universitat Politècnica de Catalunya, Barcelona, Spain

This series publishes monographs and carefully edited books inspired by the thematic conferences of ECCOMAS, the European Committee on Computational Methods in Applied Sciences. As a consequence, these volumes cover the fields of Mathematical and Computational Methods and Modelling and their applications to major areas such as Fluid Dynamics, Structural Mechanics, Semiconductor Modelling, Electromagnetics and CAD/CAM. Multidisciplinary applications of these fields to critical societal and technological problems encountered in sectors like Aerospace, Car and Ship Industry, Electronics, Energy, Finance, Chemistry, Medicine, Biosciences, Environmental sciences are of particular interest. The intent is to exchange information and to promote the transfer between the research community and industry consistent with the development and applications of computational methods in science and technology.

Book proposals are welcome at
Eugenio Oñate
International Center for Numerical Methods in Engineering (CIMNE)
Technical University of Catalunya (UPC)
Edificio C-1, Campus Norte UPC Gran Capitán
s/n08034 Barcelona, Spain
onate@cimne.upc.eduwww.cimne.com
or contact the publisher, Dr. Mayra Castro, mayra.castro@springer.com

Indexed in SCOPUS, Google Scholar and SpringerLink.

More information about this series at http://www.springer.com/series/6899

Andrés Kecskeméthy · Francisco Geu Flores
Editors

Multibody Dynamics 2019

Proceedings of the 9th ECCOMAS Thematic Conference on Multibody Dynamics

Springer

Editors
Andrés Kecskeméthy
Lehrstuhl für Mechanik und Robotik
Universität Duisburg-Essen
Duisburg, Nordrhein-Westfalen, Germany

Francisco Geu Flores
Lehrstuhl für Mechanik und Robotik
Universität Duisburg-Essen
Duisburg, Nordrhein-Westfalen, Germany

ISSN 1871-3033
Computational Methods in Applied Sciences
ISBN 978-3-030-23134-7 ISBN 978-3-030-23132-3 (eBook)
https://doi.org/10.1007/978-3-030-23132-3

This Springer imprint is published by the registered company Springer Nature Switzerland AG
The registered company address is: Gewerbestrasse 11, 6330 Cham, Switzerland

Preface

The present proceedings book collects a selection of full-papers as a subset of the submitted extended abstracts presented at the 9th ECCOMAS Thematic Conference on Multibody Dynamics, held on 15–18 July 2019, at the University of Duisburg-Essen, Duisburg, Germany.

Multibody dynamics is an exciting area of mechanics which merges various disciplines such as structural dynamics, multiphysics problems, computational mathematics, control theory and computer science in order to deliver methods and tools for the virtual prototyping of complex mechanical systems. In this setting, multibody dynamics play a central role in the modelling, analysis, simulation, and optimization of mechanical systems in a variety of fields and for a wide range of applications. As new methods and procedures are being proposed at a fast pace in academia, research laboratories and industry, it becomes important to provide researchers in multibody dynamics with appropriate venues for exchanging ideas and results. To answer these needs, the ECCOMAS Thematic Conference on Multibody Dynamics was initiated in 2003 in Lisbon. Since this event, the conference was held every two years in changing locations and with large attraction in the international scientific community: Madrid (2005), Milan (2007), Warsaw (2009), Brussels (2011), Zagreb (2013), Barcelona (2015), and Prague (2017). The objective of this conference series was to bring together researchers from different fields where multibody dynamics play a key role. This includes not only theoretical fields where multibody dynamics were traditionally established, but also in particular applications in which multibody dynamics might contribute new perspectives for new-generation practical and industrial challenges.

The present conference has attracted again a strong interest in the international community: a total of 165 contributions from 27 countries were selected for presentation after severe peer review by two independent reviewers. From the submitted full papers, 64 articles were anew selected after severe peer review by two independent reviewers. This book contains the thus identified full papers, which reflect the importance multibody dynamics research has in current and future multibody dynamics technologies from technical systems over robotic control up to human motion interactions.

The book is of interest to researchers, doctoral students, teachers and engineers specializing in multibody dynamics. It is divided into eight sections corresponding to the wide range of methodologies and applications encompassed in this thriving field: biomechanics; contact and constraints; mechatronics, robotics and control; flexible multibody dynamics; formulations and numerical methods; optimization and sensitivity analysis; efficient simulation and real-time applications; and applications in vehicle dynamics and aerospace devices.

We thank the authors for submitting their valuable contributions for this conference as well as the reviewers for performing the reviews in due time. We also thank the publisher Springer for the timely implementation of this book and the valuable advice during the production process. We are very indebted to the University of Duisburg-Essen as well as a long list of sponsors for the operational and financial support of this conference. Last but not least, we thank the European Community on Computational Methods in Applied Sciences, ECCOMAS, for the ideal support by offering its patronage for this conference.

Andrés Kecskeméthy
Francisco Geu Flores

Contents

Biomechanics

Investigation of Tympanic Membrane Influences on Middle-Ear Impedance Measurements and Simulations

Benjamin Sackmann[1,4(✉)], Birthe Warnholtz[2], Jae Hoon Sim[2], Dmitrii Burovikhin[1], Ernst Dalhoff[3], Peter Eberhard[4], and Michael Lauxmann[1]

[1] Reutlingen University, Alteburgstrasse 150, 72762 Reutlingen, Germany
benjamin.sackmann@reutlingen-university.de
[2] University Hospital Zurich, Frauenklinikstrasse 24, 8091 Zurich, Switzerland
[3] University of Tübingen, Elfriede-Aulhorn-Str. 5, 72076 Tübingen, Germany
[4] Institute of Engineering and Computational Mechanics, University of Stuttgart, Pfaffenwaldring 9, 70569 Stuttgart, Germany

Abstract. This study simulates acoustic impedance measurements in the human ear canal and investigates error influences due to improperly accounted evanescence in the probe's near field, cross-section area changes, curvature of the ear canal, and pressure inhomogeneities across the tympanic membrane, which arise mainly at frequencies above 10 kHz. Evanescence results from strongly damped modes of higher order, which can only be found in the near field of the sound source and are excited due to sharp cross-sectional changes as they occur at the transition from the probe loudspeaker to the ear canal. This means that different impedances are measured depending on the probe design. The influence of evanescence cannot be eliminated completely from measurements, however, it can be reduced by a probe design with larger distance between speaker and microphone. A completely different approach to account for the influence of evanescence is to evaluate impedance measurements with the help of a finite element model, which takes the precise arrangement of microphone and speaker in the measurement into account. The latter is shown in this study exemplary on impedance measurements at a tube terminated with a steel plate. Furthermore, the influences of shape changes of the tympanic membrane and ear canal curvature on impedance are investigated.

1 Introduction

Today's audiometric methods for the detection of hearing loss are mostly based on a comparison of measurements with standard curves representing the statistical range of normal hearing. Therefore, currently available clinical procedures for acoustic impedance measurements are often limited to a qualitative diagnosis of middle-ear pathologies with limited specificity. To resolve these limits, we previously established a model-based approach to evaluate the impedance measurements on individuals with a finite element (FE) model of the ear [5].

A. Kecskeméthy and F. Geu Flores (Eds.): ECCOMAS 2019, COMPUTMETHODS 53, pp. 3–10, 2020.
https://doi.org/10.1007/978-3-030-23132-3_1

Currently there is no standardized commercial device available for measuring acoustic impedances in the ear canal (EC). Instead, only energy absorbance is measured taking into account only amplitude, but not phase information. The main reason for this is the dependency of the measured impedances on the geometry of the EC and on the probe design, which directly influences the evanescent waves. Evanescence results from strongly damped modes of higher order, which can only be found in the near field of the sound source and are excited due to sharp cross-sectional changes as they occur at the transition from the probe speaker to the EC.

Furthermore, the middle-ear impedance which describes the averaged velocity of the tympanic membrane (TM) caused by the pressure at the TM, is derived from the EC impedance based on the assumption that the residual EC can be modelled as an 1D-transmission line neglecting spatial pressure distribution and complex vibrations of the TM, which is questioned within this study.

For the investigations, a customized impedance probe is used as shown in Fig. 1 and it is shown how it can be modelled using an FE approach to account for the important influences such as evanescence. The investigations are done first on tubes with rigid termination and a flexible steel plate on which the impedance is measured. It is shown, how the parameters needed to simulate the measurements of the tube with steel plate termination can be achieved by first identifying the damping parameters of air in rigid tubes and then fitting the other parameters in a second step. Furthermore, the need to include evanescence in the simulations is studied. Furthermore, with an FE model of the ear, validated by comparing to measurements from literature, specific changes in the TM and EC shape are introduced and it is investigated how they affect the impedance.

2 Methods

2.1 How Is Impedance Measured?

The acoustic impedance Z_{EC} is defined as the ratio of sound pressure p and sound flow $u = vA_{lsp}$ with the velocity v and cross-section A_{lsp} of the driving loudspeaker and is measured at the EC entrance with a probe consisting of a microphone (EK-3024, Knowles) and loudspeaker (ER-3A, Etymotics), as shown in Fig. 1.

The probe is characterized by the two parameters probe impedance and probe pressure with a calibration procedure based on the Thévenin equivalent. These parameters are determined from pressure measurements on four waveguides of known impedance by solving an overdetermined system of equations using a least-squares method [1]. To improve the quality of the calibration, an error metric [6] is minimized by iteratively adjusting the calibration lengths.

In impedance measurements in the ear canal, the sound source has a smaller diameter than the measured tube or EC, which produces a sound field composed of propagating plane waves and rapidly decreasing evanescent waves as seen in Fig. 2.

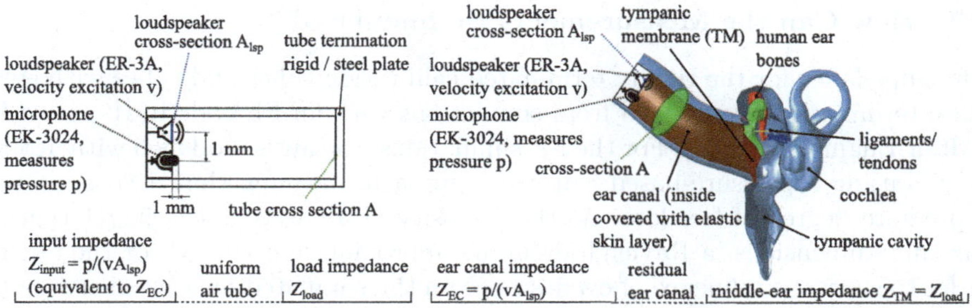

Fig. 1. Impedance probe with parameters and application in tubes (left) and ear model (right) in simulations and measurements, respectively.

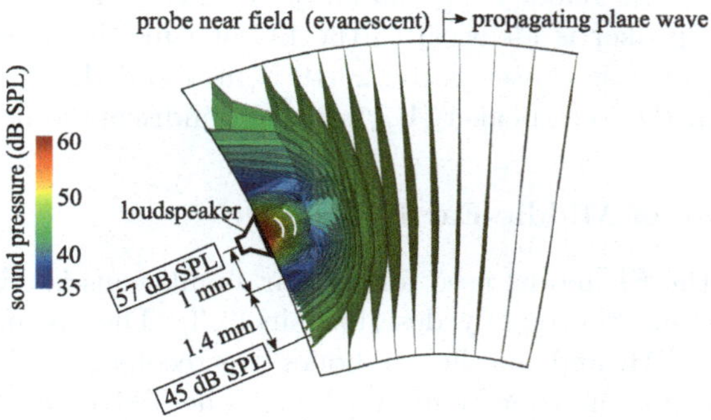

Fig. 2. Evanescent wave in the near field of the probe and propagating plane wave at 4 kHz in the EC of 7 mm diameter.

Because the microphone is often placed in the plane of the loudspeaker, measurements and calibrations are affected by the evanescent waves. Currently this effect can be eliminated or reduced from the source parameters during the calibration by bevelling the probe to increase the aperture [7], as also done here, or by improved error metrics [4]. However, even though the Thévenin parameters of the probe can be corrected, the actual measurement at the ear is still impaired by evanescent modes and, therefore, depends on the specific probe [4].

The termination impedance $Z_{TM} = p_{TM}/u_{TM}$ is often calculated from sound pressure p_{EC} and sound flow u_{EC} at the probe in the EC assuming a plane wave propagation with the transmission matrix as

$$\begin{bmatrix} p_{TM} \\ u_{TM} \end{bmatrix} = \begin{bmatrix} a_{11} & a_{12} \\ a_{21} & a_{22} \end{bmatrix} \cdot \begin{bmatrix} p_{EC} \\ u_{EC} \end{bmatrix} = \begin{bmatrix} \cosh(jkL) & -Z_c \sinh(jkL) \\ -\frac{1}{Z_c} \sinh(jkL) & \cosh(jkL) \end{bmatrix} \cdot \begin{bmatrix} p_{EC} \\ u_{EC} \end{bmatrix} \quad (1)$$

with the propagation factor k, the distance L between probe and termination and the characteristic impedance $Z_c = \rho c/A$ (density ρ, speed of sound c, probe cross-section A). The positive wave propagation direction is defined from probe input to termination.

2.2 How Can the Measurements Be Simulated?

The impedance for the rigidly terminated calibration tubes and tubes with steel plate termination are derived from simulations with an FE code in Hyperworks (Altair Engineering Inc.). For the FE simulations, the air is modelled with acoustic FE using a pressure-based Eulerian approach, meaning that each node has a pressure degree of freedom. As the viscosity of air is very low in relation to the tube dimensions, a Rayleigh damping approximation is used for the air. In order to show the influence of evanescence on the simulated measurements, both a uniform excitation and a local excitation are simulated with the FE model. The loudspeaker is modelled as an ideal sound source with a constant frequency spectrum via a velocity boundary condition at the position of the probe. For uniform excitation the velocity v is chosen to be $1 \cdot 10^{-5}$ m/s and the cylindrical diameter of the speaker is identical to the EC or tube diameter of 7 or 9 mm. For local excitation v is chosen to be $2 \cdot 10^{-3}$ m/s and thus is adapted to the excitation area of the loudspeaker A_{lsp} with a cylindrical diameter of 1 mm.

2.3 Simulation of Middle-Ear Impedance

An overview of the FE model used for the ear simulations is given on the right of Fig. 1. Details are thoroughly described in [2,3]. The geometry of the EC, tympanic cavity, TM, and middle-ear bones are reconstructed from computer tomography of a post-mortem temporal bone. The TM is modelled with shell elements and subdivided into six regions each with a constant thickness. The middle-ear bones are modelled as rigid bodies characterized by their mass and inertia. The ligaments, tendons and joints are represented by passive 6×6 spring-damper elements. For the simulation of the ear impedance, we use a uniform velocity excitation at the EC, which corresponds to an ideal probe-independent excitation and, thus, can be compared to other simulations. The length of the EC is 8 mm analogous to temporal bone measurements. For comparability of the simulations with literature measurements, the volume of the middle-ear cavity and the mastoid are selected according to an individual temporal bone (24L in [8]) to 0.9 cm^3 each, with the mastoid volume being considerably smaller compared to its natural state due to preparation.

3 Results and Discussion

3.1 Tubes, Measurements, and Simulations

By fitting the FE simulations to input-impedance measurements on tubes with rigid termination the Rayleigh damping coefficients for the air ($\alpha_{fluid} = 80$ Hz, $\beta_{fluid} = 2.55 \cdot 10^{-6}$ s) are derived. The assumption of fluid damping with Rayleigh coefficients in the simulation turned out to be suitable for prediction of the measured input impedances. For the probes with steel plate (diameter 9 mm) the damping parameters of air determined from the rigid tubes are used. Additionally, the parameters for the tube length (17.8 mm), plate thickness

(32.5 μm), Young's Modulus (200 GPa), and the Rayleigh damping for the plate ($\alpha_{structure} = 16$ Hz, $\beta_{structure} = 1.3 \cdot 10^{-7}$ s) are determined from impedance measurements by parameter fitting. The impedance simulations of the steel-plate probes (green curve in Fig. 3) are in a very good agreement with the measurements. This can only be reached when the precise geometric arrangement of speaker and microphone is taken into account in the modelling.

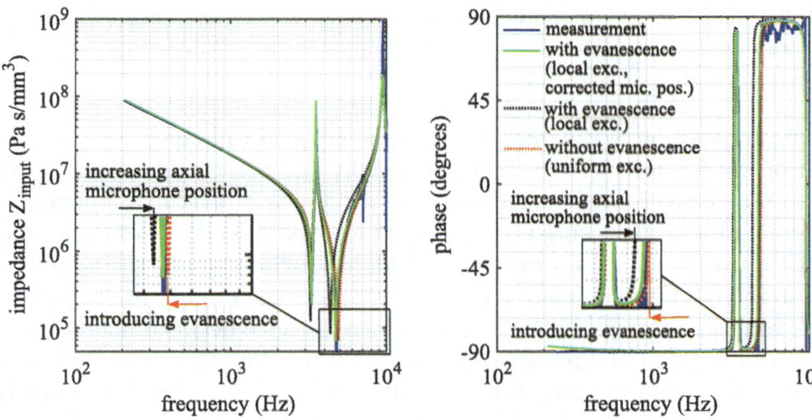

Fig. 3. Comparison of measurements and simulations at the tube with steel plate termination. The simulation is done both with uniform as well as with local excitation. The green curve shows the impedance for local excitation with corrected microphone position according to the real probe.

Evanescence in the probe's near field has an effect independent of the absolute pressure amplitude, in particular at the pressure anti-resonances where amplitude is small, while the resonances are hardly influenced. Evanescence lengthens the wavelength artificially and the anti-resonances shift to lower frequencies, which increases the frequency spacing to the uninfluenced resonances. A corresponding opposite effect is observed when the microphone is not located in the probe plane but extends axially further into the cavity, see Fig. 3. Therefore, in the model the precise geometric arrangement of microphone and loudspeaker needs to be included.

3.2 Ear Simulations

Figure 4 shows the pressure distribution in the EC, derived from an FE model with an EC length of 20 mm, similar to real patient measurements. As seen in the color plot, the EC can be divided into three regions of error sources.

In the probe's near field (region I), evanescent modes occur and lead to pressure variations up to 30 dB SPL at 20 kHz at the probe side of the EC and to more than 20 dB SPL at 4 kHz, see Fig. 2, which is far away from useful tolerances. If the evanescence is not accounted for appropriately with consideration

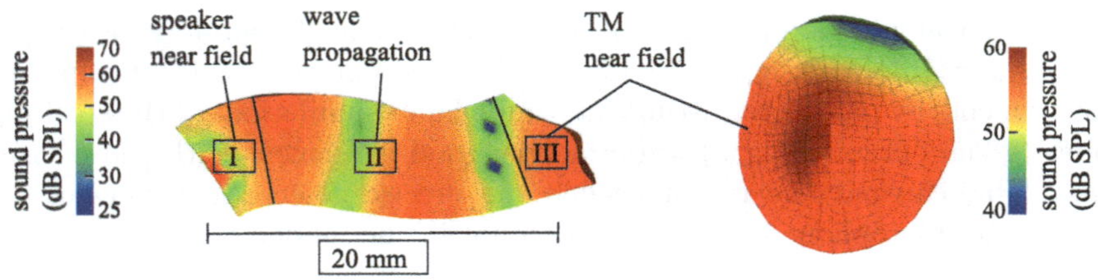

Fig. 4. Inhomogeneity of pressure distribution along the EC with classification in error source regions.

of the exact position of the microphone, this will lead to large errors in the evaluation of impedance measurements, especially when the microphone is placed close to the speaker where small position variations of the microphone lead to completely different measured pressures. The probe should be designed in a way that the microphone is out of the near field, otherwise manufacturing tolerances of the probe might be crucial. There are two ways to design a probe, one which is more robust to tolerances, by increasing the distance between microphone and loudspeaker in the probe plane as much as possible, the other by moving forward the microphone into the measured EC. By using such a probe, the evanescence can be appropriately simulated with an FE model, and be accounted for in the evaluation of measurements.

In the wave propagation region II, errors are mainly introduced by an inappropriately accounted or not accounted cross-section area change in the EC and at higher frequencies (18–20 kHz) also by changes in the EC axis direction. Not accounting for the area change leads to tolerable errors below 4 kHz, however, large errors of more than 5 dB occur above 4 kHz in the case with conical EC, where the diameter of the EC at the probe is increased by 20% over a length of 8 mm, see Fig. 5. Therefore, the area function of the EC needs to be accounted for in the simulation. There are approaches available to estimate this function from measurements, however, further microphones for multiple pressure measurements in the EC are needed.

In the TM near field (region III) significant errors can be introduced by the large pressure inhomogeneity across the TM and leads to considerable errors of more than 5 dB at 10 kHz and even more than 10 dB above 15 kHz or 20 dB at 20 kHz as seen in Fig. 4 (right). This error type is mainly influenced by the angle between TM and EC axis because it originates from differences of the sound path lengths. These pressure inhomogeneities can be accounted for by an FE simulation, however, a simple 1D transmission-line model cannot account for them and will lead to large errors in the calculation of Z_{TM} above 10 kHz.

To validate the model, two different cases are looked at, a normal case and a case with completely removed TM. Figure 6 shows the simulated Z_{TM} compared to individual and averaged measurements from literature [8]. Compared to the literature data, the model shows a valid modelling of the air in the cavities and also the curves of the normal ear match the important characteristics in

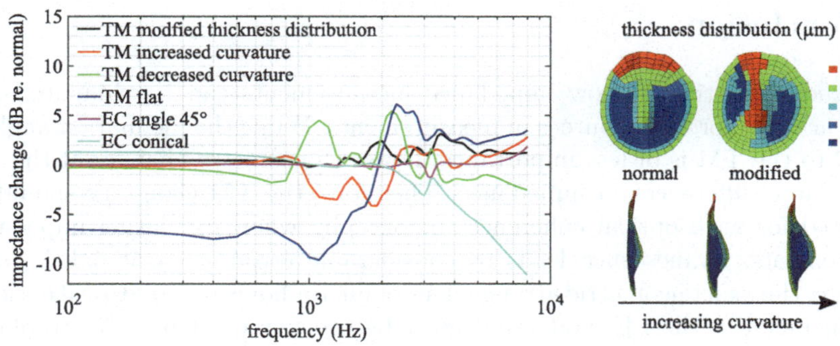

Fig. 5. TM influences on impedance compared to EC influences.

the literature data qualitatively well. In some frequency ranges the simulated impedance has magnitudes up to a factor of 2 times lower than those of temporal bone 24L, however, changes in that magnitudes are within the confidence interval of normal ears.

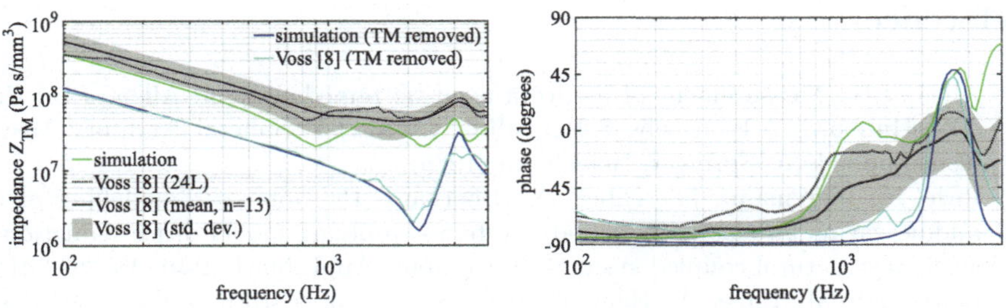

Fig. 6. Simulation of TM impedance with the FE model of the ear for a temporal bone with sealed antrum compared to literature data [8] of a normal ear and an ear with removed TM.

For an evaluation of impedance measurements on individuals with the FE model of the ear it is important to know furthermore the effects of specific shape and thickness changes of TM. Therefore, sensitivity studies seen in Fig. 5 are conducted.

Changes in the TM thickness distribution lead to significant impedance changes at about 2 and 3.5 kHz. Changes in the curvature of the TM lead to significant impedance changes from 1 to 4 kHz. A flat TM leads to changes up to 10 dB at 1 kHz and significantly changes the static stiffness. Considering the exact geometry of the TM in an FE simulation based evaluation of impedance measurements is therefore essential. Main influences of the inclination angle between EC and TM, which was changed from perpendicular to 45° are above 8 kHz and of less importance.

4 Conclusion

This study investigated how impedance measurements can be simulated and how to account for error sources in measurements. While the inclination angle of the EC to the TM is of less importance, however, geometry changes of the TM lead to large differences in impedance. Therefore, the TM geometry should be accounted for with optical coherence tomography when simulating impedance measurements. Evanescence leads to considerable errors even at 4 kHz, and, therefore, the exact geometric arrangement of microphone and loudspeaker needs to be included in the FE model and must be accounted for in order to obtain accurate impedance measurements. To be able to account for the pressure inhomogeneities above 10 kHz FE simulation is indispensable. The area function of the EC needs to be accounted for above 4 kHz otherwise in temporal bone preparations a straight EC should be used.

Acknowledgements. This work has been funded by Volkswagen Foundation (Az. 93949) and a scholarship of the Ministry of Science, Research and Art Baden-Württemberg (MWK). This support is gratefully acknowledged.

References

1. Allen, J.B.: Measurement of eardrum acoustic impedance. In: Allen, J.B., Hall, J.L., Hubbard, A.E., Neely, S.T., Tubis, A. (eds.) Peripheral Auditory Mechanisms, pp. 44–51. Springer, Heidelberg (1986)
2. Ihrle, S., Lauxmann, M., Eiber, A., Eberhard, P.: Nonlinear modelling of the middle ear as an elastic multibody system - applying model order reduction to acousto-structural coupled systems. J. Comput. Appl. Math. **246**, 18–26 (2013)
3. Lauxmann, M., Eiber, A., Haag, F., Ihrle, S.: Nonlinear stiffness characteristics of the annular ligament. J. Acoust. Soc. Am. **136**, 1756–1767 (2014)
4. Nørgaard, K.R., Fernandez-Grande, E., Laugesen, S.: Incorporating evanescent modes and flow losses into reference impedances in acoustic Thévenin calibration. J. Acoust. Soc. Am. **142**, 3013–3024 (2017)
5. Sackmann, B., Dalhoff, E., Lauxmann, M.: Model-based hearing diagnostics based on wideband tympanometry measurements utilizing fuzzy arithmetic. Hear. Res. (2019). https://doi.org/10.1016/j.heares.2019.02.011
6. Scheperle, R.A., Goodman, S.S., Neely, S.T.: Further assessment of forward pressure level for in situ calibration. J. Acoust. Soc. Am. **130**, 3882–3892 (2011)
7. Siegel, J.H., Neely, S.T.: Eartip modification greatly reduces evanescent waves. In: Proceedings of the 40th MidWinter Meeting, pp. 11–15 (2017)
8. Voss, S.E., Rosowski, J.J., Merchant, S.N., Peake, W.T.: Acoustic responses of the human middle ear. Hear. Res. **150**, 43–69 (2000)

Anterior Cruciate Ligament Injuries Alter the Kinematics and Kinetics of Knees with or Without Meniscal Deficiency

Xiaode Liu[1], Hongshi Huang[2], Shuang Ren[2], Yingfang Ao[2(✉)], and Qiguo Rong[1(✉)]

[1] Department of Mechanics and Engineering Science, College of Engineering, Peking University, Beijing 100871, China
qrong@pku.edu.cn
[2] Institute of Sports Medicine, Peking University Third Hospital, Beijing 100191, China
aoyingfang@163.com

Abstract. The purpose of this paper was to study the biomechanical behaviors of knees with anterior cruciate ligament deficient (ACLD) with or without a combined medial or/and lateral meniscal injury during level walking. The motion capture system and the modeling system (Any-Body) were applied to simulate the kinematic and kinetic properties of knees. The results show that the knees with ACLD exhibited significantly less extension than the control knees at the mid stance. A lower extension moment and adduction moment in all ACLD-affected knees were detected during the terminal stance when compared with control knees. The ACLDML group showed significantly lower proximodistal compressive forces and anteroposterior and mediolateral shear forces, while the shear forces tended to increase in the ACLD, ACLDL, and ACLDM groups.

1 Introduction

It is well known that the anterior cruciate ligament (ACL) plays an important role in knee joint stability by limiting anterior tibial translation and maintaining axial and transverse rotation of the knee [1]. Many studies have investigated ACL biomechanics under various loading conditions, especially for the patients who have ACL injuries. ACL deficient (ACLD) knees tend to exhibit abnormal biomechanical characteristics compared to the control group and often associate with an incidence of knee osteoarthritis. Majority of previous studies have investigated the characteristics of knees with isolated ACLD. ACL injuries, however, are commonly combined with meniscal tears. In the United States, approximately 40% to 80% of ACL injuries are combined with menisci injuries [2]. Bellabarba et al. [3] reported that 41% to 82% of acute ACL-injured knees and 58% to 100% of knees with chronic ACL deficiency had meniscal tears. Some studies [4–6] have showed that the location of the meniscus tear could influence kinematics in ACLD knees.

A. Kecskeméthy and F. Geu Flores (Eds.): ECCOMAS 2019, COMPUTMETHODS 53, pp. 11–17, 2020.
https://doi.org/10.1007/978-3-030-23132-3_2

However, the effect of different meniscus injury patterns on the kinematics and kinetics in ACLD knees during gait, has not been well studied. Meanwhile, accurate determination of kinetics involves however difficulties and limitations in both in vitro cadaver and in vivo imaging studies. Musculoskeletal modeling on the other could circumvent such shortcomings. However, there are few studies focusing on the musculoskeletal modeling analysis. In this study, based on a multibody dynamics software, a three-dimensional musculoskeletal modeling method was used to investigate the kinematics and kinetics of ACLD knees with or without a combined medial or lateral meniscal injuries during level walking. We hypothesized that gait mechanics would present a more abnormal pattern in ACLD knees with medial or/and lateral meniscal injury than in those without.

2 Methods

Between January 2014 and December 2016, the experimental data was collected by using an optical motion capture system during normal walking. The ethical approval was obtained from the university's ethics committee and written informed consent was attained from all participants. The ACLD knees were diagnosed by clinical examination, magnetic resonance imaging and confirmed during arthroscopic ACL reconstruction surgery. 29 patients with unilateral ACLD knees (contralateral side intact; injury time range: 6 months–4 years, age range: 18–34 years; body mass index rage, 20.52–35.65 $\mathrm{kg/m^2}$) were recruited before undergoing ACL reconstruction. Among these patients, 12 patients had isolated unilateral ACL injuries (ACLD group), 5 had combined ACL and lateral meniscal injuries (ACLDL group), 5 had combined ACL and medial meniscal injuries (ACLDM group) and 7 had combined ACL and medial/lateral meniscal injuries (ACLDML group). When evaluating the meniscal tears, we did not consider the type of tear (i.e., longitudinal root tear, horizontal cleavage tear, or complex tear) because of the limited sample size. None of the knee cartilage defects were higher than grade II according to the Outerbridge system [7]. 15 healthy male with no history of musculoskeletal or neuromuscular disorders in the lower extremities and no measurable ligamentous instability on clinical examination volunteered for this study (Control group).

The experimental data were collected by 100-Hz eight-camera motion capture system (Vicon MX; Oxford Metrics, Yarnton, Oxfordshire, UK), and ground-reaction forces were measured using two 1000-Hz embedded force plates (Advanced Mechanical Technology Inc., Watertown, MA, USA). A set of markers were attached to participants lower limbs to track segmental motion during walking. Based on the validated plug-in-gait model, anatomical markers were taped to following anatomical lower limbs locations: the anterior and posterior superior iliac spine, medial and lateral femoral epicondyles, malleoli, and medial and lateral sides of the calcaneus, the frontal and lateral aspects of the thigh and the shank, posterior part of the calcaneus, heads of the first, second, and fifth metatarsal bones, base of the first metatarsal bone, and navicular, hallux.

After performing a standing trial, participates were asked to walk from a specified point. A successful gait trial was recorded when each foot stepped on the force plates at a self-selected speed. Five successful gait trials were assessed for each participant.

A three-dimensional (3D) musculoskeletal model was employed using multi-body dynamics software (AnyBody Modeling System, version 6.0.5; AnyBody Technology A/S). This model was constructed based on the University of Twente lower extremity model (TLEM) and has been validated to calculate muscle forces and joint moments [8]. The model consists of 160 muscle-tendon actuators and 6 joint degrees of freedom: the hip joint was modeled as a spherical joint with 3 degrees of freedom; the knee joint was modeled as a hinge joint with only 1 degree of freedom for flexion-extension, and a universal joint was considered for the ankleCsubtalar complex.

As for the multi-body dynamics analysis in the TLEM, initial condition operations were firstly performed to identify the parameters of segment lengths and the (virtual) marker positions. Anatomic frames on the femur and tibia corresponding to the static standing trial were defined based on the respective tibiofemoral joint (TFJ) landmarks (Fig. 1). In order to match the outer markers (experimental data) well with the inner markers (markers attached to the TLEM model based on the same plug-in-gait model), least-squares minimization were performed between the outer markers and inner markers positions. The TLEM model was re-modeled according to our anthropometric data (body weight, body height, pelvis width, thigh, shanks, and foot length) for each subject, and the personalized model was scaled with a mass-fat scaling algorithm. Kinematics and inverse dynamics analysis were successively performed for each simulation: first, the skeletal model was employed to calculate the joint kinematic waveforms and segmental motions related to the experimental gait trials; second, the musculoskeletal model was used in the inverse dynamic simulations to calculate the joint forces and moments. Joint contact forces were derived from the inter-segment loading and muscle force that acted on the joint. For the muscle recruitment, we adopted a min/max recruitment solver in the software to solve the muscle redundancy problem. Five different gait trials were simulated for each participant, and the average values (including the sagittal joint angle, 3D reaction forces, and joint moments of the TFJ were compared between the control and ACLDs knees during level walking.

3 Results

The ACLD knees, with or without meniscal deficiency, had significant less extension (Fig. 2A) and lower extension moments (Fig. 2C) than the control knees at mid-stance, including the maximum extension moment (ACLD: -0.011 ± 0.014 Nm/(kg*m), ACLDL: -0.011 ± 0.010 Nm/(kg*m), ACLDM: -0.012 ± 0.011 Nm/(kg*m), ACLDML: -0.018 ± 0.008 Nm/(kg*m), control: -0.024 ± 0.017 Nm/(kg*m), $p < 0.05$). Compared with the control group, the ACLDL and ACLDML groups showed a decreased proximal-distal compressive

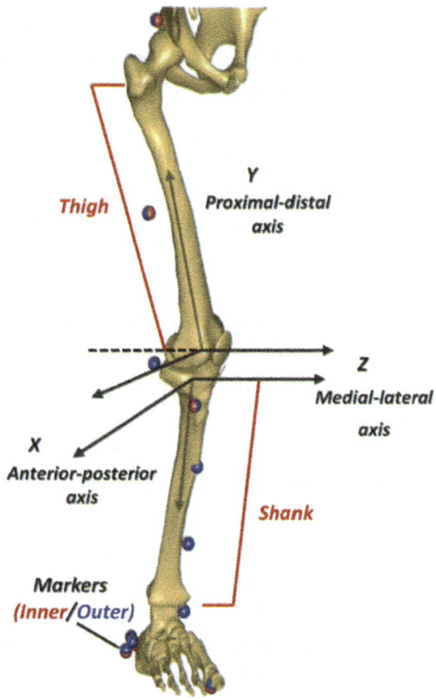

Fig. 1. Definition of local tibiofemoral coordinate systems.

forces, while ACLDM group presented an increased trend. For group ACLDML, in particular, the reduction in the maximum compressive force was significant (ACLDML: 3.45 ± 0.52 N/kg, control: 4.46 ± 1.65 N/kg, $p < 0.05$) (Fig. 2B). Meanwhile, there is no significant statistical difference at the maximum flexion moment, except for the comparison between control knees and ACLDML knees (Fig. 2C). No significant differences in the flexion moment were observed between ACLDL and ACLDM groups during the stance phase.

4 Discussion

This study investigated the sagittal kinematics and 3D kinetics of ACLD-affected knees with or without meniscal injuries during level walking. The results of our current research support the initial hypothesis that the kinematic and kinetic response of ACLD-affected knees could be dependent on the type of meniscal injuries. Some previous studies have investigated the effects of meniscal tears on the gait parameters of ACLD-affected knees [4–6]. Ren et al. [4] showed that ACLD-affected knees with medial meniscal posterior horn tears exhibited extension deficiency, more external tibial rotation, and lower extension and internal rotation moments. Zhang et al. [6] recently reported that a combined

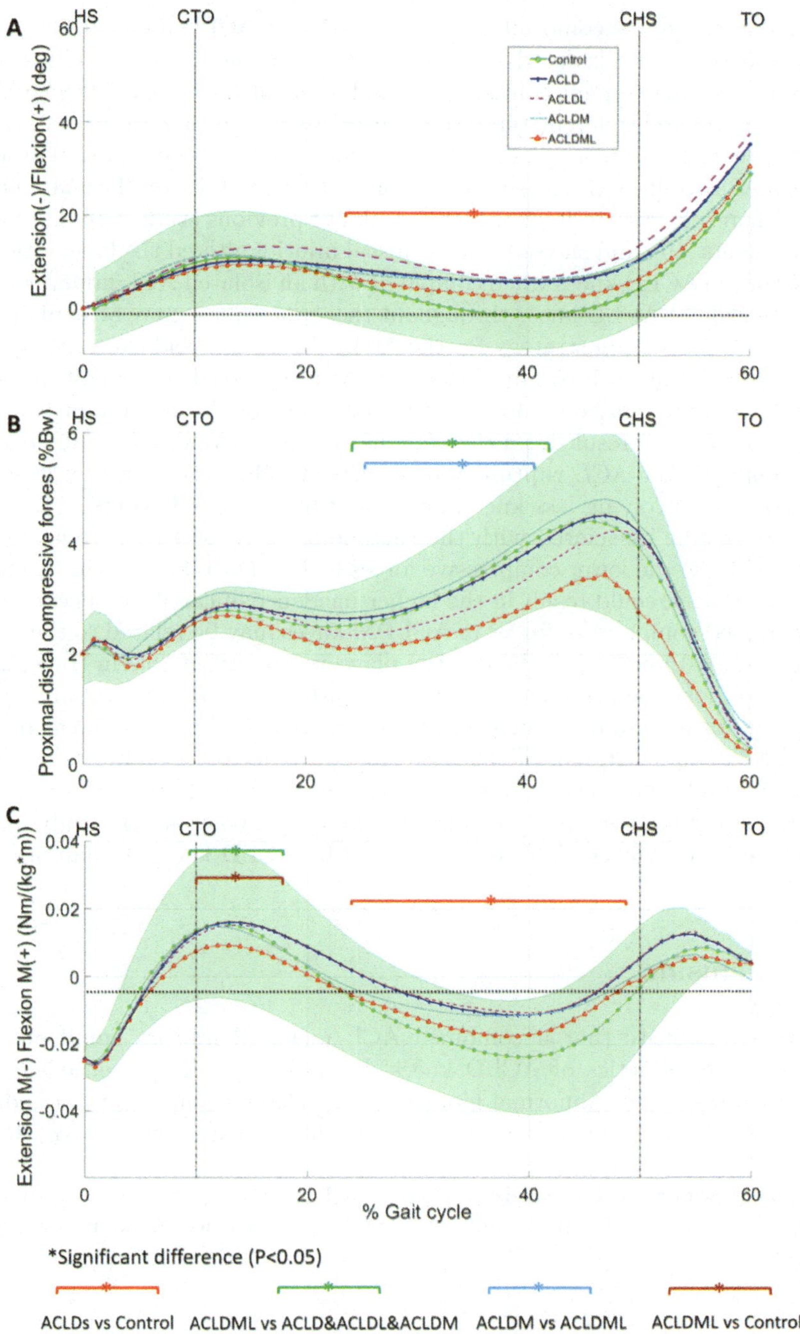

Fig. 2. TFJ kinematics and kinetics of control and patients (ACLD, ACLDL, ACLDM and ACLDML) knees. Segments with significant statistical differences between the patients and the control groups were marked with asterisks. The green shaded area represents means standard deviation of the control group. HS: heel strike; CHS: contralateral heel strike; TO: toe off; CTO: contralateral toe off. deg: degree.

ACL/meniscus injury could alter the kinematics of ACLD-affected knees compared with knees with isolated ACL injury. However, these studies did not assess the alterations in the joint contact force and moment in patients with combinations of ACLD and different types of meniscal tears (e.g., lateral or/and medial meniscal tears). As far as we know, this is the first inverse dynamical study to evaluate ACLD-affected knees with concomitant medial or/and lateral meniscal tears. Our result were in line with the results of previous studies [9] and in vivo measurements [10], and showed that meniscal injuries altered the kinematics and kinetics of the ACLD knees when compared with an isolated ACL injury. The less knee extension in ACLD knees throughout the stance phase may be explained by a protective adaptation strategy for the ACLD knees to avoid excessive anterior tibial displacement at maximum extension. As the injury knees result in a lower demand on the quadriceps muscles, the reduction in the maximum flexion of ACLDM knees may result from the nervous system to avoid pain. Subjects with knee stability after ACL rupture consistently stabilize their knee with a stiffening strategy involving less knee motion and higher muscle contraction. Joint unloading maybe associated with the cascade of early degenerative changes at the knee [11], so the lower compressive forces in ACLDML knees, with combined meniscal injuries, could result in the higher incidence of cartilage degenerations. The increased compressive forces in ACLDM knees may be related to the type of medial meniscal injuries, as the menisci plays an important role in transmitting tibiofemoral loads and reducing pressure on articular cartilage [12] and release of the anterior intermeniscal ligament results in increased peak contact pressures in the medial compartment of the knee [13]. Chronic ACLD subjects tended to present a muscle contraction adaptation strategy, which increased the activity of quadriceps and hamstrings. Therefore, the lower extension in the study may be related to the muscle co-contraction adopted by ACLD knees to maintain knee stability.

5 Conclusion

The results indicate that a combined ACL/meniscal injuries could alter the kinematics and kinetics of ACLD knees depending on the presence and type of meniscal tears. The abnormal biomechanics behavior may be associated with the cascade of early degenerative changes and subsequent onset of osteoarthritis.

Acknowledgements. This study was supported by Beijing Municipal Natural Science Foundation (No. 7172120) and National Natural Science Foundation of China (No. 11872074).

References

1. Andersen, H.N.: The anterior cruciate ligament does play a role in controlling axial rotation in the knee. J. Knee Surg. Sports Traumatol. Arthrosc. **5**(3), 145–149 (1997)
2. Granan, L.P.: Sport-specific injury pattern recorded during anterior cruciate ligament reconstruction. Am. J. Sports Med. **41**, 2814–2818 (2013)
3. Bellabarba, C., Bush-Joseph, C.A., Bach Jr., B.R.: Patterns of meniscal injury in the anterior cruciate-deficient knee: a review of the literature. Am. J. Orthop. (Belle Mead NJ) **26**(1), 18–23 (1997)
4. Shuang, R.: Three dimensional knee kinematics and kinetics in ACL-deficient patients with and without medial meniscus posterior horn tear during level walking. J. Gait Posture **66**, 26–31 (2018)
5. Ali, H.: Meniscus injuries alter the kinematics of knees with anterior cruciate ligament deficiency. Orthop. J. Sports Med. **2**, 1–8 (2014)
6. Zhang, Y.: Anterior cruciate ligament injuries alter the kinematics of knees with or without meniscus deficiency. Am. J. Sports Med. **44**, 3132–3139 (2016)
7. Outerbridge, R.E.: The etiology of chondromalacia patellae. J. Bone Joint Surg.-Br. **43**, 752–757 (1961)
8. Horsman, K.: The twente lower extremity model: consistent dynamic simulation of the human locomotor apparatus. Ph.D. thesis, University of Twente (2007)
9. Worsley, P.: Predicted knee kinematics and kinetics during functional activities using motion capture and musculoskeletal modelling in healthy older people. J. Gait Posture **33**, 268–273 (2011)
10. Kutzner, I.: Loading of the knee joint during activities of daily living measured in vivo in five subjects. J. Biomech. **43**, 2164–2173 (2010)
11. Wellsandt, E.: Decreased knee joint loading associated with early knee osteoarthritis after anterior cruciate ligament injury. J. Am. J. Sports Med. **44**(1), 143–151 (2016)
12. Guess, T.M.: A subject specific multibody model of the knee with menisci. J. Med. Eng. Phys. **32**(5), 505–515 (2010)
13. Paci, J.M.: Knee medial compartment contact pressure increases with release of the type I anterior intermeniscal ligament. J. Am. J. Sports Med. **37**(7), 1412–1416 (2009)

Investigation of Inhomogeneous Stiffness and Damping Characteristics of the Human Stapedial Annular Ligament

D. Burovikhin[1(✉)], Benjamin Sackmann[1], Merlin Schär[2], J. H. Sim[2], P. Eberhard[3], and M. Lauxmann[1]

[1] Reutlingen University, Alteburgstr. 150, 72762 Reutlingen, Germany
{dmitrii.burovikhin,benjamin.sackmann, michael.lauxmann}@reutlingen-university.de

[2] University Hospital Zürich, Frauenklinikstrasse 24, 8091 Zürich, Switzerland
{merlin.schaer,jaehoon.sim}@usz.ch

[3] Institute of Engineering and Computational Mechanics, University of Stuttgart, Pfaffenwaldring 9, 70569 Stuttgart, Germany
peter.eberhard@itm.uni-stuttgart.de

Abstract. This study describes a non-contact measuring and system identification procedure for evaluating inhomogeneous stiffness and damping characteristics of the annular ligament in the physiological amplitude and frequency range without the application of large static external forces that can cause unnatural displacements of the stapes. To verify the procedure, measurements were first conducted on a steel beam. Then, measurements on an individual human cadaveric temporal bone sample were performed. The estimated results support the inhomogeneous stiffness and damping distribution of the annular ligament and are in a good agreement with the multiphoton microscopy results which show that the posterior-inferior corner of the stapes footplate is the stiffest region of the annular ligament.

1 Introduction

The stapes is a bone found in the middle ear of humans and other mammals which is involved in the conduction of sound vibrations to the inner ear, see Fig. 1. The stapedial annular ligament (AL) is a ring of fibrous tissue that connects the base of the stapes, its footplate, to the oval window of the inner ear. The anatomical dimensions of the AL support the hypothesis of an inhomogeneous stiffness distribution. In [3] it has been revealed that the cross-section of the AL is posteriorly narrower and thicker, resulting in a higher stiffness on the posterior side, and in [1] it is stated that the properties of the AL largely determine the transfer characteristic of the middle ear in the lower frequency range.

A number of studies on the topic of the stiffness characteristics of the human stapedial AL were conducted in the past, but all of them were focused on determining the stiffness properties of the AL in the quasi-static frequency range by

A. Kecskeméthy and F. Geu Flores (Eds.): ECCOMAS 2019, COMPUTMETHODS 53, pp. 18–25, 2020.
https://doi.org/10.1007/978-3-030-23132-3_3

applying large, mostly static, external forces to the stapes, which caused displacements beyond the physiological range and, as a consequence, introduced artefacts that hid underlying physical properties of the AL [1,2,4,7]. Moreover, so far the inhomogeneous stiffness and damping properties of the AL have not been investigated in the dynamic frequency range.

The measurement and system identification procedure described here offers a non-contact way to determine the stiffness and damping characteristics of the AL within the physiological amplitude and frequency range by using the inertia forces of the stapes itself instead of large external forces. The inertia properties of the stapes were determined with the help of a micro-CT scanner. Based on that the inhomogeneous stiffness and damping characteristics of the AL were estimated within the physiological frequency range. We verified the procedure by conducting measurements on a theoretically well-known mechanical structure similar to that of the stapes-AL system, where the AL was represented by a steel beam and the stapes by a small steel cube. The introduced system identification procedure can potentially help to find correlations between the inertia properties of the stapes and the stiffness and damping properties of the AL among different human cadaveric temporal bone samples.

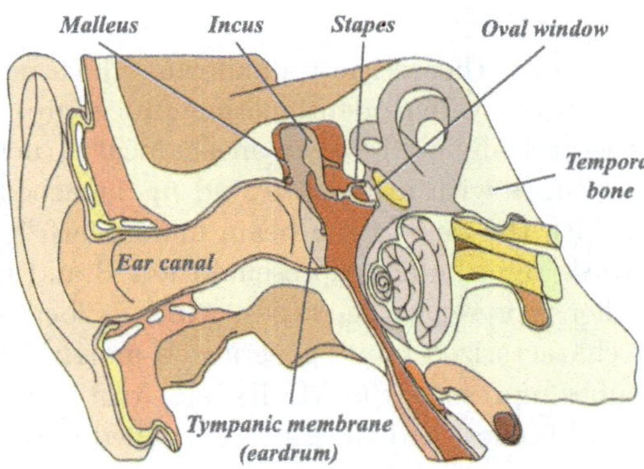

Fig. 1. Anatomical structure of the human ear

2 Identification of Structural System Parameters from Dynamic Response Data

The stapes and the peripheral bone (PB) can be modelled as two rigid bodies connected by the AL modelled as a spring/damper combination, as shown schematically in Fig. 2. The stapes is represented by its generalized mass matrix $\mathbf{M}$ which is referred to the centre of its footplate and obtained by scanning the stapes and the plastic plate, to which the temporal bone sample is glued, with a micro-CT scanner [8]. The plastic plate serves as a reference for the scanner. The mass of the AL can be neglected. The stapes is subjected to the base excitation from the PB mounted on a shaker table. The spatial motion of the stapes and the PB is measured by a 3D laser Doppler vibrometer (3D-LDV) at multiple points and the generalized coordinates for each rigid body are calculated with reference to the centre of the stapes footplate.

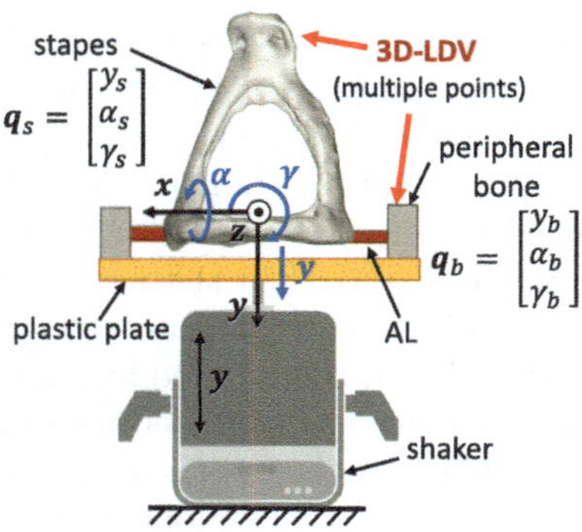

Fig. 2. Stapes - measurement setup

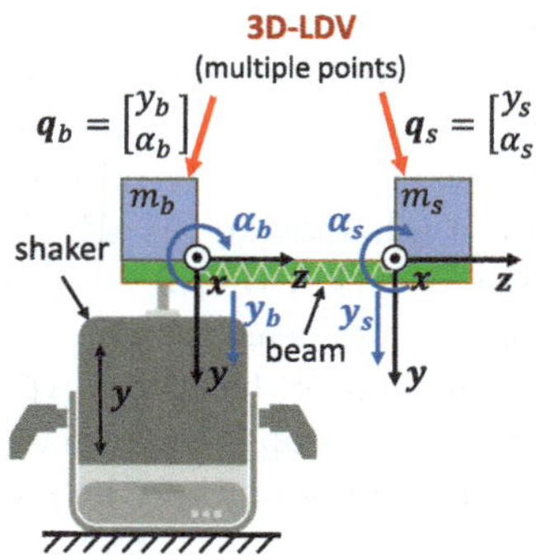

Fig. 3. Beam - measurement setup

The stapes-AL system is theoretically unknown, which makes it difficult to confirm the validity of our system identification procedure. Therefore, it was first verified on a theoretically well-known mechanical structure consisting of two 10×10 mm steel cubes with masses m_b and m_s attached to a steel beam. Each cube, together with the part of the beam underneath it, is modelled as a rigid body. One of these rigid bodies represents the PB and is mounted on the shaker table. The other represents the stapes, which is located at the free end of the beam, and is characterized by its generalized matrix $\mathbf{M}$. The part of the beam between the cubes represents the AL. Its mass can be neglected compared to the mass of the rigid bodies, and thus, it is modelled only with regard to its stiffness and damping properties. The spatial motion of the cubes is measured by the 3D-LDV system at multiple measurement points and their generalized coordinates are calculated with reference to the origin of their axes.

Both the stapes-AL system and the beam system represent a structure subjected to base excitation whose dynamic equation of motion is given by

$$\mathbf{M} \cdot \ddot{\mathbf{q}}_s(t) + \mathbf{D} \cdot \dot{\mathbf{q}}_s(t) + \mathbf{K} \cdot \mathbf{q}_s(t) = \mathbf{D} \cdot \dot{\mathbf{q}}_b(t) + \mathbf{K} \cdot \mathbf{q}_b(t), \tag{1}$$

with the generalized displacement vectors of the structure $\mathbf{q}_s(t)$ and the base $\mathbf{q}_b(t)$, generalized symmetric viscoelastic damping matrix $\mathbf{D}$, and the generalized symmetric stiffness matrix $\mathbf{K}$.

The assumption of harmonic base excitation at k number of frequencies ω_j and steady state response yields

$$\omega_j^2 \, \mathbf{M} \cdot \mathbf{q}_s(\omega_j) = (\mathbf{K} + i\omega_j \, \mathbf{D}) \cdot \Delta\mathbf{q}(\omega_j) \; , \;\; j = 1, 2, ..., k, \tag{2}$$

with the relative displacement vector $\Delta\mathbf{q}(\omega_j) = \mathbf{q}_s(\omega_j) - \mathbf{q}_b(\omega_j)$. Having in mind that $\Delta\mathbf{q}(\omega_j) = \Delta\mathbf{q}'(\omega_j) + i\Delta\mathbf{q}''(\omega_j)$ and $\mathbf{q}_s(\omega_j) = \mathbf{q}_s'(\omega_j) + i\mathbf{q}_s''(\omega_j)$, the amplitude and phase of each measured point at the j-th excitation frequency ω_j are transformed to the real and imaginary part of the steady-state response

$$\begin{aligned} \omega_j^2\,\mathbf{M}\cdot\mathbf{q}_s'(\omega_j) &= \mathbf{K}\cdot\Delta\mathbf{q}'(\omega_j) - \omega_j\,\mathbf{D}\cdot\Delta\mathbf{q}''(\omega_j)\,, \\ \omega_j^2\,\mathbf{M}\cdot\mathbf{q}_s''(\omega_j) &= \mathbf{K}\cdot\Delta\mathbf{q}''(\omega_j) + \omega_j\,\mathbf{D}\cdot\Delta\mathbf{q}'(\omega_j). \end{aligned} \tag{3}$$

Equation 3 can be written as

$$\mathbf{A}_j\cdot\mathbf{p} = \mathbf{b}_j\;,\quad j = 1, 2, ..., k, \tag{4}$$

with a design parameter vector $\mathbf{p}$ containing the unknown stiffness and damping coefficients of matrices $\mathbf{K}$ and $\mathbf{D}$. For k number of frequencies Eq. 4 yields an overdetermined system of linear equations, see Eq. 5, that can be solved with the least squares method by calculating the pseudo inverse $\mathbf{A}^I$ of matrix $\mathbf{A}$ to avoid singularity in matrix $\mathbf{A}$ caused by measurement noise

$$\underbrace{\begin{bmatrix} A_1 \\ \vdots \\ A_j \end{bmatrix}}_{\mathbf{A}}\cdot\mathbf{p} = \underbrace{\begin{bmatrix} \mathbf{b}_1 \\ \vdots \\ \mathbf{b}_j \end{bmatrix}}_{\mathbf{b}} \implies \mathbf{p} = \mathbf{A}^I\cdot\mathbf{b}. \tag{5}$$

The stapes-AL system is more complex than that of the beam, partly, because it is more damped, and partly because of larger measurement errors caused by misalignment and measurement noise. Because of that solving Eq. 5 for the stapes-AL system using an unconstrained least squares solution leads to large estimation errors, since, in this case, a local minimum found by LSQ algorithm can become the global one. Instead, we can iteratively minimize the function $|\mathbf{b} - \mathbf{A}\cdot\mathbf{p}|^2$ by solving a constrained optimization problem with the help of MATLAB's *fmincon* programming solver. From [6] the signs of the stiffness coefficients and their plausible range can be determined and the stiffness and damping coefficients can be constrained in a range where the system is stable. The stability of the system is determined by the eigenvalues criterion meaning that the real part of the eigenvalues of the system estimated with these coefficients should be negative in order for the system to be stable.

2.1 The Beam Experiment Results

The motion of the beam system has two degrees of freedom (DOF), one translational y_s along y-axis and one rotational α_s around x-axis, see Fig. 3. From the differential equation of the cantilever beam, the theoretical stiffness matrix of the beam can be derived and is given by

$$\frac{1}{EI}\begin{bmatrix} \frac{l^3}{3} & -\frac{l^2}{2} \\ -\frac{l^2}{2} & l \end{bmatrix}\cdot\begin{bmatrix} F \\ M \end{bmatrix} = \begin{bmatrix} y_s \\ \alpha_s \end{bmatrix} \implies \mathbf{K}_{theory} = \begin{bmatrix} c_{1,th} & c_{2,th} \\ c_{2,th} & c_{3,th} \end{bmatrix} = EI\begin{bmatrix} \frac{12}{l^3} & \frac{6}{l^2} \\ \frac{6}{l^2} & \frac{4}{l} \end{bmatrix} \tag{6}$$

Table 1. Beam parameters

Parameter	Value	Unit
Young's modulus, E	210000	$\frac{\mathrm{N}}{\mathrm{mm}^2}$
Moment of inertia, $I = \frac{bh^3}{12}$	0.44287	mm^4
Height of the beam, h	0.81	mm
Width of the beam, b	10	mm
Length of the beam, l	48	mm

with force F and rotational moment M applied at the free end of the beam. For the beam parameters, see Table 1.
For a beam with two DOF, Eq. 2 can be written as

$$\omega_j^2 \mathbf{M} \cdot \underbrace{\begin{bmatrix} y_s(\omega_j) \\ \alpha_s(\omega_j) \end{bmatrix}}_{\mathbf{q}_s(\omega_j)} = \left(\underbrace{\begin{bmatrix} c_1 \ c_2 \\ c_2 \ c_3 \end{bmatrix}}_{\mathbf{K}} + i\omega_j \underbrace{\begin{bmatrix} d_1 \ d_2 \\ d_2 \ d_3 \end{bmatrix}}_{\mathbf{D}} \right) \cdot \underbrace{\begin{bmatrix} y_s(\omega_j) - y_b(\omega_j) + l\alpha_b(\omega_j) \\ \alpha_s(\omega_j) - \alpha_b(\omega_j) \end{bmatrix}}_{\Delta \mathbf{q}(\omega_j)}. \quad (7)$$

Equation 7 can be written in the form of Eq. 5 and solved for the unknown stiffness and damping coefficients. The generalized mass matrix $\mathbf{M}$ is calculated from the geometric parameters given in Table 1. To verify the results, the motion of the free end of the beam is simulated using estimated $\mathbf{K}_{est}$ and $\mathbf{D}_{est}$ matrices by calculating the complex frequency response of a structure subjected to base excitation which is given by

$$\mathbf{H}_j = (-\omega_j^2 \mathbf{M} + i\omega_j \mathbf{D}_{est} + \mathbf{K}_{est})^{-1} \cdot (i\omega_j \mathbf{D}_{est} + \mathbf{K}_{est}), \quad j = 1, 2, ..., k, \quad (8)$$

with the $n \times n$ frequency response matrix $\mathbf{H}_j$ for each frequency j, where n is the number of DOF, which in the case of the beam equals two.

Figure 4 shows a comparison between the measured and simulated motions of the free end y_s and α_s related to the measured y-motion of the base $y_{b\ mes}$ in order to normalize the results. This is done because $y_{b\ mes}$ is the dominant motion, since the shaker table excites the structure primarily in the y-direction. The solid lines show the measured motion and the dashed lines show the estimated motion. To quantitatively compare the estimated and theoretical stiffness coefficients, a deviation error in % is calculated

$$\text{deviation} = \begin{bmatrix} (c_{1,est} - c_{1,th})/c_{1,th} \\ (c_{2,est} - c_{2,th})/c_{2,th} \\ (c_{3,est} - c_{3,th})/c_{3,th} \end{bmatrix} 100\% = \begin{bmatrix} 3.17 \\ 1.28 \\ 0.38 \end{bmatrix} \%. \quad (9)$$

A misalignment of the centre of mass of the free end of 100 µm in the z-direction can already explain the deviation error of about 3% in estimating the stiffness values as well as an error of about 1% in estimating the resonance frequencies of the beam, see Fig. 4. This shows the validity of the proposed system identification procedure.

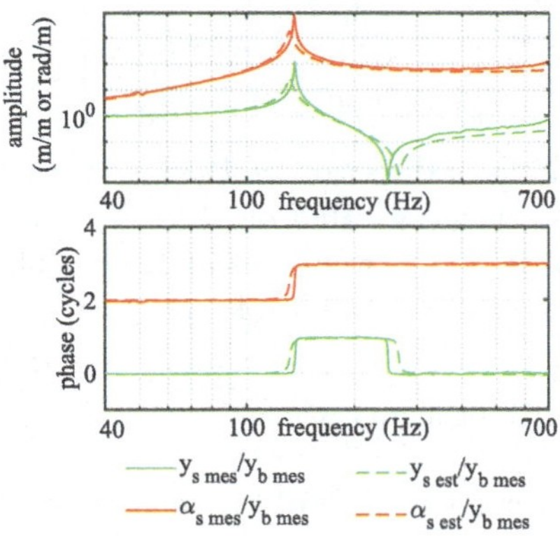

Fig. 4. Beam - comparison between measured and estimated (simulated) data

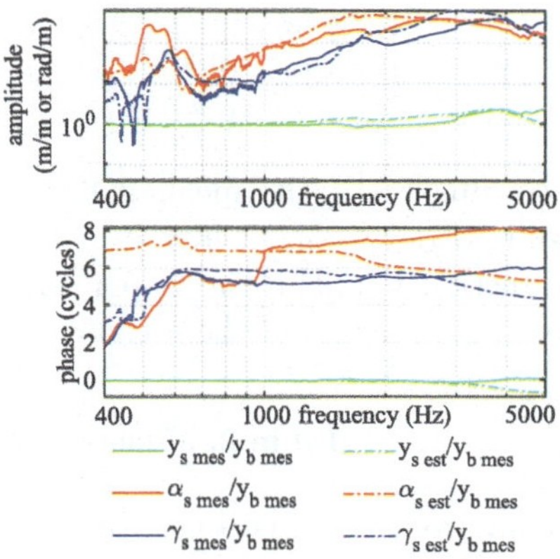

Fig. 5. Stapes - comparison between measured and estimated (simulated) data

2.2 The Stapes Experiment Results

It is generally accepted that the motion of the stapes has three DOF, one translational piston-like motion y_s along the y-axis, and two rotational motions α_s and γ_s around the x-axis and z-axis, respectively, as shown in [5].

For a stapes system with three DOF, Eq. 2 can be written as

$$\omega_j^2\,\mathbf{M}\cdot\underbrace{\begin{bmatrix} y_s(\omega_j)\\ \alpha_s(\omega_j)\\ \gamma_s(\omega_j)\end{bmatrix}}_{\mathbf{q}_s(\omega_j)} = \Bigg(\underbrace{\begin{bmatrix} c_1 & c_2 & c_3\\ c_2 & c_4 & c_5\\ c_3 & c_5 & c_6\end{bmatrix}}_{\mathbf{K}} + i\omega_j \underbrace{\begin{bmatrix} d_1 & d_2 & d_3\\ d_2 & d_4 & d_5\\ d_3 & d_5 & d_6\end{bmatrix}}_{\mathbf{D}}\Bigg)\cdot\underbrace{\begin{bmatrix} y_s(\omega_j)-y_b(\omega_j)\\ \alpha_s(\omega_j)-\alpha_b(\omega_j)\\ \gamma_s(\omega_j)-\gamma_b(\omega_j)\end{bmatrix}}_{\Delta\mathbf{q}(\omega_j)} . \quad (10)$$

As in the case of the beam, Eq. 10 can be written in the form of Eq. 5 and solved for the unknown stiffness and damping coefficients. The generalized mass matrix of the stapes $\mathbf{M}$, obtained with the help of a micro-CT scanner, reads

$$\mathbf{M} = \begin{bmatrix} 3.2\cdot 10^{-6}\,\mathrm{kg} & -434.88\cdot 10^{-12}\,\mathrm{kg\cdot m} & 143.04\cdot 10^{-12}\,\mathrm{kg\cdot m}\\ -434.88\cdot 10^{-12}\,\mathrm{kg\cdot m} & 9.77\cdot 10^{-12}\,\mathrm{kg\cdot m^2} & -29.7\cdot 10^{-15}\,\mathrm{kg\cdot m^2}\\ 143.04\cdot 10^{-12}\,\mathrm{kg\cdot m} & -29.7\cdot 10^{-15}\,\mathrm{kg\cdot m^2} & 11.14\cdot 10^{-12}\,\mathrm{kg\cdot m^2}\end{bmatrix}.$$

Just like in the beam experiment, to verify the results, the frequency response is simulated for $\mathbf{K}_{est}$ and $\mathbf{D}_{est}$, see Eq. 8. Figure 5 shows a comparison between the measured and simulated motions of the stapes y_s, α_s and γ_s related to the measured y-motion of the base $y_{b\ mes}$. The solid lines show the measured motion and the dashed lines show the estimated motion. The estimated stiffness and damping matrices read

$$\mathbf{K}_{est} = \begin{bmatrix} 1722\,\frac{\mathrm{N}}{\mathrm{m}} & 0.414\,\frac{\mathrm{N}}{\mathrm{rad}} & -0.48\,\frac{\mathrm{N}}{\mathrm{rad}}\\ 0.414\,\mathrm{N} & 1.37\cdot 10^{-3}\,\frac{\mathrm{Nm}}{\mathrm{rad}} & -6.41\cdot 10^{-5}\,\frac{\mathrm{Nm}}{\mathrm{rad}}\\ -0.48\,\mathrm{N} & -6.41\cdot 10^{-5}\,\frac{\mathrm{N\cdot m}}{\mathrm{rad}} & 3.79\cdot 10^{-3}\,\frac{\mathrm{N\cdot m}}{\mathrm{rad}}\end{bmatrix} \quad \text{and}$$

$$\mathbf{D}_{est} = \begin{bmatrix} 22.89 \cdot 10^{-3}\,\frac{\mathrm{N \cdot s}}{\mathrm{m}} & 61.42 \cdot 10^{-5}\,\frac{\mathrm{N \cdot s}}{\mathrm{rad}} & -36.79 \cdot 10^{-5}\,\frac{\mathrm{N \cdot s}}{\mathrm{rad}} \\ 61.42 \cdot 10^{-5}\,\mathrm{N \cdot s} & 30.15 \cdot 10^{-9}\,\frac{\mathrm{N \cdot s \cdot m}}{\mathrm{rad}} & -67.56 \cdot 10^{-13}\,\frac{\mathrm{N \cdot s \cdot m}}{\mathrm{rad}} \\ -36.79 \cdot 10^{-5}\,\mathrm{N \cdot s} & -67.56 \cdot 10^{-13}\,\frac{\mathrm{N \cdot s \cdot m}}{\mathrm{rad}} & 56.46 \cdot 10^{-9}\,\frac{\mathrm{N \cdot s \cdot m}}{\mathrm{rad}} \end{bmatrix}.$$

According to [7] the mean stiffness value in the y-direction is 1238 $\frac{\mathrm{N}}{\mathrm{m}}$ with a standard deviation of 973 $\frac{\mathrm{N}}{\mathrm{m}}$. The estimated stiffness value of 1722 $\frac{\mathrm{N}}{\mathrm{m}}$ in the y-direction found in the upper left corner of $\mathbf{K}_{est}$ fits well in that range. The modal parameters of the estimated system can be found in Table 2.

Table 2. Modal parameters of the estimated system

Parameters	1st Eigenmode	2nd Eigenmode	3rd Eigenmode
Eigenfrequencies	$f_1 = 1753$ Hz	$f_2 = 2783$ Hz	$f_3 = 3862$ Hz
Eigenvectors	$\phi_1 = \begin{bmatrix} 4.7\ \mathrm{nm} \\ -22.1\ \mu\mathrm{rad} \\ 0.76\ \mu\mathrm{rad} \end{bmatrix}$	$\phi_2 = \begin{bmatrix} -13.1\ \mathrm{nm} \\ -2.1\ \mu\mathrm{rad} \\ -15.6\ \mu\mathrm{rad} \end{bmatrix}$	$\phi_3 = \begin{bmatrix} 34.3\ \mathrm{nm} \\ 0.83\ \mu\mathrm{rad} \\ -10.3\ \mu\mathrm{rad} \end{bmatrix}$
Damping ratios	$\zeta_1 = 0.12$	$\zeta_2 = 0.14$	$\zeta_3 = 0.18$

To interpret the eigenvectors, we can multiply their α- and γ-components by the height of the stapes (approximately 3 mm) to derive the motion of the stapes head in the x-, y-, and z-directions. Based on this information it can be concluded that in the first eigenmode the α-motion is dominant, in the second the γ-motion is dominant, and in the third the piston-like y-motion is dominant.

Within the frequency range, where our system was estimated (400 Hz to 5 kHz), the maximum displacement amplitude in the y-direction is 148 nm, which is lower than the expected displacement amplitude of 150 nm measured with acoustic excitation at 100 dB SPL as shown in [5]. This means that the displacement amplitudes used in this study fit within the physiological range.

It was shown in [6] that applying a centric force orientated in the lateral direction to the stapes footplate moves the stapes head laterally and in the posterior-inferior direction, which means that the posterior-inferior corner of the stapes footplate is the stiffest region of the AL. This type of spatial motion corresponds to the signs of the stiffness coefficients estimated here, which supports the inhomogeneous stiffness distribution of the AL. The same results were observed in [3], where the morphometry of an intact human AL along the full boundary of the annular rim was conducted via multiphoton microscopy revealing cross sections of the AL resembling a sandwich-composite structure. For this structure, the main effect on the variation in bending stiffness originates from thickness variations of the core layer, and not the variation in thickness of the face layers. While the total thickness and core thickness are highest around the posterior and anterior poles, the AL is found to be narrowest in the posterior-inferior region, and widest in the anterior region.

There are several reasons why the measured and estimated curves in Fig. 5 do not match precisely. We assume a three DOF system, however, due to the shaker excitation, which is qualitatively different to an acoustic excitation used in [5], in-plane motions, especially x- and y-translations, can become too large to be neglected and may cause estimation errors. Thus, a further investigation of a stapes-AL system with 5 and 6 DOF may provide more accurate results. In addition, the inaccuracies in alignment of the stapes reference frame with respect to the plastic plate and to the 3D-LDV system lead to further evaluation errors. In order to reduce those uncertainties, the mentioned error sources should be systematically investigated.

3 Conclusion

In this study a non-contact measuring and system identification procedure for evaluating inhomogeneous stiffness and damping characteristics of the AL is introduced. What makes this procedure unique is that it estimates the stiffness and damping coefficients in the physiological amplitude and frequency range without the application of large static external forces that can cause unnatural displacements of the stapes. To verify the procedure, measurements were first conducted on a steel beam. Then measurements on an individual human cadaveric temporal bone sample were performed. The estimated results support the inhomogeneous stiffness and damping distribution of the AL and are in a good agreement with the multiphoton microscopy results obtained in [3] where it was concluded that the posterior-inferior corner of the stapes footplate is the stiffest region of the AL. The procedure itself, however, requires further investigation with regard to the mentioned error sources.

References

1. von Békésy, G.: Experiments in Hearing, pp. 1514–1523. McGraw-Hill, New York (1960)
2. Gan, R.Z., Yang, F., Zhang, X., Nakmali, D.: Mechanical properties of stapedial annular ligament. Med. Eng. Phys. **33**, 330–339 (2011)
3. Schär, M., Dobrev, I., Chatzimichalis, M., Röösli, C., Sim, J.H.: Multiphoton imaging for morphometry of the sandwich-beam structure of the human stapedial annular ligament. Hear. Res. (2019). https://doi.org/10.1016/j.heares.2018.11.011
4. Helmholtz, H.: Die Mechanik der Gehörknöchelchen und des Trommelfells, pp. 34–37. Max Cohen & Sohn, Bonn (1869)
5. Lauxmann, M., Sim, J.H., Chatzimichailis, M., Heckeler, C., Ihrle, S., Huber, A., Eiber, A.: In-plane motions of the stapes in human ears. J. Acoust. Soc. Am. **132**, 3280–3291 (2012)
6. Lauxmann, M., Eiber, A., Haag, F., Ihrle, S.: Nonlinear stiffness characteristics of the annular ligament. J. Acoust. Soc. Am. **136**, 1756–1767 (2014)
7. Waller, T.-S.: Dynamic measurement of the circular stapes ligamentum using electrostatic forces. Ph.D. thesis, Faculty of Medicine of the Bavarian Julius Maximilian University of Würzburg, pp. 28–29 (2002)
8. Sim, J.H., Puria, S., Steele, C.R.: Calculation of the inertial properties of the malleus-incus complex from micro-CT imaging. J. Mech. Mater. Struct. **2**(8), 1515–1524 (2007)

Comparison of Measured EMG Data with Simulated Muscle Actuations of a Biomechanical Human Arm Model in an Optimal Control Framework – Direct Vs. Muscle Synergy Actuation

Marius Obentheuer[1(✉)], Michael Roller[1], Staffan Björkenstam[2], Karsten Berns[3], and Joachim Linn[1]

[1] Fraunhofer Institute for Industrial Mathematics, Fraunhofer Platz 1, Kaiserslautern, Germany
{Marius.Obentheuer,Michael.Roller, Joachim.Linn}@itwm.fraunhofer.de
[2] Fraunhofer-Chalmers Centre, Chalmers Science Park, Gothenburg, Sweden
staffan.bjorkenstam@fcc.chalmers.se
[3] Robotics Research Lab, TUK, Kaiserslautern, Germany
berns@rhrk.uni-kl.de

Abstract. We developed a biomechanical digital human model (DHM) simulation framework that uses (synergetic) Hill type muscles as actuators and optimal control (OC) for motion generation. In this work, we start investigating the underlying actuation signals of the Hill type muscles. We have set up a *weight lifting test* ('biceps curls') in the motion lab, where we measure the muscle activation via electromyography (EMG). The via muscles actuated simulation model produces human like trajectories for different types of OC cost functions, whereas the underlying muscle actuations strongly differ from each other. Our first results indicate that a muscle synergy actuation is more robust concerning the variation of activation signals and that a specific mix of cost functions preserves the resulting motion behavior while producing more human like actuation signals.

1 Introduction

There is an increasing demand from industry for tools that includes models of humans in the simulation process, and predicts human motions and behavior. This would facilitate doing e.g. ergonomic evaluations or assembly simulations in an early stage of development, when still working with digital prototypes. Modelling and controlling a digital human model (DHM) is a challenging task, that comprises multiple sub problems. Beside finding a modelling approach which represents the human body in an appropriate way, the question of how to generate motions, fulfill a given task and control that model in general might be even more challenging. Due to the high number of joints in the human body, each task can be fulfilled by a multitude of solutions (kinematical redundancy). Additionally, the number of muscles in the human body is

A. Kecskeméthy and F. Geu Flores (Eds.): ECCOMAS 2019, COMPUTMETHODS 53, pp. 26–33, 2020.
https://doi.org/10.1007/978-3-030-23132-3_4

much larger than the number of joints, which means that the same kinematical motion can be fulfilled by different muscle actuations (anatomical redundancy), making work with biomechanical DHMs that include muscle models more complex.

One approach to overcome this complexity is to use tracked (samples) of motions of a real human and map this motion capture data on a digital manikin. This can deliver good and realistic kinematical motions when a physical prototype exists or when the digital scenario is very similar to the one used to record the motion(s). However, when it comes to motions of new scenarios, these approaches are limited to what has been measured before. Additionally, as these approaches mainly work with a pure kinematical representation, inner forces or joint torques, which are important quantities for e.g. ergonomic evaluations, have to be estimated in a post processing step, often with additional software tools. Further on, it might be dangerous to produce good looking but unrealistic motions, e.g. when loads of simulated and recorded use cases differ. Predictions, based on inner forces and strains estimated in a post-processing step, would then be wrong.

Another approach is to simply position the manikin manually, meaning that every joint angle has to be specified by the user. Generating realistic positioning in this way would require expert knowledge of ergonomists, and would be very time consuming even for simple use cases. Moreover, the result is user dependent and limited to pure postural assessment, giving no information on timings and inner loads.

We developed a biomechanical dynamic DHM modelling approach that uses optimal control (OC) for motion generation, which we already presented in earlier publications [4, 7]. The human body is modeled as a multibody system that can be actuated by joint torques and Hill type muscles, which can be activated individually or by synergetic groupings of muscles (see Sect. 2).

To validate our simulations results, we compare them to measurements taken in a motion lab (motion capture and EMG). In earlier investigations, we focused on human arm motions and could show that a muscle and muscle synergy actuation predict the characteristics of measured human reaching motions very well, and has several advantages compared to a joint torque actuation [5].

In this work, we focus on the muscle activation signals, which are important quantities for ergonomic assessments. We have set up a *weight lifting test* in the motion lab, where test persons perform "biceps curls" (elbow flexion) with different weights, hand orientations and from different start configurations (shoulder joint angles), as described in Sect. 3.

In Sect. 4, we simulate the weight lifting with the same weights and motion constraints as measured in the lab, and compare the outcome of direct muscle actuation with a muscle synergy actuation for different OC cost functions.

2 Control Approach

We model the human body as multibody system (MBS), where bones are modeled as rigid bodies, connected by joints, which can be restricted to the human range of motion. The model is actuated either by joint torques (AM-T) or by simplified Hill type muscles. The muscles are force elements with a force-length and a force-velocity dependency representing the natural characteristics of human muscle force generation. Muscles are connected to the rigid bodies and create pulling forces between their attachment points (straight line). To approximate human muscle paths, they can be led over via points.

The muscles can be actuated directly (AM-M) or by using predefined groupings of muscles (AM-S), so called time invariant muscle synergies, where each muscle is present in each synergy with a fixed amplitude that does not change over time.

This approach is derived from the muscle synergy hypothesis in neuroscience, which tries to explain how the human central nervous system (CNS) might simplify motor control [2]. A multitude of studies have investigated muscle synergies on the input space side (actuation/EMG signals) and show that a subset of synergies is capable to reproduce measured EMG signals from experiments with a variety of different motions [1, 2, 4]. Recent studies also showed that muscle synergies are suitable to control biomechanical models [6], or as an underlying control principle for humanoid robots [3].

We develop a novel combination of optimal control, dynamic biomechanical (muscle) modelling and muscle synergy actuation, which allows us to investigate and exploit the properties of the different elements of our approach in order to generate human like motions in a predictive manner. Actuation signals for motion generation are calculated by the OC framework simply by defining the start and the end configuration of the MBS. The joint torques (AM-T), muscle actuations (AM-M), and weights for the muscles synergies (AM-S), which lead to the resulting trajectories, velocities and accelerations that achieve the defined "goal" are pure outcomes of the OC framework, and depend on the minimized OC cost function. "… Optimal control models of biological movement have explained behavioral observations on multiple levels of analysis …" [9], and are a promising approach for DHM control.

In contrast to other approaches working with a reduced number of degree of freedom (DOF) or implemented muscles in their model, our base is a fully dynamic, seven DOF human arm model where each human muscle (head) is modelled as Hill type muscle. Additionally, our simulation models are adapted to the anthropometry of the test persons, to allow a better comparison of measured and simulated data.

3 The Weight Lift Test – Motion Lab Experimental Setup

The weight lift test is designed to investigate EMG signals of (fore)arm muscles at elbow flexions with different loads and arm orientations, with a reduced set of kinematical DOF's, to focus on anatomical redundancy. The test persons perform a flexion of the elbow joint at three different shoulder angles Ω (Test Scenario TS1- 3, see Fig. 1), each with two forearm orientations (neutral and supinated *-N/S*). In TS 1, one additional measurement, without restriction of the forearm orientation is measured (free *-f*). Each task (Shoulder angle Ω and forearm orientation) is performed with five different weights *(W0 = 0 kg, W1 = 1,75 kg, W2 = 4 kg, W3 10 kg and W4 = 14 kg)* and repeated three times. Weights are released in between repetitions. Between tasks performed with high loads (*W2-W4*), the test subject takes a longer time to rest to avoid fatigue. At the start position (and before each repetition), the arm is stretched and muscles are relaxed (see e.g. Fig. 1 *right*). The test subject is advised not to change forearm orientation (supination/pronation) and to keep the shoulder and elbow joint in position while executing the motion.

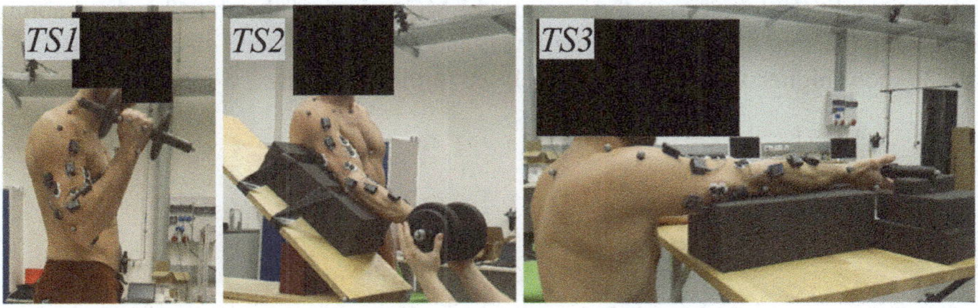

Fig. 1. The three different test scenarios TS1-3 of the weight lift test: Shoulder angle Ω 0° (left), 45° (middle) and 90° (right).

Motions should be done in a "controlled" manner, with no further specification for motion speed given. EMG Sensors are placed as depicted in Fig. 2, with a special focus on the dominant elbow flexors (Biceps, Brachialis, Brachioradialis), where sensors are positioned according to [8]. One field of attention here is the correlation of biceps short head and long head. To what extent the CNS can actuate different muscle heads individually is an important question when it comes to biomechanical modelling (and control).

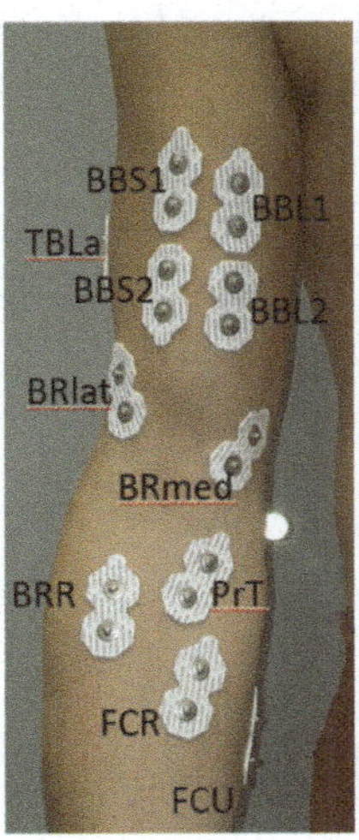

Short name	Muscle
BBS1	M. biceps brachii (short head)
BBS2	M. biceps brachii (short head)
BBL1	M. biceps brachii (long head)
BBL2	M. biceps brachii (long head)
BRlat	M. brachialis lateral
BRmed	M. brachialis medial
BRR	M. brachioradialis
PrT	M. pronator teres
FCR	M. flexor carpi radialis
FCU	M. flexor carpi ulnaris
TBLo	M. triceps brachii long (not depicted) []
TBLa	M. triceps brachii lateral

Fig. 2. EMG sensor placement, short names and muscles

When each muscle head is modelled as single Hill type muscle (as done in our model), the controller can independently exploit the mechanical properties of each head. This might not be the case in humans and could therefore lead to "wrong" muscle actuation values. If a clear correlation of motor unit recruitment of short and long head exists (and can be measured via surface EMG), this could be considered by the controller, e.g. by using muscle synergies. In Fig. 3, the EMG values of the short- and the long head of the biceps muscle of *TS1-f* for different loads are depicted. At motions without weight (*W0*, first row) the measured EMG values of the short head are about a factor 3 higher than those of the long head, and the lower placed electrodes (BBS2 & BBL2) show higher values than those closer to the shoulder.

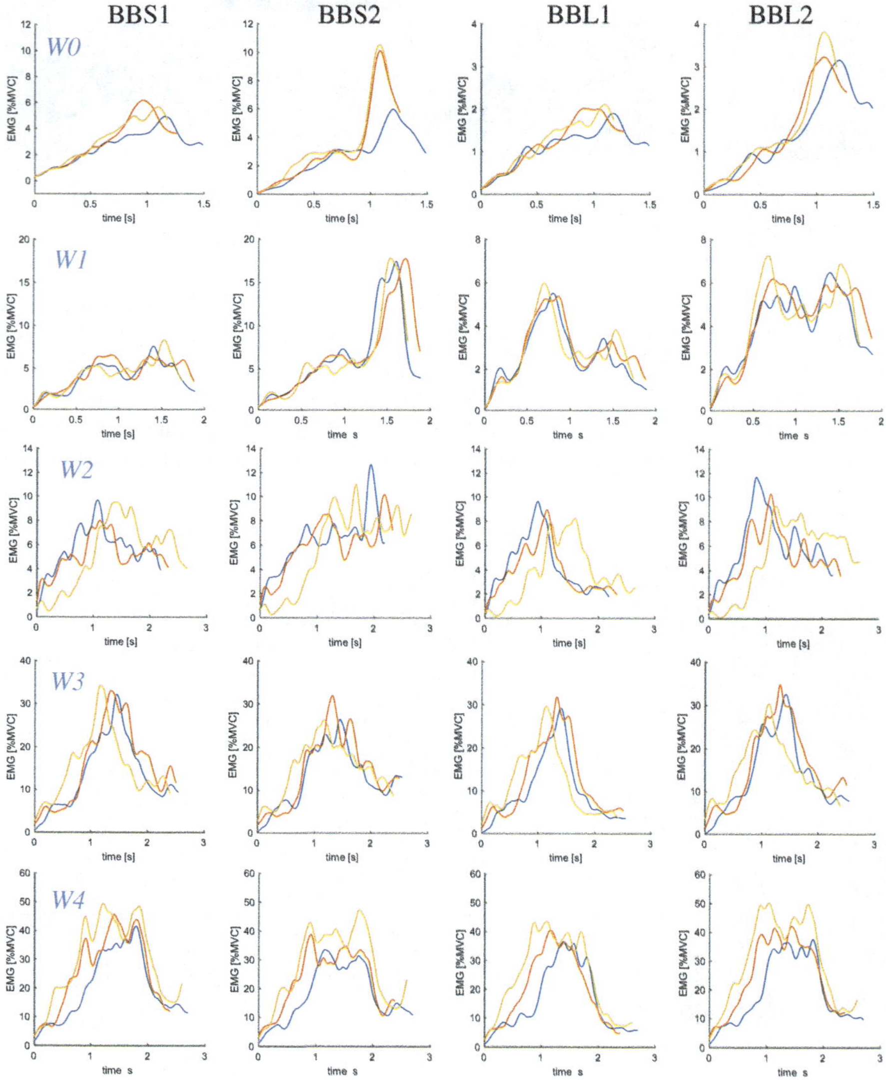

Fig. 3. EMG values (filtered and normalized to MVC) of TS1-f of the short head (BBS column 1&2) and the long head (BBL, column 3&4) of the biceps muscle (2 electrodes each). Each row shows the measured values for a certain weight (w0-w4), and in each plot, the values for the three repetitions are depicted (blue, red and yellow lines), including electromechanical delay (EMD) and grasping the handle.

This observation is preserved for *W1* in moderated form (Note that y-axis values of BBS and BBL differ in rows one and two (*W0* and *W1*)). For higher loads (*W2*-*W4)*, the measured maximum values and the characteristic of the EMG signal become very similar for each muscle head and electrode.

4 The Weight Lift Test – Simulation Setup

In Fig. 4, the simulated motions for Test scenario *TS1-f-W0* are depicted. The start configuration of the MBS is given, and the goal is to lift the wrist over a certain height. The via muscles actuated model produces motions which are very similar to those measured in the lab, confirming our observations made in earlier tests [5].

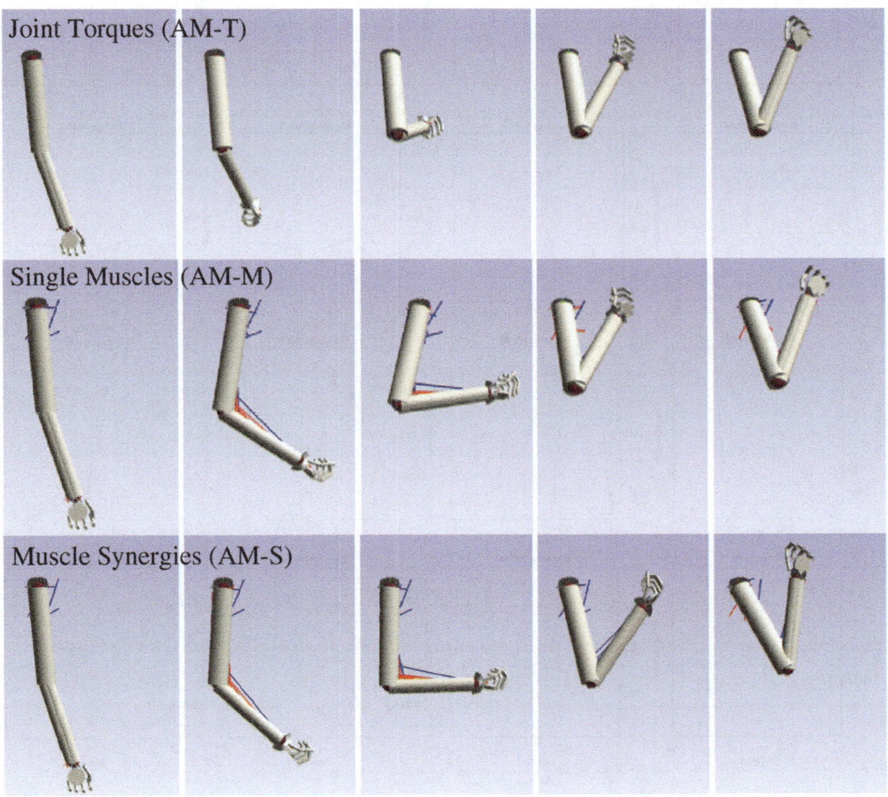

Fig. 4. Simulated motion sequence of TS1-f-W0 (forearm orientation not restricted).

In AM-T, the forearm is moved inwards and pronated. In AM-M and AM-S in contrast, the forearm is supinated during the motion (as observed in the motion lab) and both motions look very "human like", whereas the via AM-S produce motion looks a bit smoother and the elbow joint is turned outwards less. However, both motions are very similar to the measured motions, which is a necessary condition to compare measured and simulated muscle actuations. Note that the muscle synergies used in these simulations were extracted at the *basic reaching test* [4], which indicates them a certain generality. In Fig. 5, the simulated muscle activations ($a \in [01]$) for *TS1-f-W0* and different OC cost functions are depicted. In all cases, the resulting trajectories are similar, with different velocity profiles. When time is minimized (first row), the controller makes use of the maximal muscle forces, which leads to a "bang-bang" actuation in AM-M which is moderated in AM-S, but still the actuation reaches 1 (100%).

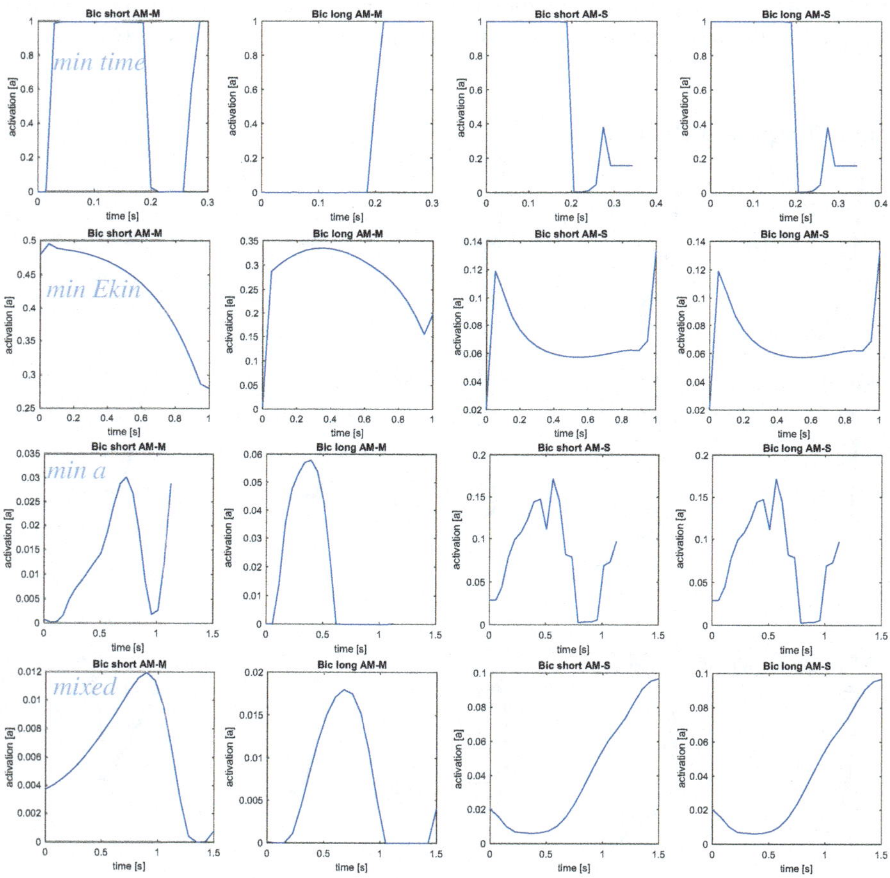

Fig. 5. Activation level a ($a \in [01]$) of the Hill type muscles of the biceps long -and short head for a direct muscle actuation (AM-M, column 1&2) and muscle synergy actuation (column 3&4) for simulation of TS1-f-W0. Different OC cost functions (rows).

Minimizing kinetic energy (second row) leads to smother activation curves with a maximum activation of 50% in AM-M and about 14% in AM-S, which is getting closer to the measured maximum values (12% MVC, compare Fig. 3 BBS2). Minimizing the activation a (third row) shows smooth curves in AM-M with a good match of max values for the long head of the biceps (around 5%, Bic long AM-M vs BBL Fig. 3), but short head values are very low. The maximum activations in AM-M are less influenced and remain almost constant (14% min Ekin vs.16% min a). The last row shows the resulting muscle activations when minimizing a certain mixed OC cost function that deliver human like motions and speed profiles in earlier tests. Activation levels in AM-M are smooth but very low, whereas maximum activation levels in AM-S (10%), as well as the characteristics of the activation curve, resemble measured values quite well.

In Conclusion, predicting EMG signals remains a challenging tasks, especially if the motion framework or controller should be valid for a wide range of different motions. First results of our framework show that the OC cost function heavily influences the muscle activation signals, although the resulting trajectories remain similar. Muscle synergy actuation appears to be more robust to these variations. Measurements of biceps long- and short head support observations of other studies that the CNS does not activate both heads in a one to one correlation, as implemented in our muscle synergy model (but also not completely independent, as in AM-M). This will be investigated in further studies, as well as load cases with higher weights.

Acknowledgments. This work was supported by the Fraunhofer Int. Programs under Grant N. MAVO 828 424. We acknowledge the use of the facilities and the support at the Chair of Applied Dynamics motion lab at FAU Erlangen.

References

1. d'Avella, A., Lacquaniti, F.: Control of reaching movements by muscle synergy combinations. Front. Comput. Neurosci. **7**(42), 2013 (2013). https://doi.org/10.3389/fncom.2013.00042
2. Bizzi, E., Cheung, V.C.K., d'Avella, A., Saltiel, P., Tresch, M.: Combining modules for movement. Brain Res. Rev. **57**(1), 1–270 (2008)
3. Cimolato, A., Piovanelli, E., Bortoletto, R., Menegatti, E., Pagello, E.: Muscle synergies for reliable NAO arm motion control: an online simulation with real-time constraints. In: 2016 IEEE International Conference on Simulation, Modeling, and Programming for Autonomous Robots (SIMPAR), San Francisco, CA, pp. 191–196 (2016)
4. Obentheuer, M., Roller, M., Björkenstam, S., Berns, K., Linn, J.: Human like motion generation for ergonomic assessment - a muscle driven digital human model using muscle synergies. In: ECCOMAS Multibody Dynamics, Prague (2017)
5. Obentheuer, M., Roller, M., Björkenstam, S., Berns, K., Linn, J.: Comparison of different actuation modes of a biomechanical human arm model in an optimal control framework. In: IMSD Joint International Conference on Multibody System Dynamics, Lisbon (2018)
6. Razavian, R., Ghannadi, B., McPhee, J.: Control of Musculoskeletal Arm Model Using Muscle Synergy. International Society of Biomechanics, Glasgow, UK (2015)
7. Roller, M., Björkenstam, S., Linn, J., Leyendecker, S.: Optimal control of a biomechanical multibody model for the dynamic simulation of working tasks. In: ECCOMAS, Prague (2017)
8. Staudemann, D., Taube, W.: Brachialis muscle activity can be assessed with surface electromyography. J. Electromyogr. Kinesiol. **25**, 199–204 (2014)
9. Todorov, E.: Optimality principles in sensorimotor control. Nat. Neurosci. **7**(9), 907–915 (2004)

A Detailed Kinematic Multibody Model of the Shoulder Complex After Total Shoulder Replacement

Sven Herrmann[1], Märuan Kebbach[2], Robert Grawe[3], Kelsey Kubiak[1], Katrin Ingr[1], Rainer Bader[2], and Christoph Woernle[3](✉)

[1] Institute for Biomechanics, Trauma Center Murnau, Prof.-Küntscher-Str. 8, 82418 Murnau, Germany
sven.herrmann@bgu-murnau.de

[2] Department of Orthopedics, University Medicine Rostock, Doberaner Strasse 142, 18057 Rostock, Germany
{maeruan.kebbach,rainer.bader}@med.uni-rostock.de

[3] Chair of Technical Dynamics, University of Rostock, Justus-von-Liebig-Weg 6, 18059 Rostock, Germany
{robert.grawe,woernle}@uni-rostock.de

Abstract. Multibody modeling allows reproducible and comparative analyses of shoulder dynamics after total shoulder replacement (TSR). For inverse dynamics an accurate representation of the musculoskeletal kinematics from motion capture is fundamental. Although current optimization-based approaches effectively identify the underlying skeletal motion in terms of position, velocities and accelerations are not consistently provided. The purpose was therefore to introduce a multibody model of the shoulder complex after TSR implantation with detailed representation of the muscle apparatus that directly generates the musculoskeletal kinematics from motion capture data. The inherent inverse kinematics problem is resolved by implementation of a potential field method. Sensitivity analysis was performed to determine the model's tracking capability. Scapular motion showed overall good agreement with measurements from the literature. The approach yields equivalent results to current approaches with the benefit of directly computing accelerations without formulation of an optimization problem. The multibody model presented will be used for further inverse dynamics analyses regarding various loading conditions of different TSR designs.

1 Introduction

The main reasons for failure of total shoulder replacement (TSR) were found to be instability until joint dislocation, rotator cuff or deltoid deficiency, joint stiffness and implant loosening [5]. These clinical implications are directly related to TSR dynamics. While faulty positioning and misalignment of the implant components as well as the condition of muscle structures were considered to play a

A. Kecskeméthy and F. Geu Flores (Eds.): ECCOMAS 2019, COMPUTMETHODS 53, pp. 34–42, 2020.
https://doi.org/10.1007/978-3-030-23132-3_5

key role in joint stability [16], the underlying causes have still not been identified. In this context, inverse dynamics approaches based on multibody modeling of the shoulder [15] were shown to be capable of reproducing physiological TSR loading conditions [10, 14].

For inverse dynamics analyses an accurate representation of the musculoskeletal geometry and kinematics is fundamental. Kinematic multibody models have been frequently used to translate measured skin marker data from motion capture into joint angles by optimization techniques [1,2,7,8]. This approach was shown to effectively identify the underlying skeletal motion at the position level while reducing inherent errors such as soft tissue artifacts [3,4,13]. However, velocities and accelerations are not provided impeding a consistent inverse dynamics analysis. Furthermore, the representation of the shoulder muscle apparatus is frequently excluded although its functionality could already be assessed at the kinematics level.

The purpose of this work is therefore to introduce a multibody model of the shoulder complex after TSR implantation that directly generates the musculoskeletal kinematics from motion capture data. In contrast to previous optimization-based approaches, the inherent inverse kinematic problem is resolved by a potential field method [18]. By a detailed representation of the muscle apparatus of upper body and extremity the model can be directly extended for subsequent dynamic analyses.

2 Materials and Methods

The multibody model is presented including the formulation of the mapping of motion capture data with the potential field method. The computational implementation is described, and the model is assessed with respect to its tracking capabilities.

2.1 Model Topology

The model consists of a multibody system with four moving rigid segments and overall 13 joint coordinates arrayed in vector $\boldsymbol{q} \in \mathbb{R}^{13}$. The kinematic chain begins at the thorax, assumed to be ground-fixed, followed by the clavicula, the scapula, the humerus and the lower arm modeled as one rigid segment (Fig. 1). The humeroulnar joint is modeled by a revolute joint, and both the sternoclavicular and acromioclavicular joints by spherical joints [4]. The glenohumeral joint is represented by a kinematic sub-chain consisting of three revolute joints with co-intersecting axes and three in pairs orthogonal prismatic joints.

The kinematic chain is supported between the thorax and scapula by means of two holonomic constraints allowing the scapula to glide over the thorax approximated by an ellipsoid [15]. Each constraint ensures that a scapular point P_j remains on the ellipsoidal surface with coordinates given in vector $\boldsymbol{r}_{e_j}$. The constraints are formulated in terms of $\boldsymbol{q}$ at the position, velocity and acceleration levels ($j = 1, 2$),

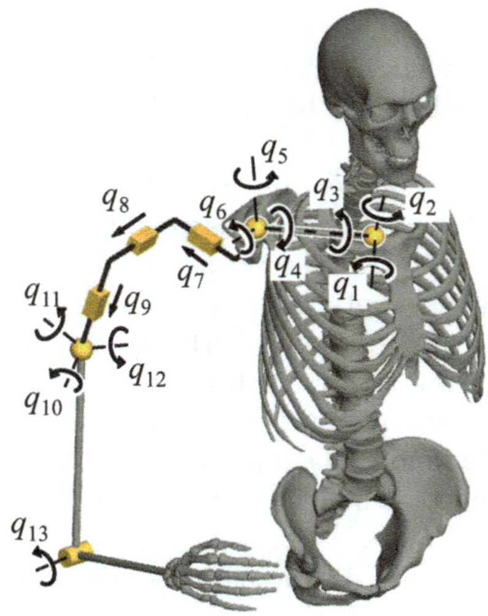

Fig. 1. Multibody model topology with illustration of all relevant segments and the joint coordinates $\boldsymbol{q}$.

$$g_j \equiv \boldsymbol{r}_{e_j}^{\mathrm{T}} \boldsymbol{A} \boldsymbol{r}_{e_j} - 1 = 0 \quad \text{with} \quad \boldsymbol{r}_{e_j} = \boldsymbol{r}_{P_j}(\boldsymbol{q}) - \boldsymbol{r}_E, \tag{1}$$

$$\dot{g}_j \equiv \boldsymbol{G}_j \dot{\boldsymbol{q}} = 0 \quad \text{with} \quad \boldsymbol{G}_j = 2 \boldsymbol{r}_{e_j}^{\mathrm{T}} \boldsymbol{A} \boldsymbol{J}_{P_j}, \tag{2}$$

$$\ddot{g}_j \equiv \boldsymbol{G}_j \ddot{\boldsymbol{q}} + \bar{\eta}_j = 0 \quad \text{with} \quad \bar{\eta}_j = 2 \left(\dot{\boldsymbol{q}}^{\mathrm{T}} \boldsymbol{J}_{P_j}^{\mathrm{T}} \boldsymbol{A} \boldsymbol{J}_{P_j} + \boldsymbol{r}_{e_j}^{\mathrm{T}} \boldsymbol{A} \dot{\boldsymbol{J}}_{P_j} \right) \dot{\boldsymbol{q}}, \tag{3}$$

where $\boldsymbol{A} \in \mathbb{R}^{3,3}$ denotes the ellipsoidal semi-axes summarized in a diagonal matrix, $\boldsymbol{r}_E \in \mathbb{R}^3$ the position vector of the ellipsoid, $\boldsymbol{J}_{P_j} \in \mathbb{R}^{3,13}$ the Jacobian matrix of point P_j and the row vector $\boldsymbol{G}_j \in \mathbb{R}^{1,13}$ the constrained direction at point P_j. With 13 joint degrees of freedom and two kinematic loops, the overall number of degrees of freedom is 11 according to the Chebychev-Grübler-Kutzbach criterion.

2.2 Mapping Motion Capture Data to the Multibody Model

The inherent inverse kinematics problem is defined by finding consistent joint coordinates $\boldsymbol{q}(t)$ for a given set of skin marker trajectories such that the model kinematics replicates the recorded subject motion. In practice, it is sought after to match measured skin markers M_i at each time step with corresponding model points R_i. The present distance s_i and distance rate $\dot{s}_i$ can then be expressed in terms of the joint coordinates $\boldsymbol{q}$ and marker trajectories $\boldsymbol{r}_{M_i}(t)$ by

$$s_i = \sqrt{\boldsymbol{s}_i^{\mathrm{T}} \boldsymbol{s}_i} \quad \text{with} \quad \boldsymbol{s}_i = \boldsymbol{r}_{R_i}(\boldsymbol{q}) - \boldsymbol{r}_{M_i}(t), \tag{4}$$

$$\dot{s}_i = \frac{1}{s_i} \boldsymbol{s}_i^{\mathrm{T}} \dot{\boldsymbol{s}}_i \quad \text{with} \quad \dot{\boldsymbol{s}}_i = \boldsymbol{J}_{R_i} \dot{\boldsymbol{q}} - \dot{\boldsymbol{r}}_{M_i}(t), \tag{5}$$

where matrix $\boldsymbol{J}_{R_i} \in \mathbb{R}^{3,13}$ represents the Jacobian matrix of reference point R_i. In case of ideal replication, both distance and distance rate vanish, i.e. $s_i = \dot{s}_i = 0$. However, tracking deviations normally occur due to soft tissue artifacts [3,11] and model assumptions concerning segment scaling and skin marker placement [2].

Skeletal kinematics is generated by a potential field that transfers measured skin marker trajectories onto the model. The potential field is represented by spring elements connecting the markers M_i with the segment-fixed reference points R_i, complemented by dampers for asymptotically stable tracking behavior. Likewise, glenohumeral translations are restrained by a bushing element connecting the humeral head with an additional marker defined in the global reference frame that mimics the spatial motion of the glenoid rotation center.

Given n spring-dampers including the bushing element, the generalized spring-damper forces related to the joint coordinates $\boldsymbol{q}$ can be expressed as

$$\boldsymbol{\tau}^{\mathrm{s}} = \boldsymbol{J}^{\mathrm{T}}\boldsymbol{B}\left[\boldsymbol{C}\,\boldsymbol{s} + \boldsymbol{D}\,\dot{\boldsymbol{s}}\right] \tag{6}$$

with the transpose of the global Jacobian $\boldsymbol{J}^{\mathrm{T}} = [\boldsymbol{J}_{R_1}^{\mathrm{T}} \ldots \boldsymbol{J}_{R_n}^{\mathrm{T}}] \in \mathbb{R}^{13,3n}$, the unit direction vectors of the spring-damper forces described in matrix $\boldsymbol{B} \in \mathbb{R}^{3n,n}$ and spring and damping constants provided in diagonal matrices $\boldsymbol{C} \in \mathbb{R}^{n,n}$ and $\boldsymbol{D} \in \mathbb{R}^{n,n}$, respectively. Spring deflections and deflection rates obtained from Eqs. (4) and (5) are summarized in vectors $\boldsymbol{s} \in \mathbb{R}^{n}$ and $\dot{\boldsymbol{s}} \in \mathbb{R}^{n}$, respectively. The equations of motion are then given as a set of differential algebraic equations with the 13 joint coordinates $\boldsymbol{q}$ and two reaction force coordinates $\boldsymbol{\lambda}$,

$$\begin{bmatrix} \boldsymbol{M} & -\boldsymbol{G}^{\mathrm{T}} \\ -\boldsymbol{G} & \boldsymbol{0} \end{bmatrix} \begin{bmatrix} \ddot{\boldsymbol{q}} \\ \boldsymbol{\lambda} \end{bmatrix} = \begin{bmatrix} \boldsymbol{\tau}^{\mathrm{s}} + \boldsymbol{\tau}^{\mathrm{c}} \\ \bar{\boldsymbol{\eta}} \end{bmatrix}, \tag{7}$$

where $\boldsymbol{M} \in \mathbb{R}^{13,13}$ denotes the mass matrix, $\boldsymbol{G} = [\boldsymbol{G}_1\ \boldsymbol{G}_2]^{\mathrm{T}} \in \mathbb{R}^{2,13}$ the matrix of the two constraints (2) and $\boldsymbol{\tau}^{\mathrm{c}} \in \mathbb{R}^{13}$ the generalized centrifugal and Coriolis forces.

2.3 Model Implementation and Model Assessment

Marker trajectories were obtained from motion analyses of one healthy subject (male, 27 a, 177 cm, 68 kg) with 21 attached skin markers. The recordings included a static posture at neutral position as well as five motion cycles of full arm flexion and abduction. The measured trajectories were processed using a second-order Butterworth filter [1] with a cut-off frequency of 2 Hz and cubic spline interpolation. After motion capture, MRI data of the subject's upper extremity including the attached skin markers were recorded.

The model topology was set up in the multibody software SIMPACK v9.7 (Dassault Systèmes Deutschland GmbH, Gilching, Germany). Skeletal geometries were derived from the male Visible Human dataset [17] using reconstructing techniques. Joint rotation centers and axes were obtained by fitting spheres or cylinders into articulating surfaces. Local reference frames as well as rotation sequences were defined according to [19]. The bone segments were uniformly scaled onto the subject according to the recorded MRI scans. Inertia properties of the bones were set to unit values ($m = 10^{-3}$ kg, $I_{xx} = I_{yy} = I_{zz} = 10^{-6}$ kg m^2) to accomplish the considered task of marker tracking. The position of the reference points R_i were calibrated using the MRI data and the static posture scenario. Overall 22 point-to-point force elements represent the spring-dampers and the bushing element. An anatomic TSR was virtually implanted. For calculation of the muscle paths the attachment sites of 48 muscles were derived from anatomic descriptions and further discretized into uniform point distributions using GEOMAGIC v12 (3D Systems, Rock Hill, SC, USA). Overall 152 muscle paths were modeled as straight line elements. Muscle wrapping was incorporated either as segment-fixed points or using spherical deflection where needed.

Sensitivity analysis was performed to determine the impact of the spring-damper elements on the model's tracking capability. Spring constants c_i were uniformly altered between 0.01 and 100 N/m with damper-spring ratios $k = d_i/c_i$ varying from 0 to 10 s. The spring-damper values of the bushing element were set as a thousandfold higher throughout this study. The overall mean of spring deflections and the mean norm of the joint accelerations $\ddot{\boldsymbol{q}}$ served as output parameters. After finding adequate spring-damper values, the tracking deviations between skin marker trajectories and segment-fixed reference points were evaluated for each scenario [2]. Finally, scapular motion was compared to measurements based on electromagnetic tracking [12] and transcortical bone pins [9]. In each simulation test, the model of Eq. (7) was integrated by means of a RADAU5 integrator.

3 Results and Discussion

3.1 Sensitivity Analysis

Spring deflections consistently reached convergence with spring constants beyond 0.5 N/m indicating the greatest achievable tracking quality (Fig. 2). The level of convergence started to rise for damper-spring ratios above 0.1. The mean norm of joint accelerations reached minimum values around $c_i = 2.5$ N/m and

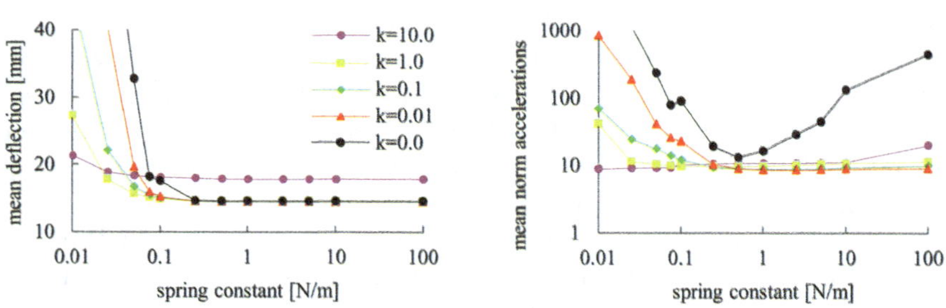

Fig. 2. Sensitivity analysis of spring constants c_i and damper-spring ratio k on overall mean spring deflection (left) and mean norm of joint accelerations (right).

$k = 0.01$ s showing no benefit in terms of continuous accelerations for lower or higher damping. Hence, uniform constants of $c_i = 2.5$ N/m and $d_i = 0.025$ Ns/m were considered to be sufficient for subsequent analyses.

By using small damping values, the model can be interpreted to minimize the potential energy of the virtual springs. By this it is equivalent to standard approaches using least-squares optimization with uniform weighting factors in the objective function [1,2,7,8]. It was mentioned that non-uniform weighting may achieve better tracking results when emphasizing markers near the M. deltoid insertion and the humeral epicondyles [3]. This option was here not tested but can be taken into account by increasing the spring constants connected to the relevant markers.

3.2 Tracking Deviations

For all scenarios, the bone segments consistently followed the given marker trajectories with minor tracking deviations (Fig. 3). The mean deviations were at 7.1 ± 2.2 mm in the static neutral posture which increased to 14.5 ± 7.6 mm for full arm flexion and 14.6 ± 7.7 mm for abduction. Maximum deviations were recorded with values of 24.1 ± 11.8 mm for flexion and 21.2 ± 10.6 mm for abduction.

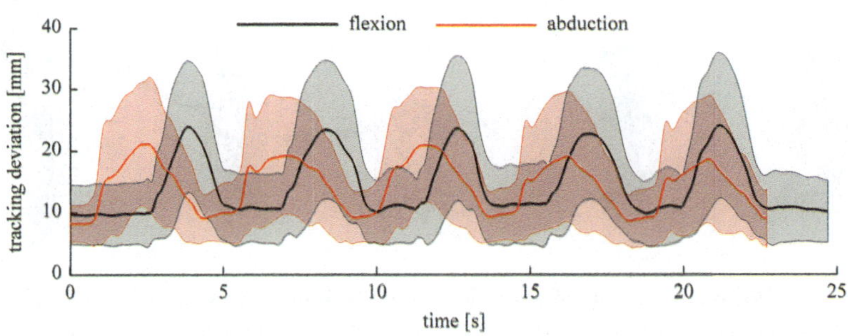

Fig. 3. Tracking deviations (mean ± standard deviation) between marker trajectories and segment-fixed reference points during five motion cycles of full arm flexion and abduction.

As the segment-fixed reference points were also calibrated based on the static posture, the tracking deviations observed at neutral could be related to scaling and marker placement errors [2]. The considerable increase during flexion and abduction scenarios, however, can be mainly attributed to soft tissue artifacts. This is in line with the notion that these artifacts are highly task-specific [3]. Comparison between marker placement defined at neutral and reevaluated at full arm elevation even revealed that mean tracking deviations can rise up to 87 mm with respect to the actual scapula positioning [11].

3.3 Kinematic Analysis

For both flexion and abduction scenarios, the incorporated muscle apparatus demonstrated realistic muscle paths over the entire ranges of motion (Fig. 4). Individual paths were shown to be effectively deflected by the implemented muscle wrapping. Hence, penetration with bones or adjacent muscles was avoided. Simulated motion reproduced scapular positioning well following major trends of published data [9,12] throughout both full arm scenarios (Fig. 5). The simulation slightly underestimated internal and downward rotation showing a smaller curvature for internal rotation. These differences can result from discrepancies in the established local reference frames and variations in the execution of the scenarios.

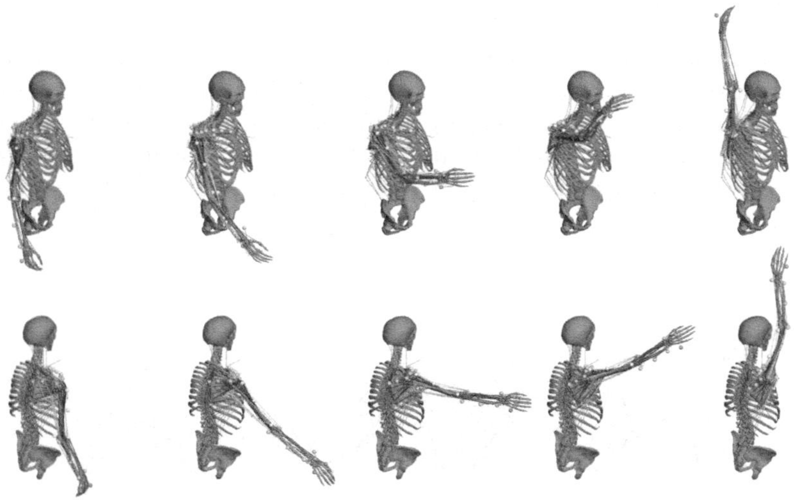

Fig. 4. Multibody simulation of shoulder complex kinematics after TSR implantation during full arm flexion (top) and abduction (bottom). The muscle apparatus is represented by straight line elements including muscle wrapping. Small spheres indicate current marker positions.

Optimization-based approaches were shown to replicate comparable kinematics at the position level to measurements based on intracortical bone pins [3,4,13]. The presented comparison implies that the implemented potential field method generates equivalent results to these approaches. However, an advantage is seen in the fact that accelerations, needed for inverse dynamics simulation [10,14,15], are directly computed without formulation of an optimization problem.

Furthermore, glenohumeral translations were restrained by the implemented bushing element to values below 0.2 mm. Hence, the modeled glenohumeral joint approximates a spherical joint which is frequently assumed in inverse dynamics analysis [10,14,15]. However, this assumption does not reflect physiological

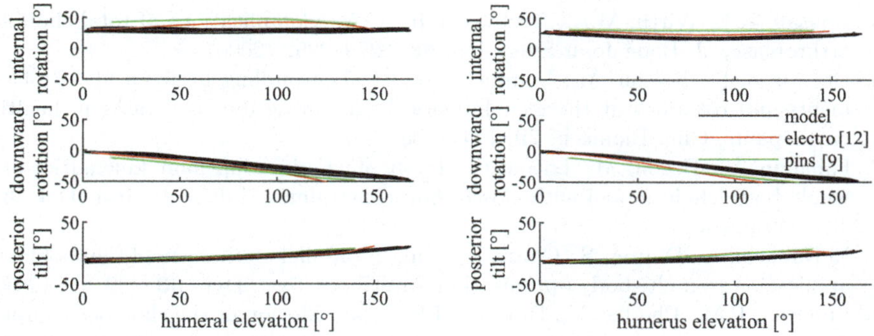

Fig. 5. Simulated scapular positioning with respect to the thorax for full arm flexion (left) and abduction (right) plotted against humeral elevation in comparison to measurements based on electromagnetic tracking [12] and transcortical bone pins [9]. Rotations were defined according to [19].

measures in vivo [6]. A more accurate description may be achieved by replacing the bushing element with adequate kinematic constraints [8] in future studies.

4 Conclusion

Our presented multibody model reliably simulated the musculoskeletal kinematics for full arm flexion and abduction based on motion capture data. This will be used for further inverse dynamics analyses regarding various loading conditions of different TSR designs.

Acknowledgments. The authors would like to thank the Deutsche Forschungsgemeinschaft (WO 452/11-1, BA 3347/14-1 and HE 7885/1-1) for supporting the research work presented, and Mr. Hannes Wackerle, Trauma Center Murnau, for providing the motion capture data.

References

1. Andersen, M.S., Damsgaard, M., Rasmussen, J.: Kinematic analysis of over-determinate biomechanical systems. Comput. Methods Biomech. Biomed. Eng. **12**, 371–384 (2009)
2. Andersen, M.S., Damsgaard, M., MacWilliams, B., et al.: A computationally efficient optimisation-based method for parameter identification of kinematically determinate and over-determinate biomechanical systems. Comput. Methods Biomech. Biomed. Eng. **13**, 171–183 (2010)
3. Begon, M., Dal Maso, F., Arndt, A., et al.: Can optimal marker weightings improve thoracohumeral kinematics accuracy? J. Biomech. **48**, 2019–2025 (2015)
4. Begon, M., Bélaise, C., Naaim, A., et al.: Multibody kinematics optimization with marker projection improves the accuracy of the humerus rotational kinematics. J. Biomech. **62**, 117–123 (2017)

5. Bohsali, K.I., Wirth, M.A., Rockwood Jr., C.A.: Complications of total shoulder arthroplasty. J. Bone Joint Surg. Am. **88**, 2279–2292 (2006)
6. Dal Maso, F., Raison, M., Lundberg, A., et al.: Coupling between 3D displacements and rotations at the glenohumeral joint during dynamic tasks in healthy participants. Clin. Biomech. **29**, 1048–1055 (2014)
7. Fohanno, V., Begon, M., Lacouture, P., et al.: Estimating joint kinematics of a whole body chain model with closed-loop constraints. Multibody Syst. Dyn. **31**, 433–449 (2014)
8. El Habachi, A., Duprey, S., Chèze, L., et al.: A parallel mechanism of the shoulder - application to multi-body optimisation. Multibody Syst. Dyn. **33**, 439–451 (2015)
9. Ludewig, P.M., Phadke, V., Braman, J.P., et al.: Motion of the shoulder complex during multiplanar humeral elevation. J. Bone Joint Surg. Am. **91**(2), 378–389 (2009)
10. Martins, A., Quental, C., Folgado, J., et al.: Computational reverse shoulder prosthesis model: experimental data and verification. J. Biomech. **48**, 3242–3251 (2015)
11. Matsui, K., Shimada, K., Andrew, P.D.: Deviation of skin marker from bone target during movement of the scapula. J. Orthop. Sci. **11**(2), 180–184 (2006)
12. Meskers, C.G., Vermeulen, H.M., de Groot, J.H., et al.: 3D shoulder position measurements using a six-degree-of-freedom electromagnetic tracking device. Clin. Biomech. **13**, 280–292 (1998)
13. Naaim, A., Moissenet, F., Duprey, S., et al.: Effect of various upper limb multibody models on soft tissue artefact correction: a case study. J. Biomech. **62**, 102–109 (2017)
14. Nikooyan, A.A., Veeger, H.E., Westerhoff, P., et al.: Validation of the Delft Shoulder and Elbow Model using in-vivo glenohumeral joint contact forces. J. Biomech. **43**, 3007–3014 (2010)
15. Quental, C., Folgado, J., Ambròsio, J., et al.: A multibody biomechanical model of the upper limb including the shoulder girdle. Multibody Syst. Dyn. **28**, 83–108 (2012)
16. Sperling, J.W., Hawkins, R.J., Walch, G., et al.: Complications in total shoulder arthroplasty. J. Bone Joint Surg. Am. **95**(6), 563–569 (2013)
17. Spitzer, V., Ackerman, M.J., Scherzinger, A.L., et al.: The visible human male: a technical report. J. Am. Med. Inform. Assoc. **3**, 118–130 (1996)
18. Wallrapp, O., Grund, T., Böhm, H.: Human motion analysis and dynamic simulation of rowing. In: Proceedings of 2005 ECCOMAS Thematic Conference Multibody Dynamics, Madrid, Spain (2005)
19. Wu, G., van der Helm, F.C., Veeger, H.E., et al.: ISB recommendation on definitions of joint coordinate systems of various joints for the reporting of human joint motion-Part II: shoulder, elbow, wrist and hand. J. Biomech. **38**, 981–992 (2005)

Multibody Analysis of a 3D Human Model with Trunk Exoskeleton for Industrial Applications

Elisa Panero[1(✉)], Giovanni Gerardo Muscolo[1], Laura Gastaldi[2], and Stefano Pastorelli[1]

[1] DIMEAS-Department of Mechanical and Aerospace Engineering, Politecnico di Torino, c.so Duca degli Abruzzi 24, Turin, Italy
{elisa.panero,giovanni.muscolo, stefano.pastorelli}@polito.it
[2] DISMA-Department of Mathematical Sciences, Politecnico di Torino, c.so Duca degli Abruzzi 24, Turin, Italy
laura.gastaldi@polito.it

Abstract. This paper presents a multibody analysis of a 3D human model with a wearable exoskeleton for trunk support in industrial applications. Multibody computational models reveal to be a suitable solution for investigating the human-exoskeleton interaction, for properly developing the mechanical design and for analyzing the effect on human joints loads. With the final aim of developing a proper wearable device for the support of human trunk during manual lifting tasks, a 3D multibody model of human body interacting with an exoskeleton has been implemented. Different position of exoskeleton assistance joints and two assistance levels are investigated in terms of human waist and hip joints loads reduction and interface forces at contact points. The results comparison allows defining maps of most suitable and critical areas for exoskeleton joints positioning.

1 Introduction

During last decades, robotics and wearable technologies have reached a fundamental role in several fields. In industry, due to the numerous cases of work-related musculoskeletal diseases and risks, the interest on the maintenance and the promotion of workers' well-being has become a crucial objective [1].

Collaborative robots and exoskeletons are promising solutions for the emergent challenges in workers' safety and health. Nevertheless, the developing and the application of these technologies need particular attention, in order to investigate, design and characterize the human robot interaction [15]. Concerning industrial exoskeletons, several passive and powered devices [2, 5] have been proposed both as commercial product [18] and as research prototype [11], to support human trunk, lower and upper body segments. However, many challenges are still open.

A. Kecskeméthy and F. Geu Flores (Eds.): ECCOMAS 2019, COMPUTMETHODS 53, pp. 43–51, 2020.
https://doi.org/10.1007/978-3-030-23132-3_6

One of the main problems deals with the correct alignment of rigid exoskeleton structure with the human body. In addition, the fact that usually devices can be used by several workers and can be adjusted to different size increases the design difficulty. To overcome the misalignment of the rigid structure during movements, some mechanical arrangements and combination of passive degree of freedoms (DOFs) have been proposed for lower and upper limb exoskeletons [3, 8].

Concerning the exoskeleton for trunk supports, the complexity of human spine and vertebras' DOFs requires more investigations. Previous studies and commercial devices align the exoskeleton assisting joints with the human hips. Moreover, the majority of human biomechanical models developed for evaluating human joints loads during flexion-extension with assisting device approximates the human trunk as one rigid segment linked to the lower legs by hip joints [7].

Computational approach seems to be an encouraging solution for investigating interaction between human and exoskeleton to support the definition of the mechanical design, the assistance strategy of the device, and for analyzing the effect on human joints loads [9]. Compared to the "*try&error*" method, the model-based method allows advantages in terms of cost, time and analysis [14].

The aim of the present research deals with the development of a 3D multibody analysis of human body wearing an external device. In particular, the development of a 3D human multibody model, the modelling of an exoskeleton system for trunk support and the investigation of their interaction is described. Starting from alignment with human hip joints as standard configuration, different positions of the exoskeleton joints are simulated. The resulting human joints loads and interface forces are analyzed.

2 The 3D Multibody Model

In the following sections, a description of the 3D multibody models of the human body and of the exoskeleton is reported. The multibody model has been implemented in Matlab Simscape Multibody software.

2.1 Human Body Model

The human body is modeled as a multibody model composed by 15 rigid links. Segments are connected by 14 joints, with a total of 34 DOFs. Model data are assumed according to the anthropometrics of a 50 percentile Italian man (body mass 75 kg, body height 1.71 m). Table 1 shows a sketch of the human manikin in frontal and sagittal planes, body segment lengths, mass distribution and center of masses positions [10, 19]. The trunk is divided in two different links: pelvis and torso, connected by a three rotational DOFs joint (waist), while hip joints link the pelvis to lower limbs with a three rotational DOFs joint. The waist approximates the last lumbar vertebra of the human spine. This is the point of human spine mainly involved during flexion-extension motion. It also corresponds to the local zone most affected by low back pain caused by excessive joint loads. In addition to the biomechanical importance, the introduction of waist joint allows differentiating lifting motion strategies of stoop and squat. The joints range of motion and the orientation of local reference axes are defined according to ISB standard [17]. Standing posture with upper limbs aligned to the trunk is assumed as neutral reference position (images in Table 1), while flexion angles are positive and extension ones negative. The model has been validated in a previous study [12].

Table 1. Human body anthropometry and inertial parameters

Human body 3D model	Segment name		Segment length (m)	Segment breadth (m)	% Distance of COM from proximal end	% Total mass
body mass= 75 Kg body height= 1.71 m 	Head + Neck		0.19	0.15	56	8.4
	Torso	Lumbar	0.41	0.22	42	22
		Thorax				15
	Pelvis		0.10	0.29	74	13
	Upperarm		0.32	/	45	2.8
	Forearm		0.25	/	42	1.7
	Hand		0.19	0.09	40	0.6
	Thigh		0.48	0.18	43	10
	Shank		0.41	0.12	41	4.3
	Foot		length 0.26 height 0.05	0.10	43	1.4

2.2 Exoskeleton and Interface Model

The exoskeleton is modeled as a structure with a total mass of 2.5 kg, distributed in four rigid elements: trunk support, pelvis belt and two thigh supports (Fig. 1a). Hinge joints J_1 connect pelvis belt and each thigh support, hinge joints J_2 connect each thigh support to the trunk support. The assistance of the exoskeleton is modeled as torque applied in joints J_2, named assistance joints. The human-exoskeleton interfaces are modeled as kinematic connections at trunk and thighs, they all allow translation in the vertical direction and rotation in the sagittal plane. Figure 1a depicts the human body wearing the exoskeleton with J_1-J_2 aligned to human hip joints.

3 Simulation Procedure and Analysis

The simulation reproduces a trunk flexion-extension motion of 6 s and it consists of flexion (2s), grasping of 15 kg external mass (2s), and lifting (2s) phases. Kinematics of human joints describes the inputs, while dynamics at human joints and contact points represent the outputs. The Matlab solver was selected as variable-step ode15s, based on differential algebraic equations (DAEs). Two strategies are simulated: stoop (bending with extended legs) and squat (bending with flexed legs). Table 2 shows the graphical representation of motions in these two cases and the maximum joint angles for lower limbs and trunk. The arms kinematics is equal for both cases (80° flexed shoulders in grasp and 60° flexed elbows in lifting).

The exoskeleton joint position and two levels of assistance were selected as the changing variables. As suggested by literature and commercial devices, the alignment of exoskeleton joints to the human hip joints is selected as standard configuration. From the starting co-axial solution (X = 0, Y = 0), a parameter set of 200 joints configuration is defined, as depicted in Fig. 1b [13]. The translation is delimited in a circle (0.19 m radius) and it is mapped with specific and homogeneous order. The simulation starts from the point in the first circle, with coordinates: X = 0.01 m, Y = 0 m and continues in the same circle with a counterclockwise direction. The 21st point has the same Y coordinate as the first point but translated in the second circle (X = 0.03 m, Y = 0 m), until the last ones (X = 0.19 m, Y = 0 m). Two constant levels of exoskeleton assistance were considered: 20 Nm and 40 Nm for each side (total assistance of 40 Nm and 80 Nm, respectively). These values correspond to 20% and 40% of maximum extension waist torque during stoop motion, lifting an external load of 15 kg without the exoskeleton (~ 210 Nm).

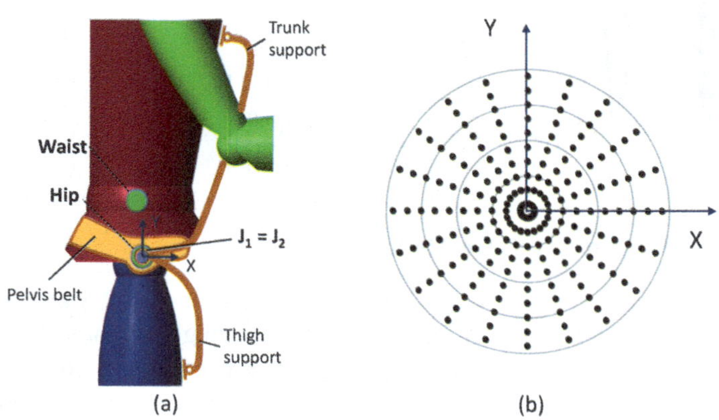

Fig. 1. (a) Sketch of the human model with exoskeleton: waist and hip joints, exoskeleton joints J_1 and J_2 coaxial to hip joints. Reference frame XY; (b) maps of exoskeleton joints positions.

The dynamic results of maximum forces at interface points (trunk and thigh) and maximum torques at human joints (waist and hip) have been considered as an overall performance index. To evaluate human effort, the maximum torques from simulation with exoskeleton in different configurations were compared to ones obtained in simulation of human body without exoskeleton. Some researchers are working on the definition of standard guidelines for industrial exoskeleton development [6, 16], but there is not a specific limit for pressure due to the device. Considering that the limit of skin capillarity pressure of 32 mmHg (4.5 kPa) established in clinics and more recently updated to 47 mmHg (6 kPa) and the results of a recent study on wearable robotics [4], the maximum value of 15 kPa has been considered to avoid discomfort. A surface of 250 cm^2 was hypothesized for trunk and 150 cm^2 for thigh supports, identifying thresholds of 350 N for trunk and 200 N for thigh contact.

Table 2. Simulation input kinematics

Stoop lifting strategy				
Flexion	Grasping	Extension	Maximum joint angles in grasping phase	
			Waist	+100°
			Hip	-5°
			Knee	0°
			Ankle	-10°
Squat lifting strategy				
Flexion	Grasping	Extension	Maximum joint angles in grasping phase	
			Waist	+55°
			Hip	+80°
			Knee	+110°
			Ankle	+40°

4 Results

Peaks results refer to maximum forces and torques along the simulation with the joint J_2 in a specific point, while *values* refer to local maximum and minimum peaks among several simulations. Figure 2 shows the maximum peaks of torques (Nm) at human waist (black), hip (light blue), and the maximum peaks of interface forces (N) between human trunk (dots) and thighs (purple), and the device supports respect to the several J_2 positions (x-axis). Hip torques and thigh interface forces refer to one leg. Figure 3 shows the maps of critical (red) and suitable (green) areas for J_2 translations. The red circles depict where the interface pressure exceeds the limit previously selected. Positions that minimize human torques and interface forces are stressed using green dots and green circles, respectively. In both Figs. 2 and 3, graphs A-B refer to stoop with 20 and 40 Nm of torque assistance respectively, while graphs C-D refer to squat with 20 and 40 Nm of torque assistance respectively.

5 Discussion and Conclusion

From results, several aspects can be pointed out. First, comparing the graphs in Fig. 2, the two lifting strategies and the two assistance levels highlight different maximum peaks distribution. Curves have similar trend among the simulations, showing succession of local maximum and minimum values. In squat, local maximum forces correspond to local maximum torques, while in stoop local maximum forces correspond to local minimum torques. Possible reasons could be identified in the different relative motions of human body segments, the legs motions in squat and the different

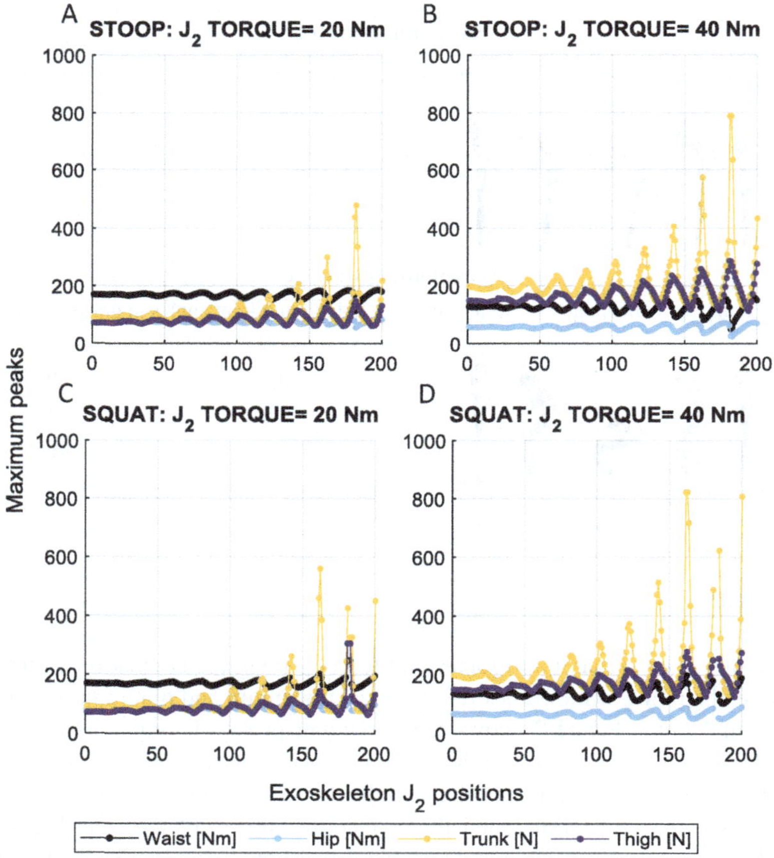

Fig. 2. Interfaces forces (trunk/thigh) and human joints torques (waist/hip) maximum peaks during stoop (A-B) and squat (C-D), related to J_2 positions. Graphs A-C refer to exo assistance of 20 Nm in each J_2 joint, graphs B-D to 40 Nm.

translations of exoskeleton contacts points. As expected, lower assistance (20 Nm for side) produces lower interface forces and only a few positions produce pressures with critical values. Nevertheless, also the reduction of human efforts is restricted. Analyzing the human torques, in some external joints conditions the maximum peaks result near to maximum extension torque calculated in simulation without exoskeleton (210 Nm for waist and 100 Nm for hip joint). The exoskeleton support is not effective in these cases.

With 40 Nm assistance for each side, the human torques are strongly reduced, but interface pressures exceed the selected thresholds in many J_2 positions. Moreover, even if some external positions reveal greater minimization for both forces and torques, the

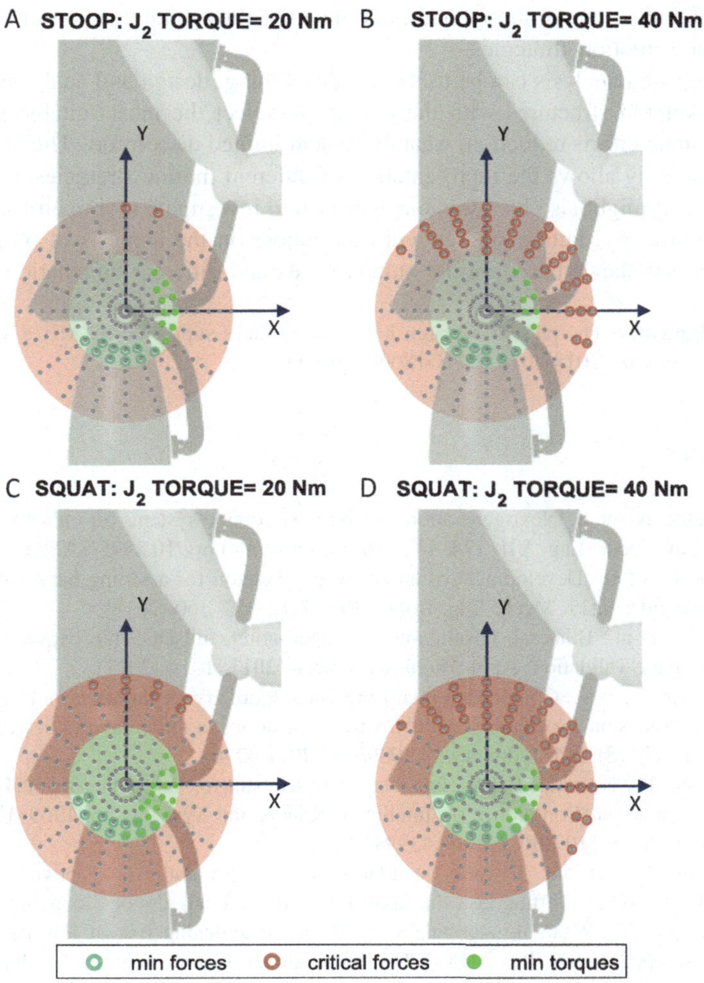

Fig. 3. Maps of optimal and critical zones based on variables in stoop (A-B) and squat (C-D). Green areas depict the suitable zone for the exo joint; the red ones underline the critical area. Graphs A-C refer to device assistance of 20 Nm in each J_2 joints, B-D to 40 Nm in each J_2 joints.

nearest local maximum results higher. Small movements of the device during task execution may move the joints in critical configurations. For all these reasons, external area of translations with radius greater than 0.1 m has been discarded, while the internal area results suitable for joint positions (Fig. 3).

Considering the exoskeleton joints translations inside the green area, all the solutions are suitable because they reduce the human joint efforts requiring lower hip and waist torques. Among them, the positions that allow greater global minimization of forces and torques are highlighted as green circles and green dots respectively. The global minimization has been defined as global minimum value plus 2.5% of that minimum for lower assistance level, while global minimum value plus 5% of that minimum for higher assistance level. Some possible solutions are shared between squat and stoop also considering different levels of assistance. Positioning the exoskeleton assistance joints in one of these configurations may allow minimization of torques and forces in both motion strategies.

The proposed analysis can be useful for developing, design and analysing different trunk exoskeleton structures with the attempt to select the most suitable solution in terms of human efforts reduction, wearability and limited discomfort. The 3D model of total human body allows the representation of different motion strategies, not only the stoop lifting. It might also be a suitable approach to investigate asymmetrical trunk and legs lifting strategy. Future studies will concentrate on the analysis of proper law of exoskeleton assistance based on the human needs and different kinematic motions.

Acknowledgments. The present research has been partially supported by MIUR grant Dipartimenti di Eccellenza 2018-2022 (E11G18000350001).

References

1. Deshpande, N., et al.: Next-generation collaborative robotic systems for industrial safety and health. Saf. Secur. Eng. VII. **174**, 187 (2018). https://doi.org/10.2495/SAFE170181
2. Durante, F., et al.: Development of an active exoskeleton for assisting back movements in lifting weights. Int. J. Mech. Eng. Robot. Res. **7**(4), 353–360 (2018)
3. Junius, K., et al.: Bilateral, misalignment-compensating, full-DOF hip exoskeleton: design and kinematic validation. Appl. Bionics Biomech. **2017**, 1–14 (2017)
4. Kermavnar, T., et al.: Computerized cuff pressure algometry as guidance for circumferential tissue compression for wearable soft robotic applications: a systematic review. Soft Robot. **5**(1), 1–16 (2018). https://doi.org/10.1089/soro.2017.0046
5. Kobayashi, H., Nozaki, H.: Development of muscle suit for supporting manual worker. In: IEEE International Conference on Intelligent Robots and Systems, pp. 1769–1774 (2007). https://doi.org/10.1109/IROS.2007.4399412
6. Lazzaroni, M., et al.: Towards standards for the evaluation of active back-support exoskeletons to assist lifting task. In: Sixth International Congress of Bioengineering (2018)
7. Luo, Z., Yu, Y.: Wearable stooping-assist device in reducing risk of low back disorders during stooped work. In: 2013 IEEE International Conference on Mechatronics and Automation, IEEE ICMA 2013, pp. 230–236 (2013). https://doi.org/10.1109/ICMA.2013.6617923

8. Malosio, M., et al.: Analysis of elbow-joints misalignment in upper-limb exoskeleton. In: 2011 IEEE International Conference on Rehabilitation Robotics, pp. 1–6. IEEE (2011)
9. Mombaur, K., et al.: Parameter optimization for passive spinal exoskeletons based on experimental data and optimal control. In: 2017 IEEE-RAS 17th International Conference Humanoid Robotics (Humanoids), pp. 535–540 (2017)
10. Muscolo, G.G., et al.: Biomechanics of human locomotion with constraints to design flexible-wheeled biped robots. In: 2017 IEEE International Conference on Advanced Intelligent Mechatronics (AIM), pp. 1273–1278. IEEE (2017). https://doi.org/10.1109/AIM.2017.8014193
11. Naf, M.B., et al.: Towards low back support with a passive biomimetic exo-spine. In: 2017 International Conference on Rehabilitation Robotics (ICORR), pp. 1165–1170. IEEE (2017)
12. Panero, E., et al.: Influence of hinge positioning on human joint torque in industrial trunk exoskeleton. In: 15h IFToMM World Congress (2019)
13. Panero, E., et al.: Model based analysis of trunk exoskeleton for human efforts reduction. In: International Conference on Robotics in Alpe-Adria Danube Region (2019)
14. Sartori, M., et al.: Neural data-driven musculoskeletal modeling for personalized neurorehabilitation technologies. IEEE Trans. Biomed. Eng. **63**(5), 879–893 (2016)
15. Schnieders, T.M., Stone, R.T.: Current work in the human-machine interface for ergonomic intervention with exoskeletons. Int. J. Robot. Appl. Technol. **5**(1), 1–19 (2017)
16. Toxiri, S., et al.: Towards standard specifications for back-support exoskeletons. In: Biosystems and Biorobotics (2019). https://doi.org/10.1007/978-3-030-01887-0_42
17. Wu, G., et al.: ISB recommendation on definitions of joint coordinate system of various joints for the reporting of human joint motion—part I: ankle, hip, and spine. J. Biomech. **35**(4), 543–548 (2002). https://doi.org/10.1016/S0021-9290(01)00222-6
18. http://en.laevo.nl/. Accessed 21 Nov 2018
19. ISO/TR 7250–2: Basic human body measurements for technological design. Part 2: Statistical summaries of body measurements from individual ISO populations (2010)

A Hill Muscle Actuated Arm Model with Dynamic Muscle Paths

Johann Penner(✉) and Sigrid Leyendecker

Chair of Applied Dynamics, University of Erlangen-Nuremberg,
Immerwahrstrasse 1, 91058 Erlangen, Germany
{johann.penner,sigrid.leyendecker}@fau.de

Abstract. This contribution presents the optimal control of a musculoskeletal multibody model with Hill muscle actuation and dynamic muscle paths. In particular, the motion of a human arm and its muscle paths is described via constrained variational dynamics. The optimal control problem in this work is based on the direct transcription method DMOCC [4], where the optimal control problem is discretised in time, and the resulting nonlinear constrained finite dimensional optimisation problem is solved. To take a step towards finding global or multiple minima, we utilize the MATLAB multistart framework for global optimisation. With the help of an example, we outline a framework to find feasible solutions and analyse several minima to which the nonlinear programming solver converges.

1 Introduction

In this work, we consider applications in which a musculoskeletal model is expected to perform certain motion tasks, while information about the corresponding required muscle activities, forces, and joint moments are of interest. Modelling this as an optimal control problem and approximating its solution numerically is a well suited simulation procedure to obtain such information. The optimal control problem comprises the minimisation of an objective function (e.g. related to the control trajectory or the motion of the system) subject to the fulfilment of constraints (e.g. the differential equation of motion or algebraic path constraints). For the numerical solution, we use the direct transcription method DMOCC to transcribe the optimal control problem into a general nonlinear program that adopts the properties of structure preserving integrators [3,4]. The corresponding large-scale nonlinear constrained optimisation problem is solved with deterministic algorithms like interior point (IP) or sequential quadratic programming (SQP) provided by `GlobalSearch`, `MultiStart` and `fmincon` from MATLAB. Furthermore, we use the open-source software tool CasADi for automatic differentiation [1].

In order to draw physiological conclusions from the simulations of the biomechanical model, the actuation is realized by the Hill-type muscle force. The essential idea of the Hill muscle model is to represent the force-length and force-velocity relations of real muscles. In our model, muscles are attached to the

A. Kecskeméthy and F. Geu Flores (Eds.): ECCOMAS 2019, COMPUTMETHODS 53, pp. 52–59, 2020.
https://doi.org/10.1007/978-3-030-23132-3_7

skeletal at the muscle's origin and insertion points and are connected by the shortest path without intersecting bones. Hence, if the positions and velocities of the skeletal bodies are known, the muscle's length and contraction velocity can be determined via a muscle wrapping model [6–8].

2 Biomechanical Model

All modelling in this work is based on Lagrangian mechanics using a variational principle. We begin with the application of the Lagrange-d'Alembert principle to constrained systems, which yields the constrained forced Euler-Lagrange equations. In addition, we use the nullspace method to reduce the system to minimum possible size [2,3].

The assumption that muscles and tendons are always under tension leads to the conclusion that they follow the path of minimum distance between origin and insertion point. This allows us to use variational calculus to describe the muscle path as a G1-continuous combination of geodesics on adjacent obstacle surfaces [7,8].

To actuate the multibody model representing the skeleton, a simple Hill type muscle model is used, see [6]. Here, the three dimensional muscle's force is characterised by the scalar muscle force value and the muscles path's direction at the muscle origin and insertion point, see Sect. 2.2.

2.1 Biomechanical Multibody System

Consider a b dimensional mechanical system in a configuration manifold with configuration vector $\boldsymbol{q}(t) \in \mathbb{R}^b$ and velocity vector $\dot{\boldsymbol{q}}(t) \in \mathbb{R}^b$, where t denotes the time variable in the interval $[t_0, t_N]$. The Lagrangian $L : \mathbb{R}^b \times \mathbb{R}^b \to \mathbb{R}$ for this system is composed of the difference between kinetic and the potential energy. Let the motion be restricted by c holonomic constraints $\boldsymbol{g} : \mathbb{R}^b \to \mathbb{R}^c$ requiring $\boldsymbol{g}(\boldsymbol{q}) = \boldsymbol{0}$. This means that the degrees of freedom of the system are reduced to $b - c$. Corresponding constraint forces then enforce the fulfilment of the constraints. Additionally, a non-conservative external force $\boldsymbol{f}(\boldsymbol{q}, \dot{\boldsymbol{q}}, \boldsymbol{A}) \in \mathbb{R}^b$ is acting on the system, where $\boldsymbol{A} : [t_0, t_N] \to [0, 1]^m$ is an external activation of m muscles. Here, the Lagrange-d'Alembert principle states

$$\delta S[\boldsymbol{q}, \boldsymbol{\lambda}] = \delta \int_{t_0}^{t_N} L(\boldsymbol{q}, \dot{\boldsymbol{q}}) - \boldsymbol{g}(\boldsymbol{q}) \cdot \boldsymbol{\lambda} \,\mathrm{d}t + \int_{t_0}^{t_N} \boldsymbol{f}(\boldsymbol{q}, \dot{\boldsymbol{q}}, \boldsymbol{A}) \cdot \delta \boldsymbol{q} \,\mathrm{d}t = \boldsymbol{0}$$

where the augmented action $S[\boldsymbol{q}, \boldsymbol{\lambda}]$ is a functional of $\boldsymbol{q}$ and the Lagrange multipliers $\boldsymbol{\lambda} \in \mathbb{R}^c$. The premultiplication of the resulting constrained forced Euler-Lagrange equations with the nullspace matrix $\boldsymbol{P}(\boldsymbol{q}) \in \mathbb{R}^{b \times (b-c)}$ eliminates the constraint forces including the Lagrange multipliers from the system. The resulting equations of motion read

$$\begin{aligned} \boldsymbol{P}^T(\boldsymbol{q}) \cdot \left[\frac{\partial L(\boldsymbol{q}, \dot{\boldsymbol{q}})}{\partial \boldsymbol{q}} - \frac{\mathrm{d}}{\mathrm{d}t} \left(\frac{\partial L(\boldsymbol{q}, \dot{\boldsymbol{q}})}{\partial \dot{\boldsymbol{q}}} \right) + \boldsymbol{f}(\boldsymbol{q}, \dot{\boldsymbol{q}}, \boldsymbol{A}) \right] &= \boldsymbol{0} \\ \boldsymbol{g}(\boldsymbol{q}) &= \boldsymbol{0} \end{aligned} \qquad (1)$$

For many applications it is possible to find a reparametrisation $\boldsymbol{q} = \boldsymbol{F}(\boldsymbol{u}) \in \mathbb{R}^b$ in terms of independent coordinates $\boldsymbol{u} \in \mathbb{R}^{b-c}$. Then the Jacobian $\mathrm{D}\boldsymbol{F}(\boldsymbol{u}) \in \mathbb{R}^{b\times(b-c)}$ of the coordinate transformation plays the role of a nullspace matrix [2,3].

2.2 Muscle Actuation

The Hill model in this work consists of a contractile component and a parallel linear elastic component, for further information and parameters see [6]. Here, the scalar muscle force $F^m \in \mathbb{R}$ of a single muscle m is calculated as a function of the activity of the muscle $A^m \in [0,1]$, the muscle length $\ell^m \in \mathbb{R}$ and the contraction velocity $v^m \in \mathbb{R}$, both given by the dynamic muscle path. Finally, the actuation force at the origin $\boldsymbol{f}_O^m = F^m \cdot \tilde{\boldsymbol{r}}_O^m$ and at the insertion $\boldsymbol{f}_I^m = F^m \cdot \tilde{\boldsymbol{r}}_I^m$ point is defined in the direction of the muscle path's (unit) tangent directions $\tilde{\boldsymbol{r}}_O^m, \tilde{\boldsymbol{r}}_I^m \in \mathbb{R}^3$, see Fig. 1.

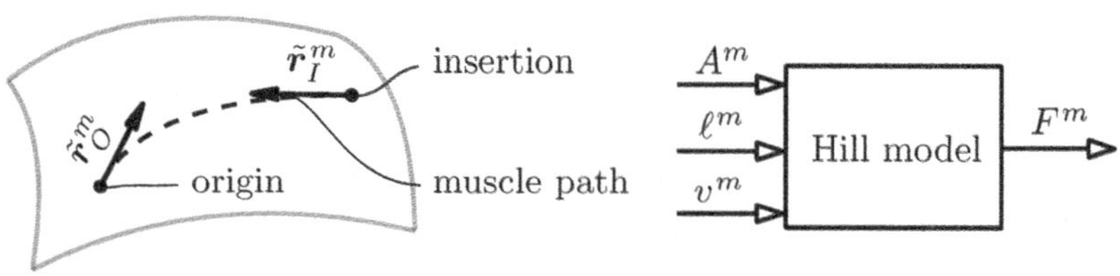

Fig. 1. Definition of the Hill muscle force and the muscle force direction.

2.2.1 Hill Muscle Model

The scalar force amount can be calculated as

$$F^m = f_l(l^m) \cdot f_v(v^m) \cdot A^m \cdot F^m_{max} + k_p \cdot (l^m - l_0) \tag{2}$$

where $k_p \in \mathbb{R}$ is the proportionality constant, $F^m_{max} \in \mathbb{R}$ is the maximal possible muscle force and $l_0 \in \mathbb{R}$ is the muscle slack length. In this model, $f_l(l^m) \in [0,1]$ is a function describing the force-length relation of the muscle and $f_v(v^m) \in [0,1.4]$ represents the Hill-hyperbola like force-velocity behaviour, see [6].

2.2.2 Dynamic Muscle Path

In order to describe the shortest connection between two points on a surface that represents a muscle passing smoothly over obstacle surfaces, we define the action

$$E[\gamma] = \int_{s_0}^{s_K} \left(\frac{1}{2} \left\| \gamma'(s) \right\|^2 - \phi(\gamma) \cdot \lambda \right) \mathrm{d}s$$

of a curve $\gamma \in C^2([s_0, s_K], \mathbb{R}^3)$ from the geodesic start point $\gamma(s_0)$ to the end point $\gamma(s_K)$. Therefore, we assume that the muscle completely touches the surface, thus the Lagrangian is augmented by a scalar valued function of holonomic surface constraints $\phi(\gamma) = 0 \in \mathbb{R}$ and Lagrange multipliers $\boldsymbol{\lambda} \in \mathbb{R}$. The local minimizers of E, that also locally minimize the length are the so called geodesics [9]. The variational principle $\delta E = 0$ yields that for γ to be a stationary point of E, the corresponding Euler-Lagrange equation

$$\begin{aligned} \gamma'' + \boldsymbol{\Phi}(\gamma)^T \cdot \lambda &= \mathbf{0} \\ \phi(\gamma) &= 0 \end{aligned} \tag{3}$$

has to hold, where $\boldsymbol{\Phi}(\gamma) = {}^{\partial\phi(\gamma)}/_{\partial\gamma} \in \mathbb{R}^{1\times 3}$ is the surface constraint Jacobian. For the sake of completeness, it should be noted that the Jacobian of a non-singular differentiable parametrization $\gamma = \boldsymbol{F}(\boldsymbol{\nu})$ with respect to the surface coordinates $\boldsymbol{\nu} \in \mathbb{R}^2$ can also be used to project (3) into the tangent space of the manifold defined by the surface constraint, compare with Eq. (1). This will be included in future work.

3 Optimal Control

The optimal control problem in this work is discretised with by direct transcription method DMOCC, using a variational integrator to constrain the optimisation problem. Within this formulation, all continuous quantities have to be approximated with discrete counterparts, see [3,4].

It is well known that the set of optimisation algorithms in `fmincon` from MATLAB does not guarantee that a global minimum is found. Also, the initial guess has a great influence on to the local minimum to which the algorithm converges. For these reasons, we first generate a feasible solution as initial guess for the subsequent multistart framework for global optimisation from MATLAB[1] [10]. This heuristic algorithm to find global minima yields several minima for the provided optimisation problem. While there is no guarantee of global optimality, this does increase the information content of the solution compared to `fmincon`.

3.1 Resulting Constrained Nonlinear Optimisation Problem

In the following, the discrete configuration vector $\boldsymbol{q}_d = \{\boldsymbol{q}_n\}_{n=0}^N$ is an approximation of the continuous configuration on a discrete time grid with constant time step $\Delta t \in \mathbb{R}$ and $N \in \mathbb{N}$ time nodes. Similarly, the muscle path is defined on an discrete arc length grid with fixed arc length fraction $\Delta s \in \mathbb{R}$ with $K \in \mathbb{N}$ nodes, leading to the discrete muscle path vector $\gamma_n = \{\gamma_n^k\}_{k=0}^K$ at the n-th time step. Moreover, the midpoint and trapezoidal rule is used to approximate the integrals in the variational principle. Altogether, this leads to the constrained

[1] https://de.mathworks.com/help/gads/how-globalsearch-and-multistart-work.html.

nonlinear optimisation problem

$$\min_{\boldsymbol{u}_d,\gamma_d,\boldsymbol{A}_d,\Delta t} J(\boldsymbol{u}_d,\gamma_d,\boldsymbol{A}_d,\Delta t) = \min_{\boldsymbol{u}_d,\gamma_d,\boldsymbol{A}_d,\Delta t} \sum_{n=0}^{N-1} C_d(\boldsymbol{u}_n,\boldsymbol{u}_{n+1},\gamma_n,\boldsymbol{A}_n,\Delta t) \quad \text{(4a)}$$

$$\text{subject to} \quad \begin{array}{l} \cdot \text{ discrete version of (1)} \\ \cdot \text{ discrete version of (3)} \\ \cdot \text{ initial and final condition} \\ \cdot \text{ path constraints} \end{array} \quad \text{(4b)}$$

where discrete Euler-Lagrange equations are used to contain the optimisation problem. Here, the discrete local coordinate $\boldsymbol{u}_d = \{\boldsymbol{u}_n\}_{n=1}^{N}$ from the nodal reparametrisation $\boldsymbol{q}_{n+1} = \boldsymbol{F}(\boldsymbol{u}_{n+1},\boldsymbol{q}_n)$, the discrete muscle path $\gamma_d = \{\{\gamma_n^k\}_{k=0}^{K}\}_{n=0}^{N}$, the sequence of muscle activities $\boldsymbol{A}_d = \{\boldsymbol{A}_n\}_{n=0}^{N-1}$ and the time step Δt are the optimisation variables.

3.2 Objectives

Within this work, the following objective functions are examined

$$J_{const} = 0 \qquad J_A(\boldsymbol{A}_d) = \sum_{n=0}^{N-1} \|\boldsymbol{A}_n\|^2 \qquad J_t(\Delta t) = N\Delta t \qquad (5)$$

which represent a constant function value, the so-called control effort and the manoeuvre time. While the control effort and the manoeuvre time are typical and widely used criteria for optimal motion, the constant function is used to find feasible solutions that serve as a initial guess.

3.3 Results

In the following example, the multibody system consists of two rigid bodies and a revolute joint, which represent the upper and lower arm as well as the

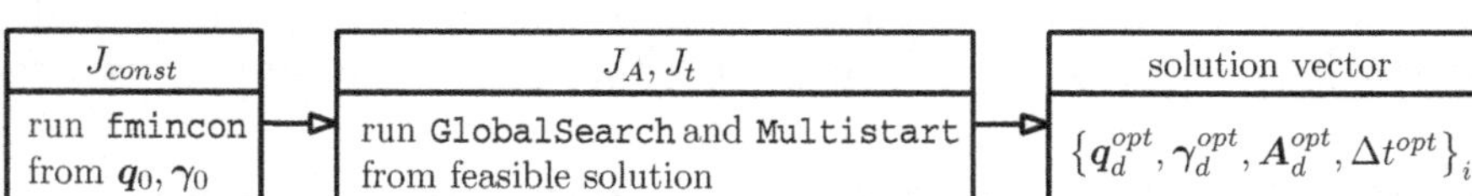

Fig. 2. Simulation procedure to finding global or multiple minima

Table 1. Optimisation results calculated using sequential quadratic programming (SQP) and interior point (IP) algorithms from MATLAB

Objective	fmincon		MultiStart		GlobalSearch	
	SQP	IP	SQP	IP	SQP	IP
J_A	0.0193	0.0193	0.0193	0.0193	0.0193	0.0069, 0.0193, 0.0391
J_t	0.4639	0.4639	0.4639	0.4639	0.4639, 0.6980	0.4639, 0.9220

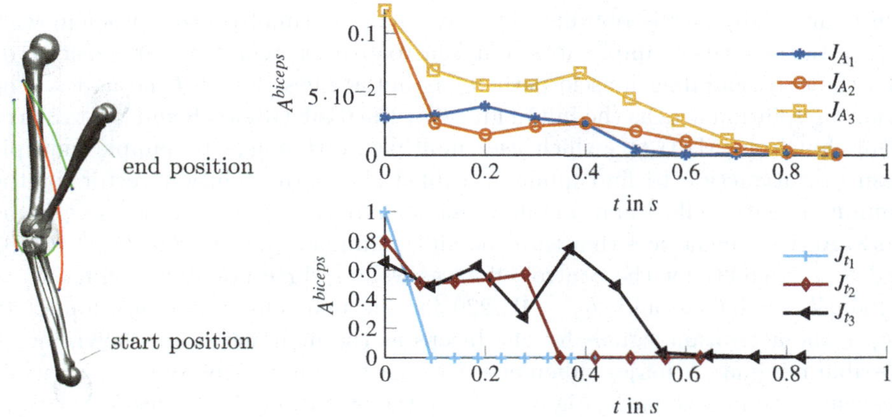

Fig. 3. Left: Start and end position of the rest-to-rest manoeuvre, **right top**: Activation sequence of the biceps for minima of J_A, **right bottom**: Activation sequence of the biceps for minima of J_t (For the 3d bone model see https://www.thingiverse.com/thing:1543880.)

elbow joint. A Hill muscle model as in [6] is used to model the musculus triceps and biceps. Moreover, the muscle path of the triceps is modeled around an ellipsoid and the biceps around two cylinders via a G1-continuous combination of geodesics, see [7]. The system is discretized with $K = 30$ arc length nodes and $N = 10$ time nodes, which leads to 1285 optimisation variables. We solve an optimal control problem for a rest-to-rest manoeuvre from a given initial state to a predefined final state from multiple start points. Figure 2 shows the used approach to find global or multiple minima. The first step is to find a feasible solution, i.e. a sequence of configurations and activations that satisfy the equality

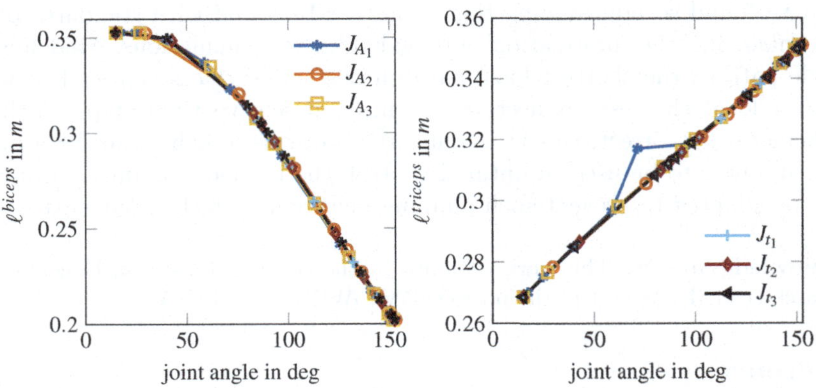

Fig. 4. Left: Muscle length evolution of the biceps for all possible minima, **right**: Muscle length evolution of the triceps for all possible minima

constraints (4b) and is obtained by solving the optimal control problem with J_{const}. In the given example, it is sufficient to give `fmincon` the start position $\boldsymbol{q}_0$ and the corresponding muscle path $\boldsymbol{\gamma}_0$ as initial guess for all time nodes. The resulting solution acts as the initial guess for the `GlobalSearch` and `MultiStart` algorithm from MATLAB, which uses multiple start points to sample multiple basins of attraction to find minima. Table 1 shows the solution vector for the minimum control effort and minimum manoeuvre time produced by these frame work. At first one notices that three possible minima $J_{A_1} = 0.0069, J_{A_2} = 0.0193$ and $J_{A_3} = 0.0391$ for the control effort as well as three possible minima $J_{t_1} = 0.4639, J_{t_2} = 0.6980$ and $J_{t_3} = 0.9220$ for the manoeuvre time are found. In Fig. 3, the activation signals for the biceps as the main actor are shown for all possible minima. A major difference between the minima becomes clear when analysing the muscle path. Figure 4 shows the evolution of the muscle length of the biceps and triceps plotted over the elbow joint angle. For the same rest-to-rest manoeuvre, the algorithm found different possible G1-continuous combinations of geodesics. Although they are not very different in length, the muscle force direction changes and has a big impact on the result. Some of the solutions also show an unrealistic behaviour of the muscle path. The muscle path vector $\boldsymbol{\gamma}_n$ can jump between local shortest connections (for example the solution vector of J_{A_1} in Fig. 4) during two consecutive time steps. Since we have described the muscle path as the shortest connection between two points, such a unphysical behaviour is not prevented by the solver and is part of the solution. When looking for global minima, this issue has to be addressed by improving the wrapping surfaces and the muscle path formulation.

4 Conclusion

In this work an optimal control simulation method for biomechanical systems with dynamic muscle paths is presented. A simple rest-to-rest manoeuvre is solved as an example. The multistart framework for global optimisation from MATLAB found several minima that could not be found with standard solvers, thus enhancing the information obtained from the simulations. Modeling the muscle path as the shortest local path on obstacle surfaces can lead to jumps between local shortest connections during consecutive time steps. Although mathematically correct, this is a completely unrealistic behaviour for a muscle path. In order to be used in optimal control, the muscle wrapping formulation must be adapted to prevent such jumping behaviour, which is our next step.

Acknowledgements. This work is funded by the Federal Ministry of Education and Research (BMBF) as part of the project 05M16WEB - DYMARA.

References

1. Andersson, J.A., Gillis, J., Horn, G., Rawlings, J.B., Diehl, M.: CasADi: a software framework for nonlinear optimization and optimal control. Math. Program. Comput. **11**(1), 1–36 (2019)

2. Betsch, P., Leyendecker, S.: The discrete null space method for the energy consistent integration of constrained mechanical systems. Part II: multibody dynamics. Int. J. Numer. Meth. Eng. **67**(4), 499–552 (2006)
3. Leyendecker, S., Marsden, J.E., Ortiz, M.: Variational integrators for constrained dynamical systems. ZAMM J. Appl. Math. Mech. **88**(9), 677–708 (2008)
4. Leyendecker, S., Ober-Blöbaum, S., Marsden, J.E., Ortiz, M.: Discrete mechanics and optimal control for constrained systems. Optim. Control Appl. Meth. **31**(6), 505–528 (2010)
5. Marsden, J.E., West, M.: Discrete mechanics and variational integrators. Acta Numerica **10**, 357–514 (2001)
6. Maas, R., Leyendecker, S.: Biomechanical optimal control of human arm motion. Proc. Inst. Mech. Eng. Part K J. Multi-body Dyn. **227**(4), 375–389 (2013)
7. Penner, J., Leyendecker, S.: Multi-obstacle muscle wrapping based on a discrete variational principle. In: Proceedings of the European Consortium for Mathematics in Industry (ECMI) Conference, Budapest, Hungary (2018, in review)
8. Scholz, A., Sherman, M., Stavness, I., Delp, S., Kecskeméthy, A.: A fast multi-obstacle muscle wrapping method using natural geodesic variations. Multibody Syst. Dyn. **36**(2), 195–219 (2016)
9. Thielhelm, H., Vais, A., Brandes, D., Wolte, F.E.: Connecting geodesics on smooth surfaces. Vis. Comput. **28**(6–8), 529–539 (2012)
10. Ugray, Z., Lasdon, L., Plummer, J., Glover, F., Kelly, J., Martí, R.: Scatter search and local NLP solvers: a multistart framework for global optimization. INFORMS J. Comput. **19**(3), 328–340 (2007)

Optimal Control Simulations of Two-Finger Precision Grasps

Uday Phutane[1(✉)], Michael Roller[2], Staffan Björkenstam[3], and Sigrid Leyendecker[1]

[1] Chair of Applied Dynamics, Friedrich-Alexander-Universität Erlangen-Nürnberg, 1 Immerwahrstrasse, 91058 Erlangen, Germany
{uday.phutane,sigrid.leyendecker}@fau.de
[2] Fraunhofer ITWM, Fraunhofer Platz 1, 67663 Kaiserslautern, Germany
michael.roller@itwm.fraunhofer.de
[3] Fraunhofer Chalmers Centre, Chalmers Science Park, 412 88 Gothenburg, Sweden
staffan.bjorkenstam@fcc.chalmers.se

Abstract. Grasping is a complex human activity performed with readiness through a complicated mechanical system as an end effector, i.e. the human hand. Here, we apply a direct transcription method of discrete mechanics and optimal control with constraints (DMOCC) to reproduce human-level grasping of an object with a three-dimensional model of the hand, actuated through joint control torques. The equations of motions describing the hand dynamics are derived from a discrete variational principle based on a discrete action functional, which gives the time integrator structure-preserving properties. The grasping action is achieved through a series of constraints, which generate a hybrid dynamical system with a given switching sequence and unknown switching times. To determine a favourable trajectory for grasping action, we solve an optimal control problem (ocp) with an objective involving either the contact polygon centroid or the control torques subject to discrete Euler-Lagrange equations, boundary conditions and path constraints.

1 Introduction

Humans have developed the skills and dexterity over years to perform grasping very naturally. However, to replicate it in simulation and practice with human precision is a challenging task given the highly sophisticated coordination between the fingers. Although first treated in the literature as a purely kinematic problem, the dynamics concerning grasping simulations are crucial. Using the paradigm of discrete mechanics and optimal control for constrained systems (DMOCC), see [1], we simulate a two-finger multibody system to predict motions for precision grasps for two objective functions.

2 Hand Model

We consider a two-finger model, as shown in Fig. 1, composed of the thumb and index finger, hereafter both referred to as only 'finger', with the wrist and forearm.

A. Kecskeméthy and F. Geu Flores (Eds.): ECCOMAS 2019, COMPUTMETHODS 53, pp. 60–67, 2020.
https://doi.org/10.1007/978-3-030-23132-3_8

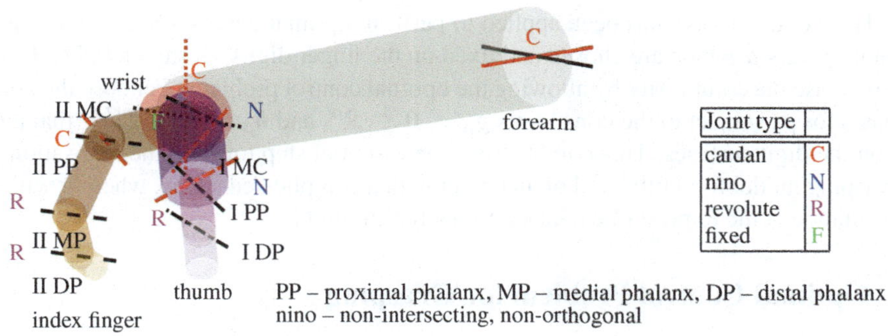

Fig. 1. The two finger model with the joint axes for different joints. The adjoining table shows the joint types. The thumb and index finger are denoted with roman numerals I and II.

The multibody system is modelled with $\boldsymbol{q} \in \mathbb{R}^{108}$ redundant coordinates. With $\boldsymbol{g}_{int} \in \mathbb{R}^{54}$ internal constraints and a combination of revolute, cardan, nino and fixed joints set up through external constraints $\boldsymbol{g}_{ext} \in \mathbb{R}^{41}$, see [3,6], the model is reduced to its minimal coordinates $\boldsymbol{u} \in \mathbb{R}^{13}$, through the null-space method and a nodal reparameterisation, see [3]. The model is actuated using joint torques $\boldsymbol{\tau} \in \mathbb{R}^{13}$ to give the redundant forces $\boldsymbol{f} \in \mathbb{R}^{108}$. The dynamics of the object to be grasped, with redundant configuration $\boldsymbol{q}^O \in \mathbb{R}^{12}$, is included in the model. The discrete Euler-Lagrange (DEL) equations of motion are derived through a discrete variational principle. This gives a symplectic time stepping scheme with structure preserving properties.

3 Contact Model

The contact between the fingers and the object is modelled through holonomic constraint functions $\boldsymbol{g}_c\left(\boldsymbol{q}, \boldsymbol{q}^O\right) = \boldsymbol{0}$. The spatial derivatives $\boldsymbol{G}_c$ and $\boldsymbol{G}_c^O$ transposed multiplied with Lagrangian multipliers $\boldsymbol{\lambda}_c$ form the contact force, for the finger and object configurations, respectively. The grasping action is composed of two steps, namely reaching and then grasping and moving the object. These are modelled by a sequence of two different sets of equations of motion and constraints. We define n_c contact points $\boldsymbol{\rho} \in \mathbb{R}^{3 \cdot n_c}$ in the local coordinate-frames of the finger digits, which can be either prescribed or determined by the optimiser. On the object, the contact points are generally free on a given surface and to be chosen by the optimiser. We use gap functions $\boldsymbol{g}_{c1} \geq \boldsymbol{0} \in \mathbb{R}^{n_c}$ between the contact points on the fingers and the corresponding object surfaces. Further, inequality constraints $\boldsymbol{h}_{\boldsymbol{\rho}^O} \leq \boldsymbol{0} \in \mathbb{R}^{4 \cdot n_c}$, for box (and $\boldsymbol{h}_{\boldsymbol{\rho}^O} \leq \boldsymbol{0} \in \mathbb{R}^{2 \cdot n_c}$ for cylinder) to keep the contact points on the object $\boldsymbol{\rho}^O$ within its dimensions. At gap closure, the contact points $\boldsymbol{\rho}^O$ on the object (and on the fingers if they have not been prescribed) are determined. We then use spherical joint constraints $\boldsymbol{g}_{c2} = \boldsymbol{0} \in \mathbb{R}^{3 \cdot n_c}$, see [4], which restrict relative translation between the contact points. Additionally, we constrain the normal component of the contact forces to be pushing on the object $\boldsymbol{h}_{normal} \leq \boldsymbol{0} \in \mathbb{R}^{n_c}$ and to lie in the friction cone $\boldsymbol{h}_{fric} \leq \boldsymbol{0} \in \mathbb{R}^{n_c}$, using Coulomb's static friction law.

This contact model has been applied to perform optimal control for grasping using contact points $\boldsymbol{\rho}$ which are chosen and fixed on the finger digits, as shown in [7]. Here, we increase the complexity by allowing the optimal control problem to choose the coordinates for $\boldsymbol{\rho}$, subject to the constraints $\boldsymbol{g}_{\rho} = \mathbf{0} \in \mathbb{R}^{n_c}$ and $\boldsymbol{h}_{\rho} \leq \mathbf{0} \in \mathbb{R}^{2 \cdot n_c}$ that they lie on the digit surfaces. This complexity is an essential step towards the realisation of a grasping model for a full hand of an object with a complicated shape, where it will be very cumbersome to prescribe contact points for all digits.

4 Optimal Control Problem for Grasping

As the grasping action is composed of two stages with different dynamics, the optimal control problem (ocp) is composed of two phases, as shown in Fig. 2, see [4]. Here, we define a fixed number of time nodes N_k and N_m for the reaching and grasping phases, respectively. The ocp is solved to determine discrete optimal trajectories $\boldsymbol{u}_d, \boldsymbol{q}_d^O$ for the two systems, control torques $\boldsymbol{\tau}_d$ for the hand, Lagrangian multipliers $\boldsymbol{\lambda}_{c,d}$, contact points $\boldsymbol{\rho}$ on the finger digits, and thus on the object, and optimal durations T_k, T_m for the reaching and grasping phases, respectively, with respect to a particular objective. We approximate the solution of the ocp through a direct transcription method, i.e. we transform it into a nonlinear constrained optimisation problem, see [1]. We define a discrete objective function

$$J_d\left(\boldsymbol{u}_d, \boldsymbol{\tau}_d, \boldsymbol{q}_d^O, \boldsymbol{\lambda}_{c,d}, \boldsymbol{\rho}, T_k, T_m\right) = \sum_{n=0}^{N-1} B_d\left(\boldsymbol{u}_n, \boldsymbol{u}_{n+1}, \boldsymbol{\tau}_n, \boldsymbol{q}_n^O, \boldsymbol{q}_{n+1}^O, \boldsymbol{\lambda}_{c,n}, \boldsymbol{\rho}, T_k, T_m\right) \tag{1}$$

as a sum of a cost function B_d, which has to be minimised. As side constraints, the DEL equations of motion, initial and final configuration and momentum conditions, and discrete path constraints have to be fulfilled.

For the reaching phase, we solve the dynamics of the hand and the object independently through the DEL equations for the two systems. After closing the contact at node N_k using the gap functions, the two systems are coupled through the spherical joints with constraints on the contact forces as described in Sect. 3. Also, we use discrete path constraints such as joint angle limits as inequality constraints $\boldsymbol{h}_{limits} \leq \mathbf{0} \in \mathbb{R}^{2 \cdot n_c \cdot N}$. Finally, we define the initial configuration, initial and final momentum for the complete system, and the final configuration for the object.

4.1 Objective Functions

To solve the grasping ocp, we select two objective functions, first of which involves the contact points on the object $J_{1,d}$ and second of which involves the control torques $J_{2,d}$.

Grasp Contact Polygon Centroid $\left(J_{1,d}\right)$. This objective minimises the distance between the object center of mass $\boldsymbol{\varphi}^O$ and the contact polygon centroid $\boldsymbol{\rho}_{cen}$, see [5]. This achieves a good spread of the contact points around the object, thereby ensuring an even distribution of the contact forces.

$$J_{1,d}\left(\boldsymbol{u}_d, \boldsymbol{q}_d^O\right) = \frac{1}{2} ||\boldsymbol{\rho}_{cen} - \boldsymbol{\varphi}^O||^2, \quad \text{where } \boldsymbol{\rho}_{cen} = \frac{1}{n_c} \sum_{i=1}^{n_c} \boldsymbol{\rho}_i \tag{2}$$

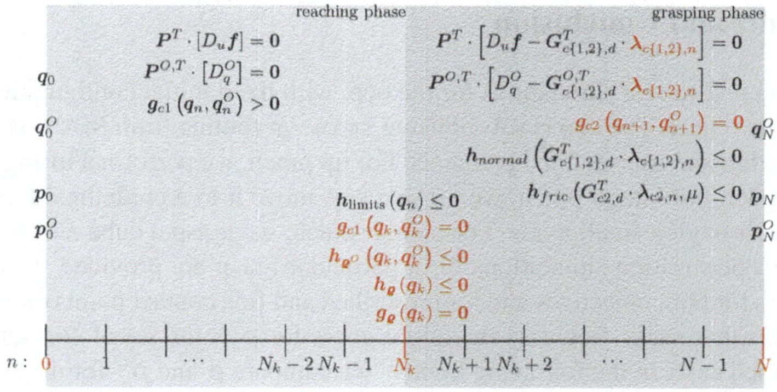

Fig. 2. The grasping ocp setup with the time grid of the reaching and the grasping phase and the equations that form the equality and inequality constraints in the two phases. The placeholders $D_u \boldsymbol{f} = D_1 L_d (\boldsymbol{q}_n, \boldsymbol{q}_{n-1}) + D_2 L_d (\boldsymbol{F}_d(\boldsymbol{u}_{n+1}, \boldsymbol{q}_n), \boldsymbol{q}_n) + \boldsymbol{f}_{n-1}^{+} + \boldsymbol{f}_n^{-}$ and $D_q^O = D_1 L_d (\boldsymbol{q}_n^O, \boldsymbol{q}_{n-1}^O) + D_2 L_d (\boldsymbol{q}_{n+1}^O, \boldsymbol{q}_n^O)$, see [1], are the DEL.

Rate of Change of Control Torques $(J_{2,d})$. This objective minimises changes in the control torques to ensure a smooth movement of the fingers for the complete simulation duration and is a common biomechanical measure, see [3,4].

$$J_{2,d}(\boldsymbol{\tau}_d, T_k, T_m) = \frac{1}{2} \sum_{n=0}^{N-2} (t_{n+1} - t_n) \left(\frac{\boldsymbol{\tau}_{n+1} - \boldsymbol{\tau}_n}{t_{n+1} - t_n} \right)^2 \tag{3}$$

4.2 Two Finger Grasp Taxonomy

We simulate three two-finger precision grasps, as per grasp types defined in [2]. The tip pinch grasp, see Fig. 3(a), holds thin cylindrical objects such as a toothpick. This grasp is performed with two contact points for the distal phalanges of the two fingers, see Fig. 3(d). The lateral pinch grasp, see Fig. 3(b), holds thin objects with flat faces such as a key or a credit card. This is simulated with two contact points on the index finger medial phalanx and one contact point on the thumb distal phalanx, see Fig. 3(e). The palmar pinch grasp, see Fig. 3(c), holds thin or thick objects such as a card, a dice or a ball. This is simulated with two contact points each on the index finger and the thumb distal phalanges, see Fig. 3(f).

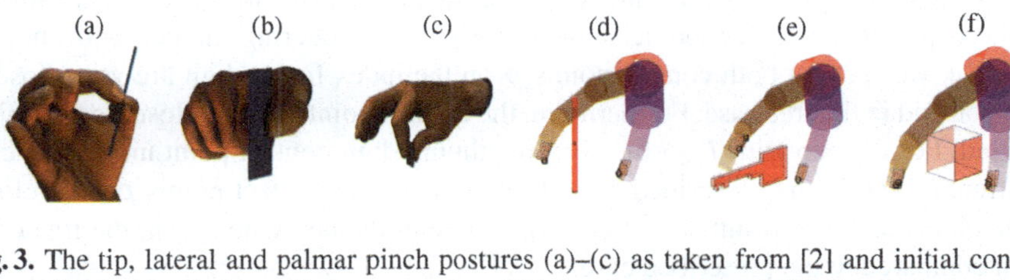

Fig. 3. The tip, lateral and palmar pinch postures (a)–(c) as taken from [2] and initial configurations for the simulation with contact points ((◇) for the thumb and (□) for the index finger) defined on the fingers for the corresponding grasps (d)–(f). The shaded areas are the limits for the contact points on the different finger digits.

5 Results and Conclusion

We perform rest-to-rest simulations for the ocp, with fixed initial configurations for the hand and the object. The ocp is solved using `fmincon` routine from MATLAB[1] with the SQP algorithm and user-defined gradients. For tip pinch, we perform a lifting action for a thin cylinder. For lateral pinch, we grasp a key, move it to a predefined location, and then rotate through a small angle. For palmar pinch, we grasp a cube and place it on a plane. The optimisation simulations for a particular grasp are provided with the same initial guess for both objectives and for prescribed and free contact points on the fingers. This implies that in the free case the initial guess for $\boldsymbol{\rho}$ is the set of contact points of the prescribed case. In the following figures, we compare $\boldsymbol{\rho}$ and $\boldsymbol{\rho}^O$ for the free and the prescribed cases.

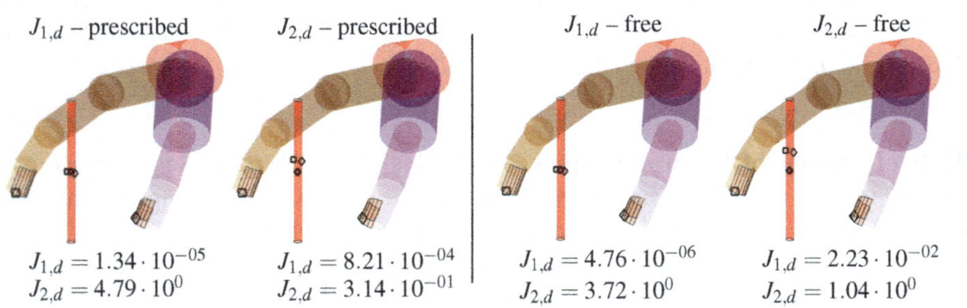

Fig. 4. The contact points for the tip pinch grasps for the simulations with prescribed and free contact points for the ocp with $J_{1,d}$ and $J_{2,d}$. The distance $J_{1,d}$ and torque change value $J_{2,d}$ from Eq. (2) and Eq. (3), respectively, is written below every picture. The contact points for the thumb and the finger are shown with ($\Diamond$) and ($\square$) markers, respectively.

For the tip pinch, the contact points $\boldsymbol{\rho}$ obtained from the ocp in the free case are not very far from the prescribed contact points, as shown in Fig. 4. With respect to the contact points $\boldsymbol{\rho}^O$ on the object, the distance $J_{1,d}$ shows a lower minimum for the ocp with $J_{1,d}$ in the free case, than in the prescribed case. For the ocp with the $J_{2,d}$, we see that the contact points are farther away from the centroid in the free case, than in the prescribed case. Interestingly, for ocp with $J_{2,d}$ objective, the $J_{2,d}$ value is lower in the prescribed case, than in the free case.

For the lateral pinch, the distances $J_{1,d}$ for the prescribed and free cases are similar for the ocp with $J_{1,d}$ objective, as shown in Fig. 5. Considering the ocp with the $J_{1,d}$ objective, we see that both contact points $\boldsymbol{\rho}$ on the index finger digit are very close to the distal end in the free case. Furthermore, the contact points $\boldsymbol{\rho}^O$ are closer to the object centroid. For the ocp with $J_{2,d}$ objective, the thumb digit contact point in the free case is different than in the prescribed case. In particular, the contact points $\boldsymbol{\rho}^O$ are closer to the object and also results in a lower $J_{1,d}$ value in the ocp with $J_{2,d}$ in the free case, when compared in the prescribed case.

[1] https://de.mathworks.com/help/optim/ug/fmincon.html.

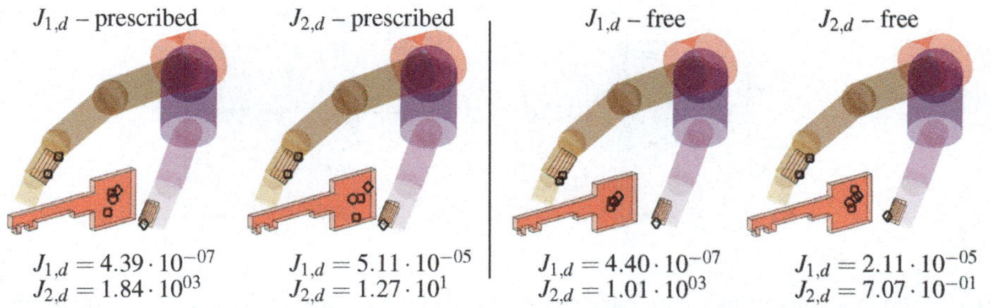

Fig. 5. The contact points for the lateral pinch grasps for the simulations with prescribed and free contact points for the ocp with $J_{1,d}$ and $J_{2,d}$. The distance $J_{1,d}$ and torque change value $J_{2,d}$ from Eq. (2) and Eq. (3), respectively, is written below every picture. The contact points for the thumb and the finger are shown with (◇) and (□) markers, respectively.

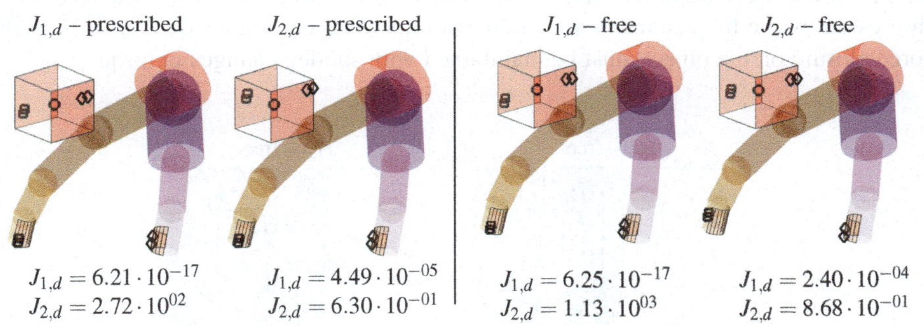

Fig. 6. The contact points for the palmar pinch grasps for the simulations with prescribed and free contact points for the ocp with $J_{1,d}$ and $J_{2,d}$. The distance $J_{1,d}$ and torque change value $J_{2,d}$ from Eq. (2) and Eq. (3), respectively, is written below every picture. The contact points for the thumb and the finger are shown with (◇) and (□) markers, respectively.

For the palmar pinch, the positions of contact points $\boldsymbol{\rho}$ on the fingers for the free case are similar across the two objectives and two cases, with a notable exception observed for the index finger digit for $J_{2,d}$ objective, as shown in Fig. 6. The distances $J_{1,d}$ for the ocp with $J_{1,d}$ are numerically zero in the prescribed and free cases. For the ocp with $J_{2,d}$, the distance $J_{1,d}$ in the free case is higher than in the prescribed case. Interestingly, the ocp with $J_{2,d}$, finds a totally different minimum to lift the object with the side of the index finger digit, instead of the pulp.

The evolution of joint control torque between $J_{1,d}$ and $J_{2,d}$ is shown in Fig. 7 for the palmar pinch in the index finger metacarpophalangeal (MCP) joint. To smooth the torque profile, the optimiser chooses a significant reduction in the torque order magnitude for the ocp with $J_{2,d}$. This, however, is not accompanied with a reduction in the contact forces, as shown for the palmar pinch in Fig. 8, suggesting that the contact forces rather depend on the object weight. Also from Fig. 8, we can observe that the

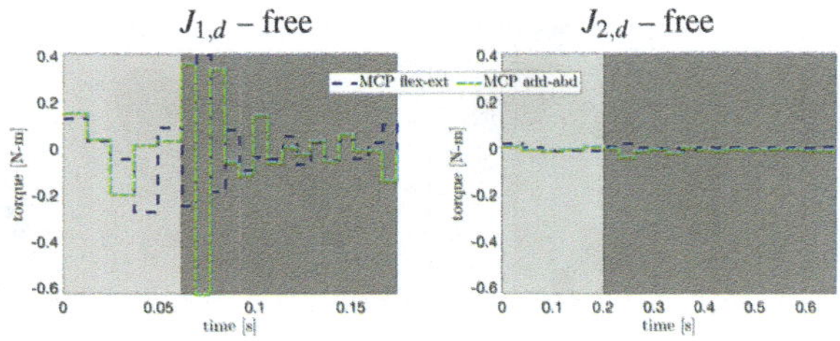

Fig. 7. The index finger metacarpophalangeal (MCP) joint control torques for the palmar pinch simulation for the simulations with free contact points on the fingers for the ocp with $J_{1,d}$ and $J_{2,d}$.

contact forces for the ocp with $J_{2,d}$ are more equal for the two fingers, when compared to the ocp with $J_{1,d}$. This also refers to the two different grasps that is shown in Fig. 6 (right). It can be interpreted that the smoother torque profiles ensure balanced force impressed by the fingers and hence, more stability. If forces are unequal, as in $J_{1,d}$, the force closure on the object must be maintained with sudden changes in torque.

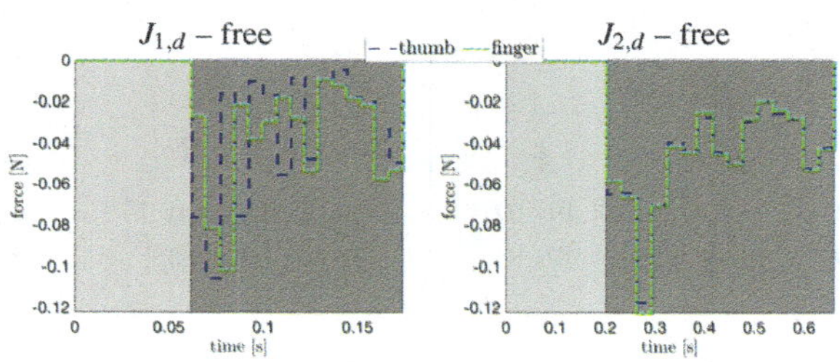

Fig. 8. The contact forces impressed by the thumb and the index finger on the object for the palmar pinch in the free case for the ocp with $J_{1,d}$ and $J_{2,d}$.

To conclude, leaving the choice of contact points on the finger as well as the object as free parameters to the optimiser adds a new dimension of intricacy to the grasping ocp. While the differences in the choice of contact points in the ocp simulations across objectives with the prescribed and free case are fairly straightforward, their implications on the control torque profiles require further analysis.

Acknowledgements. This work is supported by the Fraunhofer Internal Programs under Grant No. MAVO 828424.

References

1. Leyendecker, S., Ober-Blöbaum, S., Marsden, J.E., Ortiz, M.: Discrete mechanics and optimal control for constrained systems. Optimal Control Appl. Methods **31**, 505–528 (2010)
2. Feix, T., Romero, J., Schmiedmayer, H., Dollar, A.M., Kragic, D.: The GRASP taxonomy of human grasp types. IEEE Trans. Hum.-Mach. Syst. **46**(1), 66–77 (2016)
3. Maas, R., Leyendecker, S.: Biomechanical optimal control of human arm motion. J. Multibody Dyn. **227**(4), 375–389 (2013)
4. Koch, M., Leyendecker, S.: Structure preserving simulation of monopedal jumping. Arch. Mech. Eng. **60**(1), 127–146 (2013)
5. Roa, M.A., Suárez, R.: Grasp quality measures: review and performance. Auton. Robots **38**, 65–88 (2014)
6. Phutane, U., Roller, M., Björkenstam, S., Linn, J., Leyendecker, S.: Kinematic validation of a human thumb model. In: Proceedings of the ECCOMAS Thematic Conference on Multibody Dynamics, Prague (2017)
7. Phutane, U., Roller, M., Leyendecker, S.: Optimal control simulations of lateral and tip pinch grasping. In: Proceedings of the ECMI Conference, Budapest (2018, in review)

Reinforcement Learning Applied to a Human Arm Model

Michael Burger, Simon Gottschalk(✉), and Michael Roller

Fraunhofer ITWM, Fraunhofer-Platz 1, 67663 Kaiserslautern, Germany
{michael.burger,simon.gottschalk,michael.roller}@itwm.fraunhofer.de

Abstract. In this contribution, we focus on a muscle actuated human arm model [1] and discuss the applicability of Reinforcement Learning (RL) [2] in order to control it. The content is divided into five sections. We start with the introduction of the human arm model and continue with the optimization method the authors of the model applied. Afterwards, we bring the optimization problem into a form such that RL can handle it and introduce the RL approach we are planning to apply. Before we close with the conclusion, we have a look at the results of the techniques in the numerics section.

1 Human Arm Model

In our first section, we introduce the human arm model. The model is derived and discussed in detail in [1] and is used in [3] to study the influence of different actuation modes by comparing resulting solutions to measurements generated in a movement laboratory. The arm consists of rigid bodies representing the upper arm and the forearm. Additionally, several rigid bodies representing the hand are attached to the forearm without any degree of freedom. The upper arm is linked by a spherical joint to the ground, while the upper arm and the forearm are linked by a revolute joint for the flection and one for the rotation in the elbow. The arm is actuated by 29 Hill's muscles. A sketch of this arm model can be seen in Fig. 1.

In [1] the authors describe how the equations of motion can be derived. They introduce the Lagrange function and derive the equations by using the Lagrangian-d'Alembert principle. All in all, a discrete model of the human arm with $N+1$ states $q^I \in \mathbb{R}^n$ and controls $u^I \in [0,1]^k$ for a fixed step size $h := \frac{t_f - t_0}{N}$ is considered:

$$\partial_2 L_d(q^{I-1}, q^I) + \partial_1 L_d(q^I, q^{I+1}) + F_-^h(q^I, q^{I+1}, u^I, u^{I+1}) \\ + F_+^h(q^{I-1}, q^I, u^{I-1}, u^I) = 0 \text{ for } I = 1, \dots N-1. \quad (1)$$

Here $L_d : \mathbb{R}^n \times \mathbb{R}^n \to \mathbb{R}$ is the discrete Lagrange function. The expression $\partial_1 L_d(q^I, q^J)$ stands for the derivative of the discrete Lagrangian function with respect to the first component. Accordingly, $\partial_2 L_d(q^I, q^J)$ is the derivative with respect to the second component. The function $F_-^h : \mathbb{R}^n \times \mathbb{R}^n \times [0,1]^k \times [0,1]^k \to \mathbb{R}^n$ and the function

A. Kecskeméthy and F. Geu Flores (Eds.): ECCOMAS 2019, COMPUTMETHODS 53, pp. 68–75, 2020.
https://doi.org/10.1007/978-3-030-23132-3_9

$F_{+}^{h} : \mathbb{R}^n \times \mathbb{R}^n \times [0,1]^k \times [0,1]^k \rightarrow \mathbb{R}^n$ are the left and right discrete external forces. For more details on the derivation of the optimal control problem we refer to [4].

Obviously, this system contains more muscles than degrees of freedom, i.e. the system is overactuated. This is very typical for biomechanical systems and complicates the optimization.

2 Numerical Solution Method for the Optimal Control Problem

After we have the discrete equations of motion, we now introduce the optimization problem and the typical way to tackle it. In order to formulate the problem, we need to define the task we are interested in, e.g. reaching a desired point in the room with the tip of the middle finger as fast as possible. Therefore, we need the function $c : \mathbb{R}^n \times \mathbb{R}^n \rightarrow \mathbb{R}^{n^2}$ defining the initial and end position. The objective function $J^h : \mathbb{R}^{n\times(N+1)} \times \mathbb{R}^{n\times(N+1)} \times [0,1]^{k\times(N+1)} \times [0,1]^{k\times(N+1)} \rightarrow \mathbb{R}$ represents the way this task shall be done. Typical objective functions are minimal effort or the mentioned minimal time. It can be defined with the functions $\Phi_d : \mathbb{R}^n \times \mathbb{R}^n \times [0,1]^k \times [0,1]^k \rightarrow \mathbb{R}$ and $\chi : \mathbb{R}^n \times \mathbb{R}^n \times \mathbb{R} \times \mathbb{R} \rightarrow \mathbb{R}$ where the former function defines the goal in each time step in the trajectory and the latter describes the task in the initial and end position.

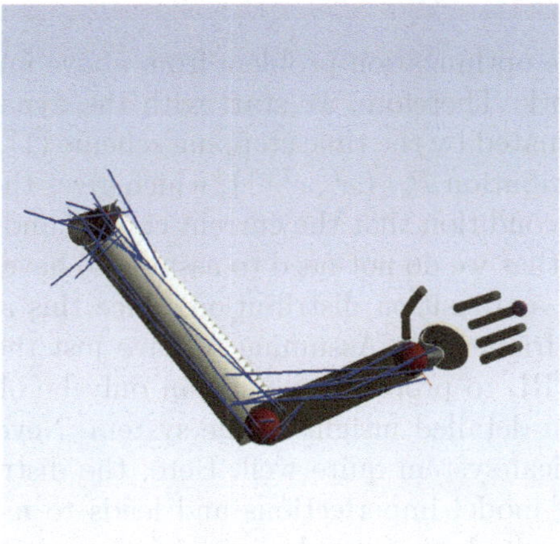

Fig. 1. Sketch of the arm model. The rigid bodies are represented by grey cylinders. The joints are located in the red balls and the muscles are visualized by blue lines. Thereby, blue stands for passive muscle. If a muscle gets active, it turns to red (see Fig. 3). A small ball is attached at the end of the middle finger. This is a marker which is used to define the reaching task by setting the desired end-position of this marker.

The discretized optimal control problem we are considering is (cf. [1,5]):

$$\min_{q,u,\lambda} \quad J^h := \sum_{I=0}^{N-1} \Phi_d(q^I, q^{I+1}, u^I, u^{I+1}) + \chi\left(q^0, q^N, t_0, t_N\right), \tag{2}$$

$$\begin{aligned} \text{subject to } \; & \partial_2 L_d(q^{I-1}, q^I) + \partial_1 L_d(q^I, q^{I+1}) + F_-^h(q^I, q^{I+1}, u^I, u^{I+1}) \\ & + F_+^h(q^{I-1}, q^I, u^{I-1}, u^I) = 0, \text{for } I = 1, \ldots N-1, \qquad (3) \\ & c(q^0, q^N) = 0. \qquad (4) \end{aligned}$$

The applied technique in order to get this discretized optimal control problem is called DMOC [6]. This technique is based on the idea to discretize the continuous Lagrange-d'Alembert principle by approximating the occurring integrals. Then, the discrete optimization problem can be solved by an interior point method for example. In the numerics this is done by the optimizer IPOPT [7]. At this point, it is important to notice that this is an open-loop optimization. This means, that we optimize the control for one specific problem. For each further reaching point we need to solve this optimization problem again. In the upcoming approach, we will see that a controller is trained. The application scope of this controller depends, among other things, on the training data.

For the next section, we introduce $x^I := \left(q^{I-1}, q^I\right)$ which will be the observation for the RL approach. In case of $I = 0$, we set $x^0 := \left(q^0, q^0\right)$. This notation is for the sake of overview and is no restriction.

3 Reinforcement Learning

Now we can bring the optimization problem from above into the Reinforcement Learning [2] framework. Therefore, we start with the dynamical system, whose solution was approximated by the time stepping scheme (1), which is now hidden in the transition distribution $P_{u^I}\left(x^I, x^{I+1}\right)$, which gives the distribution for the next state under the condition that the current control and state is given. It is a big advantage of RL that we do not need to assume to have the detailed dynamical system but only a transition distribution, since this assumption is weaker than in the classical framework. Assuming to have just the transition distribution allows to apply RL to problems which can only be observed from outside and without having a detailed insight in the system. Nevertheless, in our case we know the dynamical system quite well. Here, the distribution only enables the considerations of model imperfections and leads to more robust solutions, which means that small changes in the initial value or the target position or other small disturbances can be handled. The reward function $r(x^I, u^I)$ takes the role of the objective function. In our example where we are aiming to reach a target point y_{target} with a marker position y_{marker} attached at the hand, we define the reward as ten times minus the euclidean distance between y_{marker} and y_{target}. Additionally, a huge reward is added in the end, if the task is fulfilled. In summary, we assume to have a Markov decision process [8]. The goal of RL is to find a suitable controller. There are several possible ways to do this. We want to focus on a policy-based, model-free RL approach. This means, we omit doing

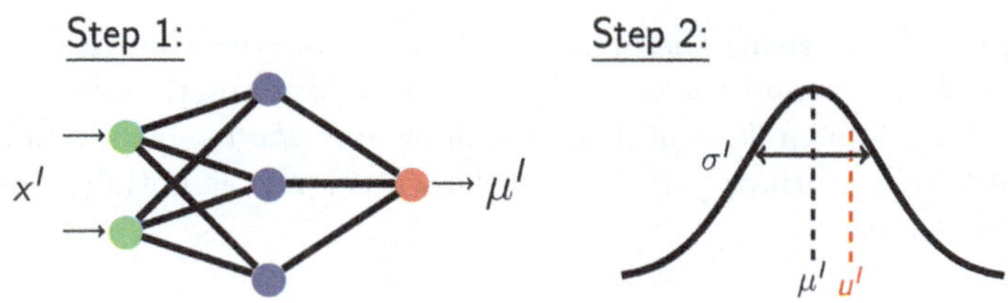

Fig. 2. Sketch of the policy. In the first step the current state x^I goes into the neural network, which is defined by its parameters θ. In a second step, the control u^I is sampled from a Gaussian distribution which is defined by an exploration variance σ^I and the output of the neural network, which is used as the mean value.

system identification (i.e. model-free) and train an approach to get a distribution for the next control (policy) depending on the current state.

To be more precise, we assume to have the policy $\pi(u^I|x^I)$, which gives the distribution from that we can sample the next action u^I under the condition that the current state x^I is known. In our case, we use a Gaussian distribution as policy and compute its parameters with a neural network, which gets the current state as input (Fig. 2). In this way, we derive a parametrized policy, since the policy does now only depend on the finitely many weights and biases of the neural network. Since we are interested in a controller, suitable parameters (abbreviated by θ) need to be found. Here suitable means we search for parameters which maximizes the expected reward over a whole trajectory $\tau = \{x^0, u^0, x^1, u^1, \dots\}$:

$$\max_{\theta} \mathbb{E}\left[\sum_{I=0}^{N} r\left(x^I, u^I\right)\right]. \tag{5}$$

In the beginning θ is initialized randomly and is improved step by step by iterative approaches like a gradient method [9]. In this contribution, we prefer a *proximal policy optimization* technique [10] which is based on the *Trust Region Policy optimization* [11]. The general idea of Trust Region optimization is to approximate the objective function and to solve the resulting simplified optimization problem at least on a small neighborhood of the current parameters θ. With introducing the definition of the advantage function $A_{\pi_\theta}(x, u)$ as the expected reward when having the action u at state x compared to the reward we expect, if an average action is applied, we get in each iteration step the optimization problem (see [11]):

$$\begin{aligned} \max_{\tilde{\theta}} \quad & \mathbb{E}_{x\sim\rho_{\pi_\theta}, u\sim\pi_\theta}\left[\frac{\pi_{\tilde{\theta}}(u|x)}{\pi_\theta(u|x)} A_{\pi_\theta}(x, u)\right] \\ s.t. \quad & \mathbb{E}_{x\sim\rho_{\pi_\theta}}\left[D_{KL}\left[\pi_\theta(\cdot, x), \pi_{\tilde{\theta}}(\cdot, x)\right]\right] \leq \delta. \end{aligned} \tag{6}$$

The parameter δ defines the size of the trust region. The constraint is defined by the Kullback-Leibler divergence $D_{KL}\left[\pi_\theta(\cdot, x), \pi_{\tilde{\theta}}(\cdot, x)\right] = \sum_u \pi_\theta(u|x) \log \frac{\pi_\theta(u|x)}{\pi_{\tilde{\theta}}(u|q)}$ (see [12] and [13]) and with ρ_{π_θ} we denote the occurrence distribution of states during one trajectory using the policy π_θ. In the numerics the constraint is used as a penalty objective and the weight of this term is adapted in each step as it is discussed in [10]. The implementations are done in *Python* and the neural network is performed by Tensorflow [14]. Our implementation is essentially based on the algorithm of Coady [15].

4 Numerical Results

The task for the human arm is to reach a certain point. Therefore, we have a marker attached at the end of the middle finger and the related end position is defined. The initial position can be seen in Fig. 3(a) while the desired end position of the finger tip is defined and represented by a small ball in Fig. 3. The Fig. 3 shows snapshots of the solution of the RL approach. It shows the initial posture, the posture after six (respectively ten) time steps and the end position. We can see that the marker at the finger tip and the desired end position overlap and thus, that the task is fulfilled and the movement itself looks plausible from a human point of view. The movement can be described as reaching out the hand in order to present something in the hand. It can be observed that the upper arm is pulled back a little bit and the hand is turned such that the palm points to the ceiling. This makes sense, since in this way the important muscles are not twisted and have a strong influence. In Fig. 4 we see the increase of the summed up reward over the iterations. At the beginning we have a huge increase. Later on, the improvement becomes smaller between the iterations. Nevertheless, on average the reward still increase. But the variance from the average is bigger than in the beginning, because now some trajectories reach the desired end position and a huge reward is activated. To be more precise,

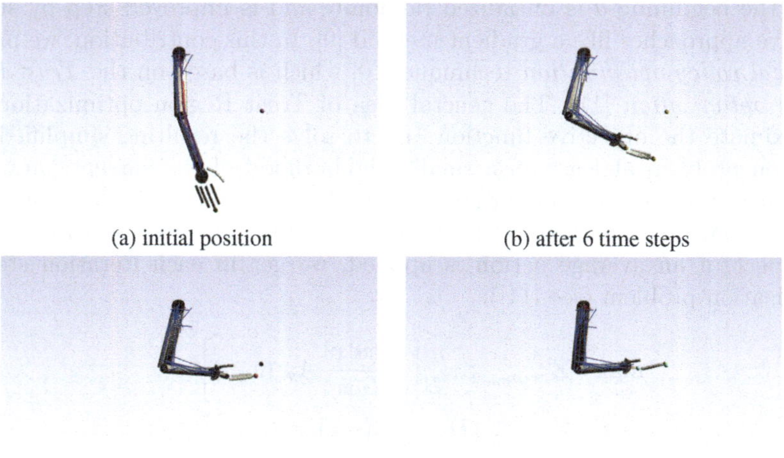

(a) initial position (b) after 6 time steps (c) after 10 time steps (d) final position

Fig. 3. Trajectory

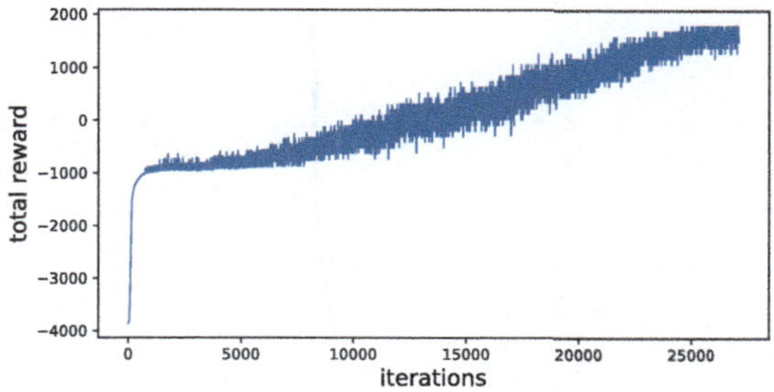

Fig. 4. Total reward during training

the discontinuity of the reward function leads to this observed behaviour. In the end, the average summed up rewards does not increase anymore and the maximum seems to be reached. At this time the learning is completed. The trajectory we have seen in Fig. 3 is generated with the trained controller. The related control, which the controller generated during the trajectory can be seen in Fig. 5. These controls are the activations of the Hill's muscle models. The activation is a value between zero (not active) and one (full active). It can be seen that the end position is reached after 16 time steps. The dominating muscles are the *musculus teres minor* acting in the shoulder, the *musculus brachioradialis* responsible for the supination of the forearm and the *biceps* for flexing the arm. In Fig. 6 we consider the movement of a marker attached at the palm of the hand during one trajectory. While the yellow line is the solution of the optimal control problem (Sect. 2) searching for the controls in order to reach the point as fast as possible, the blue lines represent trajectories generated with the trained controller. The controller generates a smooth s-shaped movement of the hand, which looks plausible from a human point of view.

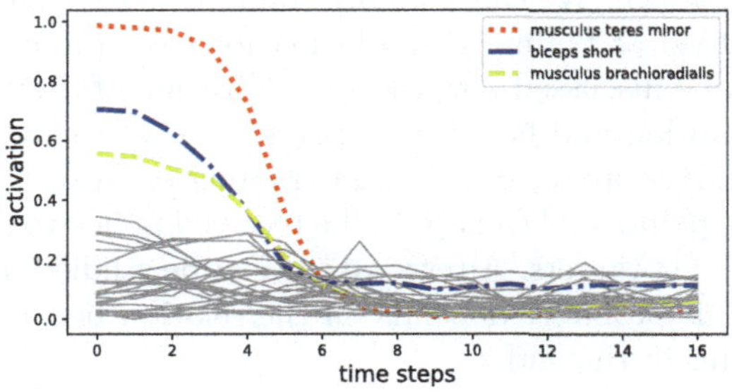

Fig. 5. Activations of the muscle models involved in the human arm model. The activation is a value between zero (not active) and one (fully actuated).

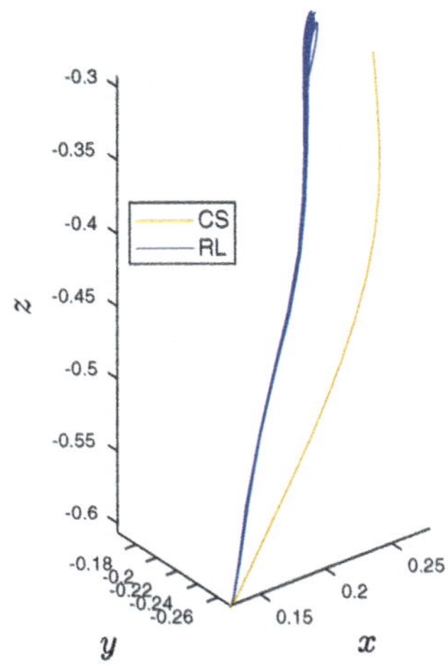

Fig. 6. Motion of a marker attached at the pal of hand. Classical optimal control trajectory (CS). RL approach (RL).

5 Conclusion

We considered a muscle actuated model of a human arm and tackled the control problem with a Reinforcement Learning approach. We introduced the used RL approach based on the trust region idea and successfully applied it to the arm model. We receive a human-like movement which reaches the desired end position and documented the RL training.

In future works, we will focus on more complex tasks. One possible application scenario could be a variation of the task described in [3]. In this work, a human in front of a wall moves its finger tip to points, which are marked in a certain pattern on a wall. We can possibly train one controller in order to reach several points instead of finding the solution for each target point separately. Afterwards, it can be discussed how the generalization of the task influences the accuracy of the solution and how this depends on the size and structure of the neural network. Furthermore, in a second training one can train the controller with several target points and can apply the received policy to a point, which was not one of the training targets. Afterwards, it can be studied, in which cases the task succeeded. This probably depends on the relation between training points and the target point in the end.

Acknowledgements. The authors are grateful for the funding by the Federal Ministry of Education and Research of Germany (BMBF), project number 05M16UKD.

References

1. Roller, M., Björkenstam, S., Linn, J., Leyendecker, S.: Optimal control of a biomechanical multibody model for the dynamic simulation of working tasks. In: Proceedings of the 8th ECCOMAS Thematic Conference on Multibody Dynamics, Prague, pp. 817–826 (2017)
2. Sutton, R.S., Barto, A.G.: Introduction to Reinforcement Learning, 1st edn. MIT Press, Cambridge (1998)
3. Obentheuer, M., Roller, M., Björkenstam, S., Berns, K., Linn, J.: Comparison of different actuation modes of a biomechanical human arm model in an optimal control framework. In: Proceedings of the 5th Joint International Conference on Multibody System Dynamics IMSD, Lisbon (2018)
4. Björkenstam, S., Leyendecker, S., Linn, J., Carlson, J.S., Lennartson, B.: Inverse dynamics for discrete geometric mechanics of multibody systems with application to direct optimal control. J. Comput. Nonlinear Dyn. **13**(10), 101001 (2018)
5. Gerdts, M.: Optimal Control of ODEs and DAEs. De Gruyter Textbook. De Gruyter, Berlin (2012)
6. Ober-Blöbaum, S., Junge, O., Marsden, J.: Discrete mechanics and optimal control: an analysis. ESAIM Control Optim. Calc. Var. **17**, 322–352 (2011)
7. Wächter, A., Biegler, L.T.: On the implementation of an interior-point filter line-search algorithm for large-scale nonlinear programming. Math. Program. **106**, 25–57 (2006)
8. Puterman, M.L.: Markov Decision Processes: Discrete Stochastic Dynamic Programming, 1st edn. Wiley, New York (1994)
9. Williams, R.J.: Simple statistical gradient-following algorithms for connectionist reinforcement learning. Mach. Learn. **8**, 229–256 (1992)
10. Schulman, J., Wolski, F., Dhariwal, P., Radford, A., Klimov, O.: Proximal policy optimization algorithms. CoRR, vol. abs/1707.06347 (2017)
11. Schulman, J., Levine, S., Abbeel, P., Jordan, M.I., Moritz, P.: Trust region policy optimization. In: ICML. JMLR Workshop and Conference Proceedings, Lille, France, vol. 37, pp. 1889–1897. JMLR.org (2015)
12. Kullback, S.: Information Theory and Statistics. Wiley, New York (1959)
13. Kullback, S., Leibler, R.A.: On information and sufficiency. Ann. Math. Statist. **22**, 79–86 (1951)
14. Abadi, M., Agarwal, A., Barham, P., Brevdo, E., Chen, Z., Citro, C., Corrado, G.S., Davis, A., Dean, J., Devin, M., Ghemawat, S., Goodfellow, I., Harp, A., Irving, G., Isard, M., Jia, Y., Jozefowicz, R., Kaiser, L., Kudlur, M., Levenberg, J., Mané, D., Monga, R., Moore, S., Murray, D., Olah, C., Schuster, M., Shlens, J., Steiner, B., Sutskever, I., Talwar, K., Tucker, P., Vanhoucke, V., Vasudevan, V., Viégas, F., Vinyals, O., Warden, P., Wattenberg, M., Wicke, M., Yu, Y., Zheng, X.: TensorFlow: large-scale machine learning on heterogeneous systems (2015). Software tensorflow.org
15. Coady, P.: AI gym workout (2017). https://learningai.io/projects/2017/07/28/ai-gym-workout.html. 26 Oct 2018

Contact and Constraints

Dynamic Modeling and Analysis of Pool Balls Interaction

Eduardo Corral[1(✉)], Raúl Gismeros[1], Filipe Marques[2], Paulo Flores[2], María Jesús Gómez García[1], and Cristina Castejon[1]

[1] MaqLab Research Group, Universidad Carlos III de Madrid, Madrid, Spain
{ecorral, rgismero, mjggarci, castejon}@ing.uc3m.es
[2] CMEMS-UMinho, Department of Mechanical Engineering, University of Minho, Braga, Portugal
{fmarques, pflores}@dem.uminho.pt

Abstract. This work presents the development, test and verification of a pool billiard model in the context of Multibody System Dynamics (MSD) methodologies. This study aims at examining and comparing the effects of considering several different contact forces models when dealing with impact and frictional phenomena. Thus, during the contact events two kinds of forces are considered, i.e., normal contact forces and frictional forces.

1 Introduction

The field of MSD has experienced a significant growth in recent decades thanks to the increase in computing capacity. This methodology can be applied in a broad variety of fields, namely vehicle dynamics [1], biomechanical studies (orthopaedics, accident recreations), robotics [2], among others [3–5]. This wide range of applications, as well as the current research activity with constant innovations, justify the fact that MSD is an interesting and active research topic.

The main purpose of this work is to create a realistic model that can predict properly the behaviour of the different bodies (pool balls) given the force and direction of the shot. Unlike many of the projects developed so far [6, 7], a contact force approach has been considered to define the interactions between bodies [8]. In this approach, the normal contact forces are expressed as continuous functions of the pseudo-penetration between the contacting bodies, as well as their geometric and material characteristics [9]. The presence of frictional forces on the interaction between billiard balls and between the balls and the table has also been considered [10]. All these forces have been defined using parameters obtained from previous experiments [11].

A MATLAB® program for dynamic analysis of spatial multibody systems called MUBODYNA (MUltiBOdy DYNAmics) has been utilized to carry out this work [12]. This program allows a great degree of customization from the user, with the possibility of defining multiple multibody formulations and modifying the model properties with little effort.

Thus, several cue stick directions and positions of impact on the cue ball have been chosen as simulation scenarios. The most relevant issues related with the suitable

A. Kecskeméthy and F. Geu Flores (Eds.): ECCOMAS 2019, COMPUTMETHODS 53, pp. 79–86, 2020.
https://doi.org/10.1007/978-3-030-23132-3_10

selection of normal, tangential contact force models for the different interactions (ball-cloth, ball-ball and ball-cushion) are subsequently discussed, having in mind the computational efficiency, and the choice of proper values for the force models parameters [13]. The friction forces are computed utilizing different models with the purpose to appraise the most relevant and appropriate options [10].

2 Pool Table Model

Two different approaches are considered when dealing with contact problems: "non-smooth" dynamics and "compliant" or contact forces approach [13]. The first one, also known as "momentum based" formulation, assumes the bodies truly rigid and works with concept of impulse [14], while the second expresses the contact forces as continuous functions of penetration between contacting bodies, using position and velocity information [15]. Rigid models allows a simple, computationally efficient modeling, whereas one of the main drawbacks of compliant modelling is the computational efficiency. Furthermore, another weakness of this kind of modelling is the complexity of selection of appropriate contact parameters, such as the equivalent stiffness or the degree of nonlinearity of the penetration, critical in complex contact scenarios and non-metallic materials [16].

A quick search on the subject shows that most of the works made so far about pool ball/ball interaction are "non-smooth" models [6, 7, 17, 18], and contact force approach only focusses on two ball interaction [8]. The main aim of our research is to develop a complete contact forces model that considers not only ball interactions, but also ball-cloth and ball-cushion interactions. Within the spirit of this methodology, special attention is paid to the contact detection itself, in terms of both computational accuracy and efficiency [13].

A multibody dynamic system can be formulated and solved in many ways. In this paper, the equations are implemented in MUBODYNA, a MATLAB® program used to perform forward dynamic simulations for spatial multibody systems, using several different multibody formulations [12].

For this algorithm, relative coordinates are used. Every body has their local coordinate system, and to describe their rotation the Euler parameters are used. The translational motion is described in terms of Cartesian coordinates, while rotational motion is specified using the technique of Euler parameters.

2.1 Geometric Considerations of the Model

An Eight-ball table, the most popular on billiard game, has been considered to develop the model. It consists of 15 coloured balls (seven striped balls, seven solid-colored balls and the black 8 ball) and a white ball, which will be hit by a cue. Following the WPA (World Pool Association), a regulatory 9-Foot table has been selected, with a dimension of 2.54 x 1.57 m [19]. The cushions should be triangular, but the ball-cushion contact has been modeled and implemented as a sphere-plane contact. No pockets have been considered in this model. The balls are also regulatory (57.15 mm diameter, 170 g mass, solid and made of Phenol-formaldehyde resin) [19]. Table 1 shows the values of elastic properties used on the model.

Table 1. Material properties of the bodies

Body	Material	Young modulus [MPa]	Poisson ratio [-]
Cloth	Wool	3100[1]	0.38[2]
Cushion	Rubber	7[3]	0.5[4]
Ball	PF resin	3800[5]	0.375[6]

[1]Value obtained from *Animal fibers sheet* [20].
[2]Value obtained from *On the Poisson's ratios of a woven fabric* [21].
[3]Value obtained from *A Practical Guide to Vertebrate Mechanics* [22].
[4]Value obtained from *Materials for engineering* [23].
[5]Value obtained from *MakeItFrom.com* [24].
[6]Value obtained from *Delaware composites design encyclopedia* [25].

2.2 Normal and Friction Force of the Model

Three interactions can be identified on our model: ball-ball, ball-cloth and ball-cushion. For each one, a suitable contact and frictions forces set has been chosen based on the properties of the contacting bodies and looking for computational efficiency.

For the interaction between balls, a nonlinear Lankarani and Nikravesh model has been selected [9]. This contact force model gives good outcomes for general mechanical contacts, in particular for the cases in which the energy dissipated during

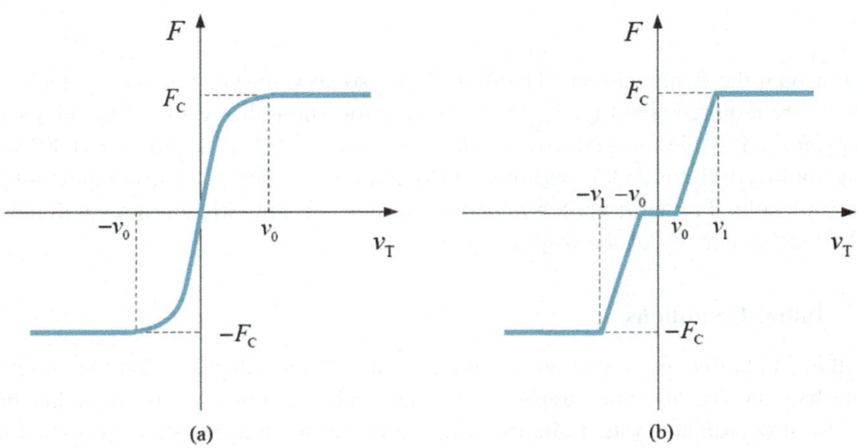

Fig. 1. a) Threlfall friction model; b) Ambrosio friction model

the contact is relatively small when compared to the maximum absorbed elastic energy, i.e., for the values of the coefficient of restitution close to unit. About the friction force, a Threlfall model has been chosen (see Fig. 1a) [10], as friction is not a matter of concern in this interaction.

On the other hand, for the ball-cloth contact, a Hertz contact model has turned out to be the best choice, due to its simplicity and computational efficiency [9, 15]. This last aspect has been critical, as it will be discussed later. An Ambrosio friction model has been selected (see Fig. 1b) [10], as stick–slip motion is not required in this model: when the white ball is hit, the friction between the balls and the cloth is so low that they tend to slip immediately.

Finally, for the ball-cushion interaction, a Lankarani and Nikravesh model has been chosen, as the cushion absorbs most of the impact energy through elastic deformation. Friction is minimal in this interaction so it has been neglected.

The values of both friction and restitution coefficients have been taken from the work of Alciatore [11]. Specific values for every interaction are shown on Table 2.

Table 2. Restitution and Friction coefficients

Case	Contact force model	Friction force model	Friction coefficient	Restitution coefficient
Ball-ball	Lankarani and Nikravesh[1]	Threlfall	0.06	0.93
Ball-cloth	Hertz[1]	Ambrosio	0.2	-
Ball-cushion	Lankarani and Nikravesh[1]	-	-	0.85

[1]For a value of the nonlinear power exponent n = 1.5 [9]

For both the Ambrosio and Threlfall model, some values of tolerance velocity (v_0 and v_1) are required (see Fig. 1). In this model, for Threlfall a value of $v_0 = 0.01$ m/s has been used, while for Ambrosio model, values of $v_0 = 0.0001$ and $v_1 = 0.001$ have been employed. Both models resemble to Coulomb's law, and provide computationally efficient results. Furthermore, for velocities close to zero, the friction force will always be low independently of the displacement.

2.3 Initial Conditions

For the simulations presented on this paper, four balls are deployed on the table: the white ball and the first three upper balls of the rack (see Fig. 2). All the bodies have their local coordinate system aligned with the global reference system, so their Euler parameters are (1, 0, 0, 0). The hit has been implemented as an initial velocity of the white ball of value 10.729 m/s (24 MPH), correspondent to professional break shot [26], in the positive direction of x-axis. The inertia of the balls is J = 0.000055 kg·m^2. The center of the table corresponds to the position (0, 0, 0), so the initial position of the white ball is (−0.635, 0, 0.028576) m and the rack of the coloured balls starts on (0.635, 0, 0.028575) m.

3 Numerical Results

The dynamic equations of motion have been solved using four different methods: the standard Lagrange multipliers [27], the Augmented Lagrangian formulation [28], the Baumgarte stabilization method [29] and the direct correction method [30], all of them accompanied by the Ordinary Differential Equations (ODE) integrator. An analysis time of 3 s, with a 0.01 s time differential has been fixed. Figure 2 shows the evolution of the simulation for the standard Lagrange multipliers method.

As it can be seen on Fig. 3, in the early stages of the simulation, when the white ball is hit, it slides until it collides with the coloured balls, then starting to roll without losing energy (with constant angular and linear velocity).

The shot implemented on the simulation has only x-axis velocity, and the balls are perfectly aligned, so the cue ball should not have y-axis linear acceleration. It has been proven that, for time differentials greater than 0.00001 s, the Augmented and direct correction methods does not keep the symmetry of the model (the cue ball exhibits y-axis linear acceleration, so it has non-zero y-coordinate). Thus, it can be said that both Augmented and direct correction methods are not computationally efficient for this model.

Although it is necessary to validate the model with a real situation, so parameters such as the friction or the restitution coefficients can be adjusted, the tests carried out until now confirms that the model provides realistic behaviour for the different methods of resolution. Figure 4 shows the computing times for the different methods and time differentials used on this work. These values also consider the post processor time, which for smaller time differential can be greater than the analysis time.

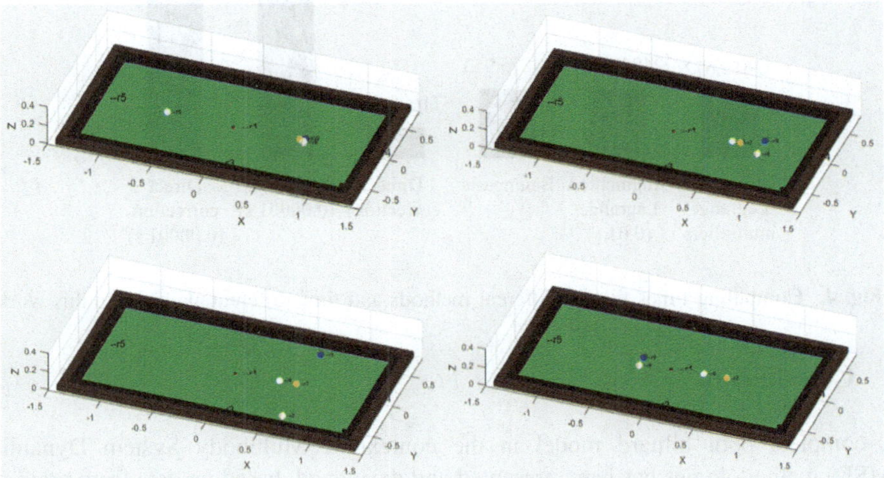

Fig. 2. Evolution of the simulation, for the standard Lagrange multipliers method

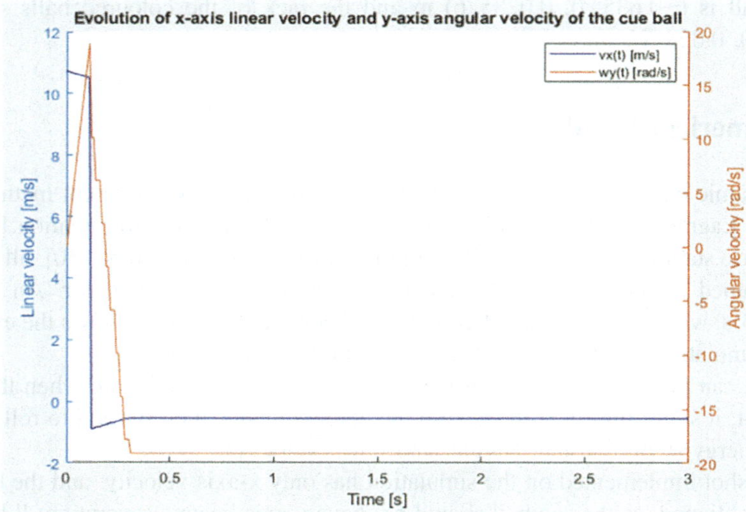

Fig. 3. Evolution of linear velocity on x-axis direction and angular velocity on y-axis direction, for the standard Lagrange multipliers method

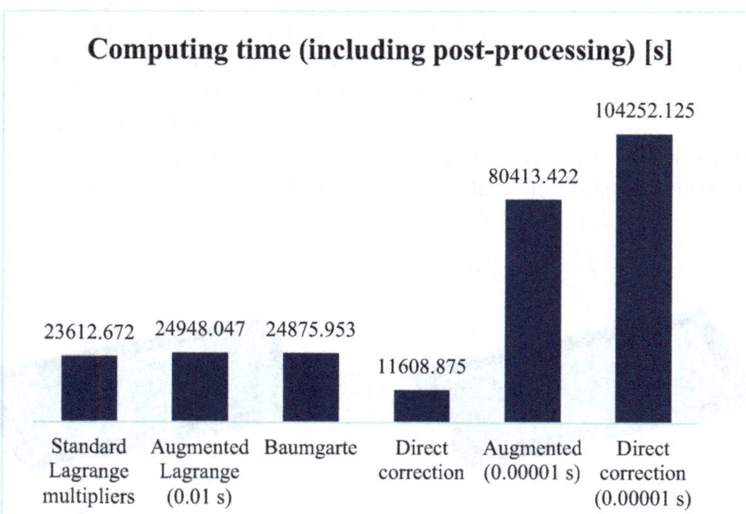

Fig. 4. Computing times for the different methods and time differentials used on this work

4 Conclusions

A complete pool billiard model in the context of Multibody System Dynamics (MSD) methodologies has been presented and developed, based on data from previous works and considering all the interactions between the different bodies. The results show which resolution method is more convenient to use in this model, in order to get a balance between computing time and results accuracy.

Knowing that the model works properly, an adjustment based on the measure of the data obtained from real shots could be done to get as much resemblance to reality as possible.

Acknowledgements. The authors would like to thank the financing of the University Carlos III of Madrid and the collaboration between the University of Minho and the University Carlos III of Madrid, with which it has been possible to carry out this research.

References

1. Abad, E.C., Alonso, J.M., García, M.J.G., García-Prada, J.C.: Methodology for the navigation optimization of a terrain-adaptive unmanned ground vehicle. Int. J. Adv. Robot. Syst. **15**, 1–11 (2018). https://doi.org/10.1177/1729881417752726
2. Corral, E., Meneses, J., Castejón, C., García-Prada, J.C.: Forward and inverse dynamics of the biped PASIBOT. Int. J. Adv. Rob. Syst. **11**, 1–9 (2014)
3. Al Nazer, R., Rantalainen, T., Heinonen, A., Sievänen, H., Mikkola, A.: Flexible multibody simulation approach in the analysis of tibial strain during walking. J. Biomech. **41**, 1036–1043 (2008)
4. O'Riordain, K., Thomas, P., Phillips, J., Gilchrist, M.: Reconstruction of real world head injury accidents resulting from falls using multibody dynamics. Clin. Biomech. **18**, 590–600 (2003)
5. Zhu, W., Piedboeuf, J., Gonthier, Y.: A dynamics formulation of general constrained robots. Multibody Sys. Dyn. **16**, 37–54 (2006)
6. Jia, Y.B., Mason, M., Erdmann, M.: A state transition diagram for simultaneous collisions with application in billiard shooting. In: Chirikjian, G.S., Choset, H., Morales, M., Murphey, T. (eds.) Algorithmic Foundation of Robotics VIII. Springer Tracts in Advanced Robotics, vol. 57, pp. 135–150. Springer, Berlin, Heidelberg (2009)
7. Ivanov, A.: Theorem for change of the rigid body generalized impulse. In: XVII Anniversary International Scientific Conference by Construction and Architecture VSU'2018, Sofia, Bulgaria (2018)
8. Hu, S., Guo, X.: A dissipative contact force model for impact analysis in multibody dynamics. Multibody Sys. Dyn. **35**, 131–151 (2015)
9. Machado, M., Moreira, P., Flores, P., Lankarani, H.M.: Compliant contact force models in multibody dynamics: evolution of the Hertz contact theory. Mech. Mach. Theory **53**, 99–121 (2012)
10. Marques, F., Flores, P., Pimenta Claro, J.C., Lankarani, H.M.: A survey and comparison of several friction force models for dynamic analysis of multibody mechanical systems. Nonlinear Dyn. **86**, 1407–1443 (2016)
11. Alciatore, D.: Billiards and pool physics FAQs. Billiards.colostate.edu. http://billiards.colostate.edu/threads/physics.html#properties
12. Flores, P.: MUBODYNA -A MATLAB program for dynamic analysis of spatial multibody systems. University of Minho, Guimarães, Portugal (2012)
13. Flores, P., Ambrósio, J.: On the contact detection for contact-impact analysis in multibody systems. Multibody Sys. Dyn. **24**, 103–122 (2010)
14. Pfeiffer, F., Glocker, C.: Multibody Dynamics with Unilateral Contacts. Wiley, New York (1996)

15. Flores, P., Ambrósio, J., Claro, J., Lankarani, H.: Dynamic behaviour of planar rigid multi-body systems including revolute joints with clearance. In: Proceedings of the Institution of Mechanical Engineers, Part K: J. Multi-Body Dyn. **221**(2), 161–174 (2007). https://doi.org/10.1243/14644193jmbd96
16. Askari, E., Flores, P., Dabirrahmani, D., Appleyard, R.: Nonlinear vibration and dynamics of ceramic on ceramic artificial hip joints: a spatial multibody modelling. Nonlinear Dyn. **76**(2), 1365–1377 (2014). https://doi.org/10.1007/s11071-013-1215-y
17. Leckie, W., Greenspan, M.: Pool physics simulation by event prediction 1: motion transitions. ICGA J. **28**(4), 214–222 (2005). https://doi.org/10.3233/icg-2005-28403
18. Stewart, D.: Rigid-body dynamics with friction and impact. SIAM Rev. **42**(1), 3–39 (2000). https://doi.org/10.1137/s0036144599360110
19. Equipment Specifications - WPA Pool: WPA Pool (2001). https://wpapool.com/equipment-specifications/
20. Zimniewska, M., Kicińska-Jakubowska, A.: Animal fibers sheet, p. 1. Poznan (2012). http://dnfi.org/wp-content/uploads/2012/01/fact-sheet-animal-fibers.pdf
21. Sun, H., Pan, N., Postle, R.: On the Poisson's ratios of a woven fabric. Compos. Struct. **68** (4), 505–510 (2005). https://doi.org/10.1016/j.compstruct.2004.05.017
22. McGowan, C.: A Practical Guide to Vertebrate Mechanics. Cambridge University Press, Cambridge, UK (1999)
23. Ashby, M., Jones, D.: Engineering Materials 1: An Introduction to Properties, Applications and Design, 3rd edn. Butterworth-Heinemann, Oxford (2005)
24. Phenol Formaldehyde (PF, Phenolic):: MakeItFrom.com: Makeitfrom.com. https://www.makeitfrom.com/material-properties/Phenol-Formaldehyde-PF-Phenolic
25. Carlson, L., Gillespie, J.: Delaware Composites Design Encyclopedia - 3: Processing and Fabrication Technology, p. 83. Lancaster, Pa (1990)
26. Onoda, G.: Faster than a speeding bullet? Billiards Digest, p. 34 (1989). http://www.sfbilliards.com/Misc/onoda_all_txt.pdf
27. Nikravesh, P.: Computer-Aided Analysis of Mechanical Systems. Prentice-Hall, London (1988)
28. Garcia De Jalon, J., Bayo, E.: Kinematic and Dynamic Simulation of Multibody Systems. Springer, New York (1994)
29. Baumgarte, J.: Stabilization of constraints and integrals of motion in dynamical systems. Comput. Methods Appl. Mech. Eng. **1**(1), 1–16 (1972). https://doi.org/10.1016/0045-7825(72)90018-7
30. Marques, F., Souto, A., Flores, P.: On the constraints violation in forward dynamics of multibody systems. Multibody Syst. Dyn. **39**(4), 385–419 (2017). https://doi.org/10.1007/s11044-016-9530-y

Dynamics of Machine-Process Combinations

Friedrich Pfeiffer(✉)

Lehrstuhl fuer Angewandte Mechanik, TU-Muenchen,
Boltzmannstrasse 15, 85748 Garching, Germany
friedrich.pfeiffer@tum.de

Abstract. Machines and mechanisms realize processes, from the shaping process of a milling machine to the motion process of an automotive system. The dynamics of a machine generated by a properly chosen set of constraints in combination with an appropriate drive system is designed to meet the prescribed requirements of some process, which is done by projecting the machine equations of motion on the process dynamics and thus by looking at the process as a system of additional constraints. The presentation presents a corresponding approach applying multibody system theory in combination with transformations from the machine side to the process side and vice versa. Practical aspects are discussed and an example given.

1 Machine-Process Dynamics

We consider a machine or mechanism on the one side and a process to be performed by this machine on the other side. Both systems possess a dynamical character and must be treated accordingly. We assume, that the process motion is given in some way, and we look for the motion of the machine necessary to realize the process requirements.

We start with a classical formulation of Lagrange I for multibody systems, see [1]:

$$\begin{aligned}
\boldsymbol{M}\ddot{\boldsymbol{z}} + \boldsymbol{f}^g - \boldsymbol{f}^e - \boldsymbol{f}^a - \boldsymbol{f}^p - \boldsymbol{f}^c = \boldsymbol{0}, &\quad \in \mathbb{R}^{6n} \\
\dot{\boldsymbol{\Phi}} = \boldsymbol{W}^T\dot{\boldsymbol{z}} + \bar{\boldsymbol{w}} = \boldsymbol{0}, &\quad \in \mathbb{R}^{p} \\
\ddot{\boldsymbol{\Phi}} = \boldsymbol{W}^T\ddot{\boldsymbol{z}} + [(\frac{d\boldsymbol{W}^T}{dt})\dot{\boldsymbol{z}} + (\frac{d\bar{\boldsymbol{w}}}{dt})] = \boldsymbol{W}^T\ddot{\boldsymbol{z}} + \hat{\boldsymbol{w}} = \boldsymbol{0}, & \\
\boldsymbol{f}^c = -\boldsymbol{W}(\boldsymbol{z}, t)\boldsymbol{\lambda}, & \qquad (1)
\end{aligned}$$

$\boldsymbol{M}$ is the symmetric mass matrix, $\boldsymbol{f}^g, \boldsymbol{f}^e, \boldsymbol{f}^a, \boldsymbol{f}^p, \boldsymbol{f}^c$ gyroscopic, external, driving, process and constraint forces or torques, $\boldsymbol{z}$ a set of non-minimal machine coordinates, $\boldsymbol{\Phi}$ the combined constraints of the machine and the process and $\boldsymbol{W}$ the constraint matrix. These equations can be resolved for $\ddot{\boldsymbol{z}}$ and $\boldsymbol{\lambda}$, resulting in

$$\begin{aligned}
\ddot{\boldsymbol{z}} &= -\boldsymbol{M}^{-1}(\boldsymbol{f}^g - \boldsymbol{f}^e - \boldsymbol{f}^a - \boldsymbol{f}^p + \boldsymbol{W}\boldsymbol{\lambda}) && \in \mathbb{R}^{6n} \\
\boldsymbol{\lambda} &= -(\boldsymbol{W}^T\boldsymbol{M}^{-1}\boldsymbol{W})^{-1}\left[\boldsymbol{W}^T\boldsymbol{M}^{-1}(\boldsymbol{f}^g - \boldsymbol{f}^e - \boldsymbol{f}^a - \boldsymbol{f}^p) - \hat{\boldsymbol{w}}\right] && \in \mathbb{R}^{p} \qquad (2)
\end{aligned}$$

A. Kecskeméthy and F. Geu Flores (Eds.): ECCOMAS 2019, COMPUTMETHODS 53, pp. 87–95, 2020.
https://doi.org/10.1007/978-3-030-23132-3_11

Projecting these equations from the machine $\boldsymbol{z}$-space to the process $\boldsymbol{w}$-space with the help of the transformation gradients $\dot{\boldsymbol{z}} = (\frac{\partial \boldsymbol{z}}{\partial \boldsymbol{w}})\dot{\boldsymbol{w}}$, $\ddot{\boldsymbol{z}} = \frac{d}{dt}(\frac{\partial \boldsymbol{z}}{\partial \boldsymbol{w}})\dot{\boldsymbol{w}} + (\frac{\partial \boldsymbol{z}}{\partial \boldsymbol{w}})\ddot{\boldsymbol{w}} = [\frac{\partial}{\partial \boldsymbol{w}}(\frac{\partial \boldsymbol{z}}{\partial \boldsymbol{w}})\dot{\boldsymbol{w}}]\dot{\boldsymbol{w}} + (\frac{\partial \boldsymbol{z}}{\partial \boldsymbol{w}})\ddot{\boldsymbol{w}}$ and regarding a quadratic relationship of the gyroscopic forces and the additional constraint term $\boldsymbol{f}^g = (\boldsymbol{\omega}\boldsymbol{\omega}^T)\boldsymbol{f}_{g0}$, $\hat{\boldsymbol{w}} = (\boldsymbol{\omega}\boldsymbol{\omega}^T)\hat{\boldsymbol{w}}_0$ the following equations can be generated

$$\boldsymbol{M}\{(\frac{\partial \boldsymbol{z}}{\partial \boldsymbol{w}})\ddot{\boldsymbol{w}} + [\frac{\partial}{\partial \boldsymbol{w}}(\frac{\partial \boldsymbol{z}}{\partial \boldsymbol{w}})\dot{\boldsymbol{w}}]\dot{\boldsymbol{w}}\} + (\boldsymbol{f}^g - \boldsymbol{f}^e - \boldsymbol{f}^a - \boldsymbol{f}^p + \boldsymbol{W}\boldsymbol{\lambda}) = \boldsymbol{0}. \quad (3)$$

Introducing $\boldsymbol{\lambda}$ from Eq. (2), regarding the transformation gradients and also taking into account $\boldsymbol{\omega} = \boldsymbol{H}\dot{\boldsymbol{q}}$, $\boldsymbol{f}^g = [(\boldsymbol{H}\frac{\partial \boldsymbol{q}}{\partial \boldsymbol{w}})(\dot{\boldsymbol{w}}\dot{\boldsymbol{w}}^T)(\boldsymbol{H}\frac{\partial \boldsymbol{q}}{\partial \boldsymbol{w}})^T]\boldsymbol{f}_{g0} = (\boldsymbol{H}^*(\dot{\boldsymbol{w}}\dot{\boldsymbol{w}}^T)\boldsymbol{H}^{*,T})\boldsymbol{f}_{g0}$, q being for example minimal coordinates, results in

$$\begin{aligned}
&(\frac{\partial \boldsymbol{z}}{\partial \boldsymbol{w}})(\frac{\partial \dot{\boldsymbol{w}}}{\partial \boldsymbol{w}})\dot{\boldsymbol{w}} + \left[\frac{\partial}{\partial \boldsymbol{w}}(\frac{\partial \boldsymbol{z}}{\partial \boldsymbol{w}})\dot{\boldsymbol{w}}\right]\dot{\boldsymbol{w}} + \left[\boldsymbol{G}(\boldsymbol{H}^*(\dot{\boldsymbol{w}}\dot{\boldsymbol{w}}^T)\boldsymbol{H}^{*,T})\boldsymbol{f}_{g0} + \right.\\
&\quad \left.\boldsymbol{M}^{-1}\boldsymbol{W}(\boldsymbol{W}^T\boldsymbol{M}^{-1}\boldsymbol{W})^{-1}(\boldsymbol{H}^*(\dot{\boldsymbol{w}}\dot{\boldsymbol{w}}^T)\boldsymbol{H}^{*,T})\hat{\boldsymbol{w}}_0\right] - \boldsymbol{G}(\boldsymbol{f}^e - \boldsymbol{f}^a - \boldsymbol{f}^p) = \boldsymbol{0},\\
&\qquad \boldsymbol{G} = \boldsymbol{M}^{-1}\left[\boldsymbol{E} - \boldsymbol{W}(\boldsymbol{W}^T\boldsymbol{M}^{-1}\boldsymbol{W})^{-1}\boldsymbol{W}^T\boldsymbol{M}^{-1}\right] \quad \in \mathbb{R}^{6n,6n},\\
&\boldsymbol{f}^c = -\boldsymbol{W}\boldsymbol{\lambda} = \boldsymbol{W}(\boldsymbol{W}^T\boldsymbol{M}^{-1}\boldsymbol{W})^{-1}\left[\boldsymbol{W}^T\boldsymbol{M}^{-1}(\boldsymbol{f}^g - \boldsymbol{f}^e - \boldsymbol{f}^a - \boldsymbol{f}^p) - \hat{\boldsymbol{w}}\right] \quad (4)
\end{aligned}$$

Whenever the constraint forces of the machine and the process are needed, this set will be an appropriate one.

Not considering the constraint forces and for this case generating the equations of motion can be achieved by a direct projection from machine to process coordinates by applying the orthogonality relation included in the Eq. (1), namely $\boldsymbol{W}^T(\frac{\partial \dot{\boldsymbol{z}}}{\partial \dot{\boldsymbol{w}}}) = \boldsymbol{W}^T(\frac{\partial \boldsymbol{z}}{\partial \boldsymbol{w}}) = \boldsymbol{0}$ or $(\frac{\partial \boldsymbol{z}}{\partial \boldsymbol{w}})^T\boldsymbol{W} = \boldsymbol{0}$. Multiplying Eq. (3) from the left with $(\frac{\partial \boldsymbol{z}}{\partial \boldsymbol{w}})^T$ gives after some rearrangements

$$\boldsymbol{M}_C\ddot{\boldsymbol{w}} + (\frac{\partial \boldsymbol{z}}{\partial \boldsymbol{w}})^T\{\boldsymbol{M}[\frac{\partial}{\partial \boldsymbol{w}}(\frac{\partial \boldsymbol{z}}{\partial \boldsymbol{w}})\dot{\boldsymbol{w}}]\dot{\boldsymbol{w}} + (\boldsymbol{f}^g - \boldsymbol{f}^e - \boldsymbol{f}^a - \boldsymbol{f}^p)\} = \boldsymbol{0} \quad \in \mathbb{R}^{6n}, \quad (5)$$

with $[\boldsymbol{M}_C = (\frac{\partial \boldsymbol{z}}{\partial \boldsymbol{w}})^T\boldsymbol{M}(\frac{\partial \boldsymbol{z}}{\partial \boldsymbol{w}})]$. This can be brought into the form

$$\begin{aligned}
\ddot{\boldsymbol{w}} + \{(\frac{\partial \boldsymbol{w}}{\partial \boldsymbol{z}})[\frac{\partial}{\partial \boldsymbol{w}}(\frac{\partial \boldsymbol{z}}{\partial \boldsymbol{w}})\dot{\boldsymbol{w}}]\dot{\boldsymbol{w}} + \boldsymbol{M}_C^{-1}(\frac{\partial \boldsymbol{z}}{\partial \boldsymbol{w}})^T(\boldsymbol{H}^*(\dot{\boldsymbol{w}}\dot{\boldsymbol{w}}^T)\boldsymbol{H}^{*,T})\boldsymbol{f}_{g0}\}\\
- \boldsymbol{M}_C^{-1}(\frac{\partial \boldsymbol{z}}{\partial \boldsymbol{w}})^T(\boldsymbol{f}^e + \boldsymbol{f}^a + \boldsymbol{f}^p) = \boldsymbol{0} \quad \in \mathbb{R}^p, \quad (6)
\end{aligned}$$

with the pseudo-inverse $[(\frac{\partial \boldsymbol{w}}{\partial \boldsymbol{z}}) = \left[(\frac{\partial \boldsymbol{z}}{\partial \boldsymbol{w}})^T(\frac{\partial \boldsymbol{z}}{\partial \boldsymbol{w}})\right]^{-1}(\frac{\partial \boldsymbol{z}}{\partial \boldsymbol{w}})^T]$. Expressing these tensors by sums gives the following result, where the gyroscopic term is lengthy but not evaluated here

$$\begin{aligned}
&\ddot{w}_\mu + \sum_{j,k}^{p} b_{\mu jk}\dot{w}_k\dot{w}_j + c_\mu = 0,\\
&b_{\mu jk} = \sum_{i=1}^{6n}\{(\frac{\partial w_\mu}{\partial z_i})\frac{\partial}{\partial w_j}(\frac{\partial z_i}{\partial w_k}) + [\boldsymbol{M}_C^{-1}(\frac{\partial \boldsymbol{z}}{\partial \boldsymbol{w}})^T(\boldsymbol{H}^*\boldsymbol{H}^{*,T})\boldsymbol{f}_{g0}]_{\mu jk}\}\dot{w}_j\dot{w}_k\\
&c_\mu = \left[\boldsymbol{M}_C^{-1}(\frac{\partial \boldsymbol{z}}{\partial \boldsymbol{w}})^T(\boldsymbol{f}^e + \boldsymbol{f}^a + \boldsymbol{f}^p)\right]_\mu. \quad (7)
\end{aligned}$$

This is a quadratic form similar to geodesics [2]. The symmetric coefficients $b_{\mu jk}$ contain all information of the machine and the process, for example geometry, topology, all constraints and of course all connections, whatsoever. The coefficient c_μ includes all external influence as expressed by forces.

Things become much simpler when considering only a one-dimensional process, a trajectory of any kind, with the property $p = 1$, $\boldsymbol{w} \rightarrow w = s$. The derivatives then reduce to $(\frac{\partial \boldsymbol{z}}{\partial w}) = (\frac{\partial \boldsymbol{z}}{\partial s}) = \boldsymbol{z}'$, $\dot{\boldsymbol{w}} = \dot{s}$, $\ddot{\boldsymbol{w}} = \ddot{s} = \frac{1}{2}(\dot{s}^2)'$, and the rotation term writes $(\boldsymbol{H}^*(\dot{\boldsymbol{w}}\dot{\boldsymbol{w}}^T)\boldsymbol{H}^{*,T})\boldsymbol{f}_{g0} = \boldsymbol{H}^2\boldsymbol{f}_{g0}(\dot{s}^2)$ with $\boldsymbol{H}^2 = (\boldsymbol{H}^*\boldsymbol{H}^{*,T})$. Equation (5) then writes

$$\frac{1}{2}[(\boldsymbol{z}')^T\boldsymbol{M}(\boldsymbol{z}')](\dot{s}^2)' + (\boldsymbol{z}')^T[\boldsymbol{M}\boldsymbol{z}'' + \boldsymbol{H}^2\boldsymbol{f}^{g0}](\dot{s}^2) - (\boldsymbol{z}')^T(\boldsymbol{f}^e + \boldsymbol{f}^a + \boldsymbol{f}^p) = \boldsymbol{0}, \quad (8)$$

which for further considerations will be put into the form

$$\begin{aligned} & A_i(s)(\dot{s}^2)' + B_i(s)(\dot{s}^2) + C_i(s) = T_i, \qquad (i = 1, 2, ...6n), \\ & A_i = \frac{1}{2}[(\boldsymbol{z}')^T\boldsymbol{M}(\boldsymbol{z}')]_i, \qquad B_i = (\boldsymbol{z}')_i^T[\boldsymbol{M}\boldsymbol{z}'' + \boldsymbol{H}^2\boldsymbol{f}^{g0}]_i, \\ & C_i = -(\boldsymbol{z}')_i^T(\boldsymbol{f}^e + \boldsymbol{f}^p)_i, \qquad T_i = ((\boldsymbol{z}')_i^T\boldsymbol{f}^a)_i. \end{aligned} \quad (9)$$

The constraint forces follow from the Eq. (4) in a comparable form

$$\begin{aligned} & f_i^c + \beta_i(s)(\dot{s}^2) + \gamma_i(s) = \tau_i(s), \qquad (i = 1, 2, ...6n), \\ & \beta_i = \{-\hat{\boldsymbol{W}}[\boldsymbol{W}^T\boldsymbol{M}^{-1}\boldsymbol{H}^2\boldsymbol{f}^{g0} + \boldsymbol{H}^2\dot{\boldsymbol{w}}_0]\}_i, \\ & \gamma_i = \{+\hat{\boldsymbol{W}}[\boldsymbol{W}^T\boldsymbol{M}^{-1}(\boldsymbol{f}^e + \boldsymbol{f}^p)]\}_i, \\ & \tau_i = \{-\hat{\boldsymbol{W}}[\boldsymbol{W}^T\boldsymbol{M}^{-1}\boldsymbol{f}^a]\}_i, \qquad \hat{\boldsymbol{W}} = \boldsymbol{W}(\boldsymbol{W}^T\boldsymbol{M}^{-1}\boldsymbol{W})^{-1} \end{aligned} \quad (10)$$

These are remarkable equations, especially Eq. (9). They represent a linear set of differential equations, linear in $(\dot{s}^2)$. The coefficients are highly nonlinear functions of the path coordinate s including the whole structure of the machine-process-combination (see above). Each machine component with its coordinates $\boldsymbol{z}_i$ is described by one of these equations. For design and layout purposes it is helpful to go the following way: A machine component may be equipped with some kind of drive ($\boldsymbol{f}^a \neq 0$), there may be some external forces ($\boldsymbol{f}^e$) or some additional process forces ($\boldsymbol{f}^p$). Any of these cases means $C_i \neq 0$. And even if $C_i = 0$, the following procedure can go as presented. All types of forces possess usually some maximum and some minimum value. Regarding this in defining C_i in more detail comes out with two equations for each i, the number (6n) is doubled. All (6n) equations then possess in the $(\dot{s}^2)'$ – $(\dot{s}^2)$ - plane two-times (6n) straight lines which altogether intersect to form a polygon. The properties of these polygons can be used for any design problems and purposes.

2 Motion Space

A motion space can be constructed by considering the Eq. (9) together with a vector $T_i = ((\boldsymbol{z}')_i^T \boldsymbol{f}^a)_i$ of machine driving torques or forces. We have 6n equations for p process coordinates $\boldsymbol{w}$, which are usually prescribed. In addition $6n > p$, so that we may use the set of 6n relations to build a motion space limited by the extreme drive torques or forces [5]. Taking into account these possible extreme values of $T_i \rightarrow (T_{i,min}, T_{i,max})$, we arrive at the following condition

$$T_{i,min} \leq [A_i(s)(\dot{s}^2)' + B_i(s)(\dot{s}^2) + C_i(s)] \leq T_{i,max}, \tag{11}$$

which is one of the limitations for the machine structure. Further limitation might be the following: The angular or translational velocities of constraining elements may be limited due to some maximum speeds of the drive train components, $(0 \leq [q_i'\dot{s}] \leq \dot{q}_{i,max})$, the path velocity itself might become constrained by some manufacturing process, $(0 \leq |r'|\dot{s} \leq +v_{max})$, the coefficient A_i being zero gives a kind of constraint considering Eq. (9), $T_{i,min} \leq B_i(s)(\dot{s}^2) + C_i(s)] \leq T_{i,max} \Longrightarrow (\dot{s}^2)_{A=0} = \frac{-C_i(s)+T_{i,min,max}}{B_i(s)}$, which are two vertical lines constraining the velocity $(\dot{s}^2)$. These examples, and possibly additional ones, define together a maximum velocity $\dot{s}_G$ along the path which must not be exceeded:

$$0 \leq (\dot{s}^2) \leq (\dot{s}^2)_G, \quad (\dot{s}^2)_G = \min\left[(\frac{\dot{q}_{i,max}}{q_i'})^2, \ (\frac{v_{max}}{|r'|})^2, \ (\dot{s}^2)_{A=0} \right]. \tag{12}$$

Considering the limiting cases $(T_{i,min}, T_{i,max})$ in connection with Eq. (9) we write

$$\begin{aligned} (\dot{s}^2)'_{max} &= \frac{-B_i(s)(\dot{s}^2) - C_i(s) + T_{i,max}}{A_i(s)}, \\ (\dot{s}^2)'_{min} &= \frac{-B_i(s)(\dot{s}^2) - C_i(s) - T_{i,min}}{A_i(s)}, \qquad (i = 1, 2, \cdots, n_D), \end{aligned} \tag{13}$$

which define two parallel straight lines in the $[(\dot{s}^2)', (\dot{s}^2)]$-plane, each for constant s. The number n_D indicates the number of driven components. It is usually

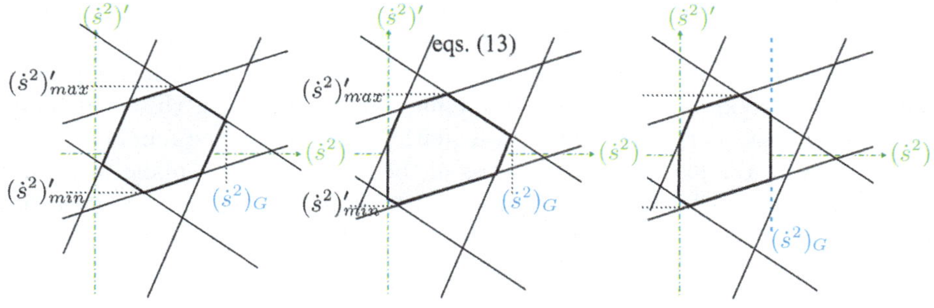

Fig. 1. Polygons of possible motion for s = constant [5]

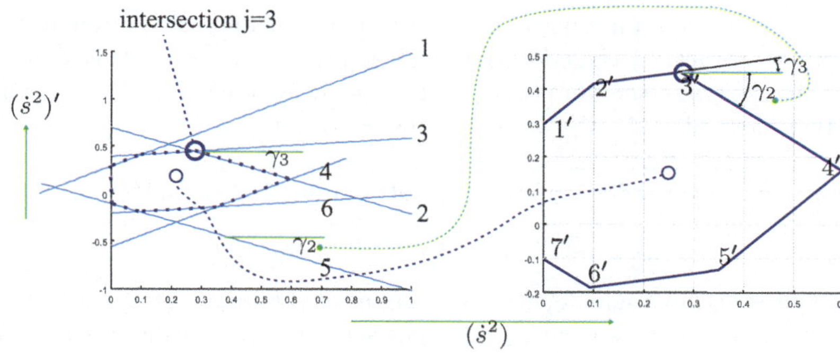

Fig. 2. Polygon evaluation ($\gamma_3 - \gamma_2 > 0$, $\gamma_2 - \gamma_4 < 0$, $\gamma_4 - \gamma_6 > 0$ with indices corresponding to straight line numbers)

smaller than the total number of components ($n_D \leq 6n$). The altogether ($2n_D$) straight lines form a polygon confined at the left side by the axis $(\dot{s}^2) = (\dot{s}^2)_{min} \geq 0$ and at the right side by the $(T_{i,min}, T_{i,max})$-straight-lines or by the constraint $(\dot{s}^2)_G$ given by Eq. (12). Without violating the constraints, motion can take place only within or on the polygons as shown in Fig. 1, where the situation for one path point s is illustrated. These graphs contain the following information:

Firstly, we immediately obtain the maximum possible velocity $\dot{s}^2_{max}$ for any process point s, from an intersection or from a $\dot{s}^2_G$-limit.

Secondly, for every path velocity $\dot{s}^2$ smaller than the maximum velocity $\dot{s}^2_{max}$ we can select two values of the derivative $(\dot{s}^2)'$ from the boundaries of the polygon, $((\dot{s}^2)'_{min}, (\dot{s}^2)'_{max})$. We call these points extremal points [4].

Combining all polygons for all path points s we obtain a constrained phase space bounded by ruled surfaces due to the straight line characteristics of the polygons [3]. We shall call this phase space mobility or motion space.

For the ***polygon evaluation*** we need the intersections of the straight lines as defined by the Eq. (13) and the sequence of points forming finally a polygon (Fig. 1). Starting with Eq. (13) we may write them in the form

$$(\dot{s}^2)' = a_{i0} + a_{i1}(\dot{s}^2), \qquad \text{with} \qquad a_{i0} = -\frac{C_i(s) + T_i}{A_i(s)}, \quad a_{i1} = -\frac{B_i(s)}{A_i(s)}, \tag{14}$$

where T_i stands for $T_{i,max}, T_{i,min}$ according to the line under consideration. An intersection of two lines (i, k) requires $(\dot{s}^2_i)' = (\dot{s}^2_k)'$ and $(\dot{s}^2_i) = (\dot{s}^2_k)$. From Eq. (14) we then come out with

$$(\dot{s}^2_i) = (\dot{s}^2_k) = -(\frac{a_{k0} - a_{i0}}{a_{k1} - a_{i1}}) \geq 0. \tag{15}$$

As any $(\dot{s}^2)$ must not be negative, $(\dot{s}^2) \geq 0$, we put $(\dot{s}^2) = 0$ and determine the relevant intersection points with $(\dot{s}^2) = 0$ in the case, that Eq. (15) results in a negative value (for example the middle situation of Fig. 1).

For starting the evaluation of the polygon for s = const. we have to prescribe a starting point number one, for example the intersection of two lines i and k, usually at the left minimum side of the polygon and very often just zero, depending on practical considerations. Then progressing from this point j we look for the next intersection point (j + 1) as near as possible to the known one, by applying formula (15) and watching the condition

$$\Delta r = \sqrt{[(\dot{s}_j^2)' - (\dot{s}_{j+1}^2)']^2 + [(\dot{s}_j^2) - (\dot{s}_{j+1}^2)]^2} = \min[\Delta r_{j \in 2n_D}; \Delta r > 0, \Delta\gamma > 0],$$
$$\Delta\gamma = \text{atan}(a_{j1}) - \text{atan}(a_{j+1,1}). \quad (16)$$

with the intersection angle $\Delta\gamma$ from the slopes of the intersecting lines (Eq. 14):

We proceed that way until $\Delta\gamma$ changes sign thus indicating the right maximum side of the polygon. Then going back to the starting point number one requires a change of sign in the relations (16). As a result we come out with as much polygon points as we have lines ($2n_D$) augmented by one. Knowing these polygon corners we are able to evaluate the area covered by the polygon. According to the Gauss trapeze rule we have

$$A_{polygon} = \frac{1}{2}\sum_{j=1}^{n_D}[(\dot{s}_{j+1}^2) - (\dot{s}_j^2)][(\dot{s}_{j+1}^2)' + (\dot{s}_j^2)'], \quad (17)$$

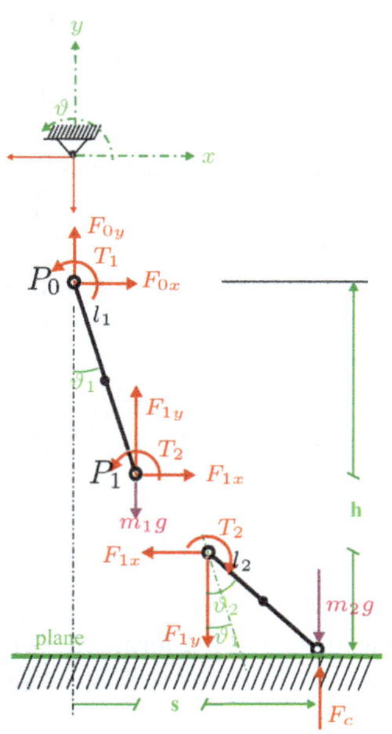

Fig. 3. Double pendulum sliding on a plane

where the meaning is obvious. The algorithm is illustrated by Fig. 2, where we have three pairs of parallel straight lines numbered 1 to 6, which intersect. We start with the point intersecting the ordinate, which corresponds to the negative intersection point of the two straight lines 1 and 5. But as $(\dot{s}^2) \geq 0$, always, we must put back any negative intersection to the ordinate or to $(\dot{s}^2) = 0$. We then step forward according to the relation (16), until we have reached point 4, which corresponds to $(\dot{s}^2) = (\dot{s}_{max}^2)$. This point is indicated by a negative angle $\Delta\gamma = \text{atan}(a_{j1}) - \text{atan}(a_{j+1,1})$. We then go backward to point 7 and from there to point 1, the starting point. (The numbers on the left graph of Fig. 2 are the numbers of the straight lines, whereas the numbers on the right graph are those of the polygon corners.)

Classical problems in the area of robots and mechanisms are minimum time along a path or time restrictions together with force limitations [4,7]. This might be done directly or by adapting the size if some mobility space with respect to the problem. Example is polishing of surfaces. It makes sense that for such cases and beyond the given drives, the design parameters of a machine are optimized. The necessary information is contained in the motion space. Its size is mainly given by the prescribed torques, and its form by the machine parameters. Therefore also its volume follows the given torques, but to a certain extent also the machine parameters. This allows an optimization with torques fixed and the machine parameters defined by design limits.

3 Example Double Pendulum

The double pendulum case is illustrated by Fig. 3 being self-explaining. Working with three coordinates $[\boldsymbol{z} = (x_1\ y_1\ x_2\ y_2)^T,\ \boldsymbol{q} = (\vartheta_1\ \vartheta_2)^T,\ s = x_2]$, where the $\boldsymbol{q}$'s are minimal coordinates, we have to watch the sequence $\boldsymbol{z}' = (\frac{\partial \boldsymbol{z}}{\partial \boldsymbol{q}})\boldsymbol{q}'$. The pendulum description and the constraints are

$$\begin{pmatrix} x_1 \\ y_1 \end{pmatrix} = l_1 \begin{pmatrix} +\sin\vartheta_1 \\ -\cos\vartheta_1 \end{pmatrix}, \qquad \begin{pmatrix} x_2 \\ y_2 \end{pmatrix} = \begin{pmatrix} +l_1\sin\vartheta_1 + l_2\sin\vartheta_2 \\ -l_1\cos\vartheta_1 - l_2\cos\vartheta_2 \end{pmatrix} \tag{18}$$

and with the pendulum constraints $(\Delta x = (x_2 - x_1),\ \Delta y = (y_2 - y_1))$

$$\ddot{\boldsymbol{\Phi}} = \boldsymbol{W}^T \ddot{\boldsymbol{z}} + \hat{\boldsymbol{w}} = \begin{pmatrix} x_1 & y_1 & 0 & 0 \\ -\Delta x & -\Delta y & +\Delta x & +\Delta y \\ 0 & 0 & 0 & 1 \end{pmatrix} \cdot \begin{pmatrix} \ddot{x}_1 \\ \ddot{y}_1 \\ \ddot{x}_2 \\ \ddot{y}_2 \end{pmatrix} + \begin{pmatrix} (\dot{x}_1^2 + \dot{y}_1^2) \\ (\dot{x}_2 - \dot{x}_1)^2 + (\dot{y}_2 - \dot{y}_1)^2 \\ 0 \end{pmatrix} = \boldsymbol{0}. \tag{19}$$

The constraint include the "machine-constraints" in form of the pendulum relations for the link length and also the "process-constraints" in form of the horizontal line in Fig. 3. Some numerical results are shown below.

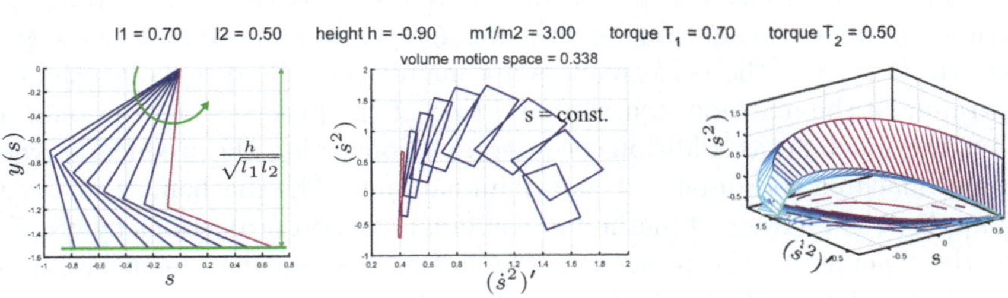

Fig. 4. Typical results for a driven double pendulum with relevant motion space

The results presented here are a selection of many simulations on the basis of above relations. Figure 4 depicts on the left side the configuration of a double pendulum sliding along a given path with the coordinate $y = -h = constant$. It is assumed, that this represents a bilateral constraint with no unilateral effects like contact or detachment. As a consequence, the constraint forces might become negative. At each path point s we get a limited area of accelerations and velocities enclosed by a parallelepiped with four corners according to the two degrees of freedom of the double pendulum, see middle graph of Fig. 4. These parallelepipeds are composed to a volume by connecting them along the path coordinate s, right chart of Fig. 4. The resulting structure represents the limited space, where motion for the chosen parameters can take place.

Figure 5 shows the relevant constraint forces for the data given in Fig. 4. We have selected the forces of joint 1 (left graph) and the constraint y = −h, right and middle, for the last case also some of the quadrangles in the middle graph. The constraint forces are evaluated under the assumption of a velocity range $(0 \leq (\dot{s}^2) \leq (\dot{s}^2)_{max})$, where $(\dot{s}^2)_{max}$ follows from the quadrangle structure of Fig. 4. This makes sense physically.

The results provide us at least with an indication, that mobility spaces represent a powerful tool for considering machine-process-combinations.

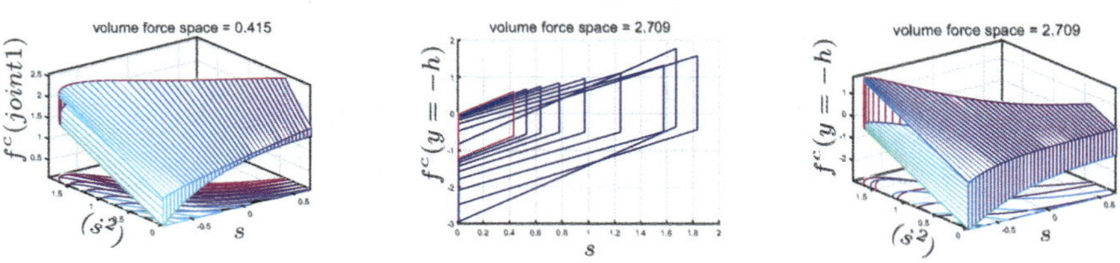

Fig. 5. Constraint forces (data of Fig. 4)

4 Summary

A combination of a machine performing a process is considered, where the geometric structure of the machine and the kinematics (dynamics) of the process are given. In both cases the performances are mainly influenced by constraints, the constraints of the machine structure and that of the process. The last ones may act on the machine as additional constraints converting the frequently known minimal machine coordinates to non-minimal ones. The mathematical description applies the Lagrange I method in combination with projections by gradients from the machine to the process or vice versa. The equations can be reduced to quadratic forms, which allow an interpretation by motion spaces, for general cases of quite complex structure, for the one-dimensional case representing trajectories of rather clear form. The method offers a new design tool for machines and mechanisms. A double pendulum example is given.

References

1. Bremer, H.: Elastic Multibody Dynamics. Springer Science+Business Media B.V., Dordrecht (2008)
2. Duschek, A., Hochrainer, A.: Grundzüge der Tensorrechnung in analytischer Darstellung. Springer, Wien (1960)
3. Hoschek, J.: Liniengeometrie. B. I. Hochschulskripten, Bibliographisches Institut Zürich (1971)
4. Johanni, R.: Optimale Bahnplanung bei Robotern. Fortschritt-Berichte VDI, Reihe 18, Nr. 51, VDI-Verlag Düsseldorf (1988)
5. Pfeiffer, F.: Optimal trajectory planning for manipulators. In: Systems and Control Encyclopedia. Pergamon Press, Oxford, New York (1990)
6. Pfeiffer, F.: Mechanical System Dynamics. Springer, Heidelberg (2008)
7. Richter, K.: Kraftregelung elastischer Roboter. Fortschritt-Berichte VDI, Reihe 8, Nr. 259, VDI-Verlag Düsseldorf (1991)

Modeling of Elastic Cages in the Rolling Bearing Multi-Body Tool CABA3D

Dmitry Vlasenko(✉) and Bodo Hahn(✉)

Schaeffler Technologies AG & Co. KG, Industriestr. 1-3, 91074 Herzogenaurach, Germany
{vlasedit,hahnbdo}@schaeffler.com

Abstract. The paper is concerned with the dynamical simulation of rolling bearings with elastic cages in the simulation software CABA3D (Computer Aided Bearing Analyzer 3D). The modeling process (import of finite element models, model reduction, model verification) is considered in detail. Two different contact simulation methods (slice contact and node-to-surface contact approaches) are shown and compared.

1 Introduction

Simulation of rolling bearing dynamics is extremely helpful for the understanding of processes acting in bearings. The information about kinematics, load distribution and friction forces of bearing elements (rings, cages, rolling elements, etc.) provided by simulation software is needed for the cost-effective optimization of bearing design as well as prediction and prevention of bearing failures.

Nowadays CABA3D (Computer Aided Bearing Analyzer 3D) developed by Schaeffler Technologies is one of the most powerful industrial software tools for the simulation of dynamic processes in rolling bearings. CABA3D, in contrast to commercial, general-purpose multi-body software, has elasto-hydrodynamic friction and contact models specially designed for rolling bearings [2].

Lightweight cages are widely used in distinct types of rolling bearings. The advantages of the low cage weight include the reduction of inertial and centrifugal forces while maximizing the lubricant effectiveness. Often, lightweight cages cannot be simulated as rigid bodies since the cages' elasticity has considerable influence on the dynamic behavior. Investigation of many effects like deformation of cage pockets, cage-instability, the response on impacts or stresses and fatigue is only possible if the deformation of the cages is considered.

2 Model Order Reduction Techniques in CABA3D

Finite element (FE)-models of elastic cages usually have several thousand degrees of freedom and extremely high eigenfrequencies. This entails the need for reduction of the models before the model dynamics is simulated.

A. Kecskeméthy and F. Geu Flores (Eds.): ECCOMAS 2019, COMPUTMETHODS 53, pp. 96–103, 2020.
https://doi.org/10.1007/978-3-030-23132-3_12

Since the deformations of cages are small compared to their rigid body motion in rotational and translational direction, the most suited reduction methods in this case are methods based on the floating frame formulation [7, 8, 11]. According to the method, the total motion of an elastic body is divided into two parts: rigid body motion represented by the motion of the body reference frame and deformations with respect to this frame. In this method, the position d^i of the i-th node in the global inertial frame over time t is expressed as

$$\mathbf{d}^i(t) = \mathbf{x}(t) + \mathbf{A}(t) \cdot \left(\bar{\mathbf{r}}_0^i + \bar{\mathbf{u}}^i(t)\right) \quad (1)$$

where $\mathbf{x}$ denotes a position for the origin of the body reference frame in the global coordinate system, $\mathbf{A}$ is a body rotation matrix, $\bar{\mathbf{r}}_0$ is the undeformed local position of the node in the body frame, and $\bar{\mathbf{u}}^i$ is the local nodal displacement.

The model reduction techniques assume that the deformations vector $\bar{\mathbf{u}} = \left(\bar{\mathbf{u}}^1 \quad \ldots \quad \bar{\mathbf{u}}^N\right)$ can be approximated by linear combinations of several generalized deformation modes (Eq. 2).

$$\bar{\mathbf{u}} = \mathbf{S} \cdot \mathbf{q} \quad (2)$$

$\mathbf{S}$ is the modal matrix, consisting on deformation modes, and $\mathbf{q}$ is the vector of modal coordinates. It is assumed that the count of modes is much less than the count of nodes in the FE-model, e.g. the size of $\mathbf{q}$ is much less than the size of $\bar{\mathbf{u}}$.

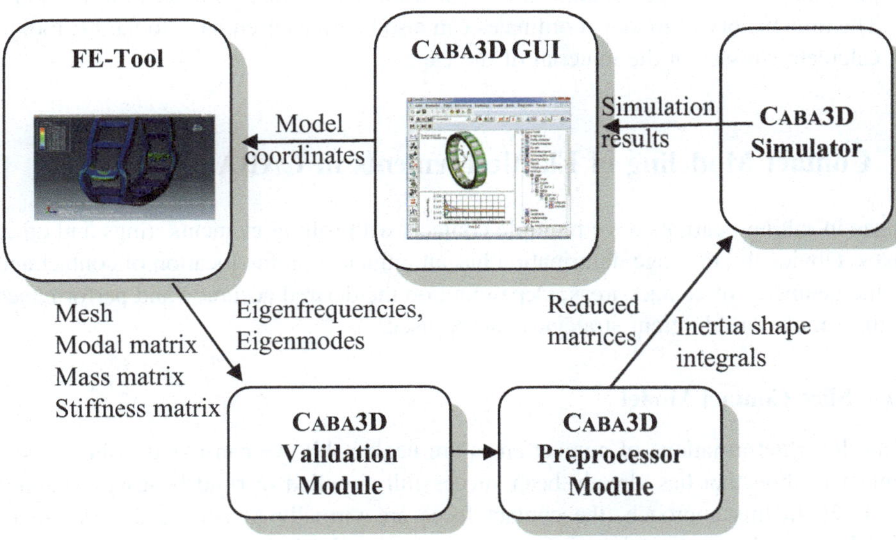

Fig. 1. Simulation scheme of elastic cages

The advantages of reduced models are low computational cost and small simulation results file size compared to FE simulations. At the same time, reduced models can also create an accurate simulation of macro- and micro-elasticity of cages (e.g. ovalization of cages, deformation contacting faces, etc.) [10].

The standard methods of modal reduction (e.g. Craig-Bampton [3]), are already implemented in FEA tools (e.g. ABAQUS, ANSYS) and can be used for the simulation of elastic cages in CABA3D.

The simulation scheme of elastic cages is shown in Fig. 1:

1. On the first stage modal matrix, mass matrix and stiffness matrix, together with the eigenfrequencies and eigenmodes, are imported from the FE-tool to the CABA3D Validation Module where the quality of model reduction is validated using different methods: Normalized Relative Eigenfrequency Difference (NRED), Modal Assurance Criterion (MAC), comparison of frequency response, etc. [5, 7].
2. If the quality of the reduction is sufficient, the next modeling step is started. The model parameters (modal matrix, mass matrix, stiffness matrix, mesh data) are imported to the CABA3D Preprocessor Module. The module generates the reduced mass, stiffness and damping matrices together with the time-constant inertia shape integrals of the reduced model [7, 11].
3. During the third step, the dynamics of a bearing model including the elastic cage is simulated over time. The simulation results (coordinates, velocities, forces, pressures, etc.) are calculated.
4. The simulation results are analyzed in CABA3D GUI. Using motion data, application engineers generate 2D and 3D diagrams showing time changes of needed parameters (forces, deformations, etc.) or make the animation of bearing motion.
5. The time history of modal coordinates can also be transferred back to the FE-tool to calculate stresses in the material of the cage.

3 Contact Modeling of Elastic Elements in CABA3D

Cages in rolling bearings have multiple contacts with rolling elements, rings and other cages. Obviously, the cage deformation has an influence on the location of contact and on the geometry of contact areas. Depending on the desired accuracy and performance of the simulation, different strategies can be used.

3.1 Slice Contact Model

If the local deformations of contact areas are neglectable we can use the slice model contact method that has already been successfully applied in rigid bearing dynamics [2, 4, 9]. In this approach, the contact faces are virtually treated as discrete slices (circular or polygonal) and analytical surfaces, as it is shown in Fig. 2. To find the intersection points between slices and surfaces the analytical root finding algorithm is used [9]. The contact forces and torques acting on contacting bodies are calculated as a sum of forces and torques acting on each slice. The method is accurate and numerically efficient in the case of simple geometries (e.g. cylinder, torus).

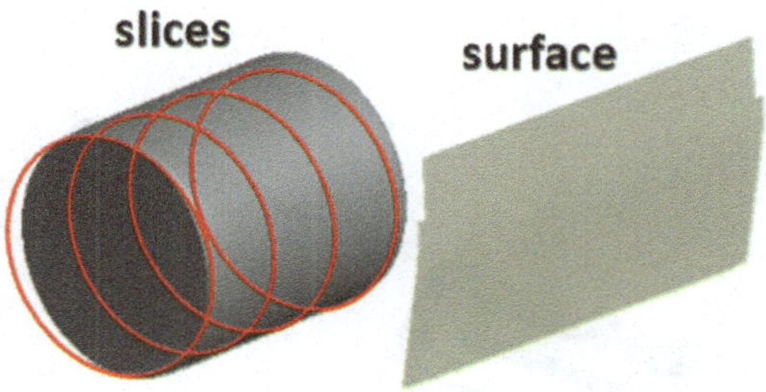

Fig. 2. Slice contact model

To account for the influence of elastic deformations of contact areas, the distributed coupling algorithm [1] can be used. In this algorithm, a contact surface is connected to a group of coupling FE-nodes, as it is shown in Fig. 3.

Fig. 3. Coupling of a contact area with FE-Nodes

The position and the velocity of the area coordinate system with respect to the body frame are calculated from the displacements of coupling nodes. The contact forces and moments acting on the surface are distributed at each coupling node.

3.2 Node-to-Surface Contact

The node-to-surface contact model is used if local deformations are important. In this case, every FE-node on the surface of the cage is considered and checked whether it has contact with other bearing elements, as it is shown in Fig. 4.

The distance from FE-nodes to the rigid surface of the contact partner is calculated analytically [6]. In comparison to the slice contact model, this approach is more accurate and can be implemented for all surface types. The disadvantage of the method is a higher computational cost because of the big number of nodes.

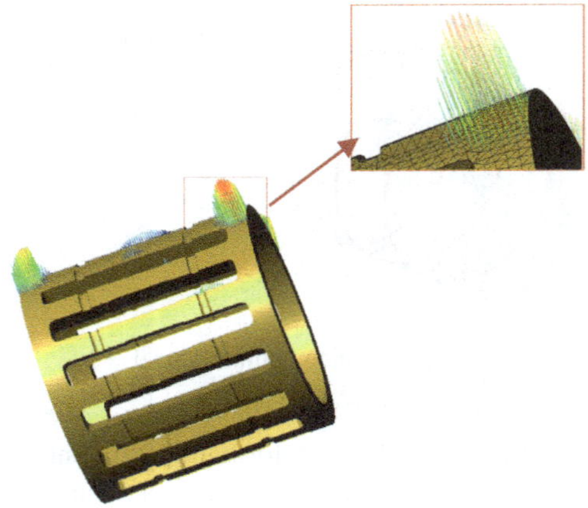

Fig. 4. Pressure distribution in the node-face contact

4 Simulation Results

4.1 Validation

One important step before the models are used in real applications is validation. In the following, two examples are presented.

The first one shows a comparison of the deformation and stress calculation using CABA3D with the results of a finite element analysis. A simple static load case is used for this comparison. Figure 5 shows very good correlation for both stresses and deformation. Even stress concentrations are calculated accurately which is important for a potential strength assessment in a postprocessing step.

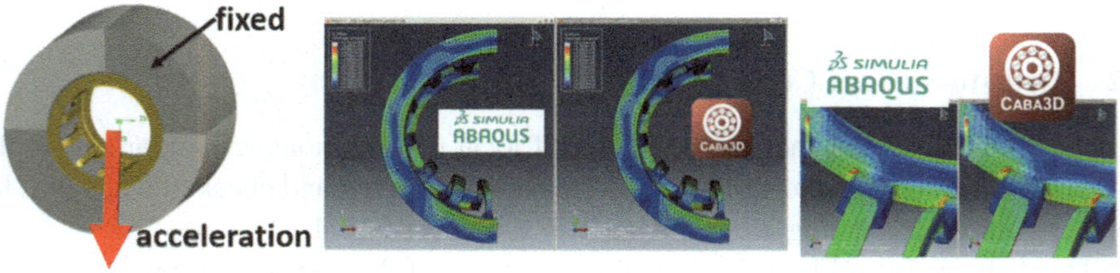

Fig. 5. Comparison of stresses and deformations

The second example (Fig. 6) shows the contact pressure distribution between the cage and the outer ring due to constant acceleration. It can be seen clearly that the slice model cannot represent such a complex shape. Some contact points are not even located on the cage. However, if we look at the node-to-surface model the pressure distribution is almost identical to the finite element analysis.

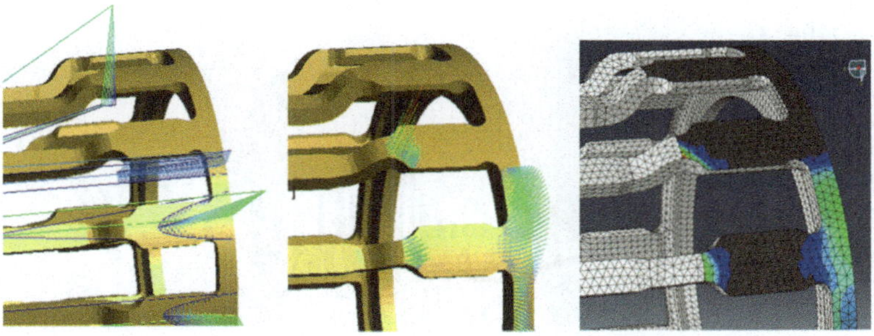

Fig. 6. Contact pressure: slice model, node-to-surface model, FE analysis

4.2 Application Example

Figure 7 shows a common application example of the elastic cage model: A two-stroke crank drive that contains four bearings. Especially the cages at the crank pin and at the piston pin experience high loads due to acceleration. These two cages are modeled with the described elastic cage model. All contacts between rollers and cages are modeled with the slice model due to simple cage pocket shapes. The contacts between cages and their outer guidance surface use the node-to-surface model.

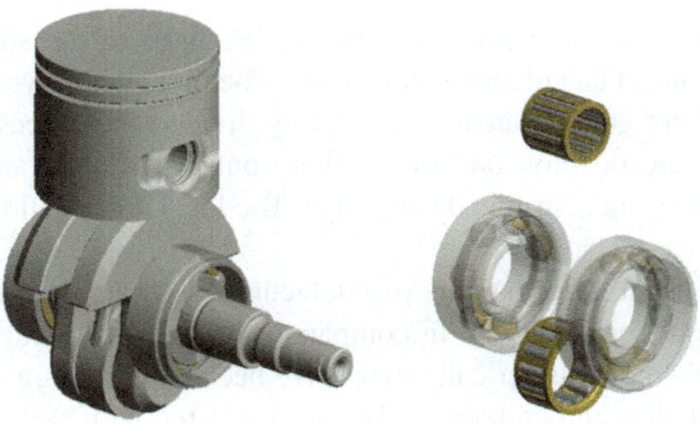

Fig. 7. Crank drive model with four bearings

Figure 8 depicts one example that shows that the modeling of the cage can have a significant influence on the simulation results. The diagram shows the forces between the cage and the guidance surface at the conrod. An impact occurs due to the contact alteration of the cage between the top and bottom dead center of the piston. When using a rigid cage model, the forces would be highly overestimated during the impact phase. However, there is good agreement between the two models during the time span with constant acceleration.

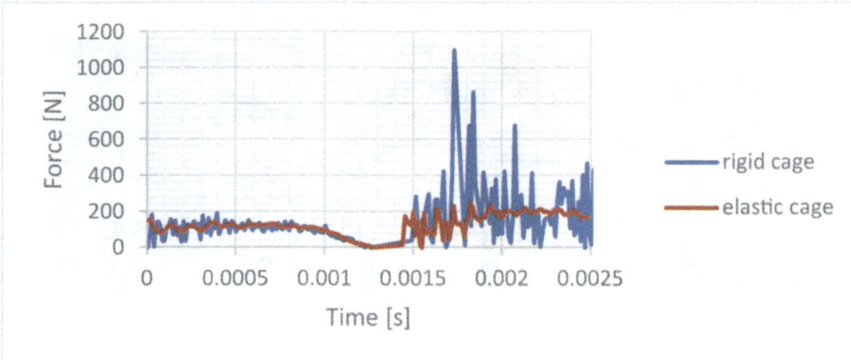

Fig. 8. Contact force: cage to conrod

5 Conclusions

In this article, the modeling of elastic cages in the rolling bearing multi-body tool CABA3D is described. It is shown that the Craig-Bampton reduction method is well suited for the simulation of cage dynamics. The advantages of the approach are the minor computational costs, small simulation results file size and accurate simulation of elasticity effects.

Depending on the desired accuracy and performance of the simulation, different strategies for the contact calculation can be used. The first alternative is the slice-to-face contact model, where contact areas are virtually divided into slices. The cage deformations influence the position and orientation contact surfaces, since they are connected to FE-nodes using distributed couplings. The local deformations of contact areas are neglected.

In contrary, if local deformations of contacting faces are important, the node-to-surface contact model can be used. In comparison to the slice model, this approach is more accurate but is more numerically expensive because of the high number of nodes.

Reduction strategies and contact models are validated with high-level FE methods. The example of the crank drive shows that the developed elastic cage model can be used for practical applications. It enables the development and improvement of new cage designs.

References

1. Abaqus Theory Manual, Abaqus Version 6.13 Edition
2. Bakolas, V., Koch, O.: Bearing optimization using a 3D-dynamic simulation tool. In: Proceedings of STLE Annual Meeting & Exhibition. Florida (2009)
3. Craig, R.R.: Coupling of substructures for dynamic analyses: an overview. In: Structures, Structural Dynamics and Material Conference, 41st AIAA/ASME/ASCE/AHS/ASC. Atlanta (2000)

4. Grillenberger, H., Hahn, B., Koch, O.: Elastische Käfiginstabilität in Wälzlagern – Simulation und Versuch. Tagungsband 16. Antriebstechnisches Kolloquium, 1, pp. 153–164. Auflage (2015)
5. Hahn, B., Gaile, S., Vlasenko, D.: Detailed cage analysis using CABA3D. In: Proceedings of the 73rd STLE Annual Meeting and Exhibition. Minneapolis (2018)
6. Konyukhov, A., Schweizerhof, K.: Computational Contact Mechanics. Springer, Heidelberg (2013)
7. Makhavikou, V.: Line-fitting method of model order reduction for elastic multibody systems. unv. Diss., Universität Magdeburg (2015)
8. Shabana, A.: Dynamics of Multibody Systems, 2nd edn. Cambridge University Press, Cambridge (1998)
9. Vesselinov, V.: Dreidimensionale Simulation der Dynamik von Wälzlagern. Karlsruhe (2003)
10. Vlasenko, D.: Dynamical simulation of rolling bearings with elastic parts in CABA3D. In: Ambrósio, J.A.C.; European Mechanics Society: Rolling Contact Mechanics for Multibody System Dynamics: EUROMECH Colloquium 578, Funchal, Madeira, Portugal, 10–13 April 2017, Funchal (2017)
11. Wallrapp, O., Schwertassek, R.: Dynamik flexibler Mehrkörpersysteme. Braunschweig/Wiesbaden (1998)

Analysis of the Influence of the Links' Flexibility and Clearance Effects on the Dynamics of the RUSP Linkage

Krzysztof Augustynek(✉) and Andrzej Urbaś

University of Bielsko-Biala, Bielsko-Biala 43-309, Poland
{kaugustynek, aurbas}@ath.bielsko.pl

Abstract. A mathematical model of the RUSP linkage with a flexible link and clearance in a revolute joint is presented. In the proposed model it is assumed that the clearance exists only in the cut-joint. A new spatial model of the revolute joint with a radial and axial clearance is proposed. In this approach, the clearance joint is discretized by means of the contact elements located on the cylindrical and frontal surfaces of the journal and bearing which can automatically detect collisions between contacting bodies. The normal contact force is modelled using the Nikravesh-Lankarani formula. The LuGre friction model is applied to model tangent friction force. Flexible link is discretized by means of the Rigid Finite Element method. In simulations, the influence of the link's flexibility and clearance in the revolute joint on the dynamic response of the linkage is analyzed.

1 Introduction

The clearance exists in real joints of linkages and it is the result of a wear in joints, machining and assembly errors. Spatial models of the clearance in revolute joints are subject of analysis of many papers [2, 3, 5–7]. Some of the papers take into account only radial clearance [2, 3] whilst other ones [5–7] analyse radial and axial clearances together. The significant contribution to the development of mathematical models of the clearance in spatial linkages with flexible components is the paper [3].

The mathematical model of the spatial RUSP linkage with the clearance in the revolute joint is developed in the paper. A novel spatial model of the clearance in the revolute joint is formulated. This model takes into account the radial and axial clearances and allows to analyse different scenarios of the contact between the journal and bearing without the need to formulate additional conditions. Due to application of the joint coordinates in description of kinematics of the linkage it is assumed that clearance exists only in the cut-joint. In numerical simulations the influence of the clearance and link's flexibility on the dynamics of the RUSP linkage are analysed.

A. Kecskeméthy and F. Geu Flores (Eds.): ECCOMAS 2019, COMPUTMETHODS 53, pp. 104–111, 2020.
https://doi.org/10.1007/978-3-030-23132-3_13

2 Mathematical Model of the RUSP Linkage with the Clearance

The analyzed RUSP linkage is presented in Fig. 1. The linkage is built of four links and loaded by means of driving ($\mathbf{t}_{dr}^{(1,1)}$) and resistance ($\mathbf{t}_{res}^{(1,1)}$) torques. The cut-joint method is used to transform the linkage with the closed-loop into the open-loop structure. The linkage is cut at the universal joint(U) and as a result the two open-loop kinematics chains ($c \in \{1,2\}$) are obtained. Each chain contains two links ($n_l^{(c)}\Big|_{c=1,2} = 2$) which are denoted $(c,l)\big|_{\substack{c=1,2 \\ l=1,\ldots,n_l^{(c)}}}$.

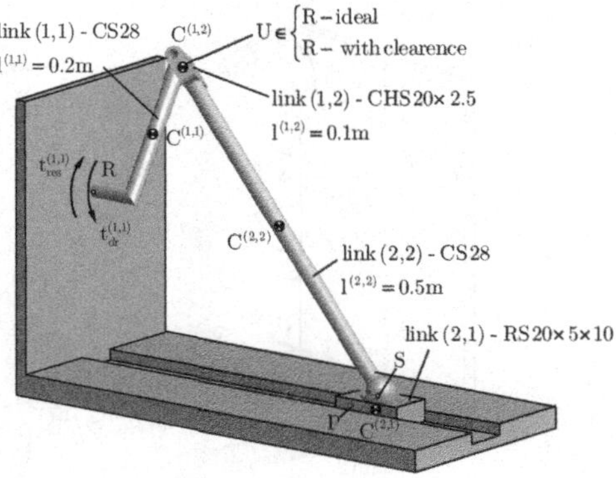

Fig. 1. Model of RUSP linkage.

It is assumed that the friction and clearance are considered only in the cut-joint and other joints are treated as ideal. The formalism of joint coordinates and homogeneous transformation matrices are applied to describe the kinematics of the linkage considered [8]. Joint coordinates vectors describing the motion of the particular links with respect to previous ones can be written as follows:

- subchain 1:

$$\mathbf{q}^{(1)} = \left(q_i^{(1)}\right)_{i=1,\ldots,n_{dof}^{(1)}} = \left[\psi^{(1,1)} \quad \psi^{(1,2)}\right]^T, \tag{1}$$

- subchain 2:

$$\mathbf{q}^{(2)} = \left(q_i^{(2)}\right)_{i=1,\ldots,n_{dof}^{(2)}} = \left[z^{(2,1)} \quad \psi^{(2,2,0)} \quad \theta^{(2,2,0)} \quad \varphi^{(2,2,0)} \quad \tilde{\mathbf{q}}_f^{(2,2)^T}\right]^T, \tag{2}$$

where $\tilde{\mathbf{q}}_f^{(2,2)^T}$ is vector describing components of the generalized coordinates vector due to link (2,2) flexibility.

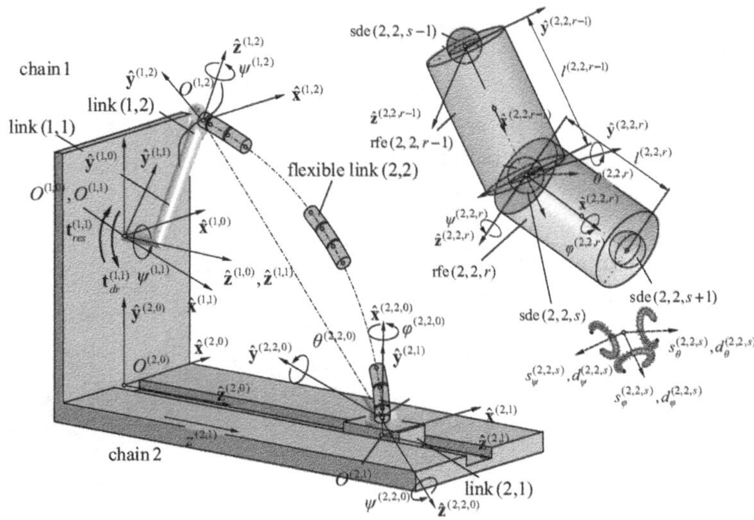

Fig. 2. Coordinate systems assignment and discrete model of the coupler.

It is assumed that the coupler can be treated as a rigid or a flexible. The Rigid Finite Element Method is applied to discretize the flexible coupler [2, 8]. In this method flexible link is replaced by the set of rigid finite elements (rfe) interconnected by means of the massless and dimensionless spring damping elements (sde) (Fig. 2). Generalized forces vector resulting from the elastic deformations of the flexible link can be expressed in the following form:

$$\mathbf{Q}_f^{(2,2)} = \mathbf{S}^{(2,2)}\tilde{\mathbf{q}}_f^{(2,2)} + \mathbf{D}^{(2,2)}\dot{\tilde{\mathbf{q}}}_f^{(2,2)}, \tag{3}$$

where $\tilde{\mathbf{q}}_f^{(2,2)} = \left[\tilde{\mathbf{q}}^{(2,2,1)^T} \quad \ldots \quad \tilde{\mathbf{q}}^{(2,2,r)^T} \quad \ldots \quad \tilde{\mathbf{q}}^{(2,2,n_{rfe}^{(2,2)}-1)^T}\right]^T$, $\tilde{\mathbf{q}}^{(2,2,r)} = \left[\psi^{(2,2,r)} \quad \theta^{(2,2,r)} \quad \varphi^{(2,2,r)}\right]^T$ is generalized coordinates vector of rfe(2,2, r), $\mathbf{S}^{(2,2)}$, $\mathbf{D}^{(2,2)}$ are diagonal matrices containing stiffness and damping coefficients of sdes.

The proposed model of a spatial revolute joint with radial and axial clearances is shown in Fig. 3. In the presented approach the bearing and journal are connected by means of $n_{ce}^{(r)}$ radial contact elements (ce$_r$) located around perimeter of the bearing on n_r levels measured along $\hat{\mathbf{z}}^{(b)}$ axis and $n_{ce}^{(a)} \times n_a$ axial contact elements (ce$_a$) located on the frontal surfaces of the journal and bearing. It is assumed that the clearance is small and rotations of $\hat{\mathbf{z}}^{(j)}$ axis with respect to $\hat{\mathbf{z}}^{(b)}$ are also small.

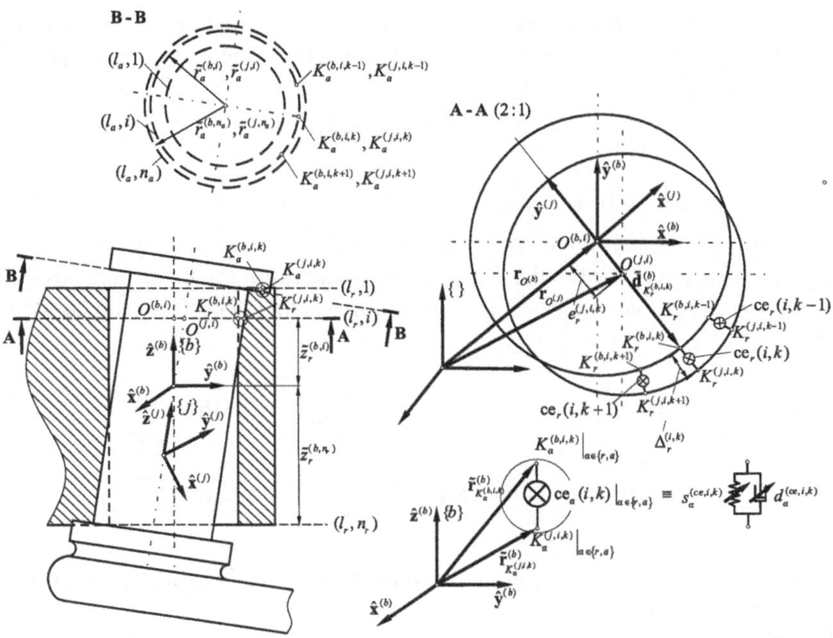

Fig. 3. Spatial model of the revolute joint.

The contact elements are active when the penetration depth due to impact between the bearing and journal is nonzero. Radial and axial penetrations for $\mathrm{ce}_r(i,k)$ and $\mathrm{ce}_a(i,k)$ can be calculated as follows:

$$\Delta_r^{(i,k)} = \left\| \tilde{\mathbf{r}}^{(b)}_{K_r^{(j,i,k)}} - \tilde{\mathbf{r}}^{(b)}_{K_r^{(b,i,k)}} \right\|, \tag{4.1}$$

$$\Delta_a^{(i,k)} = \left\| \tilde{z}^{(b)}_{K_a^{(j,i,k)}} - \tilde{z}^{(b)}_{K_a^{(b,i,k)}} \right\|, \tag{4.2}$$

where $\tilde{\mathbf{r}}^{(b)}_{K_r^{(j,i,k)}}$, $\tilde{\mathbf{r}}^{(b)}_{K_r^{(b,i,k)}}$ are position vectors of the radial contact points expressed in bearing frame $\{b\}$, $\tilde{z}^{(b)}_{K_a^{(j,i,k)}}$, $\tilde{z}^{(b)}_{K_a^{(b,i,k)}}$ are z components of the axial contact points expressed in frame $\{b\}$. Normal contact force acting at contact points is modelled by means of the Lankarani and Nikravesh formulae given by [4]:

$$f_{n,\alpha}^{(i,k)}\Big|_{\alpha\in\{r,a\}} = d_\alpha^{(ce,i,k)} \dot{\Delta}_\alpha^{(i,k)} + s_\alpha^{(ce,i,k)} \Delta_\alpha^{(i,k)}, \tag{5}$$

where $d_\alpha^{(ce,i,k)} = \frac{3\, s_\alpha^{(ce,i,k)} \Delta_\alpha^{(i,k)^n} \left(1-k_e^2\right)}{4 \cdot \dot{\Delta}_\alpha^{(i,k)}}$ are damping coefficients depending on the restitution coefficient k_e, the penetration velocity before the impact ${}^{-}\dot{\Delta}_\alpha^{(i,k)}$ and exponent for metallic bodies $n = 1.5$, $s_\alpha^{(ce,i,k)}$ are generalized stiffness coefficients depending on a material and geometric properties of contacting bodies defined in [4]. Tangent friction force is modelled by the LuGre friction model which can be expressed as [1]:

$$f_{t,\alpha}^{(i,k)}\Big|_{\alpha\in\{r,a\}} = \mu_{\alpha}^{(i,k)} f_{n,\alpha}^{(i,k)} = \left(\sigma_0 z_{\alpha}^{(i,k)} + \sigma_1 \dot{z}_{\alpha}^{(i,k)} + \sigma_2\, v_{t,\alpha}^{(i,k)}\right) f_{n,\alpha}^{(i,k)}, \tag{6}$$

where σ_0, σ_1, σ_2 are stiffness, damping and viscous friction coefficients of the bristle, $v_{t,\alpha}^{(i,k)}$ is a tangent velocity at the contact point and $z_{\alpha}^{(i,k)}$ is state variable representing the average deformation of the bristle. The average deformation can be obtained as solution of the additional state equations. Total contact force acting at $\mathrm{ce}_{\alpha}(i,k)$ can be written as follows:

$$\mathbf{f}_{c,\alpha}^{(b,i,k)}\Big|_{\alpha\in\{r,a\}} = f_{n,\alpha}^{(i,k)} \mathbf{n}_{\alpha}^{(i,k)} + f_{t,\alpha}^{(i,k)} \mathbf{t}_{\alpha}^{(i,k)} \quad - \text{ bearing}, \tag{7.1}$$

$$\mathbf{f}_{c,\alpha}^{(j,i,k)}\Big|_{\alpha\in\{r,a\}} = -{}_{b}^{j}\mathbf{R}\,\mathbf{f}_{c,\alpha}^{(b,i,k)} \quad - \text{journal}, \tag{7.2}$$

where $\mathbf{n}_{\alpha}^{(i,k)}$, $\mathbf{t}_{\alpha}^{(i,k)}$ are tangent and normal vectors to the colliding surfaces, ${}_{b}^{j}\mathbf{R}$ is rotation matrix from frame $\{b\}$ to $\{j\}$.

Generalized forces resulting from forces generated by the contact elements can determined in the following way:

$$\mathbf{c}^{(1)} = \sum_{i=1}^{n_r}\sum_{k=1}^{n_{ce}^{(r)}}\left(\mathbf{f}_{c,r}^{(b,i,k)} \frac{\partial \mathbf{r}_{K_r^{(b,i,k)}}}{\partial \mathbf{q}^{(1)}} + \mathbf{f}_{c,r}^{(j,i,k)} \frac{\partial \mathbf{r}_{K_r^{(j,i,k)}}}{\partial \mathbf{q}^{(1)}}\right) + \sum_{i=1}^{n_a}\sum_{k=1}^{n_{ce}^{(a)}}\left(\mathbf{f}_{c,a}^{(b,i,k)} \frac{\partial \mathbf{r}_{K_a^{(b,i,k)}}}{\partial \mathbf{q}^{(1)}} + \mathbf{f}_{c,a}^{(j,i,k)} \frac{\partial \mathbf{r}_{K_a^{(j,i,k)}}}{\partial \mathbf{q}^{(1)}}\right), \tag{8.1}$$

$$\mathbf{c}^{(2)} = \sum_{i=1}^{n_r}\sum_{k=1}^{n_{ce}^{(r)}}\left(\mathbf{f}_{c,r}^{(j,i,k)} \frac{\partial \mathbf{r}_{K_r^{(j,i,k)}}}{\partial \mathbf{q}^{(2)}}\right) + \sum_{i=1}^{n_a}\sum_{k=1}^{n_{ce}^{(a)}}\left(\mathbf{f}_{c,a}^{(j,i,k)} \frac{\partial \mathbf{r}_{K_a^{(j,i,k)}}}{\partial \mathbf{q}^{(2)}}\right). \tag{8.2}$$

Dynamics' equations of motion of the linkage are derived using the Lagrange equations of the second kind. The computational procedure described in [2, 8] is applied to determine components of the equations of motion. Due to existing clearance phenomenon at the cut-joint closing constraint equations can be neglected. Finally, the dynamics' equations together with the state equations [1] resulting from the LuGre friction model can be written as follows:

$$\dot{\mathbf{z}} = \mathbf{LuGre}(t, \mathbf{v}, \mathbf{z}), \tag{9.1}$$

$$\begin{bmatrix} \mathbf{M}^{(1)} & \mathbf{0} \\ \mathbf{0} & \mathbf{M}^{(2)} \end{bmatrix} \begin{bmatrix} \ddot{\mathbf{q}}^{(1)} \\ \ddot{\mathbf{q}}^{(2)} \end{bmatrix} = \begin{bmatrix} \mathbf{t}_{dr}^{(1)} - \mathbf{h}^{(1)} - \mathbf{g}^{(1)} + \mathbf{c}^{(1)} \\ -\mathbf{h}^{(2)} - \mathbf{g}^{(2)} + \mathbf{c}^{(2)} \end{bmatrix}, \tag{9.2}$$

where $\mathbf{M}^{(c)}$ is mass matrix, $\mathbf{t}_{dr}^{(1)}$ is vector containing components resulting from the driving and resistance torques, $\mathbf{h}^{(c)}$ is vector of the dynamic forces, $\mathbf{g}^{(c)}$ is vector of the gravity forces and $\mathbf{c}^{(p)}$ vector containing forces from the contact elements.

3 Numerical Simulations

The parameters of the linkage applied in numerical studies are gathered in Table 1. The crank is loaded by means of the driving and resistance torque which parameters are taken from [2]. The initial configuration of the linkage is as follows: $\psi^{(1,1)} = 90°$, $\psi^{(1,2)} = 0°$, $z^{(1,2)} = l^{(2,2)} \cos 60°$, $\psi^{(2,2)} = 30°$, $\theta^{(2,2,0)} = \varphi^{(2,2,0)} = 0°$. The numerical experiments show that the division of the crank into 4 rfes gives an acceptable accuracy of the results.

Table 1. Parameters of the linkage.

Parameter	Value	Parameter	Value
Young modulus E	2.1×10^{11} Pa	Stiffness coefficient σ_0	$100\ \mathrm{m}^{-1}$
Shear modulus G	0.8×10^{11} Pa	Damping coefficient σ_1	$\sqrt{100}\,\mathrm{s\,m}^{-1}$
Static coefficient of the friction μ_s	0.1	Coefficient of viscosity σ_2	$0\,\mathrm{s\,m}^{-1}$
Kinetic coefficient of friction μ_k	0.2	Coefficient of restitution k_r	0.9
Stribeck velocity v_s	$10^{-3}\,\mathrm{m\,s}^{-1}$	Radius of the bearing $r^{(b)}$	0.005 m

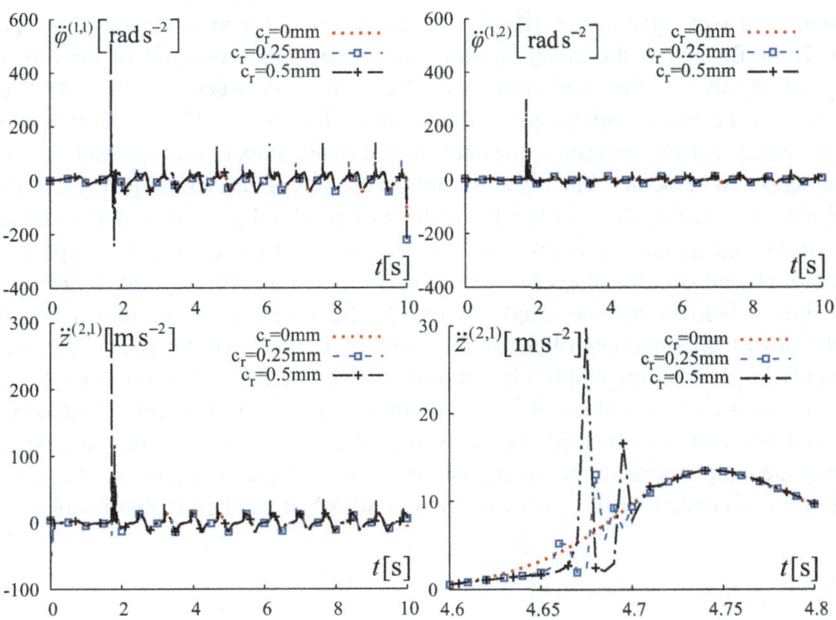

Fig. 4. Time course of the acceleration of the joint coordinates obtained for the linkage with rigid links.

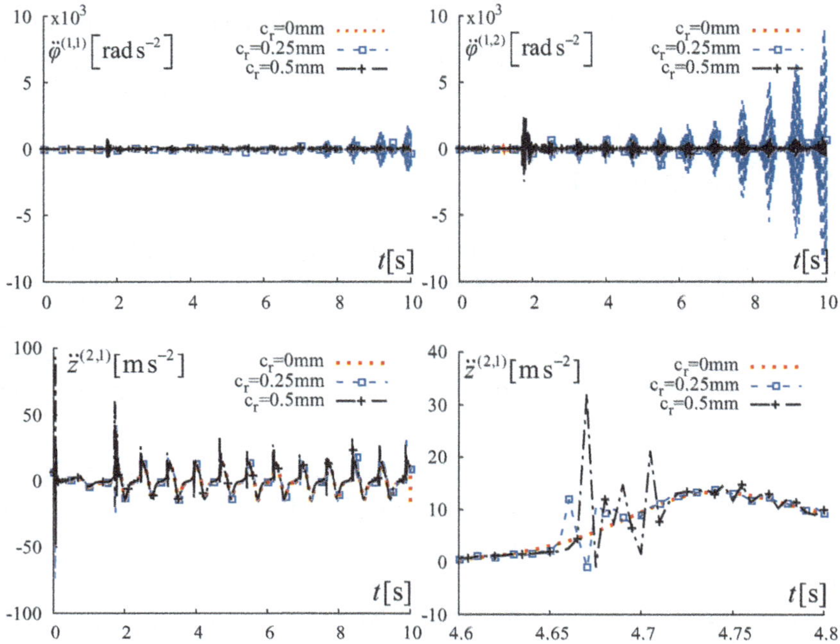

Fig. 5. Time course of the acceleration of the joint coordinates obtained for the linkage with the flexible coupler.

Equations of motions are integrated using the 4th order Runge-Kutta scheme with the constant step-size equal to 5×10^{-6} s. It is assumed that the axial clearance is equal to zero. The influence of the radial clearance on the dynamic response of the linkage is analyzed. Analyzing the time courses of the joint coordinates and their derivatives obtained for the model without and with the radial clearance (0.25 mm and 0.50 mm), it can be stated that the clearance does not have a major impact on time courses of displacements and velocities, but it is noticeable in courses of accelerations (Figs. 4 and 5).

Analyzing results obtained for the model with rigid links it can be noted that as the size of the clearance increases, changes in accelerations caused by impacts also increase. The maximum value of acceleration obtained for revolute joint (1,2) is smaller than those obtained for the crank. When the flexibility is taken into account the amplitudes of accelerations obtained for crank (1,1) and revolute joint (1,2) are significantly larger than the amplitudes obtained for the model with rigid links. It can be also observed that the values of these amplitudes arise with the time. Analyzing the accelerations courses obtained for the slider, it can be observed that peaks in the courses are smaller than in the model with rigid links. It can be explained that a part of the energy generated by the clearance joint is absorbed by the flexible coupler.

4 Conclusions

The mathematical model of the spatial RUSP linkage with flexible link and clearance in the cut-joint has been presented. The novelty of the presented paper is the new model of the revolute joints which takes into account the radial and axial clearance. The clearance joint is discretized using contact elements located on its cylindrical and frontal surfaces. Such approach allows us to analyze different collisions scenarios which can occur during the motion between the journal and bearing.

References

1. Åström, K.J., Canudas-de-Witt, C.: Revisiting the LuGre model. IEEE Control Syst. Mag. Inst. Electr. Electron. Mag. **28**(6), 101–114 (2008)
2. Augustynek, K., Urbaś, A.: Mathematical modelling of spatial linkages with clearance, friction and links' flexibility effects. In: Proceedings of the 5th Joint International Conference on Multibody System Dynamics, Lisbon, Portugal, June 24–28 (2018)
3. Dubowsky, S., Deck, J.F., Costello, H.: The dynamic modeling of flexible spatial machine systems with clearance connections. J. Mech. Transm. Autom. Des. **109**, 87–94 (1987)
4. Lankarani, H.M., Nikravesh, P.E.: A contact force model with hysteresis damping for impact analysis of multibody systems. J. Mech. Des. **112**, 369–376 (1990)
5. Liu, C., Tian, Q., Hu, H.: Dynamics and control a spatial rigid-flexible multibody system with multiple cylindrical clearance joints. Mech. Mach. Theor. **52**, 106–129 (2012)
6. Marques, F., Isaac, F., Dourado, N., Flores, P.: An enhanced formulation to model spatial revolute joints with radial and axial clearance. Mech. Mach. Theory **116**, 123–144 (2017)
7. Tian, Q., Flores, P., Lankarani, H.M.: A comprehensive survey of the analytical, numerical and experimental methodologies for dynamics of multibody mechanical systems with clearance or imperfect joints. Mech. Mach. Theory **116**, 123–144 (2017)
8. Wittbrodt, E., Szczotka, M., Maczyński, A., Wojciech, S.: Rigid Finite Element Method in Analysis of Dynamics of Offshore Structures. Ocean Engineering & Oceanography. Springer, Heidelberg (2013)

Mechatronics, Robotics and Control

Multibody Analysis and Design of an Electromechanical System Simulating Hyperelastic Membranes

Valentina Franchi[1], Gianpietro Di Rito[1], Roberto Galatolo[1], Ferdinando Cannella[2], Darwin Caldwell[2], and Giovanni Gerardo Muscolo[3](✉)

[1] Department of Civil and Industrial Engineering, University of Pisa, Largo Lucio Lazzarino, Pisa, Italy
valentinafranchi93@gmail.com,
{g.dirito, r.galatolo}@dia.unipi.it
[2] Advanced Robotics Line, Istituto Italiano di Tecnologia, Viale Morego, 33, 16163 Genoa, Italy
{ferdinando.cannella, darwin.caldwell}@iit.it
[3] Department of Mechanical and Aerospace Engineering, Politecnico di Torino, Corso Duca degli Abruzzi, 24, 10129 Turin, Italy
giovanni.muscolo@polito.it

Abstract. This work presents the multibody analysis and design of an electromechanical system (named robotics membrane) able to replicate the non-linear behavior of hyperelastic membranes. The designed robotics membrane is able to change its stiffness reproducing the deformation of different kinds of membrane. An actuator is controlled to adapt the length of the flexible elements, inside the robotics membrane, in order to modify the stiffness of the whole system. An analytical and numerical modelization in SimMechanics™ have been developed. The model validation has permitted to simulate the robotic membrane deformation in different load conditions in order to compare the results with the behavior of real hyperelastic membranes. The obtained results validated our system and underlined that our idea is concrete and applicable in the real environment.

1 Introduction

A flexible multibody system may consist of elastic and rigid components connected by joints and/or force elements such as springs, dampers, and actuators [1, 2]. In recent years, especially in the realm of robotics, the research has moved to soft materials and elastomeric membranes, which are characterized by stiffness and mechanical properties that are completely different from those of metals or composites. These new studies allow obtaining devices that interact more "naturally" with the external environment [3].

Soft robotics is thus an emerging sector of bio-robotics, in which robots can be made up soft elements [4, 5].

Marchese, Rus et al. [6] have studied fluidic elastomer actuators and developed soft complex systems like soft manipulators and a soft robotic fish.

A. Kecskeméthy and F. Geu Flores (Eds.): ECCOMAS 2019, COMPUTMETHODS 53, pp. 115–122, 2020.
https://doi.org/10.1007/978-3-030-23132-3_14

Elastomeric membranes have received considerable attention in recent years due to their applicability in numerous engineering areas, including space applications, actuators, sensors, robotics, and bioengineering devices. Intensive research has been conducted on the development of new membrane materials, including shape memory polymers and dielectric elastomers. Another important employment of elastomeric materials are dielectric elastomers [7, 8].

In nature, it's easy to find multiple cases of animals that need gliding to move in their habitat. In animals like birds or bats, gliding is simply one of the flight phases; other animals, that are commonly terrestrial, have a particular shape that allows them to glide, for example, from one tree to another. Also marine animals e.g. gliding fishes make a sort of flight of several distances outside water, to escape predators. Creatures like sugar gliders, flying snakes or flying lizards can modify their shape to obtain a lifting surface for gliding.

A special attention has been dedicating to the flight dynamics of sugar gliders [9], which use the flexible skin between body and paws (patagium) to obtain a sort of parachute allowing them to glide. Observing their trajectories, it's clear that movements of the limbs are strongly correlated with body rotations, suggesting that sugar gliders make extensive use of limb and tail movements to control their orientation during gliding flight. Gliding mammals also have musculature within the wing membrane, muscles are positioned such that if contracted or relaxed they may allow lesser or greater billowing of the wing.

Form the observation made on small group of animals, authors noted that flying squirrels, upon launching, dropped a vertical distance before beginning to glide, and often rose a smaller vertical distance just before landing. They observed different types of glide paths depending on the environment [10].

The interest in gliding performances of sugar-glider has been the starting point for the development of a robotics membrane system simulating the behavior of elastomeric membranes, with control function enabling the variation of the stiffness of the system [11]. This may allow realizing a system able to adapt to external conditions in which it operates.

The purpose of this work is thus to realize a preliminary study of a multibody electromechanical system with modern soft materials, for applications of variable stiffness and bioinspiration.

This paper is structured as follows: Sect. 2 presents the electromechanical robotics membrane; Sect. 3 shows the modelling and analysis of the robotics membrane; Sect. 4 presents the model validation and discussion. The paper ends with conclusion.

2 Electromechanical Robotics Membrane

The objective of this work is to develop an electromechanical system simulating the behavior of hyperelastic membranes. The designed robotics membrane is able to modify its stiffness in order to replicate the deformation of various elastic membranes which can differ, for example, in type of material or in thickness.

Figure 1 shows a first prototype of the robotics membrane, six analysed skeleton structures, and a first version of the 3D printed skeleton.

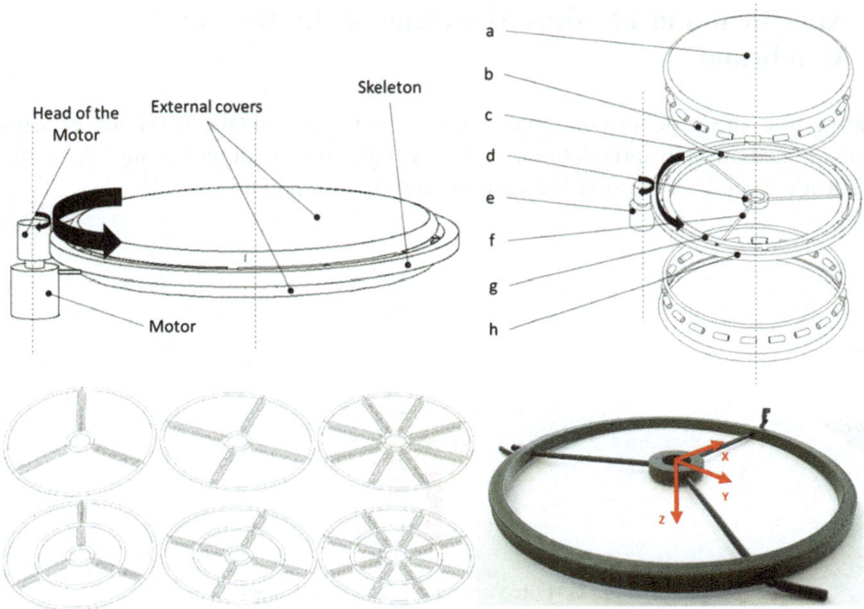

Fig. 1. Sketch of the complete robotic membrane (above); six hypothesis of skeleton (below-on the left); the realized skeleton prototype (below-on the right).

The starting idea is to create a preliminary model that replicates the displacement of the central point of a hyperelastic membrane. In order to realize it, the authors concentrated all the membrane's mass in a central element (a ring) connecting it to the edge by some elastic element. Greater is the number of the elastic element, more accurate will be the shape of deformation.

The basic structure is composed of a central little ring (d), which concentrates the mass of the membrane, and two external rings, one fixed (g) and one that can rotate (h). The inner ring (d) is connected via elastic elements (f) to the external ring (h). The structures is closed with two covers (a), each of which is a thin non-elastic membrane; in order to avoid the influence of the covers on the robotics system displacement behavior, the thin membranes may be unrolled, in fact they are fixed to cylindrical rings (b) that can rotate around fixed rings (c) thanks to torsional springs. When an external load is applied on the external cover, the force is transmitted directly on the skeleton and the inner ring (d) will be in motion. The elastic elements (f) absorb the external load and transfer a resultant force that balance the external load. If a different displacement of the central point of the system, and therefore a different membrane behavior, is required, it's necessary to vary the stiffness of the system. To obtain this result, the tensioning of the elastic elements has to be changed thanks to the motor (e) rotation. Motor rotation, in fact, produces the external ring rotation (h) that stretch the elastic elements changing their tension. Some pulleys connected to the ring (g) allow a change of direction of elements (f) from radial to tangential direction.

3 Analytical and Physical Modelling of the Robotics Membrane

Starting from the basic idea of a mass-spring system, we can take as references scheme a mass linked through springs to the wall. A central mass with two springs acted on by an outside force F is the simplest case to consider as shown in Fig. 2.

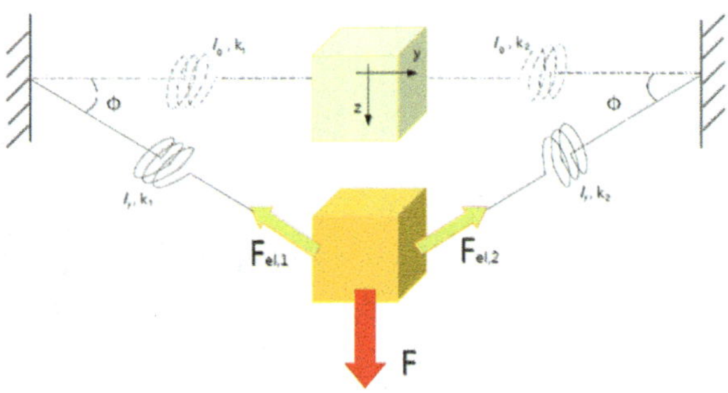

Fig. 2. Force acting on the central mass.

The external force acts in the vertical direction z, so it is possible to write the balance of forces in z and y directions. We assumed that the mass is in the middle and that each spring has initial length l_0; with these hypotheses the equation for the static equilibrium ($\dot{y} = \ddot{y} = 0, \dot{z} = \ddot{z} = 0$) in the maximum displacement point is:

$$\begin{aligned} m\ddot{y} &= -F_{el1,y} + F_{el2,y} = 0 \\ m\ddot{z} &= F - F_{el1,z} - F_{el2,z} = 0 \end{aligned} \tag{1}$$

the elastic force of each part can be expressed as the product between the elastic constant k and $\Delta l = l_f - l_0$, and so the equilibrium equations become:

$$\begin{aligned} m\ddot{y} &= -k_1 \Delta l cos\phi_1 + k_2 \Delta l cos\phi_2 = 0 \\ m\ddot{z} &= F - k_1 \Delta l sin\phi_1 - k_2 \Delta l sin\phi_2 = 0 \end{aligned} \tag{2}$$

Thanks to the symmetry of the problem and assuming $k_1 = k_2 = k$ we derive that $\phi_1 = \phi_2 = \phi$ and so the system has just one degree of freedom in z direction. We can extend the relations to a system with n springs obtaining the general equation for the static equilibrium in z-direction:

$$F - nk\Delta l sin\phi = 0 \tag{3}$$

The correlations between the lengths and the angle of rotation of the springs ϕ can be get from geometrical consideration:

$$\begin{aligned} l_0 &= l_f cos\phi \\ \Delta z &= l_f sin\phi \\ \Delta l = l_f - l_0 &= l_0\left(\tfrac{1}{cos\phi} - 1\right) \end{aligned} \tag{4}$$

If the elastic elements are subjected to length variation (s) (see Fig. 3), the dynamic behaviour of the mass, in order to obtain the desired tensioning, could be obtained as in follows [12]. The equations of motion can be derived with an energy method like the use of Lagrange equations. First, we define z as generalized coordinate q, then we can express kinetic energy K and potential energy U like:

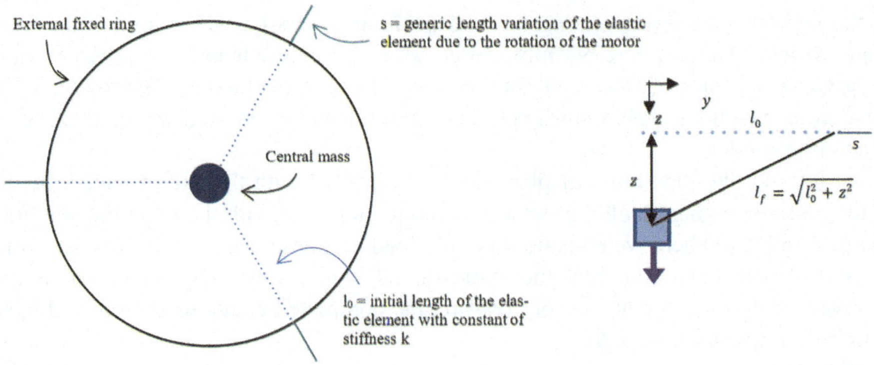

Fig. 3. Skeleton schematization considering length variation of the elastic elements.

$$\begin{aligned} K &= \tfrac{1}{2}m\dot{z}^2 \\ U = Fz + \tfrac{1}{2}\sum_{i=1}^{3} k_i(s + l_f - l_0)^2 &= Fz + \tfrac{3}{2}k\left(s + \sqrt{l_0^2 + z^2} - l_0\right)^2 \end{aligned} \tag{5}$$

The Lagrangian is defined as $L = K - U$ and consequently the Lagrange's equation is:

$$\frac{d}{dt}\frac{\delta L}{\delta \dot{q}} - \frac{\delta L}{\delta q} = F_{ext} \tag{6}$$

So the equation of motion becomes:

$$\ddot{z} + z\frac{3k}{m}\left(1 - \frac{l_0}{\sqrt{l_0^2 + z^2}} + \frac{s}{\sqrt{l_0^2 + z^2}}\right) = \frac{F}{m} \tag{7}$$

4 Model Validation and Discussion

The purpose of this section is finding confirmation of the developed model. To check the model's accuracy we need to compare experimental data available in literature with results obtained with numerical simulation.

The simulations for the validation are referred to the mechanical system with two rings and three elastic elements. For the static analysis, we want to check the position of the central mass of the membrane when subjected to an external load. In the work of Selvadurai, Shi [13], the measurements of the deflected shape of a rubber membrane for different pressures are given. The two rubber membranes used in experiments were 146 mm in diameter with thicknesses respectively of 0.794 and 1.588 mm. The resultant of the pressure acting on the membrane is considered as the external force. The system has to replicate the behaviour of the real hyperelastic membrane, so with the stiffness values previously calculated, we simulate in SimMechanics the behaviour of the central mass extrapolating its final position z. Firstly, we examined the membrane with thickness $t_1 = 0.794$ mm, under external pressure loads. The pressure inputs are given as a vector of values within the range of experimental data. We calculated the force as the resultant apply in the central point of the membrane and we applied it in the simulation model.

In Fig. 4a, the maximum displacement z associated with the external load, defined as the pressure acting on all the surface, is plotted together with the experimental graph taken from [13], where we consider only the loading phase. The two curves are almost coincident, demonstrating that the static model works well. The displacements are simulated only for a few values of pressure, the complete behaviour shown is obtained with an interpolation of data.

We repeated the same simulation for the second membrane with thickness $t_2 = 1.588$ mm. The results achieved are shown in Fig. 4b, where the comparison between simulation's results and experimental outcomes is presented. It is clear from the comparison's graphs that the curve obtained by the simulation is almost coincident with the curve of experimental data in both cases. Some divergences may be ascribed to an incorrect extrapolation of experimental data from the diagram set out in [13].

By the way, the coincidence of results is interesting and confirms that the model developed, allows to obtain a correct final displacement associated with an external load.

In conclusion, we can validate our model from a static point of view: once defined the characteristic of the robotics membrane, we can associate a particular value of the stiffness of each elastic element that permits to get a certain final displacement of the central point of the system, when subjected to a known external load. In this way we can obtain the static behaviour of hyperelastic membranes with different properties, setting the correct simulation parameters according to the case we want to replicate.

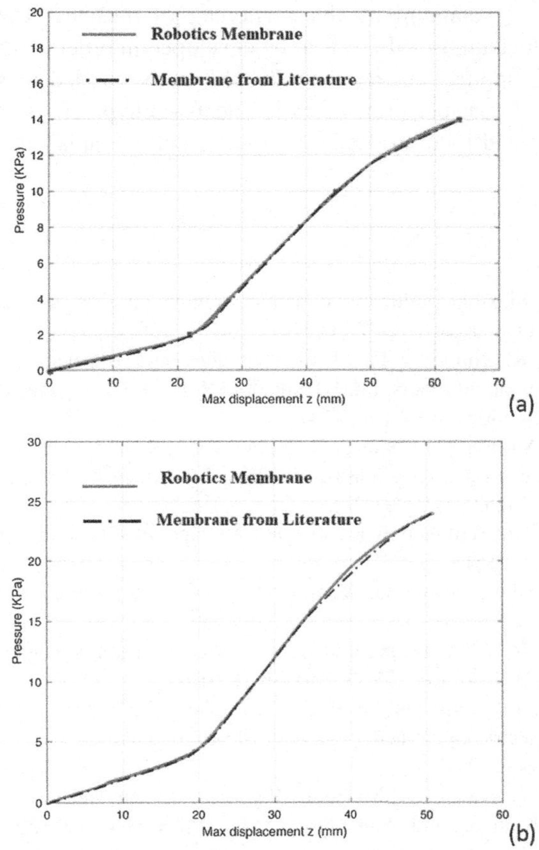

Fig. 4. Comparison between static simulation and experimental data for $t = t_1$ (a) and for $t = t_2$ (b).

5 Conclusion

This work deals with the design of a system able to modify its stiffness in order to reproduce the non-linear behaviour of hyperelastic membranes. The model of the system has been analysed in the configurations with three elastic elements, but it's possible to extend the results also to the model with more springs. The effective functioning of the system has been validated from a statical point of view. The system is able to reach the maximum z-displacement under an external load and has been compared with data form literature. From the results obtained, the authors noticed that for small loads and associated small z-displacements, the spring stiffness, that the elastic elements should have, are particularly high. This could be a problem in the physical realization of the system, but increasing the number of elastic elements, the spring stiffness request of each of them greatly decreases, improving the conditions. The research shown in this work could be set in a wider research. In fact, the work started from the bioinspired observation on "flight surfaces" of sugar glider which are able to modify their patagium by the tensioning of muscles. The presented modelization is only a first analysis that has to be developed in order to address the research on a more specific application. By the way, the study has led to significant results. Future works should consider to make experiments on hyperelastic materials, in order to have a complete and focused database of static and dynamic results.

References

1. Shabana, A.A.: Flexible multibody dynamics: review of past and recent developments. Multibody Sys. Dyn. **1**(2), 189–222 (1997)
2. Muscolo, G.G., Recchiuto, C.T.: Flexible structure and wheeled feet to simplify biped locomotion of humanoid robots. Int. J. Humanoid Rob. **14**(1), 26 pages (March 2017). http://dx.doi.org/10.1142/S0219843616500304
3. Muscolo, G.G., Moretti, G., Cannata, G.: SUAS: a novel soft underwater artificial skin with capacitive transducers and Hyperelastic membrane. Robotica **37**(4), 756–777 (2019). https://doi.org/10.1017/S0263574718001315
4. Marchese, A.D., Katzschmann, R.K., Rus, D.: A recipe for soft fluidic elastomer robots. Soft Rob. **2**(1), 7–25 (2015)
5. Rus, D., Tolley, M.T.: Design, fabrication and control of soft robots. Nature **521**(7553), 467 (2015)
6. Marchese, A.D., Rus, D.: Design, kinematics, and control of a soft spatial fluidic elastomer manipulator. Int. J. Rob. Res. **35**(7), 840–869 (2016)
7. Dubois, P., Rosset, S., Niklaus, M., Dadras, M., Shea, H.: Voltage control of the resonance frequency of dielectric electroactive polymer (deap) membranes. J. Microelectromech. Syst. **17**(5), 1072–1081 (2008)
8. O'Halloran, A., O'malley, F., McHugh, P.: A review on dielectric elastomer actuators, technology, applications, and challenges. J. Appl. Phys. **104**(7), 9 (2008)
9. Bishop, K.L.: Aerodynamic force generation, performance and control of body orientation during gliding in sugar gliders (petaurus breviceps). J. Exp. Biol. **210**(15), 2593–2606 (2007)
10. Vernes, K.: Gliding performance of the northern flying squirrel (glaucomys sabrinus) in mature mixed forest of Eastern Canada. J. Mammal. **82**(4), 1026–1033 (2001)
11. Jianbing, S., Xiang, L., Sufang, X., Wenjia, W.: Mechanical property analysis of circular polymer membrane under uniform pressure. Int. J. Polym. Sci. (2017)
12. Goncalves, P.B., Soares, R.M., Pamplona, D.: Nonlinear vibrations of a radially stretched circular hyperelastic membrane. J. Sound Vib. **327**(1–2), 231–248 (2009)
13. Selvadurai, A.P.S., Shi, M.: Fluid pressure loading of a hyperelastic membrane. Int. J. Non-Linear Mech. **47**(2), 228–239 (2012)

Haptic Simulation of Mechanisms

Jascha Norman Paris(✉), Jan-Lukas Archut, Mathias Hüsing, and Burkhard Corves

Institute of Mechanism Theory, Machine Dynamics and Robotics, RWTH Aachen University, Steinbachstraße 53B, 52074 Aachen, Germany
paris@igmr.rwth-aachen.de

Abstract. Mechanisms with non-uniform transmission ratios are used by humans in everyday and industrial environments. Often, these transfer or guidance mechanisms are hand-actuated. Approaches to develop ergonomic products with high perceived quality, such as Human-Centered Design, require prototypes. In order to provide prototypes of mechanisms to be designed regularly and at an early development stage, haptic feedback systems can be used to simulate virtual prototypes and make them haptically perceptible by the user. This contribution discusses challenges of this *haptic simulation* and introduces measures to increase overall performance. This includes an integration scheme, a passivity controller implementation, a beneficial friction model and *haptic simulation* of direct dynamics. Furthermore, adaptation of mechanisms during *haptic simulation* is presented in order to allow fast and convenient evaluation of virtual prototypes.

1 Introduction: Developing Hand-Actuated Mechanisms

Hand-actuated mechanisms are used by humans in everyday and industrial environments. These transfer or guidance mechanisms, such as car doors, furniture, machine housings, etc., represent a class of products with which humans interact directly. In the design process of these products, ergonomic requirements have to be considered. As the technical function is often solved, industry attempts to achieve perception of quality and excitement during usage as a unique selling point. Door design in the automotive industry is a good example for the effort put into the design of mechanisms to achieve good product perception. In order to focus on human expectations, behaviour and properties, the Human-Centered Design was introduced. In this process, prototypes and regular testing by users are the central elements. However, prototypes are time and cost intensive.

As a solution, Haptic Feedback Systems (HFS) can be used to simulate virtual prototypes (VP) and let the user feel the mechanism to be designed. This haptic simulation enables user tests and the evaluation of haptics at an early development stage without any hardware prototyping.

The usage of HFS as virtual prototyping tool was outlined and used systematically by Bordegoni et al. [1]. Clover et al. implemented the very first simulation of a simple mechanism using a haptic feedback system in 1997 [2]. The haptic simulation of a car door was carried out by Strolz et al. in 2008 and Kölling et al. developed a general purpose HFS for the simulation of mechanisms [4,7]. All of these papers use position controlled HFS and fixed model parameters.

A. Kecskeméthy and F. Geu Flores (Eds.): ECCOMAS 2019, COMPUTMETHODS 53, pp. 123–130, 2020.
https://doi.org/10.1007/978-3-030-23132-3_15

2 Concept: Haptic Simulation

The proposed overall system for the haptic simulation is shown in Fig. 1. This system mainly consists of the kinematics of the RePlaLink robot, it's real-time controller running the simulation, an additional PC and the users.

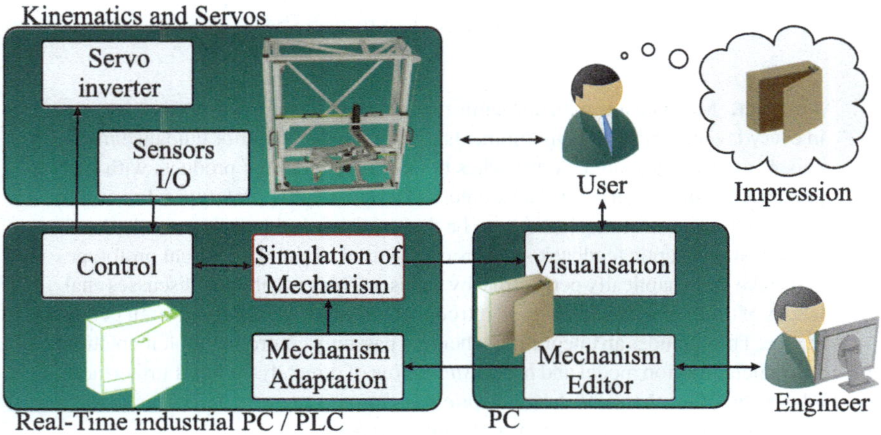

Fig. 1. The haptic feedback system RePlaLink for haptic simulation of mechanisms

Potential users are development engineers or future end-users. For the user it is possible to perceive the VP by interacting with the HFS and see the visualisation of the prototype at the same time. Furthermore, it is possible to use a mechanism editor to modify the parameters of the VP during simulation, e.g. link lengths or masses. For the visualisation and the mechanism editor an additional PC is used, which communicates via OPC UA, i.e. an industrial machine to machine communication protocol, with the real-time controller. The controller is an industrial PLC, i.e. a programmable logic controller or industry PC with real-time operating system. It simulates the VP with a cycle time of 1.2 ms based on measurements of the forces exerted by the user. Because of the average calculation power of the built-in Intel Atom Z520 with 1330 MHz, it is not appropriate to use a smaller cycle time. Though, shorter cycle times reduce integration errors and non-passivity. Another software module handles desired adaptations of the simulated mechanism. Furthermore, the PLC controls the overall system and sends target values to the servo inverters. The kinematics of the robot are composed of a five-bar mechanism for positioning in plane and an additional serial, sixth link for the rotation. These three degrees of freedom are sufficient to mimic most hand-actuated mechanisms. Industrial drive technology is used for the actuation of the HFS. Three servo drives are powered and controlled by their inverters. For the measurement of user forces a six-axis force sensor is attached between the sixth link and the handle.

In each cycle of the haptic simulation the forces exerted by the user are measured and used as input for the multi-body-simulation of the mechanism to be designed. The position output of this simulation is mapped to the coordinates of the kinematics of the REPLALINK. The robot is position-controlled to mimic the behaviour of the mechanism to be designed. Thereby, the user has the impression of grasping the VP. The REPLALINK is a kinaesthetic HFS capable of high forces and stiffness, though it uses standard industrial drive technology.

This contribution focuses on the real-time multi-body-simulation of the mechanism to be designed. In Sect. 3 the model of the VP is presented. Integration schemes, passivity and possibilities to simulate the direct dynamics of the VP are discussed in Sect. 4. The online-modification of parameters of the VP is introduced in Sect. 5. As mathematical notation arrows are used for vectors $\vec{v}$, bold symbols are used for matrices $\mathbf{A}$ and all other values are scalars.

3 Modelling

Hand-actuated mechanisms are often planar and consist only of few links. Though, for ergonomic reasons additional force elements such as support springs are used. The motion of hand-actuated mechanisms is often limited. Therefore, a four-bar mechanism with spring support and two limit positions as shown in Fig. 2 is used as exemplary VP. This could be the mechanism of a cupboard to be designed as illustrated in Fig. 1. The direct kinematics DK and inverse kinematics IK of the mechanism can be solved efficiently by solving dyads with the half-tangent-method.

The crank angle σ can be used as minimal coordinate and the closing conditions for accelerations can be formulated as Eqs. (1) to (3). The pose of the handle is $\vec{p}_H$, the rotation of all links about their centres of gravity (cog) $\vec{\varphi}_{cog}$ and the positions of these cog $\vec{x}_{cog}$. Required are the transmission ratios between the crank angle and the handle $\vec{I}$, the cog rotations $\vec{I}_R$ and the cog translation $\vec{I}_T$ together with analogous ratios for centrifugal and Coriolis accelerations $\mathbf{I}_{CC}$.

$$\ddot{\vec{p}}_H = \vec{I} \cdot \ddot{\sigma} + \mathbf{I}_{CC} \cdot \mathrm{diag}\left(\vec{I}_R\right) \cdot \vec{I}_R \cdot \dot{\sigma}^2 \tag{1}$$

$$\ddot{\vec{\varphi}}_{cog} = \vec{I}_R \cdot \ddot{\sigma} + \mathbf{I}_{CC,R} \cdot \mathrm{diag}\left(\vec{I}_R\right) \cdot \vec{I}_R \cdot \dot{\sigma}^2 \tag{2}$$

$$\ddot{\vec{x}}_{cog} = \vec{I}_T \cdot \ddot{\sigma} + \mathbf{I}_{CC,T} \cdot \mathrm{diag}\left(\vec{I}_R\right) \cdot \vec{I}_R \cdot \dot{\sigma}^2 \tag{3}$$

Together with the mass matrix $\overline{\mathbf{M}}$ the dynamics equations for the four-bar are defined with Eqs. (4) to (6). Thereby, M_k is the resulting torque of the spring support and M_{LP} is the resulting torque of the limit positions with springs k_i as well as damping d_i and is calculated with Eq. (7). The individual joint friction torques $M_{F,i}$ are calculated with an adapted Threlfall representation as proposed by [5] with a static torque $M_{Fs,i}$ and an additional viscous term, see Eq. (9) with Eq. (8). This representation has the advantage of being continuously differentiable and has shown improved

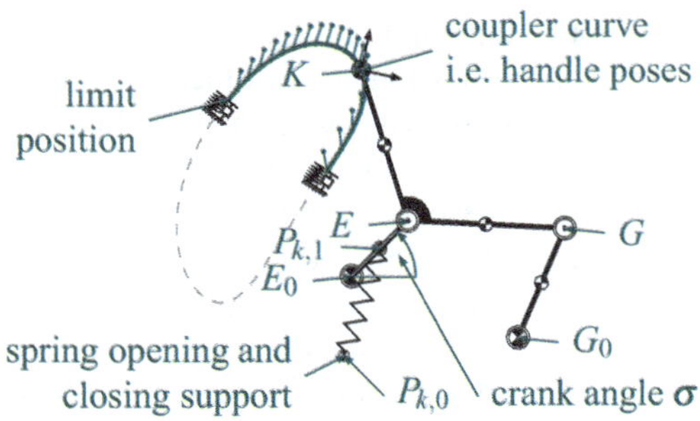

Fig. 2. Example four-bar mechanism as virtual prototype for the haptic simulation with limit positions and force support. A cupboard is a possible application. Chosen values $\overline{E_0E} = 20\,\text{mm}$, $\overline{G_0G} = 25\,\text{mm}$, $\overline{EG} = 35\,\text{mm}$, $\overline{EK} = 600\,\text{mm}$, $\angle GEK = 80°$, $E_0 = (0,0)$, $G_0 = (40\,\text{mm}, 0)$, $P_{k,0} = (-20\,\text{mm}, -100\,\text{mm})$, $m_{EE} = 0.294\,\text{kg}$, $m_{GG} = 0.368\,\text{kg}$, $m_{EG} = 0.514\,\text{kg}$, $m_{EK} = 8.823\,\text{kg}$, $J_{EE} = 9.804\,\text{kg mm}^2$, (see Fig. 3)

stability in haptic simulation. The actual friction values have to be chosen by the design engineer as no real prototype of the VP exists for measurements.

$$J_{red} = [\vec{I}_T^T \ \vec{I}_R^T] \cdot \overline{\mathbf{M}} \cdot \begin{bmatrix} \vec{I}_T \\ \vec{I}_R \end{bmatrix} \tag{4}$$

$$\frac{\mathrm{d}J_{red}}{\mathrm{d}\sigma} = [\vec{I}_T^T \ \vec{I}_R^T] \cdot \overline{\mathbf{M}} \cdot \begin{bmatrix} \mathbf{I}_{CC,T} \\ \mathbf{I}_{CC,R} \end{bmatrix} \cdot \text{diag}\left(\vec{I}_R\right) \cdot \vec{I}_R \tag{5}$$

$$0 = J_{red} \cdot \ddot{\sigma} + 0.5 \cdot \frac{\mathrm{d}J_{red}}{\mathrm{d}\sigma} \cdot \dot{\sigma}^2 + \vec{I}_F \cdot \vec{M}_F + \vec{I}^T \cdot \overrightarrow{F_{User}} + M_k + M_{LP} \tag{6}$$

$$M_{LP} = \begin{cases} -k_{\max} \cdot (\sigma - \sigma_{\max}) - d_{\max} \cdot \dot{\sigma}, & \sigma > \sigma_{\max} \\ -k_{\min} \cdot (\sigma - \sigma_{\min}) - d_{\min} \cdot \dot{\sigma}, & \sigma < \sigma_{\min} \\ 0, & \text{else} \end{cases} \tag{7}$$

$$\vec{\omega}_{\text{rel}} = \begin{bmatrix} 1 & 0 & 0 \\ -1 & 1 & 0 \\ 0 & -1 & 1 \\ 0 & 0 & 1 \end{bmatrix} \cdot \vec{I}_R^T \cdot \dot{\sigma} \tag{8}$$

$$M_{F,i} = \begin{cases} \mu_{C,i} \cdot M_{Fs,i} \cdot \frac{1-e^{\frac{-3 \cdot |\omega_{\text{rel},i}|}{\omega_{F,0}}}}{1-e^{-3}} \cdot \text{sign}\left(\omega_{\text{rel},i}\right) + \mu_{v,i} \cdot \omega_{\text{rel},i}, & \omega_{\text{rel},i} \leq \omega_{F,0} \\ \mu_{C,i} \cdot M_{Fs,i} \cdot \text{sign}\left(\omega_{\text{rel},i}\right) + \mu_{v,i} \cdot \omega_{\text{rel},i}, & \omega_{\text{rel},i} > \omega_{F,0} \end{cases} \tag{9}$$

4 Real-Time Simulation

An impedance $Z(s)$ is defined as the transfer function from velocity to force, i.e. $F(s)/v(s)$, and admittance is the inverse. Consequential, HFS are impedance-type controlled when using force-control and using position measurements for the VP. Admittance-type HFS are position-controlled and use force measurements for the VP. For the admittance-type control, as explained in Sect. 2, it is necessary to solve the inverse dynamics, i.e. integrate the accelerations calculated. Due to the limited calculation power of the PLC it is required to use efficient integration schemes. For the simulation of mechanical systems it is possible to use symplectic integrators, i.e. do the velocity integration after the acceleration integration. Under the premise to solve the dynamic equation only once per time step the scheme in Eqs. (10) to (14) based on symplectic half trapezoidal steps is proposed. Because only one force measurement is available per PLC cycle, a fixed integration step is chosen accordingly.

$$\dot{\sigma}_{n-0.5} = \dot{\sigma}_{n-1} + \frac{T_s}{2} \cdot \ddot{\sigma}_{n-1} \tag{10}$$

$$\sigma_{n-0.5} = \sigma_{n-1} + \frac{T_s}{2} \cdot \dot{\sigma}_{n-0.5} \tag{11}$$

$$\ddot{\sigma}_n = \frac{1}{J_{red}} \cdot \left(\frac{\mathrm{d}J_{red}}{\mathrm{d}\sigma} \cdot \dot{\sigma}^2 + \vec{I}_F \cdot \vec{M}_F + \vec{I}^T \cdot \vec{F}_{User,n} + M_k + M_{LP} \right) \Bigg|_{n-0.5} \tag{12}$$

$$\dot{\sigma}_n = \dot{\sigma}_{n-1} + \frac{T_s}{2} \cdot (\ddot{\sigma}_n + \ddot{\sigma}_{n-1}) \tag{13}$$

$$\sigma_n = \sigma_{n-1} + \frac{T_s}{2} \cdot (\dot{\sigma}_n + \dot{\sigma}_{n-1}) \tag{14}$$

The comparison of different integration schemes shows Fig. 3. Considering the Runge-Kutta as the most accurate scheme shown, it can be noticed that symplectic integrations schemes perform better than their counterpart. The proposed integration scheme with only one function evaluation performs better than Euler or trapezoidal methods and has proven better stability of the overall HFS in heuristic experiments.

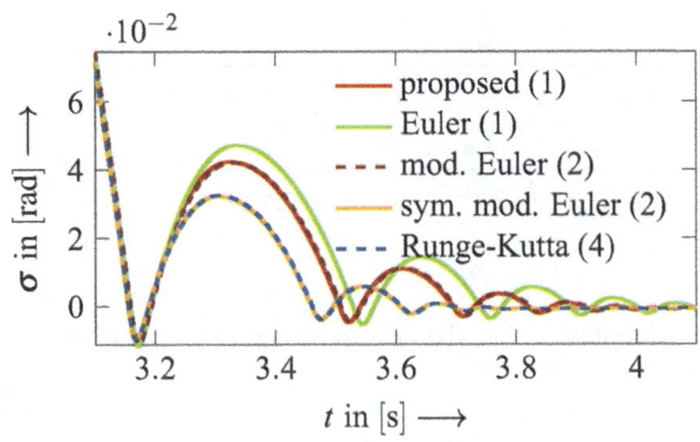

Fig. 3. Comparison of different integration schemes. The legend shows the required number of evaluations of the dynamics equations in brackets. Chosen values: $J_{GG} = 19.15\,\mathrm{kg\,mm^2}$, $J_{EG} = 52.54\,\mathrm{kg\,mm^2}$, $J_{EK} = 0.265\,\mathrm{kg\,m^2}$, $l_{k,0} = 5\,\mathrm{mm}$, $k = 4500\,\frac{\mathrm{N}}{\mathrm{m}}$, $\sigma_{\min} = 0°$, $\sigma_{\max} = 135°$, $\mu_{v,i} = 0.25$, $\mu_{C,i} \cdot M_{Fs,i} = 0.25\,\mathrm{N\,m}$, $\omega_{F,0} = 0.05\,\frac{\mathrm{rad}}{2}$, $k_{\max} = 5.0 \times 10^3\,\frac{\mathrm{N}}{\mathrm{m}}$, $k_{\min} = 1.8 \times 10^4\,\frac{\mathrm{N}}{\mathrm{m}}$, $d_{\max} = 2.3 \times 10^3\,\frac{\mathrm{N\,s}}{\mathrm{m}}$, $d_{\min} = 2.9 \times 10^4\,\frac{\mathrm{Ns}}{\mathrm{m}}$. Simulation: $\sigma_0 = 67.5°$, $\dot{\sigma}_0 = 0\,\frac{\mathrm{rad}}{\mathrm{s}}$, $\ddot{\sigma}_0 = 0\,\frac{\mathrm{rad}}{\mathrm{s}^2}$, $\vec{F}_{User} = \vec{0}$

Passivity is the property of a mechanical system to not generate energy. Mainly due to sampling, simulated mechanical systems might not maintain their passivity by contrast to real counterparts. Therefore, Hannaford et al. presented the passivity controller concept [3]. Ryu et al. [6] proposed the usage of model information as reference energy for the time domain passivity control. Based on that, a new admittance-type control passivity controller is proposed.

The passivity controller observes the energy flows in the system and whenever non-passive behaviour is identified, the calculated acceleration $\ddot{\sigma}_n$ of the VP is corrected. The observed time period lies between $n-0.5$ and $n+0.5$, which is the time where velocity and acceleration are set to $\dot{\sigma}_n$ and $\ddot{\sigma}_n$ respectively. The change of stored energy is calculated for the kinetic energy ΔE_{kin} and the potential spring energy $\Delta E_{pot,k}$. Note, in this case gravity is perpendicular to the moving plane. Applied power is calculated based on the torques and the current velocity for acceleration P_{kin}, friction and damping P_F as well as springs P_k. Afterwards, a comparison of the change of stored energy with the power applied is possible. If energy generation is detected, the corrections in Eqs. (15) to (17) are used. In summary, the simulated VP and herewith the overall system is stabilised to a certain amount. With the correction of the acceleration more damping is applied to the system based on physical relationships. The simulation behaves more like the real counterpart, though because of approximations energy generation is not completely avoided but strongly reduced.

$$\ddot{\sigma}_{corr,F} = \left\{ \left(\vec{I}_F \cdot \vec{M}_F + M_{LP,d}\right) \cdot \left(\frac{\dot{\sigma}_{n-0.5}}{\dot{\sigma}_{n-0.5} - \dot{\sigma}_{n+0.5}} - 1\right), \quad P_{\mathrm{F}} > 0 \right. \tag{15}$$

$$\ddot{\sigma}_{corr,k} = \begin{cases} (M_k + M_{LP,k}) \cdot \left(\frac{\Delta E_{\mathrm{pot,k}} + \Delta E_{\mathrm{pot,LP}}}{P_{\mathrm{k}} \cdot T_s} - 1\right), & P_{\mathrm{k}} > \frac{\Delta E_{\mathrm{pot,k}} + \Delta E_{\mathrm{pot,LP}}}{T_s} > 0 \\ (M_k + M_{LP,k}) \cdot \left(\frac{\Delta E_{\mathrm{pot,k}} + \Delta E_{\mathrm{pot,LP}}}{P_{\mathrm{k}} \cdot T_s} - 1\right), & 0 > P_{\mathrm{k}} > \frac{\Delta E_{\mathrm{pot,k}} + \Delta E_{\mathrm{pot,LP}}}{T_s} \end{cases} \tag{16}$$

$$\ddot{\sigma}_{corr,kin} = \begin{cases} -\dfrac{P_{\mathrm{kin}} - \frac{\Delta E_{\mathrm{kin}}}{T_s}}{\dot{\sigma}_{n+0.5} \cdot \left(J_{red} + \frac{\mathrm{d}J_{red}}{\mathrm{d}\sigma} \cdot \dot{\sigma}_{n+0.5} \cdot T_s\right)}, & \frac{\Delta E_{\mathrm{kin}}}{T_s} > P_{\mathrm{kin}} > 0 \\ -\dfrac{P_{\mathrm{kin}} - \frac{\Delta E_{\mathrm{kin}}}{T_s}}{\dot{\sigma}_{n+0.5} \cdot \left(J_{red} + \frac{\mathrm{d}J_{red}}{\mathrm{d}\sigma} \cdot \dot{\sigma}_{n+0.5} \cdot T_s\right)}, & 0 > \frac{\Delta E_{\mathrm{kin}}}{T_s} > P_{\mathrm{kin}} \end{cases} \tag{17}$$

(a) with spring-damper **(b)** elastic model **(c)** find nearest position

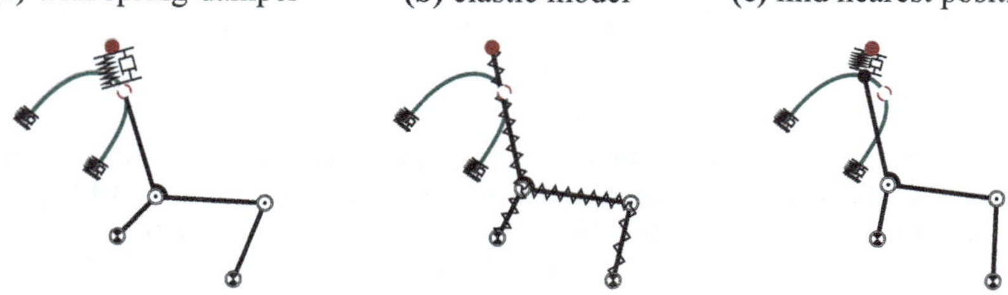

Fig. 4. Possibilities for the impedance-type simulation of the VP, i.e. solving direct dynamics

In contrast to admittance-type control of HFS as used above, impedance-type control requires to feed the VP with a position measurement. Whenever the VP has less degrees of freedom than the HFS, in most cases position measurements will be invalid.

Three possibilities to overcome this issue are shown in Fig. 4. It is possible to use a spring-damper connection as a virtual coupling between HFS and VP. Similarly the mechanism can be modelled compliantly. Though, both methods require the integration of the dynamics equation. Another possibility is to find the nearest position of the measured position on the coupler curve. A spring-damper element is used to calculate the force normal to the coupler curve. The force tangential to the coupler curve is calculated with the direct dynamics and the finite difference of the nearest position and the last position of the four-bar. This way actual impedance-type simulation is possible for VP with fewer degrees of freedom than the HFS.

5 Online-Modification

For fast and convenient virtual prototyping using haptic simulation, it is crucial to be able to change parameters quickly online and during simulation.

These changes should be applied smoothly without high accelerations or velocities. Therefore, whenever the four-bar parameters are updated via the mechanism editor and send to the PLC, a dwell trajectory for each kinematic parameter is planned with polynomials of fifth order. In order to save calculation time these polynomials are evaluated every tenth time step of the PLC. Subsequently, with a linear interpolation new set values for each time step are calculated and used for the dynamic calculations.

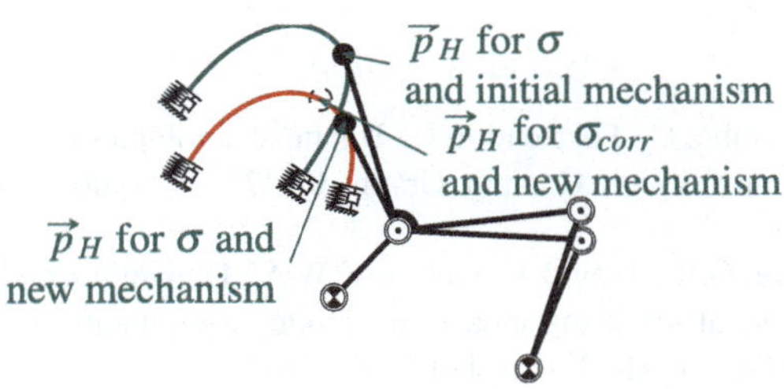

Fig. 5. State-correction for online adaptation of kinematic parameters in order to reduce unintended motion

Changes of kinematic parameters may lead to motion of the handle, though the crank angle stays constant, see handle poses in Fig. 5. Such a motion is unintended because no force was applied in direction of the trajectory. Furthermore, this may lead to high accelerations or may push the mechanism into the limit position. In order to avoid such unintended motions, a state correction is introduced. In the event of parameter changes a corrected state, i.e. crank angle, is calculated. To minimize the unintended motion of the handle, the associated crank angle of the closest position of the new mechanism to the old mechanism is determined.

$$\sigma_{n-0.5,corr.} = \min_{x} \|\mathrm{DK}_{old}\left(\sigma_{n-0.5}\right) - \mathrm{DK}_{new}\left(x\right)\|_2 \text{ with } \sigma_{\min} < x < \sigma_{\max} \quad (18)$$

The solution should be in between the limit positions and therefore a golden section search minimisation Eq. (18) can be used.

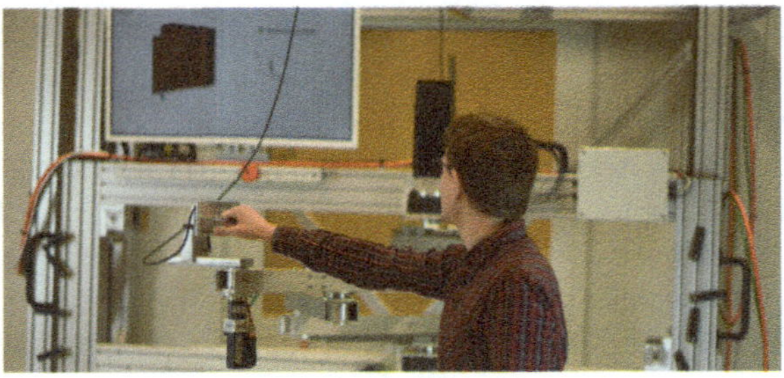

Fig. 6. Application of the HFS RePlaLink for the haptic simulation of a cupboard.

6 Conclusion

This paper presents the haptic simulation of mechanisms and discusses measures to increase performance and stability. These are the analysis of integration schemes and the proposal of an efficient scheme with only one evaluation of the dynamics equation. Passivity controller concepts are adapted to the admittance type system. Furthermore, parameter changes during simulation are implemented and realised using a state-correction approach. The application of the system is shown in Fig. 6.

References

1. Bordegoni, M., Colombo, G., Formentini, L.: Haptic technologies for the conceptual and validation phases of product design. Comput. Graph. **30**, 377–390 (2006). https://doi.org/10.1016/j.cag.2006.02.012
2. Clover, C.L., Luecke, G.R., Troy, J.J., McNeely, W.A.: Dynamic simulation of virtual mechanisms with haptic feedback using industrial robotics equipment. In: Proceedings of ICRA. IEEE (1997). https://doi.org/10.1109/robot.1997.620121
3. Hannaford, B., Ryu, J.H., Kim, Y.S.: Stable control of haptics. In: Touch in Virtual Environments, pp. 47–70 (2002). ISBN 0-13-065097-8
4. Kölling, T.G., Paris, J.N., Hüsing, M., Corves, B.: Bewerten der Ergonomie handbetätigter Mechanismen durch virtual prototyping mit haptischem Display. Getriebe aktuell, pp. 14–23 (2016). ISSN 2199-9775
5. Marques, F., Flores, P., Claro, J.C.P., Lankarani, H.M.: A survey and comparison of several friction force models for dynamic analysis of multibody mechanical systems. Nonlinear Dynam. **86**, 1407–1443 (2016). https://doi.org/10.1007/s11071-016-2999-3
6. Ryu, J.-H., Preusche, C., Hannaford, B., Hirzinger, G.: Time domain passivity control with reference energy following. IEEE Trans. Control Syst. Technol. **13**, 737–742 (2005). https://doi.org/10.1109/TCST.2005.847336. ISSN 10636536
7. Strolz, M., Buss, M.: Haptics: Perception, Devices and Scenarios, pp. 712–717. Springer, Heidelberg (2008). https://doi.org/10.1007/978-3-540-69057-3_91

Solution Techniques for Problems of Inverse Dynamics of Flexible Underactuated Systems

Timo Ströhle(✉) and Peter Betsch

Institute of Mechanics, KIT, Karlsruhe, Germany
{timo.stroehle,peter.betsch}@kit.edu

Abstract. A new approach to the inverse dynamics of flexible mechanical systems is proposed. In contrast to the commonly applied sequential discretization in space and time, a simultaneous space-time discretization of the problem at hand is proposed. In particular, two alternative approaches are investigated: (i) a space-time finite element formulation, and (ii) the method of characteristics. The focus is put on mechanical systems whose motion is governed by quasilinear hyperbolic partial differential equations. Numerical examples are presented which confirm that the two alternative methods under investigation can be successfully applied to the considered class of underactuated mechanical systems.

1 Introduction

Servo constraints have been successfully applied to solve the inverse dynamics of discrete mechanical systems. In this approach the equations governing the motion of the discrete mechanical system at hand are supplemented by algebraic servo constraints. The servo constraints serve the purpose of partially prescribing the motion of the mechanical system. The resulting set of differential algebraic equations (DAEs) has typically a differentiation index that is equal to 3 or higher. The numerical solution of high-index DAEs demands the application of index reduction techniques before specific time-stepping schemes can be applied (see [2, 11] and the references cited therein). In principle, a similar approach can also be applied to the inverse dynamics of flexible mechanical systems such as elastic strings and beams. To this end, a discretization in space needs be applied first to generate the discrete mechanical system. Then servo constraints can be appended leading again to DAEs. However, the index of the resulting DAEs can be quite high hindering their numerical solution.

Therefore, we propose an alternative approach to solve inverse dynamics problems of flexible mechanical systems. Our approach relies on a simultaneous discretization in space and time. For this purpose, we consider two alternative procedures, namely a formulation based on the method of characteristics and a space-time finite element method. Both formulations have in common the simultaneous discretization in space and time, which is in contrast to the commonly applied sequential discretization in space and time. In the present work

A. Kecskeméthy and F. Geu Flores (Eds.): ECCOMAS 2019, COMPUTMETHODS 53, pp. 131–138, 2020.
https://doi.org/10.1007/978-3-030-23132-3_16

we focus on mechanical systems whose motion is governed by quasilinear hyperbolic partial differential equations (PDEs). In particular, we consider systems whose motion is governed by

$$\begin{aligned} &\partial_t^2 r(s,t) - B(s,t)\partial_s^2 r(s,t) = c(s,t) && (s,t) \in \Omega \\ &\partial_s r(0,t) = f(t),\ \partial_s r(l,t) = 0,\ r(l,t) = g(t) && t \in T \\ &r(s,0) = r_0(s),\ \partial_t r(s,0) = v_0(s) && s \in S \end{aligned} \tag{1}$$

Here, $s \in S = [0,l]$ contains the arc-length of a reference curve in $\mathbb{R}^{n_{\mathrm{dim}}}$, $n_{\mathrm{dim}} \in \{1,2,3\}$, $t \in T = [0,t_e]$ is the time domain of interest, and $r(s,t) \in \mathbb{R}^{n_{\mathrm{dim}}}$. We further introduce the space-time domain $\Omega = S \times T$. The main task is to find $f(t) \in \mathbb{R}^{n_{\mathrm{dim}}}$ for $g(t) \in \mathbb{R}^{n_{\mathrm{dim}}}$ prescribed, such that (1) is satisfied. For simplicity of exposition, in the first part of this work we focus on $n_{\mathrm{dim}} = 1$ and $B(s,t) = b^2$, where $b \in \mathbb{R}$ is constant.

2 Semi-discrete Formulation

Problem (1) constitutes an initial boundary value problem (IBVP), the numerical solution of which is commonly attained by applying a sequential discretization in space and time. For example, to apply finite elements for the space discretization, (1) is recast in weak form. Multiplication of the first equation in (1) with a sufficiently smooth test function $w(s)$ and integrating over the domain $S = [0,l]$ yields, after integration by parts,

$$\int_S w \cdot \partial_t^2 r \, ds + \int_S \partial_s w \cdot b^2 \partial_s r \, ds = \left[w \cdot b^2 \partial_s r \right]_{s=0}^{l} + \int_S w \cdot c \, ds \tag{2}$$

Inserting the Neumann boundary conditions $\partial_s r(0,t) = f(t)$ and $\partial_s r(l,t) = 0$, the first term on the right-hand side of (2) simplifies to

$$\left[w \cdot b^2 \partial_s r \right]_{s=0}^{l} = -w(0) \cdot b^2 f(t). \tag{3}$$

Using piecewise continuous polynomials for the approximation of the test function $w(s)$ and the trial function $r(s,\cdot)$, yields the semi-discrete equations of motion which assume the form of ordinary differential equations (ODEs). In the semi-discrete formulation, the desired motion at the end-point corresponding to $s = l$ can now be imposed by an algebraic servo constraint

$$r(l,t) - g(t) = 0 \tag{4}$$

where $g(t)$ is a prescribed, sufficiently smooth, function. Thus, the resulting servo-constraint problem is governed by a system of DAEs. The (differentiation) index of the DAEs can be exceedingly high [3,8,9] hindering an efficient numerical solution.

3 Method of Characteristics

As the first alternative to the sequential discretization approach outlined above, we consider the method of characteristics which can be applied to the solution of Problem (1). To this end, we recast the PDE in (1) in the form of a system of first order PDEs by introducing $q(s,t) = \partial_t r(s,t)$ and $p(s,t) = \partial_s r(s,t)$

$$\begin{aligned} \partial_t q - b^2 \partial_s p &= c \\ \partial_t p - \partial_s q &= 0 \end{aligned} \tag{5}$$

Introducing the column vectors $x = \begin{bmatrix} q\; p \end{bmatrix}$ and $C = \begin{bmatrix} c\; 0 \end{bmatrix}$, the system of first order PDEs 5 can be written as the square matrices $A(s,t)$ and $B(s,t)$ as well as a column vector together with more compactly:

$$\partial_t x + A \partial_s x = C \tag{6}$$

with the matrix $A \in \mathbb{R}^{2\times 2}$. In the method of characteristics (cf. [1,7,10]) a curve $t = k(s)$ is called a characteristic curve if

$$\det \left(I - A \frac{d}{ds} k(s) \right) = 0 \tag{7}$$

Due to the fact that our problem is hyperbolic, (7) leads to two real-valued solutions for the characteristic curves:

$$\left(\frac{dt}{ds}\right)_1 = b^{-1} \qquad \left(\frac{dt}{ds}\right)_2 = -b^{-1} \tag{8}$$

where $(\,\cdot\,)_\alpha$, $\alpha = 1,2$, refers to the two characteristic curves. Along the characteristic curves, $q(s,t)$ and $p(s,t)$ have to satisfy the following ODEs:

$$\left(\frac{dq}{ds}\right)_1 - b\left(\frac{dp}{ds}\right)_1 = +\frac{c}{b} \qquad \left(\frac{dq}{ds}\right)_2 + b\left(\frac{dp}{ds}\right)_2 = -\frac{c}{b} \tag{9}$$

Equations (8) and (9) together form a system of ODEs, which can be solved numerically, e.g. by using finite differences. The boundary and initial conditions specified in (1) can be applied directly at the nodes of the characteristic net.

4 Space-Time Finite Elements

As the second alternative to the sequential discretization approach outlined in Sect. 2, we make use of space-time finite elements (cf. [6]). In a first step we introduce the velocity $v(s,t) = \partial_t r(s,t)$, so that the PDE in (1) can be rewritten as

$$\begin{aligned} \partial_t r - v &= 0 \\ \partial_t v - b^2 \partial_s^2 r &= c \end{aligned} \tag{10}$$

Multiplying each equation in (10) with a sufficiently smooth test function $w_1(s,t)$ and $w_2(s,t)$, respectively, and integrating over the space-time domain Ω yields

$$\int_\Omega w_1 \cdot (\partial_t r - v)\, d\Omega = 0 \tag{11}$$

$$\int_\Omega w_2 \cdot \left(\partial_t v - b^2 \partial_s^2 r\right) d\Omega = \int_\Omega w_2 \cdot c\, d\Omega \tag{12}$$

Applying integration by parts to the second term on the left-hand side of (12) leads to

$$\int_\Omega w_2 \cdot b^2 \partial_s^2 r\, d\Omega = \int_T \left[w_2 \cdot b^2 \partial_s r\right]_{s=0}^{l} dt - \int_\Omega \partial_s w_2 \cdot b^2 \partial_s r\, d\Omega \tag{13}$$

Taking into account the Neumann boundary conditions $\partial_s r(0,t) = f(t)$ and $\partial_s r(l,t) = 0$, we get

$$\int_\Omega w_2 \cdot \partial_t v\, d\Omega + \int_T w_2(0,t) \cdot b^2 f(t)\, dt + \int_\Omega \partial_s w_2 \cdot b^2 \partial_s r\, d\Omega = \int_\Omega w_2 \cdot c\, d\Omega \tag{14}$$

The servo-constraint $r(l,t) = g(t)$ which has to be satisfied for all $t \in T$ can be enforced in a weak sense by using a test function $w_3(t)$ to obtain

$$\int_T w_3(t) \cdot (r(l,t) - g(t))\, dt = 0 \tag{15}$$

Equations (11), (14), and (15) constitute the newly proposed space-time finite element formulation. The test functions $w_1(s,t)$, $w_2(s,t)$, $w_3(t)$, along with the trial functions $r(s,t)$, $v(s,t)$, $f(t)$ can now be approximated by piecewise continuous polynomials. To this end, we make use of Lagrangian shape functions. Applying standard finite element procedures yields an algebraic system of equations for the determination of the nodal degrees of freedom associated with the discrete trial functions.

5 Numerical Examples

Linearly Elastic Bar

As a first model problem, a bar with length l, cross-sectional area A, density ρ and Young's modulus E is investigated. The task is to find the force $F(t)$ which is acting on one end of the bar ($s = 0$) such that the other end ($s = l$) tracks a prescribed trajectory

$$g(t) = \begin{cases} 0 & t < 1 \\ \frac{1}{2}\sin\left(\frac{\pi}{2}t - \pi\right) + \frac{1}{2} & 1 \leq t \leq 3 \\ 1 & t > 3 \end{cases} \tag{16}$$

This model problem will be solved numerically using the methods presented in Sects. 3 and 4, namely the method of characteristics and a space-time finite element method, respectively. The results can be compared with an analytical solution. Assuming linear constitutive relations and linear kinematics, the servo-constrained longitudinal wave propagation in the bar is governed by the following IBVP:

$$\begin{aligned} &\rho A\partial_t^2 r(s,t) - EA\partial_s^2 r(s,t) = 0 && (s,t) \in \Omega \\ &EA\partial_s r(0,t) = F(t),\ EA\partial_s r(l,t) = 0,\ r(l,t) = g(t) && t \in T \\ &r(s,0) = r_0(s),\ \partial_t(s,0) = v_0(s) && s \in S \end{aligned} \tag{17}$$

This IBVP falls into the framework of Problem (1), by setting $n_{\text{dim}} = 1$, $B(s,t) = b^2$, and

$$b^2 = \frac{E}{\rho}, \qquad c = 0, \qquad f(t) = \frac{F(t)}{EA} \tag{18}$$

Semi-discrete Formulation: As outlined in Sect. 2, finite elements can be applied for the space discretization of the elastic bar leading to the semi-discrete servo-constraint problem. The semi-discrete problem can be viewed as spring-mass system in which the number of masses, say n, corresponds to the number of nodes in the finite element model. It can be shown that the index of the DAEs is $2n+1$, see [3,5]. Accordingly, using a reasonably accurate finite element discretization (in space) yields DAEs with excessively high index.

Simultaneous Space-Time Discretization: Both the method of characteristics and the space-time finite element method can be directly applied as described in Sects. 3 and 4, respectively. Both methods yield numerical results which coincide very well with the analytical reference solution (Figs. 1 and 2).

Nonlinearly Elastic String

The second example deals with large planar ($n_{\text{dim}} = 2$) deformations of an elastic string. In the undeformed stress-free reference configuration the string

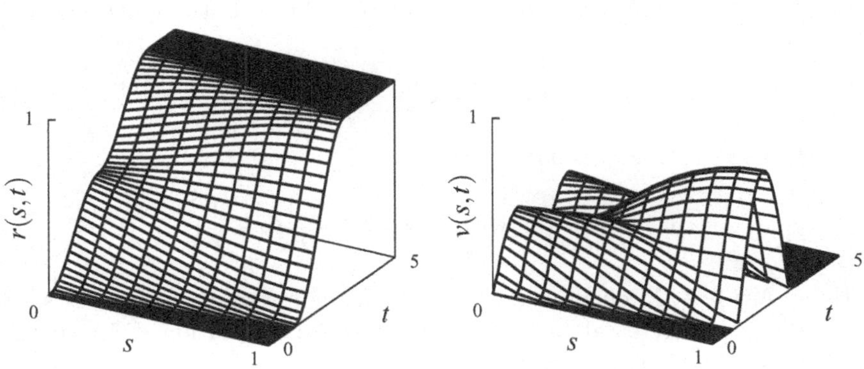

Fig. 1. Numerical solution for r and v computed using the space-time finite element method

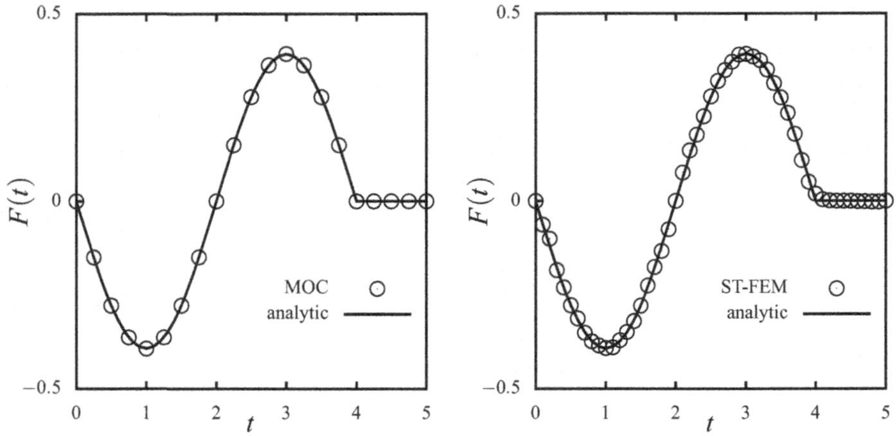

Fig. 2. Numerical solution for the actuating force $F(t)$ computed using the method of characteristics (left) and the space-time finite element method (right) compared with the analytical reference solution (solid line)

has length l, cross-sectional area A, density ρ and Young's modulus E. The task is to find the external force $F(t)$ which is acting on one end of the rope ($s = 0$) such that the other end ($s = l$) follows a prescribed trajectory $g(t) = \left[g_x(t)\; g_y(t)\right]$ given by

$$g_x(t) = g_y(t) = \begin{cases} 0 & t < 2 \\ \dfrac{1}{2}\sin\left(\dfrac{\pi}{2}(t-3)\right) + \dfrac{1}{2} & 2 \le t \le 4 \\ 1 & t > 4 \end{cases} \tag{19}$$

The motion of the string is characterized by $r(s,t) \in \mathbb{R}^2$, and the force in the extensible string is denoted by $n(s,t) \in \mathbb{R}^2$. Furthermore, $b(s,t) \in \mathbb{R}^2$ denotes a body force per unit reference length. The motion of the elastic string is governed by (cf. [4])

$$\partial_s n(s,t) + b(s,t) = \rho A \partial_t^2 r \tag{20}$$

where the constitutive relation is given by

$$n = N\frac{\partial_s r}{\|\partial_s r\|} \quad \text{with} \quad N(\nu) = \frac{EA}{2}(\nu - \frac{1}{\nu}) \tag{21}$$

Here, $N(\nu) \in \mathbb{R}$ denotes the tension force in the string and $\nu(s,t) = \|\partial_s r(s,t)\|$ is the stretch at (s,t). A straightforward calculation shows that (20) together with (21) lead to

$$\partial_t^2 r - B\partial_s^2 r = \frac{1}{\rho A} b(s,t) \quad \text{with} \quad B = \frac{E}{2\rho}\left(\left(1 - \frac{1}{\nu^2}\right)I + \frac{2}{\nu^4}(\partial_s r \otimes \partial_s r)\right) \tag{22}$$

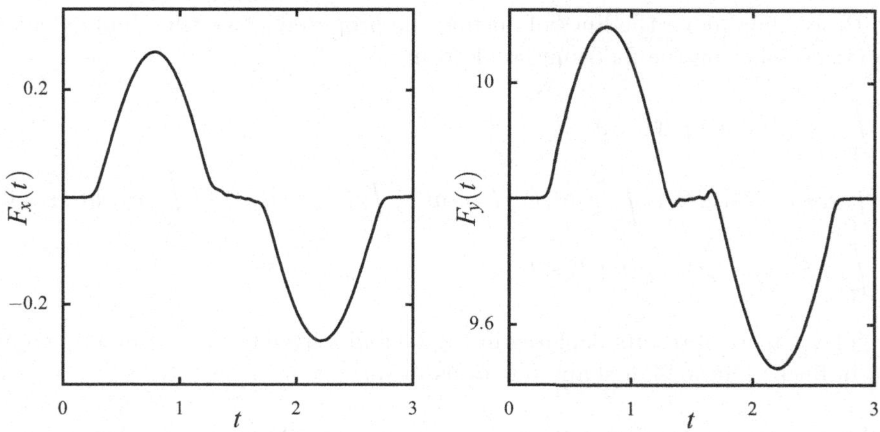

Fig. 3. Numerical solution for the components $F_x(t)$ and $F_y(t)$ of the actuating force $F(t)$

Accordingly, the motion of the extensible string is governed by a quasilinear hyperbolic PDE which again fits into the framework of Problem (1). In addition to $n_{\text{dim}} = 2$, we have

$$c = \frac{b}{\rho A}, \qquad f(t) = F(t)\frac{\nu(0,t)}{N(\nu(0,t))} \tag{23}$$

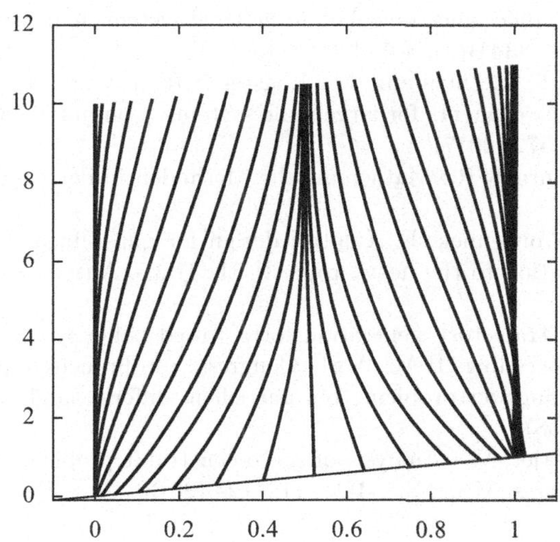

Fig. 4. Snapshots of the moving elastic string acted upon on the upper end by force $F(t)$ such that the lower end follows a prescribed straight line from the starting point $(0, 0)$ to the endpoint $(1, 1)$

Proceeding along the lines of Sect. 4, the proposed space-time finite element approach relies on the following weak form

$$\begin{aligned}
&\int_{\Omega} w_1 \cdot (\partial_t r - v)\, d\Omega = 0 \\
&\int_{\Omega} w_2 \cdot \rho A \partial_t v\, d\Omega + \int_{T} w_2(0,t) \cdot F(t)\, dt + \int_{\Omega} \partial_s w_2 \cdot n\, d\Omega - \int_{\Omega} w_2 \cdot b\, d\Omega = 0 \\
&\int_{T} w_3(t) \cdot (r(l,t) - g(t))\, dt = 0
\end{aligned}$$

The numerical results depicted in Figs. 3 and 4 have been obtained by applying bi-linear Lagrangian shape functions in space-time.

Acknowledgements. This work was supported by the Deutsche Forschungsgemeinschaft (DFG) under Grant BE 2285/12-1. This support is gratefully acknowledged.

References

1. Abbott, M.B.: An Introduction to the Method of Characteristics. Elsevier, New York (1966)
2. Altmann, R., Betsch, P., Yang, Y.: Index reduction by minimal extension for the inverse dynamics simulation of cranes. Multibody Syst. Dyn. **36**(3), 295–321 (2016)
3. Altmann, R., Heiland, J.: Simulation of multibody systems with servo constraints through optimal control. Multibody Syst. Dyn. **40**(1), 75–98 (2017)
4. Antman, S.S.: Nonlinear Problems of Elasticity, 2nd edn. Springer, New York (2005)
5. Blajer, W.: Dynamics and control of mechanical systems in partly specified motion. J. Franklin Inst. **334B**(3), 407–426 (1997)
6. Hesch, C., Schuß, S., Dittmann, M., Eugster, S.R., Favino, M., Krause, R.: Variational space-time elements for large-scale systems. Comput. Methods Appl. Mech. Eng. **326**, 541–572 (2017)
7. Hilbert, D., Courant, R.: Mathematische Methoden in der Physik. Springer, Heidelberg (1937)
8. Knüppel, T., Woittennek, F.: Control design for quasi-linear hyperbolic systems with an application to the heavy rope. IEEE Trans. Autom. Control **60**(1), 5–18 (2015)
9. Murray, R.A.: Trajectory generation for a towed cable system using differential flatness. In: Proceedings IFAC World Congress, San Francisco, pp. 395–400 (1996)
10. Sauer, R.: Anfangswertprobleme bei partiellen Differentialgleichungen. Springer, Heidelberg (1958)
11. Seifried, R., Blajer, W.: Analysis of servo-constraint problems for underactuated multibody systems. Mech. Sci. **4**(1), 113–129 (2013)

Investigation of the Behavior of Vibration-Damped Flexible Link Robots in End-Effector Contact: Simulation and Experiment

Florian Pucher(✉), Hubert Gattringer, and Andreas Müller

Institute of Robotics, Johannes Kepler University Linz,
Altenbergerstraße 69, 4040 Linz, Austria
{florian.pucher,hubert.gattringer,a.mueller}@jku.at

Abstract. Lightweight robots with flexible links are perfect candidates for applications which require contact with the environment due to their inherited compliance. However, due to the lower stiffness compared to typical industrial robots it is imperative to handle occurring link vibrations. This can be done with different vibration damping control approaches. This paper focuses on a method based on analytical estimation of the joint torques using accelerometers. The joint torques are estimated by reconstruction of the link dynamics. This is valid as long as no external forces are present. Stability of a robot with active vibration damping control in a contact scenario is essential. In this regard the switching caused by unilateral contact must be investigated carefully. In this paper the stability of the control system is verified in simulation and moreover in an experiment.

1 Introduction

The main benefits of lightweight robots are quite obvious. Small mass and inertia parameters gives rise to agile manipulators with lower power consumption than their rigid counterparts. Another interesting aspect is the inherited compliance, which results in lower contact forces when the robot is colliding with the environment. This type of enhanced passive safety obtained by combination of low mass and high compliance is desirable not only in the growing field of human-robot interaction (HRI).

However, the cost for all these advantages is the need to deal with vibrations due to low stiffness of the links and a reduced accuracy of the endpoint position. There are various methods for dealing with link oscillations. These approaches can be distinguished according to the sensor type as methods using (1) strain gauges [6], (2) accelerometers [4], and (3) vision systems [7]. The use of joint torque sensors is also worth mentioning, but while delivering good results, most industrial robots do not have this feature. When joint torque sensors are not included in the design process, mounting them afterwards can be a challenging

A. Kecskeméthy and F. Geu Flores (Eds.): ECCOMAS 2019, COMPUTMETHODS 53, pp. 139–146, 2020.
https://doi.org/10.1007/978-3-030-23132-3_17

task and economically not viable. This paper focuses on using acceleration sensors for vibration damping control. These are easier to apply on a robot system than strain gauges and require less computational power compared to image processing for a vision-based system.

Depending on the control law in combination with observers, the behavior of a flexible link robot in presence of contact forces can differ from the case of a robot in free motion. The main concern is the stability under switching conditions induced by unilateral contact. Based on the acceleration measurements the joint torques are estimated and state feedback control is applied. Simulation results for a 3-DOF robot with flexible links in end-effector contact with and without vibration damping control are shown and experimentally verified.

2 Modeling

A variety of models for elastic multibody systems exists with different accuracy, complexity, and modeled physical effects. These include finite element models (FEM), models based on a direct Ritz-approach as in [1], and lumped element models (LEM). An elastic 3 DOF serial robot as in Fig. 1 can be modeled as a rigid body linkage actuated by series elastic actuators (SEA). The corresponding lumped element model is given by

$$\mathbf{M}_M\ddot{\mathbf{q}}_M + \boldsymbol{\tau}_{Fric}(\dot{\mathbf{q}}_M) + \mathbf{K}\left(\mathbf{q}_M - \mathbf{q}_L\right) = \boldsymbol{\tau}_{Mot} \tag{1}$$

$$\mathbf{M}\left(\mathbf{q}_L\right)\ddot{\mathbf{q}}_L + \mathbf{C}(\mathbf{q}_L,\dot{\mathbf{q}}_L)\dot{\mathbf{q}}_L + \mathbf{G}(\mathbf{q}_L) - \underbrace{\mathbf{K}\left(\mathbf{q}_M - \mathbf{q}_L\right)}_{\boldsymbol{\tau}_J} = -\underbrace{\mathbf{J}^T(\mathbf{q}_L)\mathbf{F}_E}_{\boldsymbol{\tau}_E}. \tag{2}$$

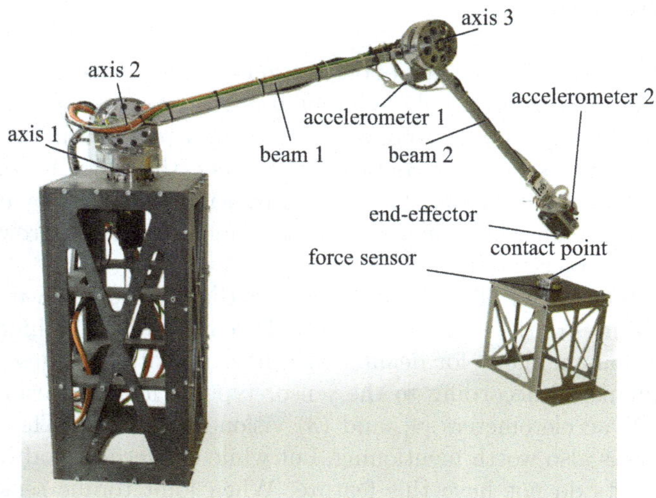

Fig. 1. Photo of the robotic system under consideration

While the Lagrange method is popular for derivation of the equations of motion, the chosen approach is the projected Newton-Euler formulation, also known as the PROJECTION EQUATION

$$\sum_{i=1}^{N_B} \left[\left(\frac{\partial_R \mathbf{v}_c}{\partial \dot{\mathbf{q}}} \right)^T \left(\frac{\partial_R \boldsymbol{\omega}_c}{\partial \dot{\mathbf{q}}} \right)^T \right]_i \begin{bmatrix} {}_R\dot{\mathbf{p}} + {}_R\tilde{\boldsymbol{\omega}}_{IR}\, {}_R\mathbf{p} - {}_R\mathbf{f}^e \\ {}_R\dot{\mathbf{L}} + {}_R\tilde{\boldsymbol{\omega}}_{IR}\, {}_R\mathbf{L} - {}_R\mathbf{M}^e \end{bmatrix}_i = \mathbf{0} \tag{3}$$

which will be also used for state estimation in the following section. For N_B bodies, this method projects the linear and angular momenta, ${}_R\mathbf{p}$ and ${}_R\mathbf{L}$ respectively, as well as the impressed forces and torques, ${}_R\mathbf{f}^e$ and ${}_R\mathbf{M}^e$, into the directions of unconstrained motion. Subscript R denotes the reference frame, ${}_R\tilde{\boldsymbol{\omega}}_{IR}$ is the angular velocity of the reference frame relative to the inertial frame I. ${}_R\mathbf{v}_c$ and ${}_R\boldsymbol{\omega}_c$ are the translational and angular velocity of the center of mass, and $\dot{\mathbf{q}}$ are the generalized velocities. $\mathbf{q}_M$ and $\mathbf{q}_L$ are the motor coordinates and the link coordinates respectively. Equation (1) describes the motor dynamics, where $\mathbf{M}_M$ represents the motor inertias, and $\boldsymbol{\tau}_{Fric}(\dot{\mathbf{q}}_M)$ and $\boldsymbol{\tau}_{Mot}$ are the friction and motor torques, respectively. The dynamics of the rigid body linkage is governed by (2), in terms of the inertia matrix $\mathbf{M}(\mathbf{q}_L)$, centrifugal and Coriolis terms $\mathbf{C}(\mathbf{q}_L, \dot{\mathbf{q}}_L)\dot{\mathbf{q}}_L$, and generalized gravity forces $\mathbf{G}(\mathbf{q}_L)$. The joint torques $\boldsymbol{\tau}_J$ due to the SEA are determined by the stiffness matrix $\mathbf{K}$, which is a diagonal matrix comprising the lumped stiffness associated to the individual joints. $\mathbf{F}_E$ is the end-effector force acting between robot and environment, and $\mathbf{J}^T(\mathbf{q}_L)$ is the end-effector Jacobian.

Ideal unilateral contact can be modeled by switching geometric constraints. This requires using a momentum consistent formulation [2], preferably in form of an efficient $O(n)$-algorithm for flexible link robots as presented in [3]. In this paper, however, since the focus is more on the results than on efficiency in implementation, a simple spring-damper contact model is used. The contact plane is defined by a position vector ${}_I\mathbf{r}_0$ and a normal vector $|{}_I\mathbf{n}| = 1$, pointing outwards the contact surface. The deformation is $s = ({}_I\mathbf{r}_0 - {}_I\mathbf{r}_E)^T {}_I\mathbf{n}$. With

$$F_{spring} = \begin{cases} cs & s > 0 \\ 0 & else \end{cases} \tag{4}$$

$$F_{damper} = \begin{cases} d\dot{s} & s > 0 \quad \wedge \quad \dot{s} > 0 \\ 0 & else \end{cases} \tag{5}$$

the contact force is computed by $\mathbf{F}_E = -(F_{spring} + F_{damper})_I\mathbf{n}$.

Furthermore, stiction in the robot joints is also modeled. Equation (1) can be written for each axis $i = 1, 2, 3$ independently with diagonal Matrices $\mathbf{M}_M$ and $\mathbf{K}$. The stiction model is shown in Fig. 2. It is implemented with a state machine for each motor. The motor velocity threshold value is denoted by ε and the critical stiction torque is $\tau_{Fric,0,i}$.

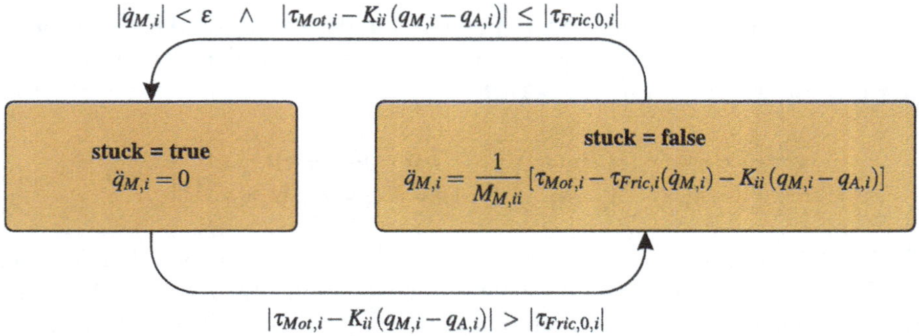

Fig. 2. Stiction switching conditions

3 Vibration Damping Control

Typically, industrial robots are controlled via joint position feedback with a PD or PID controller structure. In presence of joint elasticities or flexible beams this approach leads to undesirable link vibrations. The simplicity of an elastic joint model is beneficial for control design due to the flatness property, proposed in [5]. Without external disturbances, i.e. $\mathbf{F}_E = \mathbf{0}$, the state $\mathbf{x}^T = (\mathbf{q}_M, \mathbf{q}_L, \dot{\mathbf{q}}_M, \dot{\mathbf{q}}_L)$ and the input $\boldsymbol{\tau}_{Mot}$ can be calculated by an algebraic function of the flat output $\mathbf{q}_L$ and its time derivatives. The motor positions can be computed by rewriting Eq. (2) as

$$\begin{aligned}\mathbf{q}_M &= \mathbf{q}_L + \mathbf{K}^{-1}(\mathbf{M}_L(\mathbf{q}_L)\ddot{\mathbf{q}}_L + \mathbf{C}(\mathbf{q}_L, \dot{\mathbf{q}}_L)\dot{\mathbf{q}}_L + \mathbf{G}(\mathbf{q}_L)) \\ &= \mathbf{q}_M(\mathbf{q}_L, \dot{\mathbf{q}}_L, \ddot{\mathbf{q}}_L)\end{aligned} \tag{6}$$

and derivation of Eq. (6) twice w.r.t. time leads to $\dot{\mathbf{q}}_M$ and $\ddot{\mathbf{q}}_M$, which can be finally substituted into Eq. (1) for calculation of the motor torques $\boldsymbol{\tau}_{Mot}(\mathbf{q}_L, \dot{\mathbf{q}}_L, \ddot{\mathbf{q}}_L, \mathbf{q}_L^{(3)}, \mathbf{q}_L^{(4)})$. The control law

$$\begin{aligned}\boldsymbol{\tau}_{Mot} =& \mathbf{K}_{P,M}(\mathbf{q}_{M,d} - \mathbf{q}_M) + \mathbf{K}_{D,M}(\dot{\mathbf{q}}_{M,d} - \dot{\mathbf{q}}_M) \\ &+ \mathbf{K}_{P,\tau}(\boldsymbol{\tau}_{J,d} - \hat{\boldsymbol{\tau}}_J) + \mathbf{K}_{D,\tau}(\dot{\boldsymbol{\tau}}_{J,d} - \dot{\hat{\boldsymbol{\tau}}}_J) + \boldsymbol{\tau}_{Mot,FF}.\end{aligned} \tag{7}$$

is a full state feedback of the elastic joint model, where instead of the original state $\mathbf{x}$ a transformed state $\boldsymbol{\xi}^T = (\mathbf{q}_M, \boldsymbol{\tau}_J, \dot{\mathbf{q}}_M, \dot{\boldsymbol{\tau}}_J)$ is used. The feed-forward torques $\boldsymbol{\tau}_{Mot,FF}$ and the desired states $(\mathbf{q}_{M,d}, \boldsymbol{\tau}_{J,d}, \dot{\mathbf{q}}_{M,d}, \dot{\boldsymbol{\tau}}_{J,d})$ are calculated from the desired trajectory of the link coordinates $\mathbf{q}_{L,d}$. To this end, continuous derivatives of $\mathbf{q}_{L,d}$ up to the fourth derivative w.r.t. time are needed to assure continuous feed-forward torques, i.e. $\boldsymbol{\tau}_{Mot,FF}(\mathbf{q}_{L,d}, \dot{\mathbf{q}}_{L,d}, \ldots, \mathbf{q}_{L,d}^{(4)})$.

Estimations $\hat{\boldsymbol{\tau}}_J$ for the joint torques $\boldsymbol{\tau}_J$ (and the time derivative) are used in the control law (7). Additional sensors, accelerometers in this case, can be used for a classical state observer approach, as in [8]. However, observers cause time delays, or even instability of the closed loop system. Hence, in [4] an analytical calculation of the joint torques is used

$$\hat{\boldsymbol{\tau}}_J = \sum_{i=1}^{N_L} \left(\frac{\partial_R \mathbf{v}}{\partial \dot{\mathbf{q}}_L} \right)_i^T \left[m \underbrace{\left({}_R\dot{\mathbf{v}} + {}_R\tilde{\boldsymbol{\omega}}_{IR}\, {}_R\mathbf{v} - {}_R\mathbf{g} \right)}_{{}_R\mathbf{a}} \right]_i , \tag{8}$$

where ${}_I\mathbf{g}$ is the vector of gravitational acceleration. This is done by reconstructing the link dynamics with accelerometers, assuming a lumped mass m and an acceleration measurement ${}_R\mathbf{a}$ at the end-position of the i-th link. The estimated joint torques are then calculated with Eq. (8) where the absence of external forces $\mathbf{F}_E = \mathbf{0}$ is presumed and the influence of inertia parameters as opposed to Eq. (3) is neglected. The time-derivative of $\hat{\boldsymbol{\tau}}_J$ is computed by numerical differentiation. The complete control structure is shown in Fig. 3.

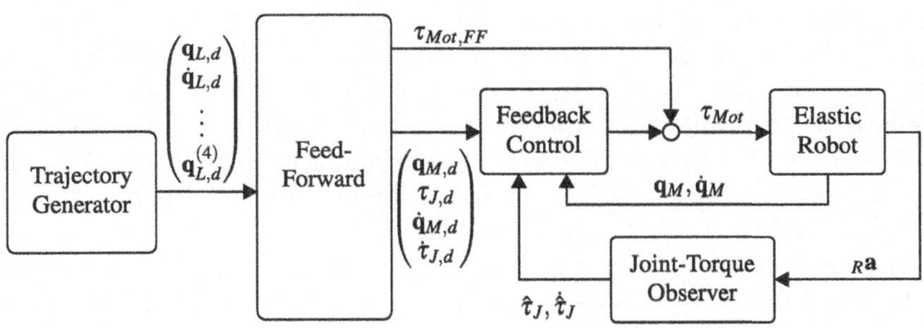

Fig. 3. Control structure

A comparison of a pure PD-motor controlled robot and a robot with additional vibration damping is shown in Fig. 4. The reaction to an external disturbance is examined. The acceleration signal of the tool center point (TCP) is successfully suppressed with vibration damping control, where pure PD-control of the motor coordinates fails to achieve a fast decreasing acceleration amplitude.

4 Vibration Damping Control During Contact

In a contact situation the aforementioned assumption of absent external forces is violated, i.e. $\mathbf{F}_E \neq \mathbf{0}$. Without additional knowledge the estimated $\hat{\boldsymbol{\tau}}_J$ cannot be separated into the joint torques $\boldsymbol{\tau}_J$ and the torques $\boldsymbol{\tau}_E$ resulting from external forces. The feedback signal for vibration damping in a contact scenario is $\hat{\boldsymbol{\tau}}_J \approx \boldsymbol{\tau}_J - \boldsymbol{\tau}_E$. When contact is modeled as switching of geometric constraints, the system becomes a non-smooth dynamical system. The theoretical stability analysis of this system is beyond the scope of this paper. Stability in this case is shown via simulation and experiment in the next section.

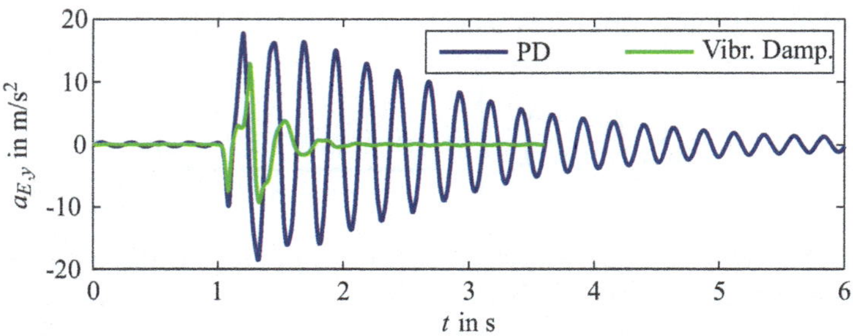

Fig. 4. Measured TCP acceleration signal (external disturbance)

5 Simulation and Experimental Results

Figure 1 shows the experimental setup. The three-axis *Elastic Laboratory Robot (ELLA)* with flexible links is equipped with synchronous motors and Harmonic-Drive gear units. The *ELLA* is operated with an industrial control system from *B&R Industrial Automation*. The PC-controller of the servo drives and the decentralized vibration damping are running on this system. Also, three-axis acceleration sensors used for vibration damping control are mounted at the elbow joint and the end-effector of the robot.

An aluminum plate is mounted via a force sensor on a table (as shown in Fig. 1). The robot is controlled to get into contact with the plate. The planned desired trajectory is a straight vertical line 0.2 m downwards. To meet the requirements for calculation of continuous feed-forward torques the trajectory is based on a $\sin^2$-jerk profile. A maximum velocity of 0.2 m/s is achieved. The trajectory endpoint is chosen to result in an absolute value of the static contact force of 50 N. The sensor force $\mathbf{F}_E^T = \begin{pmatrix} 0 \; 0 \; F_{Ext} \end{pmatrix}$ is measured. Simulation and experimental results are compared when the *ELLA* is controlled with and without ($\mathbf{K}_{P,\tau} = \mathbf{K}_{D,\tau} = \mathbf{0}$) vibration damping control.

In Fig. 5 the contact force is shown when no vibration damping is active for both simulation and experiment. Simulation and experiments are in good agreement in terms of the number of impacts and the static contact force. The main difference of simulation and measurements is the time-difference between two subsequent impacts and the peak values. This is mainly due to the shortcomings of the LEM in accurate modeling of the endpoint position. However, this model is sufficient for analyzing a contact scenario with the previously described vibration damping control.

The sensor force oscillations with changing signs, which can be seen between the impacts, can be attributed to the vibration excitation of the aluminum mass attached to the sensor due to force impacts. A more interesting aspect is the lower damping in the simulation model, which is apparent from the force oscillation during steady contact in simulated data. The contact model only considers a force perpendicular to the contact surface. No forces in the tangent plane are simulated. Therefore, the absence of friction at the robot endpoint in simulation

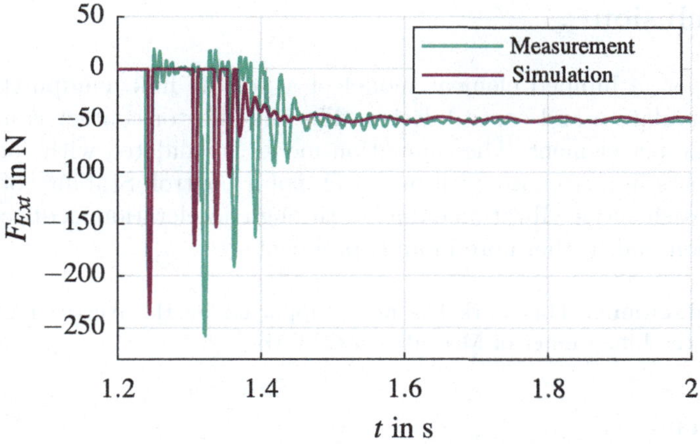

Fig. 5. Simulated and measured contact force without vibration damping control, $\mathbf{K}_{P,\tau} = \mathbf{K}_{D,\tau} = \mathbf{0}$

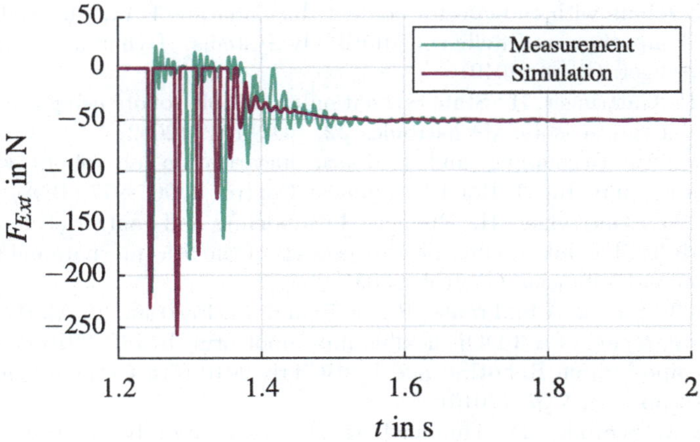

Fig. 6. Simulated and measured contact force with active vibration damping control

causes this difference. Notably is also that the frequency of this effect corresponds to the second eigenfrequency of the robot.

With the simulation model validated for the chosen contact scenario, active vibration damping is applied. In Fig. 6 the simulated contact force suggests a stable control system during impact phase. The oscillations in the contact phase are suppressed by vibration damping control compared to Fig. 5. The robotic system also remains stable in the experiment, as can be seen from the force measurement.

6 Conclusion

In this paper, a lumped element model of a flexible link manipulator is used both for simulation and control design. The end-effector contact is modeled by a spring-damper element. The simulation model is validated with contact force measurements in a scenario with pure PD motor control. Stability of the robot in contact with active vibration damping through acceleration feedback is shown in simulation and furthermore in an experiment.

Acknowledgement. This work has been supported by the Austrian COMET-K2 program of the Linz Center of Mechatronics (LCM).

References

1. Bremer, H.: Elastic Multibody Dynamics: A Direct Ritz Approach. Springer, Heidelberg (2008)
2. Pfeiffer, F.: Mechanical System Dynamics. Springer, Heidelberg (2008)
3. Gattringer, H., Müller, A., Pucher, F., Reiter, A.: O(n) algorithm for elastic link/joint robots with end-effector contact. In: Zahariev, E.V., Cuadrado, J. (eds.) IUTAM Symposium on Intelligent Multibody Systems - Dynamics, Control, Simulation. Springer, Cham (2019)
4. Staufer, P., Gattringer, H.: State estimation on flexible robots using accelerometers and angular rate sensors. Mechatronics **22**, 1042–1049 (2012)
5. De Luca, A.: Decoupling and feedback linearization of robots with mixed rigid/elastic joints. Int. J. Robust Nonlinear Control **8**, 965–977 (1998)
6. Höbarth, W., Gattringer, H., Bremer, H.: Modeling and control of an articulated robot with flexible links/joints. In: Proceedings of the 9th International Conference on Motion and Vibration Control (2008)
7. Malzahn, J., Phung, A.S., Franke, R., Hoffmann, F., Bertram, T.: Markerless visual vibration damping of a 3-DOF flexible link robot arm. In ISR 2010 (41st International Symposium on Robotics) and ROBOTIK 2010 (6th German Conference on Robotics), pp. 1–8. VDE (2010)
8. De Luca, A., Schroder, D., Thummel, M.: An acceleration-based state observer for robot manipulators with elastic joints. In: Proceedings 2007 IEEE International Conference on Robotics and Automation, pp. 3817–3823. IEEE (2007)

Possibilistic Investigation of Mechanical Control Systems Under Uncertainty

Andreas Hofmann(✉), Michael Hanss, and Peter Eberhard

Institute of Engineering and Computational Mechanics, University of Stuttgart, Pfaffenwaldring 9, 70569 Stuttgart, Germany
andreas.hofmann@itm.uni-stuttgart.de

Abstract. Uncertainty in linear control systems is often modeled by real- or complex-valued perturbations. The key idea of robust control techniques is to check whether the design specifications are satisfied even for the worst-case uncertainty. A quite intuitive and apparently natural way to model uncertainty is using fuzzy sets and arithmetic. The resulting uncertain system models do not only consider worst-case scenarios, but also the shape of uncertainty modeled by imprecise quantities. Possibilistic measures can then be employed to investigate the satisfaction of stability and performance objectives and to draw conclusions about the robustness of the resulting controller. The gradualness in the quantification of uncertainty allows for more comprehensive statements than just worst-case scenarios.

1 Introduction

Linear PID-controllers are still the most frequently used controllers for mechanical systems. Various tuning rules and guidelines exist to achieve given control objectives, worth mentioning are the well-known Ziegler-Nichols tuning rules, [12]. The idea of these rules is not only to provide stability and a desired performance for a given nominal plant model, but also to establish stability and performance margins in the case of uncertainty, [9]. More advanced techniques like the well-established $\mathscr{H}_\infty$ robust control paradigm include real- or complex-valued perturbations to model uncertainty, and the key idea is to check whether the design specifications are satisfied even for the worst-case uncertainty. A quite intuitive and apparently natural way to model uncertainty is fuzzy sets and arithmetic, [6]. It leads to comprehensive models that include uncertainty stemming from various sources, such as approximation, simplification and lack of knowledge, or variations in the operating conditions, [7]. The resulting uncertain system models do not only consider worst-case scenarios, but also the shape of uncertainty. These uncertain models are well suited for analysis and optimization in the framework of possibility theory, [8]. In the following, a mechanical control system under uncertainty is investigated using a possibilistic approach. In the presence of modeling uncertainty expressed by uncertain parameters, the stability constraint and given performance objectives lead to uncertain inequalities.

A. Kecskeméthy and F. Geu Flores (Eds.): ECCOMAS 2019, COMPUTMETHODS 53, pp. 147–154, 2020.
https://doi.org/10.1007/978-3-030-23132-3_18

In contrast to crisp inequalities, these uncertain inequalities can also be satisfied just to some degree, and possibilistic measures can be employed to quantify this degree of satisfaction. These measures can be used to assess the criteria of stability and performance, and to draw conclusions about the influence of the uncertainty on the control problem and about the guarantee of robustness.

2 Fuzzy Arithmetical Uncertainty Modeling

In general, the member elements of an ordinary set $A \subset \mathbb{R}$ can be defined by using a function $\mu_A : X \to \{0,1\}$ which indicates the membership of an element $x \in X$ if $\mu_A(x) = 1$, and its non-membership if $\mu_A(x) = 0$. A fuzzy set, instead, is a generalization of an ordinary set A where an element is not only defined by complete membership or non-membership, but by a gradual degree of membership, [11]. In this case, the characteristic function μ_A is replaced by a membership function $\mu_{\tilde{A}} : X \to [0,1]$, and the fuzzy set is defined as $\tilde{A} = \{(x, \mu_{\tilde{A}}(x)) | x \in \mathbb{R}, \mu_{\tilde{A}}(x) \in [0,1]\}$. Fuzzy numbers are a special type of fuzzy sets with some specific properties. For a fuzzy number $\tilde{p}$, the corresponding fuzzy set is convex, and the membership function $\mu_{\tilde{p}}$ satisfies $\mu_{\tilde{p}}(x) = 1$ for only one value $x = \bar{x}$. This value is called nominal value and it represents the crisp value one might preferably choose in a conventional, crisp-valued problem. Furthermore, the support of a fuzzy number is given by all elements with non-zero membership values, i.e. $\text{supp}(\tilde{p}) = \{x \in \mathbb{R} \,|\, \mu_{\tilde{p}}(x) > 0\}$. An example of a fuzzy number, which is suitable for the representation of parametric uncertainty, is the triangular fuzzy number

$$\tilde{p} = \text{tfn}(\bar{x}, \alpha_\text{l}, \alpha_\text{r}) \tag{1}$$

as shown in Fig. 1. The membership function of this fuzzy number is defined by

$$\mu_{\tilde{p}}(x) = \begin{cases} 0 & \text{for} \quad x \leq \bar{x} - \alpha_\text{l} \\ 1 + (x - \bar{x})/\alpha_\text{l} & \text{for} \quad \bar{x} - \alpha_\text{l} < x < \bar{x} \\ 1 - (x - \bar{x})/\alpha_\text{r} & \text{for} \quad \bar{x} \leq x < \bar{x} + \alpha_\text{r} \\ 0 & \text{for} \quad x \geq \bar{x} + \alpha_\text{r} \end{cases} \tag{2}$$

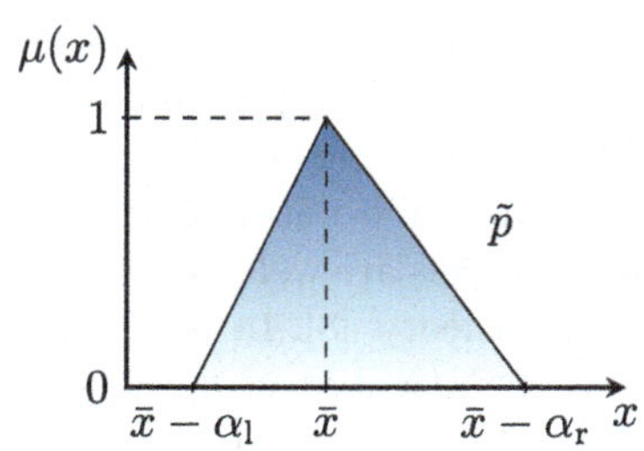

Fig. 1. Triangular fuzzy number $\tilde{p} = \text{tfn}(\bar{x}, \alpha_\text{l}, \alpha_\text{r})$.

with α_l and α_r as the left-hand and right-hand worst-case deviations from the nominal value. In general, there are no more limitations on the membership function than the above-stated properties of a fuzzy number. With regard to uncertainty modeling, fuzzy numbers of various shapes allow an appropriate description and quantification of parametric uncertainty. Fuzzy arithmetic then enables the forward propagation of the fuzzy-valued parametric uncertainties through the system. This allows conclusions about how the uncertainty in the parameters affects the overall system output.

3 Possibilistic Ranking of Fuzzy Numbers

Possibility theory provides a unified framework for a robust treatment of both aleatory and epistemic uncertainty. In a state of perfect knowledge only aleatory uncertainties remain and the occurrence of an event is a matter of chance. Partial or incomplete information is often denoted by epistemic uncertainty and can be described by means of a set of possible values, [5]. Epistemic uncertainty arises due to lack of knowledge about the true underlying probability distribution. A possibility distribution describes more or less plausible values of an uncertain variable, [1]. The theory of fuzzy sets provides a mathematical basis for the calculus of possibility distribution, [10]. Possibility theory provides two measures of occurrence of an event: the possibility Π and the necessity $\mathcal{N}$. The measures are defined through the possibility distribution $\pi : \mathbb{R} \to [0,1]$. The possibility

$$\Pi(A) = \sup_{x \in A} \pi(x)\,, \tag{3}$$

where $\Pi(A) = 0$ means impossibility and $\Pi(A) = 1$ that the event A is absolutely possible. The necessity of an event, also called certainty measure,

$$\mathcal{N}(A) = \inf_{x \notin A} (1 - \pi(x)) = 1 - \Pi(\bar{A}) \tag{4}$$

is the counterpart of possibility. The event A is not necessary if $\mathcal{N}(A) = 0$ and absolutely sure if $\mathcal{N}(A) = 1$. The measures verify

$$\forall A, B \subseteq \mathbb{R} \quad \Pi(A \cup B) = \max(\Pi(A), \Pi(B)) \tag{5}$$

and

$$\forall A, B \subseteq \mathbb{R}, \quad \mathcal{N}(A \cap B) = \min(\mathcal{N}(A), \mathcal{N}(B))\,. \tag{6}$$

As a consequence of Eq. (5), the possibility measure is not self-dual and only the inequality $\Pi(A)+\Pi(\bar{A}) \geq 1$ holds. With Eq. (4) it follows that the possibility of an event is at least as high as the necessity, $\mathcal{N}(A) \leq \Pi(A)$. Possibility and necessity provide bounds on the confidence. In this sense, they can be interpreted as upper and lower bounds of probability assignments, [4]. Possibilistic measures can be employed to quantitatively assess the relative position of a fuzzy number $\tilde{p}$ and a crisp value r, namely to evaluate the event that the one is greater or less than the other, [3]. It verifies that

$$\Pi(\tilde{p} < r) = \sup_{\substack{x \\ x \leq r}} \mu_{\tilde{p}}(x) \tag{7}$$

and

$$\mathcal{N}(\tilde{p} < r) = 1 - \Pi(\overline{\tilde{p} < r}) . \tag{8}$$

Especially the necessity (8) will be used to evaluate stability and performance criteria in the following mechanical control example.

4 Mechanical Control Example

Consider the mechanical system of a cart with a disturbance weight, as shown in Fig. 2. The cart of mass m_{K} is constrained by a spring of stiffness c_{K}. The disturbance mass m_{G} is coupled to the cart by a linear spring-damper combination with stiffness c_{G} and damping d_{G}. The coordinate x_{K} describes the absolute position of the cart, and x_{G} is the relative position of the disturbance mass against the cart. This low-order linear model can be seen as a simplified model of a truck with a moving load, e.g. a liquid. The system is driven by a driving force F. The control objective is, to stabilize the horizontal position x_{K} of the cart by a PI-controller with the transfer function

$$K(s) = k_{\mathrm{C}} \left(1 + \frac{1}{t_{\mathrm{I}} s}\right) \tag{9}$$

with the parameters k_{C} and t_{I}. The dynamics of the system with input $u = F$ and output $y = x_{\mathrm{K}}$ can be represented by a proper minimum-phase transfer function

$$G(s) = \frac{m_{\mathrm{G}} s^2 + d_{\mathrm{G}} s + c_{\mathrm{G}}}{m_{\mathrm{G}} m_{\mathrm{K}} s^4 + d_{\mathrm{G}} (m_{\mathrm{G}} + m_{\mathrm{K}}) s^3 + (c_{\mathrm{G}} (m_{\mathrm{G}} + m_{\mathrm{K}}) + c_{\mathrm{K}} m_{\mathrm{G}})\, s^2 + c_{\mathrm{K}} d_{\mathrm{G}} s + c_{\mathrm{G}} c_{\mathrm{K}}} . \tag{10}$$

The controller design has to guarantee stability of the system and has to fulfill performance requirements on the maximum available control force and the frequency response. The stability of the closed-loop system can be inspected using the Nyquist stability criterion $|L(j\omega_{180})| < 1$ of the open-loop transfer function $L = GK$, i.e. the encirclement of the critical point must not change. The performance objectives can be expressed by restrictions on the $\mathcal{H}_\infty$ norm of the sensitivity function $S = (I + GK)^{-1}$, by

$$|S(j\omega)| < \frac{1}{|w_{\mathrm{P}}(j\omega)|} \quad \forall \omega \Leftrightarrow \|w_{\mathrm{P}} S\|_\infty < 1 , \tag{11}$$

where

$$w_{\mathrm{P}}(s) = \frac{s/a + \omega_{\mathrm{B}}}{s + \omega_{\mathrm{B}} b} \tag{12}$$

is a given performance weight with a and b to restrict the high- and low-frequency amplitude and the required bandwidth frequency ω_{B}. In the same manner, the input restrictions may lead to $\|w_{\mathrm{u}} KS\|_\infty < 1$ where $1/w_{\mathrm{u}}$ is an upper bound on the magnitude of the closed-loop input signal u.

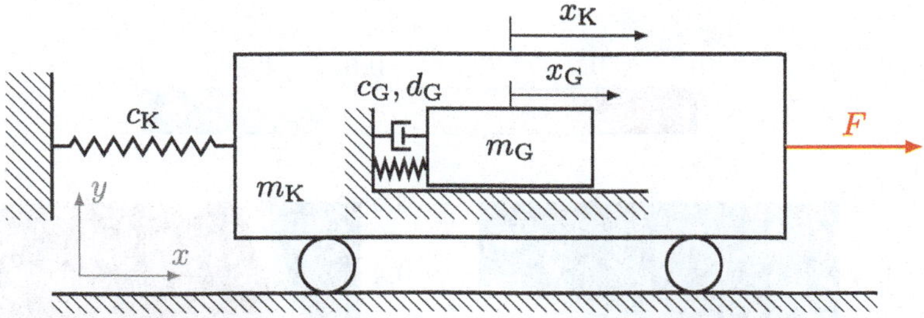

Fig. 2. Cart with disturbance weight.

5 Results

For the system given by the transfer function (10), let the load of the cart, i.e. the disturbance weight, be an uncertain mass modeled by the fuzzy value $\tilde{m}_G = \mathrm{tfn}(1, 0.3, 0.5)$ kg. The other parameters shall be considered as crisp values with $d_G = 1$ Ns/m, $c_G = 1$ N/m, $c_K = 10$ N/m and $m_K = 1$ kg. This leads to an uncertain system dynamics $\tilde{G}(s, \tilde{m}_G)$. In a first step, the stability of the closed loop is considered. The possibilistic assessment of the Nyquist stability criterion for a controller based on the Ziegler-Nichols tuning rules for the nominal system is shown in Fig. 3 (left). It can clearly be seen that the criterion is not completely fulfilled for the uncertain control system. Hence, the system response to an input disturbance is not robustly stable, as can be seen in Fig. 3 (right). While the nominal system is stable the uncertain system becomes unstable for lower levels of membership of the fuzzy values, i.e. for a higher degree of uncertainty. The Nyquist stability criterion on the open-loop amplitude is only applicable if the nominal plant is stable, i.e. the encirclement of the critical point does not change. The evaluation of the necessity measure of the uncertain stability inequality automatically satisfies this prerequisite. The necessity of the event of stability

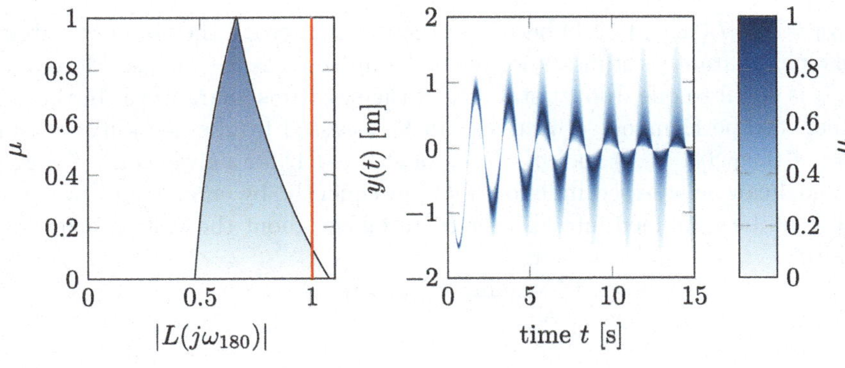

Fig. 3. Evaluation of the Nyquist stability criterion (left) and system response (right) in the presence of uncertainty.

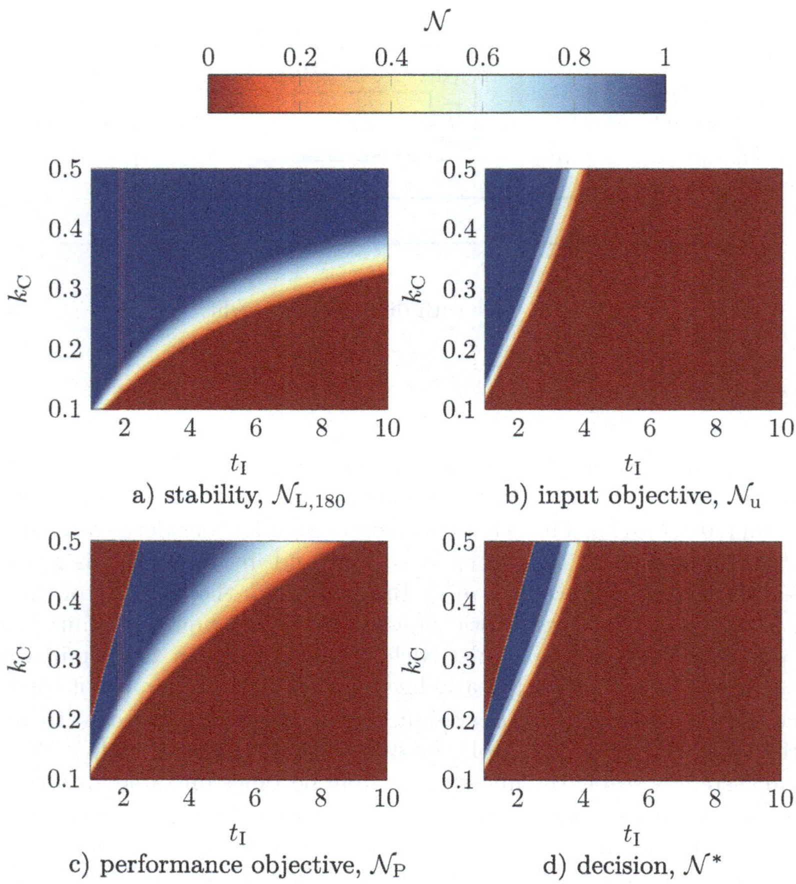

Fig. 4. Confluence of constraints and goals in a decision-making scheme.

$\mathcal{N}_{L,180} = \mathcal{N}(|\tilde{L}(j\omega_{180})| < 1)$ becomes greater than zero, if the nominal system is stable and guarantees stability for memberships $\mu_{\tilde{m}_G} > 1 - \mathcal{N}_{L,180}$. Moreover, if $\mathcal{N}_{L,180}$ is equal to one, robust stability of the system is guaranteed. In the same manner, the performance objectives can be assessed by the necessity measures $\mathcal{N}_P = \mathcal{N}(\|w_P\tilde{S}\|_\infty < 1)$ and $\mathcal{N}_u = \mathcal{N}(\|w_u\tilde{S}\|_\infty < 1)$ for a given controller $K(s)$. Each goal can be assessed individually. Subsequently, by employing Eq. (6), the results can be combined into an overall statement about the system behavior in the form

$$\mathcal{N}^* = \min(\mathcal{N}_{L,180}, \mathcal{N}_P, \mathcal{N}_u) \,. \tag{13}$$

The best possible controller is found by solving the decision making scheme in a fuzzy environment of Bellman and Zadeh [2], stating that the decision is the *"confluence of constraints and goals"* and the best controller maximizes the confidence of the decision

$$\mathscr{N}_{\max} = \max_{K(s)} \mathscr{N}^* \,. \tag{14}$$

There exists a robust controller if the joint necessity $\mathscr{N}_{\max}$ becomes 1. If this is not the case, a bound of the membership $\mu_{\tilde{m}_G} = 1 - \mathscr{N}_{\max}$ can be computed. The control objectives are guaranteed for levels of membership of the fuzzy values higher than the computed bound. The details of the decision making process are depicted in Fig. 4. It can be clearly seen, that there exists a set of possible controllers which robustly fulfill the given objectives. In addition, the gradual stability and performance maps as well as the final decision map provides information about the sensitivity of a constraint, goal or the final decision to the uncertainty. A sharp decrease of the necessity measure can indicate a bound on the controller which is not at all or insignificantly affected by the uncertainty. A slow decrease, instead, indicates the opposite. This characteristics can be used to assess the robustness of a controller. More specifically, it can be investigated to what extent the controller gains can vary before the control performance decreases significantly. For example, let us assume that the controller gains can be implemented with an accuracy of $\pm 10\%$, only. Hence, the decision should be evaluated accordingly. The possible accuracy defines a set on which the minimum necessity provides information about the robustness of the given controller.

6 Conclusions

The novelty of the presented approach consists in using comprehensive system models which include a quantification of the inherent uncertainty in combination with possibility theory and decision making. It is possible to combine constraints and performance objectives for the assessment of the uncertain control systems. The possibilistic measures allow detailed statements about the possibility that the control objectives are fulfilled in the presence of uncertainty. In the case that the control objectives are not fulfilled, a lower bound on the membership level can be computed and thus the tolerated uncertainty. In addition, the possibilistic measures can be used to conclude about the sensitivity of constraints and objectives to the uncertainty, and robustness analysis of the resulting controller. Furthermore, this work can be used as a basis for an optimization setting to find an optimal robust controller. Yet, fuzzy sets cannot only be employed to model parametric uncertainty, but also imprecise or vague performance specifications. This allows an extension of this approach to imprecise performance objectives.

References

1. Baudrit, C., Dubois, D.: Practical representations of incomplete probabilistic knowledge. Comput. Stat. Data Anal. **51**(1), 86–108 (2006)
2. Bellman, R.E., Zadeh, L.A.: Decision-making in a fuzzy environment. Manage. Sci. **17**(4), 141–164 (1970)
3. Dubois, D., Prade, H.: Ranking fuzzy numbers in the setting of possibility theory. Inf. Sci. **30**(3), 183–224 (1983)
4. Dubois, D., Prade, H.: When upper probabilities are possibility measures. Fuzzy Sets Syst. **49**(1), 65–74 (1992)
5. Dubois, D., Prade, H.: Gradualness, uncertainty and bipolarity: making sense of fuzzy sets. Fuzzy Sets Syst. **192**, 3–24 (2012)
6. Hanss, M.: Applied Fuzzy Arithmetic - An Introduction with Engineering Applications, 2nd edn. Springer, Berlin (2010)
7. Hofmann, A., Hanss, M.: Fuzzy arithmetical controller design for active road vehicle suspension in the presence of uncertainties. In: 22nd International Conference on Methods and Models in Automation and Robotics (MMAR), Miedzyzdroje, Poland, pp. 582–587 (2017)
8. Hose, D., Mäck, M., Hanss, M.: A possibilistic approach to the optimization of uncertain systems. In: ICVRAM ISUMA Uncertainties Conference, Florianopolis, SC, Brazil (2018)
9. Skogestad, S., Postlethwaite, I.: Multivariable Feedback Control: Analysis and Design, 2nd edn. Wiley, Chichester (2005)
10. Zadeh, L.A.: Fuzzy sets as a basis for a theory of possibility. Fuzzy Sets Syst. **1**, 3–28 (1978)
11. Zadeh, L.A.: Fuzzy sets. Inf. Control **8**(3), 338–353 (1965)
12. Ziegler, J.G., Nichols, N.B.: Optimum settings for automatic controllers. Trans. ASME **64**(11) (1942)

Nonlinear Position Control of a Very Flexible Parallel Robot Manipulator

Peter Eberhard(✉) and Fatemeh Ansarieshlaghi

Institute of Engineering and Computational Mechanics, University of Stuttgart, Pfaffenwaldring 9, 70569 Stuttgart, Germany
{peter.eberhard,fatemeh.ansari}@itm.uni-stuttgart.de

Abstract. In this paper, we investigate the control of a very flexible parallel robot with high accuracy. This robot has two very flexible long links and can be modeled as an underactuated multibody system since it has fewer control inputs than degrees of freedom for rigid body motion and deformation. Therefore, these flexibilities are taken into account in the control design. In order to obtain high performance in the end-effector trajectory tracking, an accurate and efficient nonlinear controller is designed. This nonlinear feedback controller is based on the Lyapunov approach using the measurable states of the system. Then, it is carefully tested on the flexible parallel robot. The simulation and experimental results show that the end-effector tracks desired trajectories with high accuracy. Also, the designed controller is compared to previous works and the results show that the controller can achieve higher tracking performance.

1 Introduction

Light-weight manipulators attract a lot of research interest because of their complementing advantages. The advantages of light-weight robots include low energy usage, less mass, and often high working speeds. However, due to the light-weight design, the bodies have a significant flexibility which yields undesired deformations and vibrations. Therefore, this manipulator is modeled as a flexible multibody system and the flexibilities must be taken into account in the control design. The flexible system with significant deformations complicates the control design because there are more generalized coordinates than control inputs. In order to obtain high performance in the end-effector trajectory tracking of a flexible manipulator, an accurate and efficient nonlinear feedforward and feedback controller are advantageous. The difficulty of designing a nonlinear feedback controller with high performance for a highly flexible system is increased, when the controller does not have access to direct measurement of the end-effector and all the system states. To overcome this problem, the system model can be reduced based on the system constraints. Finally, based on the reduced model of the system, a nonlinear feedback controller can be designed.

To investigate a flexible manipulator here, a lambda shape robot is used. In previous works on the lambda robot, some linear controllers and a nonlinear

A. Kecskeméthy and F. Geu Flores (Eds.): ECCOMAS 2019, COMPUTMETHODS 53, pp. 155–162, 2020.
https://doi.org/10.1007/978-3-030-23132-3_19

feedback controller were designed based on the system model. The linear controllers were designed using the measurable system states [6,9]. Also, a nonlinear controller was designed based on the feedback linearization approach and all the estimated states were obtained by a nonlinear observer [2].

The novelty of this work is, that a nonlinear feedback controller for high-speed trajectory tracking of a very flexible parallel lambda robot is designed. This controller is based on the reduced model of the system and the direct method of Lyapunov [8] using the measurable system states. Using the reduced model of the system and only measurable states, there is no need to observe and estimate all the system states. Therefore, in this method, the corresponding estimation error is removed.

In this paper, the nonlinear feedback controller is implemented on the simulated model of a very flexible parallel robot and tested on the real system. The results show that the end-effector tracks a trajectory with high performance and higher accuracy compared to the previous works.

The paper is organized as follows: Sect. 2 describes the robot and Sect. 3 includes the modeling of the flexible parallel lambda robot. Section 4 explains the architecture of the nonlinear controller. In Sect. 5, the proposed nonlinear controller is simulated and tested on the robot and the results are discussed.

2 Flexible Lambda Robot

The used lambda robot is a simple parallel robot manipulator. This robot has highly flexible links. The end of the short link is connected in the middle of the long link using rigid bodies. This connection creates a closed loop kinematics constraint that causes the parallel configuration of the robot. This robot has two prismatic actuators connecting the links to the ground. The links are connected using passive revolute joints to the linear actuators. Another revolute joint is used to connect the short link and the middle of the long link. An additional rigid body is attached to the free end of the long link as an end-effector. The drive positions and velocities are measured with two optical encoders. Three full Wheatstone bridge strain gauge sets are attached on the long flexible link to measure its deformation. The lambda robot configuration is shown in Fig. 1 has been built in hardware, see [6] at the Institute of Engineering and Computational Mechanics of the University of Stuttgart.

The electrical part of the hardware includes some power supplies for motors, strain gauge's amplifiers, digital/analog input-outputs boards, one *Speedgoat* target, a host computer, etc. For controlling the robot, the online control is done with a *Speedgoat* performance real-time target machine running a *Mathworks xPCtarget* kernel, which is called *Simulink* Real-Time since *Matlab R2014a*. Also, to observe the controller progress a graphical user interface is available for the input, output, safety logic of the lambda robot and the communication.

3 Modeling of the Flexible Lambda Robot

The modeling process of the flexible manipulator with λ configuration can be separated into three major steps. First, the flexible components of the system are modeled with the linear finite element method in the commercial finite element code *ANSYS* with about six hundred degrees of freedom in total. Next, in order to control the λ robot, the degrees of freedom of the flexible bodies shall be decreased. Therefore, the modal model order reduction method is utilized to reduce the order of flexible multibody model. Then, all the rigid and flexible parts are modeled as a multibody system with a kinematic loop. The equation of motion with a kinematic loop constraint of the flexible parallel manipulator using the generalized coordinates $\boldsymbol{q} \in \boldsymbol{R}^5$ is

$$\boldsymbol{M}(\boldsymbol{q})\ddot{\boldsymbol{q}} + \boldsymbol{k}(\boldsymbol{q}, \dot{\boldsymbol{q}}) = \boldsymbol{g}(\boldsymbol{q}, \dot{\boldsymbol{q}}) + \boldsymbol{B}(\boldsymbol{q})\boldsymbol{u} + \boldsymbol{C}^{\mathrm{T}}(\boldsymbol{q})\boldsymbol{\lambda}\,, \tag{1a}$$

$$\boldsymbol{c}(\boldsymbol{q}) = \boldsymbol{0}\,. \tag{1b}$$

The symmetric, positive definite mass matrix $\boldsymbol{M} \in \boldsymbol{R}^{5\times5}$ depends on the joint angles and the elastic coordinates. The vector $\boldsymbol{k} \in \boldsymbol{R}^5$ contains the generalized centrifugal, Coriolis and Euler forces, and $\boldsymbol{g} \in \boldsymbol{R}^5$ includes the vector of applied forces and inner forces due to the body elasticity. The input matrix $\boldsymbol{B} \in \boldsymbol{R}^{5\times2}$ maps the input vector $\boldsymbol{u} \in \boldsymbol{R}^2$ to the system. The constraint equations are defined by $\boldsymbol{c} \in \boldsymbol{R}^2$. The Jacobian matrix of the constraint $\boldsymbol{C} = \partial\left(\boldsymbol{c}(\boldsymbol{q})\right)/\partial\boldsymbol{q} \in \boldsymbol{R}^{2\times5}$ maps the reaction force $\boldsymbol{\lambda} \in \boldsymbol{R}^2$ due to the kinematic loop. The flexible lambda robot in hardware and its mechanical model are shown in Fig. 1 left, see also [1,3,6].

While the system has a kinematics loop as a constraint, the system coordinates $\boldsymbol{q}$ can be divided to dependent $\boldsymbol{q}_{\mathrm{d}} = [\alpha_1, \alpha_1]^T$, passive joints and independent $\boldsymbol{q}_{\mathrm{i}} = [s_1, s_2, q_e]^T$, active joints and flexible link. Therefore, the system constraints can be written as

$$\ddot{c} = \boldsymbol{C}\ddot{\boldsymbol{q}} + \boldsymbol{c}'' = \boldsymbol{C}_{\mathrm{i}}\ddot{\boldsymbol{q}}_{\mathrm{i}} + \boldsymbol{C}_{\mathrm{d}}\ddot{\boldsymbol{q}}_{\mathrm{d}} + \boldsymbol{c}'', \tag{2}$$

where $\boldsymbol{C}_{\mathrm{d}}$ and $\boldsymbol{C}_{\mathrm{i}}$ are dependent and independent parts of the Jacobian matrix of the constraint and $\boldsymbol{c}''$ presents local accelerations due to the constraints. Based on Eq. (2), the system coordinates can be formulated by

$$\ddot{\boldsymbol{q}} = \begin{bmatrix}\ddot{\boldsymbol{q}}_{\mathrm{i}}\\ \ddot{\boldsymbol{q}}_{\mathrm{d}}\end{bmatrix} = \begin{bmatrix}\boldsymbol{I}_{3\times3}\\ -\boldsymbol{C}_{\mathrm{d}}^{-1}\boldsymbol{C}_{\mathrm{i}}\end{bmatrix}\ddot{\boldsymbol{q}}_{\mathrm{i}} + \begin{bmatrix}\boldsymbol{0}_{3\times3}\\ -\boldsymbol{C}_{\mathrm{d}}^{-1}\boldsymbol{c}''\end{bmatrix} = \boldsymbol{J}_{\mathrm{c}}\ddot{\boldsymbol{q}}_{\mathrm{i}} + \boldsymbol{b}''. \tag{3}$$

Finally, by left-side multiplication with the transposed of the Jacobian matrix $\boldsymbol{J}_{\mathrm{c}}$ and replacing $\ddot{\boldsymbol{q}}$ by Eq. (3), the equation of motion (1a) can be determined, see [4,5], by

$$\bar{\boldsymbol{M}}\ddot{\boldsymbol{q}}_{\mathrm{i}} = -\bar{\boldsymbol{k}} + \bar{\boldsymbol{g}} + \bar{\boldsymbol{B}}\boldsymbol{u} = \bar{\boldsymbol{f}} + \bar{\boldsymbol{B}}\boldsymbol{u}. \tag{4}$$

The new dynamics formulation of the lambda robot, Eq. (4), based on the independent states of the system is named the reduced model.

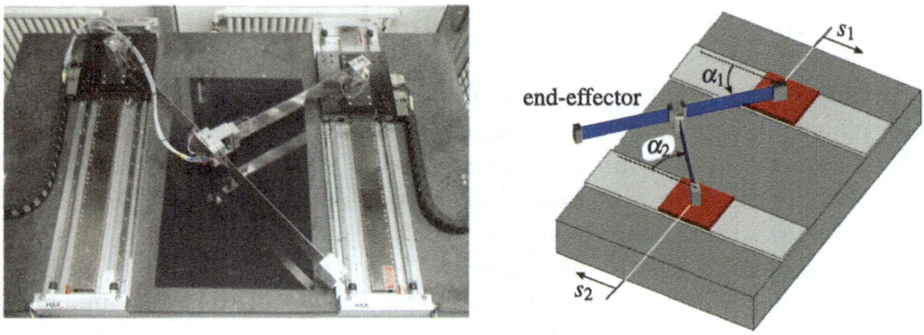

Fig. 1. Lambda robot, mechanical setup of the robot and simulation model of flexible parallel robot.

4 Control of the Flexible Lambda Robot

The lambda robot control is separated into feedforward and feedback control parts. The designed nonlinear controller structure is shown in Fig. 2.

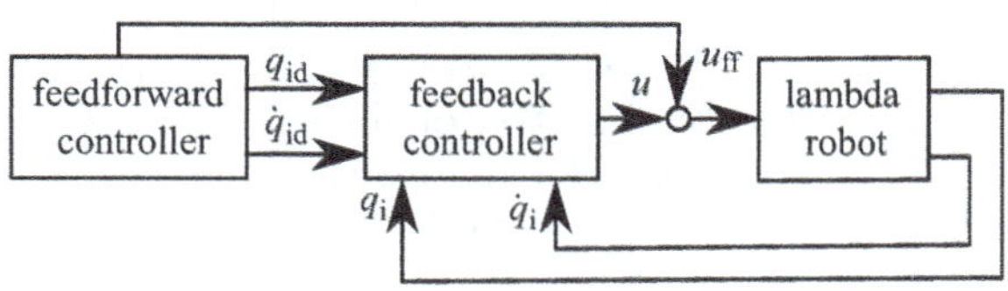

Fig. 2. Nonlinear feedback controller structure.

In the feedforward control part, the offline part, the desired trajectories of the system states ($\boldsymbol{q}_{id}$, $\dot{\boldsymbol{q}}_{id}$) and the feedforward current of the actuators $\boldsymbol{u}_{ff}$ are calculated using a two-point boundary value problem based on the desired end-effector trajectory. The two-point boundary value problem is used while the flexible multibody system is a non-minimum phase system with internal dynamics [5]. The feedback control part computes the lambda robot input ($\boldsymbol{u}$) using the reduced nonlinear dynamics of the robot in Eq. (4), the measured states ($\boldsymbol{q}_i$, $\dot{\boldsymbol{q}}_i$), and the desired states based on the direct method of Lyapunov [8] in real-time. This controller is named reduced nonlinear feedback controller.

To design a nonlinear feedback controller for the system in Eq. (4), the control law is obtained for the lambda robot as

$$\boldsymbol{u} = \bar{\boldsymbol{B}}^{-1}(-\bar{\boldsymbol{f}} + \bar{\boldsymbol{M}}(\boldsymbol{K}_{P}\boldsymbol{e} + \boldsymbol{K}_{D}\dot{\boldsymbol{e}})), \tag{5}$$

where $\boldsymbol{q}_{id}$ is the desired value for $\boldsymbol{q}_i$, consequently $\dot{\boldsymbol{q}}_{id}$ is the desired value for $\dot{\boldsymbol{q}}_i$ and the error and the dynamics of error are calculated by $\boldsymbol{e} = \boldsymbol{q}_i - \boldsymbol{q}_{id}$ and $\dot{\boldsymbol{e}} = \dot{\boldsymbol{q}}_i - \dot{\boldsymbol{q}}_{id}$. The desired values depend on the desired trajectory of the end-effector and can be computed via the feedforward part and $\boldsymbol{q}_{id}$ and $\dot{\boldsymbol{q}}_{id}$

can be set. The matrices $\boldsymbol{K}_\mathrm{P}$ and $\boldsymbol{K}_\mathrm{D}$ correspond to the weighting of feedback errors and can be designed via the LQR method or tuned by hand. Also, they should satisfy the stability conditions for nonautonomous systems as a uniform stability, based on the Lyapunov theorem. The inverse of the input matrix $\bar{\boldsymbol{B}}$ is not so straightforward to calculate, since it is not of full row rank. Therefore, the existing left-inverse is used as a pseudo-inverse to yield $\bar{\boldsymbol{B}}\bar{\boldsymbol{B}}^{-1} = \boldsymbol{I}$. The vector $\boldsymbol{u}$ presents the control input of the robot manipulator and that is the output of the designed position controller based on the Lyapunov method.

5 Simulation and Experimental Results

The controller is tested in real-time on the machine with 250 µs sampling time. Therefore, the camera with image processing is not applicable for online tracking the end-effector position and it can be used only for offline validation. To validate the designed nonlinear feedback controller, the end-effector tracks a trajectory and a camera records the movie during the trajectory tracking. Then, the recorded movie is used for offline validation.

To track the end-effector position and transfer the recorded movie from pixel to meter, two light points are attached to the end-effector with specified distance. Figure 3 shows these light points on the end-effector and the offline image processing results of the light points that are recorded by the camera during tracking a line trajectory. Also, the position of the camera is fixed and the camera view for each point of the trajectories are different. Hence, camera calibration is required. For this goal, a checkerboard is used for the calibration in [7]. The camera parameters are used to correct an image for lens distortion. Then, the corrected images are used for tracking the two light points on the end-effector.

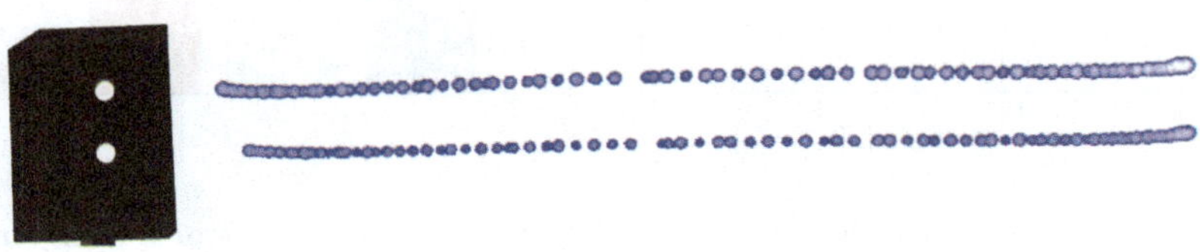

Fig. 3. End-effector and the light points (left), offline image processing result for the recorded movie of the end-effector position (right).

The CPU time to calculate the system input based on the system dynamics model, the measurements, and desired signals is less than 100 µs. That means that the total CPU time for the running controller loop and getting the measurements is less than sampling time. Thus, the controller fits for the real-time test.

In order to investigate the controller performance, the robot based on the designed nonlinear controller tracks two different trajectories with different velocities as test scenarios. Also, the tracking performance of the robot's end-effector is compared with the previous work. Hence, the presented nonlinear feedback controller in Eq. (5) (named the reduced nonlinear controller) and the feedback controller in [2] (named the nonlinear controller) are implemented on the lambda robot in simulation and tested on the real system. Also, the end-effector trajectory tracking error and the strain of the long flexible link are chosen as comparison benchmarks.

The first scenario is a line tracking task. The goal of this scenario is to validate the tracking performance in high-speed motion and investigate the end-point error. To this end, the robot end-effector shall track a line with length 0.283 m and with a maximum velocity of 1.2374 m/s. The end-effector based on the controllers tracks a line with the start point $x_s = [-0.6, -0.5]^T m$ and the end point $x_e = [-0.8, -0.3]^T m$. The movie in this scenario is recorded with approximately 300 frames per second. The simulation and experiment results are shown in Fig. 4.

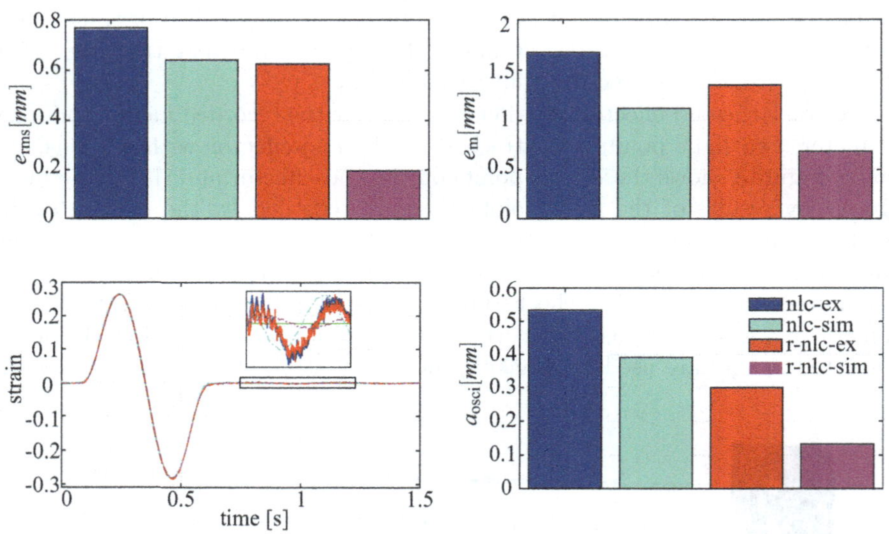

Fig. 4. Comparing root-mean-square (e_{rms}) and maximum (e_{m}) error during tracking the line trajectory as well as the strain gauge results of the long link and the maximum oscillation amplitude (a_{osci}) at the end-point of the trajectory using two controllers, i.e., the nonlinear controller (nlc) and the reduced nonlinear controller (r-nlc) in experimental test (-ex) on the robot and in simulation (-sim).

For the second scenario, the end-effector shall track an eight shape trajectory. The goal is to validate the tracking performance of the proposed controller in a long more complicated nonlinear trajectory and the trajectory corners. Therefore, the robot end-effector shall track this trajectory with a length of 0.943 m and with a maximum velocity of 1.03 m/s. In this task, the end-effector tracks the eight shape trajectory that is described with its center/start/end point $x_c = [-0.7, -0.3]^T m$. The position of the eight shape points are defined with

$[a \sin(2\phi), -a \sin(\phi)]^{\mathrm{T}}$ where $a = 0.1$ m and the angle $\phi = [\pi, 3\pi]$. For this scenario, the movie is recorded with approximately 150 frames per second since the recorded area is bigger than the first scenario. The simulation and experiment results are shown in Fig. 5.

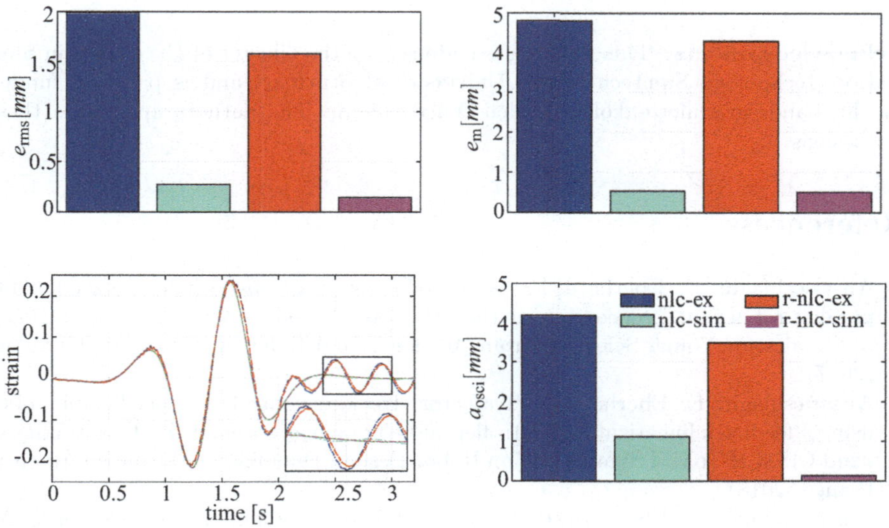

Fig. 5. Comparing the results of two controllers for tracking the eight shape trajectory.

The simulation results on the lambda robot model and the experimental test on the real system show that the presented controller in this paper is able to track a trajectory with high accuracy and high performance. Also, in comparison with the previous work, the tracking performance of the controller in simulation and the experimental test is improved. The presented reduced nonlinear controller reduces the root-mean-square (e_{rms}) and the maximum (e_{m}) trajectory tracking error at the minimum 21% and 11%, respectively. Also, at the end point, the oscillation amplitude (a_{osci}) is decreased at the minimum 17% in simulation and experimental tests that are shown in Figs. 4 and 5.

6 Conclusions

In this paper, a nonlinear feedback controller was designed based on the reduced model of a flexible parallel lambda shape robot. The controller is tested on the lambda robot in simulation and experiment. The nonlinear feedback controller obtains the robot inputs based on the measurable system states and the desired signals. The controller computes the system inputs based on the direct Lyapunov method. The experiment results for the very flexible parallel robot show that the controller successfully tracks the desired trajectories with high accuracy. Also, experimental validation results demonstrate that the tracking error and the oscillation amplitude of the presented controller are decreased in comparison with the previous works. For future work, the used method and designed controller will be utilized to control the lambda robot's end-effector interaction with an environment using force feedback.

Acknowledgements. This research is endorsed by the Cluster of Excellence in Simulation Technology SimTech at the University of Stuttgart and is partially funded by the Landesgraduiertenkolleg Baden-Württemberg. The authors appreciate these discussions.

References

1. Ansarieshlaghi, F., Eberhard, P.: Design of a nonlinear observer for a very flexible parallel robot. In: Proceedings of the 7th GACM Colloquium on Computational Mechanics for Young Scientists from Academia and Industry, Stuttgart, Germany (2017)
2. Ansarieshlaghi, F., Eberhard, P.: Trajectory tracking control of a very flexible robot using a feedback linearization controller and a nonlinear observer. In: Proceedings of 22nd CISM IFToMM Symposium on Robot Design, Dynamics and Control, Rennes, France (2018)
3. Ansarieshlaghi, F., Eberhard, P.: Experimental study on a nonlinear observer application for a very flexible parallel robot. Int. J. Dyn. Control (2018). https://doi.org/10.1007/s40435-018-0467-2
4. Ansarieshlaghi, F., Eberhard, P.: Hybrid force/position control of a very flexible parallel robot manipulator in contact with an environment. In: Proceedings of 16th International Conference on Informatics in Control, Automation and Robotics, Prague, Czech Republic (2019, submitted)
5. Burkhardt, M., Holzwarth, P., Seifried, R.: Inversion based trajectory tracking control for a parallel kinematic manipulator with flexible links. In: Proceedings of the 11th International Conference on Vibration Problems, Lisbon, Portugal (2013)
6. Burkhardt, M., Seifried, R., Eberhard, P.: Experimental studies of control concepts for a parallel manipulator with flexible links. In: Proceedings of the 3rd Joint International Conference on Multibody System Dynamics and the 7th Asian Conference on Multibody Dynamics, Busan, Korea (2014)
7. Fetić, A., Jurić, D., Osmanković, D.: The procedure of a camera calibration using Camera Calibration Toolbox for MATLAB. In: Proceedings of the 35th International Convention MIPRO, Opatija, Croatia (2012)
8. Kalil, H.K.: Nonlinear Systems, 3rd edn. Prentice-Hall, Upper Saddle River (2002)
9. Morlock, M., Burkhardt, M., Seifried, R.: Friction compensation, gain scheduling and curvature control for a flexible parallel kinematics robot. In: Proceedings of the: IEEE/RSJ International Conference on Intelligent Robots and Systems, Hamburg, Germany (2015)

A Compliant and Redundantly Actuated 2-DOF 3$\underline{R}$RR PKM: Best of Both Worlds?

Robin Cornelissen[1], Andreas Müller[2], and Ronald Aarts[1](✉)

[1] Faculty of Engineering Technology, Structural Dynamics, Acoustics and Control, University of Twente, P.O. Box 217, 7500 AE Enschede, The Netherlands
R.J.Cornelissen@student.utwente.nl, R.G.K.M.Aarts@utwente.nl
[2] Institute of Robotics, Johannes Kepler University Linz, Altenbergerstraße 69, 4040 Linz, Austria
a.mueller@jku.at

Abstract. Due to their deterministic behaviour, compliant mechanisms are well-suited for high-precision applications. In this paper the benefits of redundant links and actuation are investigated in terms of increasing support stiffness and homogenising actuator loads.

The manipulator is modelled with lumped inertia properties of the links and non-linear relations for the joint stiffnesses. The lumped parameter model allows a fast system level performance optimisation of the joint geometry simultaneously exploiting joint pre-bending and preloading, where the stiffness matrices of all joints are computed numerically efficient with non-linear flexible beam elements.

This model is applied to optimise the design of a compliant and redundantly actuated 2-DOF 3$\underline{R}$RR parallel kinematic manipulator. The improvement of support stiffness is demonstrated with an analysis of the first parasitic natural frequency. Balancing of the actuator torques is concluded from a potential energy analysis.

1 Introduction

In precision applications compliant mechanisms, or more precisely flexure-based mechanisms, are frequently used where the motion is enabled by elastic deformation of slender elements. Deterministic behaviour is ensured by the absence of friction, hysteresis and backlash. However, in spite of significant recent achievements in terms of the range of motion [6], it remains a challenge to maintain a high support stiffness, i.e. stiffness in directions in which the mechanism is not supposed to move, especially when the flexure joints in the mechanism undergo large deflections.

Redundantly actuated parallel kinematic manipulators (PKM) are researched to combine the advantages of PKM, i.e. the high stiffness, low inertia and large accelerations, with an improved handling of singularities and optimised actuator

A. Kecskeméthy and F. Geu Flores (Eds.): ECCOMAS 2019, COMPUTMETHODS 53, pp. 163–171, 2020.
https://doi.org/10.1007/978-3-030-23132-3_20

loading made possible by the redundancy [3,5]. In [4] the design of an experimental planar 2-DOF test setup with three actuators has been presented. The platform is well suited to evaluate models and control strategies.

The main goal of the present paper is to investigate whether the combination of compliant joints and redundant actuation can be exploited to further improve the dynamic properties of PKM, while circumventing the drawbacks of compliant joints. More specifically, the support stiffness from additional links is expected to reduce the decrease of this stiffness commonly observed at large joint angles. Furthermore, the redundant actuation offers a possibility to combine load balancing techniques with preloading of the compliant joints to balance the actuator effort, which is needed to position the end-effector (EE) at any position except for the equilibrium position, i.e. the centre of the workspace.

For this analysis the manipulator has to be modelled taking the non-linear stiffness of the joints into account. Such joint models are usually rather complicated and hence computationally too expensive when used to optimise the manipulator dynamic performance. For this reason many studies, including [6,7], investigate a single compliant joint. We propose to describe the complete manipulator with lumped inertia properties of the links and non-linear relations for the stiffness matrices of the joints. The lumped model allows a fast system level performance evaluation that is used to determine optimal joint orientation, pre-bending and preloading. The stiffness matrix of each joint depends on the joint geometry. It is computed throughout the full operating range of the joint where leaf springs are modelled with the non-linear flexible beam elements of the SPACAR software package [2]. Next piecewise linear interpolation is used to approximate all coefficients in this matrix as functions of the joint angle.

The main part of this paper are the analysis and optimisation methods for the support stiffness, Sect. 3, and actuator torques, Sect. 4. These are applied to the example manipulater introduced in Sect. 2. The results are presented and discussed in Sect. 5 with conclusions in Sect. 6.

2 Example Manipulator

The dynamic and kinetostatic models developed in the next sections will be demonstrated for a planar 2-DOF 3$\underline{R}$RR PKM equipped with compliant joints, Fig. 1(a). The three arms of the manipulator are similar but rotated 120° relative to each other. The shoulder joints and actuators are located at the vertices of an equilateral triangle, Fig. 1(b). This figure illustrates the basic geometric parameters like the total length L of each arm and the distance R of each actuator to the centre of the triangle. The workspace reachable by the EE is bounded by three circular arcs with radii L as indicated with the unshaded area in the figure. In order to reach all points within this workspace, the joints have to allow joint angles in some range. The ranges for the shoulder and elbow joints are indicated in the figure with the angles θ_s^r and θ_e^r, respectively. These angles depend on the ratio L/R and the division of the total arm length L into the upper arm

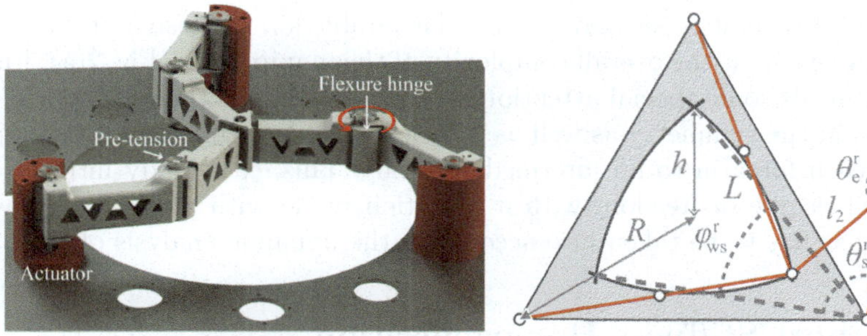

(a) Schematic manipulator overview. (b) Workspace.

Fig. 1. Planar 2-DOF 3RRR PKM similar to [4] except for the use of compliant joints, Fig. 2.

length l_1 and forearm length l_2. For the manipulator considered in this paper these dimensions are

$$R = 0.2305 \text{ m}, \quad L = 0.2536 \text{ m}, \quad l_1 = l_2 = L/2. \tag{1}$$

Then the joint angles ranges, with the lower and upper limit between brackets, are:

$$\theta_s^r = 50°(-17°...33°), \quad \theta_e^r = 83°(-49°...34°), \quad \theta_w^r = 50°(-22°...28°). \tag{2}$$

All compliant joints used in this example manipulator are butterfly joints, Fig. 2 [1]. For the shoulder joints this joint type is selected as it is known for its small pivot shift which makes it relatively easy to connect a common rotational

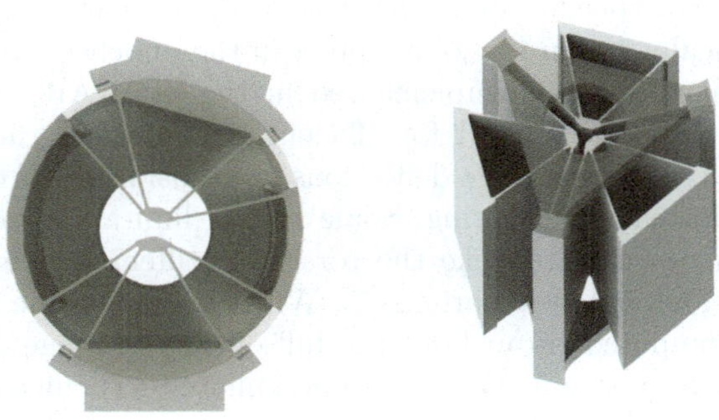

(a) Elbow joint (pre-bended). (b) Wrist joint.

Fig. 2. Two of the three compliant butterfly joint of the planar 2-DOF 3RRR PKM. The "intermediate" body that rigidly connects two or three connections between leaf springs are positioned on the top and bottom of the hinge and the top part is transparent in the images.

actuator to the upper arm. The elbow joints exhibit rather large rotation angles for which other joint types may be more favourable [6], but also a butterfly joint will be used to limit the overall complexity of the manipulator, Fig. 2(a). Finally, the wrist needs some special attention as it has to connect three links, Fig. 2(b). To maintain the symmetry as well as sufficient support stiffness, butterfly joints connect each forearm to an intermediate body. This extra body introduces an additional degree of freedom with a vibration mode with a relatively low frequency that has to be taken into account in the dynamic analysis of Sect. 3.

3 Support Stiffness Homogenisation

The support stiffness of compliant manipulators is often evaluated from the system's natural frequencies as in this way an implicit scaling of translational and rotational stifnesses is obtained. For a 2-DOF manipulator the first and second natural frequencies should be low as these are associated with the motion of the intended two degrees of freedom. The third natural frequency is the first parasitic frequency that should be as high as possible throughout the entire workspace. These natural frequencies can be computed with a detailed non-linear model of the full manipulator in which the EE position is varied in its entire workspace. For a design optimisation the evaluation of many alternative designs will be needed and such model would be too computationally intensive. Hence, a simplified lumped model is proposed.

This lumped parameter model should account for the mass properties of the rigid links and the stiffnesses of the joints. For the example manipulator of Sect. 2 seven rigid bodies can be identified, i.e. three upper arms, three forearms and the intermediate body connecting all forearms. Each body is characterised by its 6×6 mass matrix that is constant in a local frame connected to the body.

The links are connected to each other and the ground with nine joints in the shoulders, elbows and wrist. For each joint its full 6×6 stiffness matrix should be considered. For a specific joint with known dimensions this matrix can be evaluated numerically e.g. in the joint's pivot in the undeflected state. The non-linear flexible beam elements implemented in the SPACAR software package [2] have proven to be well-suited for efficient modelling of various compliant joint types [6,7]. Even when large deflections are considered three or four beam elements suffice for each leaf spring. Some known limitations arising from the beam model can be corrected like the torsional stiffening due to constrained warping in short and wide leaf springs [7]. With this model the stiffness matrix of each joint is computed throughout the full operating range of the joint and piecewise linear interpolation is used to approximate its coefficients as functions of the joint angle.

At system level all degrees of freedom of the rigid links are combined. In the example manipulator the seven links have 42 degrees of freedom in total. For a straightforward eigenvalue analysis it is proposed to define these independent coordinates in a coordinate frame located at the EE and aligned with the global coordinate frame. For the system level mass matrix the link mass matrices $\mathbf{M}_{\mathrm{upp}i}$,

$\mathbf{M}_{\mathrm{for}i}$, $\mathbf{M}_{\mathrm{int}}$ for upper arm i, forearm i and intermediate body respectively are combined to a block diagonal matrix $\mathbf{M}$. Similarly, the joint stiffness matrices are combined by adding the stiffness matrices $\mathbf{K}_{\mathrm{s}i}$, $\mathbf{K}_{\mathrm{e}i}$, $\mathbf{K}_{\mathrm{w}i}$ of shoulder i, elbow i and wrist i respectively,

$$\mathbf{K} = \begin{bmatrix} \mathbf{K}_{\mathrm{s1}} + \mathbf{K}_{\mathrm{e1}} & -\mathbf{K}_{\mathrm{e1}} & 0 & 0 & 0 & 0 & 0 \\ -\mathbf{K}_{\mathrm{e1}} & \mathbf{K}_{\mathrm{e1}} + \mathbf{K}_{\mathrm{w1}} & 0 & 0 & 0 & 0 & -\mathbf{K}_{\mathrm{w1}} \\ 0 & 0 & \mathbf{K}_{\mathrm{s2}} + \mathbf{K}_{\mathrm{e2}} & -\mathbf{K}_{\mathrm{e2}} & 0 & 0 & 0 \\ 0 & 0 & -\mathbf{K}_{\mathrm{e2}} & \mathbf{K}_{\mathrm{e2}} + \mathbf{K}_{\mathrm{w2}} & 0 & 0 & -\mathbf{K}_{\mathrm{w2}} \\ 0 & 0 & 0 & 0 & \mathbf{K}_{\mathrm{s3}} + \mathbf{K}_{\mathrm{e3}} & -\mathbf{K}_{\mathrm{e3}} & 0 \\ 0 & 0 & 0 & 0 & -\mathbf{K}_{\mathrm{e3}} & \mathbf{K}_{\mathrm{e3}} + \mathbf{K}_{\mathrm{w3}} & -\mathbf{K}_{\mathrm{w3}} \\ 0 & -\mathbf{K}_{\mathrm{w1}} & 0 & -\mathbf{K}_{\mathrm{w2}} & 0 & -\mathbf{K}_{\mathrm{w2}} & \mathbf{K}_{\mathrm{w123}} \end{bmatrix}, \tag{3}$$

where $\mathbf{K}_{\mathrm{w123}} = \mathbf{K}_{\mathrm{w1}} + \mathbf{K}_{\mathrm{w2}} + \mathbf{K}_{\mathrm{w3}}$. It should be noted that the link mass and joint stiffness matrices are initially computed in a local frame attached to the links' centre of mass or the joints' pivots, respectively. For each EE position the inverse kinematic solution for all links and joints can be found relatively easy. With these geometric data all link and joint matrices can be transformed to the EE frame with pre and post multiplication with adjoint matrices that reflect the involved sequence of translations and rotations.

Recall that for each joint the coefficients of the stiffness matrix are linearly interpolated as functions of the joint angle. In the transformation to the EE frame the joint orientations γ_{s} and γ_{e} of shoulder and elbow joints are design parameters. Additionally the shoulder and elbow joints can be manufactured with pre-bending angles β_{s} and β_{e}. These angles specify an offset between the joint angles at which the EE is in its equilibrium position and at which the joint is in the undeflected configuration, i.e. exhibits the largest support stiffness. In the computation of the actual joint stiffness, these pre-bending angles have to be added to the angles used in the linear interpolations.

4 Actuator Torque Balancing and Potential Energy

For the analysis of the support stiffness in the previous section, in particular the joint stiffnesses in the stiff directions play a role. The compliant directions are less relevant as the lowest natural frequencies are discarded. However, the actuator torques needed to position the EE depend on these joint stiffnesses and can be computed from a kinetostatic analysis. Each joint is described by a stiffness $k_{\mathrm{s}i}$, $k_{\mathrm{e}i}$ or $k_{\mathrm{w}i}$ in the driving direction, which is the lower right coefficient in the respective stiffness matrix of the previous section and is weakly dependent on the joint angle. Furthermore, for the shoulder and wrist joints pretension angles α_{s} or α_{e} are taken into account. This pretension angle is defined as the joint angle at which the joint is undeformed and hence no joint torque arises.

For any EE position the motor torques τ_i $(i = 1, 2, 3)$ acting on the shoulder joints needed to keep the system statically balanced can be determined from equilibrium equations that can be derived for all links. As the manipulator is redundantly actuated, there is no unique solution for the actuator torques.

The null space solution is used to compute the actuator torques that minimise the 2-norm of the vector $\boldsymbol{\tau}$ with the three actuator torques τ_i [4]. It appears that this actuator norm $||\boldsymbol{\tau}||$ scales proportionally to the variation of the potential energy in the compliant joints throughout the workspace, which is relatively easy to evaluate in order to optimise the joint pretension angles α_s and α_e.

5 Results

The analysis methods outlined in the previous sections will now be applied for the design of the example manipulator proposed in Sect. 2. For high performance metal links and joints would be preferred, but for a cost-effective first prototype it is chosen to use 3D printed parts made by Selective Laser Sintering (SLS) of Nylone (PA2200). Typical material properties are its Young's modulus E = 1.7 GPa and density $\rho = 930$ kg/m^3. Preloading of the joints is not possible with this material as creep will result in rather quick unloading. Hence an alternative solution for preloading should be applied like low-stiffness preloaded (metal) springs positioned parallel to the joints, as can be recognised in Fig. 1(a).

Homogenised Support Stiffness. As outlined in Sect. 2 it was chosen to use butterfly hinges for all joints in the example manipulator. Butterfly hinges have an extra internal degree of freedom due to rotation of an intermediate body which may show a relatively low frequent vibration mode. This motion can be constrained [1] when its frequency appears to be too low. As was also pointed out in Sect. 2, the EE has an intermediate body as well for which the same consideration applies. For the considered manipulator it appeared that lightweight intermediate bodies can be used, such that the natural frequencies of the internal modes were higher than some other parasitic mode and hence no precautions had to be taken to constrain these modes.

Figure 3 illustrates some steps in the optimisation procedure of which Table 1 summarises some geometric parameters. In Fig. 3(a) the variation of the support stiffness throughout the EE workspace is evaluated in terms of the first natural frequency for some initial design. For butterfly joints the reduction of the support stiffness is relatively small as is shown in Fig. 4. Nevertheless the first parasitic natural frequency varies more that a factor 2 throughout the workspace.

Table 1. Geometric parameters of optimised joints. The vertical dimension of all hinges is 40 mm. The thickness of all leaf springs is 0.4 mm. The table shows the other in-plane dimensions.

	Shoulder		Elbow		Wrist
	Height [mm]	Width [mm]	Height [mm]	Width [mm]	Height [mm]
Individual hinge optimisation	27.4	12.2	25.7	10.1	17.9
Full optimisation	20.2	19.8	29.4	17.8	18.2

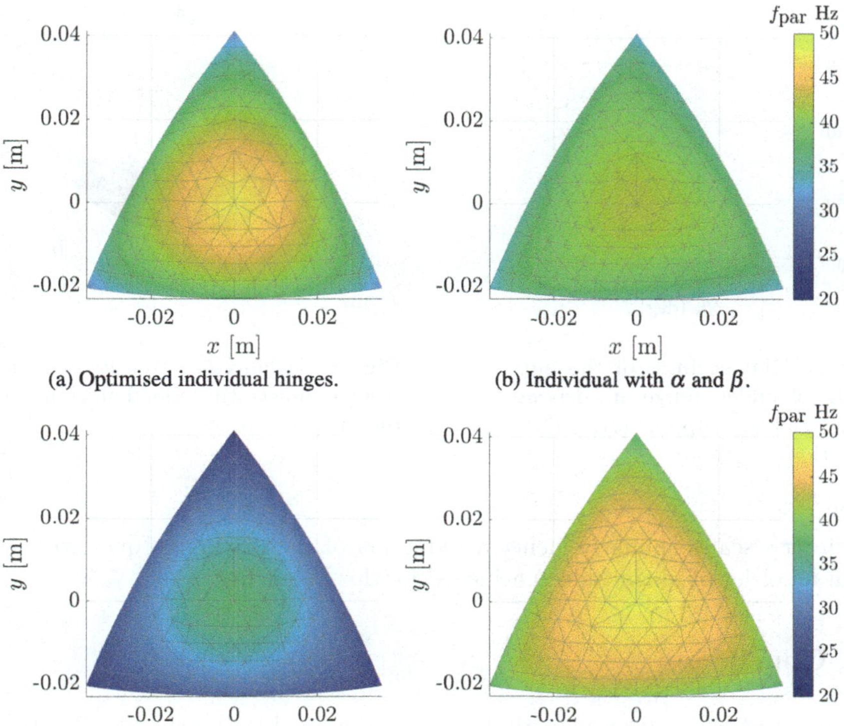

(a) Optimised individual hinges.
(b) Individual with α and β.
(c) Optimised dimensions ($\alpha = \beta = 0$).
(d) Optimised all.

Fig. 3. Optimising the support stiffness, evaluated with the first parasitic natural frequency.

With an optimal pre-bending and joint orientation this variation can be reduced as shown in Fig. 3(b), although the primary optimisation goal is increase of the lowest frequency near the corners. Probably the redundant arm also offers additional homogenisation of the stiffness.

A further improvement is possible by optimising the joint parameters, Fig. 3(c), and finally combining all optimisation efforts, Fig. 3(d). In the final result, not only the lowest parasitic natural frequency has been increased, but also its largest value near equilibrium is large again although this was not an optimisation goal.

Balanced Actuator Torques. In the design so far the actuator torques are not considered. To illustrate the actuator torque balancing, Fig. 5 shows the potential energy E stored in the compliant joints relative to the equilibrium configuration. The figure shows the steep increase in energy storage towards the extreme edges of the workspace without pretensioned compliant joints ($\alpha_s = \alpha_e = 0°$). With pretension a significant reduction of the potential energy is possible as is demonstrated in the figure with $\alpha_s = 42°$, $\alpha_e = -50°$. The torque

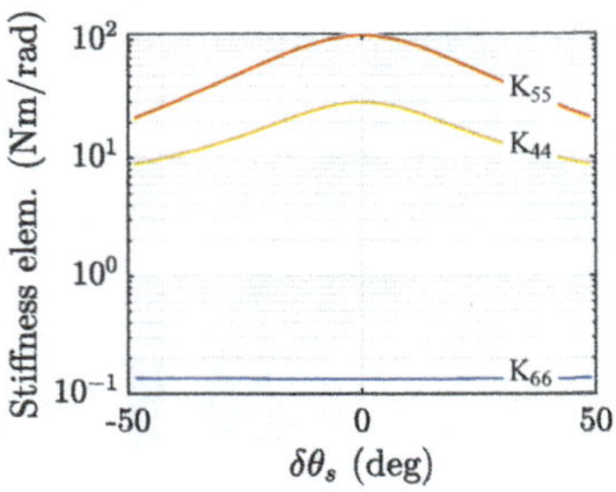

Fig. 4. Joint stiffness of the butterfly shoulder hinge in driving (K_{66}) and two support (K_{44}, K_{55}) directions.

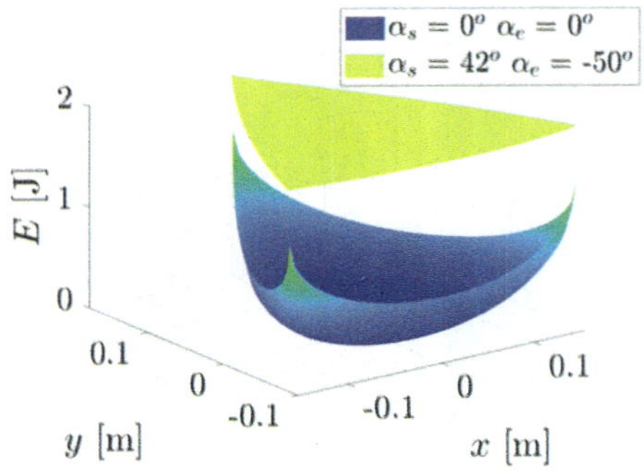

Fig. 5. Balanced potential energy storage illustrating the balancing of the driving torque.

norm $||\boldsymbol{\tau}||$ scales similarly, hence a reduction of the driving torques with more than an order of magnitude is achieved by this balancing.

6 Conclusion

In this paper we propose an analysis method that enables a system level optimisation of a parallel kinematic manipulator with redundant actuation and compliant (or flexure) joints. The goal of the optimisation is to combine the redundancy with preloading and pre-bending of the joints to obtain a high support stiffness throughout the workspace and to reduce the actuator torques. The approximate model used in this optimisation combines lumped mass properties of the rigid links with non-linear stiffness matrices of the joints computed as functions of the joint angles.

The method is applied to the design of a planar 2-DOF 3RRR parallel manipulator. It appears that orientation and pre-bending of the compliant joints can be optimised to avoid a significant decrease of the support stiffness and preloading helps to lower the required actuator torques. The proposed design has been manufactured with 3D printing to experimentally verify the expected dynamic behaviour, as will be published in the future.

References

1. Henein, S., Spanoudakis, P., Droz, S., Myklebust, L.I., Onillo, E.: Flexure pivot for aerospace mechanisms. In: Proceedings of the 10th European Space Mechanisms and Tribology Symposium, San Sebastian, Spain (2003)
2. Jonker, J.B., Meijaard, J.P.: SPACAR - computer program for dynamic analysis of flexible spatial mechanisms and manipulators. In: Schiehlen, W. (ed.) Multibody Systems Handbook, pp. 123–143. Springer, Heidelberg (1990)

3. Kock, S., Schumacher, W.: A parallel x-y manipulator with actuation redundancy for high-speed and active-stiffness applications. In: Proceedings of the 1998 IEEE International Conference on Robotics and Automation, Leuven Belgium, pp. 2295–2300 (1998)
4. Krajoski, K., Müller, A., Gattringer, H., Jörgl, M.: Design, modeling and control of an experimental redundantly actuated parallel platform. In: Niel, K., Roth, P.M. (eds.) Proceedings of the OAGM&ARW Joint Workshop 2016 "Computer Vision and Robotics", University of Applied Sciences Upper Austria, Wels Campus, pp. 209–216 (2016)
5. Müller, A., Hufnagel, T.: Model-based control of redundantly actuated parallel manipulators in redundant coordinates. Robot. Auton. Syst. **60**(4), 563–571 (2012)
6. Naves, M., Brouwer, D.M., Aarts, R.G.K.M.: Building block-based spatial topology synthesis method for large-stroke flexure hinges. J. Mech. Robot. **9**(4), 041006 (2017). 9 pages
7. Wiersma, D.H., Boer, S.E., Aarts, R.G.K.M., Brouwer, D.M.: Design and performance optimization of large stroke spatial flexures. J. Comput. Nonlinear Dyn. **9**(1), 011016 (2013). 10 pages

On the Modeling of Redundantly-Actuated Mechanical Systems

Yaojun Wang[1,2](✉), Bruno Belzile[3], Jorge Angeles[3], and Qinchuan Li[1]

[1] Faculty of Mechanical Engineering and Automation, Zhejiang Sci-Tech University, 928 NO.2 St, Hangzhou 310018, China
bluepi@126.com, lqchuan@zstu.edu.cn

[2] Department of Electrical and Electronic Engineering, Zhejiang Institute of Mechanical and Electrical Engineering, 528 Binwen Rd, Hangzhou 310053, China

[3] Department of Mechanical Engineering and Centre for Intelligent Machines, McGill University, 817 Sherbrooke St West, Montreal H3A 0C3, Canada
{bruno,angeles}@cim.mcgill.ca

Abstract. Dynamics modeling is essential in the design and control of mechanical systems, the focus of the paper being *redundantly-actuated systems,* which bring about special challenges. The authors resort to the *natural orthogonal complement* (NOC) to derive the dynamics model; benefiting from the elimination of the constraint wrenches, the NOC offers a simple and systematic alternative to the modeling of the systems of interest, which leads to low computational demands. The actuator torques/forces are found by means of a minimum-norm solution; then, by relying on the QR-decomposition, a simple, robust procedure is applied to compute the optimum solution, while obviating the explicit computation of the right Moore-Penrose generalized inverse, which requires the inverse of a product that is prone to ill-conditioning. The methodology is illustrated with an application to a redundantly-actuated parallel-kinematics machine with three degrees of freedom and four actuators.

1 Introduction

In modeling the dynamics of multibody systems, explicit equations of motion for the system of interest are to be generated using a suitable methodology. Two methods are mostly employed to construct the dynamics equations, namely, Newton-Euler (NE) and Euler-Lagrange (EL) [8,13]. The procedure described below is based on the Newton-Euler equations, cast as one single vector equation of dimension p, the dof of the system. There are less common methods, such as the virtual work principle [2,11], d'Alembert's principle [4,5], the principle of Hamilton [9], and Kane's formulation [13], among others, to cope with the aforementioned task. To simplify the mathematical modeling of multibody systems, we chose the NOC [8], based on the reciprocity relations between feasible

A. Kecskeméthy and F. Geu Flores (Eds.): ECCOMAS 2019, COMPUTMETHODS 53, pp. 172–179, 2020.
https://doi.org/10.1007/978-3-030-23132-3_21

twists and constraint wrenches. The NOC allows the elimination of the constraint wrenches upon multiplication of the unconstrained dynamics equations by an orthogonal complement of the velocity constraints. The inclusion of actuation redundancy into the NOC model, and an efficient and robust method to solve the ensuing norm-minimization problem, are first proposed and then illustrated by application to a novel PKM, as yet under development.

2 The Modeling of Redundantly-Actuated Mechanical Systems

It is assumed that the mechanical system at hand is composed of n rigid bodies. The system has r redundantly-actuated joints, its degree of freedom (dof) being $p < r$. Referring to the motion of the ith body, a few concepts are introduced. The six-dimensional *twist* $\mathbf{t}_i$ of and the *wrench* $\mathbf{w}_i$ acting on the ith body are:

$$\mathbf{t}_i \equiv [\boldsymbol{\omega}_i^T \; \dot{\mathbf{c}}_i^T]^T \quad \text{and} \quad \mathbf{w}_i \equiv [\mathbf{n}_i^T \; \mathbf{f}_i^T]^T, \qquad i = 1, \ldots, n \tag{1}$$

where $\boldsymbol{\omega}_i$ and $\dot{\mathbf{c}}_i$ are the angular velocity of the body and the velocity of its center of mass, respectively, while $\mathbf{n}_i$ and $\mathbf{f}_i$ are the moment about and the force applied at the center of mass of the body, correspondingly. Next, the wrench acting on the ith body is decomposed into an *external* wrench $\mathbf{w}_i^E$ and a *constraint* wrench $\mathbf{w}_i^C$, i.e.,

$$\mathbf{w}_i = \mathbf{w}_i^E + \mathbf{w}_i^C \tag{2}$$

Then, the 6×6 *angular velocity dyad* $\mathbf{W}_i$ and *inertia dyad* $\mathbf{M}_i$ of the ith body are defined as

$$\mathbf{W}_i \equiv \begin{bmatrix} \boldsymbol{\Omega}_i & \mathbf{O} \\ \mathbf{O} & \mathbf{O} \end{bmatrix} \quad \text{and} \quad \mathbf{M}_i \equiv \begin{bmatrix} \mathbf{I}_{Ci} & \mathbf{O} \\ \mathbf{O} & m_i \mathbf{1} \end{bmatrix} \tag{3}$$

where $\boldsymbol{\Omega}_i$ is the 3×3 cross-product matrix (CPM)[1] of vector $\boldsymbol{\omega}_i$. Moreover, $\mathbf{O}$ and $\mathbf{1}$ are the 3×3 zero and identity matrices, respectively, while $\mathbf{I}_{Ci}$ is the 3×3 *inertia tensor* about the center of mass of the ith body, and m_i the mass of the body. Thus, the NE equations of motion of the ith body are cast in a compact form:

$$\mathbf{M}_i \dot{\mathbf{t}}_i = -\mathbf{W}_i \mathbf{M}_i \mathbf{t}_i + \mathbf{w}_i^E + \mathbf{w}_i^C, \quad i = 1, \ldots, n \tag{4}$$

Moreover, the twist of the ith body is represented as

$$\mathbf{t}_i = \mathbf{T}_i \dot{\mathbf{x}}, \quad i = 1, \ldots, n \tag{5}$$

where $\mathbf{T}_i$ is the $6 \times p$ *twist-shaping matrix* of the ith body, which maps the array of *feasible Cartesian rates*, $\dot{\mathbf{x}} \equiv [\dot{x}_1 \ldots \dot{x}_p]^T$, of the system, into the twist array of the ith body. Note that the power developed by the constraint wrench $\mathbf{w}_i^C$ on the feasible twist $\mathbf{t}_i$ vanishes, i.e.,

$$\mathbf{t}_i^T \mathbf{w}_i^C = 0 \tag{6}$$

[1] The CPM $\mathbf{V} \in \mathbb{R}^{3\times3}$ of any vector $\mathbf{v} \in \mathbb{R}^3$, is defined as $\mathbf{V} = (\partial \mathbf{v} \times \mathbf{x} / \partial \mathbf{x}), \forall \mathbf{x} \in \mathbb{R}^3$.

Upon substitution of Eq. (5) into Eq. (6), we obtain

$$\dot{\mathbf{x}}^T \mathbf{T}_i^T \mathbf{w}_i^C = 0 \tag{7}$$

Since Eq. (7) must be satisfied for arbitrary values of the Cartesian rates $\dot{\mathbf{x}}$, because they are independent, the product $\mathbf{T}_i^T \mathbf{w}_i^C$ must vanish. Therefore, upon multiplying both sides of Eq. (4) by $\mathbf{T}_i^T$, we obtain

$$\mathbf{T}_i^T \mathbf{M}_i \dot{\mathbf{t}}_i = -\mathbf{T}_i^T \mathbf{W}_i \mathbf{M}_i \mathbf{t}_i + \mathbf{T}_i^T \mathbf{w}_i^E, \quad i = 1, \ldots, n \tag{8}$$

The external wrench $\mathbf{w}_i^E$ is now decomposed into actuator, gravity, and dissipation wrenches $\mathbf{w}_i^A$, $\mathbf{w}_i^G$, $\mathbf{w}_i^D$, respectively; correspondingly, $\boldsymbol{\tau}$, $\boldsymbol{\gamma}$, and $\boldsymbol{\delta}$ are the resulting generalized actuator, gravity and dissipation forces exerted on the system. Hence, upon differentiation of both sides of Eq. (5) with respect to time and substitution of the resulting expression of $\dot{\mathbf{t}}_i$ into Eq. (8), the independent constrained dynamics model of the ith body is obtained. Upon applying Eq. (8) to each rigid body, then adding the resulting equations sidewise, we obtain the mathematical model of the system as

$$\boldsymbol{\tau} = \mathbf{I}\ddot{\mathbf{x}} + \mathbf{C}\dot{\mathbf{x}} - \boldsymbol{\gamma} - \boldsymbol{\delta} \tag{9}$$

where

$$\mathbf{I} \equiv \sum_{i=1}^{n} \mathbf{T}_i^T \mathbf{M}_i \mathbf{T}_i, \ \mathbf{C} \equiv \sum_{i=1}^{n} \mathbf{T}_i^T (\mathbf{M}_i \dot{\mathbf{T}}_i + \mathbf{W}_i \mathbf{M}_i \mathbf{T}_i), \ \gamma \equiv \sum_{i=1}^{n} \mathbf{T}_i^T \mathbf{w}_i^G, \ \delta \equiv \sum_{i=1}^{n} \mathbf{T}_i^T \mathbf{w}_i^D \tag{10}$$

Considering that the angular velocity dyad $\mathbf{W}_i$ is skew-symmetric and the inertia dyad $\mathbf{M}_i$ is symmetric, it can be readily shown that: (*i*) $\mathbf{I}$ is symmetric and positive definite; and (*ii*) $\dot{\mathbf{I}} - 2\mathbf{C}$ is skew-symmetric. These are two basic results of mechanical systems of rigid bodies, leading to the controllability of a multibody system, holonomic or nonholonomic, with a linear PD controller.

3 The Minimum-Norm Solution of Actuator Torques/Forces

By virtue of the principle of virtual work [2], the power supplied by the actuators, represented in joint space $\boldsymbol{\theta}$, equals its counterpart in Cartesian space, i.e.,

$$\boldsymbol{\tau}_a^T \dot{\boldsymbol{\theta}} = \boldsymbol{\tau}^T \dot{\mathbf{x}} \tag{11}$$

where $\dot{\boldsymbol{\theta}} \equiv [\theta_1, \ldots, \theta_r]^T$ is the array of the actuated-joint rates, $\boldsymbol{\tau}_a$ being the array of the actuator torques/forces. Moreover, the $r \times p$ *rate-shaping matrix* $\mathbf{A}$, that maps $\dot{\mathbf{x}}$ into $\dot{\boldsymbol{\theta}}$, is introduced as

$$\dot{\boldsymbol{\theta}} = \mathbf{A}\dot{\mathbf{x}} \tag{12}$$

Upon substitution of Eq. (12) into Eq. (11), we obtain

$$\boldsymbol{\tau}_a^T \mathbf{A}\dot{\mathbf{x}} = \boldsymbol{\tau}^T \dot{\mathbf{x}} \tag{13}$$

Now, the Cartesian rates stored in array $\dot{\mathbf{x}}$ are independent, and hence, can be assigned arbitrarily. Therefore, $\dot{\mathbf{x}}$ can be "deleted" from Eq. (13), which leads to

$$\boldsymbol{\tau} \equiv \mathbf{A}^T \boldsymbol{\tau}_a \tag{14}$$

i.e., a system of p equations in $\boldsymbol{\tau}_a$ for $r > p$ components of $\boldsymbol{\tau}_a$. Hence, given any value of $\boldsymbol{\tau}$ in Eq. (14), to carry the robot through a given Cartesian trajectory, infinitely-many values of $\boldsymbol{\tau}_a$ are available for the task at hand. Under the foregoing conditions, the obvious strategy is to choose the actuator-torque (or force) array of *minimum cost*, which means of minimum norm. Of the various norms available, the Euclidian norm lends itself to a closed-form solution, reason why we chose it here. Therefore, we obtain the actuator-torque (or force) array $\boldsymbol{\tau}_a$ in terms of the right Moore-Penrose generalized inverse of $\mathbf{A}^T$, provided that matrix $\mathbf{A}$ is of full rank, which we will assume in this paper. Symbolically, the minimum-norm solution[2] $\boldsymbol{\tau}_{a0}$ sought: [1]

$$\boldsymbol{\tau}_{a0} = \mathbf{A}^\dagger \boldsymbol{\tau}, \quad \mathbf{A}^\dagger = \mathbf{A}(\mathbf{A}^T\mathbf{A})^{-1} \tag{15}$$

which is the *right Moore-Penrose generalized inverse* (RMPGI) of the "tall" rectangular matrix $\mathbf{A}^T$. It can be seen that the straightforward evaluation of $\mathbf{A}^\dagger$ involves the inversion of a matrix product, which is not only computationally costly, but also prone to ill-conditioning [6]. A simple, robust method to compute the desired solution relies on the QR-decomposition of $\mathbf{A}$. First, let the QR-decomposition of $\mathbf{A}$ be

$$\mathbf{A} = \mathbf{QR} = [\mathbf{Q}_L \quad \mathbf{Q}_R] \begin{bmatrix} \mathbf{U} \\ \mathbf{O}_{r'p} \end{bmatrix}, \quad r' = r - p \tag{16}$$

whence $\mathbf{A} = \mathbf{Q}_L\mathbf{U}$ with $\mathbf{Q}_L$ of $r \times p$, $\mathbf{Q}_R$ of $r \times (r-p)$, $\mathbf{U}$ of $p \times p$ and $\mathbf{O}_{r'p}$ representing the $(r-p) \times p$ zero matrix. Then, notice that, by virtue of the orthogonality of $\mathbf{Q}$,

$$\mathbf{A}^T\mathbf{A} = \mathbf{R}^T\mathbf{Q}^T\mathbf{Q}\mathbf{R} = \mathbf{R}^T\mathbf{R} = \mathbf{U}^T\mathbf{U} \tag{17}$$

Upon substitution of Eqs. (16) and (17) into Eq. (15), $\mathbf{A}^\dagger$ is obtained as

$$\mathbf{A}^\dagger = [\mathbf{Q}_L \quad \mathbf{Q}_R] \begin{bmatrix} \mathbf{U}^{-T} \\ \mathbf{O}_{r'p} \end{bmatrix} = \mathbf{Q}_L\mathbf{U}^{-T} \tag{18}$$

which means that $\mathbf{A}^\dagger$ need not be computed verbatim as in Eq. (15), once the QR-decomposition of $\mathbf{A}$ is available. The QR-decomposition, moreover, is obtained by means of *Householder reflections* [6], which is a *direct procedure*,

[2] The solution at stake is the one of minimum *Euclidean norm*.

as opposed to *iterative*, like the singular-value decomposition. The generalized inverse $\mathbf{A}^{\dagger}$ is computed by *forward substitution* from the lower-triangular system

$$\mathbf{U}^T\mathbf{A}^{\dagger} = \mathbf{Q}_L \tag{19}$$

Furthermore, upon application of the r Householder reflections $\mathbf{H} \equiv \mathbf{H}_r \ldots \mathbf{H}_2\mathbf{H}_1$, the orthogonal matrix $\mathbf{Q}$ is simply $\mathbf{H}^T$. The procedure will be illustrated with an application to a PKM with three dof and four actuators.

4 Application to a 2PUR-2RPU PKM

The machine under study is a redundantly actuated 2PUR-2RPU 2R1T[3] [14] parallel robot, as shown in Fig. 1, its geometric parameters being listed in Table 1. The ith limb for $i = 1, 2, 3, 4$, is $A_iB_iC_i$. A Cartesian coordinate frame, $\mathcal{B}(X, Y, Z)$, is attached to the fixed base at the intersection O of lines B_1B_2 and C_3C_4. A moving coordinate frame, $\mathcal{P}(U, V, W)$, is attached to the moving platform at P, the midpoint of the line segment $\overline{A_3A_4}$. The platform is capable of two rotations, through angles α and β, about corresponding skew axes, X and V, at right angles, and one translation, ζ, along vector $\mathbf{p} = \overrightarrow{OP}$, normal to V.

In the definitions below, for $i = 1, \ldots, 7$, the ith link of the system denotes, in the given order: the moving link of the P joint of the first limb; link B_1A_1; the moving link of the P joint of the second limb; link B_2A_2; link B_3A_3; link B_4A_4; and the MP. As well, for $i = 1$, 2, l_c denotes the distance from the center of mass of link B_iA_i to the center of the corresponding U joint, B_i, while q_c denotes the distance from the center of mass of link B_iA_i, for $i = 3$, 4, to the center of the corresponding U-joint, A_i.

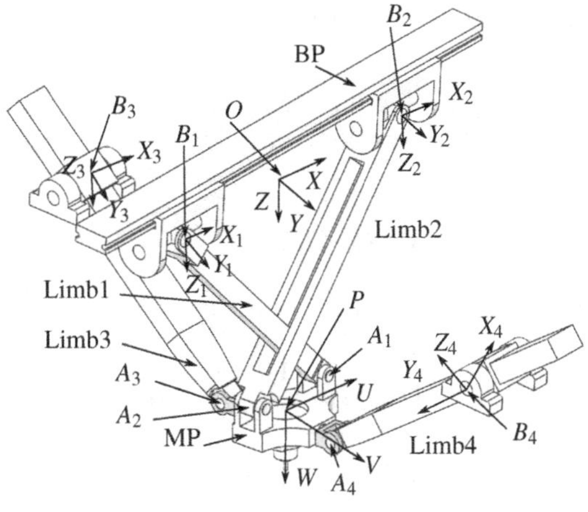

Fig. 1. 2PUR-2RPU robot

[3] R: rotation; T: translation.

Table 1. Robot geometric data

Description	Symbol	Value	Units
Limb B_iA_i length ($i = 1, 2$)	l	120	mm
OC_i length ($i = 3, 4$)	l_3	80	mm
PA_i length ($i = 1, \ldots, 4$)	f	60	mm
C_iB_i length ($i = 3, 4$)	d	20	mm
Distance between A_1A_2 and A_3A_4	e	0	mm

The helical path, which combines the line path and the circle path, is widely used in PKM dynamics analysis [12]. Therefore, the tracing of a helical-path is simulated to illustrate the dynamics modeling proposed here. The path to be traced, fixed on the base frame, is given by

$$x = r\cos\phi - r, \quad y = r\sin\phi, \quad z = z_0 + h_0\phi/(2\pi) \tag{20}$$

with the numerical values: $r = 5\,\text{mm}$; initial height $z_0 = 127.48\,\text{mm}$; pitch $h_0 = 20\,\text{mm}$; and the angle $0 \leq \phi \leq 2\pi$.

The trapezoidal velocity trajectory, consisting of three phases: acceleration, coasting, and deceleration, is known for its simplicity, hence very common in the industrial practice. Its acceleration, however, is discontinuous. In order to obtain the continuity of the acceleration profile, we resort to a modified trapezoidal trajectory [10], which is divided into six phases of either second degree polynomial or cycloidal function.

Furthermore, it is noted that, even though the mobility of the mechanism under study is 2R1T, the Cartesian coordinates of the operation point of the tool-head, attached to the moving platform (MP), can still be controlled at the sacrifice of the orientation dof, as illustrated with a case study in what follows.

Assuming that links and joints are made of 6061-T6 Alloy, with a density of $2700\,\text{kg/m}^3$, the mass of each body and their moment-of-inertia matrices at the respective centers of mass, as listed below:

$$\begin{aligned}
m_1 &= 3.261 \cdot 10^{-2}\text{kg}, [\mathbf{I}_1] = \text{diag}(4.855, 12.742, 12.795) \cdot 10^{-6}\text{kg} \cdot \text{m}^2 \\
m_2 &= 3.300 \cdot 10^{-2}\text{kg}, [\mathbf{I}_2] = \text{diag}(40.048, 39.770, 1.054) \cdot 10^{-6}\text{kg} \cdot \text{m}^2 \\
m_3 &= 3.261 \cdot 10^{-2}\text{kg}, [\mathbf{I}_3] = \text{diag}(4.855, 12.742, 12.795) \cdot 10^{-6}\text{kg} \cdot \text{m}^2 \\
m_4 &= 4.200 \cdot 10^{-2}\text{kg}, [\mathbf{I}_4] = \text{diag}(52.324, 50.786, 2.543) \cdot 10^{-6}\text{kg} \cdot \text{m}^2 \\
m_5 &= 6.082 \cdot 10^{-2}\text{kg}, [\mathbf{I}_5] = \text{diag}(109.396, 3.349, 110.827) \cdot 10^{-6}\text{kg} \cdot \text{m}^2 \\
m_6 &= 6.082 \cdot 10^{-2}\text{kg}, [\mathbf{I}_6] = \text{diag}(109.396, 3.349, 110.827) \cdot 10^{-6}\text{kg} \cdot \text{m}^2 \\
m_7 &= 8.200 \cdot 10^{-2}\text{kg}, [\mathbf{I}_7] = \text{diag}(16.631, 27.155, 37.459) \cdot 10^{-6}\text{kg} \cdot \text{m}^2
\end{aligned}$$

The initial values are chosen as $\alpha = \beta = 0\,\text{rad}$, and $\zeta = 70\,\text{mm}$. The corresponding time-histories of displacements, velocities, accelerations, and forces, of the actuated joints, are calculated and depicted in Figs. 2, 3, 4 and 5.

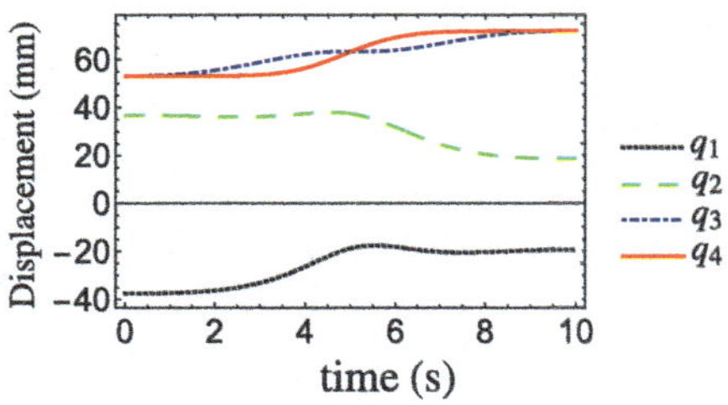

Fig. 2. Actuated-joint displacements

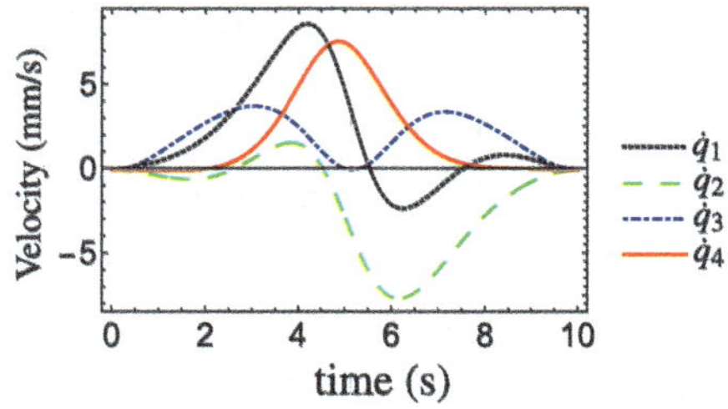

Fig. 3. Actuated-joint velocities

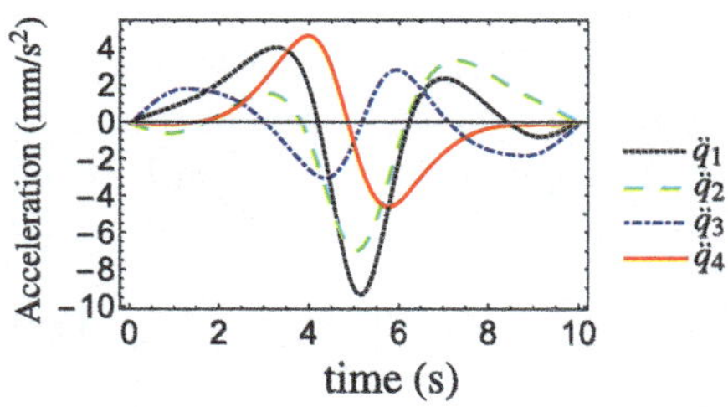

Fig. 4. Actuated-joint accelerations

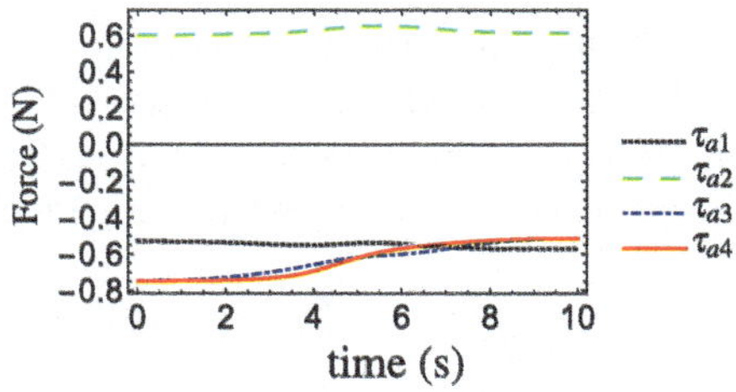

Fig. 5. Actuator-forces

5 Conclusions

A methodology applicable to redundantly-actuated systems, based on the NOC, was introduced. The method described was applied to a redundantly-actuated PKM. The simplified model applies to both forward-dynamics and inverse-dynamics. It it pointed out that internal forces can be generated under actuation redundancy [3], thereby exerting additional stress on the system. Hence, great care should be taken upon application of control schemes to redundantly-actuated systems, as unavoidable geometric errors of the systems can produce unacceptably high internal forces. Finally, the benefits of the minimization of the norm of the array of actuator torques $\boldsymbol{\tau}_{a0}$ are made apparent in the plots in Fig. 5, the torque values along the prescribed trajectory undergo only slight variations.

Acknowledgements. The work reported here was possible under: CSC (China Scholarship Council) Grant No. 201708330573; NSFC (National Science Foundation of China) Grant Nos. 51525504 and U1713202; NSERC (Natural Sciences and Engineering Research Council of Canada) Grant No. 4532-2010 and its Postdoctoral Fellowship Program.

References

1. Ben-Israel, A., Greville, T.N.: Generalized Inverses: Theory and Applications. Springer Science & Business Media, New York (2003)
2. Codourey, A., Burdet, E.: A body-oriented method for finding a linear form of the dynamic equation of fully parallel robots. In: Proceedings of the IEEE International Conference on Robotics and Automation, Albuquerque, New Mexico, pp. 1612–1618 (1997)
3. Müller, A.: Internal preload control of redundantly actuated parallel manipulators–its application to backlash avoiding control. IEEE Trans. Robot. **21**(4), 668–677 (2005)
4. Schlotter, A., Pfeiffer, F.: Modeling of a new telerobot. Multibody Sys. Dyn. **6**(4), 343–353 (2001)
5. Hui, C., Yiu, Y., Li, Z.: Dynamics and control of redundantly actuated parallel manipulators. IEEE/ASME Trans. Mechatron. **8**(4), 483–491 (2003)
6. Golub, G.H., Loan, C.F.V.: Matrix Computations. JHU Press, Baltimore (2012)
7. Angeles, J.: Fundamentals of Robotic Mechanical Systems Theory, Methods, and Algorithms. Springer, Basel (2014)
8. Angeles, J., Lee, S.: The modeling of holonomic mechanical systems using a natural orthogonal complement. Trans. Can. Soc. Mech. Eng. **13**(4), 81–89 (1989)
9. Miller, K.: Optimal design and modeling of spatial parallel manipulators. Int. J. Robot. Res. **23**(2), 127–140 (2004)
10. Biagiotti, L., Melchiorri, C.: Trajectory Planning for Automatic Machines and Robots. Springer Science & Business Media, Heidelberg (2008)
11. Tsai, L.: Solving the inverse dynamics of a Stewart-Gough manipulator by the principle of virtual work. J. Mech. Design **122**(1), 3–9 (2000)
12. Wang, L., Xu, H., Guan, L.: Kinematics and inverse dynamics analysis for a novel 3-PUU parallel mechanism. Robotica **35**(10), 2018–2035 (2017)
13. Kane, T.R., Levinson, D.A.: The use of Kane's dynamical equations in robotics. Int. J. Robot. Res. **2**(3), 3–21 (1983)
14. Wang, Y., Belzile, B., Angeles, J., Li, Q.: Kinematic analysis and optimum design of a novel 2$\underline{P}$UR-2R$\underline{P}$U Parallel Robot. Technical Report TRCIM2019-28-01-01, Centre for Intelligent Machines, McGill University, Montreal (2019)

An Individual Pitch Control Concept for Wind Turbines Based on Inertial Measurement Units

János Zierath[1](✉), Thorben Kallen[2], Dirk Machost[1], Reik Bockhahn[1], Thomas Konrad[2], Sven-Erik Rosenow[1], Uwe Jassmann[2], and Dirk Abel[2]

[1] W2E Wind to Energy GmbH, Grubenstrasse 44, 18055 Rostock, Germany
{jzierath,dmachost,rbockhahn,serosenow}@wind-to-energy.de
[2] Institute of Automatic Control, RWTH Aachen University, Campus-Boulevard 30, 52074 Aachen, Germany
{T.Kallen,T.Konrad,U.Jassmann,D.Abel}@irt.rwth-aachen.de

Abstract. Wind turbines are mainly designed by their fatigue behaviour. These fatigue behaviour can be mainly influenced by the control algorithm. So wind turbine designer struggle to enhance their control strategies to minimise the fatigue loads acting on their wind turbines. In this contribution an advanced control strategy for individual pitch control is presented, where no sensors within the blades are necessary. The loads in the blades are reconstructed by simplified models representing the characteristic dynamics of the wind turbine. The models are supported by a single Inertial Measurement Unit (IMU) within the nacelle measuring the translational accelerations and angular velocities about the three independent axis. This contribution mainly focus on the development of the concept, while practical tests will be a future task. Nevertheless, the single steps of the reconstruction and control concept are validated using a well established multibody model.

1 Introduction

Wind turbines are mechatronic systems converting mechanical energy into electrical steered by a complex closed-loop control. The quality of the control algorithms is essential for a safe operation of the wind turbine on the one hand but also for quantity of energy production on the other hand. Furthermore the control algorithms influence the construction of a wind turbine by creating more or less higher loads. So recent developments using model predictive control try to reach these three partly contrary goals for an optimal wind turbine design [1].

The actuating variables of a three-bladed horizontal axis wind turbine are the generator torque and the pitch angles of the three blades. Collective pitch control meaning that all blades are pitched synchronously is most common in

A. Kecskeméthy and F. Geu Flores (Eds.): ECCOMAS 2019, COMPUTMETHODS 53, pp. 180–187, 2020.
https://doi.org/10.1007/978-3-030-23132-3_22

industrial applications, but recent developments introduce individual pitch control to reduce the loads acting on a wind turbine. Therefore the blade loads are measured e.g. by a fibre-optic system and used as additional process variables. The concept presented in this research work does not need any sensor in the rotating system of the wind turbine circumventing technical problems transferring the signal.

This contribution is organised as follows. First the basic idea of the control concept is described in Sect. 2. In Sect. 3 the model based reconstruction supported by Inertial Measurement Units (IMUs) is presented. Hints are given on the characterising dynamics of a wind turbine. Then, the Individual Pitch Control (IPC) concept is described. Finally some results are presented and a conclusion is given.

2 Basic Idea

Wind turbine design is mainly driven by its fatigue behaviour, so more load cycles acts on a wind turbine than on a helicopter [2]. The aim of the control concept presented here is to reduce the fluctuations of the aerodynamic loads. The fluctuations are caused on the one hand by periodic effects such as tower shadow or vertical wind shear and stochastic effects caused by wind turbulence on the other hand. From basic mechanical principles it is obvious that the aerodynamic pressure distribution along the blades can be summarised to a resulting thrust force F_s acting on the rotor disc, see Fig. 1. Furthermore, the point of attack is not necessary located at the rotor disc centre. This may result in an excitation of the wind turbine structure due to the levers d_y and d_z.

The basic idea of the control concept presented in this contribution is to minimise the levers d_y and d_z by adjusting the lift of the blades individually. The lift of the blade can be steered by its pitch angle. The eccentricity of the point of attack leads to moments about the yaw and the tilt axis at the tower top. This may result into additional oscillations of the wind turbine which can be measured as a motion of the nacelle.

Thus, the control concept bases on two steps. Within the first step resulting thrust force of the rotor as well as the torsion and tilt bending moment at tower top are reconstructed. Based on this reconstruction the resulting thrust forces F_1, F_2 and F_3 of each blade are estimated, see Fig. 1.

These thrust forces are used in a second step as control variables. The pitch angles are the correcting variables controlled in that way that the levers d_y and d_z vanishes, see Fig. 1.

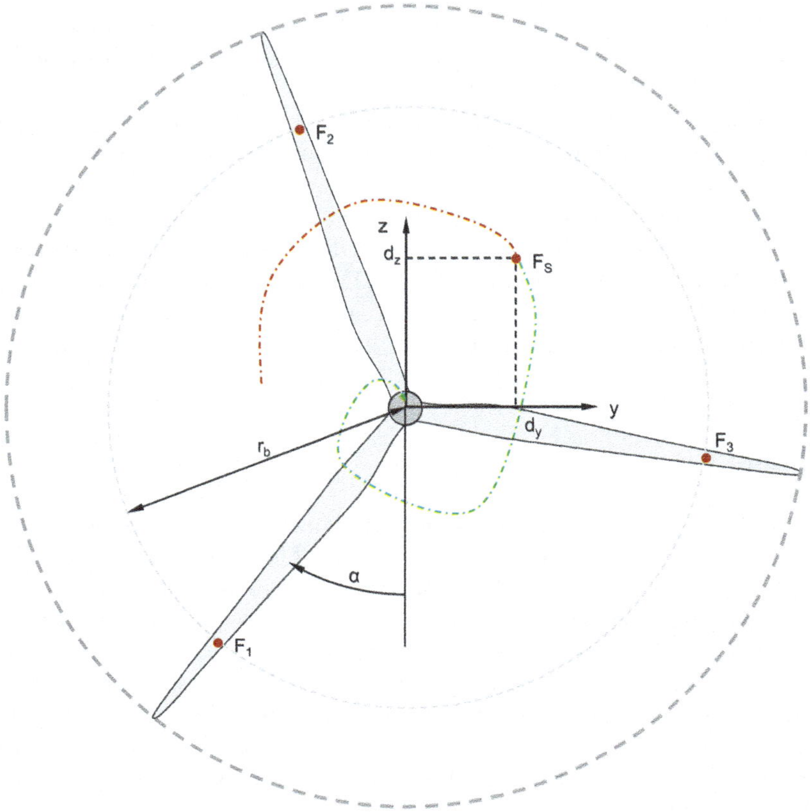

Fig. 1. Aerodynamic imbalances on the rotor disc

3 Model-Based Load Reconstruction Supported by IMUs

The main part of this contribution is the reconstruction of the blade loads. The quality of the reconstruction is essential for the control concept. So each part of the reconstruction is presented separately to minimise the deviations within the reconstruction process. The validation of the reconstruction has been carried out using a complex alaska/Wind model following the modelling principles well described in [3], [4] and [5]. Exemplary the time series for turbulent wind conditions with a mean wind speed of 21 m/s are shown in Fig. 2.

For the validation process beside load cases with steady wind conditions of 6 m/s, 12 m/s and 18 m/s also turbulent load cases with turbulent wind conditions with four different wind seeds from 4 m/s up to 25 m/s has been simulated resulting in overall 91 load cases. Virtual sensor are created and transferred to the output files. The validation process are carried out afterwards offline using MATLAB allowing a fast variation of the model parameters.

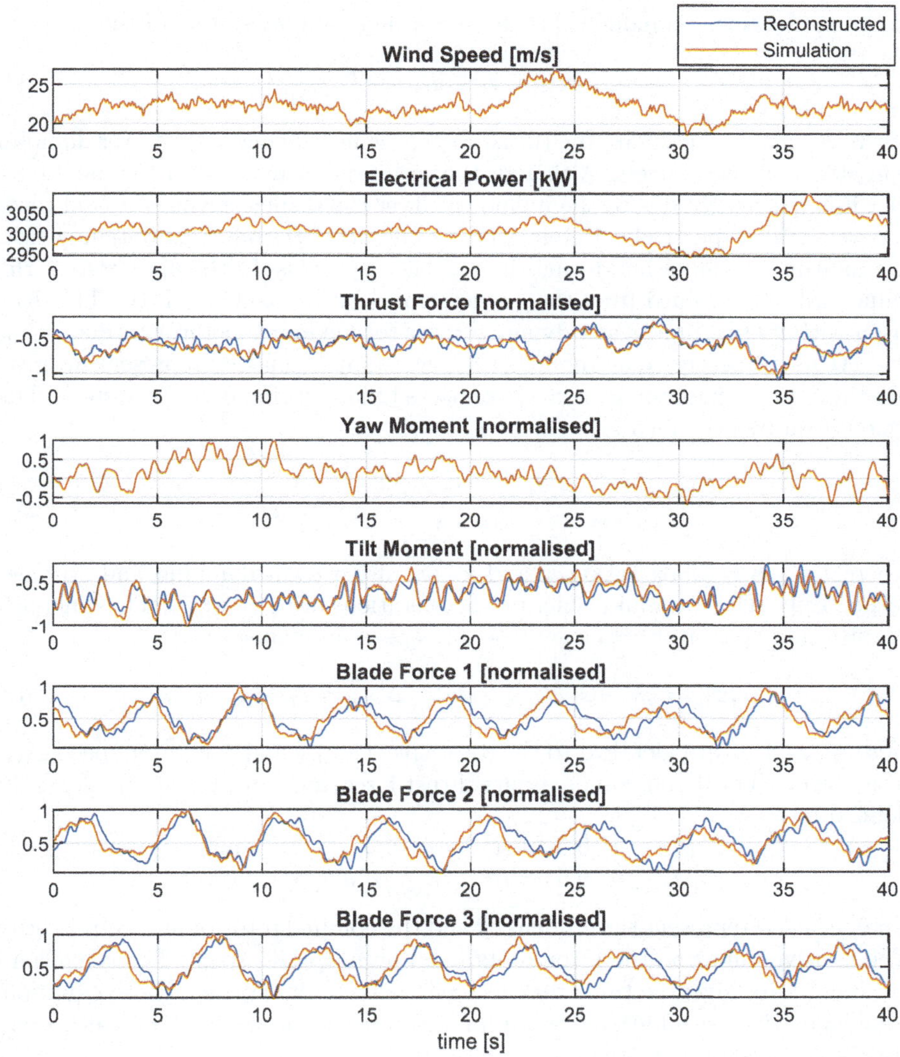

Fig. 2. Results of the reconstruction of the loads for turbulent wind conditions - Wind speed, electrical power, thrust force at yaw bearing, yaw moment, tilt moment, thrust forces at blade root for blades 1, 2, 3

3.1 Resulting Thrust-Force of the Rotor

The resulting thrust force is obtained by an Kalman filter as already used in [1], where a reduced wind turbine model has been built up by an SDOF tower model in fore-aft direction, a free-free rotational model with two degrees of freedom and a nonlinear aerodynamic model for the rotor. Other than the multibody simulation model the DOF of the reduced-order model is as small as possible.

To do so the tower is modelled as simple spring-damper system of the form

$$F_{\mathrm{s}} = m_{\mathrm{t}}\,\ddot{x}_{\mathrm{t}} + d_{\mathrm{t}}\,\dot{x}_{\mathrm{t}} + k_{\mathrm{t}}\,x_{\mathrm{t}}, \tag{1}$$

where F_{s}, m_{t}, d_{t} and k_{t} are the thrust force, the nacelle mass, the tower damping and stiffness, respectively. Although the resulting thrust force F_{s} is mentioned in 1 it is not known due to the unknown tower top displacement is x_{t} and tower top velocity, respectively. Herein, the nacelle acceleration $\ddot{x}_{\mathrm{t}}$ is obtained from the multibody simulation during the validation process. In the final process this values will be obtained from the acceleration sensor within the IMU. The drive train with its two inputs aerodynamic rotor torque and generator torque M_{rotor} and M_{gen}, respectively, is modelled as two lumped masses representing rotor inertia I_{r} and generator inertia I_{g} connected via a spring-damper-system. This system can be described as

$$\begin{aligned} M_{\mathrm{rotor}} &= I_{\mathrm{r}}\,\ddot{\varphi}_{\mathrm{r}} + d_{\mathrm{r}}\,\Delta\dot{\varphi}_{\mathrm{rg}} + k_{\mathrm{r}}\,\Delta\varphi_{\mathrm{rg}}, \\ -M_{\mathrm{gen}}\,i_{\mathrm{gear}} &= I_{\mathrm{g}}\,i_{\mathrm{gear}}\,\ddot{\varphi}_{\mathrm{g}} - d_{\mathrm{r}}\,\Delta\dot{\varphi}_{\mathrm{rg}} - k_{\mathrm{r}}\,\Delta\varphi_{\mathrm{rg}}, \end{aligned} \tag{2}$$

where i_{gear} is gearbox ratio, d_{r} and k_{r} are drive train damping and stiffness, respectively. The rotor and generator accelerations are $\ddot{\varphi}_{\mathrm{r}}$ and $\ddot{\varphi}_{\mathrm{g}}$. The difference of rotor and generator azimuth angle and speed are defined as

$$\Delta\varphi_{\mathrm{rg}} = \varphi_{\mathrm{r}} - \varphi_{\mathrm{g}}/i_{\mathrm{gear}} \quad \text{and} \quad \Delta\dot{\varphi}_{\mathrm{rg}} = \dot{\varphi}_{\mathrm{r}} - \dot{\varphi}_{\mathrm{g}}/i_{\mathrm{gear}}\,. \tag{3}$$

Here, $\dot{\varphi}_{\mathrm{r}}$ and $\dot{\varphi}_{\mathrm{g}}$ are identical to the rotor speed ω_{rotor} and generator speed ω_{gen}, respectively. The aerodynamic inputs thrust force and rotor torque are generally defined by

$$\begin{aligned} F_{\mathrm{s}} &= 0.5\,\rho\,\pi\,R^2\,v^2\,c_T(\vartheta, \omega_{\mathrm{rotor}}, v) \quad \text{and} \\ M_{\mathrm{rotor}} &= 0.5\,\rho\,\pi\,R^2\,v^3\,\omega_{\mathrm{rotor}}^{-1}\,c_P(\vartheta, \omega_{\mathrm{rotor}}, v). \end{aligned} \tag{4}$$

Here, ρ and R represent the air density and the nominal rotor radius, respectively. The relative inflow on the rotor is obtained by $v = v_{\mathrm{W}} - v_{\mathrm{rotor}}$. The coefficients c_P and c_T are given as three dimensional characteristic grids. Those equations combined with an additional state for the pitch angle ϑ result in a 6th order state space system with the state vector

$$\mathbf{x} = [\dot{\varphi}_{\mathrm{r}}, \dot{\varphi}_{\mathrm{g}}, \Delta\varphi_{\mathrm{rg}}, x_{\mathrm{t}}, \dot{x}_{\mathrm{t}}, \vartheta]^{\mathrm{T}}. \tag{5}$$

The resulting thrust force F_{s} is obtained by the interaction of the three models after estimation of the unknown states of the reduced wind turbine model as interface load between rotor and tower model. In Fig. 2 the a good agreement between the reconstructed and simulated thrust force at the yaw bearing can be seen.

3.2 Resulting Tower Torsion Moment

The tower torsion is caused by the horizontal distance d_y between the point of attack of the tower axis. In first step the assumption has been made that a SDOF torsional spring-damper is sufficient to reproduce the tower torsion,

$$M_z \equiv F_s\, d_y \approx J_z\, \ddot{\varphi}_z + b_z\, \dot{\varphi}_z + c_z\, \varphi_z \tag{6}$$

Here, the M_z is the torsion moment about the yaw axis, J_z is the mass moment of inertia of nacelle and rotor about the z-axis, b_z is the rotational damping coefficient and c_z is the torsional stiffness of the tower. The variable φ_z represents the torsion of the tower. For the validation process of that assumption the variable φ_z and its time derivatives are directly obtained from the simulation.

Figure 2 shows exemplary the according to (6) reconstructed and simulated torsion moment for two different turbulent wind speeds. An excellent agreement can be seen between the reconstructed and the simulated torsion moment. Hence the model from (6) is sufficient to reproduce the characteristic dynamics of the yaw oscillation of a wind turbine. Typically a drift can be observed on position level, which can be corrected using a moving average according to multicopter developments, see [6].

3.3 Resulting Tilt Bending Moment at Tower Top

Applying the same basic approach of the reconstruction of the yaw moment to the tilt bending moment leads to

$$M_y \equiv F_s\, dz \approx J_y\, (\ddot{\varphi}_y + \ddot{\varphi}_{\text{Model}}) + b_y\, (\dot{\varphi}_y + \dot{\varphi}_{\text{Model}}) + c_y\, (\varphi_y - \varphi_{\text{Model}})\,. \tag{7}$$

Here, it has to be pointed out that from the measured tilt angle φ_y the tilt angle corresponding to Mode 1 φ_{Model} has to be subtracted. This bases on the assumption that the thrust force F_s leads mainly to an excitation of the first fore-aft mode of the wind turbine, while a change of the point of attack mainly leads to an excitation of the second fore-aft mode.

In Fig. 2 some differences can be seen between the simulated and reconstructed tilt bending moment, which are based on the model simplification. But the overall dynamic behaviour can be reconstructed with good accuracy.

3.4 Reconstruction of Blade Thrust Forces

From the reconstructed moments at the yaw bearing the thrust forces at the blade root are estimated by a simple distribution problem, leading to the system of equations

$$\begin{aligned} F_1 \sin\alpha\, r_b + F_2 \sin(\alpha + 120°)\, r_b + F_3 \sin(\alpha + 240°)\, r_b &= -M_z \\ F_1 \cos\alpha\, r_b + F_2 \cos(\alpha + 120°)\, r_b + F_3 \cos(\alpha + 240°)\, r_b &= -M_y\,. \\ F_1 + F_2 + F_3 &= F_s \end{aligned} \tag{8}$$

Solving (8) by Gauss elimination leads to the thrust forces at the blade roots. In Fig. 2 the simulated and reconstructed blade forces at the blade root F_1, F_2 and F_3 are shown. Some differences occurs due to local dynamic effects but the overall trend agrees very well.

4 Individual Pitch Control Concept

The individual pitch control concept bases on a collective pitch angle with a superposed individual pitch angle as an offset for each blade. The collective pitch angle is controlled by a classical control concept for wind turbines, see [2] and [7], where the generator torque follows the optimal power curve below rated power and the generator speed is kept constant rated power by adjusting the pitch angle collectively. The overall pitch angle is obtained by

$$\vartheta_i = \vartheta_c + \Delta\vartheta_i \text{ with } i = 1, 2, 3, \tag{9}$$

with the collective pitch angle ϑ_c, IPC angle $\Delta\vartheta_i$ and the overall pitch angle for each blade ϑ_i. The IPC angles $\Delta\vartheta_i$ are obtained by a PD-controller, where the controlled variables are the differences of the estimated thrust forces F_i at blade root from the average thrust force $\bar{F}$ obtained by (8) and

$$\bar{F} = 1/3\,(F_1 + F_2 + F_3)\,. \tag{10}$$

5 Conclusion

This contribution presents a concept for individual pitch control of wind turbines reducing their fatigue loads without the necessity of blade sensors. The dynamic effects of aerodynamic imbalances can be measured at the nacelle. Compared to the presented concept conventional IPC concepts bases on measurements within the rotating system mainly located at the blade root. These concept addresses different problems. First a transfer of the measured signals has to be realised from the rotating system to the PLC located in the nacelle. Second the typically used fibre-optic sensors at the blade root are difficult to maintain and are much more expensive compared to the micromechanical IMUs used within this concept.

Based on the measurements in the nacelle the loads at the blade root can be reconstructed by means of mechanical models. While the agreement of the torsion moment about the yaw axis between simulation and reconstruction is very excellent the bending moment about the tilt axis agrees only within the main trends. The necessity to simplify the reconstruction to achieve real-time capabilities leads to a discrepancy of higher dynamic effects of the reconstruction.

Although the concept does not operate on a real wind turbine the concept and it components presented here are a very promising method for an advanced wind turbine controller scheme.

Acknowledgements. The authors would like to thank the German Federal Ministry for Education and Research for supporting this contribution within the programme *KMU innovativ* under the grant number 01IS17015A/B. Furthermore the authors would like to thank the European Union supporting the patent application DE 10 2017 121 082.2 of the control concept through European Regional Development Fund under the grand number TBI-V-3-012.

References

1. Jassmann, U., Zierath, J., Dickler, S., Hakenberg, M., Abel, D.: Model predictive wind turbine control for load alleviation and power leveling in extreme operation conditions. In: 2016 IEEE Multi-Conference on Systems and Control MSC 2016, Buenos Aires, Argentina (2016)
2. Hau, E.: Wind Turbines: Fundamentals, Technologies, Application, Economics. Springer, New York (2010)
3. Zierath, J., Rachholz, R., Woernle, C., Müller, A.: Load calculation on wind turbines: validation of Flex5, alaska/Wind, MSC.Adams and SIMPACK by means of Field Tests. In: Proceedings of the 2nd ASME Biennial International Conference on Dynamics for Design (DFD), Buffalo, New York, USA (2014)
4. Zierath, J., Rachholz, R., Woernle, C.: Field test validation of Flex5, MSC.Adams, alaska/Wind and SIMPACK for load calculations on wind turbines. Wind Energy **19**, 1201–1222 (2016)
5. Zierath, J., Rachholz, R., Rosenow, S.-E., Bockhahn, R., Schulze, A., Woernle, C.: Modal Testing on Wind Turbines for Validation of a Flexible Multibody Model. In: Valášek, M. (Ed.) Proceedings of the ECCOMAS Thematic Conference Multibody Dynamics, Prague, Czech Republic (2017)
6. Konrad, T., Gehrt, J.-J., Lin, J., Zweigel, R., Abel, D.: Advanced state estimation for navigation of automated vehicles. Ann. Rev. Control **46**, 181–195 (2018). https://doi.org/10.1016/j.arcontrol.2018.09.002
7. Burton, T., Jenkins, N., Sharpe, D., Bossanyi, E.: Wind Energy Handbook. Wiley, Hoboken (2011)

Flexible Multibody Dynamics

Localized Helix Configurations of Discrete Cosserat Rods

Vanessa Dörlich[1(✉)], Tomas Hermansson[2], and Joachim Linn[1]

[1] Fraunhofer ITWM, Fraunhofer–Platz 1, 67663 Kaiserslautern, Germany
{vanessa.doerlich,joachim.linn}@itwm.fraunhofer.de
[2] Fraunhofer–Chalmers Centre FCC,
Chalmers Science Park, 412 88 Gothenburg, Sweden
tomas.hermansson@fcc.chalmers.se

Abstract. Cosserat rods [1] are the prefered choice for modeling large spatial deformations of slender flexible structures at small local strains. Discrete Cosserat rod models [6,7] based on geometric finite differences preserve essential properties of the continuum theory. In previous work [9] kinetic aspects of discrete quaternionic Cosserat rods defined on a staggered grid were investigated. In particular it was shown that equilibrium configurations obtained by energy minimization correspond to solutions of finite difference type discrete balance equations for the sectional forces and moments in conservation form. The present contribution complements the numerical studies shown in [9] by considering localized helix configurations [2] of discrete Cosserat rods as more complex benchmark examples.

1 Introduction

For elastic Cosserat rods [1] the combination of slender shape and geometric nonlinearity leads to interesting effects, as the example displayed in Fig. 1 shows:

After imposing a relative torsion angle of 2π between the two terminating cross sections of an initially straight elastic rod with circular cross sections, its upper end is stepwise displaced downwards along the original cylinder axis. Due to the constant twist the configuration buckles out of plane immediately and deforms into a *localized helix configuration* in space, with torsional twist diminished in exchange of spatial bending curvature. A plane circular configuration is achieved at identical spatial positions of the two terminating cross section centers with opposite normal directions. The strip pattern indicates that at this point of the sequence torsional twist has vanished completely. The final configuration shows a further displacement step, where the configuration bends in plane with zero torsional twist.

The sequence of rod configurations visualised in Fig. 1 has been computed using a variant of the *discrete* Cosserat rod model as described in the article [7] and the more recent book chapter [8]. In a contribution [9] to the previous conference kinetic aspects of the discrete model have been discussed, with particular

A. Kecskeméthy and F. Geu Flores (Eds.): ECCOMAS 2019, COMPUTMETHODS 53, pp. 191–198, 2020.
https://doi.org/10.1007/978-3-030-23132-3_23

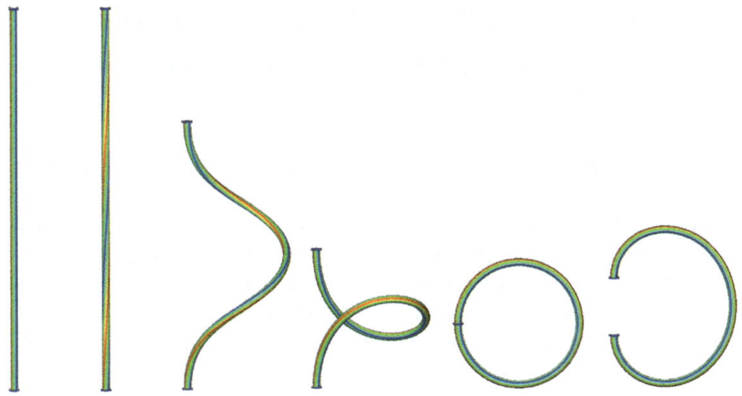

Fig. 1. Sequence of discrete elastic Cosserat rod configurations computed in *CROD*.

focus on the exact form of the discrete equilibrium equations obtained according to the Lagrangian formalism. To demonstrate the properties of the discrete Cosserat rod model for coarse discretisations of the centerline, some simple analytically solvable examples—i.e.: plane cantilever type bending of an *Elastica*, and spatial *helix* configurations, following Euler's and Kirchhoff's mathematical treatments—have been discussed, with particular emphasis on the various *first integrals* of both the continuous and the discrete model variants that govern the equilibrium configurations.

In the present contribution this work is continued by a detailed study of the more complex example of *localized helix configurations* [2] like those shown in the middle of Fig. 1. Results of numerical simulations performed with the discrete, staggered grid type quaternionic Cosserat rod model (named *CROD*, already described in [9]) will be discussed w.r.t. the findings reported in [5]. Equilibrium solutions are computed by minimizing the elastic energy of the discrete rod for given boundary conditions by an *interior point method*, using the software package *IPOPT* [10].

The results confirm the favorable properties of the discrete Cosserat rod model at coarse discretisations already found in previous work.

2 Elastic Energy and Balance Equations for Discrete Cosserat Rods

In this section we briefly summarize a few facts about our discrete rod model as described in [9] and refer to this paper for further details, including notation (see also Fig. 2).

The configuration variables of a discrete Cosserat rod consist of the sets of vertex positions $\{\mathbf{r}_j\}_{j=0,\dots,N}$ and edge centered rotational quaternions $\{\hat{\mathfrak{q}}_{j-1/2}\}_{j=1,\dots,N}$. Hamilton's principle characterizes stable static equilibrium configurations as minima of the potential energy. In the absence of external body

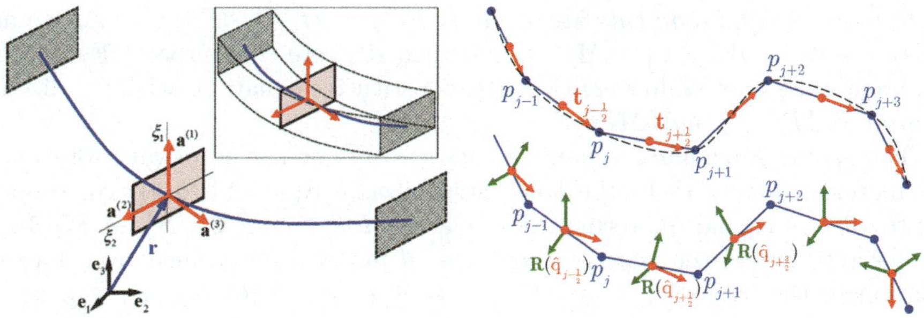

Fig. 2. *Left:* Centerline curve $\mathbf{r}(s)$ and attached moving frame $\mathsf{R}(s) = \mathbf{a}^{(k)}(s) \otimes \mathbf{e}_k$ of a *Cosserat curve*, describing the geometry of the configurations of a prismatic rod in Euclidian space. The volumetric geometry is generated by sliding the cross section spanned by the frame directors $\{\mathbf{a}^{(1)}, \mathbf{a}^{(2)}\}$ along the centerline. The position vectors of the material points in the rod volume are parametrized by: $\mathbf{x} = \mathbf{r}(s) + \xi_\alpha\, \mathbf{a}^{(\alpha)}(s)$. *Right:* Polygonal arc approximating a smooth regular geometric curve $\mathscr{C}$: The vertices $p_j \in \mathscr{C}$ located at positions $\mathbf{r}_j \in \mathbb{E}^3$ define edges $[p_{j-1}, p_j]$ of length $\ell_{j-1/2}$, with edge centered unit length tangent vectors $\mathbf{t}_{j-1/2}$. A *discrete Cosserat curve* [8] consists of a polygonal arc with edge centered quaternions $\hat{\mathsf{q}}_{j-1/2} \in S^3$, representing frames $\mathsf{R}_{j-1/2} = \mathfrak{E}(\hat{\mathsf{q}}_{j-1/2})$ via the *Euler map* $\mathfrak{E} : S^3 \to \mathsf{SO(3)}$.

forces and moments acting on a discrete Cosserat rod the static balance equations for the discrete spatial forces $\{\mathbf{f}_{j-1/2}\}_{j=1,\dots,N}$ and moments $\{\mathbf{m}_j\}_{j=0,\dots,N}$ are given by

$$\mathbf{f}_{j+1/2} - \mathbf{f}_{j-1/2} = \mathbf{0} \quad \Leftrightarrow \quad \partial \mathscr{W}^{(el)} / \partial \mathbf{r}_j = \mathbf{0}\,, \tag{1}$$

$$\mathbf{m}_{j+1} - \mathbf{m}_j + (\mathbf{r}_{j+1} - \mathbf{r}_j) \times \mathbf{f}_{j+1/2} = \mathbf{0} \quad \Leftrightarrow \quad \partial \mathscr{W}^{(el)} / \partial \hat{\mathsf{q}}_{j+1/2} = \mathbf{0}\,, \tag{2}$$

where the *discrete elastic energy*

$$\mathscr{W}^{(el)} := \sum_{j=1}^{N} h_{j-1/2}\, \frac{1}{2} \langle \Delta \boldsymbol{\Gamma}_{j-1/2}, \mathsf{C}_{\Gamma} \cdot \Delta \boldsymbol{\Gamma}_{j-1/2} \rangle + \sum_{j=0}^{N} \bar{h}_j\, \frac{1}{2} \langle \Delta \mathbf{K}_j, \mathsf{C}_K \cdot \Delta \mathbf{K}_j \rangle \tag{3}$$

approximates the stored energy $W^{(el)} = \int_0^L ds\, \mathscr{V}^{(el)}(\boldsymbol{\Gamma}, \mathbf{K})$ of the continuum model with elastic energy density $\mathscr{V}^{(el)}(\boldsymbol{\Gamma}, \mathbf{K}) = \frac{1}{2}\langle \Delta\boldsymbol{\Gamma}, \mathsf{C}_\Gamma \cdot \Delta\boldsymbol{\Gamma}\rangle + \frac{1}{2}\langle \Delta\mathbf{K}, \mathsf{C}_K \cdot \Delta\mathbf{K}\rangle$, characterized by the *effective stiffness parameters* of the local cross section assembled in the diagonal matrices $\mathsf{C}_\Gamma = \mathrm{diag}([GA_1], [GA_2], [EA])$ and $\mathsf{C}_K = \mathrm{diag}([EI_1], [EI_2], [GJ])$. The energy density $\mathscr{V}^{(el)}$ is a quadratic form of the deviations $\Delta\mathbf{K} = \mathbf{K}(s) - \mathbf{K}_0(s)$ and $\Delta\boldsymbol{\Gamma}(s) = \boldsymbol{\Gamma}(s) - \boldsymbol{\Gamma}_0$ of the material curvature $\mathbf{K} = 2\hat{\mathsf{q}}^* \circ \partial_s \hat{\mathsf{q}}$ and material tangent vector $\boldsymbol{\Gamma} = \hat{\mathsf{q}}^* \circ \partial_s \mathbf{r} \circ \hat{\mathsf{q}}$ from their reference values $\mathbf{K}_0(s)$ and $\boldsymbol{\Gamma}_0 \equiv \mathbf{e}_3$.

Discrete curvatures $\mathbf{K}_j := 2\log(\hat{\mathsf{q}}^*_{j-1/2} \circ \hat{\mathsf{q}}_{j+1/2})/\bar{h}_j$ are defined at the vertices, and $\boldsymbol{\Gamma}_{j-1/2} := \hat{\mathsf{q}}^*_{j-1/2} \circ (\mathbf{r}_j - \mathbf{r}_{j-1})/h_{j-1/2} \circ \hat{\mathsf{q}}_{j-1/2}$ are *discrete material tangent vectors* located at the edge centers, where $h_{j-1/2}$ are the edge lengths in the reference configuration, and $\bar{h}_j := \frac{1}{2}(h_{j-1/2} + h_{j+1/2})$ measures the distance of adjacent edge centers in discrete arc length. In the continuum model *material forces and moments* are given by $\mathbf{F}(s) = \partial\mathscr{V}^{(el)}/\partial\boldsymbol{\Gamma} = \mathsf{C}_\Gamma \cdot \Delta\boldsymbol{\Gamma}(s)$ and $\mathbf{M}(s) = \partial\mathscr{V}^{(el)}/\partial\mathbf{K} = \mathsf{C}_K \cdot \Delta\mathbf{K}(s)$, and their discrete counterparts $\mathbf{F}_{j-1/2}$ and $\mathbf{M}_j$ are obtained by evaluating the same constitutive equations with the discrete invariants $\Delta\boldsymbol{\Gamma}_{j-1/2}$ and $\Delta\mathbf{K}_j$.

The *spatial force and moment vectors* are obtained by a forward rotation of the material quantities to the local $SO(3)$ frame $\mathsf{R}(s) = \mathfrak{E}(\hat{\mathsf{q}}(s))$ corresponding to the rotational quaternion $\hat{\mathsf{q}}(s)$ via the *Euler map* $\mathfrak{E} : S^3 \to \mathsf{SO}(3)$. In the discrete model the *edge centered spatial forces* are obtained by a forward rotation to the frame $\mathsf{R}_{j-1/2} = \mathfrak{E}(\hat{\mathsf{q}}_{j-1/2})$ as $\mathbf{f}_{j-1/2} = \mathsf{R}_{j-1/2} \cdot \mathbf{F}_{j-1/2}$. In the present work the *vertex based spatial moments*[1] $\mathbf{m}_j \approx \mathsf{R}_j \cdot \mathbf{M}_j$ are *approximated* by a forward rotation to the SLERP–interpolated frame $\mathsf{R}_j = \mathfrak{E}(\hat{\mathsf{q}}_j)$ with $\hat{\mathsf{q}}_j := \hat{\mathsf{q}}_{j-1/2} \circ (\hat{\mathsf{q}}^*_{j-1/2} \circ \hat{\mathsf{q}}_{j+1/2})^{\frac{1}{2}}$.

The discrete balance equations (1) and (2) are conservative FD approximations of the corresponding balance equations $\partial_s\mathbf{f} = \mathbf{0}$ and $\partial_s\mathbf{m} + \partial_s\mathbf{r} \times \mathbf{f} = \mathbf{0}$ of the continuum model, which imply that in equilibrium $\mathbf{f}(s)$ and $\mathscr{M}(s) := \mathbf{m}(s) + \mathbf{r}(s) \times \mathbf{f}(s)$, as well as the scalar products $\langle \mathbf{f}, \mathscr{M} \rangle \equiv \langle \mathbf{f}, \mathbf{m} \rangle$ are necessarily *constant* along the rod, independent of its constitutive properties [3]. For uniform rods (i.e.: constant C_Γ and C_K) with a straight and untwisted reference configuration (i.e.: $\mathbf{K}_0 \equiv \mathbf{0}$) the *Hamiltonian*–like function $\mathscr{H}(s) := \frac{1}{2}\langle \mathbf{M}(s), \mathbf{K}(s) \rangle + \langle \mathbf{F}(s), \mathbf{e}_3 \rangle$ is likewise a first integral. For rods with kinetic symmetry (i.e.: $[EI_{1,2}] \equiv [EI]$) the torsional moment $M^{(3)}(s) = \langle \mathbf{m}(s), \mathbf{a}^{(3)}(s) \rangle = [GJ]\, K^{(3)}(s)$ as well as—due to constant $[GJ]$—the twist $K^{(3)}(s)$ are additional conserved quantities.

For the discrete Cosserat rod in stable equilibrium $\mathbf{f}_{j-1/2} \equiv \mathbf{f}$ is constant on all edges. The discrete balance equations $\mathbf{m}_j - \mathbf{m}_{j-1} = \mathbf{f} \times (\mathbf{r}_j - \mathbf{r}_{j-1})$ are necessarily satisfied by minimal energy configurations, such that $\mathscr{M}_j = \mathbf{m}_j + \mathbf{r}_j \times \mathbf{f}_j$ with the vertex based spatial force $\mathbf{f}_j := \frac{1}{2}(\mathbf{f}_{j-1/2} + \mathbf{f}_{j+1/2})$ and also $\langle \mathbf{f}_j, \mathscr{M}_j \rangle \equiv \langle \mathbf{f}_j, \mathbf{m}_j \rangle$ have the same values on all vertices. Also $M^{(3)}_j = [GJ]\, K^{(3)}_j$ is constant in equilibrium.

The vertex based discretization $\mathscr{H}_j := \frac{1}{2}\langle \mathbf{M}_j, \mathbf{K}_j \rangle + F^{(3)}_j$ of the first integral $\mathscr{H}(s)$ depends on the approximation of $F^{(3)}(s) = \langle \mathbf{F}(s), \mathbf{e}_3 \rangle = \langle \mathbf{f}(s), \mathbf{a}^{(3)}(s) \rangle$ on the vertices. Possible choices are $F^{(3)}(s_j) \approx \langle \mathbf{F}_j, \mathbf{e}_3 \rangle$ using averaged material forces $\mathbf{F}_j := \frac{1}{2}(\mathbf{F}_{j-1/2} + \mathbf{F}_{j+1/2})$, or $F^{(3)}(s_j) \approx \langle \mathbf{f}_j, \mathbf{a}^{(3)}_j \rangle$ using $\mathbf{f}_j$ and $\mathbf{a}^{(3)}_j = \mathsf{R}_j \cdot \mathbf{e}_3$. The approximation $F^{(3)}(s_j) \approx \langle \mathbf{f}_j, \mathbf{t}_j \rangle =: T_j$ is a proper choice for inextensible Kirchhoff rod models, obtained by assuming $\mathbf{a}^{(3)}_j \approx \mathbf{t}_j$ with vertex based unit length tangent vectors $\mathbf{t}_j := (\mathbf{t}_{j-1/2} + \mathbf{t}_{j+1/2})/\|\mathbf{t}_{j-1/2} + \mathbf{t}_{j+1/2}\|$, and is the variant chosen here.

[1] In [9] erroneously $\mathsf{R}_{j+1/2}$ was used to perform the forward rotation of $\mathbf{M}_j$ to $\mathbf{m}_j$.

3 Numerical Experiments for Localized Helices

In this section, we show and discuss numerical experiments using *CROD* which yield rod configurations in the shape of localized helices [2] like the ones depicted in Fig. 1. These experiments are based on the work of van der Heijden et al. [5] who perform real and numerical experiments on cylindric metallic wires.

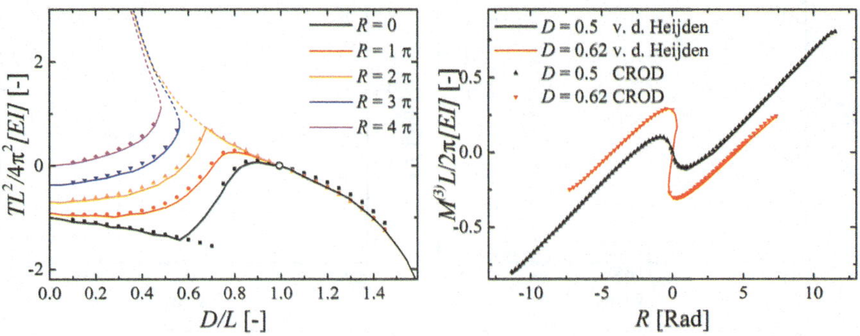

Fig. 3. Results of numerical experiments using *CROD* (symbols) in comparison to numerical results from [5] (solid for stable, dashed lines for unstable branches). Left: Experiments at fixed R. Right: Experiments at fixed D.

Two types of simulation experiments are executed under clamped boundary conditions of initially straight rods. In the first case, one end of the rod is rotated by R about the rig axis before it is translated towards the other end along the axis by the displacement D. This procedure corresponds to the sequence of configurations displayed in Fig. 1. While pretwisted rods will buckle directly into a spatial configuration once the displacement is applied, untwisted rods will buckle into a planar configuration. According to Euler's Theory of Elastica [4], the jump into the planar configuration occurs at a critical load $T_{crit} = TL^2/(4\pi^2[EI]) = -1$.

During this experiment, the rig force T (i.e. the component of the spatial force in the direction of the local end tangent) is measured as a function of the relative displacement $^D/_L$ and normalized on T_{crit}. The normalizations lead to an intersection of the curves for all prerotations at $^D/_L = 1$, where the rod is bent into a planar ring, marked with a circle in Fig. 3.

For the second type, the displacement D is applied in the first step before the end is rotated about $\pm R$. The moment about the rig axis $M^{(3)}$ (i.e. the material moment about the local end tangent) is measured in this experiment as a function of R and normalized to the length and the bending stiffness.

The numerical experiments were performed on a discrete elastic rod ($E = 5$ MPa, $\nu = 0.4$) with a radius of $\rho = 0.01\,\mathrm{m}$, length $L = 1\,\mathrm{m}$ and the number of segments $N = 15$. Although [5] used the material parameters of nitinol, the results can be compared to ours, as the ratio of torsional to bending stiffness is $\gamma = {}^5/_7$ in both cases and this determines the behavior.

We execute these experiments for different values of prerotation R and predisplacement D, respectively, and compare them to the results of [5] in Fig. 3. For both types of experiments, the results derived with *CROD* match those of [5] well. It has to be mentioned that only stable branches were simulated in these numerical experiments.

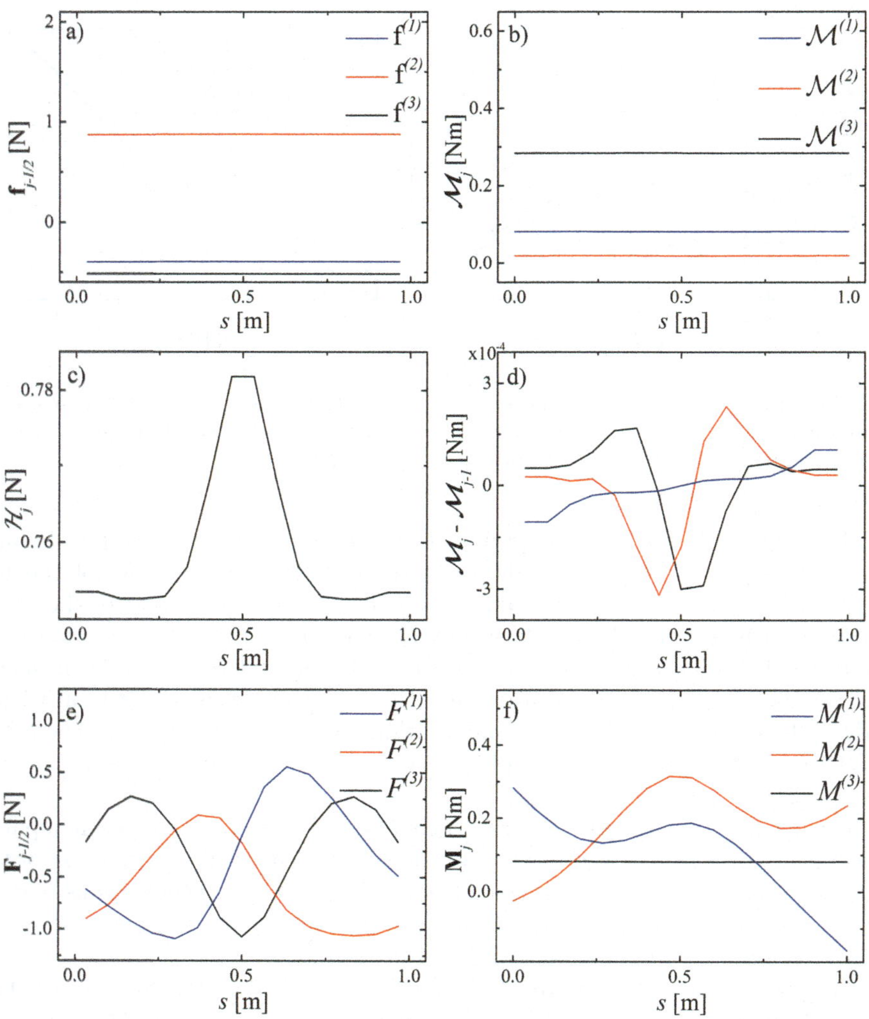

Fig. 4. Plots of *CROD* results for discrete first integrals, moment balance and material sectional quantities obtained for $D/L = 0.5$ and $R = 2\pi$.

In order to evaluate the quality of the numerical results obtained for the localized helix configurations, we take a look at the conserved quantities of the Cosserat rod model in the following. The discrete vertex based spatial force

$\mathbf{f}_{j-1/2}$ and the first integral of the spatial moment $\mathcal{M}_j$ are displayed in the plots (a) and (b) of Fig. 4. As explained in the previous section, both vector quantities should be conserved along the rod in equilibrium. From the definition of $\mathcal{M}_j$ we obtain the identity

$$\begin{aligned} \mathcal{M}_j - \mathcal{M}_{j-1} &= \mathbf{m}_j - \mathbf{m}_{j-1} + (\mathbf{r}_j - \mathbf{r}_{j-1}) \times \mathbf{f}_{j-1/2} \\ &\quad + \frac{1}{2}\mathbf{r}_{j-1} \times (\mathbf{f}_{j-1/2} - \mathbf{f}_{j-3/2}) + \frac{1}{2}\mathbf{r}_j \times (\mathbf{f}_{j+1/2} - \mathbf{f}_{j-1/2}) , \end{aligned} \tag{4}$$

which shows that the difference $\mathcal{M}_j - \mathcal{M}_{j-1}$ vanishes identically if the discrete balance equations (1) and (2) are fulfilled *exactly*, and yields a proper measure for the combined residuum of the discrete balance equations otherwise. The difference $\mathcal{M}_j - \mathcal{M}_{j-1}$ computed with the *CROD* model displayed in plot (d) of Fig. 4 shows a residual error in the range below 10^{-4}, which on the one hand can be attributed to the use of *approximate* spatial momentum vectors $\mathbf{m}_j \approx \mathsf{R}_j \cdot \mathbf{M}_j$, and on the other hand to the fact that the discrete rod configuration is only in *approximate* equilibrium, depending on the setting of the algorithmic parameters of the numerical method used for energy minimization (see also [9]).

The discrete balance equations (1) and (2) are exact criteria for configurations corresponding to stationary points of the discrete energy function (3) and therefore essentially equivalent to a *variational discretization* scheme [6] for the propagation of the frames and edge tangent vectors along the rod configuration. Different from the spatial forces $\mathbf{f}_{j-1/2}$ and the integrated moments $\mathcal{M}_j$, the discrete values $\mathcal{H}_j$ of the Hamiltonian are *not* conserved for variational schemes. The values $\mathcal{H}_j$ shown in plot (c) of Fig. 4 are computed for a coarse discretization and vary around an average value in the range of a few percent. This variation is substantially higher than the combined error of the discrete balance equations extracted from plot (d) of Fig. 4.

Plots (e) and (f) displayed in the bottom row of Fig. 4 show the variation of the material forces and moments along the (approximate) equilibrium configuration of the rod. As the spatial forces are constant, the variation of the material forces is governed by the variation of the orientation of the frame along the centerline. Furthermore, the plot (f) of the components of discrete material moment $\mathbf{M}_j$ shows that the torsional moment $M_j^{(3)}$ is (approximately) constant, as expected in equilibrium.

4 Conclusions

Discrete Cosserat rod models [6,7] based on geometric finite differences preserve essential properties of the continuum theory of Cosserat rods [1]. In previous work [9] kinetic aspects of discrete quaternionic Cosserat rods defined on a staggered grid were investigated, with particular focus on the derivation of finite difference type discrete balance equations for the sectional forces and moments in conservation form satisfied by equilibrium solutions obtained via energy minimization.

The present contribution complements the numerical studies shown in [9] by considering *localized helix configurations* of discrete Cosserat rods as more complex benchmark examples. The comparison of the numerical results obtained with the *CROD* model by energy minimization with the experimental and theoretical results of [5] shown in Fig. 3 indicates that nontrivial features of the continuum solution can be captured by our discrete rod model already at a coarse discretization (here: $N = 15$ segments). The values of the first integrals indicate that the discrete balance equations (1) and (2) are fulfilled with sufficient accuracy.

Altogether, the results of the present work underline the attractive properties of the energy minimization approach for a numerical approximation of stable equilibrium configurations of our discrete Cosserat rod model. In particular, the robustness and simple applicability of this approach provide advantages in situations where buckling occurs.

Acknowledgements. This work was funded by the *Eurostars* project *DDA-Flex* (E! 10462) under grant No. 01QE1640B (FKZ).

References

1. Antman, S.S.: Nonlinear Problems of Elasticity. Springer, New York (2005)
2. Audoly, B., Pomeau, Y.: Elasticity and Geometry - From Hair Curls to the Nonlinear Response of Shells. Oxford University Press, Oxford (2010)
3. Chouaieb, N., Maddocks, J.H.: Kirchhoff's problem of helical equilibria of uniform rods. J. Elast. **77**, 221–247 (2004)
4. Euler, L.: Additamentum I de curvis elasticis, methodus inveniendi lineas curvas maximi minimivi proprietate gaudentes. Lausanne, 1744, in Opera Omnia I, vol. 24, pp. 231–297 (1960)
5. van der Heijden, G.H.M., Neukirch, S., Goss, V.G.A., Thompson, J.M.T.: Instability and self-contact phenomena in the writhing of clamped rods. Int. J. Mech. Sci. **45**, 161–196 (2003)
6. Jung, P., Leyendecker, S., Linn, J., Ortiz, M.: A discrete mechanics approach to the Cosserat rod theory - Part 1: Static equilibria. Int. J. Num. Meth. Eng. **85**(1), 31–60 (2011)
7. Lang, H., Linn, J., Arnold, M.: Multibody dynamics simulation of geometrically exact Cosserat Rods. Multibody Sys. Dyn. **25**, 285–312 (2011)
8. Linn, J., Dreßler, K.: Discrete Cosserat rod models based on the difference geometry of framed curves for interactive simulation of flexible cables. In: Ghezzi, L., Hömberg, D., Landry, C. (eds.) Math for the Digital Factory, pp. 289–319. Springer, Cham (2017)
9. Linn, J., Hermansson, T., Andersson, F., Schneider, F.: Kinetic aspects of discrete Cosserat rods based on the difference geometry of framed curves. In: Valasek, M., et al. (Eds.) Proceedings of the ECCOMAS Thematic Conference on Multibody Dynamics 2017, pp. 163–176 (2017)
10. Wächter, A., Biegler, L.T.: On the implementation of a primal-dual interior point filter line search algorithm for large-scale nonlinear programming. Math. Program. **106**(1), 25–57 (2006)

Importance of Warping in Beams with Narrow Rectangular Cross-Sections: An Analytical, Numerical and Experimental Flexible Cross-Hinge Case Study

Marijn Nijenhuis(✉), Ben Jonker, and Dannis Brouwer

Precision Engineering, University of Twente, Enschede, The Netherlands
{m.nijenhuis,j.b.jonker,d.m.brouwer}@utwente.nl

Abstract. This paper reports on the significance of warping deformation on the stability analysis of a flexible cross-hinge mechanism, which consists of two leaf springs with rectangular cross-section. The effect of misalignments in this mechanism is studied analytically, numerically and experimentally. An analytical buckling analysis is carried out to determine the theoretical critical load of a generalized cross-hinge mechanism on the basis of first principles. A geometrically nonlinear beam finite element with a non-uniform torsion description is used to model the leaf springs numerically. The change in natural mode frequencies and stiffness as a function of the misalignment is determined by a multibody program. Measurements from a dedicated experimental set-up confirm that the inclusion of warping effects is crucial, even for narrow rectangular cross-sections: it is found that the effects of warping increase the analytical critical buckling load of the system by 55%.

1 Introduction

In precision manipulation mechanisms, the risk of indeterministic behavior is mitigated by the use flexure joints. Such joints consist of elastically deforming elements, instead of traditional hysteresis-inducing rolling or sliding components. Since the only motion is due to elastic deformation, flexure mechanisms operate without friction, backlash, stick-slip, and wear, resulting in low hysteresis and highly repeatable motion. It also means that the static performance of such mechanisms is characterized by their stiffness properties: in certain directions (typically associated with actuation) low stiffness is desired, whereas especially in load-bearing directions high stiffness is desired. Assemblies of beams with narrow rectangular cross-sections are commonly used to this end.

A. Kecskeméthy and F. Geu Flores (Eds.): ECCOMAS 2019, COMPUTMETHODS 53, pp. 199–206, 2020.
https://doi.org/10.1007/978-3-030-23132-3_24

In pursuit of higher performance (i.e. higher load-bearing stiffness), we are exploring flexure mechanisms that are termed overconstrained. This class of designs is typically avoided altogether because its stiffness properties depend strongly on the correct alignment of components (e.g. due to assembly and manufacturing tolerances). It has been shown that a flexible multibody model can predict this dependency accurately, and actually quantify the mechanism's limit of operation in terms of the allowable misalignment by means of a buckling analysis [3,4]. In this paper, we show that the effects of constrained warping have a marked influence on the critical load and therefore the allowable misalignement. An analytical lateral buckling analysis provides the theoretical critical load of the system on the basis of first principles and corroborates the measurements and simulations.

2 Experimental Set-Up

To study the phenomenon, the flexible cross-hinge in Fig. 1 serves as a case study. It consists of two leaf springs (80 mm length) with narrow rectangular cross-sections (30 mm width, 0.35 mm thickness) and a shuttle. The shuttle guides motion with respect to the base about the indicated rotation axis; in the other directions, it constrains motion. The overconstrained nature of this particular design manifests itself as a relatively high stress that occurs due to misalignment displacement v_0 (compared to misalignments in the other directions). This stress affects the stiffness properties of the mechanism, therefore its performance, and can even cause bifurcation buckling. The measurement set-up is depicted in Fig. 2.

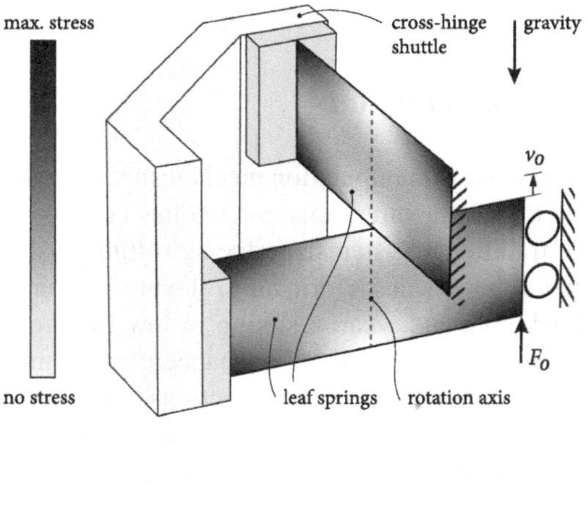

Fig. 1. Simplified illustration of the flexible cross-hinge mechanism. The Von Mises stress distribution due to misalignment displacement v_0 is indicated.

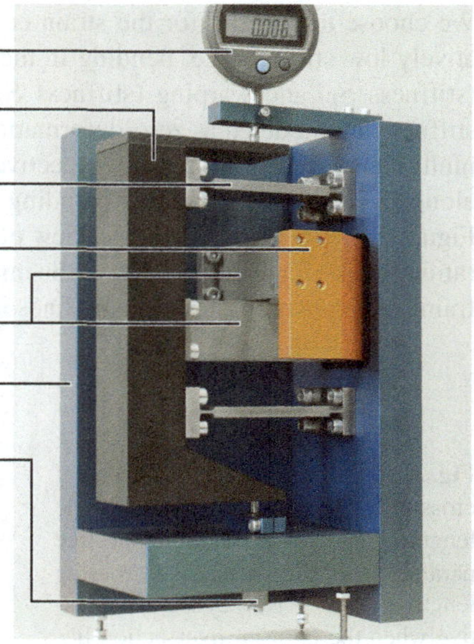

Fig. 2. Photograph of the measurement set-up of the cross-hinge mechanism. The guidance stage serves to apply a controllable misalignment to the cross-hinge mechanism. There are several design characteristics for limiting hysteresis and enabling repeatable measurements [4]. The set-up is used for measuring the mechanism's natural mode frequencies as a function of the misalignment. The natural mode frequencies serve as stiffness measures that are easier to determine experimentally.

3 Analytical Analysis

When the misalignment displacement v_0 or force F_0 exceeds a critical value, buckling occurs, support stiffness is lost and the mechanism no longer functions. To investigate the effects that warping has on the critical load of the system, an analytical buckling analysis is performed. The analysis follows the same steps and notation as Nijenhuis et al. [4], with the addition of a warping model and a change in solution procedure.

The equilibrium conditions of the system are derived using the potential energy of the system, given by

$$P_{\text{tot}} = P_{\text{leaf}}^{\text{l}} + P_{\text{leaf}}^{\text{r}} - F_0 v^{\text{l}}(0) + \begin{bmatrix} D_x & D_y & D_z \end{bmatrix} \begin{bmatrix} K_x \\ K_y \\ K_z \end{bmatrix} + \begin{bmatrix} R_\chi & R_\psi & R_\phi \end{bmatrix} \begin{bmatrix} K_\chi \\ K_\psi \\ K_\phi \end{bmatrix}. \quad (1)$$

It consists of strain energy terms, external work $F_0 v^{\text{l}}(0)$ and Lagrange multipliers $D_{x,y,z}$ and $R_{\chi,\psi,\phi}$ to account for the kinematic constraints that the shuttle imposes on the leaf spring ends. The multipliers can be interpreted as the forces and moments needed to enforce the kinematic constraints. Superscripts l and r are used to denote the left and right leaf spring. Shuttle elasticity and gravity are ignored. Using beam theory to describe the leaf springs of length L, width w and thickness t, their contribution to P_{tot} is given by

$$P_{\text{leaf}} = \frac{1}{2}\int_0^L \left[S_b \kappa_y^2 + S_t \kappa_z^2 + S_w \left(\kappa_z'\right)^2\right] \text{d}s + \int_0^L \left(N_u \gamma_x + N_v \gamma_y + N_w \gamma_z + M \kappa_x\right) \text{d}s. \quad (2)$$

We choose to account for the strain energy only due to deformation modes with relatively low stiffness, i.e. bending in the plane of lowest rigidity (stiffness S_b), torsion (stiffness S_t) and warping (stiffness S_w). The deformation modes of relatively high stiffness are modeled as zero-deformation constraints by means of additional Lagrange multipliers $N_{u,v,w}$ and M for respectively zero transverse shear strain γ_x and γ_y, zero elongation strain γ_z and zero bending curvature κ_x in the plane of highest rigidity. Figure 3 shows a schematic overview of the system. The deformed leaf spring configuration is described with respect to the initial coordinate frames x, y, z. Resolved in these frames, the position of the elastic lines is given by

$$\mathbf{r}(s) = \left[u(s)\ v(s)\ s + w(s)\right]. \tag{3}$$

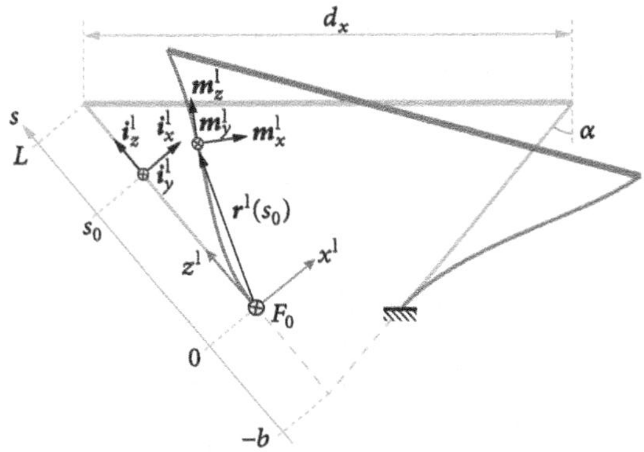

Fig. 3. For treating a wider variety of cross-hinge designs, angle α, shuttle length d_x and shuttle width d_y are free parameters in this analysis. Crossing length $b = L - (d_x/2)\csc\alpha$ is positive when the point of intersection lies on the leaf springs; negative when it lies beyond (as depicted). For the demonstrator set-up, $b = L/2$. The position vector and orientation triads are only drawn for the left leaf spring.

Also, the orientation of the orthonormal triad attached to the deformed beam cross-section $[\mathbf{m}_x\ \mathbf{m}_y\ \mathbf{m}_z]$ is described by the rotation

$$\left[\mathbf{m}_x\ \mathbf{m}_y\ \mathbf{m}_z\right] = \begin{bmatrix} 1 & 0 & 0 \\ 0 & \cos\phi & -\sin\phi \\ 0 & \sin\phi & \cos\phi \end{bmatrix} \begin{bmatrix} \cos\psi & -\sin\psi & 0 \\ \sin\psi & \cos\psi & 0 \\ 0 & 0 & 1 \end{bmatrix} \begin{bmatrix} \cos\chi & 0 & \sin\chi \\ 0 & 1 & 0 \\ -\sin\chi & 0 & \cos\chi \end{bmatrix} \left[\mathbf{i}_x\ \mathbf{i}_y\ \mathbf{i}_z\right] \tag{4}$$

with respect to the undeformed cross-section triad $[\mathbf{i}_x\ \mathbf{i}_y\ \mathbf{i}_z]$. Then, the shuttle constraints for the rotations are given by

$$\begin{aligned} K_\chi &= \chi^{\mathrm{l}} - \chi^{\mathrm{r}} = 0, \\ K_\psi &= \psi^{\mathrm{l}} - \psi^{\mathrm{r}}\cos 2\alpha + \phi^{\mathrm{r}}\sin 2\alpha = 0, \\ K_\phi &= \phi^{\mathrm{l}} - \phi^{\mathrm{r}}\cos 2\alpha - \psi^{\mathrm{r}}\sin 2\alpha = 0, \end{aligned} \tag{5}$$

and for the translations by

$$\begin{aligned} K_x &= \left(-u^{\mathrm{l}} + u^{\mathrm{r}} + d_y\psi^{\mathrm{l}}\right)\cos\alpha + \left(w^{\mathrm{l}} + w^{\mathrm{r}} + d_y\phi^{\mathrm{l}}\right)\sin\alpha = 0, \\ K_y &= -v^{\mathrm{l}} + v^{\mathrm{r}} - \left(\psi^{\mathrm{l}} + \chi^{\mathrm{l}}\phi^{\mathrm{l}}\right)d_x\cos\alpha - \left(\phi^{\mathrm{l}} - \chi^{\mathrm{l}}\psi^{\mathrm{l}}\right)d_x\sin\alpha + d_y\left(\phi^{\mathrm{l}2} + \psi^{\mathrm{l}2}\right)/2 = 0, \\ K_z &= d_x\chi^{\mathrm{l}} + \left(-w^{\mathrm{l}} + w^{\mathrm{r}} - d_y\phi^{\mathrm{l}}\right)\cos\alpha - \left(u^{\mathrm{l}} + u^{\mathrm{r}} - d_y\psi^{\mathrm{l}}\right)\sin\alpha = 0 \end{aligned} \tag{6}$$

at $s = L$. Reissner's relations for the nonlinear beam strains and curvatures energetically dual to the stress resultants are given by [5]

$$\begin{aligned} \kappa_x &= \phi' - \psi'\chi, & \gamma_x &= u' - \chi, \\ \kappa_y &= \chi', & \gamma_y &= v' + \phi(1 + w') - \psi u', \\ \kappa_z &= \psi', & \gamma_z &= w', \end{aligned} \tag{7}$$

where the prime denotes differentiation with respect to the independent coordinate s. The kinematic boundary conditions, not included in the total potential energy, are given by

$$s = 0: \qquad u^{\mathrm{l}} = u^{\mathrm{r}} = v^{\mathrm{r}} = w^{\mathrm{l}} = w^{\mathrm{r}} = \chi^{\mathrm{l}} = \chi^{\mathrm{r}} = \psi^{\mathrm{l}} = \psi^{\mathrm{r}} = \phi^{\mathrm{l}} = \phi^{\mathrm{r}} = 0 \tag{8}$$

and

$$s = 0: \qquad \psi'^{\mathrm{l}} = \psi'^{\mathrm{r}} = \psi'^{\mathrm{l}} = \psi'^{\mathrm{r}} = 0 \tag{9}$$

for the constraints on the warping deformation imposed by the mounting fillets at both ends of the leaf springs.

Application of the principle of minimum potential energy yields the equilibrium conditions for the system in terms of the configuration functions and undetermined multipliers. The fundamental solution branch is given by a zero value for all quantities, except for

$$\begin{aligned} D_y &= -F_0, & N_v^{\mathrm{l}} &= -F_0, & N_v^{\mathrm{r}} &= F_0, \\ R_\phi &= d_x F_0 \cos\alpha \cot 2\alpha, & M^{\mathrm{l}} &= -F_0\,(s - b)\,, & M^{\mathrm{r}} &= F_0\,(s - b)\,, \\ R_\psi &= -d_x F_0 \cos\alpha, \end{aligned} \tag{10}$$

indicating that load F_0 does not cause deformation in the fundamental solution, but only a "reaction" from the multipliers that enforce the idealized zero-deformation constraints. Bifurcation equilibrium conditions are obtained by requiring that the first variation of P_{tot} with respect to the *fundamental solution* vanishes. This means that only the second-order terms of P_{tot} need to be taken into account (on the basis of which the nonlinear expressions of Eqs. (5)–(7) were already truncated). The ODEs and natural boundary conditions obtained this way can be solved partially, yielding

$$\phi^{\mathrm{l}} = \phi^{\mathrm{r}} = v^{\mathrm{l}} = v^{\mathrm{r}} = w^{\mathrm{l}} = w^{\mathrm{r}} = 0. \tag{11}$$

The addition of the warping term in P_{leaf} leads to coupled variable-coefficient linear ODEs for $\psi^{\mathrm{l,r}}$ and $\chi^{\mathrm{l,r}}$ that seem to admit no closed-form solution, unlike the original analysis without a warping model [4]. The remaining boundary conditions, assuming that $0 < \alpha < \pi/2$, are

$$s = L: \qquad \begin{aligned} \psi^{\mathrm{l}} &= \psi^{\mathrm{r}} = 0, \\ \chi^{\mathrm{l}} &= \chi^{\mathrm{r}}, \\ u^{\mathrm{l}} &= u^{\mathrm{r}}, \end{aligned} \qquad \begin{aligned} d_x\chi^{\mathrm{l}} &= \left(u^{\mathrm{l}} + u^{\mathrm{r}}\right)\sin\alpha, \\ S_b\left(\chi'^{\mathrm{l}} + \chi'^{\mathrm{r}}\right) &= (b - L)\left(N_u^{\mathrm{l}} + N_u^{\mathrm{r}}\right). \end{aligned} \tag{12}$$

To obtain a solution, we reduce the complexity by splitting the solution into a symmetric part, for which $\chi^{\mathrm{l}} = -\chi^{\mathrm{r}}$, $\psi^{\mathrm{l}} = \psi^{\mathrm{r}}$, $u^{\mathrm{l}} = -u^{\mathrm{r}}$, and an anti-symmetric part, for which $\chi^{\mathrm{l}} = \chi^{\mathrm{r}}$, $\psi^{\mathrm{l}} = -\psi^{\mathrm{r}}$, $u^{\mathrm{l}} = u^{\mathrm{r}}$. For the symmetric solution, the kinematic boundary conditions and Eq. (12) together become

$$\psi(0) = \psi(L) = \psi'(0) = \psi'(L) = 0, \quad u(0) = u(L) = u'(0) = u'(L) = 0. \tag{13}$$

An approximate solution can be obtained by choosing a set of shape functions that satisfy the boundary conditions. For the symmetric solution, these can be

$$\psi(s) = \sum_{k=0}^{n} g_k \sin(\pi s/L) \sin(k\pi s/L), \quad u(s) = \sum_{k=0}^{n} h_k \sin(\pi s/L) \sin(k\pi s/L), \tag{14}$$

and $\chi(s) = u'(s)$. Substituting in P_{tot} and taking variations now yields a system of $2n$ linear algebraic equations for the undetermined coefficients g_k and h_k, whose determinant has to be zero in order to have nontrivial solutions. The smallest positive root corresponds to the critical load of the system, given by

$$F_{0,\mathrm{cr}} = \beta\left(\frac{b}{L}, \lambda\right) \frac{\sqrt{S_b S_t}}{L^2}, \qquad \lambda = \sqrt{\frac{S_t L^2}{S_w}} \approx \sqrt{\frac{24}{1+\nu} \frac{L}{w}}, \tag{15}$$

where β is a dimensionless function of the crossing ratio b/L and the spatial decay rate λ, which is a measure of the warping stiffness relative to the torsion stiffness and largely dependent on the w/L ratio of the leaf springs. The structure of $F_{0,\mathrm{cr}}$ is similar to the critical load expression in the original analysis without warping, the difference being the dependency of the geometric factor β on λ as well. Due to the length of the closed-form expression of β, numeric values of β for the symmetric buckling mode are provided in Table 1 for practically relevant values of b/L and λ.

Table 1. Values of the geometric factor β for the symmetric bucking load for $n = 15$.

b/L	$w = L/10$ ($\lambda = 42.6$)	$w = L/4$ ($\lambda = 17.2$)	$w = L/2$ ($\lambda = 8.61$)	$w = 3L/4$ ($\lambda = 5.74$)	$w = L$ ($\lambda = 4.31$)
1/2	28.61	31.41	38.32	46.84	56.41
3/4	18.02	19.49	23.34	28.33	34.01
1	11.26	12.02	14.16	17.06	20.41
5/4	7.985	8.471	9.919	11.91	14.23
3/2	6.137	6.493	7.581	9.093	10.86

4 Numerical Analysis

For simulation, a geometrically nonlinear beam finite element has been implemented in the software package SPACAR [2]. This element is based on Timoshenko's bending theory and Reissner's torsion theory in order to capture non-uniform torsion (in addition to bending, shear and elongation) of thin-walled beams with closed, symmetric cross-sections. The element is formulated in the generalized strain beam framework, in which deformation modes are defined to describe both elastic deformations and rigid-body motion of the element. This formulation has the advantages of the co-rotational formulation, while avoiding interpolation of finite rotations. The inertia properties are described with a consistent and lumped mass formulation; the latter is used to model the warping inertia of the beam cross-section [1].

5 Results and Discussion

The idealized analytical buckling analysis for the mechanism at hand ($b = L/2$) predicts $\beta = 34.62$ and a buckling load of 90.63 N, compared to a critical load of 58.25 N in the earlier analysis without warping [4].

In SPACAR, the same idealized conditions can be simulated by excluding the deformation modes of elongation and bending in the plane of highest rigidity, and disabling shear, gravity and shuttle elasticity. Figure 4 (left) shows that the behavior is the same: the critical buckling load of the idealized system increases by 55% from a converged value of 58.60 N to 90.63 N when warping is included in the SPACAR element. A qualitative explanation for the observed behavior is that the warping constraints at the ends of the leaf springs increase the torsional stiffness near the ends, effectively reducing the length over which the leaf spring twists; since the buckling mode exhibits torsion deformation, the buckling load increases.

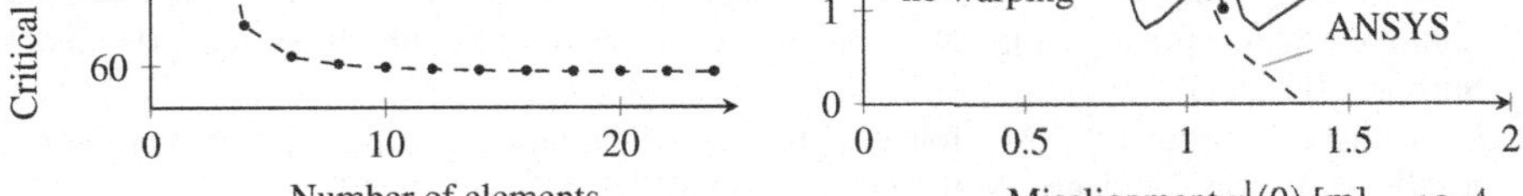

Fig. 4. Left: convergence of the critical buckling load. Right: measurement and simulation of the first natural frequency f_1.

It is observed that generally the first buckling mode is an anti-symmetric one, in which the shuttle moves and the twist angle of both leaf springs is reversed. The second mode is a symmetric one, in which the shuttle remains stationary, the bending angle of both leaf springs has the same sign and the twist angle is reversed. However, only for the case of the common doubly-symmetric embodiment of the mechanism ($b = L/2$), under the idealized conditions, and with the warping effects included, we see that the anti-symmetric mode stiffens considerably and becomes the second mode. The first mode, which now has multiplicity two, is a symmetric one in which an individual leaf spring buckles and the shuttle remains stationary. For this mode, the analytical buckling load is provided by Eq. (15) and Table 1.

A refined numerical model is obtained by including all flexible deformation modes and the effects due to the leaf spring thickness variations, shuttle elasticity, the guidance stage and gravity. Figure 4 (right) shows the first natural frequency (in a mode that is a rotation about the rotation axis) as a function of the misalignment. With increasing misalignment, the rotation stiffness and the natural frequency decrease, and buckling (mechanism failure) is observed at around 0.11 mm misalignment. The numerical SPACAR results with the effects of warping match well with the measurements (the error in critical misalignment is 5.9%) and ANSYS (with solid elements). Unexpectedly, it also shows that the effects of warping play a significant role: with warping the critical misalignment is 35 μm (i.e. 40%) larger.

6 Conclusions

An analytical buckling analysis is presented for generalized overconstrained flexible cross-hinge mechanisms. The critical load matches with simulations from a geometrically nonlinear beam finite element. It is found that the inclusion of warping deformation of beams with narrow rectangular cross-section is crucial. The results are in good agreement with measurements from a demonstrator set-up.

References

1. Jonker, J.B.: Implementation of shear deformable thin-walled beam element for flexible multibody dynamics. In: Proceedings of ECCOMAS Thematic Conference on Multibody Dynamics, Prague, Czech Republic (2017)
2. Jonker, J.B., Meijaard, J.P.: SPACAR - Computer program for dynamic analysis of flexible spatial mechanisms and manipulators. In: Schiehlen, W. (ed.) Multibody Systems Handbook. Springer, Berlin (1990)
3. Meijaard, J.P., Brouwer, D.M., Jonker, J.B.: Analytical and experimental investigation of a parallel leaf spring guidance. Multibody Syst. Dyn. **23**(1), 77–97 (2010)
4. Nijenhuis, M., Meijaard, J.P., Brouwer, D.M.: Misalignments in an overconstrained flexure mechanism: A cross-hinge stiffness investigation. Precis. Eng. (2019), Accepted
5. Reissner, E.: On one-dimensional large-displacement finite-strain beam theory. Stud. Appl. Math. **52**(2), 87–95 (1973)

Robust and Fast Simulation of Flexible Flat Cables

Michael Roller[1(✉)], Christoffer Cromvik[2], and Joachim Linn[1]

[1] Fraunhofer ITWM, Fraunhofer-Platz 1, 67663 Kaiserslautern, Germany
{michael.roller,joachim.linn}@itwm.fraunhofer.de
[2] Franhofer-Chalmers Centre, Chalmers Science Park, 41288 Gothenburg, Sweden
christoffer.cromvik@fcc.chalmers.se

Abstract. In this work we will present a novel approach to compute the potential energy and its derivatives of a shell discretized by finite elements. Afterwards a special solution strategy for quasistatic equilibrium problems with moving boundary conditions is presented. At the end numerical examples are shown, which demonstrate the benefits of this methods in simulating flexible flat cables.

1 Introduction

In almost every modern electrical devices like cellphones or cameras, flexible flat cables (**FFC**) are used. They consist of multiple thin metallic conductors, which are placed on a flat thin plastic surface strip. These cables are highly flexible because of their thinness and therefore have several advantages in comparison to classical electronic cables with circular cross sections. Because **FFC**s are highly used in manufacturing of modern computer hardware and consumer electronics, there is an increasing demand of simulating the mounting of such cables in a very early stage of product development. Because a fast simulation model of such flat cables is needed, detailed three dimensional volume models are not practical. Beam models, which are frequently used for classical cables [1], fail to work for such structures, because they are not able to capture the cross section deformation. Therefore, a shell model is a suitable choice for modeling a flexible flat cable.

Because of the two dimensional resolution of the surface, the occurring cross section deformations can be handled. In the past a variety of different shell models [2,3] have been developed for different application scenarios [4,5], where usually finite element techniques are used for discretization.

Because of their high flexibility, **FFC**s are often twisted and bent when they are mounted in the device. Therefore, different buckling phenomena can occur for which standard mathematical solution approaches fail to work. Because we are working with a displacement driven simulation, also classical methods for buckling that control the force loading fail to work.

A. Kecskeméthy and F. Geu Flores (Eds.): ECCOMAS 2019, COMPUTMETHODS 53, pp. 207–214, 2020.
https://doi.org/10.1007/978-3-030-23132-3_25

2 Shell Kinematics

In this work we use the shell model from [2], where the shell continuum is described by its two dimensional midsurface $\mathbf{x} : \Omega \to \mathbb{R}^3$ and the so called director $\mathbf{d} : \Omega \to \mathscr{S}^2$, where $\Omega \subset \mathbb{R}^2$ is the parameter domain and $\mathscr{S}^2 := \{\mathbf{d} \in \mathbb{R}^3 | \|\mathbf{d}\| = 1\}$ is the unit sphere in three dimensional space. Both functions can be combined to a mapping from the parameter domain to six dimensional vectors $\mathbf{q} = (\mathbf{x}, \mathbf{d}) \in \mathbb{R}^6$. With the help of this function the metric of surface $a_{\alpha\beta} := \mathbf{x}_{,\alpha} \cdot \mathbf{x}_{,\beta}$, the curvature $\kappa_{\alpha\beta} = \frac{1}{2}(\mathbf{x}_{,\alpha} \cdot \mathbf{d}_{,\beta} + \mathbf{d}_{,\alpha} \cdot \mathbf{x}_{,\beta})$ and the transverse shear strain $\gamma_\alpha := \gamma_\alpha = \mathbf{x}_{,\alpha} \cdot \mathbf{d}$ can be computed. Here the common short hand notation $\mathbf{x}_{,\alpha} := \frac{\partial \mathbf{x}}{\partial \eta_\alpha}$ for the partial spatial derivatives is used. Small Greek index numbers represent the numbers $1, 2$. To measure the strain in the shell a stress and strain free reference configuration $\tilde{\mathbf{q}}$ is introduced. All quantities belonging to this configuration are indicated with a tilde $(\tilde{\cdot})$. In the reference configuration it is assumed that the director is normal to the midsurface $\tilde{\mathbf{x}}_{,\alpha} \cdot \tilde{\mathbf{d}} = 0$. With this information the common strain measures of a shear deformable shell can be introduced as follows

$$m_{\alpha\beta} = \frac{1}{2} a_{\alpha\beta} - \tilde{a}_{\alpha\beta} \qquad b_{\alpha\beta} = \frac{1}{2}\kappa_{\alpha\beta} - \tilde{\kappa}_{\alpha\beta} \qquad s_\alpha = \frac{1}{2}(\gamma_\alpha - \tilde{\gamma}_\alpha) \tag{1}$$

3 Shell Elastic Energy

The elastic energy of a shell continuum can be derived from the elastic energy of a general three dimensional continuum. With the help of the variable $\zeta \in [-\frac{h}{2}, \frac{h}{2}]$, where h is the thickness of the shell continuum, the three dimensional configuration of the deformed configuration is given by

$$\Phi(\eta, \zeta) = (\mathsf{P}_x + \zeta \mathsf{P}_d)\mathbf{q}(\eta) \qquad \eta \in \Omega\,, \tag{2}$$

where $\mathsf{P}_x \in \mathbb{R}^{3\times 6}$ is the projection of $\mathbf{q}$ to the midsurface $\mathbf{x} = \mathsf{P}_x\mathbf{q}$ and $\mathsf{P}_d \in \mathbb{R}^{3\times 6}$ to the director $\mathbf{d} = \mathsf{P}_d\mathbf{q}$. In the same way also the reference shell continuum $\tilde{\Phi}$ can be parametrized. As shown in [4] the components of the Green-Lagrangain strain tensor $\mathsf{E} = \frac{1}{2}(\nabla_{\tilde{\Phi}}\Phi^T \nabla_{\tilde{\Phi}}\Phi - \mathbb{1})$ in the shell coordinate system are given by $E_{\alpha\beta} = m_{\alpha\beta} + \zeta b_{\alpha\beta} + \mathscr{O}(\zeta^2)$, $E_{\alpha 3} = s_\alpha$ and $E_{33} = 0$. As usual the higher order terms in $E_{\alpha\beta}$ are neglected. The work conjugated stress measure is the second Piola-Kirchhoff tensor S. By assuming plain stress $S^{33} = 0$ and a linear material model, which is given in the shell coordinate system by the tensor C^{ijkl}, the components of the stress tensor in the shell coordinate system can be written as $S^{\alpha\beta} = C^{\alpha\beta\iota\pi}E_{\iota\pi}$ and $S^{\alpha 3} = C^{\alpha 3\beta 3}s_\beta$. With these ingredients the elastic energy of the shell continuum with a linear material model can be written as

$$\begin{aligned} \mathscr{E}(\mathbf{q}) &= \tfrac{1}{2}\int_{\tilde{\mathfrak{B}}} \mathsf{E} : \mathsf{S}\,\mathrm{d}\bar{\mathbf{V}} \\ &= \tfrac{1}{2}\int_{\Omega} \left(m_{\alpha\beta}M^{\alpha\beta\iota\pi}m_{\iota\pi} + b_{\alpha\beta}B^{\alpha\beta\iota\pi}b_{\iota\pi} + s_\alpha S^{\alpha\beta}s_\beta\right)\mathrm{d}\bar{\eta}\,, \end{aligned} \tag{3}$$

where $\tilde{\mathfrak{B}}$ is reference shell continuum and $\mathrm{d}\bar{\eta} := \det(\tilde{a}_{\alpha\beta})\mathrm{d}\eta$ is the midsurface area measure. The material tensor splits up into three parts, one for the membrane part $M^{\alpha\beta\iota\pi} = hC^{\alpha\beta\iota\pi}$, one for the bending part $B^{\alpha\beta\iota\pi} = \frac{h^3}{12}C^{\alpha\beta\iota\pi}$ and one for the shearing part $S^{\alpha\beta} = hC^{\alpha 3\iota 3}$. All of these tensors are fully symmetric. By the derivation of (3) and the corresponding parameters some approximation has been made, which can be found in [4]. Note that all of these tensors depend on the stress free reference configuration $\tilde{\mathbf{q}}$, which is assumed to be constant. We now introduce three quad-linear functions by

$$I_m(\mathbf{q},\mathbf{p},\mathbf{u},\mathbf{v}) := \int_\Omega \mathbf{q}_{,\alpha} \cdot \mathsf{P}_m \mathbf{p}_{,\beta} M^{\alpha\beta\iota\pi} \mathbf{u}_{,\iota} \cdot \mathsf{P}_m \mathbf{v}_{,\pi} \mathrm{d}\bar{\eta} \quad \mathsf{P}_m := \mathsf{P}_x^T \mathsf{P}_x \tag{4}$$

$$I_b(\mathbf{q},\mathbf{p},\mathbf{u},\mathbf{v}) := \int_\Omega \mathbf{q}_{,\alpha} \cdot \mathsf{P}_b \mathbf{p}_{,\beta} B^{\alpha\beta\iota\pi} \mathbf{u}_{,\iota} \cdot \mathsf{P}_b \mathbf{v}_{,\pi} \mathrm{d}\bar{\eta} \quad \mathsf{P}_b := \tfrac{1}{2}(\mathsf{P}_x^T \mathsf{P}_d + \mathsf{P}_d^T \mathsf{P}_x) \tag{5}$$

$$I_s(\mathbf{q},\mathbf{p},\mathbf{u},\mathbf{v}) := \int_\Omega \mathbf{q}_{,\alpha} \cdot \mathsf{P}_s \mathbf{p} S^{\alpha\iota} \mathbf{u}_{,\iota} \cdot \mathsf{P}_s \mathbf{v} \mathrm{d}\bar{\eta} \quad \mathsf{P}_s := \mathsf{P}_x^T \mathsf{P}_d \,. \tag{6}$$

With the help of (4) the membrane energy of the shell can be computed by

$$\mathscr{E}_m(\mathbf{q}) = \tfrac{1}{8}\left(I_m(\mathbf{q},\mathbf{q},\mathbf{q},\mathbf{q}) - 2I_m(\mathbf{q},\mathbf{q},\tilde{\mathbf{q}},\tilde{\mathbf{q}}) + I_m(\tilde{\mathbf{q}},\tilde{\mathbf{q}},\tilde{\mathbf{q}},\tilde{\mathbf{q}})\right). \tag{7}$$

In the same way the bending energy $\mathscr{E}_b(\mathbf{q})$ and the shearing energy $\mathscr{E}_s(\mathbf{q})$ can be computed with (5) respectively (6). Altogether the elastic energy reads

$$\mathscr{E}(\mathbf{q}) = \mathscr{E}_m(\mathbf{q}) + \mathscr{E}_b(\mathbf{q}) + \mathscr{E}_s(\mathbf{q}) \,. \tag{8}$$

4 Shell Discretization

To discretize the continuous functions $\mathbf{q}$ and $\tilde{\mathbf{q}}$ on Ω we use the isoparametric concept. Let $\mathrm{N}^I : \Omega \to \mathbb{R}$ with $i = 1, \ldots, N$ be a suitable set of ansatz functions then the discrete versions of the configurations are given by $\mathbf{q}^d(\eta) = \mathbf{q}_I \mathrm{N}^I(\eta)$ and $\tilde{\mathbf{q}}^d(\eta) = \tilde{\mathbf{q}}_I \mathrm{N}^I(\eta)$, where $\mathbf{q}_I, \tilde{\mathbf{q}}_I \in \mathbb{R}^6$ for $i = 1, \ldots, N$. We use Einstein sum convention here to shorten the notation, where indeces which are occurring twice are contracted as sum. By using the same discretization for the arbitrary functions in (4)–(6), where the super scripted d indicates that the functions are discretized, the corresponding quad-linear forms read

$$I_m(\mathbf{q}^d,\mathbf{p}^d,\mathbf{u}^d,\mathbf{v}^d) = \mathbf{q}_I \cdot \mathsf{P}_m \mathbf{p}_J \underbrace{\int_\Omega \mathrm{N}^I_{,\alpha} \mathrm{N}^J_{,\beta} \mathrm{N}^K_{,\iota} \mathrm{N}^L_{,\pi} M^{\alpha\beta\iota\pi} \mathrm{d}\bar{\eta}}_{=:\mathrm{M}^{IJKL}} \mathbf{u}_K \cdot \mathsf{P}_m \mathbf{v}_L \tag{9}$$

$$I_b(\mathbf{q}^d,\mathbf{p}^d,\mathbf{u}^d,\mathbf{v}^d) = \mathbf{q}_I \cdot \mathsf{P}_b \mathbf{p}_J \underbrace{\int_\Omega \mathrm{N}^I_{,\alpha} \mathrm{N}^J_{,\beta} \mathrm{N}^K_{,\iota} \mathrm{N}^L_{,\pi} B^{\alpha\beta\iota\pi} \mathrm{d}\bar{\eta}}_{=:\mathrm{B}^{IJKL}} \mathbf{u}_K \cdot \mathsf{P}_b \mathbf{v}_L \tag{10}$$

$$I_s(\mathbf{q}^d,\mathbf{p}^d,\mathbf{u}^d,\mathbf{v}^d) = \mathbf{q}_I \cdot \mathsf{P}_s \mathbf{p}_J \underbrace{\int_\Omega \mathrm{N}^I_{,\alpha} \mathrm{N}^J \mathrm{N}^K_{,\iota} \mathrm{N}^L S^{\alpha\iota} \mathrm{d}\bar{\eta}}_{=:\mathrm{S}^{IJKL}} \mathbf{u}_K \cdot \mathsf{P}_s \mathbf{v}_L \,. \tag{11}$$

So the integration over the domain Ω is shifted to the functions N^I and implicitly to the constant discrete reference configuration[1], while the nodal values $\mathbf{q}_I$ are out of the integral. This is the main difference to classical FE assembly methods, where the nodal values stay inside the integral.

With the help of (9)–(11) the discrete elastic energy is given by

$$\mathscr{E}(\mathbf{q}^d) = \tfrac{1}{2}(\mathtt{m}_{IJ}\mathtt{M}^{IJKL}\mathtt{m}_{KL} + \mathtt{b}_{IJ}\mathtt{B}^{IJKL}\mathtt{b}_{KL} + \mathtt{s}_{IJ}\mathtt{S}^{IJKL}\mathtt{s}_{KL})\,. \tag{12}$$

with the following abbreviations

$$\mathtt{m}_{IJ} = \tfrac{1}{2}\left(\mathbf{q}_I \cdot \mathsf{P}_m\mathbf{q}_J - \tilde{\mathbf{q}}_I \cdot \mathsf{P}_m\tilde{\mathbf{q}}_J\right) \qquad \mathtt{b}_{IJ} = \tfrac{1}{2}\left(\mathbf{q}_I \cdot \mathsf{P}_b\mathbf{q}_J - \tilde{\mathbf{q}}_I \cdot \mathsf{P}_b\tilde{\mathbf{q}}_J\right) \tag{13}$$

$$\mathtt{s}_{IJ} = \tfrac{1}{2}\left(\mathbf{q}_I \cdot \mathsf{P}_s\mathbf{q}_J - \tilde{\mathbf{q}}_I \cdot \mathsf{P}_s\tilde{\mathbf{q}}_J\right)\,. \tag{14}$$

With this approach we are able to compute the elastic energy of the discrete shell by evaluation of the scalar products in (13)–(14) and summation as in (12). The usual assembly process over the elements is only used in pre processing to compute the entries of the tensors $\mathtt{M}^{IJKL}$, $\mathtt{B}^{IJKL}$ and $\mathtt{S}^{IJKL}$.

Note that the tensor $\mathtt{M}^{IJKL}$ is by definition a fully symmetric fourth order tensor, which means that the following relations hold $\mathtt{M}^{IJKL} = \mathtt{M}^{KLIJ}$ and $\mathtt{M}^{IJKL} = \mathtt{M}^{IJLK} = \mathtt{M}^{JIKL}$ hold. The same is true for the bending part $\mathtt{B}^{IJKL}$. The fourth order tensor representing the shear part has only the major symmetry $\mathtt{S}^{IJKL} = \mathtt{S}^{KLIJ}$, while the minor symmetries are not fulfilled in general. Also the discrete *"strain measures"* for membrane stretching and bending are symmetric second order tensors $\mathtt{m}_{IJ} = \mathtt{m}_{JI}$ $\mathtt{b}_{IJ} = \mathtt{b}_{JI}$. Because $\mathsf{P}_s \neq \mathsf{P}_s^T$ this does not hold for the shear tensor $\mathtt{s}_{IJ} \neq \mathtt{s}_{JI}$.

In this work we use quadrilateral piecewise bilinear finite elements as ansatz functions. To avoid shear locking an assumed strain approach is used as in [4]. Different from standard FE procedures no additional assembly is required.

5 Solving the Discrete Problem

The discrete configuration of the deformed shell continuum is completely determined by the vector $\mathbf{Q} = (\mathbf{q}_1, \ldots, \mathbf{q}_N) \in \mathbb{R}^{6N}$, which contains the nodal degrees of freedom of the mesh, where N is the number of nodes. To compute a static equilibrium configuration of the shell the discrete potential energy $\mathscr{E}^d(\mathbf{Q}) := \mathscr{E}(\mathbf{q}^d)$ is minimized. The unit length of the director is enforced in every node of the mesh as first described in [6], which leads to N additional side constraints. Additionally also the boundary conditions are handled as simple linear equality constraints on the specific nodes. This leads to the following discrete nonlinear constrained optimization problem

$$\min\ \mathscr{E}^h(\mathbf{Q}) \tag{15}$$

$$\text{s.t.}\quad \tfrac{1}{2}\left(\mathsf{P}_d\mathbf{q}_I \cdot \mathsf{P}_d\mathbf{q}_I - 1\right) = 0 \qquad I = 1, \ldots, N \tag{16}$$

$$\mathbf{q}_K = \bar{\mathbf{q}}_K \qquad K \in \{K \in \{1, \ldots, N\} | \mathbf{q}_K \text{ is boundary node}\}\,, \tag{17}$$

[1] This fact is hidden in the notation. Because as mentioned above the tensors $M^{\alpha\beta\iota\pi}$, $B^{\alpha\beta\iota\pi}$ and $S^{\alpha\beta}$ depend on $\tilde{q}^d$ as well as the midsurface measure $\mathrm{d}\bar{\eta}$.

where (16) is the unit length constraint and (17) represents the boundary conditions.

To solve the problem (15)–(17) usually the gradient and Hessian matrix of (15) have to be evaluated. The evaluation of these quantities can be derived from (12). The gradient of the elastic energy in direction of $\mathbf{q}_I$ is given by

$$\begin{aligned}\frac{\partial \mathscr{E}^d}{\partial \mathbf{q}_I} &= \mathsf{P}_m \mathbf{q}_J \mathtt{M}^{IJKL} \mathtt{m}_{KL} + \mathsf{P}_b \mathbf{q}_J \mathtt{B}^{IJKL} \mathtt{b}_{KL} \\ &\quad + \tfrac{1}{2}(\mathsf{P}_s \mathbf{q}_J \mathtt{S}^{IJKL} \mathtt{s}_{KL} + \mathsf{P}_s \mathbf{q}_J \mathtt{S}^{JIKL} \mathtt{s}_{KL}) \,.\end{aligned} \tag{18}$$

The Hessian matrix of (15) is also evaluated straight forward by

$$\begin{aligned}\frac{\partial^2 \mathscr{E}^d}{\partial \mathbf{q}_I \partial \mathbf{q}_J} =& 2\mathsf{P}_m \mathbf{q}_K \otimes \mathsf{P}_m \mathbf{q}_L \mathtt{M}^{IKLJ} + \mathsf{P}_m \mathtt{M}^{IJKL} \mathtt{m}_K L \\ &+ 2\mathsf{P}_b \mathbf{q}_K \otimes \mathsf{P}_b \mathbf{q}_L \mathtt{B}^{IKLJ} + \mathsf{P}_b \mathtt{B}^{IJKL} \mathtt{b}_K L \\ &+ \tfrac{1}{2}(\mathsf{P}_s \mathtt{S}^{IJKL} \mathtt{s}_{KL} + \mathsf{P}_s \mathtt{S}^{JIKL} \mathtt{s}_{KL} \\ &+ 2\mathsf{P}_s \mathbf{q}_K \otimes \mathsf{P}_s \mathbf{q}_l \mathtt{S}^{IKLJ} + 2\mathsf{P}_s \mathbf{q}_L \otimes \mathsf{P}_s \mathbf{q}_K \mathtt{S}^{JLKI}) \,.\end{aligned} \tag{19}$$

The equations (18)–(19) look quite complicated but can be computed very fast, by using the sparse structure of the tensors and their symmetries.

To solve the constrained optimization problem (15)–(17), the sequential quadratic program solver from [7] is used. This yields the advantage that no parametrization of the rotational degrees of freedom of the director is needed, because the unit length constraint is directly assured by the solver. This is in contrast to the methodology used in [4,6], where the necessary condition of (15)–(17) are solved with the help of a null space approach and a local rotational parametrization of the director.

6 Numerical Results and Discussion

In this section we will discuss the numerical behavior of the presented methods briefly. First we investigate the computational time of the evaluation method for the discretized shell presented in this work. Afterward we show an application case of the special procedure to solve an static equilibrium problem with moving boundary conditions.

6.1 Computational Time Comparison to Classical FE

We will compare the computation time of the method presented in this work, which is called tensor model from now on, with the classical assembly based FE approach from [4], which is called FE model from now on. As benchmark example a quadratic plate was discretized with a increasing number of equidistant quadratic finite elements. In Fig. 1 the computational time is plotted against the number of elements for the computation of the gradient of the discrete energy

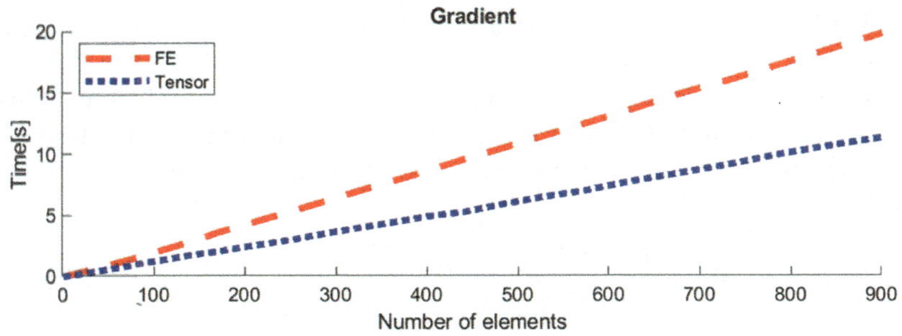

Fig. 1. Comparison of the computation time of classical FE approach to the tensor approach with application to the gradient of the energy function.

function (18) evaluated 500 times for random numbers. It can be seen that the tensor model is at least a factor of two faster than the FE model independent of the number of elements.

For the computation of the energy itself (12) and the Hessian (19) the evaluation time is essentially the same for the tensor and the FE methodology. This is a bit surprising, because for the evaluation concept of the tensor method theoretically less computations are needed. The advantage of the classical FE assembly methodology could be that the computation is always performed on a compact data set (e.g. the elements). In contrast to that in the tensor methods we have to do big jumps in storage, which could slow down the process. But this problem could perhaps be handled by a smart rearrangement of the data.

It has to be mentioned that the largest amount of computation time in the tensor model is used to evaluate the minor asymmetric shearing part, because of $\mathtt{m}^{IJ} \neq \mathtt{m}^{IJ}$ and $\mathtt{S}^{IJKL} \neq \mathtt{S}^{IJLK}$. This can be avoided by an symmetrization of the whole evaluation process, which breaks down to a system matrix A^{IJKL}, where the indices run from $1, .., 6N$.

6.2 Twisting of a Flexible Flat Cable

As benchmark example we use a rectangular flat rubber stripe with a length of $L := 300\,\text{mm}$, a width of $10\,\text{mm}$ and a thickness of $1\,\text{mm}$. The mechanical properties of the cable are a Young's modulus of $5MP$, a Poisson ratio of 0.3 and a density of $1000\,\text{kg/m}^3$. One end of the cable is fixed while the other is rotated twice and then pushed towards the other end.

In Fig. 2 the simulation results are shown. In the first row the twisting process is visualized, while in the second row the cable ends are moved towards each other up to a distance of $L/2$. In the last row the cable ends are pushed towards each other until they meet. It could be observed that the torsion of the cable has completely vanished in comparison to the row above. The transition between torsion and bending can be observed between the last image of the second row and the first of the third row. Between those pictures there is only one

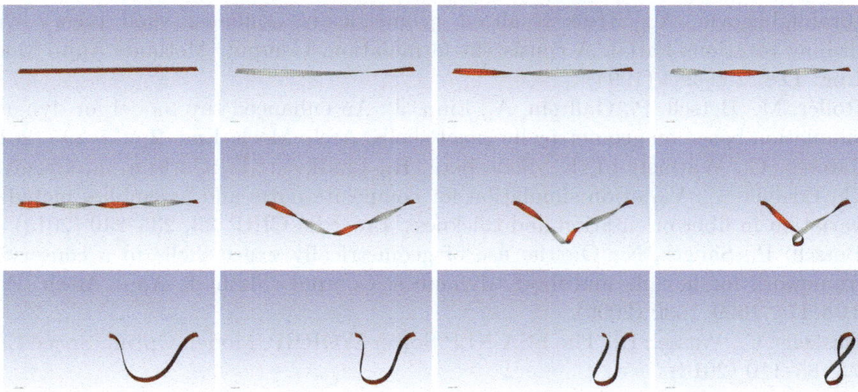

Fig. 2. Simulation results of a twice twisted and the fully compressed flat rubber cable.

simulation step. It can by seen by the big deformation between both configurations, that this is the crucial phase of the whole simulation process. This is also the point where the classical unconstrained Newton solvers fail to work, because they are not robust in such situations. Also the WORHP algorithm needs a lot of iteration (> 100), but works out of the box.

7 Conclusion

In this work we present an alternative strategy to evaluate the discrete stored energy function and its derivatives for a quadrilateral finite element shell model with linear material. This approach results in a speed up in computation time for calculating the first derivative of the stored energy. To solve a sequence of quasi-static equilibrium configurations with moving boundary conditions, we transferred the problem to a constrained optimization problem and use the solver from [7] to solve the problem. This approach seems to be more stable than the classical procedure utilizing rotational degrees of freedom and an unconstrained Newton type solver [4]. We demonstrate the benefits of our novel approach by the non-trivial example of an flexible flat cable, witch is twisted and than both ends a pushed towards each other.

References

1. Dreßler, K., Linn, J.: Discrete cosserat rod models based on the difference geometry of framed curves for interactive simulation of flexible cables. European Consortium for Mathematics in Industry, -ECMI-: Math for the Digital Factory, pp. 289–319 (2017)
2. Simo, J., Fox, D.: On a stress resultant geometrically exact shell model. Part I: Formulation and optimal parametrization. Comput. Methods Appl. Mech. Eng. **72.3**, 267–304 (1989)

3. Ibrahimbegovic, A.: Stress resultant geometrically nonlinear shell theory with drilling rotations-Part I. A consistent formulation. Comput. Methods Appl. Mech. Eng. **118**, 265–284 (1994)
4. Roller, M., Betsch, P., Gallrein, A., Linn, J.: An enhanced tire model for dynamic simulation based on geometrically exact shells. Arch. Mech. Eng. **2**, 277–295 (2016)
5. Jareteg, C., Wärmefjord, K., Söderberg, R., Lindkvist, L., Carlson, J., Cromvik, C., Edelvik, F.: Variation simulation for composite parts and assemblies including variation in fiber orientation and thickness. Procedia CIRP **23**, 235–240 (2014)
6. Betsch, P., Sänger, N.: On the use of geometrically exact shells in a conserving framework for flexible multibody dynamics. Comput. Methods Appl. Mech. Eng. **198**(17), 1609–1630 (2009)
7. Büskens, C., Wassel, D.: The ESA NLP Solver WORHP. Model. Optim. Space Eng. **73**, 85–110 (2013)

Dynamic Analysis of Compliant Mechanisms Using Absolute Nodal Coordinate Formulation and Geometrically Exact Beam Theory

Zhigang Zhang(✉), Xiang Zhou, and Zhanpeng Fang

Henan Key Laboratory of Intelligent Manufacturing of Mechanical Equipment, Zhengzhou University of Light Industry Zhengzhou, Henan, China
zhigangzhang@foxmail.com, m15358715169@163.com, 2015073@zzuli.edu.cn

Abstract. Compliant mechanism mainly relies on large deformations of compliant rods to transmit motions, forces and energy. The geometric nonlinearity of the compliant rods is one of the most serious challenges when accurate modeling and dynamic simulation are performed. The absolute nodal coordinate formulation (ANCF) and geometrically exact beam theory (GEBT) are employed to investigate the nonlinear modeling and analysis of compliant mechanisms in this paper. By taking account of deformation characteristics of the compliant rod at the external connection, a new ANCF beam element with one nodal deformation constraint is proposed. Based on the locking alleviation technique, strain split method, the effect of the locking phenomenon of the ANCF beam element on the dynamic simulation of compliant mechanism is investigated. In comparison, the Euler-Bernoulli beam element, which is a kind of locking free GEBT element, is also used for the dynamic analysis of the compliant mechanism. Finally, the numerical example of a partial compliant four-bar mechanism is presented to illustrate the accuracy and effectiveness of ANCF and GEBT beam elements for the dynamic problems of compliant mechanisms.

Keywords: Compliant mechanism · Large deformation · ANCF · Geometrically exact beam theory · Dynamic analysis

1 Introduction

The compliant mechanism is designed to use the large elastic deformation of flexible components instead of traditional joints to accomplish the transmission of motion, force and energy, and has many potential advantages such as part-count reduction, reduced assembly time and reduced wear, friction and noise [1]. One of the main challenges in the design and analysis of compliant mechanisms is dueling with large deformation of the compliant rods.

In order to simplify the modeling of the compliant rods, the Pseudo-rigid-body (PRB) model is proposed and widely adopted in the research area of compliant mechanisms [2–6]. However, because of the low accuracy of the PRB model, it is not suitable for the dynamic and fatigue analysis.

A. Kecskeméthy and F. Geu Flores (Eds.): ECCOMAS 2019, COMPUTMETHODS 53, pp. 215–222, 2020.
https://doi.org/10.1007/978-3-030-23132-3_26

It should be pointed out that the compliant mechanism which is accompanied by large deformation of flexible rods belongs to a kind of large deformation flexible multibody systems. Therefore, the modeling methods developed in the community of flexible multibody system dynamics for large deformation beams should be available for modeling and analysis of compliant mechanisms. Among these important modeling methods, absolute nodal coordinate formulation (ANCF) [7] and geometrically exact beam theory (GEBT) [8, 9] are the most used in flexible multibody dynamics. GEBT elements preserve the rigid cross-section assumption, employ global position vector and rotation vector as nodal coordinates, and allow formulating problems involving arbitrarily large displacements and rotations by using the new defined objective strain [10]. On the contrast, ANCF beam elements drop the rigid cross-section assumption, employ the global position and position gradient vectors as nodal coordinates, define a unique displacement field, and lead to a constant mass matrix and zero centrifugal and Coriolis inertia forces. By using general continuum-mechanics approaches, ANCF beam elements impose no restrictions on the amount of rotations or deformation except for the restrictions imposed by the element interpolation [11, 12].

This research focuses on the modeling of compliant mechanisms by using the ANCF beam element [13, 14] and GEBT beam element [15], respectively. Based on strain split method (SSM) [16], the locking effects of ANCF beam elements on dynamics of the compliant mechanisms are also investigated.

2 3D ANCF Beam Element

ANCF was firstly proposed by Shabana to model large deformation beams with arbitrary overall motion, and has been widely accepted as an important progress in the development of flexible multibody system dynamics. The compliant rod undergoing large deformation can be meshed by using ANCF beam elements. The position vector of an arbitrary point on the 3D ANCF shear beam element can be written as $\boldsymbol{r}(\boldsymbol{x},t) = \boldsymbol{S}(\boldsymbol{x})\boldsymbol{e}(t)$, where $\boldsymbol{x} = [x\ y\ z]^T$ is a set of material coordinates, t is time, $\boldsymbol{e} = [\boldsymbol{e}_1^T \quad \boldsymbol{e}_2^T]^T$ is the vector of element nodal coordinates which consists of the position and gradient coordinates $\boldsymbol{r}^i$, $\boldsymbol{r}_x^i$, $\boldsymbol{r}_y^i$ and $\boldsymbol{r}_z^i$ $i = 1, 2$, and $\boldsymbol{S}$ is the shape function matrix [13, 14].

The elastic forces of the ANCF element can be formulated in terms of the general continuum mechanics (GCM). Using the ANCF beam kinematic equations, the Green-Lagrange strain sensor $\boldsymbol{\varepsilon}$ can be written as $\boldsymbol{\varepsilon} = (1/2)(\boldsymbol{J}^T\boldsymbol{J} - \boldsymbol{I})$, where the matrix of position vector gradients $\boldsymbol{J} = \partial \boldsymbol{r}/\partial \boldsymbol{x} = [\boldsymbol{r}_x \quad \boldsymbol{r}_y \quad \boldsymbol{r}_z] = [\boldsymbol{S}_x\boldsymbol{e} \quad \boldsymbol{S}_y\boldsymbol{e} \quad \boldsymbol{S}_z\boldsymbol{e}]$. It is clear that the Green-Lagrange strain tensor $\boldsymbol{\varepsilon}$ is symmetric, and can be written in Voigt form as $\boldsymbol{\varepsilon}_v = [\varepsilon_{11} \quad \varepsilon_{22} \quad \varepsilon_{33} \quad 2\varepsilon_{12} \quad 2\varepsilon_{13} \quad 2\varepsilon_{23}]^T$. If the linear elastic material is employed, the Voigt form of the second Piola-Kirchhoff stress can be written in as $\boldsymbol{\sigma}_v = \boldsymbol{E}\boldsymbol{\varepsilon}_v$, where $\boldsymbol{E}$ is the matrix of elastic coefficients.

The virtual power of the element elastic forces can be written as $\delta p_s = \int_V \boldsymbol{\sigma}_v^T \delta\dot{\boldsymbol{\varepsilon}}_v dV = \left(\int_V (\partial\boldsymbol{\varepsilon}_v/\partial\boldsymbol{e})^T \boldsymbol{E}\boldsymbol{\varepsilon}_v dV\right)^T \delta\dot{\boldsymbol{e}} = \boldsymbol{Q}_s^T \delta\dot{\boldsymbol{e}}$, where V is the element volume, and $\boldsymbol{Q}_s$ is the nonlinear vector of the element generalized elastic forces which can be expressed as $\boldsymbol{Q}_s = \int_V (\partial\boldsymbol{\varepsilon}_v/\partial\boldsymbol{e})^T \boldsymbol{E}\boldsymbol{\varepsilon}_v dV$.

The element equations of motion can be formulated using the principle of virtual power. Substituting the virtual powers of inertia forces, elastic forces and external forces into the equation of virtual power, one can obtain

$$(\boldsymbol{M}\ddot{\boldsymbol{e}}+\boldsymbol{F})^T\delta\dot{\boldsymbol{e}}=0 \tag{1}$$

where $\boldsymbol{M}=\int_V \rho\boldsymbol{S}^T\boldsymbol{S}dV$ is the constant, symmetric, and positive definite mass matrix, $\boldsymbol{F}=\boldsymbol{Q}_s-\boldsymbol{Q}_e$ is the vector of nodal forces including the elastic force $\boldsymbol{Q}_s$ and external forces $\boldsymbol{Q}_e$. According to the independence of the vector of element coordinates, the element equations of motion can be written as $\boldsymbol{M}\ddot{\boldsymbol{e}}+\boldsymbol{F}=\boldsymbol{0}$.

3 New ANCF Beam Element Including One Boundary Node

In general, the columns of the matrix of the position vector gradients $\boldsymbol{J}=[\boldsymbol{r}_x \quad \boldsymbol{r}_y \quad \boldsymbol{r}_z]$ are not orthogonal unit vectors. For compliant rods which are connected with other parts by joints or fixed hinge, the local deformation at the end nodes will be constrained. The local deformation of this nodal point will be constrained because of the stiffness enhancement of the joint connection. Therefore, the effects of stretch and shear deformations of the boundary node can be neglected, and the nodal gradient vectors will become dependent and can be written as $[\boldsymbol{r}_x^i \quad \boldsymbol{r}_y^i \quad \boldsymbol{r}_z^i]=\boldsymbol{A}^i(\boldsymbol{\theta}^i)$, where $\boldsymbol{A}^i(\boldsymbol{\theta}^i)$ is an orthogonal matrix that defines the orientation of the nodal coordinate system in terms of a selected rotation parameters $\boldsymbol{\theta}^i$. Thus, a new set of independent nodal coordinate parameters can be chosen as $\boldsymbol{p}_i=[\boldsymbol{r}^{iT} \quad \boldsymbol{\theta}^{iT}]^T$.

The independence vector of element nodal coordinates of this new ANCF beam element with one boundary node can be defined as $\underline{\boldsymbol{e}}=[\boldsymbol{p}^{1T} \quad \boldsymbol{e}^{2T}]^T$, and the time derivatives of the dependent vector of element nodal coordinates $\boldsymbol{e}=[\boldsymbol{e}^{1T} \quad \boldsymbol{e}^{2T}]^T$ can be rewritten in terms of the time derivatives of the new vector of element parameters as $\dot{\boldsymbol{e}}=\boldsymbol{B}\dot{\underline{\boldsymbol{e}}}$, where $\boldsymbol{B}$ is the element velocity transformation matrix. Substituting the element velocity transformation $\dot{\boldsymbol{e}}=\boldsymbol{B}\dot{\underline{\boldsymbol{e}}}$ into the ANCF beam element equation of virtual power (1), one can obtain equations of motion of this new ANCF beam element as

$$\underline{\boldsymbol{M}}\ddot{\underline{\boldsymbol{e}}}+\underline{\boldsymbol{F}}=0 \tag{2}$$

where $\underline{\boldsymbol{M}}=\boldsymbol{B}^T\boldsymbol{M}\boldsymbol{B}$ is the element mass matrix, and $\underline{\boldsymbol{F}}=\boldsymbol{B}^T\boldsymbol{M}\dot{\boldsymbol{B}}\dot{\underline{\boldsymbol{e}}}+\boldsymbol{B}^T\boldsymbol{F}$ is the generalized element forces.

4 Locking Alleviation: Strain Split Method

As in the case of conventional elements, ANCF elements are not locking free. In this investigation, the effect of locking for the dynamic simulation of compliant mechanisms is discussed based on the new proposed locking alleviation technique, SSM approach [16]. In terms of the analyzing of the element kinematics, the basic SSM idea is to separate higher-order terms to define new Green-Lagrange strain tensors and matrices of elastic coefficients that allow alleviating the locking. In the case of 3D ANCF beam element, the position field can be expressed as $\boldsymbol{r} = \boldsymbol{r}^c + y\boldsymbol{r}_y + z\boldsymbol{r}_z$. Accordingly, the matrix of position vector gradients can be written as $\boldsymbol{J} = \boldsymbol{J}^c + \boldsymbol{J}^k$, where $\boldsymbol{J}^c = [\boldsymbol{r}_x^c \quad \boldsymbol{r}_y \quad \boldsymbol{r}_z]$ and $\boldsymbol{J}^k = [y\boldsymbol{r}_{yx} + z\boldsymbol{r}_{zx} \quad \boldsymbol{0} \quad \boldsymbol{0}]$.

Using this split of the matrix of position vector gradients, the Green–Lagrange strain tensor can be written as $\boldsymbol{\varepsilon} = (1/2)(\boldsymbol{J}^T\boldsymbol{J} - \boldsymbol{I}) = \boldsymbol{\varepsilon}^c + \boldsymbol{\varepsilon}^k$, where $\boldsymbol{\varepsilon}^c = (1/2)(\boldsymbol{J}^{cT}\boldsymbol{J}^c - \boldsymbol{I})$ is associated with the beam centerline and $\boldsymbol{\varepsilon}^c = (1/2)(\boldsymbol{J}^{cT}\boldsymbol{J}^k + \boldsymbol{J}^{kT}\boldsymbol{J}^c + \boldsymbol{J}^{kT}\boldsymbol{J}^k)$ is associated with the higher-order terms that contribute to cross-sectional deformation, bending, and curvature. The second Piola-Kirchhoff stresses can be written in Voigt form as $\boldsymbol{\sigma}_v = \boldsymbol{E}^c\boldsymbol{\varepsilon}_v^c + \boldsymbol{E}^k\boldsymbol{\varepsilon}_v^k$, where $\boldsymbol{E}^c$ and $\boldsymbol{E}^k$ are the new defined matrices of elastic coefficients [16].

5 Euler-Bernoulli Beam Element Base on GEBT

It is different from ANCF beam that the rigid cross-section assumption is employed in the development of GEBT beam element [8, 9]. As comparison, a GEBT Euler-Bernoulli beam element with the centroid position and the cross-section orientation coupled interpolated by a special approach [15] is employed to mesh the compliant rods.

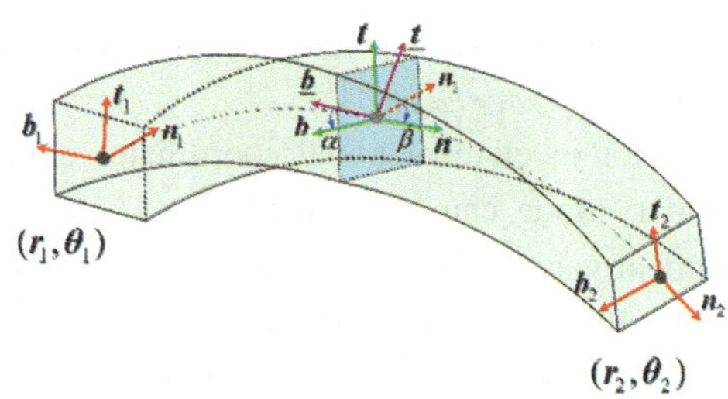

Fig. 1. GEBT Euler-Bernoulli beam element

The nodal displacement vector $\boldsymbol{r}_i$ and rotation vector $\boldsymbol{\theta}_i$ are selected as the nodal coordinate parameters as shown in Fig. 1. The configuration of the beam element is described by the beam centerline and the rotation of the beam cross-section. The global position vector of beam centerline $\boldsymbol{r}$ can be approximated by the following Hermite interpolation [15]

$$\boldsymbol{r}(s) = N_1^0\boldsymbol{r}_1 + N_1^1\boldsymbol{n}_1 + N_2^0\boldsymbol{r}_2 + N_2^1\boldsymbol{n}_2 \tag{3}$$

where $N_i^0, N_i^1, i = 1, 2$ are the basis function of Hermite interpolation.

In terms of the hypothesis of Euler-Bernoulli beam, the unit normal vector of the cross-section $\boldsymbol{n}$ can be obtained using the interpolation of the beam centerline as $\boldsymbol{n} = \boldsymbol{r}'/|\boldsymbol{r}'|$, where $\boldsymbol{r}' = \partial\boldsymbol{r}/\partial s$. The other two basis vectors $\boldsymbol{t}$ and $\boldsymbol{b}$ can be determined by the successive application of two rotations $\boldsymbol{\alpha}$ and $\boldsymbol{\beta}$ respected to the left cross-section basis $(\boldsymbol{n}_1, \boldsymbol{t}_1, \boldsymbol{b}_1)$ as shown in Fig. 1. Thus, the rotaion matrix of the cross-section can be wirtten as $\boldsymbol{R}(\boldsymbol{\theta}) = \boldsymbol{R}(\boldsymbol{\alpha})\boldsymbol{R}(\boldsymbol{\beta})\boldsymbol{R}_1$ [15]. According to GEBT, two strian vectors defined in the frame of cross-section are $\bar{\gamma} = \boldsymbol{R}^T(\boldsymbol{r}' - \boldsymbol{n})$ and $\bar{\boldsymbol{\kappa}} = \boldsymbol{R}^T\boldsymbol{T}(\boldsymbol{\theta})\boldsymbol{\theta}'$, and one can write the virtual power of the elastic forces as $\delta p_s = \int_0^L (\delta\dot{\bar{\gamma}})^{\mathrm{T}}\boldsymbol{C}_\gamma\bar{\gamma} + (\delta\dot{\bar{\boldsymbol{\kappa}}})^{\mathrm{T}}\boldsymbol{C}_{\boldsymbol{\kappa}}\bar{\boldsymbol{\kappa}}\, ds$, where the material elasticity matrixes can be written as $\boldsymbol{C}_\gamma = diag(EA, GA_t, GA_b)$ and $\boldsymbol{C}_\kappa = diag(GJ, EI_t, EI_b)$.

6 Numerical Example: Partially Compliant Four-Bar Mechanism

In order to demonstrate the effect of ANCF beam element and GEBT beam element on the performance of modeling compliant rods, a numerical example of a compliant mechanism is considered in this section.

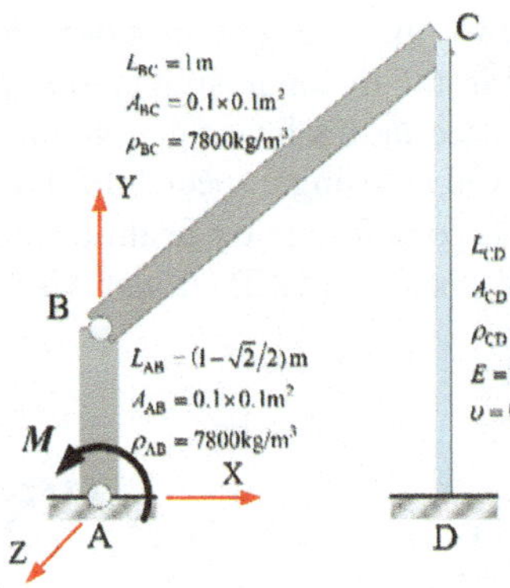

Fig. 2. Partially compliant four-bar mechanism

A partially compliant four-bar mechanism consists of two rigid rods, AB and BC, and one compliant rod CD, and the dimension and material parameters are as shown in Fig. 2. This compliant mechanism is driven by a time-varying torque $M = 50\sin(\pi t)\,\mathrm{N \cdot m}$ acting on joint A and the gravity is $\boldsymbol{g} = [-9.81 \quad 0 \quad 0]^T\,\mathrm{m/s^2}$. The flexible rod CD is meshed using ANCF beam elements and GEBT Euler-Bernoulli beam elements respectively in this problem. For the numerical results obtained by ANCF elements, it is denoted as ANCF/GCM when GCM approach is used to calculate the element elastic forces directly. And the numerical results in which the locking alleviate technique SSM approach is used are denoted as ANCF/GCM.

The GEBT Euler-Bernoulli beam element employed in this paper is locking free and the corresponding numerical results are referred as GEBT/Euler. In this section, a MATLAB computer program was developed in order to obtain the numerical results presented for ANCF and GEBT beam elements, and the system equations of motion are numerically integrated using ode15 s, the stiff ODE solver implemented in MATLAB. The simulation time is set to 5 s.

For comparison, the dynamic model of this partially compliant four-bar mechanism is also established in commercial software MSC.ADAMS and the FE_Part which is suitable for modeling large deformation cases of beam-like structures is used to mesh the compliant rod CD. The compliant rod CD is meshed with 5 elements in all simulations, and the X component of position vector of point C on the compliant rod are shown in Fig. 3.

It can be seen from Fig. 3 that the numerical results obtained using ANCF and GEBT beam elements can match well with that of commercial software ADAMS in general. One can also find that the numerical result of ANCF beam element based on GCM approach is not as good as that of ANCF beam elements based on SSM approach. This is expected due to the Poisson locking of the standard fully parameterized ANCF beam elements when the GCM approach is used. The Poisson locking is the result of using different orders of displacement interpolation in the longitudinal and transverse directions [17], which makes the ANCF beam element based on GCM appears stiffer. This example shows that the new proposed locking alleviation technique SSM approach works well in the dynamic simulation of compliant mechanism.

For the GEBT beam element used in this investigation, the deformation hypothesis of Euler-Bernoulli beam is preserved by using a special displacement-rotation coupling interpolation, which makes this large deformation beam element locking free. Therefore, very good numerical results labeled as GEBT/Euler can be obtained as shown in Fig. 3.

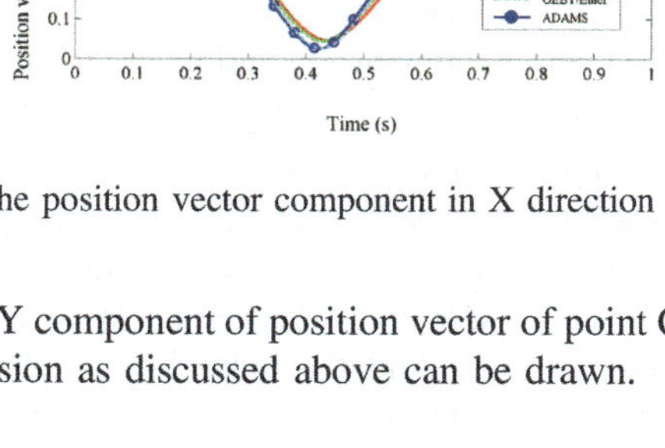

Fig. 3. The position vector component in X direction of point C

Figure 4 gives the Y component of position vector of point C on the compliant rod, and the similar conclusion as discussed above can be drawn.

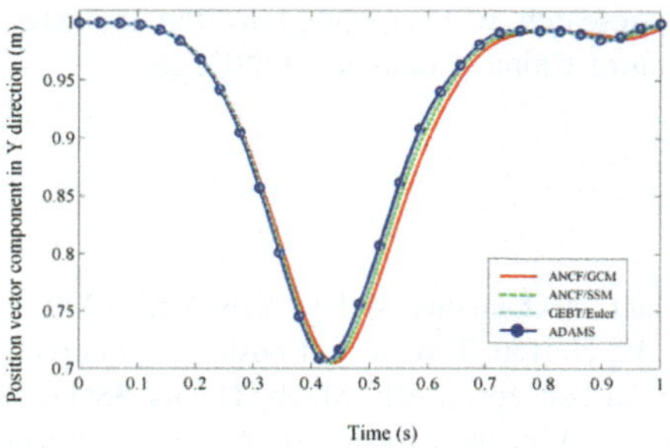

Fig. 4. The position vector component in Y direction of point C

Because of the character of the shape function of ANCF element, the high precision strian and stress analysis can be performed based on general continuum mechanics. Figure 5 shows the configuration of the compliant mechanism at different time and the distribution of normal strain ε_{11} of the compliant rod.

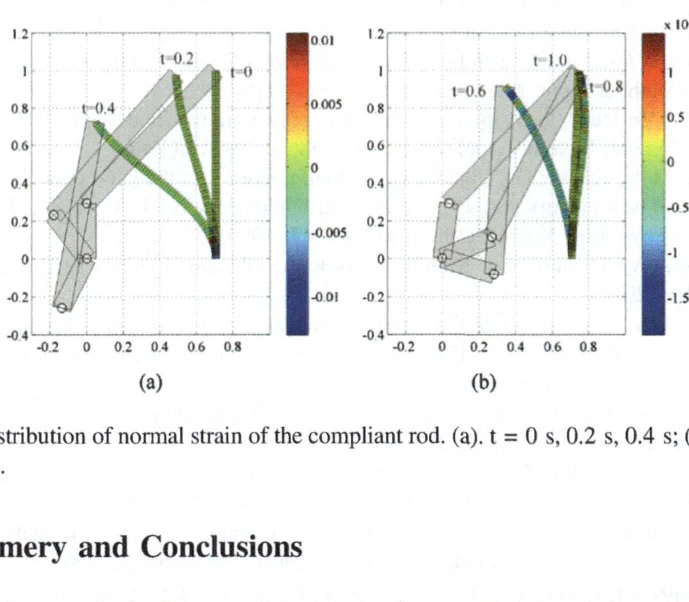

Fig. 5. Distribution of normal strain of the compliant rod. (a). t = 0 s, 0.2 s, 0.4 s; (b). t = 0.6 s, 0.8 s, 1.0 s.

7 Summery and Conclusions

The modeling methods for compliant mechanisms with large deformations are discussed in this research. The ANCF and GEBT beam elements are used to mesh the compliant rod. In order to model the compliant rod with joint, a new ANCF beam element which consists of one nodal constrain is developed. The locking alleviation technique SSM approach is also employed to demonstrate effect of locking on the deformation of compliant rod when ANCF element is used. This research provides the basis for the dynamic and fatigue analysis of compliant mechanisms.

Acknowledgments. This research was supported by the financial support of the National Natural Science Foundation of China (Grants no. 11602228).

References

1. Howell, L.L.: Compliant Mechanisms. Wiley, New York (2001)
2. Midha, A., Howell, L.L., Norton, T.W.: Limit positions of compliant mechanisms using the pseudo-rigid-body model concept. Mech. Mach. Theory **35**(1), 99–115 (2000)
3. Howell, L.L., Midha, A.: A method for the design of compliant mechanisms with small-length flexural pivots. J. Mech. Des. **116**, 280–290 (2008)
4. Howell, L.L., Midha, A., Norton, T.W.: Evaluation of equivalent spring stiffness for use in a pseudo-rigid-body model of large-deflection compliant mechanisms. J. Mech. Des. **118**, 126–131 (2008)
5. Wang, W., Bi, S., Zhang, L.G.: Dynamic modeling of compliant mechanisms based on 2R pseudo-rigid-body model. Appl. Mech. Mater. **163**, 277–280 (2012)
6. Yu, Y., Zhu, S.: 3R1H pseudo-rigid-body model for compliant mechanisms with inflection beams, pp. 39–47 (2018)
7. Shabana, A.A.: An absolute nodal coordinate formulation for the large rotation and deformation analysis of flexible bodies. Tech. Rep. MBS96-1-UIC, Dept of Mechanical Engineering, Univ of Illinois at Chicago (1996)
8. Simo, J.C.: A finite strain beam formulation. The three-dimensional dynamic problem. Part I. Comput. Methods Appl. Mech. Eng. **49**(1), 55–70 (1985)
9. Simo, J.C., Vu-Quoc, L.: A three-dimensional finite-strain rod model. Part II: computational aspects. Comput. Methods Appl. Mech. Eng. **58**(1), 79–116 (1986)
10. Crisfield, M.A., Jelenic, G.: Objectivity of strain measures in the geometrically exact three-dimensional beam theory and its finite-element implementation. Proc. R. Soc. London. Ser. A: Math., Phys. Eng. Sci. **455**(1983), 1125–1147 (1999)
11. Shabana, A.A.: Dynamics of Multibody Systems, 4th edn. Cambridge University Press, Cambridge (2013)
12. Shabana, A.A.: Computational Continuum Mechanics, 3rd edn. Wiley, Chichester (2018)
13. Yakoub, R.Y., Shabana, A.A.: Three dimensional absolute nodal coordinate formulation for beam elements: implementation and applications. J. Mech. Des. **123**(4), 614 (2002)
14. Shabana, A.A., Yakoub, R.Y.: Three dimensional absolute nodal coordinate formulation for beam elements: theory. J. Mech. Des. **123**(4), 606 (2002)
15. Zhang, Z., Qi, Z., Wu, Z., et al.: A spatial Euler-Bernoulli beam element for rigid-flexible coupling dynamic analysis of flexible structures. Shock. Vib. **2015**, 1–15 (2015)
16. Patel, M., Shabana, A.A.: Locking alleviation in the large displacement analysis of beam elements: the strain split method. Acta Mech. **229**(7), 2923–2946 (2018)
17. Nachbagauer, K.: State of the art of ANCF elements regarding geometric description, interpolation strategies, definition of elastic forces, validation and the locking phenomenon in comparison with proposed beam finite elements. Arch. Comput. Methods Eng. **21**(3), 293–319 (2014)

Dynamic Performance of Flexible Composite Structures with Dielectric Elastomer Actuators via Absolute Nodal Coordinate Formulation

Haidong Yu(✉), Yunyong Li, Aolin Chen, and Hao Wang

Key Laboratory of Digital Manufacture for Thin-walled Structure, Shanghai Jiao Tong University, Shanghai 200240, China
{hdyu, lyysjytu, d-yang, blues13, wanghao}@sjtu.edu.cn

Abstract. The composite structure with the dielectric elastomer (DE) and soft materials is the main form in the actuators in so\ft robots. The accurate dynamic model for this composite structure is important for the design and the manipulating control of soft robots. In this paper, the high-order shell elements are developed to descript the mechanical performance of composite two-layer plate based on the absolute nodal coordinate formulation (ANCF). The constitutive model for DE material and Mooney-Rivlin constitutive model for soft material are introduced. The consistent condition of deformation for composite structure is proposed when the two-layer composite shells are united together. The effects of various parameters such as voltages and material parameters on the dynamic performance of composite structure are studied. The results show that the oscillation amplitude and frequency of the free end of the composite structure increase nonlinearly with the increase of the excitation voltage. The increase of structural stiffness for the DE plate may improve the periodicity and stability of the deformation. The results may provide guidance for the dynamic modelling and actuation design for soft robots.

Keywords: Flexible composite structure · Dielectric elastomer actuator · Hyperelastic constitutive model · ANCF · Dynamic performance

1 Introduction

Flexible composite actuators have been widely applied in soft robots, which consist of soft electric actuating material and flexible medium. The accurate dynamic model of the composite flexible structure is of great difficulty due to the geometrical and material nonlinearity and the multi-element coupling problem. Chattopadhyay et al. [1] numerically studied the modeling of stiffened plates and analyzed the effect of material properties and boundary conditions on structural dynamic performance. Barik et al. [2] proposed a high-order isoparametric model, where shear-locking phenomena is considered and the stiffener can be located anywhere with arbitrary orientation. Zhao et al. [3] developed the composite stiffened plates and the coordinate transformation matrix between different nodes that may be used to greatly reduce degree-of-freedoms and improves the computing accuracy as well.

A. Kecskeméthy and F. Geu Flores (Eds.): ECCOMAS 2019, COMPUTMETHODS 53, pp. 223–230, 2020.
https://doi.org/10.1007/978-3-030-23132-3_27

The large deformation and nonlinear mechanical behavior are main problems for the flexible composite actuators in soft robots. The absolute nodal coordinate formulation (ANCF) was first proposed by Shabana [4] and has been extensively studied in recent years, which is an effective method to solve these two problems. Kubler et al. [5] proposed a three dimensional hexagonal element, and introduced slope coordinates to alleviate Poisson locking problem in numerical simulating, which realized accurate description of three-dimensional solid motion and deformation process. Zhao et al. [6] developed higher order shell elements based on ANCF to study dynamic behavior of rotating thin-walled structure. However, the real application on the engineering problem based on ANCF is rarely studied.

Hyperelastic material and dielectric elastomer material are usually used in composite actuator of soft robot. Three typical constitutive models are used for hyperelastic material such as Heo-Hookean model, Mooney-Rivlin model and Ogden model. Dielectric elastomer materials' constitutive model is mainly based on Maxwell's theory. Suo et al. [7] proposed the system free energy function of combined elastic strain energy and electric field energy. The simple linear superposition of the two kinds of energy was adopted in terms of their response time. Wisselr et al. [8, 9] proposed a new electromechanical coupling model for the viscous behavior of elastomers. The finite element analysis of circular actuators was carried out by using different constitutive models.

In this paper, a two-layer thin plate structure model is established based on ANCF. Mooney-Rivlin hyper-elastic constitutive model and the DE material constitutive equations are introduced to derive the dynamic equations with the deformation compatibility conditions. The dynamic performance of the composite structure are investigated with various DE actuating voltages and material parameters.

2 High-Order Shell Element of Composite Structure

The two-layer composite plate structure is employed to stimulate the structure of the satellite gripper, as shown in Fig. 1. The DE plate element is bonded to the upper surface of the driven plate element made of flexible material. The actuator is subjected to a certain voltage. The large deformation is observed when the DE plate actuates the composite structure, which leads to a vibration of the flexible structure. The equivalent three-dimensional model of the two-layer plate shown in Fig. 1(b). The deformation on the interface of two shell has to be consistent because they are bonded together.

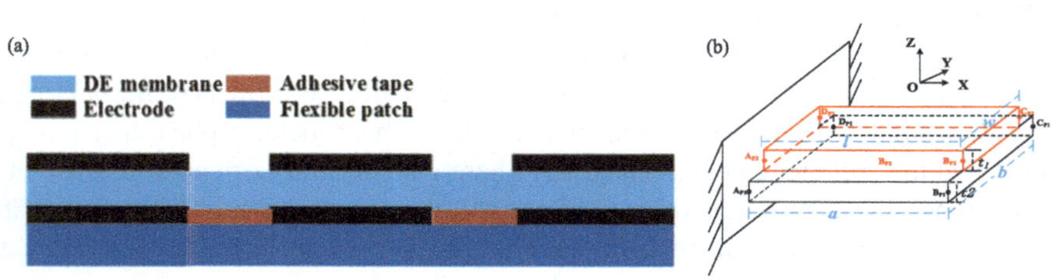

Fig. 1. The sketch of composite structures with DE actuators.

In this study, a three-dimensional plate element is presented as shown in Fig. 2. The length, width and thickness of the plate element are a, b and t, respectively. All coordinates are defined in the global coordinate system O-XYZ. And O-xyz is the local coordinate system fixed on the element, which is used to describe the relative position of an arbitrary point within the element.

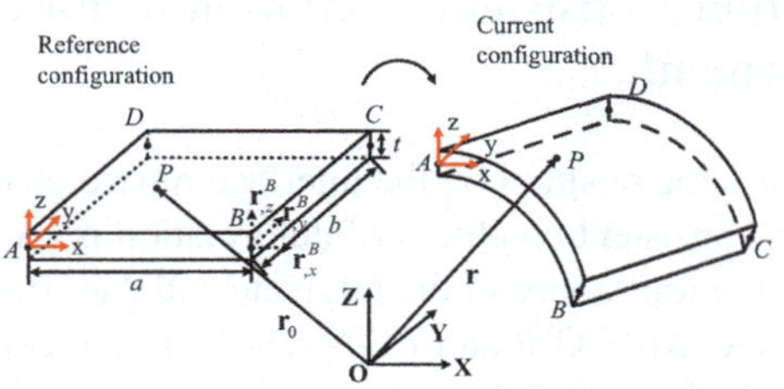

Fig. 2. Three-dimensional four-node plate element.

Based on ANCF, the definition of the nodal coordinates is denoted as

$$\mathbf{e}_i = \begin{bmatrix} \mathbf{r}_i^{\mathrm{T}} & \frac{\partial \mathbf{r}_i^{\mathrm{T}}}{\partial x} & \frac{\partial \mathbf{r}_i^{\mathrm{T}}}{\partial y} & \frac{\partial \mathbf{r}_i^{\mathrm{T}}}{\partial z} \end{bmatrix}^{\mathrm{T}} \tag{1}$$

where $\mathbf{r}_i$ denotes the global position vector of the ith node and $\partial\mathbf{r}_i/\partial k(k = x, y, z)$ represents the position gradients vector with respect to k axis. The coordinates of one plate element has 48 coordinates and is given as

$$\mathbf{e} = \begin{bmatrix} \mathbf{e}_{\mathrm{A}}^T & \mathbf{e}_{\mathrm{B}}^T & \mathbf{e}_{\mathrm{C}}^T & \mathbf{e}_{\mathrm{D}}^T \end{bmatrix}^T \tag{2}$$

By using the high-order shape function, the coordinates of the nodes in the plate element is presented as

$$\mathbf{r} = \mathbf{S}^P\mathbf{e} \tag{3}$$

where $\mathbf{S}^P$ is the shape function matrix of plate element and can be calculated as

$$\mathbf{S}^P = [S_1\mathbf{I} \quad S_2\mathbf{I} \quad S_3\mathbf{I} \quad S_4\mathbf{I} \quad \ldots \quad S_{14}\mathbf{I} \quad S_{15}\mathbf{I} \quad S_{16}\mathbf{I}] \tag{4}$$

where $\mathbf{I}$ is the 3×3 identity matrix and the shape functions S_i is defined as

$$\begin{aligned}
&S_1 = (2\xi+1)(\xi-1)^2(2\eta+1)(\eta-1)^2, && S_2 = a\xi(\xi-1)^2(2\eta+1)(\eta-1)^2,\\
&S_3 = b\eta(\xi-1)^2(2\xi+1)(\eta-1)^2, && S_4 = t\zeta(\xi-1)(\eta-1),\\
&S_5 = -\xi^2(2\xi-3)(2\eta+1)(\eta-1)^2, && S_6 = a\xi^2(\xi-1)(2\eta+1)(\eta-1)^2,\\
&S_7 = -b\eta\xi^2(2\xi-3)(\eta-1)^2, && S_8 = -t\zeta\xi(\eta-1),\\
&S_9 = \eta^2\xi^2(2\xi-3)(2\eta-3), && S_{10} = -a\eta^2\xi^2(\xi-1)(2\eta-3),\\
&S_{11} = -b\eta^2\xi^2(\eta-1)(2\xi-3), && S_{12} = t\zeta\xi\eta,
\end{aligned}$$

$$S_{13} = -\eta^2(2\xi+1)(\xi-1)^2(2\eta-3), \quad S_{14} = -a\xi\eta^2(\xi-1)^2(2\eta-3),$$
$$S_{15} = b\eta^2(\xi-1)^2(2\xi+1)(\eta-1), \quad S_{16} = -t\eta\zeta(\xi-1). \tag{5}$$

here, $\xi = x/a$, $\eta = y/b$, $\zeta = z/t$.

3 The Deformation Consistent Condition of Flexible Composite Element

The deformation at the same position on the interface of two plates should be identical. The sketch map of the consistent condition of deformation is presented in Fig. 3. Here, $A_{P1}, B_{P1}, C_{P1}, D_{P1}$ are the four nodes of the DE plate and $A_{P2}, B_{P2}, C_{P2}, D_{P2}$ are the four nodes of the driven plate. M-N-O-P and M′-N′-O′-P′ are the contact points of the two plate elements, respectively. The deformation of two plates should satisfy three conditions. (1) The tangential planes of the upper and lower plates are always coincident before and after the deformation. (2) The nodes at the same position, remain the identical deformation during the deformation process. (3) There is no relative slide on the contact surface of the two plates.

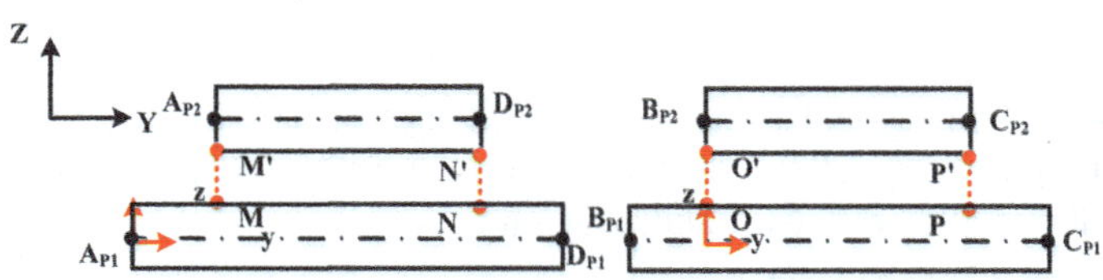

Fig. 3. The identical deformation of composite structure.

To satisfy the consistent conditions of deformation, the constraint equations of the composite structure at points M′ and M are written as

$$[\mathbf{r}_M]_{z=t/2} = [\mathbf{r}_{M'}]_{z=-t'/2}, \quad \left[\frac{\partial \mathbf{r}_M^{\mathrm{T}}}{\partial x_i}\right]_{z=t/2} = \left[\frac{\partial \mathbf{r}_{M'}^{\mathrm{T}}}{\partial x_i}\right]_{z=-t'/2}, \quad i = x, y, z \tag{6}$$

Substituting Eq. (6) into Eq. (3), it can be derived as

$$\begin{bmatrix} \mathbf{r}_M \\ \frac{\partial \mathbf{r}_M}{\partial x} \\ \frac{\partial \mathbf{r}_M}{\partial y} \\ \frac{\partial \mathbf{r}_M}{\partial z} \end{bmatrix} = \begin{bmatrix} \mathbf{S}^{P1}(0,0,\frac{t}{2})\mathbf{e}^{P1} \\ \mathbf{S}_x^{P1}(0,0,\frac{t}{2})\mathbf{e}^{P1} \\ \mathbf{S}_y^{P1}(0,0,\frac{t}{2})\mathbf{e}^{P1} \\ \mathbf{S}_z^{P1}(0,0,\frac{t}{2})\mathbf{e}^{P1} \end{bmatrix} = \mathbf{T}_M^{P1}\mathbf{e}^{P1}, \quad \begin{bmatrix} \mathbf{r}_{M'} \\ \frac{\partial \mathbf{r}_{M'}}{\partial x} \\ \frac{\partial \mathbf{r}_{M'}}{\partial y} \\ \frac{\partial \mathbf{r}_{M'}}{\partial z} \end{bmatrix} = \begin{bmatrix} \mathbf{S}^{P2}(0,0,-\frac{t'}{2})\mathbf{e}^{P2} \\ \mathbf{S}_x^{P2}(0,0,-\frac{t'}{2})\mathbf{e}^{P2} \\ \mathbf{S}_y^{P2}(0,0,-\frac{t'}{2})\mathbf{e}^{P2} \\ \mathbf{S}_z^{P2}(0,0,-\frac{t'}{2})\mathbf{e}^{P2} \end{bmatrix} = \mathbf{T}_{M'}^{P2}\mathbf{e}^{P2} \tag{7}$$

where $\mathbf{S}^{P1}$ and $\mathbf{S}^{P2}$ are the shape function of plate element 1 and 2, $\mathbf{S}_k^{P1}=\partial\mathbf{S}^{P1}/\partial k$, $\mathbf{S}_k^{P2}=\partial\mathbf{S}^{P2}/\partial k(k=x,y,z)$, respectively, $\mathbf{T}_M^{P1}$ and $\mathbf{T}_{M'}^{P2}$ are the coefficient matrix and it can be derived as

$$\mathbf{T}_M^{P1}\mathbf{e}^{P1}=\mathbf{T}_{M'}^{P2}\mathbf{e}^{P2} \tag{8}$$

Similarly, the consistent conditions of deformation at points N-N′, O-O′ and P-P′ can be obtained. They can be combined as the matrix form and denoted as

$$\begin{bmatrix}\mathbf{T}_M^{P1} & \mathbf{T}_N^{P1} & \mathbf{T}_O^{P1} & \mathbf{T}_P^{P1}\end{bmatrix}^T\mathbf{e}^{P1}=\begin{bmatrix}\mathbf{T}_{M'}^{P2} & \mathbf{T}_{N'}^{P2} & \mathbf{T}_{O'}^{P2} & \mathbf{T}_{P'}^{P2}\end{bmatrix}^T\mathbf{e}^{P2} \tag{9}$$

where $\begin{bmatrix}\mathbf{T}_{M'}^{P2} & \mathbf{T}_{N'}^{P2} & \mathbf{T}_{O'}^{P2} & \mathbf{T}_{P'}^{P2}\end{bmatrix}^{\mathrm{T}}$ is a non-singular matrix and Eq. (9) can be rewritten as

$$\mathbf{e}^{P2}=\mathbf{T}^{21}\mathbf{e}^{P1} \tag{10}$$

here, $\mathbf{T}^{21}$ is the coordinate transformation matrix of shell element 1 and 2, with which the nodal coordinates of plate 2 can be expressed with the same coordinate in plate 1. Consequently, the number of nodal coordinates of the composite structure is reduced and the calculation efficiency is obviously improved.

4 Equations of Motion of the Composite Structure

The dynamic equation of the two-layer composite unit can be derived as

$$\mathbf{M}^{P12}\ddot{e}+\mathbf{K}^{P12}\mathbf{e}=\mathbf{Q}_a \tag{11}$$

where $\mathbf{M}^{P12}$ is the mass matrix of the composite structure, $\mathbf{K}^{P12}$ is the generalized stiffness matrix of the composite structure, $\mathbf{Q}_a$ is the generalized external force and can be derived by the virtual work principle as $\mathbf{Q}_a^{\mathrm{T}}=\mathbf{F}^{\mathrm{T}}\mathbf{S}$.

According to the kinetic energy principle, the mass matrix is defined as

$$\mathbf{M}^{P1}=\int_V\rho\left(\mathbf{S}^{P1}\right)^{\mathrm{T}}\mathbf{S}^{P1}dV,\ \mathbf{M}^{P2}=\int_V\rho\left(\mathbf{S}^{P2}\right)^{\mathrm{T}}\mathbf{S}^{P2}dV \tag{12}$$

where ρ is the mass density of the element, By employing the transformation matrix, the mass matrix of the coupled element can be expressed as

$$\mathbf{M}^{P12}=\mathbf{M}^{P1}+(\mathbf{T}^{12})^{\mathrm{T}}\mathbf{M}^{P2}\mathbf{T}^{12} \tag{13}$$

The generalized stiffness matrix of the element is computed as

$$\mathbf{K}^{P}=\frac{\partial Q^{P}}{\partial\mathbf{e}}=\frac{\partial}{\partial\mathbf{e}}\left(\frac{\partial U^{P}}{\partial\mathbf{e}}\right) \tag{14}$$

where Q^P is the generalized elastic force, U^P is the strain energy of plate element. The constitutive model for DE material is generally described as

$$U^{P2} = \mathrm{U}(\lambda_1, \lambda_2, \lambda_3) + \mathrm{V}(\lambda_1, \lambda_2, \lambda_3, \tilde{D}) \tag{15}$$

where $\tilde{D}$ is the nominal electric displacement before deformation, $\lambda_1, \lambda_2, \lambda_3$ are the tensile ratios in three principal directions. So the stiffness matrix can be written as

$$\mathbf{K}^{\mathbf{P2}} = \frac{\partial}{\partial \mathbf{e}}\left(\frac{\partial \mathrm{U}(\lambda_1, \lambda_2, \lambda_3)}{\partial \mathbf{e}}\right) + \frac{\partial}{\partial \mathbf{e}}\left(\frac{\partial \mathrm{V}(\lambda_1, \lambda_2, \lambda_3, \tilde{D})}{\partial \mathbf{e}}\right) = \mathbf{K}_{\mathbf{MR}} + \mathbf{K}_{\mathbf{E}} \tag{16}$$

where $\mathbf{K}_{\mathbf{MR}}$ represents the stiffness matrix of the contribution of the elastic force, and $\mathbf{K}_{\mathbf{E}}$ represents that of the electric field force.

By introducing the Mooney-Rivlin material model into shell element, $\mathbf{K}_{\mathbf{MR}}$ can be derived as

$$\mathbf{K}_{\mathbf{MR}} = \partial(\frac{\partial W_{MR}}{\partial \mathbf{e}})/\partial \mathbf{e} \tag{17}$$

where W_{MR} is the strain energy density function given by

$$W_{MR} = \sum_{i=0}^{N}\sum_{j=0}^{N} C_{ij}(I_1 - 3)^i (I_2 - 3)^j + \sum_{k=1}^{N} \frac{1}{d_k}(I_3^2 - 1)^{2k} \tag{18}$$

By coupling the joint effects of polarization and stretching, the electric field energy density function is written as

$$\mathrm{V}(\lambda_1, \lambda_2, \lambda_3, \tilde{D}) = \frac{\tilde{D}^2}{2\varepsilon}\lambda_1^{-1}\lambda_2^{-1}\lambda_3 \tag{19}$$

where ε is the dielectric constant and $\varepsilon = \varepsilon_0 \varepsilon_r$, ε_0 is the dielectric constant in vacuum and $\varepsilon_0 = 8.85 \times 10^{-12}$ F/m, ε_r is the relative dielectric constant and $\varepsilon_r = 4$.

Based on the incompressibility condition and Green-Lagrange strain tensor, $\mathbf{K}_{\mathbf{E}}$ can be deduced as

$$\mathbf{K}_{\mathbf{E}} = \frac{\varepsilon \tilde{E}}{2}(\mathbf{S}_c + \mathbf{S}_c^T) \tag{20}$$

where $S_c = (S_{1z})^T S_{1z} + (S_{2z})^T S_{2z} + (S_{3z})^T S_{3z}$.

Similarly, the stiffness matrix for plate 1 is calculated Eq. (17) because the material is hyperelastic. Thus the generalized stiffness matrix of the coupled element can be described with the transformation matrix as

$$\mathbf{K}^{P12} = \mathbf{K}^{P1} + (\mathbf{T}^{12})^{\mathrm{T}}\mathbf{K}^{P2}\mathbf{T}^{12} = \mathbf{K}^{P1} + (\mathbf{T}^{12})^{\mathrm{T}}(\mathbf{K}_{MR}^{P2} + \mathbf{K}_{E}^{P2})\mathbf{T}^{12} \tag{21}$$

5 Discussion

Figure 4 shows the deformation of composite plate structure under DE actuation. The warp deformation is obvious because the materials are soft. The effects of various parameters on the nonlinear dynamic behaviors of composite structure are studied. Figure 5 shows the displacements of mid-point P at different excitation voltages (U = 500 V, 2 kV, and 6 kV). It is observed that the displacement of mid-point P in the Z direction is larger than that in the X and Y directions, and presents periodic change. Moreover, as excitation voltages increases, the composite structure tends to undergo a swing of greater amplitude at a higher frequency.

The displacement of the mid-point P with different hyperelastic material coefficient ratios of the actuated plates (*E*, 1.5*E*, 2*E*, and 2.5*E*) at the same excitation voltage (U = 2 kV) is shown in Fig. 6. The vibration amplitude decreases as the coefficient ratio increases. And the deformation of the composite structure is periodic. The nonlinear deformation characteristics become weaker due to the increase of the structural stiffness.

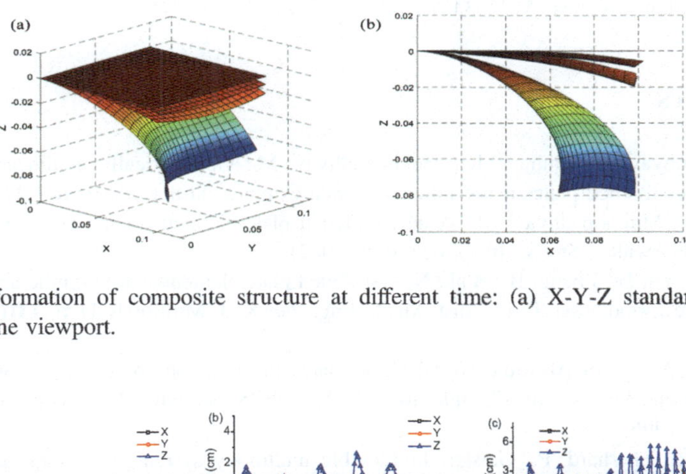

Fig. 4. Deformation of composite structure at different time: (a) X-Y-Z standard viewport; (b) X-Z plane viewport.

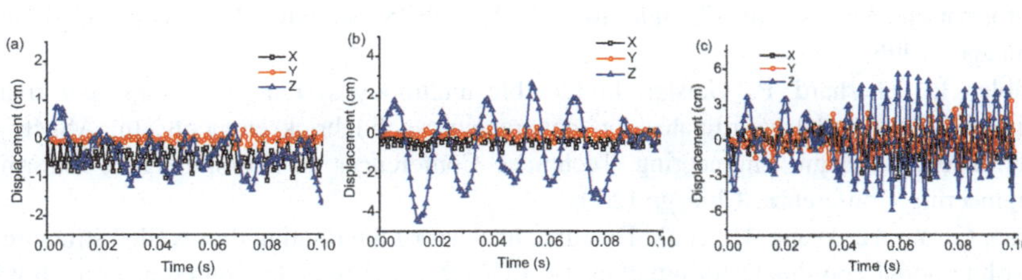

Fig. 5. Displacement of mid-point P: (a) U = 0.5 kV; (b) U = 1 kV; (c) U = 6 kV.

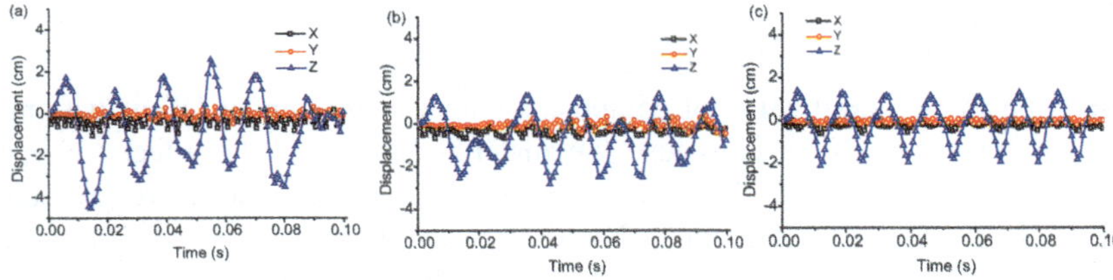

Fig. 6. Displacement of node P with different material parameters: (a) C_1 = 4 MPa, C_2 = 1 MPa; (b) C_1 = 6 MPa, C_2 = 1.5 MPa; (c) C_1 = 8 MPa, C_2 = 2MP.

6 Conclusion

The study illustrates that the consistent deformation on the interface of the composite structures may be realized due to the usage of high-order element. The dynamic model may be used to study the mechanical performance of soft robots.

Acknowledgments. The authors appreciate the financial support by the National Natural Science Foundation of China (51775345).

References

1. Chattopadhyay, B., Sinha, P.K., Mukhopadhyay, M.: Geometrically nonlinear analysis of composite stiffened plates using finite elements. Compos. Struct. **31**(2), 107–118 (1995)
2. Barik, M., Mukhopadhyay, M.: A new stiffened plate element for the analysis of arbitrary plates. Thin-walled Struct. **40**(7), 625–639 (2002)
3. Zhao, C., Yu, H., Zheng, B., et al.: New stiffened plate elements based on the absolute nodal coordinate formulation. Proce. Inst. Mech. Eng, Part K: J. Multibody Dyn. **231**(1), 213–229 (2016)
4. Shabana, A.A.: An Absolute Nodal Coordinates Formulation for the Large Rotation and Deformation Analysis of Flexible Bodies. No. MBS96-1-UIC, University of Illinois at Chicago (1996)
5. Kübler, L., Eberhard, P., Geisler, J.: Flexible multibody systems with large deformations using absolute nodal coordinates for isoparametric solid brick elements. In: ASME 2003 International Design Engineering Technical Conferences & Computers & Information Engineering Conference, Chicago (2003)
6. Zhao, C., Yu, H., Wang, H., et al.: Dynamic analysis of the rotating thin-walled structure with absolute nodal coordinate formulation. In: ECCOMAS Thematic Conference on Multibody Dynamics. Barcelona, Catalonia, Spain (2015)
7. Suo, Z.G., Zhao, X.H., Greene, W.H.: A nonlinear field theory of deformable dielectrics. J. Mech. Phys. Solid **56**, 467–486 (2008)
8. Wissler, M.: Modeling Dielectric Elastomer Actuators. Swiss Federal Institute of Technology, Zurich (2007)
9. Michel, S., Durager, C., Zobel, M., et al.: Electro active polymers as a novel actuator technology for lighter than-air vehicles. In: SPIE. Symposium on Smart Structures and Materials: Electroactive Polymer Actuators and Devices, March 18–22, 2007, San Diego, California, 65241: 1–11. SPIE, Bellingham (2007)

Approaches to Fibre Modelling in the Model of an Experimental Laboratory Mechanical System

Pavel Polach[1(✉)], Michal Hajžman[1], and Radek Bulín[2]

[1] Research and Testing Institute Pilsen,
Tylova 1581/46, 301 00 Pilsen, Czech Republic
{polach, hajzman}@vzuplzen.cz
[2] University of West Bohemia,
Univerzitní 8, 306 14 Pilsen, Czech Republic
rbulin@ntis.zcu.cz

Abstract. In this paper there are presented some possible approaches suitable for the modelling of the fibre and cable dynamics in the framework of various mechanical systems: force representation of a fibre, a point-mass model and an absolute nodal coordinate formulation. Experimental measurements focused on the investigation of the fibre behaviour were performed on an assembled weight-fibre-pulley-drive laboratory mechanical system. This mechanical system was modeled using all of the mentioned methods (newly using point-mass model) and simulations of the experimental measurements were performed. Results obtained using simulations and experimental measurements are compared and discussed.

1 Introduction

Fibres and cables play an important role in the design of many machines. One of their applications is the replacing of the chosen rigid elements of a manipulator or a mechanism with fibres. The main advantage of this design is the achievement of a lower moving inertia, which leads to a higher mechanism speed. In this paper there are presented some possible approaches suitable for the modelling of the fibres dynamics in the framework of mechanical systems: force representation of a fibre, a point-mass model and an absolute nodal coordinate formulation (ANCF).

Fibres and cables are used especially in the parallel kinematic mechanisms (e.g. [7, 14]). Cable-driven variants of the parallel kinematic mechanisms have further advantages, viz. a large range of motion, the possibility of antibacklash property [1] and easy reconfiguration. Typical disadvantages of the fibre-driven parallel kinematic mechanisms are a relatively narrow frequency bandwidth of their feedback motion control and problems with the accurate positioning of the end-effector.

The promising research direction for solution of these problems is the concept of multi-level mechanisms with a hierarchic structure composed of a parallel cable-driven mechanism for large and slow motions and an active structure connected to the mechanism platform for low and high frequency motions [5, 10].

A. Kecskeméthy and F. Geu Flores (Eds.): ECCOMAS 2019, COMPUTMETHODS 53, pp. 231–238, 2020.
https://doi.org/10.1007/978-3-030-23132-3_28

The motivation for the presented work is the development of a fibre model, which could be efficient for the usage in a mechatronic model of a manipulator consisting of fibres and an end-effector whose motion is driven by fibres – particularly for the usage in the model of QuadroSphere (see Fig. 2 left). The QuadroSphere is a tilting mechanism with a spherical motion of a platform and an accurate measurement of its position. The platform position is controlled by four fibres; each fibre is guided by a pulley from linear guidance to the platform. One of the presumptions of the high-quality QuadroSphere model is creating of the optimal dynamic model of the fibre that is supported by the experimental results.

Experimental measurements focused on the investigation of the fibre behaviour were performed on an assembled weight-fibre-pulley-drive mechanical system (e.g. [10]; see Fig. 1). The weight can be considered the end-effector.

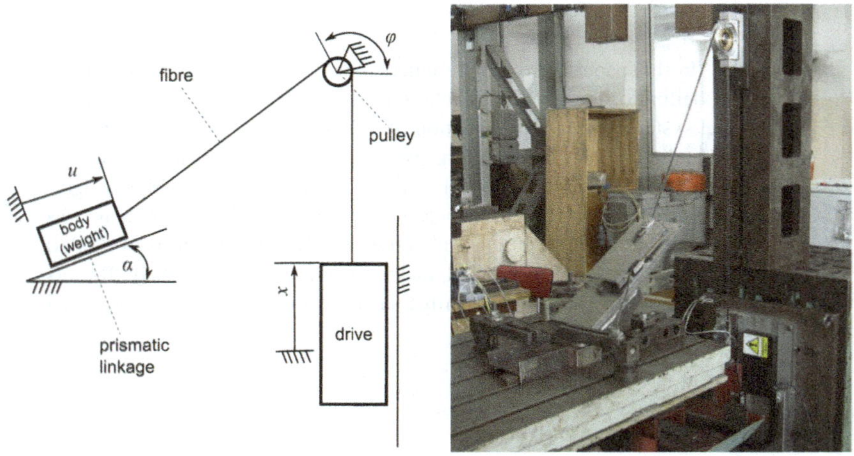

Fig. 1. Scheme and a real weight-fibre-pulley-drive mechanical system

The fibre modelling should be based on considering the fibre flexibility and suitable approaches can be based on the flexible multibody dynamics (see e.g. [6, 13]). There are many approaches to the modelling of flexible bodies in the framework of multibody systems. Comprehensive reviews of these approaches can be found in [13] or [15]. Further development together with other multibody dynamics trends is introduced e.g. in [2].

The simplest way how to incorporate fibre in the equations of motion of a mechanism is the force representation of a fibre (e.g. [4, 12]). It is supposed that the mass of fibres is small to such an extent comparing to the other moving parts that the inertia of fibres is negligible. The fibre is represented by the force dependent on the fibre deformation and its stiffness and damping properties. A variable length of the fibre due to wiring can be easily described using the force approach. The main advantage of using the force representation of the fibre consists in a very short computational time.

A more accurate approach is based on the representation of the fibre using a point-mass model (e.g. [8, 9]). It has the advantage of a lumped point-mass model. Point-masses can be connected by forces or constraints. In the case of the manipulator mechatronic model consisting of fibres and an end-effector whose motion is driven by fibres (e.g. in the case of the QuadroSphere model) utilization of the point-mass model of a fibre proved to be very prospective.

An advanced approach usable for the fibre modelling is a so called absolute nodal coordinate formulation (ANCF), which is based on the discretization of a fibre to nonlinear finite elements [6, 13]. The formulation leads to a constant mass matrix and highly nonlinear stiffness matrix.

2 Point-Mass Representation of the Fibre

Model of the mechanical system with the end-effector attached to the frame using two fibres is given as an example for introducing the point-mass representation of the fibre (see Fig. 2 right; model of this system for introducing the force representation of the cable is presented e.g. in [10]). The free length of the left fibre is l_1, the free length of the right fibre is l_2 and abbreviation EF denotes the end-effector (the weight).

For each node of the system from Fig. 2 (right) the following equation of motion can be written

$$m_i\,\ddot{\mathbf{r}}_i + b_i\frac{\mathbf{dl}_i}{dt} + b_{i+1}\frac{\mathbf{dl}_{i+1}}{dt} + k_i\,\mathbf{dl}_i + k_{i+1}\,\mathbf{dl}_{i+1} = \mathbf{f}_i, \quad i = 1, 2, \ldots, n, \tag{1}$$

where n is number of point-masses (the end-effector is considered as the point-mass too), $\mathbf{r}_i$ is the position vector of node i (or end-effector), $\ddot{\mathbf{r}}_i$ is the vector of acceleration vector of node i, $\mathbf{r}_{i-1}$ is the position vector of node $i-1$, $\mathbf{r}_{i+1}$ is the position vector of node $i+1$, $\mathbf{r}_{i,\,i-1} = \mathbf{r}_i - \mathbf{r}_{i-1}$ is the vector from node $i-1$ to node i, $\mathbf{r}_{i,\,i+1} = \mathbf{r}_i - \mathbf{r}_{i+1}$ is the vector from node $i+1$ to node i, l_i is the free length of the part of the fibre between node $i-1$ and node i, l_{i+1} is the free length of the part of the fibre between node i and node $i+1$, $\mathbf{dl}_i = -\left(\left|\mathbf{r}_{i,\,i-1}\right| - l_i\right)\frac{\mathbf{r}_{i,\,i-1}}{\left|\mathbf{r}_{i,\,i-1}\right|}\,H\left(\left|\mathbf{r}_{i,\,i-1}\right| - l_i\right)$ is the vector of deformation of the part of the fibre between node $i-1$ and node i, $\left|\mathbf{r}_{i,\,i-1}\right| = \sqrt{\mathbf{r}_{i,\,i-1}^T\,\mathbf{r}_{i,\,i-1}}$, $\mathbf{dl}_{i+1} = -\left(\left|\mathbf{r}_{i,\,i+1}\right| - l_{i+1}\right)\frac{\mathbf{r}_{i,\,i+1}}{\left|\mathbf{r}_{i,\,i+1}\right|}\,H\left(\left|\mathbf{r}_{i,\,i+1}\right| - l_{i+1}\right)$ is the vector of deformation of the part of the fibre between node i and node $i+1$, $\left|\mathbf{r}_{i,\,i+1}\right| = \sqrt{\mathbf{r}_{i,\,i+1}^T\,\mathbf{r}_{i,\,i+1}}$, $\mathbf{f}_i$ is the vector of force acting on node i, m_i is the mass of node i, k_i is the stiffness of the part of the fibre between node $i-1$ and node i, k_{i+1} is the stiffness of the part of the fibre between the node i and node $i+1$, b_i is the damping coefficient of the part of the fibre between node $i-1$ and node i, b_{i+1} is damping coefficient of the part of the fibre between node i and node $i+1$.

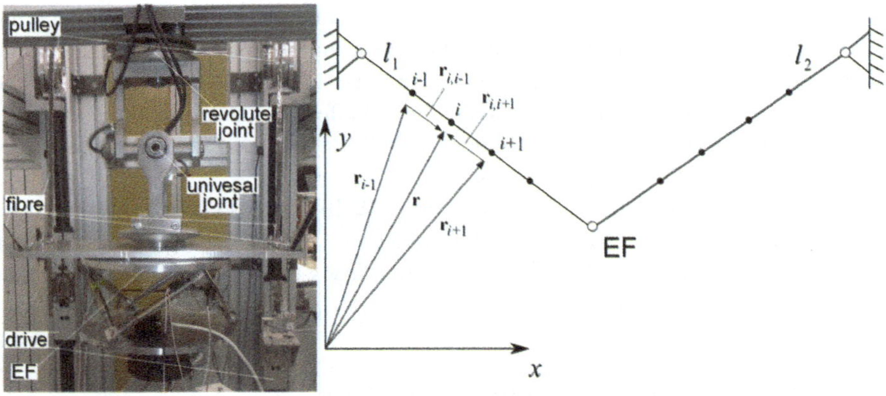

Fig. 2. The QuadroSphere tilting mechanism (left) and model of the mechanical system with the end-effector attached to the frame using two fibres with point-mass representation of the fibre in the absolute coordinate system (right)

Resultant equations of motion of all nodes and the end-effector lead to the system of differential equations of the second order in a matrix form. Note: The equation of motion for the force representation of the fibre can be obtained from Eq. (1) for $n = 1$.

3 Application

As it was already mentioned, experimental measurements focused on the investigation of the fibre behaviour were performed on an assembled weight-fibre-pulley-drive mechanical system (e.g. [10]; see Fig. 1). A carbon fibre with a silicone coating is driven with one drive and is led over a pulley. The fibre length is 1.82 m (fibre weight is 4.95 g), the pulley diameter is 80 mm. At the drive the fibre is fixed on a force gauge. At the other end of the fibre there is a prism-shaped steel weight (weight of 3.096 kg in the presented case), which moves in a prismatic linkage on an inclined plane. The angle of inclination of the inclined plane can be changed. In the presented case the angle is $\alpha = 30°$ and the pulley-fibre angle is $\varphi = 150°$. Drive periodic exciting signals can be of a rectangular, a trapezoidal (see Fig. 5 left) and a quasi-sinusoidal (see Fig. 4 left) shape and there is a possibility of variation of a signal rate. Time histories of weight position u (in direction of the inclined plane; measured by means of a dial gauge), of drive position x (in vertical direction) and of the force acting in the fibre (measured by a force gauge at drive) were recorded using a sample rate of 2 kHz. Altogether 13 different situations were tested at this configuration of assembled laboratory mechanical system.

The same system is numerically investigated using multibody models. The same quantities as at experimental measurements, i.e. time histories of the weight position and of the force acting in the fibre, were monitored. In [11, 12] results of experimental measurements and computational results obtained using the force model of the fibre are given. In [3, 10] a result obtained using the ANCF model of the fibre is given in addition. In this paper, results obtained using the newly created point-mass model of the fibre are presented (the advantage of this approach is supposed to consist in a precise physical interpretation of the problem and in a relatively short computational time).

Firstly, the phenomenological model of the massless fibre model (considering e.g. influences of fibre transversal vibration, etc.) suitable for simulations with the weigh-fibre-pulley-drive mechanical system was created [11]. The number of degrees of freedom in kinematic joints of multibody model of the weight-fibre-pulley-drive system is 5. The weight, the pulley and the drive are considered to be rigid bodies. A planar joint between the weight and the base (prismatic linkage), a revolute joint between the pulley and the base and a prismatic joint between the drive and the base (the movement of the drive is kinematically prescribed) are considered. The fibre stiffness ($34 \cdot 10^3$ N/m) and the fibre damping coefficient (27.5 N $\cdot$ s/m) were considered to be system parameters of the phenomenological model (e.g. [11]). At simulating the experimental measurements for a "quicker" drive motion it was necessary to consider the velocity-dependent stiffness and the velocity-dependent damping "coefficient" in the fibre model for the calculation of dynamic response of the system (e.g. [11]). Frequencies of drive motion – i.e. frequencies of drive periodic excitation signal – higher than 1 Hz are designated as "quicker" drive motions, frequencies of periodic drive motion lower than 1 Hz are designated as "slower" drive motions.

In the multibody model of the weight-fibre-pulley-drive using the ANCF fibre model, the pulley is modelled as a rigid body with one degree of freedom (rotation), the fibre is modelled as a deformable body and is discretized using ANCF planar L2T2 beam elements [3]. In total, 18 ANCF elements were used for the fibre discretization, from which 10 elements are used to discretize the fibre section wrapped around the pulley in order to properly model contact interaction. The rest 8 elements were used to discretized two long straight fibre sections between the weight and the pulley and between the pulley and the drive. An in-house modelling tool in the MATLAB system based on the proposed modelling methodology was created. The fibre Young's modulus is considered to be constant and is determined based on the fibre stiffness of the phenomenological model and the fibre cross-Sect. (0.001×0.01 m). Proportional damping was used.

As it has already been stated the point-mass fibre model is newly used in the model of the investigated system. Except the fibre, the model of the weigh-fibre-pulley-drive mechanical system is the same as the model with the massless fibre model. The point-masses in the fibre model are unconstrained and are connected by spring-damper elements. The fibre stiffness and fibre damping coefficient between two point-masses are considered to be constant. At present, the fibre is modelled using 7 point-masses.

At the "slower" drive motion simulations (see Figs. 3 and 4) performed with the point-mass model the results are in better coincidence with the results of experimental measurements than the results obtained using the massless model considering any phenomenological model. The coincidence of results obtained at using the ANCF method with the experimental results is comparable with results obtained at using point-mass model.

A particular signal defining the motion of the drive and time history of the measured position of the weight at the "quicker" drive motion is in Fig. 5 (the measured motion of the drive served as an input signal for the numerical simulations). At simulating the "quicker" drive motion (see Fig. 5) with the point-mass fibre model, the results are predominantly in worse coincidence with the results of experimental

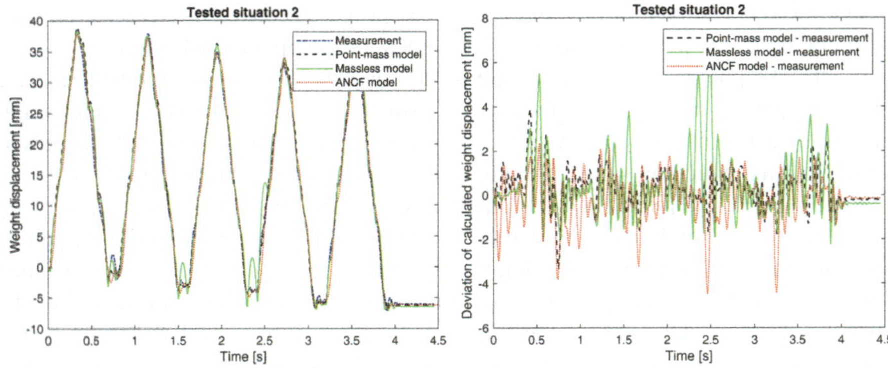

Fig. 3. Time histories of the weight displacement at a "slower" tested situation (left) and deviation of the calculated weight displacements from measured displacement (right)

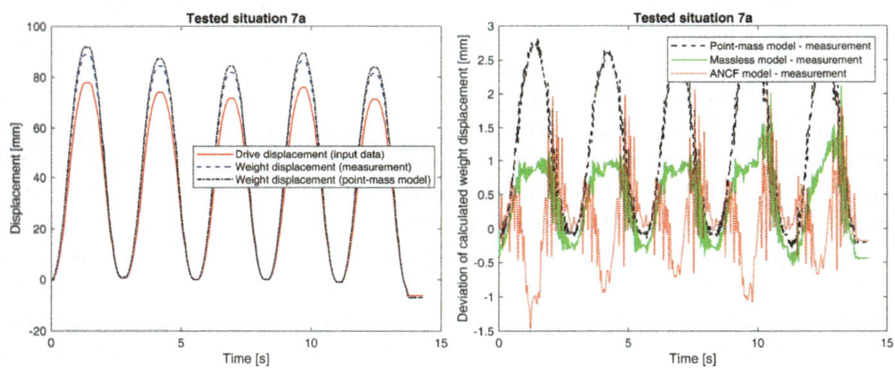

Fig. 4. Time histories of the drive and the weight displacements at a "very" "slower" tested situation (left) and deviation of the calculated weight displacements from measured displacement (right)

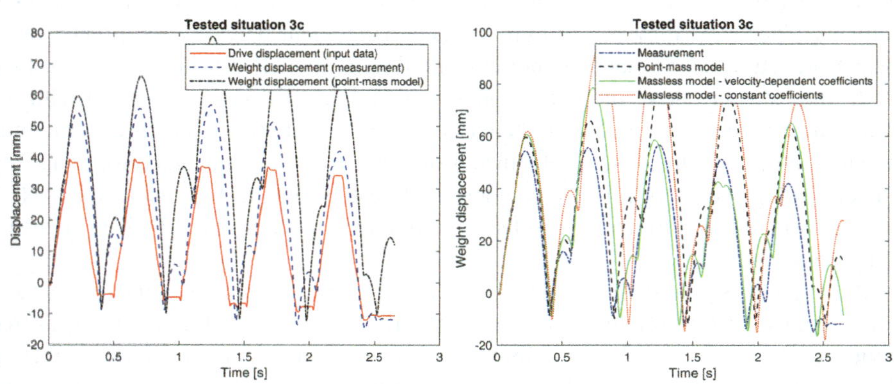

Fig. 5. Time histories of the drive and the weight displacements at a "quicker" tested situation

measurements than the results obtained using the massless model with the velocity-dependent stiffness and the velocity-dependent damping "coefficient".

4 Conclusions

This paper summarizes some possible approaches suitable for the modelling of fibre and cable dynamics in the framework of various mechanical systems. Force representation of the fibre, a point-mass model and absolute nodal coordinate formulation (ANCF).

The weight-fibre-pulley-drive mechanical system was chosen as an example of the application of above-mentioned numerical methods. This system was chosen for the reason that experimental measurements focused on the investigation of the fibre behaviour were performed on it. Newly using point-mass fibre model in the model of the weigh-fibre-pulley-drive mechanical system proved to be suitable for simulating the "slower" drive motion and still unsuitable for simulating the "quicker" drive motion. It is evident that future work will be focused on improving the point-masses model, i.e. on finding the optimum phenomenological model of stiffness and damping "coefficient" and the optimum number of point-masses used for fibre modelling. Further, it will be focused on the model of a point-mass fibre model interaction with the pulley.

Acknowledgement. The paper has originated in the framework of institutional support for the long-time conception development of the research institution provided by the Ministry of Industry and Trade of the Czech Republic to Research and Testing Institute Pilsen. The third author was supported by the project LO1506 of the Czech Ministry of Education, Youth and Sports.

References

1. Agrawal, V., Peine, W.J., Yao, B., Choi, S.-W.: Control of cable actuated devices using smooth backlash inverse. In: Proceedings of the 2010 IEEE International Conference on Robotics and Automation, pp. 1074–1079. IEEE, Anchorage (2010)
2. Bauchau, O.A.: Flexible Multibody Dynamics. Springer, Netherlands, Dordrecht (2011)
3. Bulín, R., Hajžman, M., Polach, P.: Nonlinear dynamics of a cable-pulley system using the absolute nodal coordinate formulation. Mech. Res. Commun. **82**, 21–28 (2017)
4. Diao, X., Ma, O.: Vibration analysis of cable-driven parallel manipulators. Multibody Sys. Dyn. **21**, 347–360 (2009)
5. Duan, X.-C., Qiu, Y., Du, J.-L., Zhao, Z., Duan, Q.-J.: Real-time motion planning for the macro-micro parallel manipulator system. In: Proceedings of the 2011 IEEE International Conference on Robotics and Automation (ICRA 2011), pp. 4214–4219. IEEE, Shanghai (2011)
6. Gerstmayr, J., Sugiyama, H., Mikkola, A.: Developments and future outlook of the absolute nodal coordinate formulation. In: Eberhard, P., Ziegler, P. (eds.) Proceedings of the 2nd Joint International Conference on Multibody System Dynamics, USB flash drive. University of Stuttgart, Stuttgart (2012)

7. Fahham, H.R., Farid, M., Khooran, M.: Time optimal trajectory tracking of redundant planar cable-suspended robots considering both tension and velocity constraints. J. Dyn. Syst. Meas. Contr. **133**, 011004 (2010)
8. Kamman, J.W., Huston, R.L.: Multibody dynamics modeling of variable length cable systems. Multibody Sys. Dyn. **5**, 211–221 (2001)
9. Ottaviano, E., Gattulli, V., Potenza, F., Rea, P.: Modeling a planar point mass sagged cable—suspended manipulator. In: Chang, S.-H., Ceccarelli, M., Sung, Ch.-K., Chang, J.-Y., Liu, T. (eds.) Proceedings of 14th IFToMM World Congress, pp. 2696–2702. Chinese Society of Mechanism and Machine Theory, Taipei (2015)
10. Polach, P., Bulín, R., Hajžman, M.: Approaches to the fibre modelling in a simple laboratory mechanical system. In: Marović, P., Krstulović-Opara, L., Galić, M. (eds.) Proceedings of the 9th International Congress of Croatian Society of Mechanics, USB flash drive. Croatian Society of Mechanics, Split (2018)
11. Polach, P., Byrtus, M., Šika, Z., Hajžman, M.: Fibre spring-damper computational models in a laboratory mechanical system and validation with experimental measurement. Interdisc. J. Discontinuity Nonlinearity Complexity **6**, 513–523 (2017)
12. Polach, P., Hajžman, M.: Influence of the fibre spring-damper model in a simple laboratory mechanical system on the coincidence with the experimental results. In: Font-Llagunes, J. M. (ed.) Proceedings of ECCOMAS Thematic Conference on Multibody Dynamics 2015, pp. 356–365. Universitat Politècnica de Catalunya, Barcelona (2015)
13. Shabana, A.A.: Flexible multibody dynamics: review of past and recent developments. Multibody Sys. Dyn. **1**, 189–222 (1997)
14. Taghirad, H.D., Bedoustani, Y.B.: An analytic-iterative redundancy resolution scheme for cable-driven redundant parallel manipulators. IEEE Trans. Rob. **27**, 1137–1143 (2011)
15. Wasfy, T.M., Noor, A.K.: Computational strategies for flexible multibody systems. Appl. Mech. Rev. **56**, 553–613 (2003)

Body-Fluid-Structure Interaction Simulation for a Trailing-Edge Flexible Stabilizer

Abolfazl Kiani and Meisam Mohammadi-Amin(✉)

Aerospace Research Institute, 1465774111 Tehran, Iran
mmohammadi@ari.ac.ir

Abstract. Flight stability is an interesting subject for the researchers and engineers working on design and development of flying objects. This subject will be more complicated when the body free oscillation and its mutual influence are added to it. In present investigation, a three-dimensional free oscillating body (unstable inherently) with a flexible strip attached to its trailing edge which utilized for stabilizing, is studied in a viscous subsonic flow. Navier–Stokes equations are used for fluid flow analysis and Euler–Bernoulli cantilevered beam is implemented for structure deflection modeling. For analyzing Fluid-Structure Interaction, an iterative partitioned coupling algorithm is utilized. With combining a dynamic simulation tool for body dynamics, the Body-Fluid-Structure Interaction (BFSI) is studied and analyzed. Consequently, an efficient framework is developed for multi-disciplinary analysis of highly flexible stabilizers.

1 Introduction

One of the typical means for stabilization purpose is using some kinds of stabilizers ranging from the rigid types like fins to the flexible ones e.g. ribbons or strips in various shapes and configurations. The flexible stabilizers are preferred especially in the case that mass and volume limitations are critical. However, the design process of flexible stabilizers needs several analyses including fluid-structure interaction investigation, body dynamic simulation and stability analyses in addition to different parametric surveys on the configuration, materials and also body-strip connection type. Thus, different analytic tools are essential for the multidisciplinary design study and they must be implemented in an efficient computational framework to interchange data and collaborate robustly towards the ultimate multi-physics solution.

Due to lack of such frameworks, most of similar problems have been studied experimentally during last decades. Levine and Dasser, conducted an experimental study on the characteristics of ribbons in the subsonic wind tunnel. This study was carried out based on the variation of the aspect ratio and rectangular ribbon material [1]. Auman and Dalek, Aumon and Wilkes and Lamar and Wain, conducted some empirical tests to determine the aerodynamic characteristics of ribbon stabilizers [2–4].

Wang et al. (2012), used numerical solutions for 3D page movement [5]. In one related study, Gomez and Leinheart, have used and combined commercial software to investigate body-fluid-structure interactions in the water that have got good results compared to empirical testing [6]. Xiao and Wang gave a numerical simulation of the

A. Kecskeméthy and F. Geu Flores (Eds.): ECCOMAS 2019, COMPUTMETHODS 53, pp. 239–246, 2020.
https://doi.org/10.1007/978-3-030-23132-3_29

vortex-induced vibrations of vertical raisers in linear and uniform shear flows, and the predicted numerical results are in good agreement with experimental results [7].

Dobrucali and Kinase, have concluded in their paper about prediction of vortex-induced vibration for circular cylinders using URANS, that vibration caused by vortex due to three different frequencies is highly nonlinear: the frequency of the flow, vortex fluctuation and oscillation [8]. Stabbil et al., have also developed a reduced order modeling for the analysis of flexible long cylinders [9].

The coupled simulation of flow-induced rigid body oscillations and attached flexible structure deformation has been studies rarely. In the present work, a hybrid approach, including a computer code and simultaneous use of existing solvers capabilities, is adopted to simulate the complex problem of Body-Fluid-Structure Interactions. Expected goals and achievements can be summarized as: (1) Developing a suitable computational framework for analyzing body-fluid-structure interaction. (2) Dynamic analysis of rigid body with flexible stabilizer attached to its end.

The main features of the present work in compare to the previous studies are, three-dimensionality of the solution, emphasize on the stabilization process, and developing an iterative partitioned framework for Body-Fluid-Structure interaction (BFSI) analysis. Hence, the first aim of study is to construct a proper computational framework including different solvers and routines that inter-relate various disciplines (dynamics, unsteady flow, structural) for analysis of BFSI phenomenon. The final goal is to analyze the stabilization effect of trailing edge flexible strip attached to 3D flying object and to study the effect of parameters like body center of gravity (c.g.) location and strip length and width, using the developed tool.

2 Governing Equations

2.1 Fluid Flow Governing Equations

The governing equations for fluid flow are Navier-Stokes equations, which are defined as below:

a- Continuity equation

$$\frac{\partial \rho}{\partial t} + \frac{\partial}{\partial x_j}(\rho U_j) = 0 \tag{1}$$

b- Momentum equation

$$\frac{\partial \rho U_i}{\partial t} + \frac{\partial}{\partial x_j}(\rho U_i U_j) = -\frac{\partial p}{\partial x_i} + \frac{\partial}{\partial x_j}\left[\mu_{eff}\left(\frac{\partial U_i}{\partial x_j} + \frac{\partial U_j}{\partial x_i}\right)\right] \tag{2}$$

c- Energy equation

$$\rho c\mu \frac{\partial T}{\partial t} + \rho c\mu U_i \frac{\partial T}{\partial x_i} = -P\frac{\partial U_i}{\partial x_i} + \lambda \frac{\partial^2 T}{\partial x_i^2} - \tau_{ij}\frac{\partial U_j}{\partial x_i} \tag{3}$$

where ρ is density, U is velocity, P is pressure and T is temperature of fluid. In order to solve these equations in viscous fluid mode and high Reynolds numbers, a turbulent modeling is needed, where the k-ε model is used here.

2.2 Flexible Structure Deformation Model

Considering the fact that the strip is considered thin, with a negligible width and the type of connection (clamped) to the body is such that its main oscillations are in the *XY* plane, such as a cantilevered beam with a distribution of extensive load, we use the relation of Euler-Bernoulli beam. So the final transverse motion equation under the extensive transverse force is:

$$w(x,t) = \sum_{i=1}^{\infty}\left[A_i \cos \omega_i t + B_i \sin \omega_i t + \frac{1}{\omega_i}\int_0^t Q_i(\tau)\sin \omega_i (t-\tau)d\tau\right]W_i(x) \quad (4)$$

where w is transverse displacement, W_i is natural modals and $Q_i\,(t)$ is the generalized force. The first two expressions represent free oscillation and the third term indicates the forced oscillation of the beam. A_i and B_i are fixed numbers and also are obtained using the initial conditions.

2.3 Rigid Body Dynamic Simulation

To simulate the dynamic of body, following equations are used. It is noticeable that moments are considered about the body center of mass.

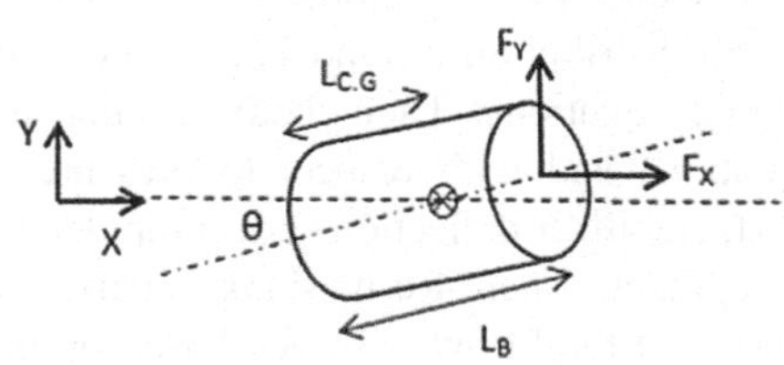

Fig. 1. Body and inserted forces from stabilizer

The equations of forces exerted on stabilizer concerning Fig. 1 (up) are determined as follows:

$$\begin{cases} F_x = \sum_{i=1}^{n}\left[P_i (A_x)_i + \tau_i (A_y)_i\right] \\ F_y = \sum_{i=1}^{n}\left[P_i (A_y)_i + \tau_i (A_x)_i\right] \end{cases} \quad (5)$$

where F_x is axial force, F_y is normal force, P_i is fluid pressure on surface element, τ_i is shear stress on surface element, A_x is projection of surface element on longitudinal direction, A_y is projection of surface element on transverse direction and i is the number of elements. The forces imposed on body from the fluid around it are obtained from inserting Eq. 5 on body elements.

The equation of moment exerted on body from the fluid around it, is as follows:

$$M_{Body} = \sum_{i=1}^{n} \left(\left[P_i (A_x)_i + \tau_i (A_y)_i \right] y_i + \left[P_i (A_y)_i + \tau_i (A_x)_i \right] x_i \right) \tag{6}$$

The equation of moment exerted on body from stabilizer concerning Fig. 1 (down) is as follows:

$$M_{Stablizer} = F_{x_{Stab}} * (L_B - L_{C.G}) sin(\theta) + F_{y_{Stab}} * (L_B - L_{C.G}) cos(\theta) \tag{7}$$

where $F_{x_{Stab}}$ and $F_{y_{Stab}}$ are respectively, axial and normal forces exerted on body from stabilizer in the contact point, L_B is body length, $L_{C.G}$ is the distance of body center of mass from its tip and θ is the angle of body longitudinal axis from horizon. Finally from relations (6) and (7) total moment for body stabilizing is obtained from below equation:

$$M_{Total} = M_{Body} + M_{Stablizer} \tag{8}$$

3 Methodology

The flowchart of the iterative partitioned coupling algorithm used in this paper is shown in Fig. 2. At first, pressure distribution around body, forces and moments are determined using fluid dynamic equations. Then, body attitude is determined using six degrees of freedom equations and with respect to pressure distribution and forces exerted on the body. Next, stabilizer deflections are computed using FSI equations and new specified points are updated. Then, Re-meshing is performed with respect to new structure points coordinate and fluid flow is solved based on new meshes. At the next step, a solution convergence criterion is considered and if verified, the new step is done. This process will be continued until the desired criterion is met.

In the staggered (parallel staggered) method, at every time step, the fluid-structure data are exchanged, and then the equations are solved simultaneously but independently. In the iterative (serial staggered) method, the fluid equations are solved first, and then the data are entered into the structural part and the structure equations are solved based on the new information of the fluid and its data are fed into the fluid section. The fluid equations are again solved based on the new structure data and this repeating cycle continues until the problem is completely solved. In this paper, iterative partitioned coupling algorithm is used for FSI analysis.

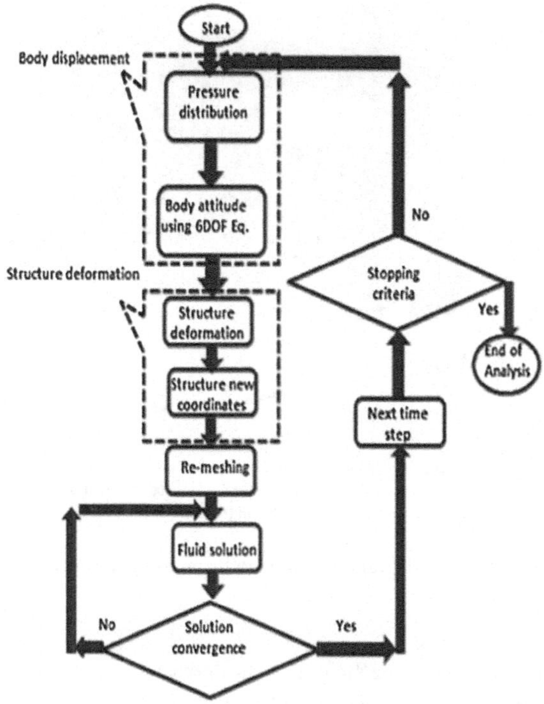

Fig. 2. BFSI solution flowchart

4 Statement of the Problem

In this section the problem is defined and grid study is performed. A three-dimensional cylinder (in the direction of longitudinal axis), along with a strip (flexible thin plate), at its end is intended to investigate the problem of fluid-structure interaction for a flow condition with Mach number 0.4 and an angle of attack 10°. This velocity is considered because the main phase of flying body mission is in this condition. Other flight conditions are not so important and we neglect them. In Fig. 3, geometry of body and strip and grids are shown.

Concerning that fluid flow analysis is in an unsteady state and the motion of the body and strip cause the grid of fluid domain to collapse, the moving mesh and Re-meshing are used at any time step. Because the geometry and the angles of cells are changed at any instant and are collapsed, therefore using of unstructured mesh is better and more appropriate.

However, because the body is rigid and only rotates, and in addition the grid of strip does not collapse; as a result, the structured mesh is used for both of them. In Fig. 3 (up), grids of body with strip and domain around it are presented.

In order to ensure the accuracy of the analysis and the independency of the results to the number of grid cells, some grids with different cell numbers were modeled which the numbers of cells were 78000, 120000, 330000, 590000 and 760000 cells. In Fig. 3

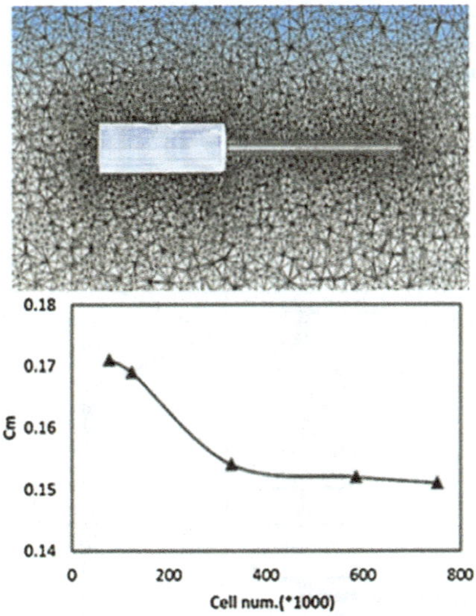

Fig. 3. Body/strip geometry and grid (up) and body pitching moment coefficient v.s grid cell number (down)

(down), the pitching moment coefficient is plotted versus the number of cells in each grid. It was found that for a model with 330000 cells, the pitching moment coefficient differs about 3% compared with 760000 cells model, Which is negligible, so it was used for next analysis.

5 Results and Discussions

First of all, we analyzed an elastic plate as a cantilevered beam which is attached to the end of a square, as presented in the work of Bazilevs [10]. As shown in Fig. 4, the transverse displacement of the end point of the plate is in good agreement in compare with the reference work results.

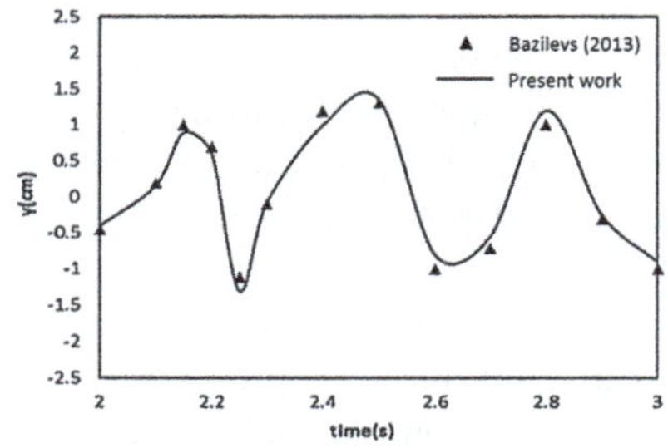

Fig. 4. Comparing of transverse displacement of end point of plate with the results of [10].

In the following, for all studies and investigations, Mach number 0.4 and angle of attack 10° are considered as the flight condition. In this investigation, finite volume method is used to solve the fluid flow. Since the flow conditions are within the compressible range, here a density based algorithm is used to associate the equations. Time discretization is performed implicitly. The standard $k - \varepsilon$ model is used for modeling the turbulent flow. The second-order accuracy has been used for spatial discretization, and the gradients have been calculated based on the least squares of the cell base. In Fig. 5 the oscillation of the body alone and body with strip are presented. The angle of body alone is increased gradually because it is unstable inherently. However, when the strip is added to the end of body, prevents from increasing the amplitude of oscillation by creating a restoring moment and changing the pressure distribution (Fig. 6).

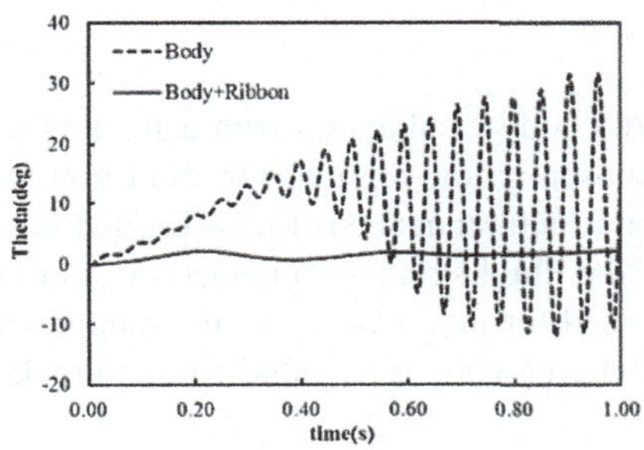

Fig. 5. Oscillations of body with and without strip v.s time

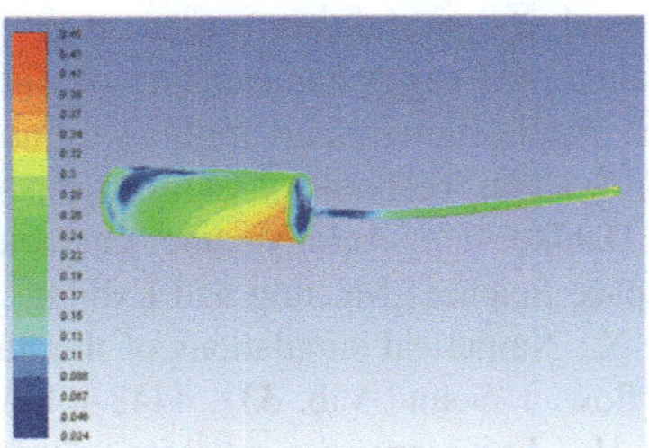

Fig. 6. Snapshot of pressure contours at t = 0.005 s

Concerning that the location of body center of mass affects its oscillations, two locations were considered for body center of mass that one of them is in the middle of body and the other one is in the front of body. It was resulted that oscillation intensity is higher when its center of mass is more backward. Finally, we studied variations of length of strip for three different lengths (L, 0.5L, 1.5L).The angle of body oscillation in Fig. 7 is presented for different strip lengths. As could be seen, the more length of strip prevents from the more oscillation angle of body.

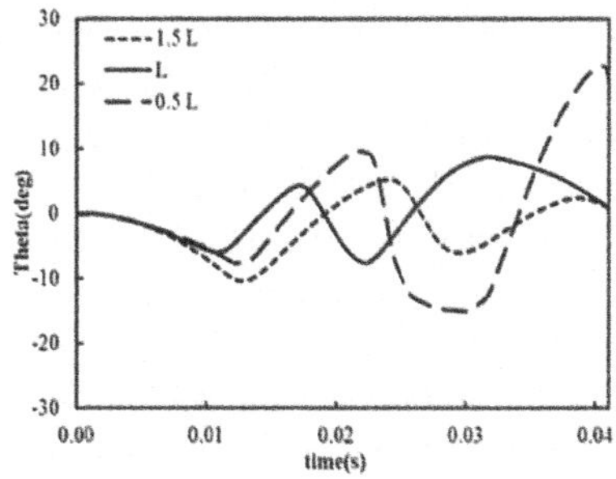

Fig. 7. Body oscillation angle vs. time for different lengths of strip

6 Conclusions

Numerical analysis of free body oscillations with a flexible stabilizer at trailing edge was done in a viscous subsonic flow regime. For fluid flow simulation, finite volume method was used and for structural analysis Euler–Bernoulli cantilevered beam theory was implemented. For the fluid-structure interaction solution, iterative partitioned coupling algorithm was used for interrelation. Combining simulation relations of body dynamics, the proposed framework is capable to capture the Body-Fluid-Structure Interactions (BFSI).

References

1. Levin, D., Daser, G., Shpund, Z.: On the aerodynamic drag of ribbons. AIAA-97-1525
2. Auman, L.M., Wayne Dahlke, C.: Drag characteristics of ribbons. AIAA 2001-2011
3. Auman, L.M., Wilks, B.L.: Application of fabric ribbons for drag and stabilization. In: AIAA Aerodynamic Decelerator Systems Technology Conference, AIAA 2005-161 (2005)
4. Auman, L.M., Wayne Dahlke, C.: Aerodynamic characteristics of ribbon stabilized grenades. In: 38th Aerospace Sciences Meeting and Exhibit, AIAA-2000-270 (2000)
5. Wang, Y., Yu, Z., Shao, X.: Numerical simulations of the flapping of a three-dimensional flexible plate in uniform flow. J. Sound Vib. **331**, 4448–4463 (2012)
6. Gomes, J.P., Leinhart, H.: Fluid-structure interaction-induced oscillation of flexible structures in laminar turbulent flows. J. Fluid Mech. **715**, 537–572 (2013)
7. Xiao, Q., Wang, E.: Numerical simulation of vortex-induced vibration of a vertical riser in uniform and linearly sheared currents. J. Ocean Eng. **121**, 492–515 (2016)
8. Dobrucali, E., Kinaci, O.K.: URANS-based prediction of vortex induced vibrations of circular cylinders. J. Appl. Fluid Mech. **10**(3), 957–970 (2017)
9. Stabile, G., Matthies, H.G., Borri, C.: A novel reduced order model for vortex induced vibrations of long flexible cylinders. J. Ocean Eng. **156**, 191–207 (2018)
10. Bazilevs, Y., Takizava, K.: Computational Fluid-Structure Interaction: Methods and Applications. Wiley, New York (2013)

Investigation of a Model Update Technique for Flexible Multibody Simulation

Andreas Schulze[1](✉), Johannes Luthe[1], János Zierath[2], and Christoph Woernle[1]

[1] Chair of Technical Dynamics, University of Rostock, Justus-von-Liebig-Weg 6, 18059 Rostock, Germany
{andreas.schulze2,johannes.luthe,woernle}@uni-rostock.de
[2] W2E Wind to Energy GmbH, Grubenstrasse 44, 18055 Rostock, Germany
jzierath@wind-to-energy.de

Abstract. In modern engineering applications, the build up of representative numerical models may be not feasible or involves considerable effort. As a result, the need for experimental investigation on the real structures and the implementation of the measured results into the numerical model arises. Within this contribution, the integration of experimentally derived data into the model build up process of a flexible body within multibody dynamics is investigated. In contrast to the conventional modelling approach, the experimental based approach omits the need of an underlying finite element model of the flexible structure. Both approaches are investigated and compared on a simple experimental test case of a flexible beam-like structure.

1 Introduction

Within the last two decades, flexible multibody dynamics has established itself as a standard method for the efficient and reliable numerical simulation of complex mechanical systems. However, with the constant growth of system complexity accompanied with modern simulation tasks, mass, stiffness and damping distribution of a flexible structure are either not known sufficiently or cannot be modelled satisfactorily without considerable effort. A solution to this problem may lie in the field of model update (MU) [6]. In MU, an underlying numerical model is matched to modal data derived from experimental identification of the real structure. In general, the numerous MU methods at hand are applied to FE models. Updating a multibody model therefore requires the MU of the flexible parts in the finite element domain and involves considerable computational effort. The direct application of MU to the complete multibody model on the other hand is subject to considerable limitations which inevitably lead to the neglect of the mostly strong non-linear transient behaviour of the multibody systems. As a result, very few research activities were carried out in the direct

A. Kecskeméthy and F. Geu Flores (Eds.): ECCOMAS 2019, COMPUTMETHODS 53, pp. 247–254, 2020.
https://doi.org/10.1007/978-3-030-23132-3_30

application of MU to multibody models, see [5]. A promising method to circumvent this issue was developed in [9] during the early days of flexible multibody simulation. That method enables modelling of the flexible system components directly through experimentally modal data under the assumption that the mass distribution of the system is known. However, a thorough investigation of this method was not carried out in later research activities and was only partially resumed in [10]. Nevertheless, in light of the complexity of modern systems and the very powerful and easily accessible modern measurement systems, the importance of the method is redefined, and it resurfaced recently in [4], where attention was paid to spatial incomplete measurement models.

The aim of this contribution is a thorough investigation and further refinement of the experimental approach to establish the method in the field of modern flexible multibody dynamics. First, the dynamics of flexible multi body systems are briefly described and important aspects for the proposed method are worked out. In Sect. 3 each individual step of the proposed method is reconstructed and compared to the conventional modelling approach. Finally, both methods are investigated on a simple beam-like structure in Sect. 4.

2 Dynamics of a Flexible Body in Multibody Simulation

A standard approach to describe the motion of a flexible body in multibody dynamics is given by the floating frame of reference formulation [7,8]. Here, the rigid body motion is described relative to the global frame, while the deformation is described relative to the body fixed floating frame of reference (FFR), see Fig. 1.

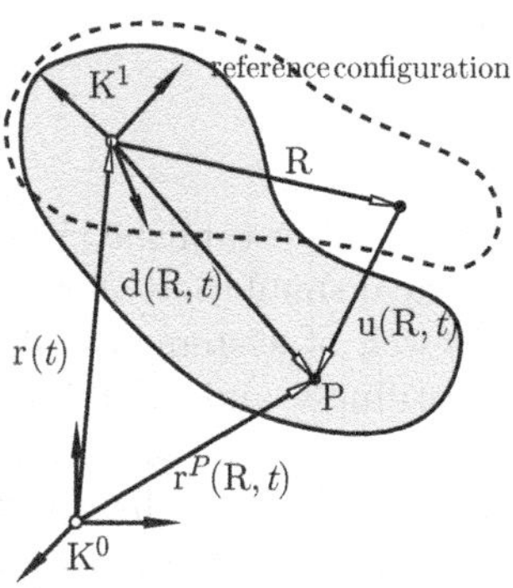

$$\mathrm{r}^P(\mathrm{R},t) = \mathrm{r}(t) + \mathrm{d}(\mathrm{R},t) = \mathrm{r}(t) + \mathrm{R} + \mathrm{u}(\mathrm{R},t)$$

$$\mathrm{u}(\mathrm{R},t) = \Phi(\mathrm{R})\,\mathrm{q}(t)$$

$$\dot{\mathrm{r}}^P(\mathrm{R},t) = \dot{\mathrm{r}}(t) + \tilde{\omega}\,\overbrace{(\mathrm{R} + \Phi(\mathrm{R})\,\mathrm{q}(t))}^{\mathrm{d}(\mathrm{R},\mathrm{q})} + \Phi(\mathrm{R})\dot{\mathrm{q}}(t) = \left[\mathrm{E} \;\; -\tilde{\mathrm{d}}(\mathrm{R},\mathrm{q}) \;\; \Phi(\mathrm{R})\right]\left[\dot{\mathrm{r}}(\mathrm{t}) \;\; \omega(t) \;\; \dot{\mathrm{q}}(t)\right]^{\top}$$

Fig. 1. Kinematics of a flexible body in floating frame of reference formulation

The equations of motion of a single, unconstrained body have the form [7,8]

$$\underbrace{\begin{bmatrix} \mathbf{M}^{tt} & \mathbf{M}^{tr} & \mathbf{M}^{tf} \\ & \mathbf{M}^{rr} & \mathbf{M}^{rf} \\ sym. & & \mathbf{M}^{ff} \end{bmatrix}}_{\mathbf{M}} \underbrace{\begin{bmatrix} \ddot{\mathrm{r}} \\ \dot{\omega} \\ \ddot{\mathrm{q}} \end{bmatrix}}_{\mathbf{b}} + \underbrace{\begin{bmatrix} \mathbf{0} \\ \mathbf{0} \\ \mathbf{K}^f\mathbf{q} + \mathbf{D}^f\dot{\mathbf{q}} \end{bmatrix}}_{\mathbf{k}^i} + \underbrace{\begin{bmatrix} \mathbf{k}^t \\ \mathbf{k}^r \\ \mathbf{k}^f \end{bmatrix}}_{\mathbf{k}^c} - \underbrace{\begin{bmatrix} \mathbf{h}^t \\ \mathbf{h}^r \\ \mathbf{h}^f \end{bmatrix}}_{\mathbf{h}} = \begin{bmatrix} \mathbf{0} \\ \mathbf{0} \\ \mathbf{0} \end{bmatrix} \quad (1)$$

with the total mass matrix $\mathbf{M}$, the vector of accelerations $\mathbf{b}$ and the vectors of generalized Coriolis and centrifugal forces $\mathbf{k}^c$, generalized internal forces $\mathbf{k}^i$ and generalized external forces $\mathbf{h}$. The superscripts t, r, f denote the quantities regarding to rigid body motion (translation t, rotation r) and flexible deformation (f) and corresponding coupling terms, respectively.

After linearisation of the equations of motion in the flexible coordinates $\mathbf{q}$, the unknown quantities in the total mass matrix $\mathbf{M}$ are given by

$$\begin{aligned}
\mathbf{M}^{tt} &= \int_V \mathbf{E}\,\rho \mathrm{d}V && \cong m\,\mathbf{E}\,,\\
\mathbf{M}^{tr} &= \int_V \tilde{\mathbf{d}}\,\rho\,\mathrm{d}V && \cong \mathbf{C0} + (\mathbf{C1q})^{\tilde{}},\\
\mathbf{M}^{tf} &= \int_V \boldsymbol{\Phi}^\top \rho\,\mathrm{d}V && \cong \mathbf{C1}\,,\\
\mathbf{M}^{ff} &= \int_V \boldsymbol{\Phi}^\top \boldsymbol{\Phi}\,\rho\,\mathrm{d}V && \cong \mathbf{C3}\,,\\
\mathbf{M}^{rr} &= \int_V \tilde{\mathbf{d}}\tilde{\mathbf{d}}^\top \rho\,\mathrm{d}V && \cong \mathbf{J0} - \textstyle\sum_{n=1}^{n_q} (\mathbf{C4}_n + \mathbf{C4}_n^\top)\, q_n\,,\\
\mathbf{M}^{rf} &= \int_V \boldsymbol{\Phi}^\top \tilde{\mathbf{d}}^\top \rho\,\mathrm{d}V && \cong \mathbf{C2} + \textstyle\sum_{n=1}^{n_q} \mathbf{C5}_n\, q_n\,.
\end{aligned} \tag{2}$$

Similarly, the generalized Coriolis and centrifugal forces $\mathbf{k}^c$ follow from

$$\begin{aligned}
\mathbf{k}^t &= \int_V \tilde{\boldsymbol{\omega}}\,(\dot{\mathbf{r}} + 2\boldsymbol{\Phi}\,\dot{\mathbf{q}} + \tilde{\boldsymbol{\omega}}\,\mathbf{d})\,\rho\,\mathrm{d}V && \cong m\tilde{\boldsymbol{\omega}}\,\dot{\mathbf{r}} + 2\tilde{\boldsymbol{\omega}}\,\mathbf{C1}\,\dot{\mathbf{q}} + \tilde{\boldsymbol{\omega}}\,\tilde{\boldsymbol{\omega}}\,(\mathbf{C0} + \mathbf{C1}\,\mathbf{q})\,,\\
\mathbf{k}^r &= \int_V \tilde{\mathbf{d}}\,\tilde{\boldsymbol{\omega}}\,(\dot{\mathbf{r}} + 2\boldsymbol{\Phi}\,\dot{\mathbf{q}} + \tilde{\boldsymbol{\omega}}\,\mathbf{d})\,\rho \mathrm{d}V && \cong (\mathbf{C0} + \mathbf{C1}\,\mathbf{q})^{\tilde{}}\,\tilde{\boldsymbol{\omega}}\,\dot{\mathbf{r}}\\
& && \quad -2\textstyle\sum_{n=1}^{n_q} \left(\mathbf{C4}_n + \sum_{m=1}^{n_q} \mathbf{C6}_{mn}\, q_m\right) \dot{q}_n \boldsymbol{\omega}\\
& && \quad +\tilde{\boldsymbol{\omega}}\,\mathbf{M}^{rr}\,\boldsymbol{\omega}\,,\\
\mathbf{k}^f &= \int_V \boldsymbol{\Phi}^\top \tilde{\boldsymbol{\omega}}\,(\dot{\mathbf{r}} + 2\boldsymbol{\Phi}\,\dot{\mathbf{q}} + \tilde{\boldsymbol{\omega}}\,\mathbf{d})\,\rho\,\mathrm{d}V && \cong \mathbf{C1}\tilde{\boldsymbol{\omega}}\,\dot{\mathbf{r}} + 2\textstyle\sum_{n=1}^{n_q} \mathbf{C5}_n^\top \dot{q}_n\,\tilde{\boldsymbol{\omega}}\\
& && \quad +\boldsymbol{\omega}^\top \left[\mathbf{C4}_k + \textstyle\sum_{n=1}^{n_q} \mathbf{C6}_{kn}\, q_m\right] \boldsymbol{\omega}\,.
\end{aligned} \tag{3}$$

In (2), (3) and hereafter the notation of [7] is adopted where n^q is the number of flexible coordinates, and $m\,\mathbf{E}$, $\mathbf{J0}$ and $\mathbf{C0}$-$\mathbf{C6}$ are time-independent volume integrals with $\mathbf{E}$ being the identity matrix. The vector of the generalized external forces $\mathbf{h}$ can also be calculated from the constant volume integrals [7].

Finally, the structural stiffness matrix $\mathbf{K}^f$ and structural damping matrix $\mathbf{D}^f$ in the vector of internal forces $\mathbf{k}^i$ can be calculated from

$$\begin{aligned}
\mathbf{K}^f &= \int_V \mathbf{B}^\top(\mathbf{R})\,\mathbf{H}(\mathbf{R})\,\mathbf{B}(\mathbf{R})\,\mathrm{d}V && \cong \boldsymbol{\lambda}\,,\\
\mathbf{D}^f &= \alpha\,\mathbf{M}^{ff} + \beta\,\mathbf{K}^f && \cong \alpha\,\mathbf{C3} + \beta\,\boldsymbol{\lambda}\,,
\end{aligned} \tag{4}$$

where $\mathbf{H}$ is the material tensor, and $\mathbf{B}$ is the strain tensor, see [7].

From Eqs. (2)–(4) it is evident that the calculation of the time-independent volume integrals $m\,\mathbf{E}$, $\mathbf{J0}$, $\mathbf{C0}$-$\mathbf{C6}$ and $\boldsymbol{\lambda}$ is mandatory for the construction of the equations of motion (1). This study aims to construct these integrals based on experimental data and compares first results to a multibody model derived with the conventional approach. Once the terms are built up correctly, further coupling of the flexible body to a multibody system is a straight forward task, see e.g. [7,8], and is therefore not included within this contribution.

3 Conventional Modelling Approach vs. Experimental Synthesis

The conventional approach to build up a flexible body in multibody dynamics starts with discretisation of the body by means of the finite element method (FEM). Here, the distribution of mass/inertia and stiffness of the body is represented by the mass matrix $\mathbf{M}^{FE}$ and stiffness matrix $\mathbf{K}^{FE}$, respectively. The damping behaviour of the body is often assumed to be of proportional type, which is also adopted in (4). Generally, the FE discretisation results in a large number of flexible degrees of freedom, and model order reduction (MOR) schemes of the type

$$\mathbf{q} = \boldsymbol{\Phi}^f(\mathbf{R})\,\mathbf{q}^f \tag{5}$$

are applied to the FE model. Here, a wide range of efficient MOR strategies is available for the build up of the reduction base $\boldsymbol{\Phi}^f$, see e.g. [3].

In a next step, rigid body motion is imposed to the FE body using FFR formulation. The rigid body motion is described by the three rigid body modes for translation $\boldsymbol{\Phi}^t$ and the three rigid body modes for rotation $\boldsymbol{\Phi}^r$ which are simply constructed from the FE mesh by

$$\boldsymbol{\Phi}^t = \left[\cdots \left[\mathbf{E}_n\ \mathbf{0}\right]^\top \cdots\right]^\top, \ \boldsymbol{\Phi}^r = \left[\cdots \left[\tilde{\mathbf{R}}_n\ \mathbf{E}\right]^\top \cdots\right]^\top, \ n = 1, \ldots, n^{FE}. \tag{6}$$

In (6) n^{FE} is the number of FE nodes and $\mathbf{R}_n$ is the location of a FE node relative to the FFR.

As described in detail in [7], the volume integrals $m\mathbf{E}$, $\mathbf{J0}$, $\mathbf{C0}$-$\mathbf{C6}$, $\boldsymbol{\lambda}$, necessary for flexible multibody simulation, can be calculated completely from the FE matrices $\mathbf{M}^{FE}$, $\mathbf{K}^{FE}$, $\boldsymbol{\Phi}^f$ and the rigid body modes $\boldsymbol{\Phi}^t$, $\boldsymbol{\Phi}^r$. The integrals required to derive the $\mathcal{O}(\mathbf{q}^0)$ terms in (1) are calculated from

$$\begin{aligned} &m\mathbf{E} = \boldsymbol{\Phi}^{t\top}\mathbf{M}^{FE}\boldsymbol{\Phi}^t, \ \mathbf{J0} = \boldsymbol{\Phi}^{r\top}\mathbf{M}^{FE}\boldsymbol{\Phi}^r, \ \mathbf{C0} = \boldsymbol{\Phi}^{r\top}\mathbf{M}^{FE}\boldsymbol{\Phi}^t, \ \mathbf{C1} = \boldsymbol{\Phi}^{t\top}\mathbf{M}^{FE}\boldsymbol{\Phi}^f, \\ &\mathbf{C2} = \boldsymbol{\Phi}^{r\top}\mathbf{M}^{FE}\boldsymbol{\Phi}^f, \ \mathbf{C3} = \boldsymbol{\Phi}^{f^\top}\mathbf{M}^{FE}\boldsymbol{\Phi}^f, \quad \boldsymbol{\lambda} = \boldsymbol{\Phi}^{f^\top}\mathbf{K}^{FE}\boldsymbol{\Phi}^f, \end{aligned} \tag{7}$$

and the integrals necessary for the calculation of the $\mathcal{O}(\mathbf{q}^1)$ terms follow from

$$\begin{aligned} \mathbf{C4}_i &= \left[\tfrac{1}{2}\boldsymbol{\Phi}^{r\top}_\alpha \left(\mathbf{M}^{FE}\widehat{\omega}_\beta + \widehat{\omega}_\beta \mathbf{M}^{FE}\right)\boldsymbol{\Phi}^f_i\right] \text{ with } \widehat{\omega}_\beta = \tfrac{\partial}{\partial \omega_\beta}\mathrm{diag}\left(\tilde{\omega}_n\right), \\ \mathbf{C6}_{ij} &= \left[\boldsymbol{\Phi}^{f^\top}_i \widehat{\omega}_\alpha \mathbf{M}^{FE}\widehat{\omega}_\beta \boldsymbol{\Phi}^f_j\right], \ \mathbf{C5}_\alpha = \tfrac{1}{2}\boldsymbol{\Phi}^{f^\top}\left(\mathbf{M}^{FE}\widehat{\omega}_\alpha + \widehat{\omega}_\alpha \mathbf{M}^{FE}\right)\boldsymbol{\Phi}^f \end{aligned} \tag{8}$$

where the subscripts α, β are cyclic permutations $\alpha, \beta = \{1, 2, 3\}$, the subscript n is the FE node index and i, j are the flexible coordinate indices.

With the FE model and (5) to (8) the conventional modelling approach is completed, and the equations of motion (1) of the flexible body can be built up.

Now the objective is to build up a dynamically equivalent multibody model based on measurements taken on the body. This process will be referred to as experimental synthesis (ES) within the scope of this paper. In contrast to the conventional approach, in ES a modal model

$$\hat{\boldsymbol{\Phi}}^f = \left[\boldsymbol{\Phi}_i\right], \quad \boldsymbol{\Lambda} = \mathrm{diag}\left(f_i^2(1 + \mathrm{i}\eta_i)\right), \quad i = 1, \ldots, i^{ES} \tag{9}$$

derived from an experimental modal analysis on the body serves as the basis of the investigation, with i^{ES} being the number of modes, $\boldsymbol{\Phi}_i$ being the ith flexible modeshape of the body scaled to unity modal mass and f_i, η_i being the corresponding natural frequency and modal damping. A comprehensive analysis on the experimental identification of a modal model is outside the scope of this contribution, instead reference is made to the classical literature in [1].

The key concept of ES is to exploit the MOR step (5) in the conventional approach to integrate the flexible body into the equations of motion (1) as a modally reduced structure. This implies the reduction base from (5) to be equal to the experimental modeshape matrix from (9). Although modal truncation is a rather poor method compared to modern MOR schemes, which means that more mode shapes may be required, it leads to a powerful advantage for the ES as the structural mass, stiffness and damping matrix can be calculated directly from measurement results

$$\mathbf{M}^{ff} = \mathbf{C3} = \mathbf{E}\,, \quad \mathbf{K}^{f} = \boldsymbol{\lambda} = \mathrm{diag}\left(\overline{m}_i\,\overline{f}_i^2\right), \quad \mathbf{D}^{f} = \mathrm{diag}\left(\overline{\eta}_i\right), \tag{10}$$

and with this, the need for an underlying FE model is omitted.

Similar to the conventional approach, rigid body motion of the flexible body is accounted for by calculating the rigid body modes from

$$\hat{\boldsymbol{\Phi}}^t = \left[\cdots\left[\mathbf{E}_n\;\mathbf{0}\right]^\top\cdots\right]^\top, \hat{\boldsymbol{\Phi}}^r = \left[\cdots\left[\tilde{\mathbf{R}}_n\;\mathbf{E}\right]^\top\cdots\right]^\top, n = 1, \ldots, n^{ES}. \tag{11}$$

In (11) n^{ES} is the number of measurement points on the body, and $\mathbf{R}_n$ describes the location of a measurement point n relative to the FFR.

To this point, two out of the nine necessary volume integrals in (7) and (8), namely $\mathbf{C3}$ and $\boldsymbol{\lambda}$ as well as the matrices $\hat{\boldsymbol{\Phi}}^f$, $\hat{\boldsymbol{\Phi}}^t$ and $\hat{\boldsymbol{\Phi}}^r$, can be calculated. Additionally the physical damping properties of the body is accounted for with (10). From (7), (8) it is obvious that for the calculation of the remaining volume integrals the description of the body mass distribution ($\mathbf{M}^{FE}$ in the conventional approach) is indispensable. As pointed out by [9] this information is not identified from experimental modal analysis, and a lumped mass formulation is proposed. To solve this issue, [4] suggests using a CAD model of the body or an underlying FE mass matrix to lump the bodies mass and inertia to the grid of measurement points using static condensation followed by a matrix update.

Within this study, the authors suggest a different approach to distribute the body mass and inertia to the to the grid of measurement points. The following prerequisites are necessary for this: The total body mass m, center of mass and inertia tensor $\mathbf{J}$ are required, which can be either derived experimentally, see e.g. [2], or from a CAD model of the body. The FFR is positioned at the body center of mass which also implies modeshapes of the body in free-free condition. These modeshapes can then be summarized to $\bar{\boldsymbol{\Phi}} = [\hat{\boldsymbol{\Phi}}^t\;\hat{\boldsymbol{\Phi}}^r\;\hat{\boldsymbol{\Phi}}^f]$, and modal reduction of the flexible body results in

$$\bar{\mathbf{M}} = \bar{\boldsymbol{\Phi}}^\top \mathbf{M}^{ES} \bar{\boldsymbol{\Phi}} = \begin{bmatrix} \hat{\boldsymbol{\Phi}}^{t\top}\mathbf{M}^{ES}\hat{\boldsymbol{\Phi}}^t & \hat{\boldsymbol{\Phi}}^{t\top}\mathbf{M}^{ES}\hat{\boldsymbol{\Phi}}^r & \hat{\boldsymbol{\Phi}}^{t\top}\mathbf{M}^{ES}\hat{\boldsymbol{\Phi}}^f \\ & \hat{\boldsymbol{\Phi}}^{r\top}\mathbf{M}^{ES}\hat{\boldsymbol{\Phi}}^r & \hat{\boldsymbol{\Phi}}^{r\top}\mathbf{M}^{ES}\hat{\boldsymbol{\Phi}}^f \\ sym. & & \hat{\boldsymbol{\Phi}}^{f\top}\mathbf{M}^{ES}\hat{\boldsymbol{\Phi}}^f \end{bmatrix} = \begin{bmatrix} m\mathbf{E} & \mathbf{0} & \mathbf{0} \\ & \mathbf{J} & \mathbf{0} \\ sym. & & \mathbf{E} \end{bmatrix} \tag{12}$$

where the only unknown quantity $\mathbf{M}^{ES}$ comprises the body mass and inertia lumped to the grid of measurement points. From Eq. (12) the equivalent total mass matrix can be derived as a least squares approximation of the form

$$\mathbf{M}^{ES} = \bar{\boldsymbol{\Phi}}^{+\top} \bar{\mathbf{M}} \bar{\boldsymbol{\Phi}}^{+} = \bar{\boldsymbol{\Phi}}^{+\top} \begin{bmatrix} m\mathbf{E} & \mathbf{0} & \mathbf{0} \\ & \mathbf{J} & \mathbf{0} \\ sym. & & \bar{\mathbf{m}} \end{bmatrix} \bar{\boldsymbol{\Phi}}^{+}. \tag{13}$$

With equations (9)–(13) a complete set of dynamically equivalent matrices $\mathbf{M}^{ES}$ (equivalent to $\mathbf{M}^{FE}$), $\boldsymbol{\Lambda}$ (equivalent to $\mathbf{K}^{FE}$ and $\mathbf{D}^{FE}$), $\hat{\boldsymbol{\Phi}}^{f}$ (equivalent to $\boldsymbol{\Phi}^{f}$) and $\hat{\boldsymbol{\Phi}}^{t}$,$\hat{\boldsymbol{\Phi}}^{r}$ (equivalent to $\boldsymbol{\Phi}^{t}$, $\boldsymbol{\Phi}^{r}$) is introduced. Based on these matrices, the volume integrals necessary for the equations of motion of the flexible body can be calculated with (7), (8) solely derived from measurements on the real physical body without the need for a preceding FE analysis.

4 Experimental Test Case

The ES and the conventional approach are investigated on a simple beam-like structure in free-free condition, with the corresponding measurement-setup shown in Fig. 2. The measurements on the structure are performed using uni-axial accelerometers and an impulse hammer. To sufficiently capture the dynamic response of the structure, a set of 57 measurement points is chosen. Figure 3 shows the corresponding FE model of the structure built up in MSC.Nastran. The identification of modeshapes, eigenvalues and modal damping from the measured frequency response is carried out using the methods of Experimental Modal Analysis (EMA) [1]. A comparison of measured and numerically derived modal data is given in Table 1 and Fig. 4 where an overall good correlation is observed.

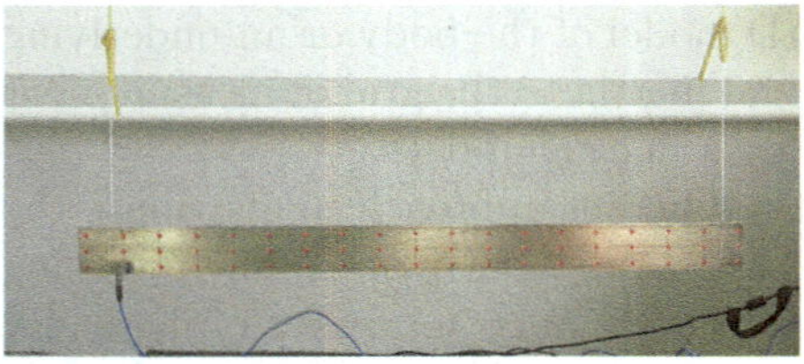

Fig. 2. Measurement setup

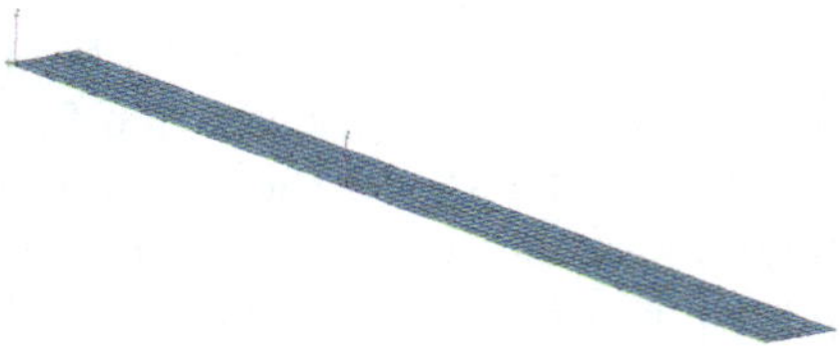

Fig. 3. Finite Element model

In a next step, the time-independent volume integrals of the structure are calculated according to Sect. 3 for the two approaches. In both approaches modal truncation was performed as MOR scheme based on the first 6 modeshapes in transversal direction of the systems corresponding to the results in Table 1 and Fig. 4. The mass information to build up the equivalent mass matrix $\mathbf{M}^{ES}$ in (12) is drawn from the FE model which is not necessary as described in Sect. 3 but

Table 1. Natural frequencies (FEM and EMA) and damping ratio of (EMA only)

Mode Nr.	Natural Freq. [Hz]		Damp. [-]
	FEM	EMA	EMA
1	100.63	99.03	5.564e−4
2	277.34	273.45	2.155e−4
3	543.55	536.48	1.275e−4
4	813.20	806.62	1.733e−3
5	898.15	886.56	2.718e−4
6	1340.80	1322.40	4.356e−4

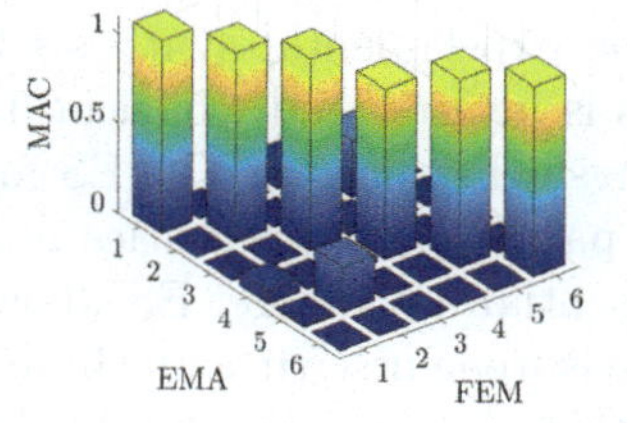

Fig. 4. Mode shape correlation

facilitates the comparison of the two approaches. As shown in Sect. 2, the time-independent volume integrals determine the equations of motions of the flexible multibody body. A deep insight into the quality of the two approaches may therefore be drawn from a direct comparison of these quantities, see Fig. 5. In Fig. 5 the differences of the terms $m\mathbf{E}, \mathbf{I0}, \mathbf{C0}$-$\mathbf{C2}$ corresponding to the first-order rigid body properties as well as modal masses $\mathbf{C3}$ are shown to be negligible. This underlines the dynamic equivalence of the two approaches since the same total body mass m and inertia $\mathbf{J}$ were chosen for $\mathbf{M}^{ES}$. While, Fig. 5 exhibits the strong advantage of ES, a difference is observed in the structural stiffness and damping matrix $\mathbf{K}^f, \mathbf{D}^f$, which results from the difference between experimental results of the real structure and the FE model. The volume integrals corresponding to the second order terms, represented in Fig. 5 by $\mathbf{C4}_4$, $\mathbf{C5}_1$ and $\mathbf{C6}_{14}$, are strongly influenced by the deformation modeshapes, and differences arise caused by differences of the numerically derived modeshapes and the experimental modeshapes.

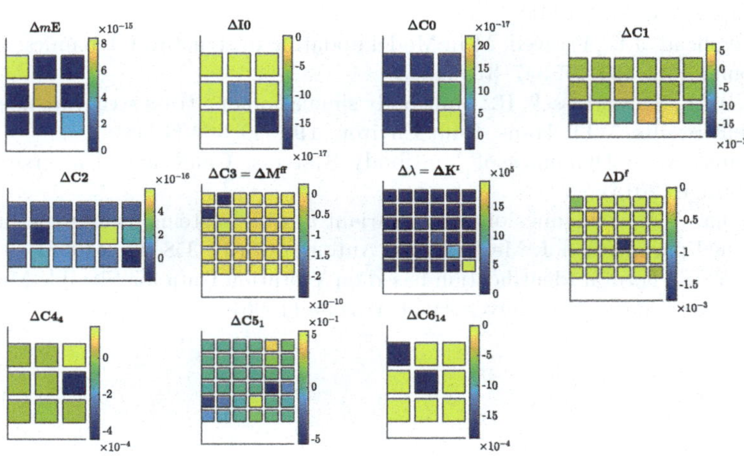

Fig. 5. Difference between volume integrals

5 Conclusion

The experimental synthesis to build up a flexible body in multibody dynamics is investigated which utilizes modal data derived from measurements on the physical structure. The ES method is elaborated in detail, and special attention is paid to the similarities and differences to the conventional approach. As an essential advantage, ES allows the flexible body to be constructed solely from measurements without the need for an underlying FE model. Herein, two conditions have to be met by ES. First the total body mass, center of mass and inertia tensor are known, either from measurements or from a CAD model of the structure and second, the floating frame of reference is attached to the center of mass of the body. Otherwise a lumped mass matrix of the body has to be constructed for ES. Both approaches are investigated on a simple experimental test case of a beam-like structure and first results underline the advantage of ES. Future research will focus on transient analysis of the derived multibody model as well as the application of ES to complex mechanical structures.

References

1. Ewins, D.J.: Modal Testing: Theory. Practice and Application. Research Studies Press Ltd., Hertfordshire (2000)
2. Doniselli, C., Gobbi, M., Mastinu, G.: Measuring the inertia tensor of vehicles. Veh. Syst. Dyn. **37**, 311–313 (2002)
3. Fehr, J., Eberhard, P.: Error-controlled model reduction in flexible multibody dynamics. J. Comput. Nonlinear Dyn. **5**(3), 031005–031005-8 (2010)
4. Lein, C., Woller, J., Hopf, H., Beitelschmidt, M.: Approach for modelling flexible bodies based on experimental data with utilization in elastic multibody simulation. In: ECCOMAS Thematic Conference on Multibody Dynamics, Prag (2017)
5. Manzato, S., Peeters, B., Toso, A., van der Auweraer, H., Osgood, R.: Model updating methodologies for multibody simulation models: application to a full-scale wind turbine. In: Proulx, T. (ed.) Linking Models and Experiments, vol. 2. Springer, New York (2011)
6. Mottershead, J.E., Friswell, M.I.: Model updating in structural dynamics: a survey. J. Sound Vib. **167**(2), 347–375 (1993)
7. Sachau, D., Schwertassek, R.: Multibody simulation of a truss structure using finite element results. WIT Trans. Built Environ. **19**, 349–364 (1996)
8. Shabana, A.A.: Dynamics of Multibody Systems. Cambridge University Press, Cambridge (2014)
9. Shabana, A.A.: Dynamics of inertia-variant flexible systems using experimentally identified parameters. J. Mech. Trans. Autom. **108**(3), 358–366 (1986)
10. Yi, T.Y.: Structural Identification based on Vibration Data for Flexible Multibody System Dynamics. The University of Arizona (1996)

Extension of the Iterative Improved Reduced System Technique to Flexible Mechanisms

Alessandro Cammarata(✉), Rosario Sinatra, and Pietro Davide Maddio

Dipartimento di Ingegneria Civile e Architettura,
University of Catania, Via. S. Sofia 64, 95125 Catania, Italy
alessandro.cammarata@unict.it

Abstract. In the present work the iterative Improved Reduced System technique (IRS) is extended to flexible mechanisms to include the joint constraints. Starting from the two-step method described in [1] by the same authors, we create an iterative method similar to the iterative IRS used in structural mechanics. Finally, the method is applied to a RSCR spatial mechanism to find the natural frequencies.

1 Introduction

Iterative dynamic condensation techniques capture the inertia effect in the dynamic condensation matrix. These techniques were invented to improve the accuracy of Guyan static condensation [2]. The first attempt was a predictor-corrector scheme called the *two-step method* [3,4]. The predictor step employs the static condensation to find the frequencies and mode shapes at masters nodes. Then, in the corrector step, the mode shapes at slave nodes are computed using partial or full mode expansion techniques as Kidder's [5] or Miller's [6] expansions. Finally, the two sets of mode shapes are used to build a new subspace in which a new reduced model is defined. Some of the two-step methods that have been proposed in the literature need for expensive computational effort because the eigenvectors are not simultaneously evaluated. Besides, the accuracy of the results depends of the static prediction and is strongly influenced by bad choices on masters definition. To cope with this issues, the two-step methods were extended using iterative techniques. Sauer [7] proposed an iterative method in which the mode shapes at slaves are updated during subsequent iterations. Sometimes, this method proved not to converge because the static modes of the masters are not updated during the iterations. Some attempts to obtain convergent iterative methods were proposed by Blair *et al.* [8] but the first convergent iterative method, referred to as *iterative IRS* or *IRS* (*Improved Reduced System*), was proposed by Friswell, Garvey and Penny in 1995 [9]. In spite of the efforts of researchers the original IRS suffered from low convergence speed and high demand for computational resources. In [10] Xia and Lin modified the

A. Kecskeméthy and F. Geu Flores (Eds.): ECCOMAS 2019, COMPUTMETHODS 53, pp. 255–263, 2020.
https://doi.org/10.1007/978-3-030-23132-3_31

iterative formula of the transformation matrix to increase the convergence speed to make the IRS comparable to the faster *subspace iteration method* (SIM). Choi et al. [11] combined the IRS with sub-structuring techniques to reduce the computational effort and to avoid the dependency of the IRS on the masters choice. In order to provide efficient solutions for large-scale eigenvalue problems, Weng et al. [12,13] extended the IRS to calculate the structural responses and response sensitivities.

Most of the dynamic condensation methods were developed in the realm of the structural mechanics. In recent years, researchers are drawing on Matrix Structural Analysis to extend the formulation to mechanisms [14]. These linear methods are well-suited for optimization tasks and design. In [15] Cammarata used condensed matrices to study complex mechanical systems. The condensation matrices were built by means of the static condensation. Then, in [1] the authors extended the static condensation method developing a two-step method for mechanisms. Here, the two-step method is extended to an iterative IRS to provide some improvements of results in the high frequency range. The iterative IRS technique will be based on the Miller's expansion while the predictor scheme will be based on the static condensation developed in [15]. Finally, the extended IRS will be applied to a spatial mechanism of type RSCR (Revolute-Spherical-Cylindrical-Revolute).

2 Extension of the Iterative Improved Reduced System

In [1] the same authors demonstrated that considering a mechanical system partitioned into inner and joint nodes it is possible to derive a transformation matrix $\mathbf{L}$ that statically reduce the joint node displacements $\mathbf{u}_j$ to the boundary node displacements $\mathbf{u}_b$ nodes. Moreover, the inner node displacements $\mathbf{u}_i$ are reduced to the joint node displacements $\mathbf{u}_j$ using the static Guyan-Iron method. The final transformation matrix Ψ_L is defined as

$$\begin{bmatrix} \mathbf{u}_i \\ \mathbf{u}_j \end{bmatrix} = \Psi_L \mathbf{u}_b \Rightarrow \Psi_L = \begin{bmatrix} \mathbf{R}_L \\ \mathbf{L} \end{bmatrix} \tag{1}$$

where $\mathbf{R}_L = \mathbf{R}_G \mathbf{L}$ and $\mathbf{R}_G$ is the classic Guyan-Iron transformation matrix. The matrix Ψ_L makes it possible to obtain the following reduced eigensystem

$$\mathbf{M}_L \ddot{\mathbf{u}}_b + \mathbf{K}_L \mathbf{u}_b = \mathbf{0} \tag{2}$$

where $\mathbf{M}_L = {\Psi_L}^T \mathbf{M} \Psi_L$ and $\mathbf{K}_L = {\Psi_L}^T \mathbf{K} \Psi_L$ are the reduced inertia and stiffness matrix, respectively. Solving the corresponding eigenvalue problem the boundary modes ϕ_{bk} can be obtained. Then, using the Miller's expansion, appropriately updated to take into account the joint-nodes, the inner modes ϕ_{ik} can be expressed in terms of the joint modes ϕ_{jk}:

$$\phi_{ik} \approx -[\mathbf{K}_{ii}^{-1}\mathbf{K}_{ij} + \lambda_k(-\mathbf{K}_{ii}^{-1}\mathbf{M}_{ij} + \mathbf{K}_{ii}^{-1}\mathbf{M}_{ii}\mathbf{K}_{ii}^{-1}\mathbf{K}_{ij})]\phi_{jk}, \quad k = 1, \ldots, m \tag{3}$$

Considering that $\phi_{jk} = \mathbf{L}\phi_{bk}$ the Miller's expansion becomes $\phi_{ik} = \mathbf{R}\phi_{bk}$, $(k = 1, \ldots, m)$, where $\mathbf{R}$ is referred to as the *dynamic condensation matrix* and it is defined as

$$\mathbf{R} = -[\mathbf{K}_{ii}^{-1}\mathbf{K}_{ij}\mathbf{L} - \mathbf{K}_{ii}^{-1}(\mathbf{M}_{ij} - \mathbf{M}_{ii}\mathbf{K}_{ii}^{-1}\mathbf{K}_{ij})\mathbf{L}\mathbf{M}_{L}^{-1}\mathbf{K}_{L}] \tag{4}$$

It can be demonstrated that $\mathbf{R}$ can be expressed as

$$\mathbf{R} = \mathbf{R}_L + \mathbf{K}_{ii}^{-1}(\mathbf{M}_{ij}\mathbf{L} + \mathbf{M}_{ii}\mathbf{R}_L)\mathbf{M}_{L}^{-1}\mathbf{K}_{L} \tag{5}$$

in which the static transformation $\mathbf{R}_L$ is the constant term of the power series expansion. In [1] we described that starting from $\mathbf{R}_L$ it is possible to define a two-step strategy to find $\mathbf{R}$ [16]. Here, we iterate this strategy to create a method that resembles the iterative IRS used in structural mechanics. To obtain an iterative method the previous Eq. (5) is rewritten as

$$\mathbf{R}^{(i)} = \mathbf{R}^{(0)} + \mathbf{K}_{ii}^{-1}(\mathbf{M}_{ij}\mathbf{L} + \mathbf{M}_{ii}\mathbf{R}^{(i-1)}){\mathbf{M}_R^{(i-1)}}^{-1}{\mathbf{K}_R}^{(i-1)} \tag{6}$$

in which the following positions are assumed: $\mathbf{R}^{(0)} = \mathbf{R}_L$, ${\mathbf{M}_R}^{(0)} = \mathbf{M}_L$, and ${\mathbf{K}_R}^{(0)} = \mathbf{K}_L$. The reduced stiffness and mass matrices ${\mathbf{K}_R}^{(i)}$ and ${\mathbf{M}_R}^{(i)}$ are obtained considering the transformation matrix

$$ {\Psi_R}^{(i)} = \begin{bmatrix} \mathbf{R}^{(i)} \\ \mathbf{L} \end{bmatrix}, \quad \text{with} \quad {\Psi_R}^{(0)} = \Psi_L \tag{7}$$

The iterative Improved Reduced System technique is summarized in the Algorithm 1. The procedures ends when a given criterion defined by the user is satisfied.

Algorithm 1. Iterative Improved Reduced System technique

1: Initialize $i \leftarrow 0$
2: find $\mathbf{L} \leftarrow$ given $\mathbf{K}$, $\mathbf{H}$, $\mathbf{G}$
3: obtain $\mathbf{R}_L = \mathbf{R}^{(0)}$, ${\mathbf{K}_R}^{(0)} = \mathbf{K}_L$, ${\mathbf{M}_R}^{(0)} = \mathbf{M}_L$
4: find $\mathbf{R}^{(i)}$ from Eq. (6)
5: build ${\Psi_R}^{(i)}$ from Eq. (7)
6: update ${\mathbf{K}_R}^{(i)}$ and ${\mathbf{M}_R}^{(i)}$ using ${\Psi_R}^{(i)}$
7: Update $i \leftarrow i + 1$
8: Go to step 4

3 Spatial Mechanism: RSCR

In this section the case study of the RSCR spatial mechanism is discussed. The RSCR is a one degree of freedom spatial mechanism that forms a closed kinematic chain in which three links are joined by means of a revolute joint (R), a spherical joint (S), a cylindrical joint (C), and another revolute joint (R), respectively. For the kinematic description of the RSCR mechanism we used the natural coordinates, as described in [17]. Referring to the Fig. 1 the natural coordinates of the three moving points 1, 2 and 3 and the three components of the unit vector $\mathbf{u}_1$ have been chosen as unknowns for the positioning problem. Other basic points and vectors, displayed in the same figure, have been considered fixed and the corresponding geometric parameters are reported in Table 1.

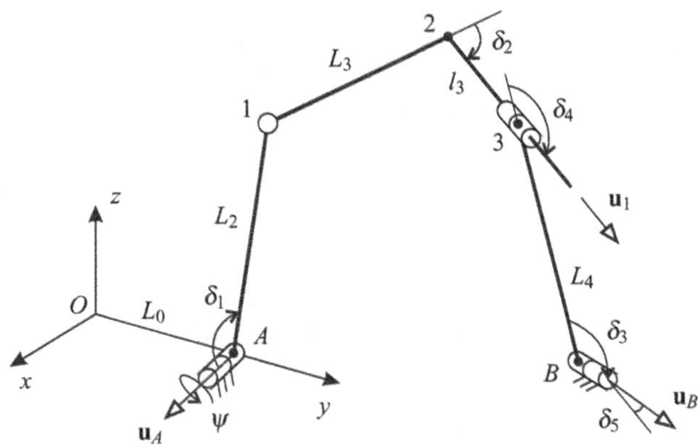

Fig. 1. Layout of the RSCR mechanism.

Table 1. Geometric parameters of the RSCR mechanism.

Notation	Description	Value	Unit
L_0	Length	0.1	(m)
L_2	Length	0.5	(m)
L_3	Length	0.3	(m)
L_4	Length	0.6	(m)
l_3	Length	$\|\mathbf{r}_3 - \mathbf{r}_2\|$	(m)
δ_1	Angle	$\pi/2$	(rad)
δ_2	Angle	$\pi/2$	(rad)
δ_3	Angle	$\pi/2$	(rad)
δ_4	Angle	$\pi/2$	(rad)
δ_5	Angle	$\pi/4$	(rad)
$\mathbf{u}_A$	Fixed unit vector	$[1,0,0]^T$	(m)
$\mathbf{u}_B$	Fixed unit vector	$[0,1,0]^T$	(m)
$\mathbf{r}_O$	Reference point	$[0,0,0]^T$	(m)
$\mathbf{r}_A$	Reference point	$[0,L_0,0]^T$	(m)
$\mathbf{r}_B$	Reference point	$[-0.1,0.8,0]^T$	(m)

3.1 Positioning Problem

The positioning problem is the first step of the proposed method. This problem is usually non-linear because involves quadratic terms and trigonometrical functions. Natural coordinates simplify the positioning problem reducing the complexity of the constraint equations that include only linear and quadratic terms in the considered unknowns. The system of constraint equations $\Psi = 0$ is recalled as below:

$$\begin{aligned}
\Psi_1 &\equiv (\mathbf{r}_1 - \mathbf{r}_A)^T(\mathbf{r}_1 - \mathbf{r}_A) - L_2^2 = 0 \\
\Psi_2 &\equiv (\mathbf{r}_1 - \mathbf{r}_A)^T(\mathbf{r}_O - \mathbf{r}_A) - L_O L_2 \cos(\psi) = 0 \\
\Psi_3 &\equiv (\mathbf{r}_1 - \mathbf{r}_A)^T \mathbf{u}_A - L_2 \cos(\delta_1) = 0 \\
\Psi_4 &\equiv (\mathbf{r}_2 - \mathbf{r}_1)^T(\mathbf{r}_2 - \mathbf{r}_1) - L_3^2 = 0 \\
\Psi_5 &\equiv (\mathbf{r}_2 - \mathbf{r}_1)^T \mathbf{u}_1 - L_3 \cos(\delta_2) = 0 \\
\Psi_6 &\equiv \mathbf{u}_1^T \mathbf{u}_1 - 1 = 0 \\
\Psi_{7-9} &\equiv (\mathbf{r}_3 - \mathbf{r}_2) \wedge \mathbf{u}_1 = 0 \\
\Psi_{10} &\equiv (\mathbf{r}_3 - \mathbf{r}_B)^T(\mathbf{r}_3 - \mathbf{r}_B) - L_4^2 = 0 \\
\Psi_{11} &\equiv (\mathbf{r}_3 - \mathbf{r}_B)^T \mathbf{u}_B - L_4 \cos(\delta_3) = 0 \\
\Psi_{12} &\equiv (\mathbf{r}_3 - \mathbf{r}_B)^T \mathbf{u}_1 - L_4 \cos(\delta_4) = 0 \\
\Psi_{13} &\equiv \mathbf{u}_B^T \mathbf{u}_1 - \cos(\delta_5) = 0
\end{aligned}$$

The system $\Psi = 0$ is solved using the Newton-Raphson algorithm to obtain $\mathbf{r}_1$, $\mathbf{r}_2$, $\mathbf{r}_3$, and $\mathbf{u}_1$. The rotation angle ψ of the revolute joint at point A is used as input parameter for the positioning problem.

3.2 Static Transformation Matrix

Once the positioning problem has been solved the mechanism is divided into three dimensional beam elements. The first link of length L_2 has been divided into eight finite elements. The second link has eight elements for the segment of length L_3 and six for the segment of variable length l_3. Finally, the link of length L_4 is divided into eight elements. The mass and stiffness matrices of each element are obtained using the structural and inertial parameters reported in Table 2.

To apply the iterative IRS all node displacements must be expressed in terms of boundary node displacements. For the case considered, the set of joint nodes includes six nodes, in fact the end nodes of each moving link. The boundary nodes that we selected for the RSCR are located at points 1, 2, and 3 of Fig. 1. Finally, all remaining are inner nodes.

Table 2. Inertial and structural parameters of the RSCR mechanism.

Notation	Description	Value	Unit
D	Outer diameter	0.016	(m)
d	Inner diameter	0.014	(m)
E	Young's modulus	110	(GPa)
G	Shear modulus	42.3	(GPa)
ρ	Material density	1600	(kg/m^3)

To build the static transformation matrix $\mathbf{L}$ introduced in Sect. 2 we used the following expressions for the joint matrix $\mathbf{H}_j$ and the rigid-body transformation matrix $\mathbf{G}_j$ [1,15]:

$$\mathbf{H}_j = \begin{bmatrix} \mathbf{h}_A^R & \mathbf{O} & \mathbf{O} & \mathbf{0} \\ \mathbf{0} & \mathbf{O} & \mathbf{O} & \mathbf{0} \\ \mathbf{0} & \mathbf{H}^S & \mathbf{O} & \mathbf{0} \\ \mathbf{0} & \mathbf{O} & \mathbf{O} & \mathbf{0} \\ \mathbf{0} & \mathbf{O} & \mathbf{O} & \mathbf{0} \\ \mathbf{0} & \mathbf{O} & \mathbf{H}^C & \mathbf{0} \\ \mathbf{0} & \mathbf{O} & \mathbf{O} & \mathbf{0} \\ \mathbf{0} & \mathbf{O} & \mathbf{O} & \mathbf{h}_B^R \end{bmatrix}, \quad \mathbf{G}_j = \begin{bmatrix} \mathbf{O} & \mathbf{O} & \mathbf{O} \\ \mathbf{1} & \mathbf{O} & \mathbf{O} \\ \mathbf{1} & \mathbf{O} & \mathbf{O} \\ \mathbf{O} & \mathbf{1} & \mathbf{O} \\ \mathbf{O} & \mathbf{1} & \mathbf{O} \\ \mathbf{O} & \mathbf{O} & \mathbf{1} \\ \mathbf{O} & \mathbf{O} & \mathbf{1} \\ \mathbf{O} & \mathbf{O} & \mathbf{O} \end{bmatrix}$$

where

$$\mathbf{h}_A^R = \begin{bmatrix} \mathbf{0} \\ \mathbf{u}_A \end{bmatrix}, \quad \mathbf{H}^S = \begin{bmatrix} \mathbf{O} \\ \mathbf{1} \end{bmatrix}, \quad \mathbf{H}^C = \begin{bmatrix} \mathbf{0} & \mathbf{u}_1 \\ \mathbf{u}_1 & \mathbf{0} \end{bmatrix}, \quad \mathbf{h}_B^R = \begin{bmatrix} \mathbf{0} \\ \mathbf{u}_B \end{bmatrix}$$

Finally, by introducing $\Delta_j = \mathbf{H}_j{}^T\mathbf{K}_{jj}\mathbf{H}_j$, the static transformation matrix $\mathbf{L}$ is written as

$$\mathbf{L} = (\mathbf{H}_j\Delta_j{}^{-1}\mathbf{H}_j{}^T\mathbf{K}_{ji}\mathbf{K}_{ii}{}^{-1}\mathbf{K}_{ij} - \mathbf{1}_j)^{-1}(\mathbf{H}_j\Delta_j{}^{-1}\mathbf{H}_j{}^T\mathbf{K}_{jj}\mathbf{G}_j - \mathbf{G}_j).$$

3.3 Numerical Results

We studied the RSCR mechanism considering an input angle $\psi = 1.86$ (rad). Moreover, an actuation stiffness $k_\theta = 10,000$ (N.m/rad) has been added to the revolute joint at point A. We applied the proposed iterative IRS method to obtain the natural frequencies of the system. Table 3 reports the first eighteen natural frequencies in function of the IRS iterations. The full model results have been also added for comparison. Figure 2 shows the high frequency range varying the iterations of the IRS methods.

Observing the numerical results it is clear that the IRS method becomes important to investigate the high frequency range. For the first ten frequencies only two steps are needed to get values with an error below 1 (Hz). For higher frequencies the number of iterations to obtain convergence grows. Particularly,

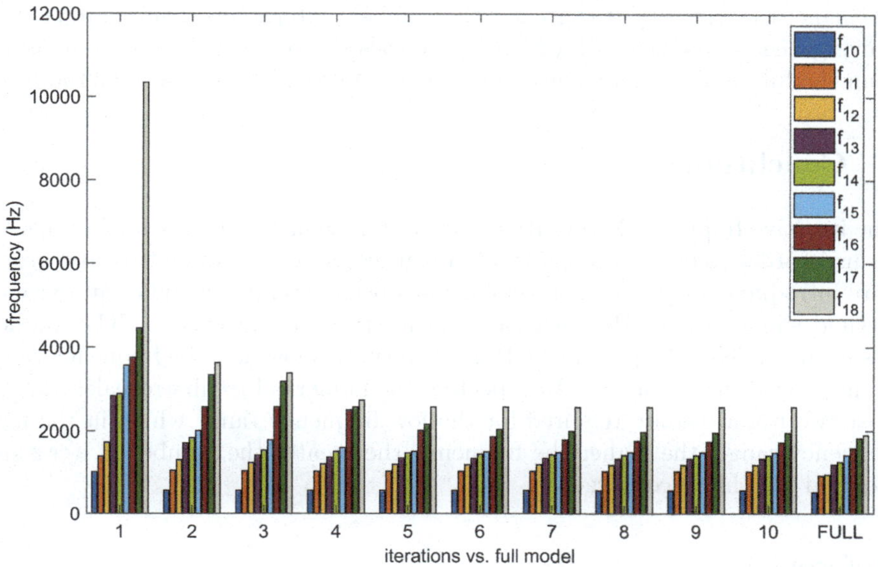

Fig. 2. High frequency range variation in function of the IRS method iterations.

Table 3. Natural frequencies of the RSCR mechanism.

Iter	f_1	f_2	f_3	f_4	f_5	f_6	f_7	f_8	f_9
1	28.861	35.125	48.919	81.703	142.495	231.254	458.180	628.889	725.867
2	28.851	35.095	48.870	81.497	140.830	219.236	374.340	469.788	528.013
3	28.851	35.095	48.870	81.497	140.830	219.233	374.146	469.534	526.874
4	28.851	35.095	48.870	81.497	140.830	219.232	374.130	469.515	526.775
5	28.851	35.095	48.870	81.497	140.830	219.232	374.123	469.502	526.738
6	28.851	35.095	48.870	81.497	140.830	219.232	374.119	469.494	526.718
7	28.851	35.095	48.870	81.497	140.830	219.232	374.116	469.487	526.704
8	28.851	35.095	48.870	81.497	140.830	219.232	374.114	469.481	526.695
9	28.851	35.095	48.870	81.497	140.830	219.232	374.113	469.477	526.687
10	28.851	35.095	48.870	81.497	140.830	219.232	374.112	469.473	526.681
Full	28.894	36.074	49.309	81.705	140.051	214.374	341.232	466.011	491.637
Iter	f_{10}	f_{11}	f_{12}	f_{13}	f_{14}	f_{15}	f_{16}	f_{17}	f_{18}
1	1011.865	1387.363	1728.544	2835.125	2889.086	3567.989	3750.863	4449.354	10339.722
2	567.904	1048.493	1301.186	1702.736	1834.823	2007.681	2576.927	3334.532	3628.284
3	564.387	1033.364	1232.855	1412.785	1538.478	1780.449	2575.381	3180.741	3379.694
4	564.115	1029.003	1205.037	1363.062	1507.143	1596.098	2501.751	2575.311	2735.058
5	563.970	1027.178	1194.653	1346.622	1469.983	1504.612	2013.062	2159.093	2575.307
6	563.867	1025.997	1189.172	1338.299	1440.932	1494.229	1854.829	2038.072	2575.307
7	563.790	1025.117	1185.583	1333.104	1426.051	1488.294	1790.942	1995.784	2575.307
8	563.731	1024.431	1183.016	1329.424	1416.778	1483.865	1755.850	1972.882	2575.307
9	563.685	1023.881	1181.080	1326.637	1410.369	1480.348	1733.280	1958.327	2575.307
10	563.647	1023.431	1179.563	1324.437	1405.639	1477.467	1717.550	1948.228	2575.307
Full	518.063	920.135	946.531	1189.928	1254.589	1413.341	1580.419	1818.826	1894.718

the higher the frequency the greater the number of iterations required to achieve convergence. Nevertheless, for high frequencies the converged values remain far from the full model results confirming the limitations of any static methods.

4 Conclusions

The iterative Improved Reduced System technique has been extended to mechanisms in order to derive the natural frequencies. Starting from a two-step procedure introduced in [1] the methodology has been extended to build an iterative method similar to the IRS method used in structural mechanics. The method has been applied to a spatial RSCR mechanism to show how the frequency range changes with the iterations. As expected, the numerical results revealed that at least two iterations are required for the low frequency range while, in the high frequency range, the higher the frequency the greater the number of iterations required to achieve convergence.

References

1. Cammarata, A., Sinatra, R., Maddio, P.: A two-step algorithm for the dynamic reduction of flexible mechanisms. In: IFToMM Symposium on Mechanism Design for Robotics, pp. 25–32. Springer (2018)
2. Guyan, R.J.: Reduction of stiffness and mass matrices. AIAA J. **3**(2), 380 (1965)
3. Papadopoulos, M., Garcia, E.: Improvement in model reduction schemes using the system equivalent reduction expansion process. AIAA J. **34**(10), 2217–2219 (1996)
4. Qu, Z.Q., Pannee, R.: Two-step methods for dynamic condensation. In: 19th AIAA Applied Aerodynamics Conference, p. 1230 (2001)
5. Kidder, R.L.: Reduction of structural frequency equations. AIAA J. **11**(6), 892–892 (1973)
6. Miller, C.A.: Dynamic reduction of structural models. J. Struct. Div. **106**(10), 2097–2108 (1980)
7. Sauer, G.: Iterative improvement of eigensolutions from reduced matrices. Commun. Appl. Numer. Methods **5**(5), 329–335 (1989)
8. Blair, M.A., Camino, T.S., Dickens, J.M.: An iterative approach to a reduced mass matrix. In: 9th Conference International Modal Analysis Conference (IMAC), vol. 1, pp. 621–626 (1991)
9. Friswell, M., Garvey, S., Penny, J.: Model reduction using dynamic and iterated IRS techniques. J. Sound Vib. **186**(2), 311–323 (1995)
10. Xia, Y., Lin, R.: Improvement on the iterated IRS method for structural eigensolutions. J. Sound Vib. **270**(4–5), 713–727 (2004)
11. Choi, D., Kim, H., Cho, M.: Iterative method for dynamic condensation combined with substructuring scheme. J. Sound Vib. **317**(1–2), 199–218 (2008)
12. Weng, S., Xia, Y., Xu, Y.L., Zhu, H.P.: An iterative substructuring approach to the calculation of eigensolution and eigensensitivity. J. Sound Vib. **330**(14), 3368–3380 (2011)
13. Weng, S., et al.: Dynamic condensation approach to calculation of structural responses and response sensitivities. Mech. Syst. Signal Process. **88**, 302–317 (2017)

14. Klimchik, A., Pashkevich, A., Chablat, D.: Fundamentals of manipulator stiffness modeling using matrix structural analysis. Mech. Mach. Theory **133**, 365–394 (2019)
15. Cammarata, A.: Unified formulation for the stiffness analysis of spatial mechanisms. Mech. Mach. Theory **105**, 272–284 (2016)
16. O'Callahan, J.C.: A procedure for an improved reduced system (IRS) model. In: Proceedings of the 7th International Modal Analysis Conference, vol. 1, pp. 17–21, Las Vegas (1989)
17. De Jalon, J.G., Bayo, E.: Kinematic and Dynamic Simulation of Multibody Systems: The Real-time Challenge. Springer, New York (2012)

Updating of Finite Element Models for Controlled Multibody Flexible Systems Through Modal Analysis

Dario Richiedei(✉) and Alberto Trevisani

Università degli Studi di Padova, DTG Vicenza, Stradella San Nicola 3, 36100 Vicenza, Italy
{dario.richiedei,alberto.trevisani}@unipd.it

Abstract. Model updating is widely used for the estimation of the correct parameters of finite element models. Indeed, model accuracy should be as high as possible to synthesize controllers and observers, as well as for fault detection. Model updating based on experimental measurements is hence necessary to ensure model correctness.

A procedure tailored for updating dynamic models of flexible-link multibody systems (FLMSs) modelled through ordinary differential equations is proposed in this paper. The aim is to correct mass, stiffness matrices for modeling accurately the main vibrational modes in the frequency range of interest. Once consistency between the model coordinates and the modal data is obtained through a suitable transformation, model updating will be solved an optimization problem that also accounts for bounds on the feasible values. The method exploits model linearization, since it allows for modal analysis.

The paper also discusses some issues in performing experimental analysis in the presence of FLMSs that are not in an asymptotically stable equilibrium configuration. First of all, it is proposed, and also validated experimentally, the use of balancing springs. Then, it is discussed theoretically and numerically the use of active control to set the mechanism in an asymptotically stable equilibrium configuration, by tackling the issues of spillover due to the controller.

1 Introduction

Model updating is often used in structural dynamics for the identification of the parameters of finite element models [4, 5, 9]. Indeed, the availability of correctly tuned models is an essential step for state estimation, control synthesis, fault diagnosis. Most of the approaches developed in the literature are based on modal analysis by comparing the natural frequencies, and less frequently the mode shapes too, of the predicted and the experimental vibrational modes.

When dealing with flexible mechanisms, model updating based on modal analysis should relies on the investigation of the local vibrational modes, i.e. on performing experimental modal analysis in a neighborhood of an asymptotically-stable equilibrium configuration. Hence, model updating techniques could be successfully extended to flexible multibody systems, whenever their specific issues are properly addressed. For example, when the mechanism lies in the vertical plane, gravity forces should be

A. Kecskeméthy and F. Geu Flores (Eds.): ECCOMAS 2019, COMPUTMETHODS 53, pp. 264–271, 2020.
https://doi.org/10.1007/978-3-030-23132-3_32

compensated in the experimental modal analysis to keep the mechanism in the original position. In contrast, the absence of an asymptotically-stable equilibrium would make the mechanism diverge from the initial configuration after the excitation applied and therefore the system dynamics would not meet the basic assumption beneath the linearized model. Asymptotical stability of the equilibrium can be achieved either by passive control, such as supporting the mechanism with springs, or by active control, i.e. controlling it through a closed-loop control scheme that regulates the motor torques according to the sensed displacements. In both cases, the effect of the springs or the control on the vibrational modes should be properly tackled in the inverse eigenvalue problem solved for model updating, to account for frequency shift and changing in mode shapes. In contrast, the use of brakes, such as those usually employed in electric motors, changes the boundary conditions. Hence, model updating leads to less reliable results. This paper proposes some preliminary results on model updating for controlled multibody systems, by discussing the effect of the control on the modal properties and by proposing an approach for model updating that handles this issue. A parametric technique is proposed, which is based on the model of the controlled system and on an objective function that represents the difference between analytical and experimental results. Experimental validation is also proposed.

2 General Description of the Updating Method

2.1 System Model

Let us consider the dynamic model of an undamped flexible multibody systems with scleronomous and holonomic constraints. It is assumed that the system equations are linearized about an equilibrium configuration, that can be either stable or unstable for the mechanism alone. The use of linearized models has been proved to be very satisfactory in the case of small displacements about the equilibrium points and hence they can be successfully used in the synthesis of control schemes and state observers [2, 3]. Additionally, eigenvectors and eigenvalues can be easily computed to identify mode shapes and natural frequencies and therefore performing model updating through linearization. A minimal set of ODE is adopted, since it leads to the exact spectrum. Indeed, it has been proved that models employing second-order ordinary differential equations coupled to a set of algebraic constraint equations, does not allow for the straightforward use of such a kind of analysis and the spectrum of the linearized model is strongly affected by the linearization method. The total motion of the links is separated into the large rigid-body motion of an equivalent rigid-link system (ERLS) and the small elastic displacements $\mathbf{u}$ with respect to the ERLS [2, 3]. The ERLS is the moving reference configuration from which elastic displacements are measured through finite element representations. Let us define $\mathbf{q}$ as the vector of the ERLS generalized coordinates, and u the vector of the elastic displacements of the nodes of the FE model with respect to the ERLS. The linearized model is the first-order terms of the Taylor's expansion computed about the equilibrium configuration of the ERLS $\mathbf{q_e}$, and recalls the one usually adopted in structural dynamics with locally constant mass, stiffness and input matrices (named respectively $\mathbf{M}$, $\mathbf{K}$, $\mathbf{B}$):

$$\begin{bmatrix} \mathcal{M}_e & \mathcal{M}_e \mathbf{S} \\ \mathbf{S}^T \mathcal{M}_e & \mathbf{S}^T \mathcal{M}_e \mathbf{S} \end{bmatrix}_{\mathbf{q}=\mathbf{q}_e} \begin{Bmatrix} \ddot{\mathbf{u}}(t) \\ \ddot{\mathbf{q}}(t) \end{Bmatrix} + \begin{bmatrix} \mathcal{K}_e & \mathbf{0} \\ \mathbf{0} & \mathcal{K}_g \end{bmatrix}_{\mathbf{q}=\mathbf{q}_e} \begin{Bmatrix} \mathbf{u}(t) \\ \mathbf{q}(t) \end{Bmatrix} = \begin{bmatrix} \mathbf{I} \\ \mathbf{S}^T \end{bmatrix}_{\mathbf{q}=\mathbf{q}_e} \{\mathbf{f}\} \quad (1)$$

$\mathcal{M}_e$ and $\mathcal{K}_e$ are the mass and stiffness matrices of all the finite elements and are computed about the equilibrium; $\mathbf{g}$ is the vector of gravity acceleration. $\mathbf{S} = \mathbf{S}(\mathbf{q})$ is the ERLS sensitivity coefficient matrix for all the nodes, that relates the velocities of the ERLS generalized coordinates to the ones of all the nodes of the ERLS. $\mathcal{K}_g$ represents the effect of gravity. $\mathbf{f}$ collects the external nodal forces.

2.2 General Issues of Model Updating

The displacement vector of the model is usually not compatible with the experimental data since the measured dofs are usually less than those of the model. In contrast, model updating needs one-to-one correspondence between the measured mode shapes and the model dofs. This correspondence can be achieved either by expanding the experimental modes [1] or by reducing the model [6, 7]. The first approach is usually the most effective one.

Another issue to tackle is to transformation the measured mode shapes from the physical reference adopted for experimental analysis (i.e. the mechanism in its static configuration), to the fictitious one defined by the ERLS. Indeed, the displacement vector of the model includes the elastic displacements with respect to the ERLS and the displacement of the ERLS itself. Hence, transformation should be performed through the kinematic constraint equations of the ERLS.

Finally, since the mode shape and the natural frequencies change with the configuration [2, 8], more equilibrium points can be studied.

2.3 A Method for Model Updating

In this work, model updating is performed on just the parameters of the mass and stiffness matrices, by exploiting the undamped model [1]. Damping is neglected in the first phase since its identification, e.g. through the Rayleigh coefficients, can be performed separately as often done in the literature. Hence, a first order problem with real eigenvectors and eigenvalues is formulated. Let $(\omega_i^2, \boldsymbol{\eta}_i)$, $i = 1, \ldots, n$ be the i-th eigenpair, where $n \leq N$ is the number of measured experimental modes (N is the number of model dofs, including both rigid and elastic coordinates), ω_i is the natural frequency and $\boldsymbol{\eta}_i$ is the mode shape.

Following a widespread approach, updating of mass and stiffness matrices is modeled through the additive corrections $\Delta\mathbf{M} \in R^{N \times N}$ and $\Delta\mathbf{K} \in R^{N \times N}$. The resulting updated matrices are $\mathbf{M}^O + \Delta\mathbf{M}$ and $\mathbf{K}^O + \Delta\mathbf{K}$, where $\mathbf{M}^O \in R^{N \times N}$ and $\mathbf{K}^O \in R^{N \times N}$ are the nominal system matrices, based on the nominal parameters.

Therefore, the following eigenvalue problem must hold for all the measured eigenpairs:

$$\mathbf{0} = \omega_i^2 \left[\mathbf{M}^O + \Delta\mathbf{M}\right]\boldsymbol{\eta}_i - \left[\mathbf{K}^O + \Delta\mathbf{K}\right]\boldsymbol{\eta}_i \tag{2}$$

By introducing the following Jacobian matrices $\mathbf{J}_{m_h} = \left[\partial\mathbf{M}^O / \partial m_h^O\right]$ and $\mathbf{J}_{k_h} = \left[\partial\mathbf{K}^O / \partial k_h^O\right]$, the updating matrices can be written in the following forms

$$\Delta\mathbf{M} = \sum_{h=1}^{N_m} m_h^O \mathbf{J}_{m_h} \bar{m}_j \quad \Delta\mathbf{K} = \sum_{h=1}^{N_k} k_h^O \mathbf{J}_{k_h} \bar{k}_h \tag{3}$$

After some transformation of the eigenvalue problem, that are here omitted for brevity, model updating is finally cast as the following constrained minimization problem, that is formulated in a pre-conditioned form to ensure good numerical conditioning:

$$\min_{\chi \in \Gamma} \left\| \chi - \left(\mathbf{A}^T\mathbf{W}^T\mathbf{W}\mathbf{A} + \lambda\boldsymbol{\Omega}\right)^{-1}\mathbf{A}^T\mathbf{W}^T\mathbf{W}\mathbf{d} \right\|_2^2 \tag{4}$$

In Eq. (4) all the unknown parameters to be updated (i.e. the entries of $\Delta\mathbf{M}$ and $\Delta\mathbf{K}$) are normalized and collected in the unknown vector $\boldsymbol{\chi}$, whose values are constrained to belong to a feasible domain Γ. Matrix $\mathbf{A}$ collects the known terms of each eigenvalue problem in Eq. (2) (i.e. those not related to $\Delta\mathbf{M}$ and $\Delta\mathbf{K}$) The positive-definite matrix $\boldsymbol{\Omega}$ is the regularization that sets different weight between the updates of the model parameters, since some of them might be more uncertain than the others. Finally, the diagonal matrix $\mathbf{W}$ weights the importance of each eigenvalue. More configuration of the mechanism can be also accounted for and weighed through $\mathbf{W}$. Overall, a convex problem is obtained.

3 Extension in the Presence of Passive or Active Control

3.1 Passive Control Through Balancing Springs

The presence of a spring balancing system, with an arbitrary spring arrangement, imposes the modification of the eigenvalue problem of the supported system due to presence of the spring stiffness and mass matrices, denoted $\mathbf{K}^S$ and $\mathbf{M}^S$ respectively (whose structure depend on the spring arrangement adopted):

$$\mathbf{0} = \omega_i^2 \left[\mathbf{M}^O + \mathbf{M}^S + \Delta\mathbf{M}\right]\boldsymbol{\eta}_i - \left[\mathbf{K}^O + \mathbf{K}^S + \Delta\mathbf{K}\right]\boldsymbol{\eta}_i \tag{5}$$

The spring stiffness, and eventually also the masses too, can be included in the vector of the problem unknowns and treated in the same way of the other updating parameters. Hence, the method proposed in Sect. 2 can be applied.

3.2 Feedback Control

A wise selection of the control architecture should be made to keep the multibody system in the correct configuration under analysis while keeping reasonable the unavoidable perturbation of the vibrational modes due to the closed-loop controller [2]. Let us assume, for ease of representation, rank-one control of a flexible mechanism, whose equivalent rigid link system has one degree of freedom. In the case of systems with more dofs, a decoupled control approach is suggested, with single-input single-output controllers that recall the rank-one control. First of all, co-located control of the joint coordinate (i.e. the joint motion is measured and controlled to track the reference position) should be chosen since it allows for a more effective and stable control, that preserves pole-zero interlacing and provides symmetric modifications of the receptance matrix. Hence, symmetry of the eigenvalue problem is ensured. As for the control law $f(t)$, it should include a proportional term, to ensure short transients, an integral term, to ensure no steady-state error, and current-feedforward $f_{FFW}(t)$ to compensate for gravity and friction forces with less effort for the integral term and hence less spillover:

$$f(t) = k_p(q_e - q(t)) + k_i \int (q_e - q(t))dt + f_{FFW}(t) \tag{6}$$

The equilibrium configuration under analysis is denoted q_e and is the reference position to track through the controller. No, or just a small, derivative action should be included since it can damp the vibrational modes and introduces derivation noise, thus reducing signal-to-noise ratio. The dynamic equations in descriptor form of the closed-loop, linearized model is therefore the following one:

$$\begin{bmatrix} -\mathbf{K}_{CL} & \mathbf{0} & \mathbf{0} \\ \mathbf{0} & \mathbf{M} & \mathbf{0} \\ \mathbf{0} & \mathbf{0} & 1 \end{bmatrix} \begin{Bmatrix} \dot{\mathbf{x}} \\ \ddot{\mathbf{x}} \\ q \end{Bmatrix} = \begin{bmatrix} \mathbf{0} & -\mathbf{K}_{CL} & \mathbf{0} \\ -\mathbf{K}_{CL} & \mathbf{0} & -k_i\mathbf{e_q} \\ \mathbf{e_q}^T & \mathbf{0} & 0 \end{bmatrix} \begin{Bmatrix} \mathbf{x} \\ \dot{\mathbf{x}} \\ \gamma \end{Bmatrix} + \begin{bmatrix} \mathbf{0} \\ \mathbf{B} \\ 0 \end{bmatrix} \{f_e\} \tag{7}$$

Matrix $\mathbf{K}_{CL} = \mathbf{K} + k_p\mathbf{e_q}\mathbf{e_q}^T$ is the stiffness matrix of the closed-loop system, i.e. the physical one "modified" through the proportional gain. f_e includes all the terms of $f(t)$ not related to the actual state and hence not affecting the vibrational modes, $\gamma(t) = \int q(t)dt$ is the integral state, $\mathbf{e_q}$ is a null column vector except at the entry related to q where it is 1. The effect of the proportional term $k_p\mathbf{e_q}\mathbf{e_q}^T$ is not critical: although it modifies both the natural frequencies and the mode shapes, the effect can be easily predicted by treating the gain as a fictitious spring, an active spring, whose stiffness is exactly known, and hence is not included among the uncertain parameters. As for the integral term, Eq. (7) reveals that the integral term $k_i \int q(t)dt$ makes more cumbersome modal analysis. On the one hand, it perturbs the poles, together with the proportional gain. On the other one, it modifies the degree of the eigenvalue problem and creates complex eigenpairs. Indeed, the eigenvalue of the controlled system (7) can be written as the solution of the following eigenvalue problem:

$$\left(\mathbf{M}\lambda^2 + \mathbf{K}_{CL} + \frac{1}{\lambda}k_i\mathbf{e_q}\mathbf{e_q}^T\right)\boldsymbol{\eta} = \mathbf{0} \tag{8}$$

where λ is the complex eigenvalue and $\boldsymbol{\eta}$ the complex eigenvector. The perturbation due to the integral term is therefore related to $\frac{1}{\lambda}k_i\mathbf{e_q}\mathbf{e_q}^T$. Hence, a wise selection of the controller, that trades between the requirement of a prompt control and small spillover, allows neglecting the complex terms and hence solving model updating for real eigenvalues and eigenvectors. The use of the model in (7) is an useful tool for predicting, by using the nominal parameters, the amount of spillover and hence tuning the suitable controller that ensure negligible modification of the eigenstructure (e.g. a spillover that is less than the frequency accuracy or resolution of the experimental modal analysis, as well as the acceptable accuracy in frequency estimation). Then, once the model has been updated by using real eigenpairs and the method proposed in Sect. 2.3, verification of the hypothesis can be done either through the model in (7) or through some matrix conditions, such as those provided by the Bauer-Fike perturbation theorem.

4 Test Case: Experimental System Supported Through Balancing Springs

An experimental test case is briefly discussed in this Section: the experimental application of the discussed model updating method to a planar six-bar linkage made by five flexible links and a rigid frame [1], where stable equilibrium is achieved through a pair of balancing spring. In this test, all the issues related to the modeling and to the representation of the model updating problem are handled. The rigid-body mechanism has 3 dofs (the absolute rotations of link 1 and link 2, and the relative rotation between link 5 and 4), while 27 elastic dofs have been adopted in the analytical model.

Experimental modal analysis has been allowed by the presence of a pair of balancing springs that set a stable equilibrium configuration, where the mechanism has been placed for impact excitations. The springs have been included in the model and modeled as one lumped spring connecting one node of the FE model to the frame, and whose mass is represented as a nodal lumped mass. Impact test has been adopted to identify the natural frequencies and the mode shapes while coordinate expansion through the shape function of the beam element and some redundant translational measurements has been adopted to extrapolate the rotational coordinates that are included in the model.

In the proposed test case, the value of the stiffness is included among the unknown parameters and constrained to range about a small interval about its nominal value. The mass has been instead assumed as known. The other parameters to be updated include the nodal masses and moments of inertia, the link mass density of the links and the Young's modulus.

Updating has been performed on the 7 lowest frequency modes within the frequency range from 0 to 200 Hz (for brevity, only the results on the first four modes are given in Table 1). The results in Table 1 shows that the method correctly updates the model by reducing the frequency error (the average percentage error decreases from 5.7% to 1.3%) and by making the model assurance criteria between the experimental and the updated modes approach the target 1 (average value for the updated model: 0.93) (Fig. 1).

Table 1. Comparison of the experimental and analytical modal properties of the 4 rightmost vibrational modes.

Frequency (Hz)			MAC	
Experimental	Nominal model	Updated model	Nominal model	Updated model
13.4	13.7	13.3	0.82	0.87
43.5	43.9	43.3	0.97	0.99
64.7	67.0	64.8	0.96	0.97
113.0	127.3	124.2	0.82	0.83

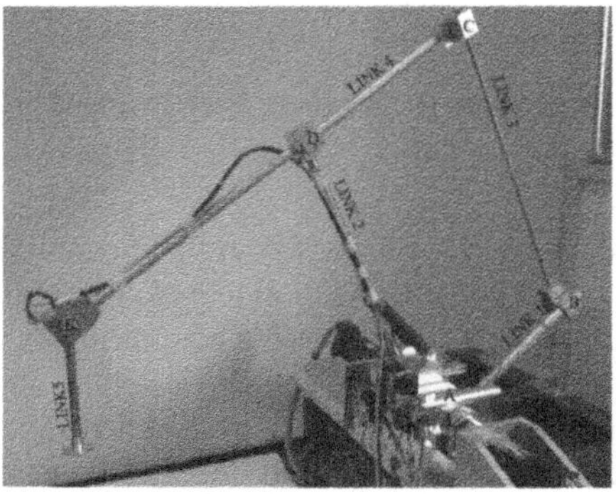

Fig. 1. Picture of the experimental setup

5 Conclusions

This paper discusses the issue of model updating in flexible multibody systems by exploiting linearized models and experimental modal analysis. First of all, a method for computing the model parameters is proposed, by formulating a constrained least square minimization problem and by addressing some issues typical of the representation of this kind of systems. Then, the issue of achieving an asymptotic stable equilibrium configuration, to perform the experimental impact tests, is discussed and two approaches are proposed to control the mechanism position: passive control through balancing springs and active feedback control. An experimental example is proposed with reference to a system balanced through springs. The results prove the method effectiveness. The future extensions of this paper will include the application of the proposed approach to model updating to the case of a servo controlled system, by exploiting the theory discussed.

References

1. Belotti, R., Caracciolo, R., Palomba, I., Richiedei, D., Trevisani, A.: An updating method for finite element models of flexible-link mechanisms based on an equivalent rigid-link system. Shock Vib. (2018)
2. Caracciolo, R., Richiedei, D., Trevisani, A.: Robust piecewise-linear state observers for flexible link mechanisms. J. Dyn. Syst. Meas. Control **130**(3), 031011 (2008)
3. Caracciolo, R., Richiedei, D., & Trevisani, A.: Deformation control in rest-to-rest motion of mechanisms with flexible links. Shock Vib. (2018)
4. Jiang, J., Yuan, Y.: Updating stiffness and hysteretic damping matrices using measured modal data. Shock Vib. (2018)
5. Mottershead, J.E., Friswell, M.I.: Model updating in structural dynamics: a survey. J. Sound Vib. **167**, 347–375 (1993)
6. Palomba, I., Richiedei, D., Trevisani, A.: Energy-based optimal ranking of the interior modes for reduced-order models under periodic excitation. Shock Vib. (2015)
7. Palomba, I., Richiedei, D., Trevisani, A.: A reduction strategy at system level for flexible link multibody systems. Int. J. Mech. Control **18**(2), 59–68 (2017)
8. Palomba, I., Vidoni, R., Wehrle, E.: Application of a parametric modal analysis approach to flexible-multibody systems. In: IFToMM Symposium on Mechanism Design for Robotics, pp. 386–394. Springer, Cham (2018)
9. Sehgal, S., Kumar, H.: Structural dynamic model updating techniques: a state of the art review. Arch. Comput. Methods Eng. **23**, 515–533 (2016)

Formulations and Numerical Methods

Modelling Rigid and Flexible Bodies with Truss Elements

Jacob Philippus Meijaard(✉)

Faculty of Engineering Technology, University of Twente,
P.O. Box 217, 7500 AE Enschede, The Netherlands
j.p.meijaard@utwente.nl

Abstract. The truss element, due to its simplicity, can fulfill the need to model multibody systems in a way that reduces the size of the problems or improves the efficiency of calculations. The truss element can be used to model rigid and flexible bodies as well as several joints with a single truss element or with aggregates built up from a number of truss elements. With an extended mass description, planar binary rigid links or links that can undergo a uniform dilatation with pin joints can be modelled by a single truss element. Planar ternary elements can likewise be modelled by three truss elements. In three dimensions, a rigid body can be modelled by six truss elements along the edges of a tetrahedron, but also three truss elements can be combined to form a triangular membrane element or six truss elements to form a constant-strain finite solid element. Applications to two benchmark problems and a Delta robot are given.

1 Introduction

There is still a need to model multibody systems in a simple way, so that the size of the problems can be made as small as possible and the efficiency of calculations is high. The truss element, due to its ultimate simplicity, is a good candidate to contribute to alleviate this need. The truss element is indeed very versatile and can be used in much wider classes of systems than just trusses; also general rigid and flexible bodies can be modelled, as well as several joints, with truss elements only, although some additional elements are useful in special cases.

In this chapter, an extended mass description for a planar truss element will be described, which makes it possible to model binary rigid links with pin joints by a single truss element. This element can also be used if the body can undergo uniform dilatation as in auxetic bodies with a Poisson ratio close to -1. Furthermore, it is shown how rigid bodies and their joints can be modelled in two and three dimensions. The modelling techniques are illustrated by the application to the two benchmark problems from the handbook [1] and to a Delta robot.

A. Kecskeméthy and F. Geu Flores (Eds.): ECCOMAS 2019, COMPUTMETHODS 53, pp. 275–282, 2020.
https://doi.org/10.1007/978-3-030-23132-3_33

2 Planar Truss Element with Extended Mass Description

The well-known planar truss element, whose configuration is described by the Cartesian coordinates of its two end-points and has its extension or compression as its only deformation, has been introduced in multibody system dynamics by van der Werff [2]. Its nodal coordinates are the four Cartesian coordinates of its two end-points p and q, see Fig. 1(a), with position vectors $\boldsymbol{r}_p = [x_p, y_p]^{\mathrm{T}}$ and $\boldsymbol{r}_q = [x_q, y_q]^{\mathrm{T}}$, respectively, and hence its current length is given by $l = \sqrt{(x_q - x_p)^2 + (y_q - y_p)^2}$. The generalized strain is $\varepsilon = l - l_0$, where l_0 is the reference length. As the configuration of a rigid body attached to a planar truss element is fully determined by the four nodal coordinates, a mass matrix for the truss element which represents the complete inertia properties of the body modelled by this element can be derived. The same mass description applies to the case in which the body can undergo a uniform dilatation as in an auxetic metamaterial with a Poisson ratio close to -1 [3].

A point of the body can be identified by the dimensionless coordinates in a local coordinate frame, which has its origin in the point p, whose x-axis points towards point q and whose y-axis is perpendicular to this axis if it is rotated in the positive direction by a right angle. The local dimensionless coordinates are $\xi = x/l$ and $\eta = y/l$. A point on the body is then located at

$$\boldsymbol{r} = \begin{bmatrix} (1-\xi)x_p + \eta y_p + \xi x_q - \eta y_q \\ -\eta x_p + (1-\xi)y_p + \eta x_q + \xi y_q \end{bmatrix} \tag{1}$$

and its time derivative is

$$\dot{\boldsymbol{x}} = \begin{bmatrix} (1-\xi)\dot{x}_p + \eta \dot{y}_p + \xi \dot{x}_q - \eta \dot{y}_q \\ -\eta \dot{x}_p + (1-\xi)\dot{y}_p + \eta \dot{x}_q + \xi \dot{y}_q \end{bmatrix}. \tag{2}$$

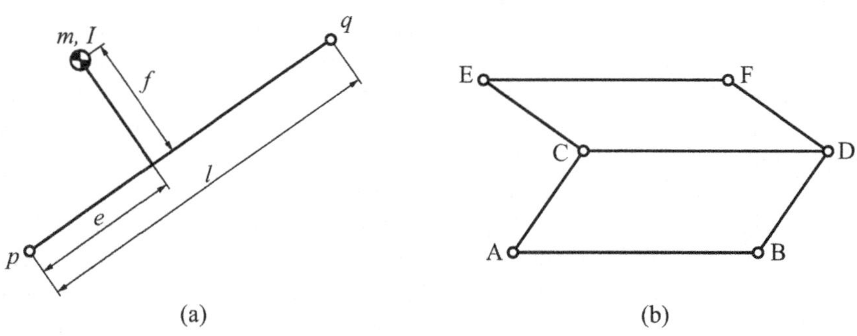

Fig. 1. Truss element with arbitrary mass distribution (a) and model of parallel joint (b)

With the mass of the body, m, the location of the centre of mass at $\xi = e/l$, $\eta = f/l$ and the moment of inertia with respect to the centre of mass I, the mass matrix associated with the nodal coordinates $[x_p, y_p, x_q, y_q]^{\mathrm{T}}$ is obtained as

$$
\begin{aligned}
\boldsymbol{M} &= \int \begin{bmatrix} 1-\xi & -\eta \\ \eta & 1-\xi \\ \xi & \eta \\ -\eta & \xi \end{bmatrix} \begin{bmatrix} 1-\xi & \eta & \xi & -\eta \\ -\eta & 1-\xi & \eta & \xi \end{bmatrix} \rho l^2 \mathrm{d}\xi \mathrm{d}\eta \\
&= \begin{bmatrix} \dfrac{I_q}{l^2} & 0 & -\dfrac{I_{\mathrm{red}}}{l^2} & -m\dfrac{f}{l} \\ 0 & \dfrac{I_q}{l^2} & m\dfrac{f}{l} & -\dfrac{I_{\mathrm{red}}}{l^2} \\ -\dfrac{I_{\mathrm{red}}}{l^2} & m\dfrac{f}{l} & \dfrac{I_p}{l^2} & 0 \\ -m\dfrac{f}{l} & -\dfrac{I_{\mathrm{red}}}{l^2} & 0 & \dfrac{I_p}{l^2} \end{bmatrix},
\end{aligned}
\tag{3}
$$

where ρ is the mass per unit of area, $I_p = I + me^2 + mf^2$ is the moment of inertia with respect to the point p, $I_q = I + m(l-e)^2 + mf^2$ is the moment of inertia with respect to the point q and $I_{\mathrm{red}} = I - me(l-e) + mf^2$ is a reduced moment of inertia. This mass matrix is constant, as m is a constant, e and f are proportional to l, and the moments of inertia are proportional to l^2. The mass matrix is also invariant under rotations of the coordinate system. This is the same mass matrix as is given for a planar body described with natural coordinates [4]. A general planar force system acting on the element in the case the body is rigid can be replaced by equivalent forces on the two nodal points. For a constant moment, these vary with the orientation.

A ternary joint connected by pin joints can simply be modelled by three truss elements in a triangle. The several planar joints can be obtained as follows. The planar revolute joint, or pin joint, can be trivially realized by connecting two elements. A fixed relative orientation can be realized by parallelograms, see Fig. 1(b), and a prismatic joint can be realized by a straight-line mechanism. It can be more convenient to model the last-mentioned joint by other elements, however.

Planar elastic bodies can be modelled by triangular finite elements modelled by three truss elements along their sides, which define constant strains within the element. Indeed, the Lagrangian strains along the directions of the sides are $[(\varepsilon_i + l_{i0})^2/l_{i0}^2 - 1]/2$, $(i = 1, 2, 3)$, from which the three independent component of the Green–Lagrange strain tensor can be calculated. The dual stress components from a constitutive relation can be transformed to generalized stresses for the element in a similar way, where for the sides that are common to two elements, the two contributions of the adjacent elements have to be added. This way of modelling has the advantage that elements share generalized strains and stresses, so these only have to be calculated and stored once. Higher-order elements can be obtained by combining several triangular elements, although the transformations become more complicated. In other respects, the procedure is similar to standard finite element formulations.

3 Spatial Systems

A spatial truss element can be defined in a way analogous to the planar case, with two nodal points with three Cartesian coordinates each, so $\boldsymbol{r}_p = [x_p, y_p, z_p]^{\mathrm{T}}$ and $\boldsymbol{r}_q = [x_q, y_q, z_q]^{\mathrm{T}}$; the current length is given by $l = \sqrt{(x_q - x_p)^2 + (y_q - y_p)^2 + (z_q - z_p)^2}$. The generalized strain is again the elongation of the element, $\varepsilon = l - l_0$. As a single truss element does not determine the orientation of a rigid body in space, more elements are needed. A minimal set would consist of three truss elements arranged in a triangle, but a more convenient model consists of six truss elements arranged along the edges of a tetrahedron. The mass distribution of a general rigid body can be replaced by four lumped masses at the vertices of a tetrahedron, which still results in a constant mass matrix. A further freedom arises if distributed masses for the truss elements are allowed, which still leads to a constant mass matrix, although it is no longer a diagonal matrix. With arbitrarily chosen vertices of the tetrahedron, a general mass distribution of a rigid body can be described by assigning appropriate values to the four lumped masses at the vertices and the six uniformly distributed masses, although some masses can be negative. The negative masses have no influence on the dynamics, because the resulting system mass matrix is still positive definite. The distribution of the masses can be obtained as follows. With homogeneous coordinates $[L_1, L_2, L_3, L_4]^{\mathrm{T}}$, $L_1 + L_2 + L_3 + L_4 = 1$, the position of a point in the body can be written as

$$\boldsymbol{r} = [L_1\boldsymbol{I}_3 \; L_2\boldsymbol{I}_3 \; L_3\boldsymbol{I}_3 \; L_4\boldsymbol{I}_3] \begin{bmatrix} \boldsymbol{r}_1 \\ \boldsymbol{r}_2 \\ \boldsymbol{r}_3 \\ \boldsymbol{r}_4 \end{bmatrix}, \tag{4}$$

where $\boldsymbol{I}_3$ is the 3×3 identity matrix and $\boldsymbol{r}_1$, $\boldsymbol{r}_2$, $\boldsymbol{r}_3$ and $\boldsymbol{r}_4$ are the position vectors of the vertices of the tetrahedron. The resulting mass matrix is a 4×4 block matrix with diagonal blocks given by the integrals

$$\boldsymbol{I}_3 \int L_i L_j \rho \mathrm{d}V = m_{ij}\boldsymbol{I}_3, \tag{5}$$

where ρ is the mass per unit of volume and $\mathrm{d}V$ is the volume element over which the integration takes place. Now the distributed masses of the truss elements connected between the nodes i and j are $6m_{ij}$, and the lumped masses are the diagonal elements minus one-third of the sum of the three distributed masses of the elements connected to the corresponding vertices; that is, the lumped mass at node i is $m_{ii} - 2\sum_{j \neq i} m_{ij}$. For a tetrahedron with a uniformly distributed mass m, we have $m/10$ on the diagonal and $m/20$ for the other elements of m_{ij}, which means that the distributed masses of the truss elements are all $3m/10$ and the lumped masses at the nodes are $-m/5$.

There are now six generalized strains for the body, so a constant strain can be represented. The aggregate of six truss elements for a rigid body has the same number of independent nodal coordinates as a beam, but the mass matrix is a constant.

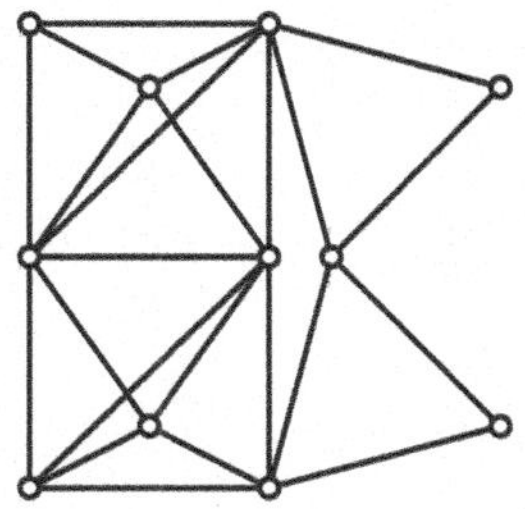

Fig. 2. Sarrus mechanism as a spatial prismatic joint, where the top can move in the vertical direction with respect to the bottom; the mechanism consists of two identical parts at an angle, in this a right angle

It is possible to combine three truss elements in a triangular membrane element, which can describe the three uniform deformations of the membrane. The mass matrix is still constant if all mass is concentrated in the plane of the membrane. The tetrahedra in space, as the triangles in the plane, can be used to construct constant-strain finite elements for general flexible bodies.

Several lower-pair joints can be modelled by truss elements only, although not all and some are complicated to realize. A spherical joint is realized by connecting two elements in a node. A revolute joint can be obtained by letting two tetrahedra have a common edge. A prismatic joint can be realized with a Sarrus mechanism, as shown in Fig. 2. A cylindrical joint is a combination of a revolute and a prismatic joint. Also a planar joint can be built up from three elementary joints. The screw pair cannot be modelled.

4 Example Problems

The usefulness of the truss element is shown in some test problems and examples. The first example is the motion of a planar seven-body mechanism used as a benchmark in [1]. The system can be modelled with ten truss elements and 18 nodal coordinates, of which eight are fixed, see Fig. 3. In order to overcome a singularity, it is better to model the system with nine truss elements and a rigid beam element, which gives an additional nodal coordinate. The original model used in [5] had eleven beam elements and one truss element, which led to a model with 31 nodal coordinates, so the new way of modelling the system leads to a reduction of the complexity. The resulting motion is exactly the same.

As a spatial example, the benchmark robot from [1] is modelled with truss elements only to show their possibilities. A schematic representation, not to scale, of the a model of the robot is shown in Fig. 4. The robot is fixed at the base point A and the points B and D can only move vertically. The truss element between the points A and B models the vertical degree of freedom, z_1, and the rotation about the vertical axis with the angle γ_1 is possible, which is the second degree of freedom of the robot. The first body is modelled by a tetrahedron of six truss elements represented by BCD. A massless intermediate tetrahedron is

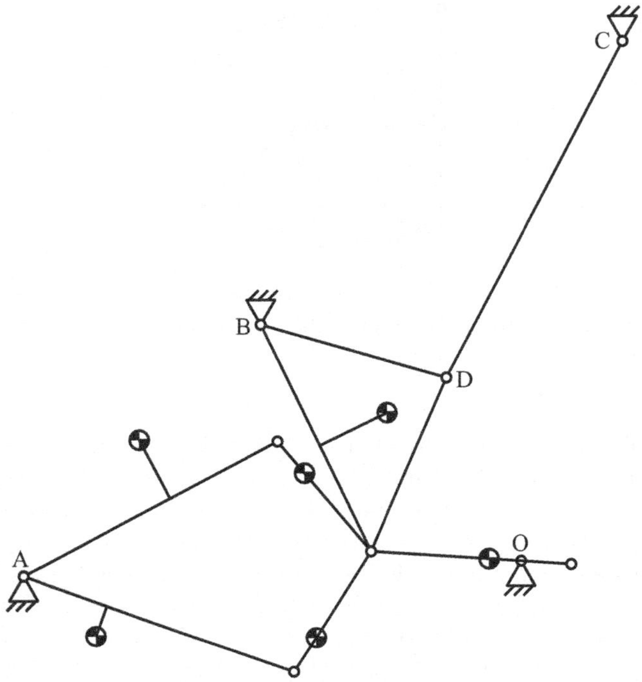

Fig. 3. Seven-body mechanism; the element between points C and D is a spring, the other elements are rigid; the centre of mass of the crank is not shown

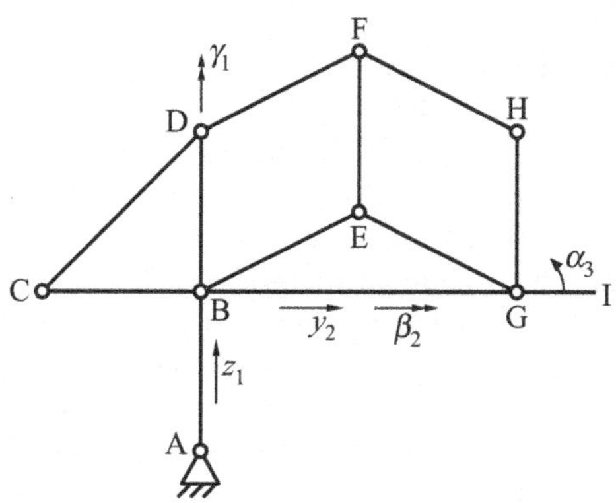

Fig. 4. Schematic representation of a model for the robotic manipulator

connected to this to make the rotation about the axis BC with the rotation angle β_2 possible. The linear motion along the axis BG with displacement y_2 is made possible by a Sarrus mechanism represented schematically by the parallelograms BDEFGH, which adds 32 truss elements to the model. The truss element BG

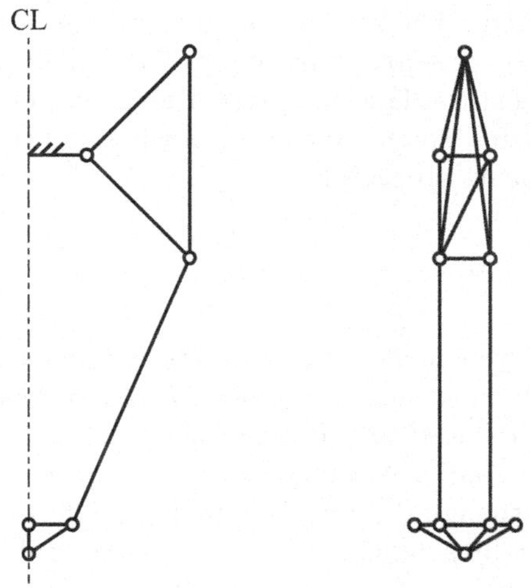

Fig. 5. Delta robot modelled with truss elements

is added as an actuator to model this degree of freedom. The second body is attached to GH by three additional truss elements and the end effector GI, which can rotate with an angle α_3 about the revolute joint G is modelled by five additional truss elements. In total, there are 53 truss elements and 63 nodal coordinates, of which seven are fixed. The degrees of freedom for the translational joints are the elongations of the two trusses, but as there are no angles in the model, the rotational degrees of freedom are replaced by convenient coordinates of three nodal points; this may lead to singularities during the simulation, but in the present example with limited angles, these do not occur.

The original model used in [5] only needed 43 coordinates, of which eight were fixed. The original as well as the present model are not the most compact models possible, so this gives a reasonably fair comparison. Although many elements and coordinates are needed, the presented model may still give rise to an efficient calculation procedure. Both models yield the same results, although in a different form.

Finally, a parallel manipulator, the Delta robot, is considered, see Fig. 5. It consists of three arms that connect a fixed base with a platform, of which one arm is shown in the figure. The other two arms are identical and rotated over $120\,^\circ$ about the centre line. Each arm consists of an upper arm connected to the base by a revolute actuated joint and a lower arm consisting of two rods arranged as the sides of a parallelogram connecting the upper arm to the moving platform. As each arm constrains one rotation of the platform and these rotations are independent, the platform can only translate. Each upper arm is modelled by nine truss elements arranged along the edges of a pyramid with a rectangular base, one of which edges is fixed to the base, and each lower arm is modelled by two truss elements. The platform is a regular hexagon modelled with 15 truss elements. One coordinate of each apex of the pyramids can be used as the degrees of freedom of the system. The model has 66 coordinates, of which 18 are fixed and three are degrees of freedom. The number of dependent coordinates, 45, is equal to the number of truss elements. A traditional way of modelling the system would use 85 coordinates, so the reduction in the number of coordinates is not so large, but the elements are simpler.

5 Conclusions

A general planar binary element with pin joints can be modelled by a single planar truss element with an appropriate generalization of its mass matrix. General applied forces can be reduced to equivalent nodal forces.

Quite complicated spatial mechanisms, such as robots, can be modelled by truss elements only, although this may lead to large collections of elements, which are, however, individually simple. It is nonetheless advisable to have other kinds of elements available. It has been shown that the inertial properties of a spatial rigid body can be modelled by six truss elements with uniform distributed mass arranged along the edges of a tetrahedron with nodes at general positions, with additional lumped masses at the nodes.

The modelling can be facilitated with some pre-processing tools, in particular the representation of moments and forces by equivalent nodal forces. The interpretation of results needs some post-processing, as angles are not directly available.

References

1. Schiehlen, W. (ed.): Multibody Systems Handbook. Springer, Berlin (1990)
2. van der Werff, K.: Kinematic and dynamic analysis of mechanisms, a finite element approach. Dissertation, Delft University Press (1977)
3. Lakes, R.S.: Negative-Poisson's-ratio materials: auxetic solids. Ann. Rev. Mater. Res. **47**, 63–81 (2017)
4. de Jalón, J.G., Bayo, E.: Kinematic and Dynamic Simulation of Multibody Systems: The Real-time Challenge. Springer, New York (1994)
5. Jonker, J.B., Meijaard, J.P.: SPACAR - computer program for dynamic analysis of flexible spatial mechanisms and manipulators. In: Schiehlen, W. (ed.) Multibody Systems Handbook, pp. 123–143. Springer, Berlin (1990)

State Observation in Beam-Like Structures Under Unknown Excitation

Johannes Luthe[1(✉)], Andreas Schulze[1], Roman Rachholz[1], János Zierath[2], and Christoph Woernle[1]

[1] Chair of Technical Dynamics, University of Rostock, Justus-von-Liebig-Weg 6, 18059 Rostock, Germany
{johannes.luthe,andreas.schulze2, roman.rachholz,woernle}@uni-rostock.de

[2] W2E Wind to Energy GmbH, Grubenstrasse 44, 18055 Rostock, Germany
jzierath@wind-to-energy.de

Abstract. Measuring and logging of fatigue loads is essential for individual life time estimations of dynamically loaded mechanical structures like operating wind turbines. A fundamental intermediate step is the accurate derivation of structural displacement fields at all critical spots, for which, in practical applications usually only a limited number of sensor information is available. Another challenge is the generally unknown system excitation, e.g. by wind loads. This contribution investigates fundamental requirements for the implementation of Unknown Input Observers (UIO) on flexible mechanical systems. To also account for large nonlinear motions, we propose a substructuring approach, which is demonstrated and validated by two numerical examples: (1) A finite element model of a comparatively simple cantilever beam structure, (2) A detailed flexible multibody model of a small-scale wind turbine test stand.

1 Introduction

The design of dynamically loaded mechanical structures like wind turbines is based on conservative load assumptions compared to the actual loads acting on operating structures, thus implying a discrepancy between the actual and anticipated fatigue progression. In order to asses the individual life span of wind turbines, reliable tracking of the actually endured fatigue loads at critical spots of the structure is of great interest. As a fundamental intermediate step, the accurate derivation of structural displacement fields at all critical spots is essential. To keep installation effort and maintenance costs low, in practical applications usually only a minimum number of sensors is installed on the structure under consideration, i.e. only a limited number of measurements is available for the structural displacement field approximation. Another challenge is the generally unknown system excitation, e.g. by wind loads. The aim of this contribution is to provide a comprehensive methodology for the approximation of displacement

A. Kecskeméthy and F. Geu Flores (Eds.): ECCOMAS 2019, COMPUTMETHODS 53, pp. 283–291, 2020.
https://doi.org/10.1007/978-3-030-23132-3_34

fields in beam-like structures under unknown excitation based on a limited number of measurements. While the Modal Decomposition and Expansion (MDE) approach presented in [5] already proved to yield accurate results in case the number of sensors at least equals the number of considered modes, it is desirable to further reduce the required sensor set. The special observer class of Unknown Input Observers (UIO) – well-known from control theory for state observation of systems excited by unknown disturbances – provides a promising approach in this context. However, UIO are only applicable to linear dynamic systems, see Sect. 2. In order to also account for the generally large nonlinear motion of wind turbine structures, we propose to disassemble the full system into a number of substructures mutually connected by interface definitions in a first stage, followed by a parallel implementation of individual UIOs for each substructure, see Sect. 3. In the context of this contribution, the quality of this implementation framework will be demonstrated in two steps: (1) Numerical implementation on a finite element model of a comparatively simple cantilever beam structure, (2) Implementation on a detailed flexible multibody model of a small-scale wind turbine test stand presented in [6].

2 Full-Order Observer for Linear Systems with Unknown Inputs

Unknown input observers are applied to estimate system states in the presence of not measurable system inputs. The basic concept goes back to [2,3,9] and has been adapted and further extended in various ways ever since. Relying on classical LUENBERGER formulations, the general idea of UIOs is to transform the observer coordinates such that the error dynamics is independent of the unknown system inputs. A comprehensive summary of the major UIO design steps is given in [1,4,8] and will be briefly introduced in this section. Considered is the design of full-order UIO for LTI systems which are exposed to unknown disturbances, modeled by the additional term $\boldsymbol{d}(t)$ in the dynamic equations,

$$\Sigma : \begin{cases} \dot{\boldsymbol{x}}(t) = \boldsymbol{A}\boldsymbol{x}(t) + \boldsymbol{B}\boldsymbol{u}(t) + \boldsymbol{E}\boldsymbol{d}(t), & \boldsymbol{x}(0) = \boldsymbol{x}_0\,, \\ \boldsymbol{y}(t) = \boldsymbol{C}\boldsymbol{x}(t)\,, \end{cases} \tag{1}$$

with states $\boldsymbol{x} \in \mathbb{R}^n$, known inputs $\boldsymbol{u} \in \mathbb{R}^m$, unknown disturbances $\boldsymbol{d} \in \mathbb{R}^r$ and measurements $\boldsymbol{y} \in \mathbb{R}^p$. The objective is to design a full-order LUENBERGER-type observer such that the estimated states $\hat{\boldsymbol{x}}(t)$ converge asymptotically towards $\boldsymbol{x}(t)$, i.e.

$$\lim_{t\to\infty} \|\hat{\boldsymbol{x}}(t) - \boldsymbol{x}(t)\| = 0, \qquad \forall\, \boldsymbol{x}_0, \hat{\boldsymbol{x}}_0\,. \tag{2}$$

This observation problem can be solved by a system of the following structure

$$\Sigma_{\text{obs}} : \begin{cases} \dot{\hat{\boldsymbol{z}}}(t) = \boldsymbol{N}\hat{\boldsymbol{z}}(t) + \boldsymbol{G}\boldsymbol{u}(t) + \boldsymbol{W}\boldsymbol{d}(t), & \hat{\boldsymbol{z}}(0) = \hat{\boldsymbol{z}}_0\,, \\ \hat{\boldsymbol{x}}(t) = \hat{\boldsymbol{z}}(t) + \boldsymbol{J}\boldsymbol{y}(t)\,, \end{cases} \tag{3}$$

where $\hat{\boldsymbol{z}}(t) \in \mathbb{R}^n$ denotes the observer state vector. By analyzing the observer error dynamics,

$$\begin{aligned}\dot{\boldsymbol{e}}(t) = \dot{\hat{\boldsymbol{x}}} - \dot{\boldsymbol{x}}(t) = \boldsymbol{N}\boldsymbol{e}(t) + (\boldsymbol{N} - \boldsymbol{N}\boldsymbol{J}\boldsymbol{C} + \boldsymbol{W}\boldsymbol{C} - \boldsymbol{A} + \boldsymbol{J}\boldsymbol{C}\boldsymbol{A})\,\boldsymbol{x}(t) \\ + (\boldsymbol{J}\boldsymbol{C}\boldsymbol{E} - \boldsymbol{E})\,\boldsymbol{d}(t) + (\boldsymbol{G} - \boldsymbol{B} + \boldsymbol{J}\boldsymbol{C}\boldsymbol{B})\,\boldsymbol{u}(t)\,,\end{aligned} \tag{4}$$

a set of existence conditions can be derived for (2) to be satisfied:

$$\boldsymbol{N} \text{ is } \textsc{Hurwitz}\text{-stable}\,, \tag{5}$$

$$\boldsymbol{N} - \boldsymbol{N}\boldsymbol{J}\boldsymbol{C} + \boldsymbol{W}\boldsymbol{C} - \boldsymbol{A} + \boldsymbol{J}\boldsymbol{C}\boldsymbol{A} = \boldsymbol{0}\,, \tag{6}$$

$$\boldsymbol{J}\boldsymbol{C}\boldsymbol{E} - \boldsymbol{E} = \boldsymbol{0}\,, \tag{7}$$

$$\boldsymbol{G} - \boldsymbol{B} + \boldsymbol{J}\boldsymbol{C}\boldsymbol{B} = \boldsymbol{0}\,. \tag{8}$$

A solution for $\boldsymbol{J}$ in Eq. (7) exists only and only if

$$\operatorname{rank}\begin{bmatrix}\boldsymbol{C}\boldsymbol{E}\\ \boldsymbol{E}\end{bmatrix} = \operatorname{rank}(\boldsymbol{C}\boldsymbol{E}) = r\,, \tag{9}$$

which is a necessary condition for the general existence of UIO. A general solution for $\boldsymbol{J}$ can be obtained from Eq. (7) according to [7]

$$\boldsymbol{J} = \boldsymbol{E}\,(\boldsymbol{C}\boldsymbol{E})^{\dagger} + \boldsymbol{H}\left(\boldsymbol{I}_p - \boldsymbol{C}\boldsymbol{E}\,(\boldsymbol{C}\boldsymbol{E})^{\dagger}\right)\,, \tag{10}$$

with $\boldsymbol{H} \in \mathbb{R}^{n\times p}$ being an arbitrary matrix and $(\boldsymbol{C}\boldsymbol{E})^{\dagger}$ being the Moore-Penrose inverse of $\boldsymbol{C}\boldsymbol{E}$. Matrix $\boldsymbol{G}$ can then be expressed by Eq. (8)

$$\boldsymbol{G} = \boldsymbol{Y}\boldsymbol{B}, \quad \text{with } \boldsymbol{Y} = \boldsymbol{I}_n - \boldsymbol{J}\boldsymbol{C}\,. \tag{11}$$

Considering Eq. (6), matrix $\boldsymbol{N}$ can be obtained by

$$\boldsymbol{N} = \boldsymbol{Y}\boldsymbol{A} - \boldsymbol{L}\boldsymbol{C}, \quad \text{with } \boldsymbol{L} = (\boldsymbol{W} - \boldsymbol{N}\boldsymbol{J}) \tag{12}$$

Under the condition that the pair $(\boldsymbol{C}, \boldsymbol{Y}\boldsymbol{A})$ is at least detectable, i.e.

$$\operatorname{rank}\begin{bmatrix} s\boldsymbol{I} - \boldsymbol{Y}\boldsymbol{A}\\ \boldsymbol{C}\end{bmatrix} = n, \quad \forall\, s \in \mathbb{C},\ \mathfrak{Re}(s) \geq 0 \tag{13}$$

matrix $\boldsymbol{L}$ can be determined by any pole placement technique such that $\boldsymbol{N}$ is stable, with the eigenvalues of $\boldsymbol{N}$ being composed of the invariant zeros of the system Σ and additional eigenvalues which can be arbitrarily assigned by appropriate choice of matrix $\boldsymbol{L}$. For $\boldsymbol{N}$ to be Hurwitz-stable all invariant system zeros need to have negative real parts.

3 Implementation Framework

The basic implementation framework of the substructuring approach for flexible state estimation is illustrated by the example case of a rather simple cantilever beam structure, see Fig. 1. Its dimensions are chosen according to the rotor blade structures of the small scale wind turbine test stand analyzed in Sect. 4. Finite element modeling provides appropriate linear equations of motion, whereas the set of degrees of freedom is reduced by the classical CRAIG-BAMPTON method. We assume that the system is excited by the external force $F(t)$. The objective is to estimate the flexible states of the reference system at locations $\mathcal{K}_1^{\mathrm{I}}$ and $\mathcal{K}_1^{\mathrm{II}}$, i.e. $\boldsymbol{x}(\mathcal{K}_1^{\mathrm{I}})$ and $\boldsymbol{x}(\mathcal{K}_1^{\mathrm{II}})$, by two separate unknown input observers. Considering finite beam elements, the sought states are composed of the six nodal coordinates and their time derivatives.

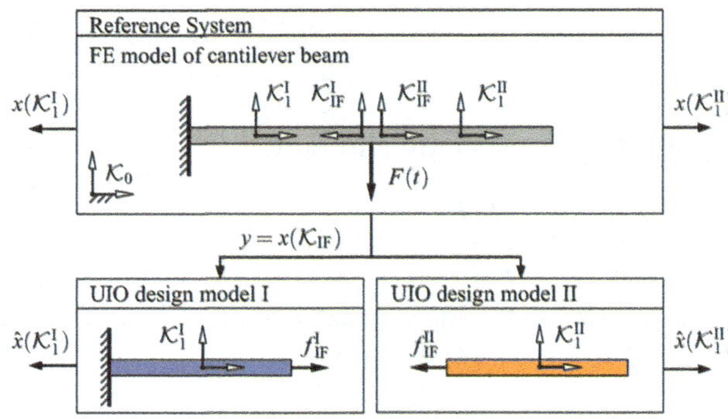

Fig. 1. Substructure approach for the UIO-based flexible state reconstruction of a cantilever beam model

3.1 UIO Design Models

In order to obtain design models for the unknown input observers, the reference structure is at first divided into the two substructures I and II by defining the interfaces $\mathcal{K}_{\mathrm{IF}}^{\mathrm{I}}$ and $\mathcal{K}_{\mathrm{IF}}^{\mathrm{II}}$ close to the attack point of $F(t)$. Now each of the two substructures is affected by unknown interface forces and moments represented by the collective vectors of reaction forces and moments $\boldsymbol{f}_{\mathrm{IF}}^{\mathrm{I}}$ and $\boldsymbol{f}_{\mathrm{IF}}^{\mathrm{II}}$, respectively, see Fig. 1. For ease of notation the upper indices are omitted hereinafter. The observer equations are then obtained on the basis of reduced linear finite element models in the form of

$$\boldsymbol{M}\ddot{\boldsymbol{q}}(t) + \boldsymbol{D}\dot{\boldsymbol{q}}(t) + \boldsymbol{K}\boldsymbol{q}(t) = \boldsymbol{f}_{\mathrm{IF}}(t)\,, \tag{14}$$

with $\boldsymbol{M}$, $\boldsymbol{D}$ and $\boldsymbol{K}$ being the mass, damping an stiffness matrices and $\boldsymbol{q}(t)$ denoting the set of generalized coordinates. By defining the state vector as $\boldsymbol{x}(t) = [\boldsymbol{q}(t), \dot{\boldsymbol{q}}(t)]^{\mathrm{T}}$, an LTI state space representation according to Eq. (1) is obtained

$$\begin{aligned} \begin{bmatrix} \dot{\boldsymbol{q}}(t) \\ \ddot{\boldsymbol{q}}(t) \end{bmatrix} &= \begin{bmatrix} \boldsymbol{0} & \boldsymbol{I} \\ -\boldsymbol{M}^{-1}\boldsymbol{K} & -\boldsymbol{M}^{-1}\boldsymbol{D} \end{bmatrix} \begin{bmatrix} \boldsymbol{q}(t) \\ \dot{\boldsymbol{q}}(t) \end{bmatrix} + \begin{bmatrix} \boldsymbol{0} \\ -\boldsymbol{M}^{-1} \end{bmatrix} \boldsymbol{f}_{\mathrm{IF}}(t) \\ \dot{\boldsymbol{x}}(t) &= \qquad\qquad \boldsymbol{A} \qquad\qquad\quad \boldsymbol{x}(t) \;+\; \quad \boldsymbol{E} \qquad \boldsymbol{d}(t) \end{aligned} \tag{15}$$

with $\boldsymbol{f}_{\mathrm{IF}}(t)$ representing the unknown disturbances $\boldsymbol{d}(t)$. Accounting for the existence conditions (5) to (8), a number of remarks on the practical implementation can be pointed out:

- The number of system outputs, i.e. measurements, p must be at least equal to the number of unknown system inputs r.
- The coordinates of the system outputs must be defined such that detectability of instable states is ensured; unobservable states need to be stable.
- In the case of interface forces being the unknown system inputs, velocity measurements $\dot{\boldsymbol{q}}(\mathcal{K}_{\mathrm{IF}})$ at the interface are necessary.
- To avoid marginal stability of matrix $\boldsymbol{N}$, additional displacement measurements are necessary.

To satisfy the existence conditions for the two UIO design models, it is sufficient to measure the system states $\boldsymbol{x}(\mathcal{K}_{\mathrm{IF}})$ at the interface of the two substructures. Within this contribution we assume perfect knowledge of the required state measurements, independent of a specific sensor concept and regardless of possible challenges connected with the practical sensor setup. The entire observer design was performed in Mathworks MATLAB on the basis of FE models built-up in MSC.NASTRAN.

3.2 Results

In order to compare time histories of reference and estimated states, the relative error $e_i(t)$ is introduced for the ith coordinate under consideration,

$$e_i(t) = \frac{q_i(t) - \hat{q}_i(t)}{\max(\hat{q}_i(t))}\,. \tag{16}$$

Figure 2 shows time histories for the reference and estimated transversal displacements $q_1^{\mathrm{I}}(t)$ and $q_1^{\mathrm{II}}(t)$ at locations $\mathcal{K}_1^{\mathrm{I}}$ and $\mathcal{K}_1^{\mathrm{II}}$, respectively. For consistent initial conditions, see Fig. 2**a**, the corresponding relative errors $e(t)$ of less than 1% confirm high agreement of the estimation with respect to the reference states. Figure 2**b** shows the results for differing initial conditions as – in contrast to the observer models – the reference system was initially statically deflected. Under these circumstances the stable error dynamics of the UIOs is especially prominent. As the convergence rate is determined by the lowest invariant system zeros, it is evident that for substructure I (smallest system zero at -8.09) the error decreases faster than for substructure II (smallest system zero at -0.30).

4 Application to a Flexible Multibody System

As experimental investigations on operating wind turbines are inherently connected to high economical expense due to necessary stand-still periods on the one hand and the need for robust measurement equipment on the other hand, a scaled test rig of a wind energy turbine has been developed (tower height 3.68 m,

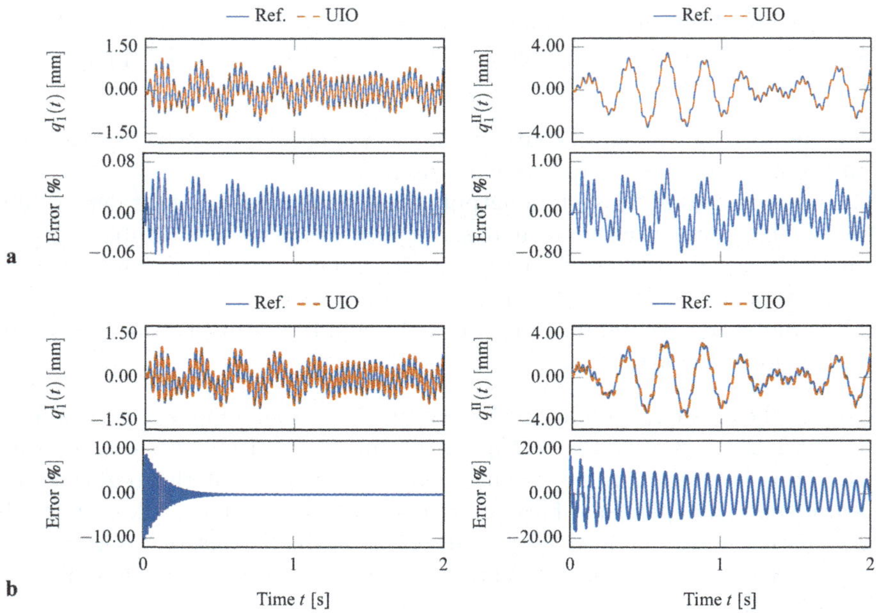

Fig. 2. Reference and estimated transversal displacements $q_1^{\mathrm{I}}(t)$ at $\mathcal{K}_1^{\mathrm{I}}$ and $q_1^{\mathrm{II}}(t)$ at $\mathcal{K}_1^{\mathrm{II}}$ following the proposed substructuring approach with **a** Consistent initial conditions of reference system and UIO models, **b** Different initial conditions of reference system and UIO models

blade length 1.81 m), which allows for fast analyses of different sensor concepts before their implementation on a real turbine. For validation purposes, also a flexible multibody model of the test rig structure has been built-up, which is utilized in this section. Due to limited amount of space, the reader is referred to [6] for further details. For the following test case we assume that the multibody system, depicted in Fig. 3, is excited by the external force $F(t)$ acting on the outer blade segments and being composed of two harmonics whose frequencies correspond to the first two blade modes. To represent operating conditions, the wind turbine model performs large nonlinear motions, i.e. the rotor angle $\alpha(t)$ and the azimuth angle $\beta(t)$ are rheonomically constrained by constant angular velocities $\dot{\alpha}$ and $\dot{\beta}$. The objective is to estimate the reference states $\boldsymbol{x}(\mathcal{K}_1^{\mathrm{TW}})$ and $\boldsymbol{x}(\mathcal{K}_1^{\mathrm{B}})$ of the multibody system at locations $\mathcal{K}_1^{\mathrm{TW}}$ (10 % tower height off the tower base) and $\mathcal{K}_1^{\mathrm{B}}$ (10 % blade length off the blade root).

4.1 UIO Design Models

As large motions are not represented by linear system equations required for the observer design, we propose to disassemble the full system into a tower and a blade substructure with unknown interface forces at $\mathcal{K}_{\mathrm{IF}}^{\mathrm{TW}}$ and $\mathcal{K}_{\mathrm{IF}}^{\mathrm{B}}$, respectively, similar to the approach proposed in Sect. 3. Under the assumption of small elastic

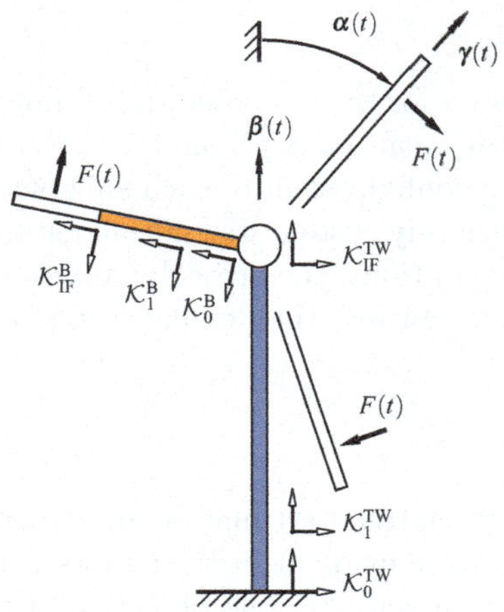

Fig. 3. Definition of substructures for parts of the flexible multibody model of a small scale wind energy turbine: 1. Tower substructure (blue, reference markers denoted by upper index TW), 2. Blade substructure (orange, reference markers denoted by upper index B).

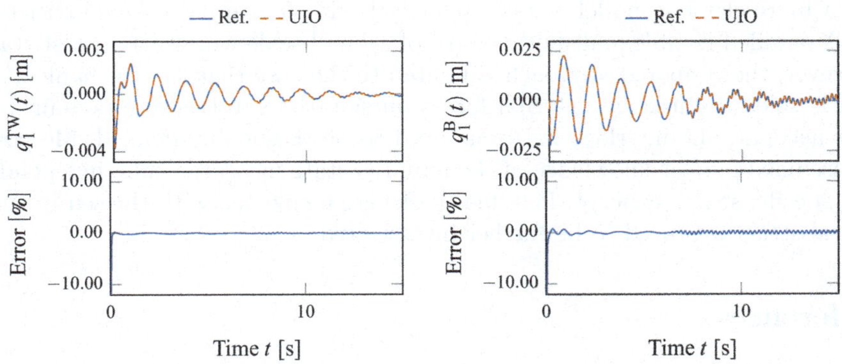

Fig. 4. Reference and estimated transversal displacements $q_1^{TW}(t)$ at $\mathcal{K}_1^{TW}$ and $q_1^{B}(t)$ at $\mathcal{K}_1^{B}$ following the proposed substructuring approach

deformations, the flexible states of the substructures can be expressed by a set of linear equations with respect to their (generally moving) base frames $\mathcal{K}_0^{TW}$ and $\mathcal{K}_0^{B}$, i.e. with respect to their undeformed reference configurations. Two separate observers are then designed in Mathworks MATLAB on the basis of system equations obtained by finite element modeling of each substructure in MSC.NASTRAN.

4.2 Results

Figure 4 shows time histories for the reference and estimated transversal displacements $q_1^{\mathrm{TW}}(t)$ and $q_1^{\mathrm{B}}(t)$ at locations $\mathcal{K}_1^{\mathrm{TW}}$ and $\mathcal{K}_1^{\mathrm{B}}$, respectively, with respect to the undeformed reference configuration of each substructure. In contrast to the observer models, the multibody system was initially statically deflected. Despite this difference in initial conditions, the error dynamics again prove a stable convergence of both observers towards the flexible states of the multibody model.

5 Conclusion

As a fundamental step for fatigue estimation in structurally loaded mechanical systems, the displacement fields at critical spots need to be derived on the basis of a limited number of sensors. The objective of this contribution was to investigate fundamental requirements for the state estimation of flexible structures by Unknown Input Observers. Since this class of observers requires linear design system, we proposed a substructuring approach for the implementation in flexible multibody systems undergoing large nonlinear motions. The high estimation quality of this framework was demonstrated by two numerical examples: (1) A finite element model of a comparatively simple cantilever beam structure, (2) A detailed flexible multibody model of a small-scale wind turbine test stand. However, the proposed approach is limited to the case that the mechanical system is solely excited by external forces outside the substructure domains. On that account, the interface locations need to be reasonably defined. Moreover, we assumed perfect knowledge of the required state measurements, independent of a specific sensor concept. Possible challenges connected with the sensor setup in real structures need to be further investigated.

References

1. Darouach, M., Zasadzinski, M., Xu, S.J.: Full-order observers for linear systems with unknown inputs. IEEE Trans. Autom. Control **39**(3), 606–609 (1994)
2. Hautus, M.: Strong detectability and observers. Linear Algebr. Appl. **50**, 353–368 (1983)
3. Kudva, P., Viswanadham, N., Ramakrishna, A.: Observers for linear systems with unknown inputs. IEEE Trans. Autom. Control **25**, 113–115 (1980)
4. Lunze, J.: Zustandsbeobachtung linearer Systeme mit unbekannten Eingängen. Automatisierungstechnik **65**, 99–114 (2017)
5. Luthe, J., Schulze, A., Rachholz, R., Zierath, J., Woernle, C.: Acceleration-based strain estimation in a beam-like structure. In: Proceedings of the 5th Joint International Conference on Multibody System Dynamics, Lisboa (2018)
6. Rachholz, R., Bartkowiak, R., Schulze, A., Luthe, J., Zierath, J., Woernle, C.: Development of a dynamically scaled wind turbine. J. Phys. Conf. Ser. **1037**, 052033 (2018). https://doi.org/10.1088/1742-6596/1037/5/052033

7. Rao, C.R., Mitra, S.K.: Generalized Inverse of Matrices and Its Applications. Wiley Series in Probability and Mathematical Statistics. Wiley, New York (1971)
8. Trinh, H., Fernando, T.: Functional Observers for Dynamical Systems. LNCIS, vol. 420. Springer, Heidelberg (2012)
9. Wang, S.H., Davison, E.: Observing partial states for systems with unmeasurable disturbances. IEEE Trans. Autom Control **23**, 481–483 (1978)

Dynamic Modelling of Lower Mobility Parallel Manipulators

Haitao Liu[1(✉)], Weifeng Chen[1], Tian Huang[1], Huafeng Ding[2], and Andres Kecskemethy[1]

[1] Tianjin University, Tianjin, China
{liuht, chenweifeng}@tju.edu.cn
[2] China University of Geosciences, No. 388 Lumo Road, Wuhan, Wuhan 430074, People's Republic of China
dinghf@cug.edu.cn

Abstract. This paper deals with the dynamic modelling of lower mobility parallel manipulators using the screw theory. The proposed approach is developed by considering both the permitted and restricted virtual displacements of a parallel manipulator, leading to the evaluation of intensities of the wrenches of actuations and constraints applied on the moving platform. By introducing the complementary part in terms of generalized constraint forces/torques, it completes the dynamic model of lower mobility parallel manipulators, and provides a feasible way to investigate the dynamic interactions between the limbs and the moving platform.

Keywords: Dynamic modelling · Parallel manipulators · Screw theory

1 Introduction

With the wide applications of parallel kinematic machines (PKM) in industry, high-speed pick-and-place operations and high-speed drilling/milling in situ for example, more and more attentions have been paid to the improvement of dynamic behaviors of the system with the goal to achieve higher accuracy and productivity. In spite of a number of commercial software that have been available on market, a general, complete and precise formulation of rigid-body dynamics of parallel kinematic machines is still in need since it is helpful to gain deep insight into dynamic performance evaluation, motor sizing, trajectory planning and control algorithm development.

In this paper, the application of dynamic model for design purpose as that stated in [1] is addressed by introducing a novel modelling approach for lower mobility parallel manipulators based on the principle of virtual work [2], which is capable of evaluating the reactions in terms of both actuation and constraint imposed on the moving platform simultaneously. The proposed method is developed drawing on the generalized Jacobian and the acceleration analysis of parallel manipulators presented in our previous works [2, 3].

Hereby, the virtual displacements related to the permitted and restricted motions can be naturally defined, and the inertial and resulting gravity wrenches of a rigid body are formulated in a body-fixed frame. Then, the virtual work of individual rigid bodies can be lumped together, resulting in analytical expressions of the actuation and constraint forces/moments.

A. Kecskeméthy and F. Geu Flores (Eds.): ECCOMAS 2019, COMPUTMETHODS 53, pp. 292–298, 2020.
https://doi.org/10.1007/978-3-030-23132-3_35

2 Kinematic Analysis

The kinematic analysis of a non-redundant actuated and non-overconstrained f-DOF $(2 \leq f \leq 6)$ parallel mechanism is briefly reviewed [2, 3].

As shown in Fig. 1, a global frame $\mathcal{K}$ is placed on the base and an instantaneous frame $\mathcal{K}'$ keeping parallel to $\mathcal{K}$ is established on the platform. In order to evaluate the tensor of inertia of the jth $(j = 1, \ldots, n_i - 1)$ body in limb i, a body-fixed frame $\mathcal{K}_{j,i}^{L}$ is attached to body j with its z-axis coincident with the $(j_a + 1)$th $(j_a = j)$ joint axis. A body-fixed frame $\mathcal{K}_{p}^{L}$ is placed on the platform. Considering the effects of constraints, $6 - n_i$ 1-DOF virtual joints are added in limb i by making an assumption that all of them are active ones. Then, referring to [2, 3], the joint velocity and the accelerator of the platform can be given as

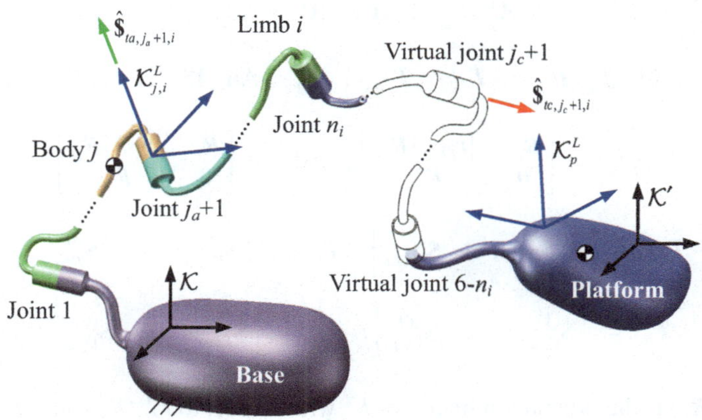

Fig. 1. Coordinate frames in the ith limb of a parallel mechanism

$$\dot{\boldsymbol{\theta}}_i = \boldsymbol{J}_i^{-1}\$_t = \boldsymbol{J}_i^{-1}\mathbf{W}^{-\mathrm{T}}\boldsymbol{\Lambda}\dot{\boldsymbol{q}} = \boldsymbol{P}_i\dot{\boldsymbol{q}}, \quad \boldsymbol{P}_i = \boldsymbol{J}_i^{-1}\mathbf{W}^{-\mathrm{T}}\boldsymbol{\Lambda}, \quad \boldsymbol{A} = \boldsymbol{J}_i\ddot{\boldsymbol{\theta}}_i + \dot{\boldsymbol{\theta}}_i^{\mathrm{T}}\boldsymbol{H}_i\dot{\boldsymbol{\theta}}_i \tag{1}$$

$$\dot{\boldsymbol{\theta}}_i = \begin{pmatrix}\dot{\boldsymbol{\theta}}_{a,i}^{\mathrm{T}} & \dot{\boldsymbol{\theta}}_{c,i}^{\mathrm{T}}\end{pmatrix}^{\mathrm{T}}, \ \dot{\boldsymbol{\theta}}_{a,i} = \begin{pmatrix}\dot{\theta}_{a,1,i} & \ldots & \dot{\theta}_{a,n_i,i}\end{pmatrix}^{\mathrm{T}}, \ \dot{\boldsymbol{\theta}}_{c,i} = \begin{pmatrix}\dot{\theta}_{c,1,i} & \ldots & \dot{\theta}_{c,6-n_i,i}\end{pmatrix}^{\mathrm{T}}$$

$$\dot{\boldsymbol{q}} = \begin{pmatrix}\dot{\boldsymbol{q}}_a^{\mathrm{T}} & \dot{\boldsymbol{q}}_c^{\mathrm{T}}\end{pmatrix}^{\mathrm{T}}, \quad \boldsymbol{q}_a = \begin{pmatrix}\dot{q}_{a,1} & \cdots & \dot{q}_{a,f}\end{pmatrix}^{\mathrm{T}}, \quad \boldsymbol{q}_c = \begin{pmatrix}\dot{q}_{c,1} & \cdots & \dot{q}_{c,6-f}\end{pmatrix}^{\mathrm{T}}$$

where $\dot{\theta}_{a,j_a,i}(j_a = 1, \ldots, n_i)/\dot{\theta}_{c,j_c,i}(j_c = 1, \ldots, 6 - n_i)$ are the joint rates of the j_ath/j_cth joint in limb i; $\dot{q}_{a,\chi}(\chi = 1, \ldots, f)/\dot{q}_{c,\vartheta}(\vartheta = 1, \ldots, 6 - f)$ denote the joint rates of the χth/ϑth actuated joint; The definition and expression of $\boldsymbol{J}_i$, $\mathbf{W}$, $\boldsymbol{\Lambda}$, $\mathbf{H}_i$ can be found in [3, 4]. From Eq. (1), the twist of permissions [2] and the accelerator of the jth body in limb i can be expressed as

$$\$_{t,j,i} = \boldsymbol{J}_{a,j,i}\boldsymbol{P}_{j,i}\dot{\boldsymbol{q}}, \quad \boldsymbol{A}_{j,i} = \boldsymbol{J}_{a,j,i}\ddot{\boldsymbol{\theta}}_{a,j,i} + \dot{\boldsymbol{\theta}}_{a,j,i}^{\mathrm{T}}\boldsymbol{H}_{a,j,i}\dot{\boldsymbol{\theta}}_{a,j,i}, \quad i = 1, 2, \ldots, l, \ j = 1, \ldots, n_i - 1 \tag{2}$$

$$\boldsymbol{J}_{a,j,i} = \begin{bmatrix} \$_{ta,1,i} & \ldots & \$_{ta,j_a,i} \end{bmatrix}, \quad \boldsymbol{P}_{j,i} = \boldsymbol{G}_{j,i}\boldsymbol{W}^{-\mathrm{T}}\boldsymbol{\Lambda}, \quad \boldsymbol{\theta}_{a,j,i} = (\theta_{1,i} \ \ldots \ \theta_{j_a,i})^{\mathrm{T}}$$

where $\boldsymbol{G}_{j,i}$ is a $(j_a \times 6)$ matrix formed by the first j_a rows of $\boldsymbol{J}_i^{-1}$; $\boldsymbol{H}_{a,j,i} \in \mathbb{R}^{6 \times j_a \times j_a}$ is the corresponding Hessian matrix extracted from $\boldsymbol{H}_i$ [3]. Since Eq. (2) is formulated in $\mathcal{K}'$, when $\$_{t,j,i}/\$_t$ and $\boldsymbol{A}_{j,i}/\boldsymbol{A}$ are evaluated with respect to $\mathcal{K}_{j,i}^L/\mathcal{K}_p^L$, the adjoint transformation should be applied.

$$\$_{t,j,i}^L = \mathrm{Ad}_{g_{j,i}}\$_{t,j,i} = \boldsymbol{T}_{j,i}\dot{\boldsymbol{q}}, \quad \$_t^L = \mathrm{Ad}_{g_p}\$_t = \boldsymbol{T}_p\dot{\boldsymbol{q}} \tag{3}$$

$$\boldsymbol{A}_{j,i}^L = \mathrm{Ad}_{g_{j,i}}\boldsymbol{A}_{j,i}, \quad \boldsymbol{A}^L = \mathrm{Ad}_{g_p}\boldsymbol{A} \tag{4}$$

$$\boldsymbol{T}_{j,i} = \mathrm{Ad}_{g_{j,i}}\boldsymbol{J}_{a,j,i}\boldsymbol{P}_{j,i} = [\boldsymbol{T}_{a,j,i} \ \ \boldsymbol{T}_{c,j,i}], \quad \boldsymbol{T}_p = \mathrm{Ad}_{g_p}\boldsymbol{W}^{-\mathrm{T}}\boldsymbol{\Lambda} = [\boldsymbol{T}_{a,p} \ \ \boldsymbol{T}_{c,p}] \tag{5}$$

$$\mathrm{Ad}_{g_{j,i}} = \begin{bmatrix} \boldsymbol{R}_{j,i} & [\boldsymbol{r}_{j,i}\times]\boldsymbol{R}_{j,i} \\ \boldsymbol{0} & \boldsymbol{R}_{j,i} \end{bmatrix}, \quad \mathrm{Ad}_{g_p} = \begin{bmatrix} \boldsymbol{R}_p & [\boldsymbol{r}_p\times]\boldsymbol{R}_p \\ \boldsymbol{0} & \boldsymbol{R}_p \end{bmatrix}$$

$$\boldsymbol{T}_{a,j,i} = \begin{bmatrix} \$_{Pa,j,i}^{(1)} & \ldots & \$_{Pa,j,i}^{(f)} \end{bmatrix}, \quad \boldsymbol{T}_{c,j,i} = \begin{bmatrix} \$_{Pc,j,i}^{(1)} & \ldots & \$_{Pc,j,i}^{(6-f)} \end{bmatrix}$$

$$\boldsymbol{T}_{a,p} = \begin{bmatrix} \$_{Pa,p}^{(1)} & \ldots & \$_{Pa,p}^{(f)} \end{bmatrix}, \quad \boldsymbol{T}_{c,p} = \begin{bmatrix} \$_{Pc,p}^{(1)} & \ldots & \$_{Pc,p}^{(6-f)} \end{bmatrix}$$

where $\boldsymbol{R}_{j,i}(\boldsymbol{R}_p)$ is the orientation matrix of $\mathcal{K}'$ with respect to $\mathcal{K}_{j,i}^L(\mathcal{K}_p^L)$ and $[\boldsymbol{r}_{j,i}\times]([\boldsymbol{r}_p\times])$ is the skew-symmetric matrix of the vector $\boldsymbol{r}_{j,i}(\boldsymbol{r}_p)$. The columns of $\boldsymbol{T}_{j,i}$ and $\boldsymbol{T}_p$ are nothing but partial screws [5].

3 Dynamic Modelling

First, the inertial wrench of a rigid body evaluated in different reference frames is investigated. The inertial wrench $\$_{wI}^C$ can be expressed in $\mathcal{K}^L$ as

$$\$_{wI}^L = \mathrm{Ad}_{g_C}^{-\mathrm{T}}\$_{wI}^C = -\boldsymbol{m}^L\boldsymbol{A}^L + \mathrm{ad}^{\mathrm{T}}(\$_t^L)\boldsymbol{m}^L\$_t^L \tag{6}$$

$$\boldsymbol{m}^L = \mathrm{Ad}_{g_C}^{-\mathrm{T}}\boldsymbol{m}^C\mathrm{Ad}_{g_C}^{-1}, \quad \mathrm{ad}^{\mathrm{T}}(\$_t^L) = \mathrm{Ad}_{g_C}^{-\mathrm{T}}\mathrm{ad}^{\mathrm{T}}(\$_t^C)\mathrm{Ad}_{g_C}^{\mathrm{T}}$$

where $\boldsymbol{m}^C$ is the tensor of inertia. Hence, the inertial wrench of the jth body in limb i, $\$_{wI,j,i}^L$, and the inertial wrench of the platform, $\$_{wI,p}^L$, can be obtained by adding subscript 'j, i' and 'p' to the corresponding terms in Eq. (6), respectively.

Suppose that the parallel mechanism is subject to the effect of a gravity field, then the gravity wrenches of the jth body in limb i, $\$^{L}_{wG,j,i}$, and the platform, $\$^{L}_{wG,p}$, evaluated in $\mathcal{K}^{L}_{j,i}$ and $\mathcal{K}^{L}_{p}$, can be obtained through adjoint transformation, respectively. Then, the overall wrench $\$^{L}_{w,p}$ ($\$^{L}_{w,j,i}$) evaluated in $\mathcal{K}^{L}_{p}$ ($\mathcal{K}^{L}_{j,i}$) can be expressed as

$$\$^{L}_{w,p} = \$^{L}_{wI,p} + \$^{L}_{wG,p}, \quad \$^{L}_{w,j,i} = \$^{L}_{wI,j,i} + \$^{L}_{wG,j,i} \tag{7}$$

Finally, the principle of virtual work states

$$\delta w = \sum_{i=1}^{l}\sum_{j=1}^{n_i-1}\left(\$^{L}_{w,j,i}\right)^{\mathrm{T}}\$^{L}_{t,j,i} + \left(\$^{L}_{w,p}\right)^{\mathrm{T}}\$^{L}_{t,p} + \mathcal{F}^{\mathrm{T}}\dot{\boldsymbol{q}} = 0 \tag{8}$$

$$\mathcal{F} = \begin{pmatrix}\mathcal{F}_a^{\mathrm{T}} & \mathcal{F}_c^{\mathrm{T}}\end{pmatrix}^{\mathrm{T}}, \quad \mathcal{F}_a = \begin{pmatrix}\mathcal{F}_{a,1} & \cdots & \mathcal{F}_{a,f}\end{pmatrix}^{\mathrm{T}}, \quad \mathcal{F}_c = \begin{pmatrix}\mathcal{F}_{c,1} & \cdots & \mathcal{F}_{c,6-f}\end{pmatrix}^{\mathrm{T}}$$

where $\mathcal{F}_{a,\chi}$ ($\chi = 1,\ldots,f$) and $\mathcal{F}_{c,\vartheta}$ ($\vartheta = 1,\ldots,6-f$) denote the intensities of the wrenches of actuations and constraints. Then, we have

$$\mathcal{F}_{a,\chi} = -\sum_{i=1}^{l}\sum_{j=1}^{n_i-1}\left(\$^{L}_{w,j,i}\right)^{\mathrm{T}}\$^{(\chi)}_{Pa,j,i} - \left(\$^{L}_{w,p}\right)^{\mathrm{T}}\$^{(\chi)}_{Pa,p}, \quad \chi = 1,\ldots,f \tag{9}$$

$$\mathcal{F}_{c,\vartheta} = -\sum_{i=1}^{l}\sum_{j=1}^{n_i-1}\left(\$^{L}_{w,j,i}\right)^{\mathrm{T}}\$^{(\vartheta)}_{Pc,j,i} - \left(\$^{L}_{w,p}\right)^{\mathrm{T}}\$^{(\vartheta)}_{Pc,p}, \quad \vartheta = 1,\ldots,6-f \tag{10}$$

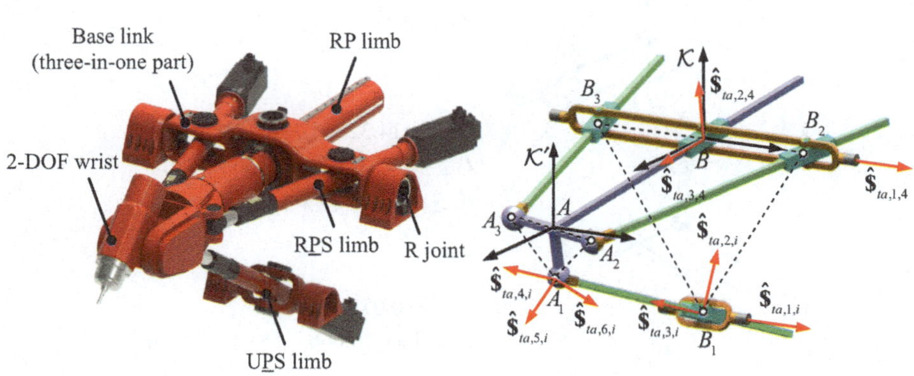

Fig. 2. Schematic diagram of the R(2RPS-RP)-UPS parallel mechanism within the TriMule robot

Table 1. Dimensional parameters (unit: m)

i	1	2	3
$\lVert\overline{AA_i}\rVert$	0.135	0.135	0.135
$\lVert\overline{BB_i}\rVert$	0.555	0.32	0.32

$\mathcal{F}_{a,\chi}$ ($\mathcal{F}_{c,\vartheta}$) can be physically interpreted as the superposition of the projection of the overall wrench of a rigid body on the direction of corresponding partial screw.

Table 2. Definitions of body-fixed frames and inertial parameters

	$\mathcal{K}^L$, $\mathcal{K}^C$	m, $\boldsymbol{I}^C$, $\boldsymbol{R}_C$, $\boldsymbol{r}_C$ (unit: kg, m, rad, s)
i	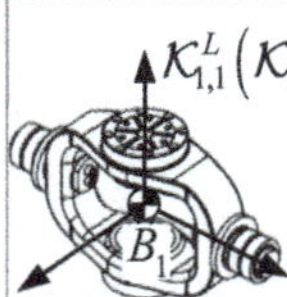	$m_{1,1} = 14.449$ $\boldsymbol{I}^C_{1,1}= \mathrm{diag}[0.059 \quad 0.123 \quad 0.168]$ $\boldsymbol{R}_{C,1,1} = \boldsymbol{I}_3,\ \boldsymbol{r}_{C,1,1} = \boldsymbol{0}$
ii	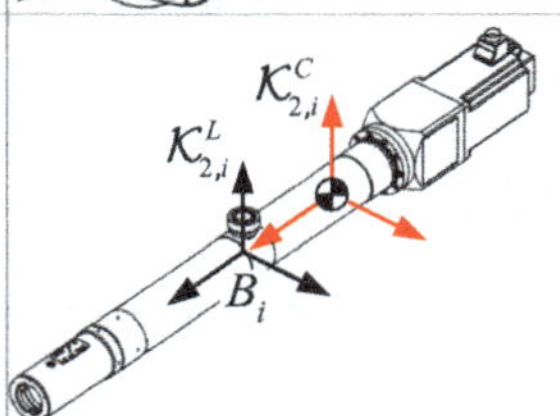	$m_{2,i} = 31.071$ $\boldsymbol{I}^C_{2,i}= \mathrm{diag}[4.541 \quad 4.538 \quad 0.062]$ $\boldsymbol{R}_{C,2,i} = \boldsymbol{I}_3,\ \boldsymbol{r}_{C,2,i} = (0 \quad 0 \quad -0.178)^{\mathrm{T}}$
iii	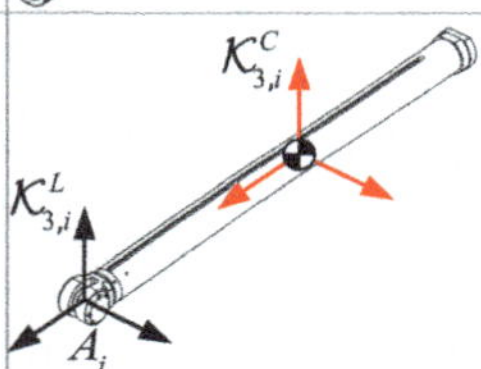	$m_{3,i} = 11.449$ $\boldsymbol{I}^C_{3,i}= \mathrm{diag}[0.865 \quad 0.865 \quad 0.007]$ $\boldsymbol{R}_{C,3,i} = \boldsymbol{I}_3,\ \boldsymbol{r}_{C,3,i} = \boldsymbol{0}$
iv	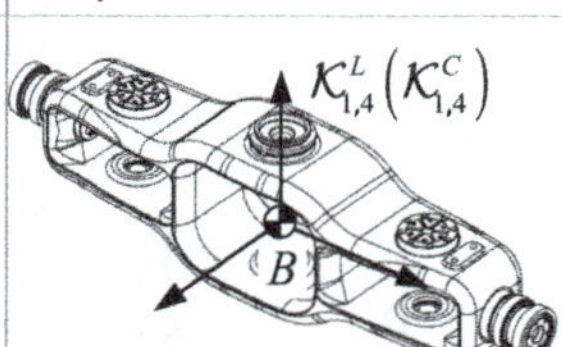	$m_{1,4} = 115.870$ $\boldsymbol{I}^C_{1,4}= \mathrm{diag}[1.172 \quad 8.106 \quad 8.628]$ $\boldsymbol{R}_{C,1,4} = \boldsymbol{I}_3,\ \boldsymbol{r}_{C,1,4} = \boldsymbol{0}$
v	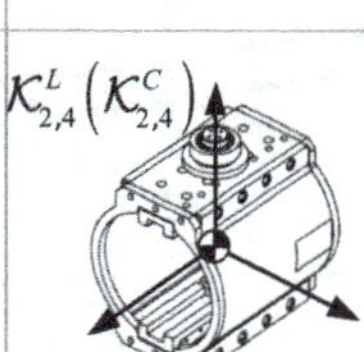	$m_{2,4} = 14.329$ $\boldsymbol{I}^C_{2,4}= \mathrm{diag}[0.126 \quad 0.086 \quad 0.115]$ $\boldsymbol{R}_{C,2,4} = \boldsymbol{I}_3,\ \boldsymbol{r}_{C,2,4} = \boldsymbol{0}$
vi	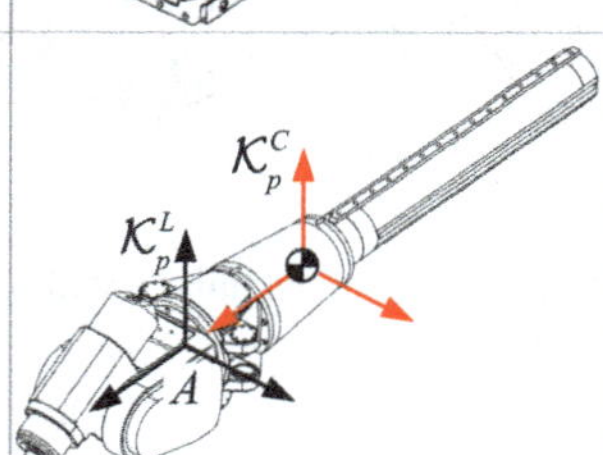	$m_p = 80.077$ $\boldsymbol{I}^C_p= \mathrm{diag}[16.066 \quad 16.043 \quad 0.638]$ $\boldsymbol{R}_{C,p} = \boldsymbol{I}_3,\ \boldsymbol{r}_{C,p} = (0 \quad 0 \quad -0.453)^{\mathrm{T}}$

4 Example

A 3-DOF R(2R<u>P</u>S-RP)-U<u>P</u>S parallel mechanism within a 5-DOF hybrid robot is taken as an example (see Fig. 2). The geometric parameters for this example are given in Table 1. While, the definitions of body-fixed frames and inertial parameters obtained from 3D software are listed in Table 2. Assume that the trajectory of the reference point A is

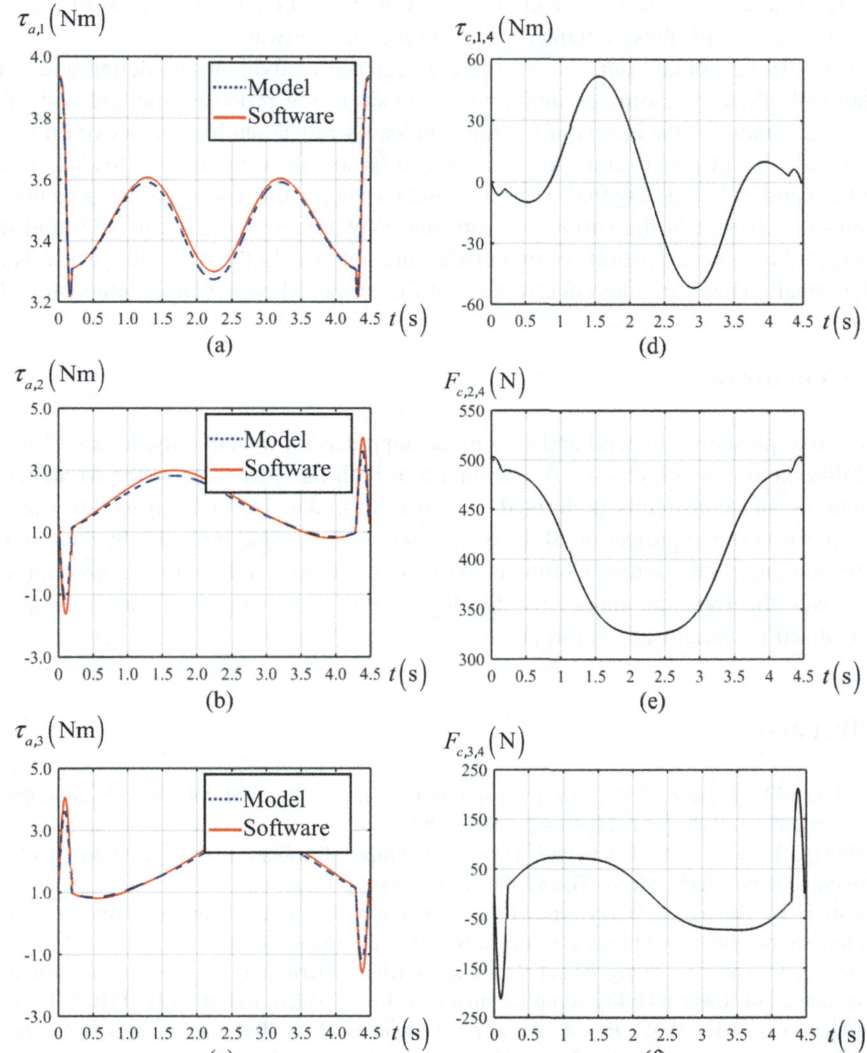

Fig. 3. Variations of the actuated torques (Figure a, b, c) versus time compared with commercial software and constrained moment and forces (Figure d, e, f) versus time

$$\begin{cases} x = -r\sin\psi(t) + x_0 \\ y = r\cos\psi(t) + y_0 \\ z = z_0 \end{cases} \tag{11}$$

$$r = 0.45\ \text{m},\ \ x_0 = 0,\ \ y_0 = -0.145\ \text{m},\ \ z_0 = 0.75\ \text{m}$$

where $(x_0 \quad y_0 \quad z_0)^{\mathrm{T}}$ is the initial value of A; $\psi(t)$ is the motion rule as the same as that given in [2]. The variations of $\mathcal{F}_{a,\chi}$ $(\chi = 1, 2, 3)$ and $\mathcal{F}_{c,\vartheta}$ $(\vartheta = 1, 2, 3)$ versus time can be obtained using Eqs. (9) and (10) (see Fig. 3). Moreover, the results of $\mathcal{F}_{a,\chi}$ are also compared with those obtained by a commercial software.

It has to be pointed out that for the convenience of dynamic modeling and computational efficiency, some assumptions are made in the proposed method with little loss of accuracy: (i) the inertia of S joints and lead-screw assemblies in active limbs are neglected; (ii) all active limbs are assumed to be axially symmetrical, leading to the inertia matrices being diagonal. However, the inertia parameters of all components are taken into account in the software. From the comparison study, it can be found that although the numerical results of two models are not exactly the same, the residuals are rather small. Therefore, the effectiveness of the proposed approach is demonstrated.

5 Conclusion

This paper presents a general and systematic approach for dynamic modelling of lower mobility parallel manipulators. A calculation of both the intensities of the wrenches of actuations and constraints is derived, which is formulated in a compact form as the superposition of the projection of the overall wrench of a rigid body on the direction of corresponding partial screw. Compared with other methods in literatures, this approach completes the dynamic model by developing the complementary part in terms of generalized constraint forces/torques.

References

1. Selig, J.M., Mcaree, P.R.: Constrained robot dynamics I: serial robots with end-effector constraints. J. Rob. Syst. **16**(9), 471–486 (1999)
2. Huang, T., Liu, H.T., Chetwynd, D.G.: Generalized Jacobian analysis of lower mobility manipulators. Mech. Mach. Theory **46**(6), 831–844 (2011)
3. Liu, H.T., Huang, T., Chetwynd, D.G.: An approach for acceleration analysis of lower mobility parallel manipulators. J. Mech. Robot. **3**(3), 150–174 (2011)
4. Huang, T., Yang, S., Wang, M., et al.: An approach to determining the unknown twist/wrench subspaces of lower mobility serial kinematic chains. J. Mech. Robot. **7**(3), 031003 (2015)
5. Gallardo, J., Rico, J.M., Frisoli, A., et al.: Dynamics of parallel manipulators by means of screw theory. Mech. Mach. Theory **38**(11), 1113–1131 (2003)

Mathematical Model of a Crane with Taking into Account Friction Phenomena in Actuators

Andrzej Urbaś(✉) and Krzysztof Augustynek

University of Bielsko-Biala, 43-309 Bielsko-Biala, Poland
{aurbas, kaugustynek}@ath.bielsko.pl

Abstract. A mathematical model of a crane with a tree structure of a kinematic chain and with closed-loop sub-chains is presented in the paper. The formulated model takes into account the flexibility of supports, link, rope and drives. Dry friction in joints is also considered. It is assumed that the clearances in joints are neglected. The formalism of joint coordinates and homogeneous transformation matrices, based on the Denavit–Hartenberg notation, are used to describe the kinematics of the crane. The equations of motion are derived using the Lagrange equations of the second kind. These equations are supplemented by the Lagrange multipliers and constraint equations formulated for each cut-joint. The flexible link is modelled using the Rigid Finite Element Method.

1 Introduction

The mathematical model for the dynamics analysis of the flexible supported crane is presented in the paper. The considered model contains the kinematic structure in the form of the open-loop kinematic chain with the closed-loop kinematic sub-chains [3]. The carrying system containing the jibs' system is treated as the main kinematic chain, whereas the actuators are connected with the main structure and they are creating the sub-chains.

The dynamics' model takes into account the different phenomena i.e. the flexibility of the supports, jib, drives and rope and friction [3, 4]. The formalism of joint coordinates and homogeneous transformation matrices is used to describe the kinematics of the crane [2]. The Rigid Finite Element Method is applied to model the flexibility of the jib [5]. The friction phenomenon in the joints of the crane is modelled by means of the LuGre friction model [1]. The dynamics equations are derived using the Lagrange equations of the second kind. Additionally, these equations are supplemented by the Lagrange multipliers and constraint equations.

The analyses performed in the paper present the influence of the changing of mass of the load, flexibility and friction on the driving systems and the trajectory of the carried load.

A. Kecskeméthy and F. Geu Flores (Eds.): ECCOMAS 2019, COMPUTMETHODS 53, pp. 299–306, 2020.
https://doi.org/10.1007/978-3-030-23132-3_36

2 Mathematical Model of the Crane

The model of crane analyzed is presented in Fig. 1. The crane is built of the main structure (m_c) mounted on the chassis (c). The chassis is connected with the ground by means of eight supports ($n_s = 8$).

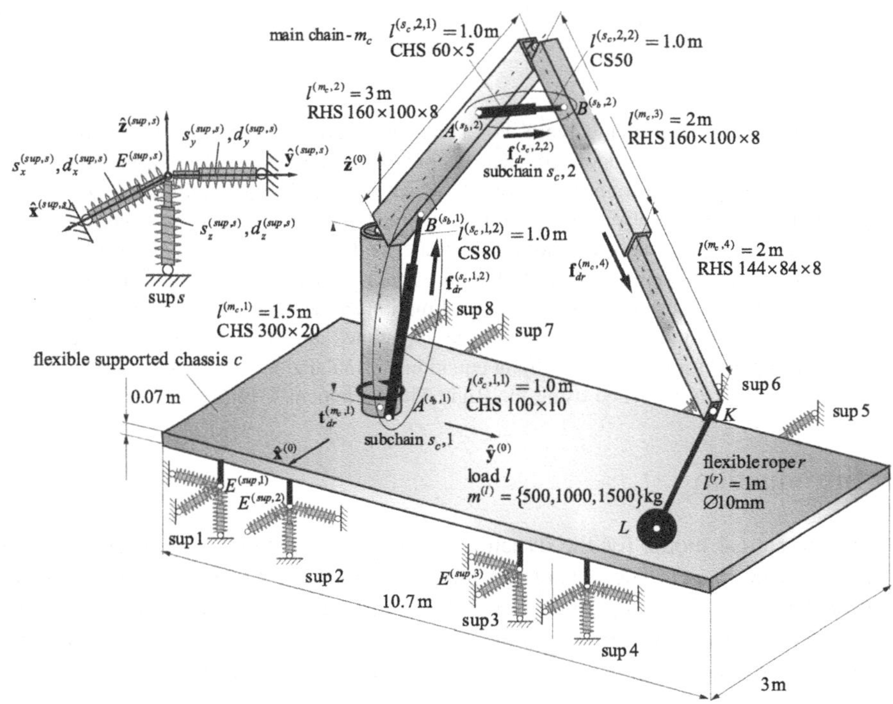

Fig. 1. Model of the crane.

The main structure forms the open-loop kinematics chain containing four links ($n_b^{(m_c)} = 4$). Two closed-loop kinematics sub-chains ($(s_c, \alpha)|_{\alpha=1,2}$) which model the actuators are connected with the main structure. Each sub-chain contains two links ($n_b^{(s_c,\alpha)}\Big|_{\alpha=1,2} = 2$). The motion of the crane is forced by means of the flexible drives (torque $\mathbf{t}_{dr}^{(m_c,1)}$ and forces $\mathbf{f}_{dr}^{(s_c,1,2)}, \mathbf{f}_{dr}^{(s_c,2,2)}, \mathbf{f}_{dr}^{(m_c,4)}$) applied to the selected links. Load (l) is treated as the concentrated mass connected with the main structure by means of flexible rope (r). It is assumed that the jib can be treated as the rigid or flexible.

The formalism of joint coordinates and homogeneous transformation matrices is used to define the kinematics of the model (Fig. 2).

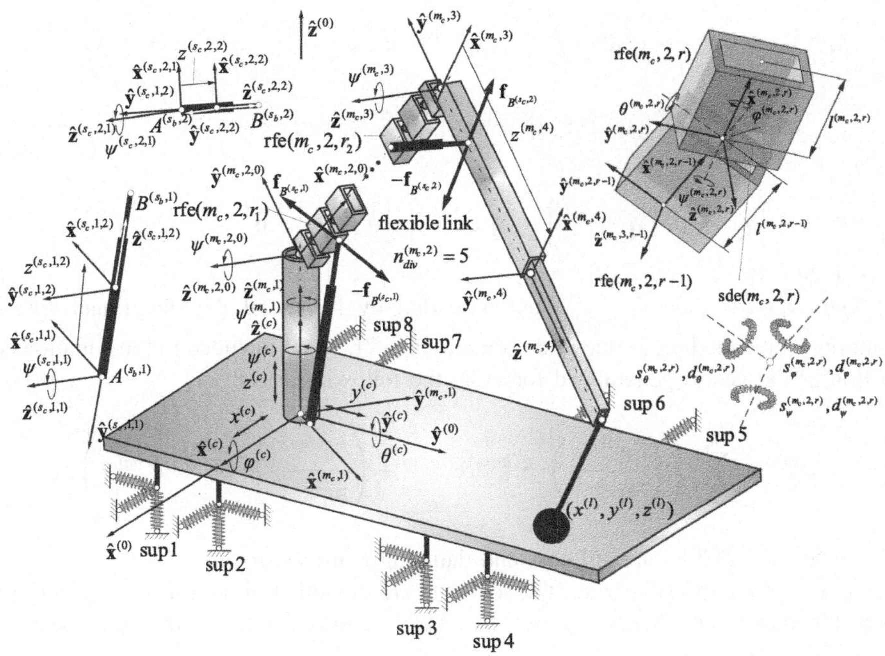

Fig. 2. Joint coordinates.

The vector of the generalised (joint) coordinates can be written in the following form:

$$\mathbf{q} = (q_j)_{j=1,\dots,n_{dof}} = \begin{bmatrix} \mathbf{q}^{(c)^T} & \bar{\mathbf{q}}^{(m_c)^T} & \bar{\mathbf{q}}^{(s_c,1)^T} & \bar{\mathbf{q}}^{(s_c,2)^T} & \mathbf{q}^{(l)^T} \end{bmatrix}^T, \tag{1}$$

where:

– chassis c : $\mathbf{q}^{(c)} = \begin{bmatrix} x^{(c)} & y^{(c)} & z^{(c)} & \psi^{(c)} & \theta^{(c)} & \varphi^{(c)} \end{bmatrix}^T$,

– main chain m_c : $\mathbf{q}^{(m_c)} = \begin{bmatrix} \mathbf{q}^{(b)^T} & \bar{\mathbf{q}}^{(m_c)^T} \end{bmatrix}^T$,

$\bar{\mathbf{q}}^{(m_c)} = \begin{bmatrix} \tilde{\mathbf{q}}^{(m_c,1)^T} & \tilde{\mathbf{q}}^{(m_c,2)^T} & \tilde{\mathbf{q}}^{(m_c,3)^T} & \tilde{\mathbf{q}}^{(m_c,4)^T} \end{bmatrix}^T$, $\tilde{\mathbf{q}}^{(m_c,1)} = \begin{bmatrix} \psi^{(m_c,1)} \end{bmatrix}$,

$\tilde{\mathbf{q}}^{(m_c,2)} = \begin{bmatrix} \tilde{\mathbf{q}}^{(m_c,2,0)^T} & \dots & \tilde{\mathbf{q}}^{(m_c,2,r)^T} & \dots & \tilde{\mathbf{q}}^{(m_c,2,n_{rfe}^{(m_c,2)})^T} \end{bmatrix}^T$, $\tilde{\mathbf{q}}^{(m_c,2,0)} = \begin{bmatrix} \psi^{(m_c,2,0)} \end{bmatrix}$,

$\tilde{\mathbf{q}}^{(m_c,2,r)}\Big|_{r=1,\dots,n_{rfe}^{(m_c,2)}} = \begin{bmatrix} \psi^{(m_c,2,r)} & \theta^{(m_c,2,r)} & \varphi^{(m_c,2,r)} \end{bmatrix}^T$, $\tilde{\mathbf{q}}^{(m_c,3)} = \begin{bmatrix} \psi^{(m_c,3)} \end{bmatrix}$,

$$\tilde{\mathbf{q}}^{(m_c,4)} = \begin{bmatrix} z^{(m_c,4)} \end{bmatrix}$$

– sub-chain $(s_c, 1)$: $\mathbf{q}^{(s_c,1)} = \begin{bmatrix} \mathbf{q}^{(b)^T} & \tilde{\mathbf{q}}^{(m_c,1)^T} & \bar{\mathbf{q}}^{(s_c,1)^T} \end{bmatrix}^T$,

$$\bar{\mathbf{q}}^{(s_c,1)} = \left[\tilde{\mathbf{q}}^{(s_c,1,1)^T} \quad \tilde{\mathbf{q}}^{(s_c,1,2)^T}\right]^T, \; \tilde{\mathbf{q}}^{(s_c,1,1)} = \left[\psi^{(s_c,1,1)}\right], \; \tilde{\mathbf{q}}^{(s_c,1,2)} = \left[z^{(s_c,1,2)}\right]$$

– sub-chain $(s_c, 2): \; \mathbf{q}^{(s_c,2)} = \left[\mathbf{q}^{(b)^T} \quad \tilde{\mathbf{q}}^{(m_c,1)^T} \quad \tilde{\mathbf{q}}^{(m_c,2,0)^T} \quad \ldots \quad \tilde{\mathbf{q}}^{(m_c,3,r_2)^T}\right.$
$\left.\bar{\mathbf{q}}^{(s_c,2)^T}\right]^T$,

$$\bar{\mathbf{q}}^{(s_c,2)} = \left[\tilde{\mathbf{q}}^{(s_c,2,1)^T} \quad \tilde{\mathbf{q}}^{(s_c,2,2)^T}\right]^T, \; \tilde{\mathbf{q}}^{(s_c,2,1)} = \left[\psi^{(s_c,2,1)}\right], \; \tilde{\mathbf{q}}^{(s_c,2,2)} = \left[z^{(s_c,2,2)}\right].$$

– load l: $\mathbf{q}^{(l)} = \left[x^{(l)} \quad y^{(l)} \quad z^{(l)}\right]^T$.

The supports ($\sup s|_{s=1,\ldots,n_s}$) are modelled by means of the directional spring-damping elements. Forces due to deformations ($\mathbf{e}^{(sup,s)}$) are included in the equations of motion as the vector generalized forces in the following form:

$$\mathbf{f}^{(sup)} = \sum_{s=1}^{n_s}\left(\left(\frac{\partial \mathbf{e}^{(sup,s)}}{\partial \tilde{\mathbf{q}}^{(m_c,1)}}\right)^T \mathbf{S}^{(sup,s)}\mathbf{e}^{(sup,s)} + \left(\frac{\partial \dot{\mathbf{e}}^{(sup,s)}}{\partial \dot{\tilde{\mathbf{q}}}^{(m_c,1)}}\right)^T \mathbf{D}^{(sup,s)}\dot{\mathbf{e}}^{(sup,s)}\right), \tag{2}$$

where: $\mathbf{S}^{(sup,s)}$, $\mathbf{D}^{(sup,s)}$ are stiffness and damping matrices of *sup s*.

As it was mentioned earlier, the actuators are modelled in the form of two closed-loop kinematic sub-chains $(s_c, \alpha)|_{\alpha=1,2}$. They require to formulate the constraint equations for the cut-joints:

$$\mathbf{C}_{B(s_c,\alpha)}\ddot{\mathbf{q}}_{B(s_c,\alpha)} = \mathbf{c}_{B(s_c,\alpha)}, \tag{3}$$

where: $\mathbf{q}_{B(s_c,1)} \in \left\{\tilde{\mathbf{q}}^{(m_c,2,0)}, \ldots, \tilde{\mathbf{q}}^{(m_c,2,r_1)}, \bar{\mathbf{q}}^{(s_c,1)}\right\}$, $\mathbf{q}_{B(s_c,2)} \in \left\{\tilde{\mathbf{q}}^{(m_c,2,r_2+1)}, \ldots, \tilde{\mathbf{q}}^{(m_c,3)},\right.$ $\left.\bar{\mathbf{q}}^{(s_c,2)}\right\}$.

The forms of matrices $\mathbf{C}_{B(s_c,\alpha)}$ and vectors $\mathbf{c}_{B(s_c,\alpha)}$ are presented in [3].

It is assumed the load is hanged to the main structure by means of the flexible rope. Generalized forces vectors (for the main structure and the load) due to the rope deformation are defined as follows:

$$\mathbf{f}^{(r,m_c)^T} = \left(f_i^{(r,m_c)}\right)_{i=1,\ldots,n_{dof}^{(m_c)}}, \tag{4.1}$$

$$\mathbf{f}^{(r,l)^T} = \left(f_i^{(r,l)}\right)_{i=1,\ldots,n_{dof}^{(l)}}, \tag{4.2}$$

where: $f_i^{(r,m_c)}, f_i^{(r,l)}$ are presented in [4].

It is assumed the selected links of the crane are driven by the flexible drives. The forces due to driving torque ($t_{dr}^{(m_c,1)}$) and driving forces ($f_{dr}^{(\alpha)}\Big|_{\alpha\in\{(m_c,4),(s_c,1,2),(s_c,2,2)\}}$) can be presented as vector generalised forces vector in the following form:

$$\mathbf{t}_{dr} = \left(t_{dr,i}^{(\alpha)}\right)_{i=1,\ldots,n_{dr}^{(\alpha)}}, \tag{5.1}$$

$$\mathbf{f}_{dr} = \left(f_{dr,i}^{(\alpha)}\right)_{i=1,\ldots,n_{dr}^{(\alpha)}}, \tag{5.2}$$

where: $t_{dr,i}^{(\alpha)}$, $f_{dr,i}^{(\alpha)}$ are presented in [3].

The Rigid Finite Element Method in the modified formulation is used to discretise $(m_c, 2)$ link. In this method, the flexible link is replaced by the system of rigid elements $\left(rfe\,(m_c, 2, r)|_{r=1,\ldots,n_{rfe}^{(m_c,2)}}\right)$ interconnected by means of spring-damping elements $\left(\text{sde}\,(m_c, 2, r)|_{r=1,\ldots,n_{sde}^{(m_c,2)}}\right)$. Generalized forces vector of forces due to the link deformation can be expressed as:

$$\mathbf{f}^{(sde)} = \mathbf{S}^{(m_c,2)}\mathbf{q}_{flex} + \mathbf{D}^{(m_c,2)}\dot{\mathbf{q}}_{flex}, \tag{6}$$

where: $\mathbf{q}_{flex} \in \tilde{\mathbf{q}}^{(m_c,2)} \backslash \left\{\tilde{\mathbf{q}}^{(m_c,2,0)}\right\}$, $\mathbf{S}^{(m_c,2)}, \mathbf{D}^{(m_c,2)}$ are stiffness and damping matrices of sdes.

The friction is taken into account in the joints of the main structure and in the actuators. The LuGre friction model is used to describe friction phenomenon. The friction coefficients are dependent on state variables which can be obtained from:

$$\dot{\mathbf{z}} = \mathbf{LuGre}\left(t, \dot{\mathbf{q}}_f, \mathbf{z}\right), \tag{7}$$

where: $\mathbf{q}_f \in \left\{\psi^{(m_c,1)}, \psi^{(m_c,2,0)}, \psi^{(m_c,3)}, z^{(m_c,4)}, z^{(s_c,1,2)}, z^{(s_c,2,2)}\right\}$.

The form of $(LuGre_i)_{i=1,\ldots,n_f}$ and the method determining of the friction coefficients are presented in [3, 4].

The recursive Newton–Euler algorithm is used to determine joint forces and torques, and subsequent friction torques (in revolute joints) and friction forces (in prismatic joints) [2].

The dynamics equations of motion with the constraint equations can be written in the following general form [5]:

$$\begin{bmatrix} \mathbf{M}(\mathbf{q}) & -\mathbf{C}(\mathbf{q}, \dot{\mathbf{q}})^T \\ \mathbf{C}(\mathbf{q}, \dot{\mathbf{q}}) & \mathbf{0} \end{bmatrix} \begin{bmatrix} \ddot{\mathbf{q}} \\ \mathbf{f}_j \end{bmatrix} = \begin{bmatrix} \mathbf{e}(\mathbf{q}, \dot{\mathbf{q}}) - \mathbf{s}(\mathbf{q}, \dot{\mathbf{q}}) - \mathbf{d}(t, \mathbf{q}, \dot{\mathbf{q}}) - \mathbf{f}(t, \mathbf{q}, \dot{\mathbf{q}}) \\ \mathbf{c}(\mathbf{q}, \dot{\mathbf{q}}) \end{bmatrix} \tag{10}$$

where: $\mathbf{M}(\mathbf{q})$ is the mass matrix, $\mathbf{C}(\mathbf{q}, \dot{\mathbf{q}})$ is the constraint matrix, $\mathbf{f}_j$ is the vector of the reaction forces in the cut-joints, $\mathbf{e}(\mathbf{q}, \dot{\mathbf{q}})$ is the vector of the Coriolis, gyroscopic and centrifugal forces, $\mathbf{s}(\mathbf{q}, \dot{\mathbf{q}})$ is the vector of the spring and damping forces formulated for the supports, link and rope, $\mathbf{d}(t, \mathbf{q}, \dot{\mathbf{q}})$ is the vector of the driving forces and torque, $\mathbf{f}(t, \mathbf{q}, \dot{\mathbf{q}})$ is the vector of the friction forces and torques, $\mathbf{c}(\mathbf{q}, \dot{\mathbf{q}})$ is the vector of the right side of constraint equations.

The dynamics of the crane is defined by the set of differential-algebraic equations with index 1. The dynamics equations of motion and state equations, are integrated using the Runge–Kutta method of the 4th order with a constant step size.

3 Numerical Simulations

In simulations the column $(m_c, 1)$ rotates around the z axis by 90° and displacements in force actuator and prismatic joint $(m_c, 4)$ also change according to the assumed function of time. The flexible link is divided into 5 rfes. The influence of the mass of the load and of link $(m_c, 2)$ flexibility on dynamics of the crane is analyzed. Figure 3 shows trajectory of the load at the final position. It can be observed that link's flexibility leads to larger oscillations of the load at the final position. Figure 4 presents how change z coordinate of the load during the motion. It can be noted that large mass of the load together with elastic deformation of the jib have large influence on position on the load during the motion.

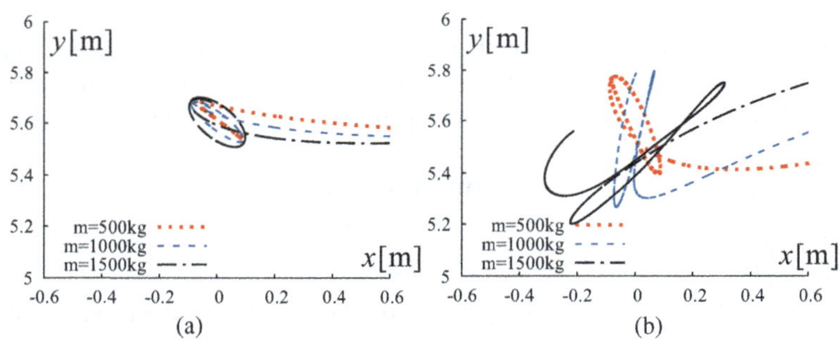

Fig. 3. Trajectory of the load in the plane $\hat{\mathbf{x}}^{(0)}\hat{\mathbf{y}}^{(0)}$ obtained for the crane with rigid links (a) and (b)

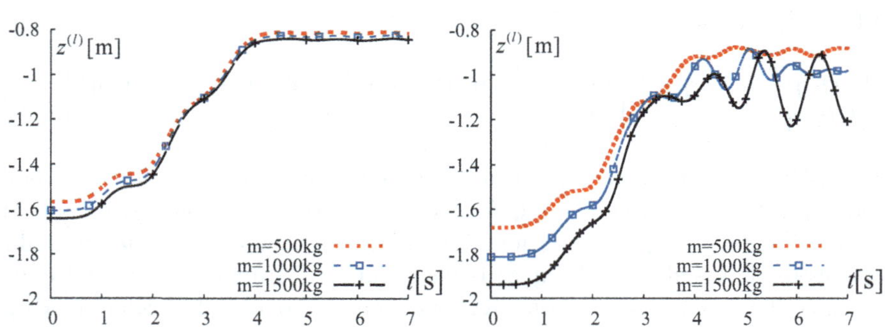

Fig. 4. Time course of the z coordinate of the load obtained for the model with rigid links (a) and flexible link (b).

The forces obtained in the force actuators realizing the assumed movement are presented in Figs. 5 and 6.

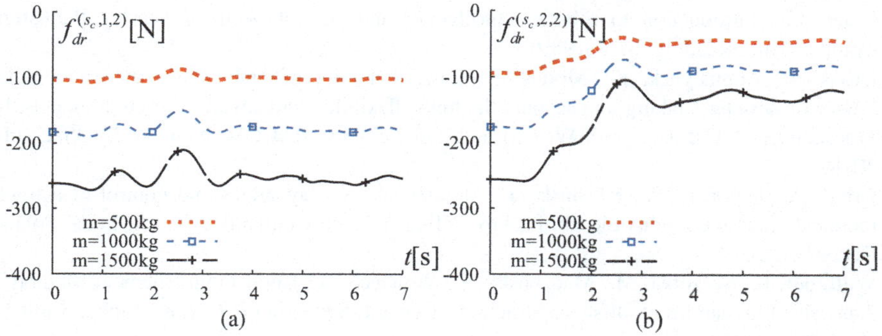

Fig. 5. Time course of the force actuator $(s_c, 1, 2)$.

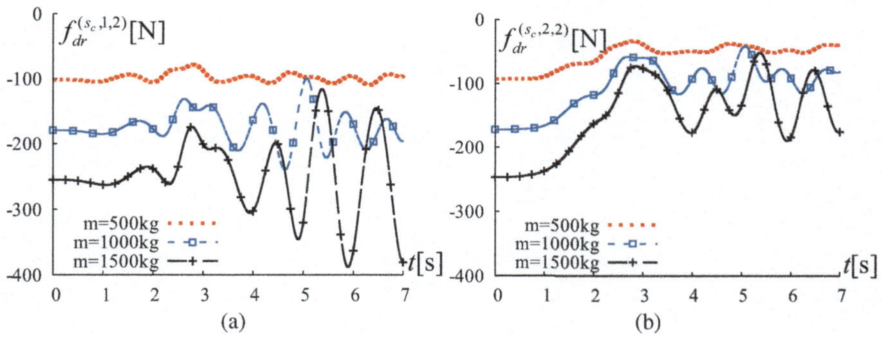

Fig. 6. Time course of the force actuator $(s_c, 2, 2)$.

The significant oscillations in forces generated by the force actuators can be observed when link $(m_c, 2)$ is treated as flexible.

4 Conclusions

The mathematical model of the crane is presented in the paper. This model was applied to the dynamics analysis taking into account such phenomena as flexibility (supports, link, drives, rope) and friction. The numerical simulations prove that these phenomena have the significant influence on the behavior of the crane. It can be seen that the jib flexibility causes significant oscillations of the courses of the driving forces in the actuators. Link flexibility has the great influence on the positioning of the load.

In the authors opinion the presented results can be useful for designers of the handling devices.

References

1. Åström, K.J., Canudas-de-Witt, C.: Revisiting the LuGre model. IEEE Control Syst. Mag. Inst. Electr. Electron. Mag. **28**(6), 101–114 (2008)
2. Craig, J.J.: Introduction to robotics: mechanics and control. Addison-Wesley Publishing Company Inc, Reading, MA (1989)
3. Urbaś, A., Augustynek, K.: Modelling of the dynamics of grab cranes with a complex kinematic structure, taking into account the links' flexibility and advanced friction models. In: Proceedings of 13th Congress WCCM and 2nd Congress PACCM, New York, USA, July, 2018
4. Urbaś, A., Jabłoński, A., Kłosiński, J., Augustynek, K.: Dynamics and control of a truck-mounted crane with a flexible jib. In: Proceedings of 14th Conference DSTA, Łódź, Poland, December, 2017
5. Wittbrodt, E., Szczotka, M., Maczyński, A., Wojciech, S.: Rigid Finite Element Method in Analysis of Dynamics of Offshore Structures. Ocean Engineering & Oceanography. Springer, Heidelberg (2013)

Closed Form of the Baker-Campbell-Hausdorff Formula for the Lie Algebra of Rigid Body Displacements

Daniel Condurache[1(✉)] and Ioan-Adrian Ciureanu[2]

[1] Technical University of Iasi, D. Mangeron Street no. 59, 700050 Iasi, Romania
daniel.condurache@tuiasi.ro
[2] "Grigore T. Popa" University of Medicine and Pharmacy Iasi, Universitatii Street no. 16, 700115 Iasi, Romania
adrian.ciureanu@umfiasi.ro

Abstract. This paper demonstrates the existence of the closed form of the Baker-Campbell-Hausdorff (BCH) formula for the Lie algebra of rigid body displacement. For this, the structure of the Lie group of the rigid body displacements $\mathbb{SE}_3$ and the properties of its algebra Lie $\mathfrak{se}_3$ are used. Also, using the isomorphism between the Lie group $\mathbb{SE}_3$ and the Lie group of the orthogonal dual tensors, a solution of this problem in dual algebra is given.

1 Introduction

The Baker–Campbell–Hausdorff (BCH) expansion formula for two noncommuting operators has wide applications in various disciplines, e.g. mathematics, theoretical physics, physical chemistry, control theory, multibody kinematics etc. Various partial results leading to what is now called the Baker–Campbell–Hausdorff formula have by now been in circulation for well over 100 years. No completely general closed formula has been produced for the BCH theorem. A variety of approaches has been taken to dealing with the series expansion directly, including calculating the higher order commutators through recurrence relations based on a deeper understanding of their algebraic properties or via combinatorial methods [1]. Thus, for given general operators **A** and **B**, the operator **C** which arises in the product exp**C** = exp**A** exp**B** is expressed as an infinite series of progressively higher order nested commutators of **A** and **B**, i.e. $\mathbf{C}(\mathbf{A},\mathbf{B}) = \mathbf{A} + \mathbf{B} + P_2(\mathbf{A},\mathbf{B}) + P_3(\mathbf{A},\mathbf{B}) + P_4(\mathbf{A},\mathbf{B}) + \ldots$, where $P_2(\mathbf{A},\mathbf{B}) = (1/2)[\mathbf{A},\mathbf{B}]$ and, in general, each higher P_n is a homogeneous Lie polynomial of order n in **A** and **B**. These higher terms may be explicitly determined by known methods, although they become rapidly more difficult to calculate.

Closed form formulae have been found in some special cases, such as when the operators belong to some particular Lie algebra so that the associated methods may be used [2–5].

However, in most applications explicit evaluation of the infinite series of the nested commutators of **A** and **B** appears to be a formidable task.

In this paper vie demonstrates the existence of the closed form of the Baker-Campbell-Hausdorff formula for the Lie algebra of rigid body displacement. For this,

A. Kecskeméthy and F. Geu Flores (Eds.): ECCOMAS 2019, COMPUTMETHODS 53, pp. 307–314, 2020.
https://doi.org/10.1007/978-3-030-23132-3_37

the structure of the Lie group of the rigid body displacements $S\mathbb{E}_3$ and the properties of its Lie algebra $s\mathbb{e}_3$ are used. Also, using the isomorphism between the Lie group $S\mathbb{E}_3$ and the Lie group of the orthogonal dual tensors, a solution of this problem in dual Lie algebra is given.

2 Closed Form BCH Formula in $s\mathbb{O}_3$

Let be $S\mathbb{O}_3 = \{\boldsymbol{R} \in \mathbb{L}(V_3, V_3) | \boldsymbol{R}\boldsymbol{R}^{\mathrm{T}} = \boldsymbol{I}, \det\boldsymbol{R} = 1\}$ the Lie group of the rotations and $s\mathbb{O}_3 = \{\tilde{\boldsymbol{\omega}} \in \mathbb{L}(\boldsymbol{V}_3, \boldsymbol{V}_3) | \tilde{\boldsymbol{\omega}}^{\mathrm{T}} + \tilde{\boldsymbol{\omega}} = \boldsymbol{O}\}$ the Lie algebra of this Lie group. In these conditions, the following remark takes place:

Remark 1: The application:

$$\begin{aligned} &\exp\colon s\mathbb{O}_3 \to S\mathbb{O}_3 \\ &\exp\tilde{\boldsymbol{\omega}} = \textstyle\sum_{k=0}^{\infty} \frac{\tilde{\boldsymbol{\omega}}^k}{k!}. \end{aligned} \tag{1}$$

is well-defined and onto.

It takes place the closed form formula:

$$\exp\tilde{\omega} = \boldsymbol{I} + \mathrm{sinc}\,\omega\tilde{\omega} + \frac{1}{2}\mathrm{sinc}^2\frac{\omega}{2}\tilde{\omega}^2, \tag{2}$$

where ω was denoted with $\omega = \sqrt{-\frac{1}{2}\mathrm{trace}\tilde{\omega}^2} = \|\mathrm{vect}\tilde{\omega}\|$ and sinc is a cardinal sin function.

The linear invariants of tensor $\boldsymbol{R} = \exp\tilde{\omega}$ it results from:

$$\mathrm{trace}\boldsymbol{R} = 1 + 2\cos\omega \tag{3}$$

$$\mathrm{vect}\boldsymbol{R} = \sin\omega\boldsymbol{\omega} \tag{4}$$

The inverse of the application (1) is a multifunction denoted with:

$$\log\colon \mathbb{S}\mathrm{O}_3 \to s\mathbb{O}_3 \tag{5}$$

$$\log\boldsymbol{R} = \alpha\tilde{\boldsymbol{u}}. \tag{6}$$

where

$$\begin{aligned} &\tilde{\boldsymbol{u}} = \pm\frac{\boldsymbol{R}-\boldsymbol{R}^T}{\sqrt{(1+\mathrm{trace}\boldsymbol{R})(3-\mathrm{trace}\boldsymbol{R})}} \\ &\alpha = 2k\pi \pm \arccos\frac{\mathrm{trace}\boldsymbol{R}-1}{2}, k \in \mathbb{Z}. \end{aligned} \tag{7}$$

For $\mathrm{trace}\boldsymbol{R} = 3$, $\log\boldsymbol{R} = 0$ and for $\mathrm{trace}\boldsymbol{R} = -1$, $\log\boldsymbol{R} \supset \{\pm\pi(2k+1)\tilde{\boldsymbol{u}}\}$ with $k \in \mathbb{Z}$ where $\mathbf{u}$ is computed as $\mathbf{u} = \frac{\boldsymbol{R}\mathbf{v}+\mathbf{v}}{\|\boldsymbol{R}\mathbf{v}+\mathbf{v}\|}$ $\forall\mathbf{v} \in \mathbf{V}_3$ that have the property $\boldsymbol{R}\mathbf{v} \neq -\mathbf{v}$.

In this case, BCH formula has a closed form that was first demonstrated in [6] and than, in an equivalent form, in [7].

Theorem 1: *Let be* $\tilde{\boldsymbol{\omega}}_1, \tilde{\boldsymbol{\omega}}_2 \in so_3$. *The bellow identity takes place:*

$$\log(\exp\tilde{\omega}_1 \exp\tilde{\omega}_1) = \tilde{\omega} \tag{8}$$

where

$$\tilde{\omega} = \alpha_1\tilde{\omega}_1 + \alpha_2\tilde{\omega}_2 + \alpha_{12}[\tilde{\omega}_1, \tilde{\omega}_2] \tag{9}$$

with:

$$\alpha_1 = \frac{\operatorname{sinc}\frac{\omega_1}{2}\cos\frac{\omega_2}{2}}{\operatorname{sinc}\frac{\omega}{2}} \tag{10}$$

$$\alpha_2 = \frac{\cos\frac{\omega_1}{2}\operatorname{sinc}\frac{\omega_2}{2}}{\operatorname{sinc}\frac{\omega}{2}} \tag{11}$$

$$\alpha_{12} = \frac{\operatorname{sinc}\frac{\omega_1}{2}\operatorname{sinc}\frac{\omega_2}{2}}{2\operatorname{sinc}\frac{\omega}{2}} \tag{12}$$

$$\omega = 2\arccos\left(\cos\frac{\omega_1}{2}\cos\frac{\omega_2}{2} - \frac{1}{4}\operatorname{sinc}\frac{\omega_1}{2}\operatorname{sinc}\frac{\omega_2}{2}\boldsymbol{\omega}_1 \cdot \boldsymbol{\omega}_2\right) \tag{13}$$

Proof: Let be $\tilde{\boldsymbol{\omega}} \in so_3$ such that:

$$\exp\tilde{\omega} = \exp\tilde{\omega}_1 \exp\tilde{\omega}_2 \tag{14}$$

Taking into account Eqs. (2), (3), (4) and (8), after some algebra, we obtain:

$$\cos\frac{\omega}{2} = \cos\frac{\omega_1}{2}\cos\frac{\omega_2}{2} - \frac{1}{4}\operatorname{sinc}\frac{\omega_1}{2}\operatorname{sinc}\frac{\omega_2}{2}\boldsymbol{\omega}_1 \cdot \boldsymbol{\omega}_2 \tag{15}$$

$$\operatorname{sinc}\frac{\omega}{2}\boldsymbol{\omega} = \operatorname{sinc}\frac{\omega_1}{2}\cos\frac{\omega_2}{2}\boldsymbol{\omega}_1 + \cos\frac{\omega_1}{2}\operatorname{sinc}\frac{\omega_2}{2}\boldsymbol{\omega}_2 + \frac{1}{2}\operatorname{sinc}\frac{\omega_1}{2}\operatorname{sinc}\frac{\omega_2}{2}\boldsymbol{\omega}_1 \times \boldsymbol{\omega}_2 \tag{16}$$

From (15) results Eq. (13). Using the identity:

$$skew(\boldsymbol{\omega}_1 \times \boldsymbol{\omega}_2) = \tilde{\omega}_1\tilde{\omega}_2 - \tilde{\omega}_2\tilde{\omega}_1 \underline{\underline{\text{def}}} [\tilde{\omega}_1, \tilde{\omega}_2], \tag{17}$$

from Eq. (16), results Eq. (9), (10), (11) and (12).

3 Closed Form BCH Formula for the Lie Algebra se$_3$

For $\boldsymbol{R} \in S\mathbb{O}_3$, $\boldsymbol{R} = \exp\widetilde{\boldsymbol{\omega}}$, the following identity takes place:

$$\dot{\boldsymbol{R}} = \widetilde{\mathrm{dexp}_{\widetilde{\boldsymbol{\omega}}}\dot{\boldsymbol{\omega}}}\boldsymbol{R} = \boldsymbol{R}\widetilde{\mathrm{dexp}_{-\widetilde{\boldsymbol{\omega}}}\dot{\boldsymbol{\omega}}} \tag{18}$$

where $\dot{\boldsymbol{\omega}}$ denotes the time derivative of $\boldsymbol{\omega}$ and $\mathrm{dexp}_{\widetilde{\boldsymbol{\omega}}}$ has the closed form [8]:

$$\mathrm{dexp}_{\widetilde{\boldsymbol{\omega}}} = \boldsymbol{I} + \frac{1}{2}\mathrm{sinc}^2\frac{\omega}{2}\widetilde{\boldsymbol{\omega}} + (1 - \mathrm{sinc}\,\omega)\frac{\widetilde{\boldsymbol{\omega}}^2}{\omega^2}. \tag{19}$$

For $\omega \neq 2k\pi, \mathrm{k} \in \mathbb{N}$, this tensor is invertible and we have:

$$\mathrm{dexp}^{-1}_{\widetilde{\boldsymbol{\omega}}} = \boldsymbol{I} - \frac{1}{2}\widetilde{\boldsymbol{\omega}} + \left(1 - \frac{\omega}{2}\cot\frac{\omega}{2}\right)\frac{\widetilde{\boldsymbol{\omega}}^2}{\omega^2}. \tag{20}$$

Let be $S\mathbb{E}_3$ the Lie group of the rigid body displacements and $s\mathbb{e}_3$ its Lie algebra. As it is known, the generic elements from $S\mathbb{E}_3$ and $s\mathbb{e}_3$ can be written in the following matrix form:

$$\begin{aligned} g &= \begin{bmatrix} \boldsymbol{R} & \mathbf{t} \\ \mathbf{0} & 1 \end{bmatrix} \in S\mathbb{E}_3 \\ \hat{\xi} &= \begin{bmatrix} \widetilde{\boldsymbol{\omega}} & \mathbf{v} \\ \mathbf{0} & 0 \end{bmatrix} \in s\mathbb{e}_3. \end{aligned} \tag{21}$$

Where $\boldsymbol{R} \in S\mathbb{O}_3, \widetilde{\boldsymbol{\omega}} \in s\mathbb{o}_3, \mathbf{t}, \mathbf{v} \in \mathbf{V}_3$. According to the author's knowledge, a closed form of the BCH formula for the Lie algebra of the rigid body displacements has not been reported in the literature.

Remark 2: The application: $\exp: s\mathbb{e}_3 \to S\mathbb{E}_3$; $\exp\hat{\xi} = \sum_{k=0}^{\infty}\frac{\hat{\xi}^k}{k!}$ is well defined and onto. Also, the below relationship takes place:

$$\exp\hat{\xi} = \begin{bmatrix} \exp\widetilde{\boldsymbol{\omega}} & \mathrm{dexp}_{\widetilde{\boldsymbol{\omega}}}\mathbf{v} \\ \mathbf{0} & 1 \end{bmatrix}. \tag{22}$$

Theorem 2: *Let be* $\hat{\xi}_1, \hat{\xi}_2 \in s\mathbb{e}_3$ *with* $\hat{\xi}_1 = \begin{bmatrix} \widetilde{\boldsymbol{\omega}}_1 & \mathbf{v}_1 \\ \mathbf{0} & 0 \end{bmatrix}$ *and* $\hat{\xi}_2 = \begin{bmatrix} \widetilde{\boldsymbol{\omega}}_2 & \mathbf{v}_2 \\ \mathbf{0} & 0 \end{bmatrix}$. *The below identity takes place:*

$$\log\left(\exp\hat{\xi}_1\exp\hat{\xi}_2\right) = \begin{bmatrix} \widetilde{\boldsymbol{\omega}} & \mathbf{v} \\ \mathbf{0} & 0 \end{bmatrix} \tag{23}$$

where

$$\tilde{\omega} = \alpha_1 \tilde{\omega}_1 + \alpha_2 \tilde{\omega}_2 + \alpha_{12} [\tilde{\omega}_1, \tilde{\omega}_2] \tag{24}$$

$$\mathbf{v} = \boldsymbol{T}_1 \mathbf{v}_1 + \boldsymbol{T}_2 \mathbf{v}_2. \tag{25}$$

The real numbers $\alpha_1, \alpha_2, \alpha_{12}$ *and the invertible tensors* $\boldsymbol{T}_1$ *and* $\boldsymbol{T}_2$ *depend solely on* $\tilde{\omega}_1$ *and on* $\tilde{\omega}_2$ *and can be determined in closed form.*

Proof: Let be $\tilde{\boldsymbol{\omega}} \in \mathbb{so}_3$ and $\mathbf{v} \in V_3$ such that

$$\exp\xi_1 \exp\xi_2 = \exp\begin{bmatrix} \tilde{\omega} & \mathbf{v} \\ \mathbf{0} & 1 \end{bmatrix}. \tag{26}$$

Taking into account Eq. (15), from Eq. (26), it follows:

$$\exp\tilde{\omega}_1 \exp\tilde{\omega}_2 = \exp\tilde{\omega} \tag{27}$$

$$\mathrm{dexp}_{\tilde{\omega}} \mathbf{v} = \exp\tilde{\omega}_1 \mathrm{dexp}_{\tilde{\omega}_2} \mathbf{v}_2 + \mathrm{dexp}_{\tilde{\omega}_1} \mathbf{v}_1. \tag{28}$$

From the Eq. (27) and Theorem 1, it follows Eq. (24). From Eq. (28), it follows the Eq. (25), where:

$$\begin{aligned} \boldsymbol{T}_1 &= \mathrm{dexp}_{\tilde{\omega}}^{-1} \mathrm{dexp}_{\tilde{\omega}_1} \\ \boldsymbol{T}_2 &= \mathrm{dexp}_{\tilde{\omega}}^{-1} \exp\tilde{\omega}_1 \mathrm{dexp}_{\tilde{\omega}_2} \end{aligned} \tag{29}$$

Another approach has a closed form solution of the BCH formula for Lie algebra of rigid body displacements can be given using the isomorphism between the Lie algebra $\mathbb{se}_3$ and Lie algebra of the dual vectors [9–11].

4 Closed Form BCH Formula in Dual Lie Algebra

Let the orthogonal dual tensor set be denoted by:

$$\underline{S\mathbb{O}}_3 = \{\underline{\boldsymbol{R}} \in \mathbf{L}(\underline{\mathbf{V}}_3, \underline{\mathbf{V}}_3) | \underline{\boldsymbol{R}\boldsymbol{R}}^T = \underline{\boldsymbol{I}}, \det\underline{\boldsymbol{R}} = 1\} \tag{30}$$

where $\underline{S\mathbb{O}}_3$ is the set of special orthogonal dual tensors and $\underline{\boldsymbol{I}}$ is the unit orthogonal dual tensor. The internal structure of any orthogonal dual tensor $\underline{\boldsymbol{R}} \in \underline{S\mathbb{O}}_3$ is illustrated in a series of results that were detailed in our previous work.

Theorem 3 (Structure Theorem) [10]. *For any* $\underline{\boldsymbol{R}} \in \underline{S\mathbb{O}}_3$ *a unique decomposition is viable*

$$\underline{\boldsymbol{R}} = (\boldsymbol{I} + \varepsilon\tilde{\rho})\boldsymbol{Q} \tag{31}$$

where $\boldsymbol{Q} \in S\mathbb{O}_3$ *and* $\boldsymbol{\rho} \in V_3$ *are called* **structural invariants**.

Taking into account the Lie group structure of $\underline{\mathbb{SO}}_3$ and the result presented in previous theorem, it can be concluded that any orthogonal dual tensor $\underline{\boldsymbol{R}} \in \underline{\mathbb{SO}}_3$ can be used globally parameterize displacements of rigid bodies.

Theorem 4 (Representation Theorem) [10]. *For any orthogonal dual tensor* $\underline{\boldsymbol{R}}$ *defined as in Eq.* (31)*, a dual number* $\underline{\alpha} = \alpha + \varepsilon d$ *and a dual unit vector* $\underline{\mathbf{u}} = \mathbf{u} + \varepsilon \mathbf{u}_0$ *can be computed to have the following equation*:

$$\underline{\boldsymbol{R}}(\underline{\alpha}, \underline{\mathbf{u}}) = \boldsymbol{I} + \sin\underline{\alpha}\,\underline{\tilde{\boldsymbol{u}}} + (1 - \cos\underline{\alpha})\underline{\tilde{\boldsymbol{u}}}^2 = \exp(\underline{\alpha}\,\underline{\tilde{\boldsymbol{u}}}) \tag{32}$$

The parameters $\underline{\alpha}$ and $\underline{\mathbf{u}}$ are called the **natural invariants** of $\underline{\boldsymbol{R}}$. The unit dual vector $\underline{\mathbf{u}}$ gives the Plücker representation of the Mozzi-Chalses axis, while the dual angle $\underline{\alpha} = \alpha + \varepsilon d$ contains the rotation angle α and the translated distance d.

The Lie algebra of the Lie group $\underline{\mathbb{SO}}_3$ is the skew-symmetric dual tensor set denoted by $\underline{\mathfrak{so}}_3 = \{\underline{\tilde{\omega}} \in \mathrm{L}(\underline{\mathbf{V}}_3, \underline{\mathbf{V}}_3) | \underline{\tilde{\omega}} = -\underline{\tilde{\omega}}^{\mathrm{T}}\}$, where the internal mapping is $\langle \underline{\tilde{\omega}}_1, \underline{\tilde{\omega}}_2 \rangle = \widetilde{\underline{\omega}_1 \underline{\omega}_2}$. The Lie algebra $\underline{\mathfrak{se}}_3$ is isomorphic to the Lie algebra of dual vectors $\underline{\mathbf{V}}_3$, having as internal operation the cross product of dual vectors.

The link between the Lie algebra $\underline{\mathfrak{se}}_3$, the Lie group $\underline{\mathbb{SO}}_3$, and the exponential map is given by the following.

Theorem 5. *The mapping*

$$\begin{gathered} \exp\colon \underline{\mathfrak{so}}_3 \to \underline{\mathbb{SO}}_3, \\ \exp(\underline{\tilde{\omega}}) = e^{\underline{\tilde{\omega}}} = \sum_{k=0}^{\infty} \frac{\underline{\tilde{\omega}}^k}{k!} \end{gathered} \tag{33}$$

is well defined and onto and takes place the below closed form formula [9]:

$$\exp\underline{\tilde{\omega}} = \boldsymbol{I} + \mathrm{sinc}\,\underline{\omega}\,\underline{\tilde{\omega}} + \frac{1}{2}\mathrm{sinc}^2\frac{\underline{\omega}}{2}\underline{\tilde{\omega}}^2. \tag{34}$$

A unit dual vector can parameterize any screw axis that embeds a rigid displacement, whereas the screw parameters (angle of rotation about the screw and the translation along the screw axis) can be structured as a dual angle. The computation of the screw axis is linked with the problem of finding the logarithm of an orthogonal dual tensor $\underline{\boldsymbol{R}}$, which is a multifunction defined by:

$$\begin{gathered} \log\colon \underline{\mathbb{SO}}_3 \to \underline{\mathfrak{so}}_3, \\ \log\underline{\boldsymbol{R}} = \{\underline{\tilde{\omega}} \in \underline{\mathfrak{so}}_3 | \exp(\underline{\tilde{\omega}}) = \underline{\boldsymbol{R}}\} \end{gathered} \tag{35}$$

and is the inverse of Eq. (33).

Based on Theorems 4 and 5, for any orthogonal dual tensor $\underline{\boldsymbol{R}}$, a dual vector $\underline{\boldsymbol{\omega}} = \underline{\alpha}\,\underline{\mathbf{u}} = \boldsymbol{\omega} + \varepsilon \mathbf{v}$ can be computed and represents the Euler dual vector, which embeds the screw axis and screw parameters. The form of $\underline{\boldsymbol{\omega}}$ implies that $\underline{\tilde{\omega}} \in \log \underline{\boldsymbol{R}}$.

Also if $\|\boldsymbol{\omega}\| < 2\pi$, Theorems 4 and 5 can be used to uniquely recover the Euler dual vector $\underline{\boldsymbol{\omega}}$, which is equivalent with computing $\log \underline{\boldsymbol{R}}$.

Theorem 6 (Isomorphism theorem) [11]: *The special Euclidean group* $(S\mathbb{E}_3, \cdot)$ *and* $(\underline{S\mathbb{O}}_3, \cdot)$ *are connected via the isomorphism of the Lie groups*

$$\Phi : S\mathbb{E}_3 \to \underline{S\mathbb{O}}_3 \tag{36}$$

$$\Phi(g) = (\boldsymbol{I} + \varepsilon \tilde{\rho})\boldsymbol{Q}$$

where $g = \begin{bmatrix} \boldsymbol{Q} & \boldsymbol{\rho} \\ \mathbf{0} & 1 \end{bmatrix}$, $\boldsymbol{Q} \in S\mathbb{O}_3$, $\boldsymbol{\rho} \in V_3$.

Remark 3: The inverse of Φ is

$$\Phi^{-1} : \underline{S\mathbb{O}}_3 \leftrightarrow S\mathbb{E}_3; \Phi^{-1}(\underline{\boldsymbol{R}}) = \begin{bmatrix} \boldsymbol{Q} & \boldsymbol{\rho} \\ \mathbf{0} & 1 \end{bmatrix} \tag{37}$$

where $\boldsymbol{Q} = Re(\underline{\boldsymbol{R}}), \boldsymbol{\rho} = vect\left(Du(\underline{\boldsymbol{R}})\boldsymbol{Q}^{\mathrm{T}}\right)$.

Using the principle of transference [12], from Theorem 1, the following theorem will be obtained.

Theorem 7: *Let be* $\underline{\boldsymbol{\omega}}_1 = \boldsymbol{\omega}_1 + \varepsilon \mathbf{v}_1$ *and* $\underline{\boldsymbol{\omega}}_2 = \boldsymbol{\omega}_2 + \varepsilon \mathbf{v}_2$ *the dual vectors that corresponds to* $\widehat{\xi}_1$ *and, respectively,* $\widehat{\xi}_2$ *from* $\underline{\mathbb{se}_3}$. *The below identities take place:*

$$\begin{gathered} \log(\exp\underline{\tilde{\omega}}_1 \log\underline{\tilde{\omega}}_2) = \underline{\tilde{\omega}} \\ \underline{\tilde{\omega}} = \underline{\alpha}_1 \underline{\tilde{\omega}}_1 + \underline{\alpha}_2 \underline{\tilde{\omega}}_2 + \underline{\alpha}_{12} [\underline{\tilde{\omega}}_1, \underline{\tilde{\omega}}_2] \end{gathered} \tag{38}$$

$$\underline{\boldsymbol{\omega}} = \underline{\alpha}_1 \underline{\boldsymbol{\omega}}_1 + \underline{\alpha}_2 \underline{\boldsymbol{\omega}}_2 + \underline{\alpha}_{12} (\underline{\boldsymbol{\omega}}_1 \times \underline{\boldsymbol{\omega}}_2) \tag{39}$$

where $\underline{\boldsymbol{\omega}} = \boldsymbol{\omega} + \varepsilon \mathbf{v}$ *and the dual numbers* $\underline{\alpha}_1, \underline{\alpha}_2, \underline{\alpha}_{12}$ *are written in closed form:*

$$\underline{\alpha}_1 = \frac{\operatorname{sinc}\frac{\underline{\omega}_1}{2} \cos\frac{\underline{\omega}_2}{2}}{\operatorname{sinc}\frac{\underline{\omega}}{2}} \tag{40}$$

$$\underline{\alpha}_2 = \frac{\cos\frac{\underline{\omega}_1}{2} \operatorname{sinc}\frac{\underline{\omega}_2}{2}}{\operatorname{sinc}\frac{\underline{\omega}}{2}} \tag{41}$$

$$\underline{\alpha}_{12} = \frac{\operatorname{sinc}\frac{\underline{\omega}_1}{2} \operatorname{sinc}\frac{\underline{\omega}_2}{2}}{2\operatorname{sinc}\frac{\underline{\omega}}{2}} \tag{42}$$

$$\underline{\omega} = 2\arccos\left(\cos\frac{\underline{\omega}_1}{2} \cos\frac{\underline{\omega}_2}{2} - \frac{1}{4}\operatorname{sinc}\frac{\underline{\omega}_1}{2} \operatorname{sinc}\frac{\underline{\omega}_2}{2} \underline{\boldsymbol{\omega}}_1 \cdot \underline{\boldsymbol{\omega}}_2\right) \tag{43}$$

Remark 4: By separating the real part and the dual part of Eq. (38), the relations (24) and (25) of the Theorem 2 are obtained. The relation (39) represents the Euler dual vector corresponding to the rigid body displacement that results from the successive composition of two rigid body displacements corresponding to the Euler dual vectors $\underline{\boldsymbol{\omega}}_2$, respectively $\underline{\boldsymbol{\omega}}_1$.

5 Conclusions

In this paper we have found a local closed-form expression for the Baker-Campbell-Hausdorff formula for the Lie algebra of rigid body displacements. For this, the structure of the Lie group of the rigid body displacements and the properties of its Lie algebra are used. Also, using the isomorphism between the Lie group of rigid body displacement and the Lie group of the orthogonal dual tensors, a solution of this problem in dual Lie algebra is given. Applications of the obtained results regarding the composition of the rigid body displacements (Euler dual vector determination) are presented. Direct and inverse kinematics of the general 3C manipulator using BCH formula will be the subject of the future works.

References

1. Iserles, A., Munthe-Kaas, Z.H., Norsett, S.P., Zanna, A.: Lie-group methods. Acta Numerica **9**, 215–365 (2000)
2. Foulis, D.L.: The algebra of complex 2 × 2 matrices and a general closed Baker–Campbell–Hausdorff formula. J. Phys. A: Math. Theor. **50**(30), 305204 (2017)
3. Lo, C.F.: Comment on 'special-case closed form of the Baker–Campbell–Hausdorff formula'. J. Phys. A: Math. Theor. **49**(21), 218001 (2016)
4. Matone, M.: Closed form of the Baker–Campbell–Hausdorff formula for the generators of semisimple complex Lie algebras. Eur. Phys. J. C **76**, 610 (2016)
5. Van-Brunt, A., Visser, M.: Special-case closed form of the Baker-Campbell-Hausdorff formula. J. Phys. A: Math. Theor. **48**(22), 225207 (2015)
6. Kenth, E.-M.: On the BCH-formula in so(3). BIT Numer. Math. **41**(3), 629–632 (2001)
7. Chirikjian, G.S., Kyatkin, A.B.: Harmonic Analysis for Engineers and Applied Scientists, Updated and Expanded Edition Edition, p. 255. Courier Dover Publications, New York (2016)
8. Park, J., Chung, W.-K.: Geometric integration on Euclidean group with application to articulated multibody systems. IEEE Trans. Rob. **21**(5), 850–863 (2005)
9. Condurache, D., Ciureanu, I.-A.: Higher-order Cayley transforms for SE(3). New Adv. Mech. Mach. Sci. **57**, 331–339 (2018)
10. Condurache, D., Burlacu, A.: Dual tensors based solutions for rigid body motion parameterization. Mech. Mach. Theory **74**, 390–412 (2014)
11. Condurache, D., Burlacu, A.: Orthogonal dual tensor method for solving the AX = XB sensor calibration problem. Mech. Mach. Theory **104**, 382–404 (2016)
12. Pennestrì, E., Valentini, P.P.: Linear dual algebra algorithms and their application to kinematics. Multibody Dyn.: Comput. Methods Appl. **12**, 207–229 (2009)

Alternative Integration Schemes for Constrained Mechanical Systems

Tobias Meyer[1(✉)], Pu Li[2], and Bernhard Schweizer[1]

[1] Institute of Applied Dynamics, Technische Universität Darmstadt, 64287 Darmstadt, Germany
{meyer,schweizer}@ad.tu-darmstadt.de
[2] Institute of Turbomachinery, School of Energy and Power Engineering, Xi'an 710049, China
lipu1215@xjtu.edu.cn

Abstract. Various methods for solving systems of differential-algebraic equations (DAE), e.g. constrained mechanical systems, are known from literature. Here, an alternative approach is suggested, which is called collocated constraints approach (CCA). The idea of the method is inspired by a co-simulation technique recently published in [11] and is based on the usage of intermediate time points. The approach is very general and can basically be applied for arbitrary DAE systems. In the paper at hand, implementations of this approach are presented for Newmark-type integration schemes [1,9,12]. We discuss index-2 formulations with one intermediate time point and index-1 implementations based on two intermediate time points. A direct application of the approach for Newmark-type integrators yields a system of discretized equations with larger dimensions. Roughly speaking, the system increases by factor 2 for the index-2 and by factor 3 in case of the index-1 formulation. It is, however, straightforward to reduce the size of the discretized DAE system by using simple interpolation techniques. Numerical examples will demonstrate the straightforward application of the approach.

1 Introduction

In order to solve DAE systems, different approaches have been developed. Applying the *Baumgarte* formulation [3], for example, the constraints are replaced by a weighted sum of the constraints and the corresponding hidden constraints. Alternatively, the index-1 formulation with drift–off effect may be used in combination with a projection step [5]. The coordinate partitioning method [14] makes use of a special set of minimal coordinates. However, the choice of the minimal coordinates may have to be switched during the simulation. In another approach, additional variables are defined, which are called dummy derivatives [10].

A. Kecskeméthy and F. Geu Flores (Eds.): ECCOMAS 2019, COMPUTMETHODS 53, pp. 315–322, 2020.
https://doi.org/10.1007/978-3-030-23132-3_38

Furthermore, there exist several approaches especially tailored for solving constrained mechanical systems of the form

$$\underline{\underline{M}}(t,\underline{q})\,\underline{\ddot{q}} = \underline{f}(t,\underline{q},\underline{\dot{q}}) - \underline{g}_{\underline{q}}^{\mathrm{T}}(t,\underline{q})\,\underline{\lambda}\,, \tag{1a}$$

$$\underline{0} = \underline{g}(t,\underline{q})\,, \tag{1b}$$

where $\underline{q}$ is an υ-dimensional vector containing the displacement variables and the rotation parameters. Equation (1a) describes *Newton*'s law and *Euler*'s equations. Equation (1b) represents the constraint equations. *Gear*, *Gupta*, and *Leimkuhler* [7] developed the stabilized index-2 formulation with an extra *Lagrange* multiplier for each constraint, so that the constraints can be solved on position and on velocity level simultaneously. Using the index-1 formulation, which was suggested by *Führer* and *Leimkuhler* [6], there are two additional *Lagrange* multipliers for each constraint, so that the constraints can be solved on position, velocity, and on acceleration level simultaneously. *Hiller* [8] modifies the stabilized index-2 formulation by introducing antiderivatives of the *Lagrange* multipliers in order to obtain an index-1 formulation. Replacing the constraints by penalty regularization [2], the DAE system is approximated by an ODE system.

The manuscript is structured as follows: Sect. 2 contains some remarks on the hidden constraints on velocity and acceleration level of the constrained mechanical system (1). In Sect. 3, the collocated constraints approach will be explained with the help of the generalized-α integration method. In Sect. 4, a numerical example is presented.

2 Hidden Constraints of Multibody Systems

A central problem in solving differential algebraic equations of higher index is that the derivatives of the algebraic equations are hidden constraints. In the equations of motion of multibody systems, the algebraic Eq. (1b) are constraints on position level. Differentiating the constraints once or twice yields the hidden constraints on velocity and acceleration level, respectively. The hidden constraints on velocity level are given by

$$\underline{\dot{g}}(t,\underline{q},\underline{\dot{q}}) := \underline{g}_t(t,\underline{q}) + \underline{g}_{\underline{q}}(t,\underline{q})\,\underline{\dot{q}} = \underline{0} \tag{2}$$

and on acceleration level by

$$\underline{\ddot{g}}(t,\underline{q},\underline{\dot{q}},\underline{\ddot{q}}) := \underline{g}_{\mathrm{II}}(t,\underline{q},\underline{\dot{q}}) + \underline{g}_{\underline{q}}(t,\underline{q})\,\underline{\ddot{q}} = \underline{0} \tag{3}$$

with

$$\underline{g}_{\mathrm{II}}(t,\underline{q},\underline{\dot{q}}) := \underline{\dot{g}}_t(t,\underline{q},\underline{\dot{q}}) + \underline{\dot{g}}_{\underline{q}}(t,\underline{q},\underline{\dot{q}})\,\underline{\dot{q}}. \tag{4}$$

Further, the initial values are given by $\underline{q}(t_0) = \underline{q}_0$ and $\underline{\dot{q}}(t_0) = \underline{v}_0$, which satisfy the constraints $\underline{g}\left(t_0,\underline{q}_0\right) = \underline{0}$ and the hidden constraints $\underline{\dot{g}}\left(t_0,\underline{q}_0,\underline{v}_0\right) = \underline{0}$.

To guarantee that the constrained mechanical system (1) is solvable, it is necessary that the matrix

$$\underline{\underline{\mathscr{M}}}(t,\underline{q}) := \begin{bmatrix} \underline{\underline{M}}(t,\underline{q}) & \underline{g}_{\underline{q}}^{\mathrm{T}}(t,\underline{q}) \\ \underline{g}_{\underline{q}}(t,\underline{q}) & \underline{\underline{0}} \end{bmatrix} \tag{5}$$

is non-singular for all $t \geq t_0$ and $\underline{q} \in \mathbb{R}^{\upsilon}$ with $\underline{g}(t,\underline{q}) = \underline{0}$. Then, consistent initial values of the *Lagrange* multipliers $\underline{\lambda}$ can be computed by solving the linear system

$$\underline{\underline{\mathscr{M}}}\left(t_0,\underline{q}_0\right)\begin{bmatrix}\underline{a}_0 \\ \underline{\lambda}_0\end{bmatrix} = \begin{bmatrix} \underline{f}\left(t_0,\underline{q}_0,\underline{v}_0\right) \\ -\underline{g}_{\mathrm{II}}\left(t_0,\underline{q}_0,\underline{v}_0\right)\end{bmatrix}. \tag{6}$$

3 Collocated Constraints Approach for the Generalized-α Method

The general idea of the collocated constraints approach (CCA) will be explained with the help of the generalized-α integration method, which is tailored for second order differential equations. The classical index-3 formulation for the multibody system (1) reads as

$$(1-\alpha_{\mathrm{m}})\,\underline{F}^{\mathrm{m}}_{n+1-\alpha,h_n} + \alpha_{\mathrm{m}}\underline{F}^{\mathrm{m}}_{n-\alpha,h_n} = (1-\alpha_{\mathrm{f}})\,\underline{F}^{\mathrm{f}}_{n+1} + \alpha_{\mathrm{f}}\underline{F}^{\mathrm{f}}_{n}, \tag{7a}$$

$$\underline{0} = \underline{g}\left(t_{n+1},\underline{q}_{n+1}\right) \tag{7b}$$

with

$$\underline{F}^{\mathrm{m}}_{\eta-\alpha,h} = \underline{\underline{M}}\left(t_\eta - \alpha h, \underline{q}_\eta - \alpha h \underline{v}_\eta\right)\underline{a}_{\eta-\alpha}, \qquad (\eta = n, n+1) \tag{8a}$$

$$\underline{F}^{\mathrm{f}}_{\eta} = \underline{f}\left(t_\eta,\underline{q}_\eta,\underline{v}_\eta\right) - \underline{g}_{\underline{q}}^{\mathrm{T}}\left(t_\eta,\underline{q}_\eta\right)\underline{\lambda}_\eta, \qquad (\eta = n, n+1) \tag{8b}$$

$$\underline{q}_{n+1} = \underline{q}_n + h_n\underline{v}_n + h_n^2\left(\left(\frac{1}{2}-\beta\right)\underline{a}_{n-\alpha} + \beta\underline{a}_{n+1-\alpha}\right), \tag{8c}$$

$$\underline{v}_{n+1} = \underline{v}_n + h_n\left((1-\gamma)\,\underline{a}_{n-\alpha} + \gamma\underline{a}_{n+1-\alpha}\right), \tag{8d}$$

which is solved for the unknown accelerations $\underline{a}_{n+1-\alpha}$ and the *Lagrange* multipliers $\underline{\lambda}_{n+1}$. The choice of the parameters α_{m}, α_{f}, β, and γ was analyzed by *Chung* and *Hulbert* [4]. It should be mentioned that $\alpha = \alpha_{\mathrm{f}} - \alpha_{\mathrm{m}}$.

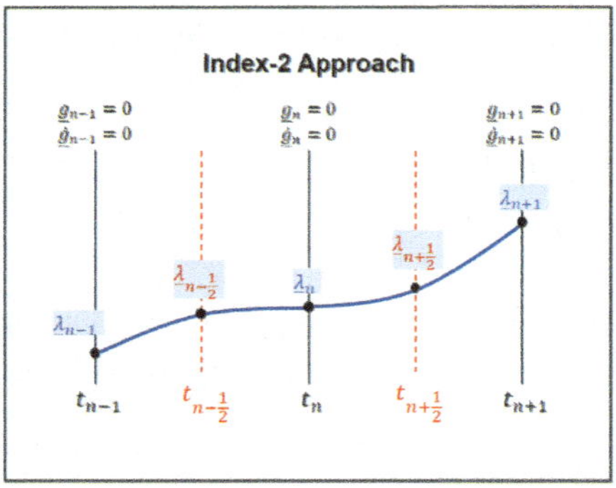

Fig. 1. Extended time grid by one intermediate time point

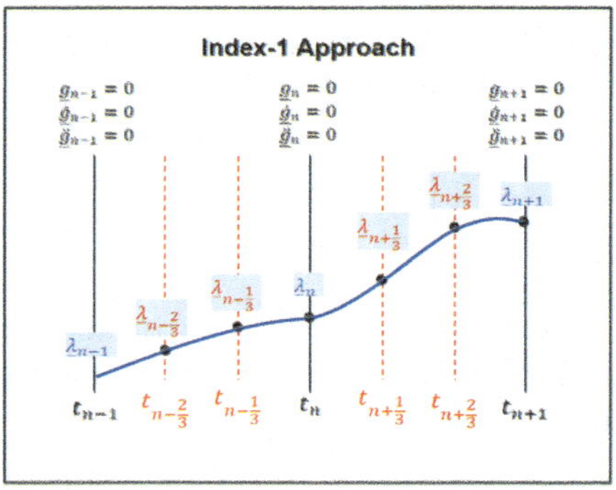

Fig. 2. Extended time grid by two intermediate time points

3.1 Index-2 Integration Scheme

Within the CCA, the time grid will be extended by one intermediate time point $t_{n+1/2}$, see Fig. 1. By ignoring the constraints on position level at the intermediate time point $t = t_{n+\frac{1}{2}}$ and by instead considering the constraints on velocity level at the integration time point $t = t_{n+1}$, yields the discretized system

$$(1-\alpha_{\mathrm{m}})\,\underline{F}^{\mathrm{m}}_{n+\frac{1-\alpha}{2},\frac{1}{2}h_n} + \alpha_{\mathrm{m}}\underline{F}^{\mathrm{m}}_{n-\frac{\alpha}{2},\frac{1}{2}h_n} = (1-\alpha_{\mathrm{f}})\,\underline{F}^{\mathrm{f}}_{n+\frac{1}{2}} + \alpha_{\mathrm{f}}\underline{F}^{\mathrm{f}}_{n}, \tag{9a}$$

$$(1-\alpha_{\mathrm{m}})\,\underline{F}^{\mathrm{m}}_{n+1-\frac{\alpha}{2},\frac{1}{2}h_n} + \alpha_{\mathrm{m}}\underline{F}^{\mathrm{m}}_{n+\frac{1-\alpha}{2},\frac{1}{2}h_n} = (1-\alpha_{\mathrm{f}})\,\underline{F}^{\mathrm{f}}_{n+1} + \alpha_{\mathrm{f}}\underline{F}^{\mathrm{f}}_{n+\frac{1}{2}}, \tag{9b}$$

$$\underline{0} = \underline{g}\left(t_{n+1}, \underline{q}_{n+1}\right) \tag{9c}$$

$$\underline{0} = \dot{\underline{g}}\left(t_{n+1}, \underline{q}_{n+1}, \underline{v}_{n+1}\right). \tag{9d}$$

Note that the inertial forces/torques $\underline{F}^{\mathrm{m}}_{\eta-\frac{\alpha}{2},\frac{1}{2}h_n}$ for $\eta = n, n+\frac{1}{2}, n+1$ are defined according to Eq. (8a). The forces/torques on the right hand side are calculated as expressed in Eq. (8b) for $\eta = n, n+\frac{1}{2}, n+1$. The position and velocity variables are determined according to Eqs. (8c) and (8d), but with step size $h_n/2$. The system defined by Eq. (9) is solved for the unknown accelerations $\underline{a}_{n+\frac{i-\alpha}{2}}$ and *Lagrange* multipliers $\underline{\lambda}_{n+\frac{i}{2}}$ for $i = 1, 2$.

3.2 Index-1 Integration Scheme

Now, the time grid will be extended by two intermediate time points $t_{n+1/3}$ and $t_{n+2/3}$, see Fig. 2. Replacing the corresponding constraints on position level at the intermediate time points $t = t_{n+\frac{1}{3}}$ and $t = t_{n+\frac{2}{3}}$ by the constraints on velocity and acceleration level at the time point $t = t_{n+1}$ results in the discretized system

$$(1-\alpha_{\mathrm{m}})\,\underline{F}^{\mathrm{m}}_{n+\frac{1-\alpha}{3},\frac{1}{3}h_n} + \alpha_{\mathrm{m}}\underline{F}^{\mathrm{m}}_{n-\frac{\alpha}{3},\frac{1}{3}h_n} = (1-\alpha_{\mathrm{f}})\,\underline{F}^{\mathrm{f}}_{n+\frac{1}{3}} + \alpha_{\mathrm{f}}\underline{F}^{\mathrm{f}}_{n}, \quad (10a)$$

$$(1-\alpha_{\mathrm{m}})\,\underline{F}^{\mathrm{m}}_{n+\frac{2-\alpha}{3},\frac{1}{3}h_n} + \alpha_{\mathrm{m}}\underline{F}^{\mathrm{m}}_{n+\frac{1-\alpha}{3},\frac{1}{3}h_n} = (1-\alpha_{\mathrm{f}})\,\underline{F}^{\mathrm{f}}_{n+\frac{2}{3}} + \alpha_{\mathrm{f}}\underline{F}^{\mathrm{f}}_{n+\frac{1}{3}}, \quad (10b)$$

$$(1-\alpha_{\mathrm{m}})\,\underline{F}^{\mathrm{m}}_{n+1-\frac{\alpha}{3},\frac{1}{3}h_n} + \alpha_{\mathrm{m}}\underline{F}^{\mathrm{m}}_{n+\frac{2-\alpha}{3},\frac{1}{3}h_n} = (1-\alpha_{\mathrm{f}})\,\underline{F}^{\mathrm{f}}_{n+1} + \alpha_{\mathrm{f}}\underline{F}^{\mathrm{f}}_{n+\frac{2}{3}}, \quad (10c)$$

$$\underline{0} = \underline{g}\left(t_{n+1}, \underline{q}_{n+1}\right) \quad (10d)$$

$$\underline{0} = \underline{\dot{g}}\left(t_{n+1}, \underline{q}_{n+1}, \underline{v}_{n+1}\right) \quad (10e)$$

$$\underline{0} = \underline{\ddot{g}}\left(t_{n+1}, \underline{q}_{n+1}, \underline{v}_{n+1}, \underline{a}_{n+1}\right), \quad (10f)$$

with the accelerations

$$\underline{a}_{n+1} = \left[\underline{E}\;\underline{0}\right] \underline{\underline{\mathcal{M}}}^{-1}\left(t_{n+1}, \underline{q}_{n+1}\right) \begin{bmatrix} \underline{f}\left(t_{n+1}, \underline{q}_{n+1}, \underline{v}_{n+1}\right) \\ -\underline{g}_{\mathrm{II}}\left(t_{n+1}, \underline{q}_{n+1}, \underline{v}_{n+1}\right) \end{bmatrix}, \quad (11)$$

where $\underline{E}$ denotes the $(\upsilon \times \upsilon)$ identity matrix and $\underline{\underline{\mathcal{M}}}(t, \underline{q})$ is defined by Eq. (5). Note that the inertial forces/torques $\underline{F}^{\mathrm{m}}_{\eta-\frac{\alpha}{3},\frac{1}{3}h_n}$ for $\eta = n, n+\frac{1}{3}, n+\frac{2}{3}, n+1$ are defined according to Eq. (8a). The forces/torques on the right hand side are calculated as in Eq. (8b) for $\eta = n, n+\frac{2}{3}, n+\frac{1}{3}, n+1$. The position and velocity variables are determined according to Eqs. (8c) and (8d), respectively, but with step size $h_n/3$. The system of Eq. (10) is solved for the unknown accelerations $\underline{a}_{n+\frac{i-\alpha}{3}}$ and *Lagrange* multipliers $\underline{\lambda}_{n+\frac{i}{3}}$ for $i = 1, 2, 3$.

4 Numerical Example

As test model, the nonlinear pendulum consisting of a rigid massless bar of length l and a point mass m is considered. The position vector $\underline{q} = \left[x\; y\right]^{\mathrm{T}}$ describes the horizontal and vertical coordinates. Gravity is artificially increased to $g_{\mathrm{grav}} = 1000$ in order to increase the natural frequency of the system. The pendulum is shown in Fig. 3. The equations of motion read as

$$m\underline{\ddot{q}} = -g_{\mathrm{grav}}\underline{e}_y - 2\underline{q}\lambda, \quad (12a)$$

$$0 = \underline{q}^{\mathrm{T}}\underline{q} - l^2, \quad (12b)$$

where $\underline{e}_y$ denotes the unit vector in y-direction. Simulations have been carried out with the parameters $m = 1$, $l = 1$ and with the initial conditions $\underline{q}(0) = \sqrt{2}/2\left(\underline{e}_x + \underline{e}_y\right)$ and $\underline{\dot{q}}(0) = \underline{0}$. The time step size is constant and has been set to $h = 1E{-}3$. The solver parameters are chosen according to

$$\beta = \frac{1}{4}(1+\alpha)^2, \quad \gamma = \frac{1}{2} + \alpha, \quad \alpha_{\mathrm{f}} = \frac{1}{2}(1-\alpha), \quad \alpha_m = \frac{1}{2}(1-3\alpha) \quad (13)$$

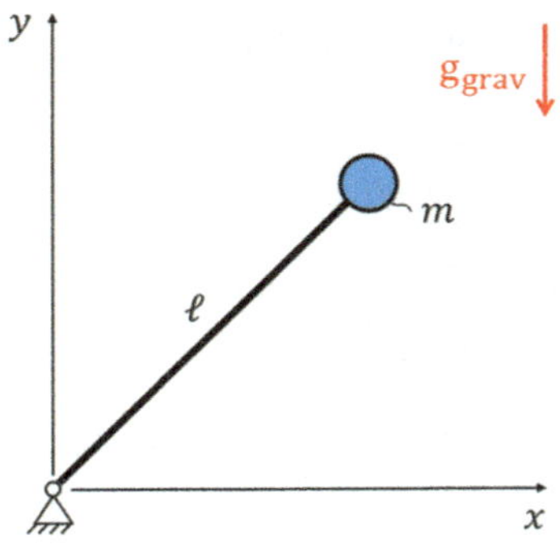

Fig. 3. Nonlinear pendulum.

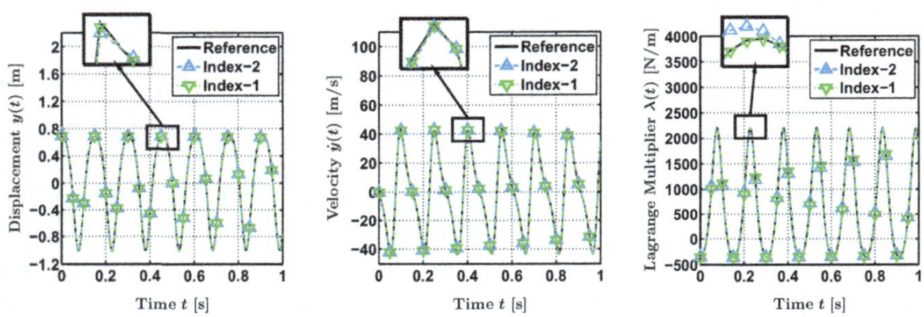

Fig. 4. Simulation results for the nonlinear pendulum with the index-2 and index-1 generalized-α approaches: displacement $y(t)$, velocity $\dot{y}(t)$, and *Lagrange* multiplier $\lambda(t)$

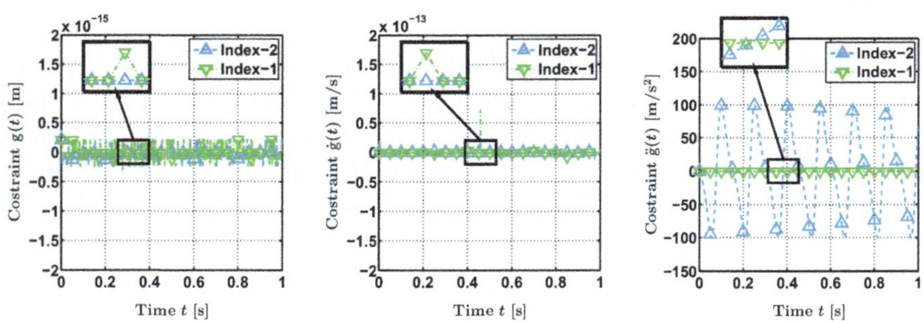

Fig. 5. Simulation results for the nonlinear pendulum with the index-2 and index-1 generalized-α approaches: residuals $g(t)$, $\dot{g}(t)$, and $\ddot{g}(t)$

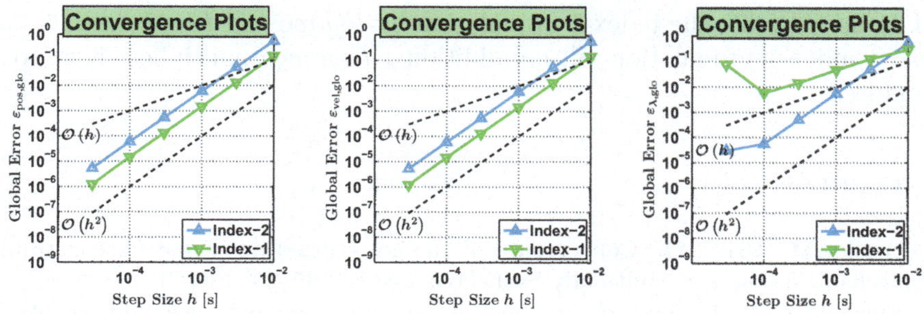

Fig. 6. Convergence plots for the nonlinear pendulum: global error over the step size h for the index-2 and index-1 generalized-α method.

with $\alpha = 3/17$. Figure 4 collects the results of the position variables in y-direction, the corresponding velocity variables $\dot{y}(t)$, and the *Lagrange* multiplier $\lambda(t)$. The residuals of the constraint Eq. (12b) on position, velocity, and acceleration level are shown in Fig. 5 interpreted as functions of time, i.e.

$$g(t) = \underline{q}^{\mathrm{T}}(t)\,\underline{q}(t) - l^2, \tag{14a}$$

$$\dot{g}(t) = 2\underline{\dot{q}}^{\mathrm{T}}(t)\,\underline{q}(t)\,, \tag{14b}$$

$$\ddot{g}(t) = 2\underline{\dot{q}}^{\mathrm{T}}(t)\,\underline{\dot{q}}(t) - \frac{2}{m}\left(\mathrm{g}_{\mathrm{grav}}\underline{e}_y + 2\underline{q}(t)\,\lambda(t)\right)^{\mathrm{T}}\underline{q}(t)\,. \tag{14c}$$

The constraints on position and on velocity level, i.e. Eqs. (14a) and (14b), are satisfied in the index-2 and in the index1-implementation. As expected, the hidden constraint on acceleration level, Eq. (14c), is only enforced by the index-1 approach. Figure 6 illustrates the convergence order of the two approaches. As can be seen, the global error of the position and velocity variables will converge with $\mathcal{O}(h^2)$ with both formulations. Applying the index-1 approach, also the *Lagrange* multipliers will converge with second order. Unfortunately, there is a loss of the convergence order using the index-2 formulation. Hence, the *Lagrange* multiplier only converges with $\mathcal{O}(h)$.

5 Conclusions

In the paper at hand, an alternative approach for solving constrained mechanical systems has been considered. The approach reduces the index of such systems from 3 to 2, if one intermediate point is introduced. Optionally, the index can be reduced to 1 by introducing two intermediate time points. However, the disadvantage is that the dimension of the discretized system is larger compared to an index-3 implementation. Strictly speaking, the size of the equation system, which has to be solved in each time step, is doubled applying the index-2 approach and tripled using the index-1 approach. The main advantage is that the method can be used for index reduction of rather general/arbitrary differential algebraic equations. In contrast to the index-2 formulation of *Gear*, *Gupta*, and *Leimkuhler* [7], the index-1 formulation by *Führer* and *Leimkuhler* [6], and *Hiller*'s index-1 formulation [8], no additional *Lagrange* multipliers have to be introduced.

References

1. Arnold, M., Brüls, O.: Convergence of the generalized-α scheme for constrained mechanical systems. Multibody Syst. Dyn. **18**(2), 185–202 (2007)
2. Bauchau, O.A., Laulusa, A.: Review of contemporary approaches for constraint enforcement in multibody systems. J. Comput. Nonlinear Dyn. **3**(1), 011005 (2008)
3. Baumgarte, J.: Stabilization of constraints and integrals of motion in dynamical systems. Comput. Methods Appl. Mech. Eng. **1**(1), 1–16 (1972)
4. Chung, J., Hulbert, G.M.: A time integration algorithm for structural dynamics with improved numerical dissipation: the generalized- method. J. Appl. Mech. **60**(2), 371–375 (1993)
5. Eich, E.: Convergence results for a coordinate projection method applied to mechanical systems with algebraic constraints. SIAM J. Numer. Anal. **30**(5), 1467–1482 (1993)
6. Führer, C., Leimkuhler, B.J.: Numerical solution of differential-algebraic equations for constrained mechanical motion. Numer. Math. **59**(1), 55–69 (1991)
7. Gear, C.W., Leimkuhler, B.J., Gupta, G.K.: Automatic integration of Euler-Lagrange equations with constraints. J. Comput. Appl. Math. **12**, 77–90 (1985)
8. Anantharaman, M., Hiller, M.: Dynamic Analysis of Complex Multibody Systems Using Methods for Differential-Algebraic Equations. Advanced Multibody System Dynamics, pp. 173–194. Springer, Dordrecht (1993)
9. Lunk, C., Simeon, B.: Solving constrained mechanical systems by the family of Newmark and α-methods. ZAMM J. Appl. Math. Mech./Zeitschrift für Angewandte Mathematik und Mechanik Appl. Math. Mech. **86**(10), 772–784 (2006)
10. Mattsson, S.E., Söderlind, G.: Index reduction in differential-algebraic equations using dummy derivatives. SIAM J. Sci. Comput. **14**(3), 677–692 (1993)
11. Meyer, T., Li, P., Lu, D., Schweizer, B.: Implicit co-simulation method for constraint coupling with improved stability behavior. Multibody Syst. Dyn. (2018). https://doi.org/10.1007/s11044-018-9632-9
12. Negrut, D., Rampalli, R., Ottarsson, G., Sajdak, A.: On the use of the HHT method in the context of index 3 differential algebraic equations of multibody dynamics. In: ASME 2005 International Design Engineering Technical Conferences and Computers and Information in Engineering Conference, pp. 207–218. American Society of Mechanical Engineers, January 2005
13. Schweizer, B., Li, P.: Solving differential-algebraic equation systems: alternative Index-2 and Index-1 approaches for constrained mechanical systems. J. Comput. Nonlinear Dyn. **11**(4), 044501 (2016)
14. Wehage, R.A., Haug, E.J.: Generalized coordinate partitioning for dimension reduction in analysis of constrained dynamic systems. J. Mech. Des. **104**(1), 247–255 (1982)

Implementation of Linear Springs and Dampers in a Newmark Second Order Direct Integration Method for 2D Multibody Dynamics

Haritz Uriarte(✉), Igor Fernández de Bustos(✉), and Gorka Urkullu(✉)

Department of Mechanical Engineering, School of Engineering of Bilbao, University of the Basque Country, Paseo Rafael Moreno "Pitxitxi", 3, 48013 Bilbao, Spain
{haritz.uriarte, igor.fernandezdebustos, gorka.urkullu}@ehu.eus

Abstract. This paper presents the mathematical developments required to introduce both linear springs and dampers into the second order Newmark method for the integration of Multibody System Dynamics (MBSD) for bidimensional problems. The advantage of the Newmark approach is that it integrates directly the second order differential equations found in MBSD, thus not duplicating variables and reducing computational cost [1]. The use of Newmark approach for MBSD is not new, but it is solved usually in a quasi-Newton procedure [2], which is easier to implement, but has worse convergence than a full Newton approach. For the full Newton approach to be achieved, however, all derivatives have to be computed analytically. In a previous work [3], the analytical derivatives needed for simple mechanisms including Bodies, Revolute and Prismatic joints were presented. The novelty presented in this document is the development of the derivatives needed for the introduction of linear springs and dampers. The resultant Karush-Kuhn-Tucker system is solved by means of the null space method ([4, 5]). The method has been developed and tested using Newton-Euler formalism and Cartesian coordinates to solve several 2D problems. Some examples are included, which have been contrasted with ADAMS. This is a preliminary work in order to afterwards develop the method for three dimensional problems.

Keywords: Multibody dynamics · Newmark integration · Newton-Euler formalism · Cartesian coordinates

1 The Newmark Method for 2D Multibody Dynamics

In this paper the required equations to implement springs and dampers in 2D Multibody Dynamics are developed. This allows the problem to be integrated afterwards using a second order Newmark method.

Since Multibody Dynamics problems are heavily non-linear, an appropriated equation solver is required; for this purpose the Newton method with exact derivatives

A. Kecskeméthy and F. Geu Flores (Eds.): ECCOMAS 2019, COMPUTMETHODS 53, pp. 323–331, 2020.
https://doi.org/10.1007/978-3-030-23132-3_39

has been chosen. Analytical derivatives of the equation are required in order to get the best performance.

The analytical derivatives can be obtained by two ways: Newmark can be applied first and afterwards be linearized, or linearizations can be done first and then Newmark be applied. In this work the second approach is used.

The linearized equilibrium equation evaluated in $t + \Delta t$, as developed in [3], is as follows:

$$M\ddot{x} + C\dot{x} + Kx = f + H^T\lambda \tag{1}$$

Newmark's equations can be introduced in this equation:

$$x(t + \Delta t) = x(t) + \Delta t\dot{x}(t) + \Delta t^2((0.5 - \beta)\ddot{x}(t) + \beta\ddot{x}(t + \Delta t)) \tag{2}$$

$$\dot{x}(t + \Delta t) = \dot{x}(t) + \Delta t((1 - \alpha)\ddot{x}(t) + \alpha\ddot{x}(t + \Delta t)) \tag{3}$$

Obtaining the recursive equation needed for solving the system, as found in [3].

2 Mathematical Development of Linear Springs and Dampers

In order to implement the linear springs and dampers, their C and K matrices have to be obtained. These are derived form a linearization of the forces they produce, and thus additional complementary forces have to be computed. The linearized equations depend on each time step's solution, and as such have to be recalculated from the displacement and (in the case of the C matrix) velocity vectors every time along with the null space.

2.1 Linear Springs

In a general case, two solids connected with a spring can be represented as follows (Fig. 1):

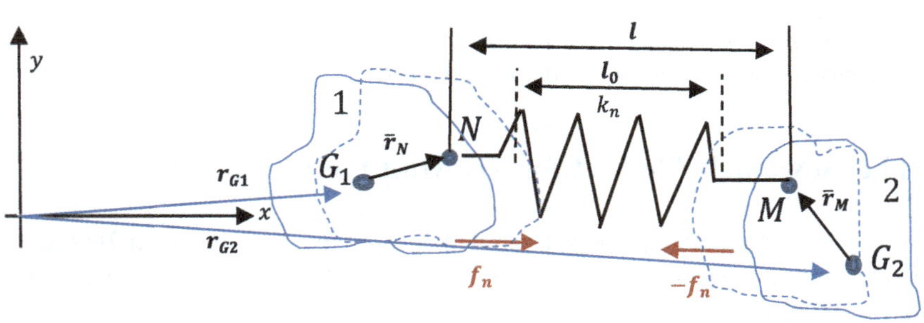

Fig. 1. Linear spring connecting two solids

The force produced by a spring is given by the following formula:

$$f_n = -k\left(1 - \frac{l_0}{l}\right)\boldsymbol{l} \tag{4}$$

Being k the stiffness constant, l_0 the unloaded length, l the length vector modulus and $\boldsymbol{l}$ the length vector.

The length vector can be expressed as follows:

$$\boldsymbol{l} = \boldsymbol{r}_N - \boldsymbol{r}_M = \boldsymbol{r}_{G1} + R_1\bar{\boldsymbol{r}}_N - \boldsymbol{r}_{G2} - R_2\bar{\boldsymbol{r}}_M \tag{5}$$

which depends on each element's rotation matrix. Thus, it has to be linearized in order to apply full Newton. The result is the following equation:

$$\boldsymbol{l} = \boldsymbol{r}_{G1} + \left(\theta_1 - \theta_1^0\right)R_1'^0\bar{r}_N + R_1^0\bar{r}_N - \boldsymbol{r}_{G2} - \left(\theta_2 - \theta_2^0\right)R_2'^0\bar{r}_M - R_2^0\bar{r}_M \tag{6}$$

Being θ each element's rotation angle. This equation can be represented in compact form as follows:

$$\boldsymbol{l} = Z\left(\boldsymbol{x}^0\right)\boldsymbol{x} - J\left(\boldsymbol{x}^0\right)\boldsymbol{x}^0 + R_1^0\bar{r}_N - R_2^0\bar{r}_M \tag{7}$$

Where $= \{x, y, \theta\}^T$, representing the CG coordinates, and:

$$Z\left(\boldsymbol{x}^0\right) = \begin{bmatrix} 1 & 0 & \left|R_1'^0\bar{r}_N\right|_x & -1 & 0 & -\left|R_2'^0\bar{r}_M\right|_x \\ 0 & 1 & \left|R_1'^0\bar{r}_N\right|_y & 0 & -1 & -\left|R_2'^0\bar{r}_M\right|_y \end{bmatrix} \tag{8}$$

$$J\left(\boldsymbol{x}^0\right) = \begin{bmatrix} 0 & 0 & \left|R_1'^0\bar{r}_N\right|_x & 0 & 0 & -\left|R_2'^0\bar{r}_M\right|_x \\ 0 & 0 & \left|R_1'^0\bar{r}_N\right|_y & 0 & 0 & -\left|R_2'^0\bar{r}_M\right|_y \end{bmatrix} \tag{9}$$

Equation (4) can be represented as follows:

$$\begin{Bmatrix} f_N \\ \tau_N \\ f_M \\ \tau_M \end{Bmatrix} = K_K\boldsymbol{x} + \boldsymbol{f}_K^0 \tag{10}$$

Being:

$$K = -k\left(1 - \frac{l_0}{l}\right)\begin{bmatrix} Z(\boldsymbol{x}^0) \\ -\left|Z(\boldsymbol{x}^0)\right|_x\left|\boldsymbol{r}_{NG}\right|_y + \left|Z(\boldsymbol{x}^0)\right|_y\left|\boldsymbol{r}_{NG}\right|_x \\ -Z(\boldsymbol{x}^0) \\ \left|Z(\boldsymbol{x}^0)\right|_x\left|\boldsymbol{r}_{MG}\right|_y + \left|Z(\boldsymbol{x}^0)\right|_y\left|\boldsymbol{r}_{MG}\right|_x \end{bmatrix} \tag{11}$$

$$f_K^0 = -k\left(1 - \frac{l_0}{l}\right)\left\{\begin{array}{c} s \\ -|s|_x|r_{NG}|_y + |s|_y|r_{NG}|_x \\ -s \\ |s|_x|r_{MG}|_y - |s|_y|r_{MG}|_x \end{array}\right\} \tag{12}$$

Where:

$$s = -J(x^0)x^0 - R_2^0\bar{r}_M + R_1^0\bar{r}_N \tag{13}$$

2.2 Linear Dampers

For linear dampers a similar representation can be done, which is shown in the next page.

The force given by a damper is expressed by the following formula (Fig. 2):

$$f_N = \frac{-c}{l^2}\left((r_N - r_M)(r_N - r_M)^T\right)(v_N - v_M) \tag{14}$$

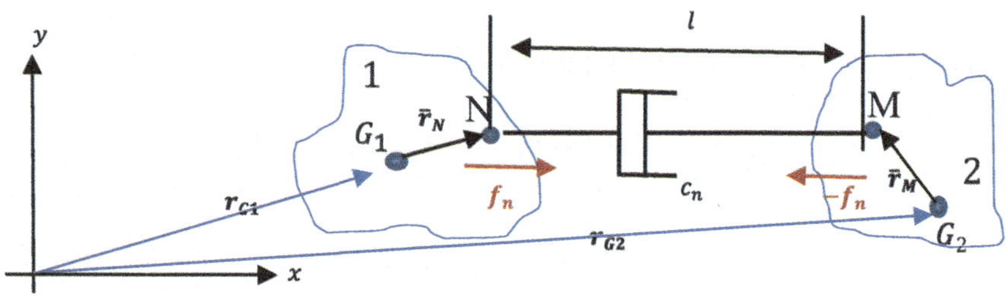

Fig. 2. Linear damper

Some variables can be regrouped as follows:

$$\begin{aligned} d &= r_N - r_M = r_{G1} + R_1\bar{r}_N - r_{G2} - R_2\bar{r}_M \\ v &= v_N - v_M = v_{G1} + \omega_1 R_1'\bar{r}_N - v_{G2} - \omega_2 R_2'\bar{r}_M \end{aligned} \tag{15}$$

This makes Eq. (14) look like this:

$$f_N = \frac{-c}{l^2}(dd^T)v \tag{16}$$

The linearization of $\boldsymbol{v}$ is shown below. The previously developed $Z(\boldsymbol{x}^0)$ matrix has also been used here.

$$\begin{aligned} \boldsymbol{v} &= \boldsymbol{v}_N - \boldsymbol{v}_M = \boldsymbol{v}_{G1} + \omega_1 R_1' \bar{\boldsymbol{r}}_N - \boldsymbol{v}_{G2} - \omega_2 R_2' \bar{\boldsymbol{r}}_M = \boldsymbol{v}(\boldsymbol{x}, \dot{\boldsymbol{x}}) = \\ &= Z(\boldsymbol{x})\dot{\boldsymbol{x}} \approx Z(\boldsymbol{x}^0)\dot{\boldsymbol{x}}^0 + \left(Z(\boldsymbol{x}) - Z(\boldsymbol{x}^0)\right)\dot{\boldsymbol{x}}^0 + Z(\boldsymbol{x}^0)\left(\dot{\boldsymbol{x}} - \dot{\boldsymbol{x}}^0\right) = \\ &= -Z(\boldsymbol{x}^0)\dot{\boldsymbol{x}}^0 + Z(\boldsymbol{x})\dot{\boldsymbol{x}}^0 + Z(\boldsymbol{x}^0)\dot{\boldsymbol{x}} \end{aligned} \tag{17}$$

In this last formula, the product $Z(\boldsymbol{x})\dot{\boldsymbol{x}}^0$ can be developed as follows using vector derivation properties:

$$Z(\boldsymbol{x})\dot{\boldsymbol{x}}^0 = Z(\boldsymbol{x}^0)\dot{\boldsymbol{x}}^0 + \left|\frac{dZ(\boldsymbol{x})\dot{\boldsymbol{x}}^0}{d\boldsymbol{x}}\right|_{\boldsymbol{x}^0} (\boldsymbol{x} - \boldsymbol{x}^0) \tag{18}$$

Once introduced in the previous formula, the linearized velocity difference vector has the following expression:

$$\boldsymbol{v} = \left|\frac{dZ(\boldsymbol{x})\dot{\boldsymbol{x}}^0}{d\boldsymbol{x}}\right|_{\boldsymbol{x}^0} (\boldsymbol{x} - \boldsymbol{x}^0) + Z(\boldsymbol{x}^0)\dot{\boldsymbol{x}} \tag{19}$$

Since Eq. (16) has a triple product of these already linearized variables, it is mandatory to linearize the triple product in order to operate with it:

$$\begin{aligned} \boldsymbol{f}_N &= \frac{-c}{l^2}\left(\boldsymbol{d}\boldsymbol{d}^T\right)\boldsymbol{v} \approx \frac{-c}{l^2}\left(\left((\boldsymbol{d} - \boldsymbol{d}^0)(\boldsymbol{d}^0)^T\right)\boldsymbol{v}^0 + \left(\boldsymbol{d}^0\left(\boldsymbol{d}^T - (\boldsymbol{d}^0)^T\right)\right)\boldsymbol{v}^0 + \left(\boldsymbol{d}^0(\boldsymbol{d}^0)^T\right)(\boldsymbol{v} - \boldsymbol{v}^0)\right) \\ &= \frac{-c}{l^2}\left(-2\left(\boldsymbol{d}^0(\boldsymbol{d}^0)^T\right)\boldsymbol{v}^0 + \left(\boldsymbol{d}^0(\boldsymbol{v}^0)^T\right)\boldsymbol{d} + \left((\boldsymbol{v}^0)^T\boldsymbol{d}^0\right)\boldsymbol{d} + \left(\boldsymbol{d}^0(\boldsymbol{d}^0)^T\right)\boldsymbol{v}\right) \end{aligned} \tag{20}$$

Introducing all linearizations and reordering items, a formula with the following scheme is achieved:

$$\begin{Bmatrix} \boldsymbol{f}_N \\ \tau_N \\ \boldsymbol{f}_M \\ \tau_M \end{Bmatrix} = \boldsymbol{f}_D^0 + K_D\boldsymbol{x} + C\dot{\boldsymbol{x}} \tag{21}$$

Being:

$$\boldsymbol{f}_D^0 = \begin{Bmatrix} \boldsymbol{g}^0 \\ -\left|\boldsymbol{g}^0\right|_y \left|\boldsymbol{r}_{NG}\right|_x + \left|\boldsymbol{g}^0\right|_x \left|\boldsymbol{r}_{NG}\right|_y \\ -\boldsymbol{g}^0 \\ \left|\boldsymbol{g}^0\right|_y \left|\boldsymbol{r}_{MG}\right|_x - \left|\boldsymbol{g}^0\right|_x \left|\boldsymbol{r}_{MG}\right|_y \end{Bmatrix} \tag{22}$$

$$K_D = \frac{-c}{l^2}\left(\left(\left(v^0\right)^T d^0\right)Q(x^0) + \left(d^0\left(v^0\right)^T\right)Q(x^0) - \left(d^0\left(d^0\right)^T\right)\left|\frac{dQ(x)\dot{x}^0}{dx}\right|_{x^0}\right) \quad (23)$$

$$C = \frac{-c}{l^2}\left(d^0\left(d^0\right)^T Q(x^0)\right) \quad (24)$$

Where:

$$g^0 = \frac{-c}{l^2}\left(\begin{array}{l} -2d^0\left(d^0\right)^T v^0 + \left(\left(v^0\right)^T d^0\right)\left(-J(x^0)x^0 + R_1^0\bar{r}_N - R_2^0\bar{r}_M\right) + \\ \left(d^0\left(v^0\right)^T\right)\left(-J(x^0)x^0 + R_1^0\bar{r}_N - R_2^0\bar{r}_M\right) + \left(d^0\left(d^0\right)^T\right)\left|\frac{dZ(x)}{dx}\dot{x}^0\right|_{x^0} x^0 \end{array}\right) \quad (25)$$

$$Q(x^0) = \left\{\begin{array}{c} Z(x^0) \\ -\left|Z(x^0)\right|_y \left|r_{NG}\right|_x + \left|Z(x^0)\right|_x \left|r_{NG}\right|_y \\ -Z(x^0) \\ \left|Z(x^0)\right|_y \left|r_{MG}\right|_x - \left|Z(x^0)\right|_x \left|r_{MG}\right|_y \end{array}\right\} \quad (26)$$

3 Examples of Application

In this section two examples of application with linear springs and dampers are presented. For each example, results in ADAMS of the same system are also included in order to prove the developed method's accuracy.

3.1 Double Pendulum with Linear Springs and Dampers

This example evolves from the one found in [2], but with completely self-defined variables, making it a whole new example. Both solids have been modeled as bars (Tables 1, and 2; Fig. 3).

Table 1. Spring and damper data

Spring data	K (N/m)	C(N sec/m)	L_0 (m)	Ground	Solid 1	Solid 2
1	15	5	0.75	(1.5,0)	(0.1,0)	–
2	6	2	0.5	–	(−0.1,0)	(−0.5,0)

Table 2. Solid data

Solid data	M (kg)	L (m)	X_0	Y_0	Θ_0	V_{x0}	V_{y0}	Ω_0
N	0.5	3	1.5	0	0	0	0	0
M	0.7	4	5	0	0	0	0	0

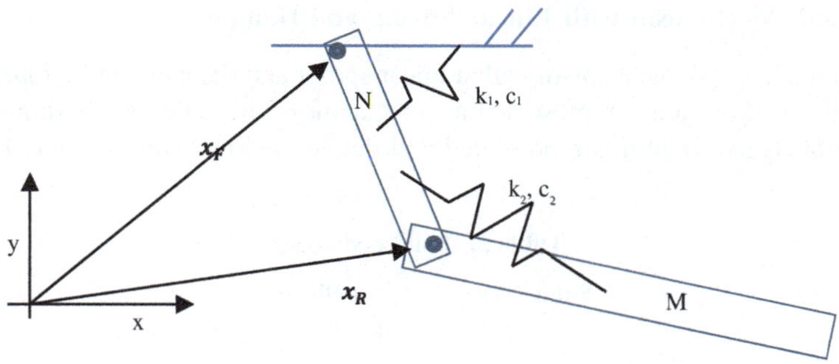

Fig. 3. Double pendulum with linear springs and dampers

This system, when simulated both in ADAMS and with the new method, gives the following results (Figs. 4, 5, 6, and 7):

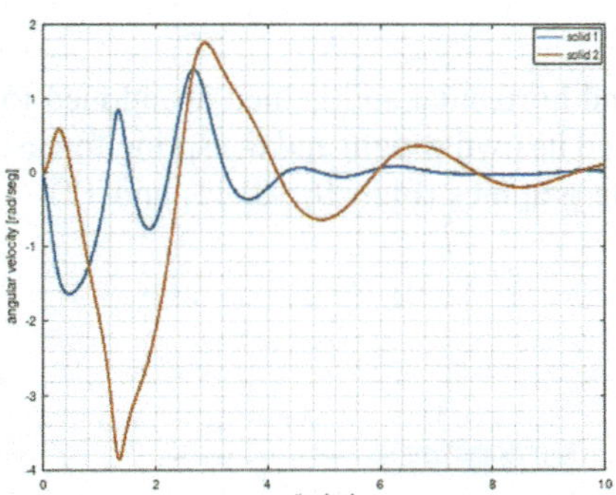

Fig. 4. Angular velocity (new method)

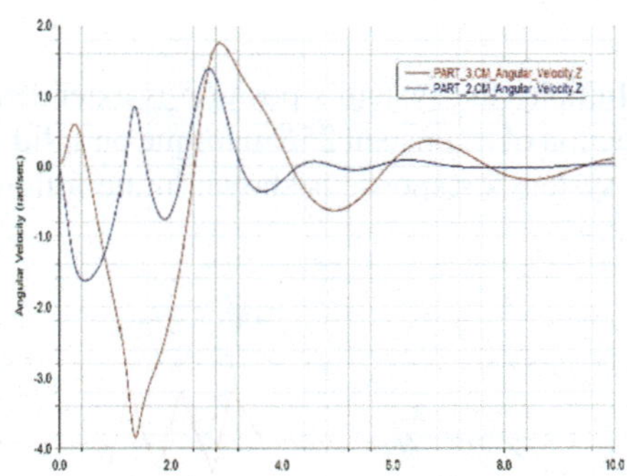

Fig. 5. Angular velocity (ADAMS)

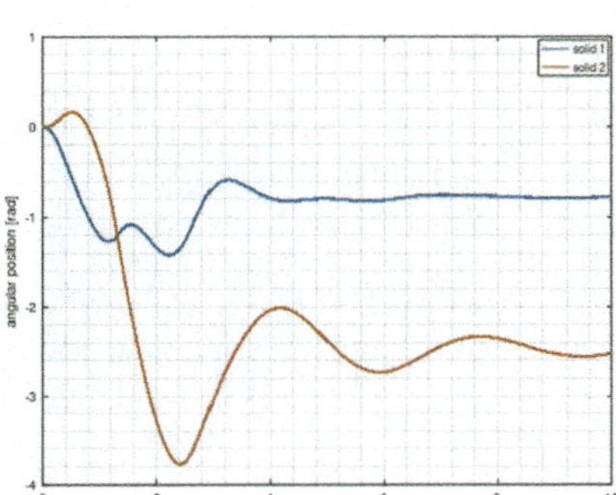

Fig. 6. Angle (new method)

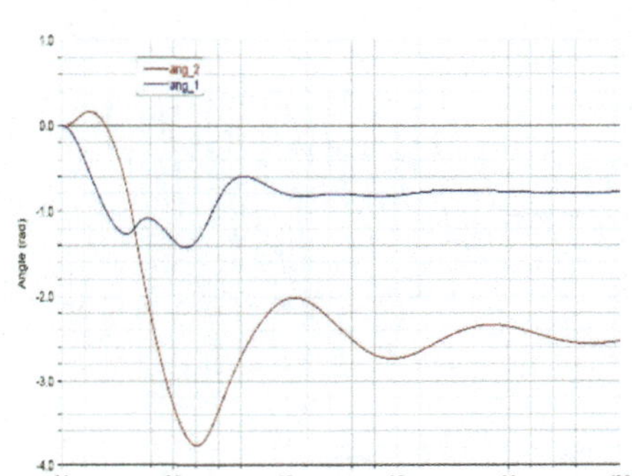

Fig. 7. Angle (ADAMS)

3.2 Crank Mechanism with Linear Spring and Damper

In this section a crank mechanism with a linear spring and damper can be found, which also features a slider joint (represented as a rectangle). The solids are both modeled as bars. The analyzed system is represented and defined below (Tables 3 and 4; Fig. 8):

Table 3. Solid body data

Solid	Mass (kg)	Length (m)
1	2	1
2	3	2

Table 4. Spring and damper data

Spring data	K (N/m)	C (N sec/m)	l_0 (m)
AC	5	5	2

Initially the system's position is extended and horizontal and is then abandoned to the action of a constant 25Nm torque on solid 1 and its own weight and the spring forces. The system's response is shown in the following graphics (Figs. 9, 10, 11, and 12):

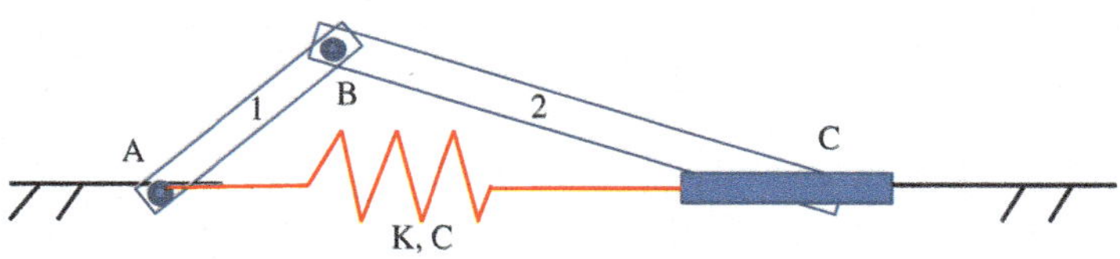

Fig. 8. Crank with linear spring and damper

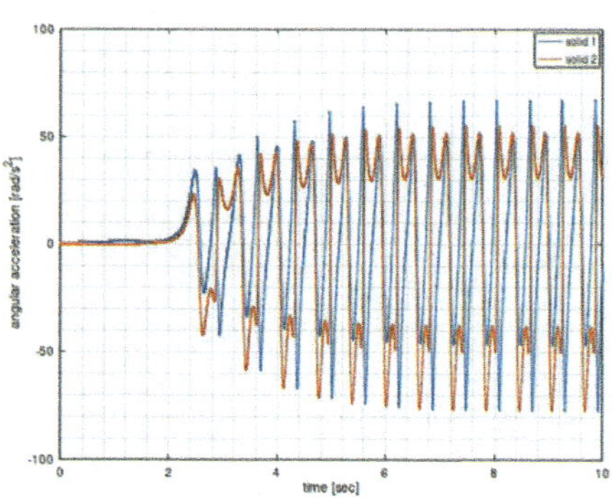

Fig. 9. Angular acceleration (new method)

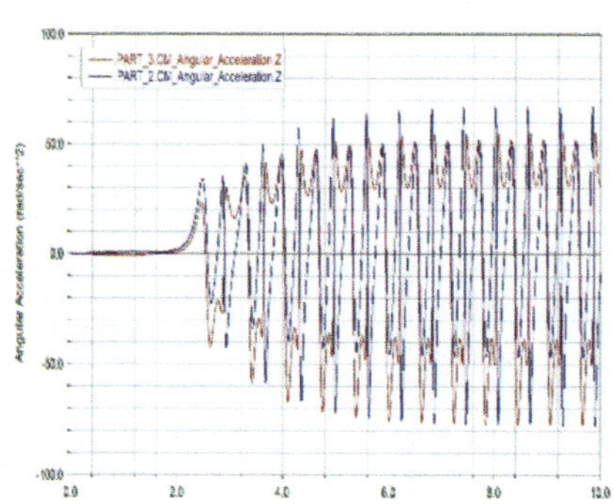

Fig. 10. Angular acceleration (ADAMS)

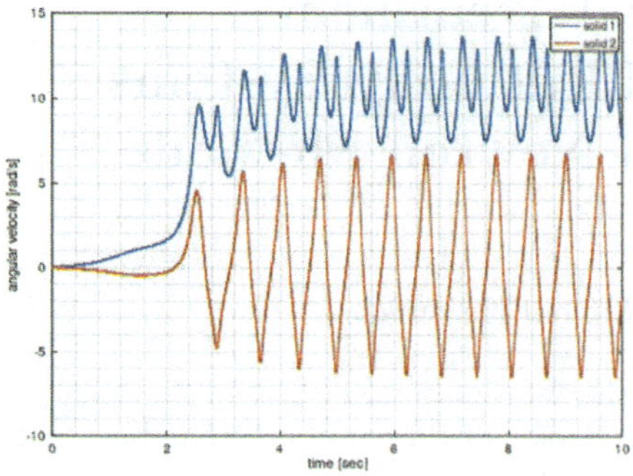

Fig. 11. Angular velocity (new method)

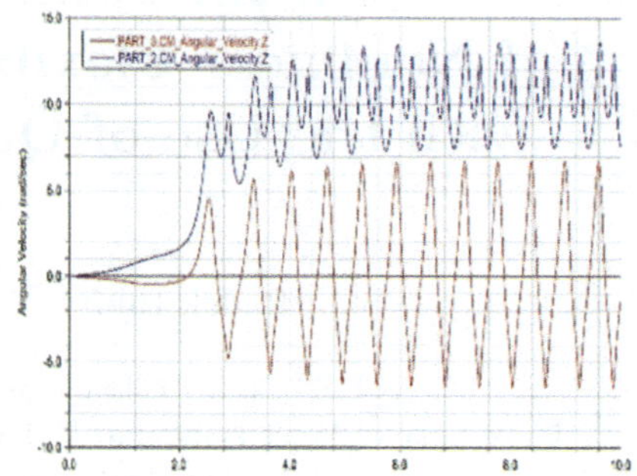

Fig. 12. Angular velocity (ADAMS)

4 Conclusions

After presenting a method for 2nd order multibody dynamics equations direct integration, which has proven to be valid, linear spring and dampers equations have been developed ad hoc for it. Because the method itself is consistent, the implementation of springs and dampers expands the possibilities of different systems to be analyzed, hence further proving the system's accuracy and precision.

For future developments, and since the 2D system has been fully characterized with this last addition of linear springs and forces, a 3D method will be developed. Furthermore, an optimized implementation should be also developed to test its computational efficiency.

Acknowledgments. The authors wish to acknowledge the Spanish and the Basque Government for the funding of this project through project DPI2016-80372-R, and IT947-16.

References

1. Shabana, A.A.: Dynamics of Multibody Systems. Cambridge University Press, Cambridge (2013)
2. Gavrea, B., Negrut, D., Potra, F.A.: The Newmark integration method for simulation of multibody systems: analytical considerations. In: ASME 2005 International Mechanical Engineering Congress and Exposition, pp. 1079–1092 (2005)
3. Uriarte, H., de Bustos, I.F., Urkullu, G., Olabarrieta, A.: On classical newmark integration of multibody dynamics. In: European Conference on Mechanism Science, pp. 97–105 (2018)
4. Cuadrado, J., Cardenal, J., Bayo, E.: Modeling and solution methods for efficient real-time simulation of multibody dynamics. Multibody Syst. Dyn. **1**(3), 259–280 (1997)
5. Urkullu, G., de Bustos, I.F., García-Marina, V., Uriarte, H.: Direct integration of the equations of multibody dynamics using central differences and linearization. Mech. Mach. Theory **133**, 432–458 (2019)

On the Numerical Treatment of Nonlinear Flexible Multibody Systems with the Use of Quasi-Newton Methods

Radek Bulín(✉) and Michal Hajžman

NTIS - New Technologies for the Information Society,
University of West Bohemia, Univerzitni 8, 306 14 Pilsen, Czech Republic
{rbulin,mhajzman}@ntis.zcu.cz

Abstract. This paper deals with the use of quasi-Newton methods for dynamical simulations of nonlinear flexible multibody systems which are modelled using absolute nodal coordinate formulation (ANCF). Three ANCF beam elements are briefly reminded and implemented. The Newmark integration method for index 3 differential-algebraic equations is coupled with the iterative quasi-Newton method in order to reduce computational time. The described algorithm is implemented and tested on the benchmark problem of a flexible pendulum.

1 Introduction

The modelling of flexible bodies as parts of multibody systems is a common issue that is a point of interest of many researchers. Several suitable methods for describing the flexibility together with the large displacements and rotations of flexible bodies were developed [12] and in most cases they lead to nonlinear formulations of motion equations. Therefore it is appropriate to deal with the numerical methods to solve these equations in reasonable computational times and accuracy.

The iterative refinement of the solution in discrete time step may be performed by using the quasi-Newton method. The basic advantage of this method is its computational effectiveness in comparison to classical the Newton method, where in most cases the Jacobi matrix needs to be evaluated numerically by finite differences. The drawback of the quasi-Newton method is the lost of convergence in case of inappropriately large time steps. This paper is dedicated to the usage of the Newmark integration method together with the quasi-Newton method for the simulations of flexible bodies modelled by absolute nodal coordinate formulation.

2 Modelling of Flexible Bodies

Since multibody systems may consist of flexible bodies, it is important to have proper methods to describe their elastic and inertia properties. One of the most

A. Kecskeméthy and F. Geu Flores (Eds.): ECCOMAS 2019, COMPUTMETHODS 53, pp. 332–339, 2020.
https://doi.org/10.1007/978-3-030-23132-3_40

promising methods in last two decades is the absolute nodal coordinate formulation (ANCF) of finite elements, that was originated in 1996 by Shabana, see [11]. Compared with the classical formulation of finite elements, ANCF uses absolute positions of nodes and partial derivatives of these positions with respect to local element parameters as nodal coordinates. This can be also advantageous from the viewpoint of possible rigid-flexible bodies connections when the natural coordinates formulation for describing rigid bodies is used [7]. In this work, several ANCF beam elements are used for flexible bodies modelling. However, the ANCF approach has been successfully applied to various types of elements such as plates, shells [8] or solid bricks [9].

The first tested element implemented in the in-house software in MATLAB is so called original ANCF beam element [5]. This type of element has two nodes and each node has 12 degrees of freedom that correspond with four vectors - the global position vector of the node and three tangential vectors that are obtained as derivatives of the global position vector of the node with respect to beam coordinates x (axial beam coordinate), y and z (two cross-sectional coordinates). Derivation of the elastic forces of the element can be found in [10], specifically the elastic line approach was used in this work. The resultant nonlinear vector of elastic forces was obtained by using symbolic operations in MATLAB software.

The second implemented element is the ANCF thin cable element, in [5] denoted as lower order cable element. It has two nodes and each node has six degrees of freedom, from which three are the components of the node position vector and three are the components of the beam center-line tangent (slope) vector. The ANCF cable element uses axial strain and beam center-line curvature to define the elastic forces. This type of element is advantageous in terms of reducing the number of degrees of freedom, but the minor disadvantage is the inability to describe torsional flexibility. The closed form of nonlinear vector of elastic forces was not obtainable by symbolic operations in MATLAB, so the Gaussian quadrature was used to its numerical evaluation.

The last tested element is denoted as L2T2 and was developed in article [1]. This type of element is a planar version of the ANCF thin cable element, it has two nodes with four degrees of freedom in each node and it uses several assumptions that lead to the simplification of the nonlinear vector of elastic forces.

One of the major advantages of used ANCF beam elements is the formulation of mass matrix which is constant, so it is computed only once before the numerical simulation. The motion equations of one unconstrained flexible body described by ANCF elements can be written in the similar form for all three tested elements

$$\mathbf{M}\ddot{\mathbf{q}} + \mathbf{Q} = \mathbf{f}, \tag{1}$$

where $\mathbf{M}$ is the mass matrix, $\mathbf{q} = \mathbf{q}(t)$ is the vector of all nodal coordinates, $\mathbf{Q} = \mathbf{Q}(\mathbf{q})$ is the vector of elastic forces that is dependent on the nodal coordinates and $\mathbf{f} = \mathbf{f}(\mathbf{q}, \dot{\mathbf{q}}, t)$ is the vector of external forces. Equation (1) can be extended by damping forces in the form of proportional damping. For this purpose, the vector of elastic forces can be written in the form $\mathbf{Q} = \mathbf{K}(\mathbf{q})\mathbf{q}$, where

$\mathbf{K}(\mathbf{q})$ is the nonlinear stiffness matrix. In this work, the damping is not considered.

The common form of motion equations of a multibody system is

$$\begin{cases} \mathbf{M}\ddot{\mathbf{q}} + \mathbf{\Phi}_{,\mathbf{q}}^T \boldsymbol{\lambda} = \mathbf{g}, \\ \mathbf{\Phi} = \mathbf{0}, \end{cases} \tag{2}$$

where $\mathbf{q}$ now contains all degrees of freedom of all bodies and $\mathbf{M}$ is the corresponding mass matrix, $\mathbf{\Phi} = \mathbf{\Phi}(\mathbf{q},\ t)$ represents holonomic constraints, $\mathbf{\Phi}_{,\mathbf{q}}$ is the Jacobi matrix of the constraints and $\boldsymbol{\lambda}$ is the vector of Lagrange multipliers. In case of use of the ANCF approach for describing a flexible body, the right side of the first equation in (2) is of the form

$$\mathbf{g}(\mathbf{q},\ \dot{\mathbf{q}},\ t) = \mathbf{f}(\mathbf{q},\ \dot{\mathbf{q}},\ t) - \mathbf{Q}(\mathbf{q}). \tag{3}$$

3 Numerical Solution of Equations of Motion

There are several publications dedicated to implicit numerical methods suitable for the solution of Index 3 differential-algebraic equations (DAE I3) of multibody systems described by Eq. (2) such as [3,6]. For example, paper [6] applies the Hilbert-Hughes-Taylor (HHT) method on the flexible bodies described by ANCF approach, but the resultant algebraic equations for the solution corrections are not further discussed in detail. In this paper, the Newmark integration method is used to obtain the algebraic equations for the solution corrections and these equations are solved by quasi-Newton method.

3.1 Newmark Integration Method for DAE I3

Based on [3], the Newmark formulas are used in the form

$$\begin{aligned} \mathbf{q}_{n+1} &= \mathbf{q}_n + h\dot{\mathbf{q}}_n + \tfrac{h^2}{2}\left[(1-2\beta)\ddot{\mathbf{q}}_n + 2\beta\ddot{\mathbf{q}}_{n+1}\right], \\ \dot{\mathbf{q}}_{n+1} &= \dot{\mathbf{q}}_n + h\left[(1-\gamma)\ddot{\mathbf{q}}_n + \gamma\ddot{\mathbf{q}}_{n+1}\right], \end{aligned} \tag{4}$$

where h is the time step, $\gamma = \frac{1}{2} + \alpha$, $\beta = \frac{1}{4}\left(\gamma + \frac{1}{2}\right)^2$ and α is the parameter of numerical damping. The equations of motion with the constraint equations (2) are formerly adjusted in order to define residuum vector $\mathbf{r}_{res}$ as

$$\begin{aligned} \mathbf{r}_{res}\left(\mathbf{q},\ \dot{\mathbf{q}},\ \ddot{\mathbf{q}},\ \boldsymbol{\lambda}\right) = \mathbf{M}\ddot{\mathbf{q}} + \mathbf{\Phi}_{,\mathbf{q}}^T\boldsymbol{\lambda} - \mathbf{g}\left(\mathbf{q},\ \dot{\mathbf{q}},\ t\right) &= \mathbf{0}, \\ \mathbf{\Phi}(\mathbf{q},\ t) &= \mathbf{0}. \end{aligned} \tag{5}$$

Equation (5) can be expressed in discrete time step $n+1$ and after the substitution of the Newmark formulas (4) inversion, see [3], into (5), following equations are obtained

$$\begin{aligned} \mathbf{r}_{res}\left(\mathbf{q}_{n+1},\ \boldsymbol{\lambda}_{n+1}\right) &= \mathbf{0}, \\ \mathbf{\Phi}\left(\mathbf{q}_{n+1}\right) &= \mathbf{0}. \end{aligned} \tag{6}$$

After the linearisation of (6) using Taylor expansion, following equations are obtained

$$\begin{aligned} \mathbf{r}_{res}\left(\mathbf{q}_{n+1}^{k+1},\ \boldsymbol{\lambda}_{n+1}^{k+1}\right) = \mathbf{0} \approx \mathbf{r}_{res}\left(\mathbf{q}_{n+1}^{k},\ \boldsymbol{\lambda}_{n+1}^{k}\right) + \mathbf{S}_t\left(\mathbf{q}_{n+1}^{k},\ \boldsymbol{\lambda}_{n+1}^{k}\right)\Delta\mathbf{q}^k + \boldsymbol{\Phi}_{,\mathbf{q}}^T\Delta\boldsymbol{\lambda}^k, \\ \boldsymbol{\Phi}(\mathbf{q}_{n+1}^{k+1}) = \mathbf{0} \approx \boldsymbol{\Phi}(\mathbf{q}_{n+1}^{k}) + \boldsymbol{\Phi}_{,\mathbf{q}}\Delta\mathbf{q}^k, \end{aligned} \tag{7}$$

where $\mathbf{S}_t$ is the Jacobi matrix of the residuum vector and has a form

$$\mathbf{S}_t = \frac{\partial \mathbf{r}_{res}}{\partial \mathbf{q}} = \frac{1}{\beta h^2}\mathbf{M} + \frac{\partial}{\partial \mathbf{q}}\left(\boldsymbol{\Phi}_{,\mathbf{q}}^T\boldsymbol{\lambda}\right) - \frac{\partial \mathbf{g}}{\partial \mathbf{q}} - \frac{\gamma}{\beta h}\frac{\partial \mathbf{g}}{\partial \dot{\mathbf{q}}}. \tag{8}$$

Now the expression (7) can be rewritten in the matrix form

$$\begin{bmatrix} \mathbf{S}_t & \boldsymbol{\Phi}_{,\mathbf{q}}^T \\ \boldsymbol{\Phi}_{,\mathbf{q}} & \mathbf{0} \end{bmatrix} \cdot \begin{bmatrix} \Delta\mathbf{q}^k \\ \Delta\boldsymbol{\lambda}^k \end{bmatrix} = -\begin{bmatrix} \mathbf{r}_{res}\left(\mathbf{q}_{n+1}^{k},\ \boldsymbol{\lambda}_{n+1}^{k}\right) \\ \boldsymbol{\Phi}(\mathbf{q}_{n+1}^{k}) \end{bmatrix}. \tag{9}$$

Expression (9) represents the system of algebraic equations for the corrections of positions $\Delta\mathbf{q}$ and corrections of Lagrange multipliers $\Delta\boldsymbol{\lambda}$ that are then used for the solution correction as follows

$$\begin{aligned} \ddot{\mathbf{q}}_{n+1}^{k+1} &= \ddot{\mathbf{q}}_{n+1}^{k} + \Delta\ddot{\mathbf{q}}^k = \ddot{\mathbf{q}}_{n+1}^{k} + \frac{1}{\beta h^2}\Delta\mathbf{q}^k, \\ \dot{\mathbf{q}}_{n+1}^{k+1} &= \dot{\mathbf{q}}_{n+1}^{k} + \Delta\dot{\mathbf{q}}^k = \dot{\mathbf{q}}_{n+1}^{k} + \frac{\gamma}{\beta h}\Delta\mathbf{q}^k, \\ \mathbf{q}_{n+1}^{k+1} &= \mathbf{q}_{n+1}^{k} + \Delta\mathbf{q}^k, \\ \boldsymbol{\lambda}_{n+1}^{k+1} &= \boldsymbol{\lambda}_{n+1}^{k} + \Delta\boldsymbol{\lambda}^k. \end{aligned} \tag{10}$$

The iterative process (10) is started with the prediction of the acceleration vector.

Equation (9) can be rewritten to the simple compact form

$$\mathbf{J}^r(\mathbf{x}^k)\Delta\mathbf{x}^k = -\tilde{\mathbf{r}}(\mathbf{x}^k), \tag{11}$$

where $\mathbf{J}^r$ is the iteration matrix, which may differ in each iteration k. The analytical expression of the iteration matrix is nearly impossible for the most of nonlinear problems, e.g. in case of a thin cable element, where the beam centerline curvature is used for the derivation of bending elastic forces, is the analytical expression very complicated. Therefore, the iteration matrix is often approximated by using numerical finite differences. Since this operation is very time consuming, it is not recommended to update the iteration matrix within every iteration [6]. In this paper, the application of the quasi-Newton method for direct construction and updating of the approximation of the iteration matrix inversion is used in order to reduce computational time needed for reaching the force equilibrium state at time step n.

3.2 Quasi-Newton Method

The quasi-Newton method is based on the approximation of iteration matrix $\mathbf{J}^r$ by a matrix $\mathbf{G}$ [4]. For each iteration k, the matrix $\mathbf{G}$ is computed based on the previously evaluated quantities, so the form of this matrix in iteration $k+1$ is

$$\mathbf{G}^{k+1} = \mathbf{G}^k + \Delta\mathbf{G}^k, \tag{12}$$

where $\Delta\mathbf{G}^k$ is the increment of the approximation matrix and several quasi-Newtonian methods are distinguished based on its form, such as Davidon's, Broyden's, DPF or BFGS method. The derivation of the increment $\Delta\mathbf{G}^k$ is based on the secant equation, for more information see [2]. The algorithm of quasi-Newton method starts with the first approximation matrix $\mathbf{G}^0$, that can be determined by using numerical finite difference method for $\mathbf{J}^r(\mathbf{x}^0)$ at the very start of the simulation. For $k = 0, 1, 2, \ldots$, following algorithm is then applied:

1. Calculation of increments $\Delta\mathbf{x}^k$ from equation $\mathbf{G}^k \Delta\mathbf{x}^k = -\tilde{\mathbf{r}}(\mathbf{x}^k)$.
2. Evaluation of the solution in step $k+1$ as $\mathbf{x}^{k+1} = \mathbf{x}^k + \Delta\mathbf{x}^k$.
3. Determination of the difference between two consecutive residual vectors as $\mathbf{y}^k = \tilde{\mathbf{r}}(\mathbf{x}^{k+1}) - \tilde{\mathbf{r}}(\mathbf{x}^k)$.
4. Calculation of $\Delta\mathbf{G}^k$ based on the chosen quasi-Newtonian method and the use of Eq. (12).

One of the benefits of the quasi-Newton method is that after few mathematical operations the approximation of the iteration matrix $\mathbf{J}^r$ inversion can be directly written in the closed form. With the denotion $\mathbf{H}^k = \left(\mathbf{G}^k\right)^{-1}$, the vector of increments can be easily evaluated as $\Delta\mathbf{x}^k = -\mathbf{H}^k \tilde{\mathbf{r}}(\mathbf{x}^k)$. Then updated matrix $\mathbf{H}^{k+1}$ can be evaluated based on the selected quasi-Newtonian method. The list of the most used methods follows [4]:

- Broyden's method

$$\mathbf{H}^{k+1} = \mathbf{H}^k + \frac{\left(\Delta\mathbf{x}^k - \mathbf{H}^k\mathbf{y}^k\right)\left(\Delta\mathbf{x}^k\right)^T \mathbf{H}^k}{\left(\Delta\mathbf{x}^k\right)^T \mathbf{H}^k \mathbf{y}^k}, \tag{13}$$

- Davidon's method

$$\mathbf{H}^{k+1} = \mathbf{H}^k + \frac{\left(\Delta\mathbf{x}^k - \mathbf{H}^k\mathbf{y}^k\right)\left(\Delta\mathbf{x}^k - \mathbf{H}^k\mathbf{y}^k\right)^T}{\left(\Delta\mathbf{x}^k - \mathbf{H}^k\mathbf{y}^k\right)^T \mathbf{y}^k}, \tag{14}$$

- Davidon-Powell-Fletcher (DPF) method

$$\mathbf{H}^{k+1} = \mathbf{H}^k + \frac{\Delta\mathbf{x}^k\left(\Delta\mathbf{x}^k\right)^T}{\left(\Delta\mathbf{x}^k\right)^T \mathbf{y}^k} - \frac{\mathbf{H}^k\mathbf{y}^k\left(\mathbf{y}^k\right)^T \mathbf{H}^k}{\left(\mathbf{y}^k\right)^T \mathbf{H}^k\mathbf{y}^k}, \tag{15}$$

- Broyden-Fletcher-Goldfard-Shanno (BFGS) method

$$\mathbf{H}^{k+1} = \left(\mathbf{I} - \frac{\Delta\mathbf{x}^k\left(\mathbf{y}^k\right)^T}{\left(\mathbf{y}^k\right)^T \Delta\mathbf{x}^k}\right)\mathbf{H}^k\left(\mathbf{I} - \frac{\mathbf{y}^k\left(\Delta\mathbf{x}^k\right)^T}{\left(\mathbf{y}^k\right)^T \Delta\mathbf{x}^k}\right) + \frac{\Delta\mathbf{x}^k\left(\Delta\mathbf{x}^k\right)^T}{\left(\mathbf{y}^k\right)^T \Delta\mathbf{x}^k}. \tag{16}$$

4 Numerical Simulations

The common benchmark problem of a flexible pendulum was used to compare proposed numerical method with the functions of `ode` family methods from MATLAB software. The pendulum of the 2 m length has a square cross section with the edge of 0.01 m, material density is 4000 kg·m^{-3} and the Young's modulus is 10^8 Pa. The rotation joint to the base frame is represented by the constraints applied to the end node. For the pendulum model, the original ANCF beam elements (Thick), ANCF thin cable elements (Thin) and planar L2T2 elements were used. Ten elements were used to describe the pendulum. The relative and absolute error of `ode` functions were left with implicit values ($RelTol = 10^{-3}$, $AbsTol = 10^{-6}$) and the solution of each time step using Newmark method is considered as converged if $|\mathbf{r}_{res}| < 10^{-6}$ and $|\mathbf{\Phi}| < 10^{-6}$. Numerical damping parameter α was set to zero in order to compare the results with the trapezoidal rule of the `ode23t`. The time step of Newmark method was set to $h = 0.0005$ s. The snapshots of the pendulum motion in chosen discrete times are shown in Fig. 1. In Table 1, the resultant computation times for the simulation of two seconds of the pendulum movement are summarized for various ANCF elements and numerical methods. Only three fastest MATLAB functions are shown. From the results it is apparent, that the Thick element is the slowest especially with the use of `ode` functions. This may be caused by the implementation of the Thick elements, because their elastic forces are derived in closed form during preprocessing by using symbolic operations of MATLAB. The evaluation with the use of Gaussian quadrature would be faster. The Newmark method with the quasi-Newton method was faster for all element types, especially Davidon's method is significantly faster than classical `ode` functions.

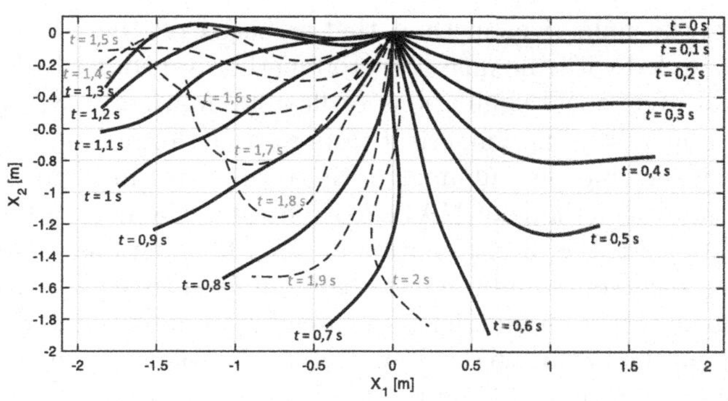

Fig. 1. Snapshots of the pendulum motion in discrete time steps.

In Fig. 2, the differences in vertical position of the flexible pendulum tip that were obtained by using various numerical methods are shown. It is apparent, that the difference between the Davidon's method and `ode23t` is in the same order of magnitude 10^{-5} as the difference between `ode23t` and `ode23`. The difference between the quasi-Newtonian methods is even smaller - order of magnitude 10^{-8}.

Table 1. Computation times in [s] of the pendulum motion simulation (2 s) for various ANCF elements and numerical integration methods.

ANCF	MATLAB functions			Newmark with quasi-Newton			
Element	ode23	ode113	ode23t	Broyden	Davidon	DPF	BFGS
L2T2	5.3	9.4	10.1	3.69	1.69	3.06	2.96
Thin	7.1	11.9	11.5	7.22	2.96	6.32	6.14
Thick	311.0	221.3	375.6	19.20	10.49	20.24	27.61

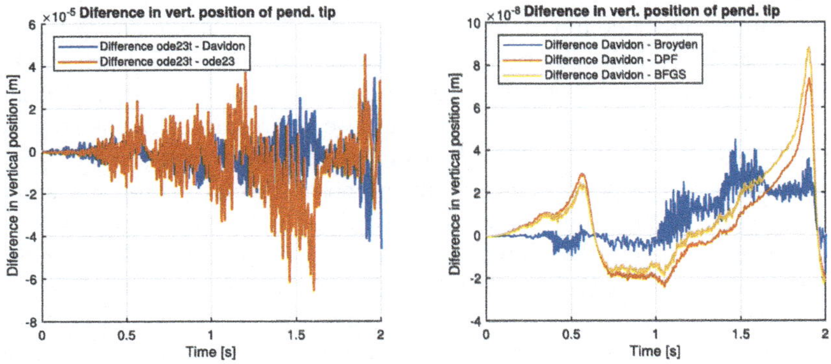

Fig. 2. Differences in vertical position of flexible pendulum tip for various numerical methods.

5 Conclusions

In this paper, the numerical method that is based on the Newmark integration scheme coupled with the quasi-Newton method was described as a suitable method for the dynamical simulations of flexible multibody systems modelled using ANCF approach. The numerical tests were performed on the benchmark problem of the flexible pendulum and the results are in a good compliance with the ode family methods of MATLAB software. The main advantage of the described approach is the direct solution of DAE I3 equations without the need of stabilization techniques and the faster computation times in comparison with MATLAB time integration functions. In case of reasonably small time steps, the quasi-Newton method converged in few steps to the solution. Currently, the proposed methods are being tested on the mechanical system of a rotating cylinder with a wrapped cable modelled by ANCF and the resultant friction and normal contact forces are in a very good compliance with the analytical equations. In future work, the quasi-Newton method will be implemented together with HHT integration method.

Acknowledgements. The work was supported from ERDF "Research and Development of Intelligent Components of Advanced Technologies for the Pilsen Metropolitan Area (InteCom)" (No.: CZ.02.1.01/0.0/0.0/17_048/0007267).

References

1. Berzeri, M., Shabana, A.A.: Development of simple models for the elastic force in the absolute nodal co-ordinate formulation. J. Sound. Vib. **235**, 539–565 (2000)
2. Dennis, J.E., Schnabel, R.B.: Numerical Methods for Unconstrained Optimization and Nonlinear Equations. SIAM, Philadelphia (1996)
3. Gradin, M., Cardona, A.: Flexible Multibody Dynamics. John Wiley & Sons, Chichester (2001)
4. Geradin, M., Idelsohn, S., Hogge, M.: Computational strategies for the solution of large nonlinear problems via Quasi-Newton methods. Comput. Struct. **13**, 73–81 (1981)
5. Gerstmayr, J., Shabana, A.A.: Analysis of thin beams and cables using absolute nodal co-ordinate formulation. Nonlinear Dyn. **45**, 109–130 (2006)
6. Hussein, B., Negrut, D., Shabana, A.A.: Implicit and explicit integration in the solution of the absolute nodal coordinate differential/algebraic equations. Nonlinear Dyn. **54**, 283–296 (2008)
7. Liu, Ch., Tian, Q., Hu, H., Garca-Vallejo, D.: Simple formulations of imposing moments and evaluating joint reaction forces for rigid-flexible multibody systems. Nonlinear Dyn. **69**, 127–147 (2012)
8. Mikkola, A.M., Shabana, A.A.: A non-incremental finite element procedure for the analysis of large deformation of plates and shells in mechanical system applications. Multibody Syst. Dyn. **9**, 283–309 (2003)
9. Olshevskiy, A., Dmitrochenko, O., Kim, Ch.-W.: Three-dimensional solid brick element using slopes in the absolute nodal coordinate formulation. J. Comput. Nonlinear Dyn. **9**(2), 021001 (2014)
10. Schwab, A.L., Meijaard, J.P.: Comparison of three-dimensional flexible beam elements for dynamic analysis: finite element method and absolute nodal coordinate formulation. In: Proceeding of IDETC/CIE, pp. 1–9 (2005)
11. Shabana, A.A.: Definition of the slopes and the finite element absolute nodal coordinate formulation. Multibody Syst. Dyn. **1**, 339–348 (1997)
12. Shabana, A.A.: Flexible multibody dynamics: review of past and recent developments. Multibody Syst. Dyn. **1**, 189–222 (1997)

Interior-Point Solver for Non-smooth Multi-Body Dynamics with Finite Elements

Dario Mangoni(✉), Alessandro Tasora, and Simone Benatti

University of Parma, Parco Area delle Scienze 181, Parma, Italy
{dario.mangoni,alessandro.tasora}@unipr.it,
simone.benatti@studenti.unipr.it

Abstract. The increasing complexity of dynamic simulations involving unilateral constraints, such as contacts, is pushing for new solvers that may address the problem of handling non-smooth impact events in a more efficient and accurate manner, especially in mixed rigid and flexible-bodies simulations. For this purpose, a new implementation of an interior-point solver for Quadratic Cone Programming is proposed. Even though the general idea of considering multibody system formulations as a parallel of an optimization/programming problem is already known in literature ([3,4,6]), still very few options are available for those problems whose complexity is due to contacts with friction *and* finite-elements at the same time. The opportunity to handle those problems in a unified numerical framework could trigger a novel interest both in scientific and applied researches.

1 Introduction

Collapsing a multibody problem with frictional contacts into an optimization problem can be easily achieved when considering the non-smooth approach, the one in which an hard constraint over interpenetration is considered. In Sect. 2 a brief overview on how to deal with the dynamic and constraint equation in order to make them compatible as an optimization problem is proposed.

Following, in Sect. 3, the interior-point algorithm, considered as the proposed solver to handle the optimization problem that stems from the previous section, is described. Because of the limitations of a short paper, only the most general case of conic constraints is considered. For specific degenerated case—such those involving no-frictional contacts—some bibliographic references are provided [5].

2 Problem Formulation

At least two main different formulations for multibody problems involving contacts can be stated. The first considers the contact as a smooth event, thus considering continuous velocity even in the case of collision. This approach aims

A. Kecskeméthy and F. Geu Flores (Eds.): ECCOMAS 2019, COMPUTMETHODS 53, pp. 340–347, 2020.
https://doi.org/10.1007/978-3-030-23132-3_41

to reuse standard linear system solvers—since this formulation boils down to a typical linear system, at most in a saddle-point shape—but has the undesired effect to require small step for the timestepper, given that the continuous variation at the speed level should be adequately caught during the time integration.

On the contrary, the DVI (Differential Variational Inequalities)/MDI (Measure Differential Inclusions) approach considers this phenomena accordingly to its non-smoothness [9]. The relative velocity between contact points is regulated and the reaction forces are computed so that the compenetration is in any case avoided. This scheme provides a different model for which the underlying time integrator has an easier task when it comes to make the method to converge i.e. longer time steps can be afforded.

The formulation of the problem, in this latter case, can be summarized as:

$$M d\boldsymbol{v} = \boldsymbol{f}(\boldsymbol{q}, \boldsymbol{v}, t)dt + C\, d\boldsymbol{\gamma} + D\, d\boldsymbol{\lambda} \tag{1a}$$

$$\Upsilon \ni d\boldsymbol{\lambda} \perp \boldsymbol{y} \in \Upsilon^* \tag{1b}$$

where the accelerations and reaction forces are here substituted by velocity variations ($d\boldsymbol{v}$) and impulses ($d\boldsymbol{\lambda}$ and $d\boldsymbol{\gamma}$) respectively. M is the mass matrix, while C and D are the constraint Jacobian matrices. $\boldsymbol{f}$ are external forces.

The variational approach together with Eq. 1b are the two main differences that clearly marks a big discrepancy with a more classical smooth formulation. Here, differently from the frictionless case [6], the orthogonality condition forces the velocities at the contact patch ($\boldsymbol{y}$) to lay on the dual of the cone described by the Coulomb friction model Υ^*. This new problem structure can be cast as a Cone QP under mild assumptions on convexity, it can be processed by an optimization solver that exploits the structure of the problem. Finite elements can be fit into this framework, without any particular restriction over the formulation or without any particular scaling on the system matrices. For example, in the tests, various kinds of elements are used: solid, beams, linear and non-linear with large displacements and deformations[1].

3 Interior-Point Solver for Cone QP

The Interior-Point solver described in the paper aims to take into account this conic constraints, in order to provide an effective solution for those problems that arise from frictional contacts. While traditional Cone QP solvers usually deal with self-dual cones (i.e. $\mu_f = 1$), the following implementation provides the correct scaling to allow to solve problems even with *arbitrary* friction coefficients. An additional Nesterov-Tood transformation [10] is added for the sake of performance.

[1] The solver has been tested with many more element types, provided by the Chrono library: linear, quadratic, non-linear, isogeometric formulations with and without plasticity, geometric and material non-linearity, beams, shells, solids, cables,...

The optimization problem that stems from the multibody problem formulated in Eq. 1 can be written as:

$$\min \quad \frac{1}{2}\boldsymbol{v}^T H \boldsymbol{v} + \boldsymbol{v}^T \boldsymbol{c} \tag{2a}$$
$$\text{s.t.} \quad C\boldsymbol{v} - \boldsymbol{b}_\gamma = \boldsymbol{0} \tag{2b}$$
$$D\boldsymbol{v} - \boldsymbol{b}_\lambda - \boldsymbol{y} = \boldsymbol{0} \tag{2c}$$
$$\boldsymbol{y} \succeq \boldsymbol{0} \tag{2d}$$

where the H matrix may include not only the mass matrix M, but also stiffness/damping matrices of finite elements. Known terms $\boldsymbol{b}_\gamma$, $\boldsymbol{b}_\lambda$, $\boldsymbol{c}$ come from the evaluation of body positions at the contact patch and from external forces. The slack variable $\boldsymbol{y}$ is introduced in ordert to formally remove the inequality condition from Eq. 2c. $\boldsymbol{\lambda}$ maps for unilateral constraints, $\boldsymbol{\gamma}$ for bilateral.

In this form, the Karush–Kuhn-Tucker conditions result in the system:

$$\boldsymbol{r}_D := H\boldsymbol{v} + \boldsymbol{c} - C^T\boldsymbol{\gamma} - D^T\boldsymbol{\lambda} = 0 \tag{3a}$$
$$\boldsymbol{r}_{P_\gamma} := C\boldsymbol{v} - \boldsymbol{b}_\gamma = 0 \tag{3b}$$
$$\boldsymbol{r}_{P_\lambda} := D\boldsymbol{v} - \boldsymbol{b}_\lambda - \boldsymbol{y} = 0 \tag{3c}$$
$$\boldsymbol{y} \circ \boldsymbol{\lambda} = 0 \tag{3d}$$
$$\boldsymbol{\lambda} \succeq 0 \tag{3e}$$
$$\boldsymbol{y} \succeq 0 \tag{3f}$$

For scalar constraints, such those arising from non-frictional contact, the $\circ$ operator has the meaning of a simple Hadamard (component-wise) product. However, for Cone QP, the constraints equations involve more than one variable, so that we have, for the i-th constraint, the *set* of variables $\boldsymbol{\lambda}_i = [\lambda_{i,0},\ \boldsymbol{\lambda}_{i,1}]^T$. Then, the $\circ$ product becomes:

$$\boldsymbol{y} \circ \boldsymbol{\lambda} = \begin{bmatrix} \boldsymbol{y}_i^T \boldsymbol{\lambda}_i \\ y_{i,0}\, \boldsymbol{\lambda}_{i,1} + \lambda_{i,0}\, \boldsymbol{y}_{i,1} \end{bmatrix}$$

The iterative process toward the solution can be described as a sequence of steps, with given *direction* and step *length.* In general, the first is obtained by the solution $\Delta\boldsymbol{q} = [\Delta\boldsymbol{v},\ \Delta\boldsymbol{\gamma},\ \Delta\boldsymbol{\lambda},\ \Delta\boldsymbol{y}]$ of the system $\boldsymbol{F}_{k+1} = \boldsymbol{F}_k + \nabla\boldsymbol{F}_k\, \Delta\boldsymbol{q} = 0$ where

$$\boldsymbol{F} = [\boldsymbol{r}_D,\ \boldsymbol{r}_{P_\gamma},\ \boldsymbol{r}_{P_\lambda},\ \boldsymbol{y} \circ \boldsymbol{\lambda}] \tag{4}$$

from which, expliciting $\nabla\boldsymbol{F}_k$ and moving known terms to the right, we have:

$$\nabla\boldsymbol{F}_k\, \Delta\boldsymbol{q} = \begin{bmatrix} H & -C^T & -D^T & 0 \\ C & 0 & 0 & 0 \\ D & 0 & 0 & -I \\ 0 & 0 & Y & \Lambda \end{bmatrix} \begin{bmatrix} \Delta\boldsymbol{v} \\ \Delta\boldsymbol{\gamma} \\ \Delta\boldsymbol{\lambda} \\ \Delta\boldsymbol{y} \end{bmatrix} = - \begin{bmatrix} \boldsymbol{r}_D \\ \boldsymbol{r}_{P_\gamma} \\ \boldsymbol{r}_{P_\lambda} \\ Y\Lambda\boldsymbol{e} + \Delta\boldsymbol{y}_p \circ \Delta\boldsymbol{\lambda}_p \end{bmatrix} + \begin{bmatrix} 0 \\ 0 \\ 0 \\ \sigma\mu_p\boldsymbol{e} \end{bmatrix} = -\boldsymbol{F}_k \tag{5}$$

where unfortunately the matrices $Y = \text{diag}(\boldsymbol{y})$ and $\Lambda = \text{diag}(\boldsymbol{\lambda})$ make the whole $\nabla \boldsymbol{F}_k$ matrix dependent on the iteration progress.

The step length is instead computed from the inequalities 3e and 3f. The interior-point algorithm, as the name may suggest, is operating inside the feasible region i.e. the one bounded by the constraints equations. In order to avoid exiting from this work space, the step length should be chosen to not exceed them.

$$\max\{\alpha_P > 0 \mid \boldsymbol{y}_p = \boldsymbol{y} + \alpha_P \, \Delta\boldsymbol{y} \succeq 0\} \tag{6a}$$

$$\max\{\alpha_D > 0 \mid \boldsymbol{\lambda}_p = \boldsymbol{\lambda} + \alpha_D \, \Delta\boldsymbol{\lambda} \succeq 0\} \tag{6b}$$

$$\alpha_p = \min\{\alpha_P, \alpha_D\} \tag{6c}$$

where the subscript $_p$ indicates a *predicted* variable, while $_P$ and $_D$ refers to primal and dual variables i.e. $\boldsymbol{y}$ and $\boldsymbol{\lambda}$ respectively. However, the generalized inequalities put an additional complexity to an otherwise simple problem.

In order to clarify the previous equations, it is required to state the equation of a generic cone. Stating that $\boldsymbol{p} \succeq 0$ or, that is the same, that $\boldsymbol{p}$ belongs to the cone $\mathscr{K}_{\mu_f}$ i.e. $\boldsymbol{p} \in \mathscr{K}_{\mu_f}$ means that $\mu_f \, p_0 \geq \|\boldsymbol{p}_1\|$.

For example, $\boldsymbol{p}$ might be the set of normal and tangential velocities at the contact patch $\boldsymbol{y}$ or the reaction forces $\boldsymbol{\lambda}$.

The step length limitation should take the conic constraint into consideration. Supposing to have to make a step in the direction $\boldsymbol{d}$, starting from the (feasible) point $\boldsymbol{p}$, then the step length α should verify

$$\mu_f (p_0 + \alpha d_0) \geq \|\boldsymbol{p}_1 + \alpha \boldsymbol{d}_1\| \tag{7}$$

In particular, we are looking for the supremum value α_M that respects the equation above and that can be found from

$$\alpha^* = \frac{-k_\alpha - \sqrt{k_\alpha^2 - a_\alpha c_\alpha}}{a_\alpha} \tag{8}$$

where

$$a_\alpha = \boldsymbol{d}^T J \boldsymbol{d}; \quad k_\alpha = \boldsymbol{p}^T J \boldsymbol{d}; \quad c_\alpha = \boldsymbol{p}^T J \boldsymbol{p}$$

where, for this general case, we have $J = \text{diag}\left\{\mu_f^2, -I_{[p-1]}\right\}$.

Naming $\alpha_b = -\frac{p_0}{d_0}$ we can have multiple cases, summarized in Table 1.

Table 1. Supremum step length α_M

α_M	$a_\alpha > 0$	$a_\alpha < 0$
$d_0 > 0$	$+\infty$	α^*
$d_0 < 0$	$\min\{\alpha^*, \alpha_b\}$	$\min\{\alpha^*, \alpha_b\}$

Table 1 is crucial in order to find the maximum length of the step α_M that is necessary to reach the border of the cone or to not exit from it. Additional considerations may be done in the case $\boldsymbol{d} = [1\ 0\ 0\ \ldots]$ or when $\mu_f = 1$. See [5].

However, this procedure is still not enough to guarantee the algorithm to be effective and some additional refinements are needed to assist the convergence. Because of this, the predictor-corrector scheme [7] is here adopted, thus introducing a double-step process. In the first step, a fastest-descent direction $\Delta \boldsymbol{q}_p$ is evaluated so to respect Eq. 5, together with a step length α_p that assures that the conic constraints are respected, as described in Eq. 6. In the second phase, thanks to information gathered during the prediction step, the tentative solution is finally updated.

It is important to notice that this predictor-corrector approach requires to solve the system twice. For direct linear solvers, however, this will not exactly double the computational cost. In fact, even though the coefficients of the matrix do change, the sparsity pattern is kept, thus allowing for optimized analysis and factorization phases.

The Nesterov-Todd scaling is introduced, together with the scaled variable $\boldsymbol{z} = W^{-T}\boldsymbol{y} = W\boldsymbol{\lambda}$ for the sake of performance. The construction of the scaling matrix W can be found in detail in [10], as well as in [5].

Finally, the addition of a custom friction coefficient μ_f impose also a slight change in the matrix D that maps for the unilateral constraints. For each conic constraint whose dimension is p_λ, in the D matrix a sub-matrix D_i of dimension $p_\lambda \times n$ (where H is $n \times n$) is added. This matrix contains the Jacobian of the constraints to which the friction coefficients should be applied. For this purpose we define

$$\bar{D} = \begin{bmatrix} \mu_f & 0 \\ 0 & I_{[p-1]} \end{bmatrix} D; \qquad \bar{\boldsymbol{b}}_\lambda = \begin{bmatrix} \mu_f \\ I_{[p-1]} \end{bmatrix} \boldsymbol{b}_\lambda; \qquad \boldsymbol{\lambda} = \begin{bmatrix} \mu_f \\ I_{[p-1]} \end{bmatrix} \bar{\boldsymbol{\lambda}} \tag{9}$$

As we may notice, for typical multibody applications, the system matrix that we will see in Eq. 10 is positive-definite (or it can be even made symmetric switching the sign of the constraint matrices C and D in the first block-row). In order to avoid breaking this pattern, it's preferable to solve for the variable $\bar{\boldsymbol{\lambda}}$ until the multibody system requires the computation of the actual value of the reaction forces $\boldsymbol{\lambda}$. Only then, the conversion expressed in Eq. 9 should be applied.

The definitive Newton-Raphson iteration that is used to evaluate the best direction along which the iteration will proceed is pursued solving the following linear system

$$\begin{bmatrix} H & -C^T & -\bar{D}^T \\ C & 0 & 0 \\ \bar{D} & 0 & W^T W \end{bmatrix} \begin{bmatrix} \Delta \boldsymbol{v} \\ \Delta \boldsymbol{\gamma} \\ \Delta \bar{\boldsymbol{\lambda}} \end{bmatrix} = - \begin{bmatrix} \boldsymbol{r}_D \\ \boldsymbol{r}_{P_\gamma} \\ \boldsymbol{r}_{P_\lambda} + W^T(\boldsymbol{z} \diamond \boldsymbol{r}_z) \end{bmatrix} \tag{10}$$

where

$$\boldsymbol{r}_z := \boldsymbol{z} \circ \boldsymbol{z} \qquad \text{for the prediction step} \tag{11}$$

$$\boldsymbol{r}_z := \boldsymbol{z} \circ \boldsymbol{z} - \sigma\mu\boldsymbol{e} + (W^{-T}\Delta \boldsymbol{y}_p \circ W\Delta\boldsymbol{\lambda}_p) \qquad \text{for the correction step} \tag{12}$$

As the reader may notice, the dimension of the system is now different from the proposed system of Eq. 5. This is achieved by the substitution $\Delta \boldsymbol{y} = D\Delta \boldsymbol{v} + \boldsymbol{r}_{P_\lambda}$.

With this framework, it is possible to formulate the whole iterative process, including the predictor-corrector scheme anticipated in the previous paragraphs. The *prediction* phase requires a temporary update of the approximate solution $[\boldsymbol{v}_p, \boldsymbol{\gamma}_p, \boldsymbol{\lambda}_p, \boldsymbol{y}_p]$. The tentative direction is provided by Eq. 10 together with Eq. 11 while the step length, as it will be for the correction step, is given so that it fulfills the indications given in Table 1. A new *predicted* solution is thus acquired and can be used to evaluate the *complementarity gap*, expressed in general as $\mu = \boldsymbol{y}^T\boldsymbol{\lambda}/n_\lambda$, and to update the matrix W^TW of Eq. 10.

The remaining unknown—the centering parameter σ—is computed in order to keep the *centrality* of the iterations path and, as often [7], is set empirically to $\sigma = (\mu_p/\mu)^3$. This additional parameter, with a negligible additional computational cost, greatly improve the convergence speed, making sure that the iteration steps may not come too close to the feasibility border and, in practice, assuring longer steps for the following iterations. In this sense, it makes the iterations follow a *central* path, equally far from all the unfeasible contour.

With these preliminary information, the final *corrected* step can then be performed in a more effective way. The system of equations for the correction step is quite similar to the previous one except very few exceptions, namely Eq. 11 is substituted by Eq. 12 and the bottom-right part of the system matrix of Eq. 10 is updated with the new variables status. The final solution can be now computed by accepting the corrected direction and by evaluating the step length α according, once again, with Table 1.

4 Results

The interior-point algorithm here proposed has been implemented in an actual working code, written in C++ language, and freely available to the public according to BSD-3 copyright license. The Chrono multibody library [8] is then taken into play to provide the multibody simulation framework required to feed the IP solver with meaningful dynamic data.

The underlying linear solver for the Chrono-Custom IP solution is MUMPS 5.1.2 [1], linked to Intel MKL BLAS library.

The proposed tests are run until the requested convergence conditions are reached.

$$\frac{\|\boldsymbol{r}_{p_\gamma}\|}{n_\gamma} < 1\times10^{-10}, \quad \frac{\|\boldsymbol{r}_{p_\lambda}\|}{n_\lambda} < 1\times10^{-10}, \quad \frac{\|\boldsymbol{r}_d\|}{n_v} < 1\times10^{-10}, \quad \mu < 1\times10^{-9}$$

The first test is proposed to show the performance of the algorithm involved in a rigid-flexible bodies simulation. The tetrahedral mesh is bent thanks to rigid bodies, as can be seen in Fig. 1a. The top cylinder is pushing against the beam, while the other two act as support. As can be seen in Fig. 1b, the performance is always slightly better compared to the CVXOPT counterpart. The convergence is here achieved for both the solvers, during the whole simulation. Pushing further the robustness of the algorithms, a more demanding benchmark is proposed. The test involves a bundle of IGA beams [2] wrapped around a

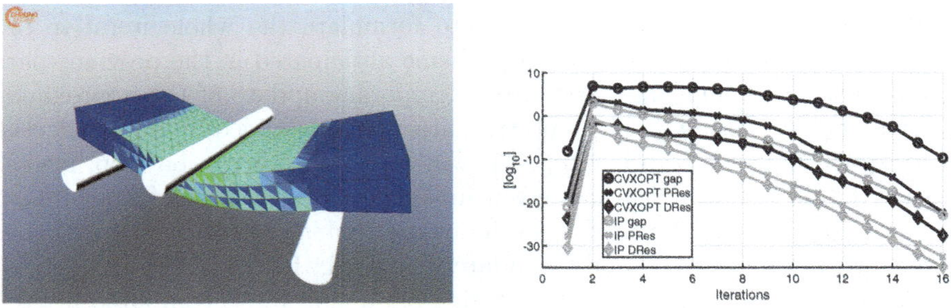

(a) Simulation setup. (b) Convergence test: CVXOPT vs Custom IP.

Fig. 1. Three-point bending test.

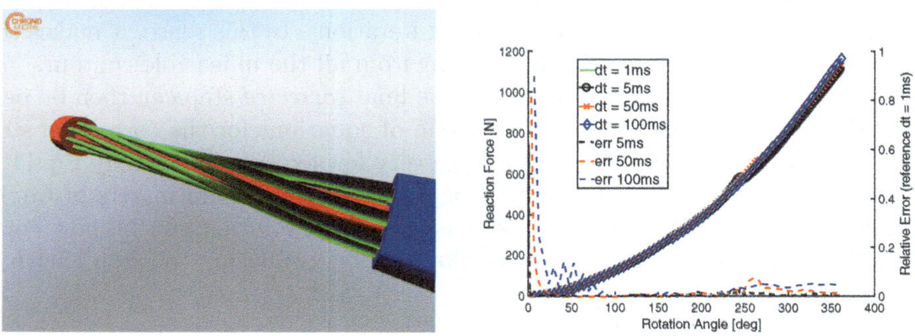

(a) Simulation setup. (b) Reaction on fixed flange

Fig. 2. Bundle of IGA beams.

central rigid cylinder (Fig. 2a). Not only the magnitude of the forces exchanged between beams and the pole are increasing, but also the overall number of reactions is augmented as far as the test proceeds. The beams, each of which made by 56 IGA elements, collide with the inner cylinder at each node point as soon as the torque applied by a rotating flange pushes them against it.

The same test has been run at different timesteps (namely 1×10^{-3} s, 5×10^{-3} s, 5×10^{-2} s and 5×10^{-1} s) showing a great robustness, showing meaningful results even at 5×10^{-1} s. The reaction forces measured at the fixed flange are reported in order to assess the stability (Fig. 2b). The simulation run at 1×10^{-3} s is taken as base reference and the reaction force, evaluated at the fixed flange (red, in the figure), is compared to other runs. As it can be seen in Fig. 2b, the error is acceptable even at the longest timestep. It has to be noticed that the error at the initial angle is high due to very low forces values, that make the relative error increase. Using smooth contacts, instead, would force the user to impose much shorter timesteps, thus heavily affecting the performance: at 1×10^{-3} s the simulation is not able to perform a full revolution.

5 Conclusion

The robustness of the IP method presented here guarantees effective results for a wide range of non-smooth multibody problems involving contacts. The frictional problem with arbitrary friction coefficient is here considered as an additional feature that turns out to be quite important for real-case scenario, in which constraints fulfill Coulomb friction models.

References

1. Amestoy, P.R., Duff, I.S., Koster, J., L'Excellent, J.-Y.: A fully asynchronous multifrontal solver using distributed dynamic scheduling. SIAM J. Matrix Anal. Appl. **23**(1), 15–41 (2001)
2. Benatti, S., Tasora, A., Mangoni, D.: An IGA-based three dimensional Cosserat rod for multibody dynamics. In: GAMM 2019 Book of Abstracts. International Association of Applied Mathematics and Mechanics (2019)
3. Kleinert, J., Simeon, B., Obermayr, M.: An inexact interior point method for the large-scale simulation of granular material. Comput. Methods Appl. Mech. Eng. **278**, 567–598 (2014)
4. Kučera, R., Machalovà, J., Netuka, H., Ženčák, P.: An interior-point algorithm for the minimization arising from 3D contact problems with friction. Optim. Methods Softw. **28**(6), 1195–1217 (2013)
5. Mangoni, D.: An interior-point method for non-smooth multibody system dynamics. Ph.D. thesis, University of Parma (2018)
6. Mangoni, D., Tasora, A., Garziera, R.: A primaldual predictorcorrector interior point method for non-smooth contact dynamics. Comput. Methods Appl. Mech. Eng. **330**, 351–367 (2018)
7. Nocedal, J., Wright, S.: Numerical Optimization. Springer Series in Operations Research and Financial Engineering. Springer, New York (2006)
8. Tasora, A., et al.: Chrono: an open source multi-physics dynamics engine, pp. 19–49. Springer, Cham (2016)
9. Tasora, A., Anitescu, M.: A matrix-free cone complementarity approach for solving large-scale, nonsmooth, rigid body dynamics. Comput. Methods Appl. Mech. Eng. **200**(5–8), 439–453 (2011)
10. Vandenberghe, L.: The CVXOPT linear and quadratic cone program solvers, March 2010

A Fast Explicit Integrator for Numerical Simulation of Multibody System Dynamics

Hui Ren(✉) and Ping Zhou

School of Astronautics, Harbin Institute of Technology, Harbin, China
renhui@hit.edu.cn, 18b918037@stu.hit.edu.cn

Abstract. A computationally efficient explicit integrator is proposed to solve the differential-algebraic equations (DAEs) in multibody system dynamics. Algebraic constraint equations in the DAEs are regularized by a simple stabilization method, yielding a set of first order ordinary differential equations (ODEs), whose large eigenvalues are located at the negative real axis. Those ODEs have specific stiff characters, and are integrated by a class of explicit integrators (the Runge-Kutta-Chebyshev family of ODE integrators) with large stability zones on the negative real axis, so as to achieve large step-sizes at the same requirement of accuracy. The integrator adopted in this work is of fourth order, verified by practical example, and compared to several popular integrators. The high efficiency of the explicit integrator renders it a good option for practical simulations of the dynamics of constraint mechanical systems.

1 Introduction

Faster integrators, that are used to solve differential-algebraic equations (DAEs) in constrained multibody systems, are always being needed in industrial practices. New integrators either with large step-sizes or minimum computations in each step of integration, are developed to solve equations faster under the required error tolerance. For stiff equations, the step-sizes in implicit integrators are usually limited by accuracy, while explicit integrators are limited by the stability. Generally speaking, to satisfy the same error criteria, the step-sizes in implicit integrators are larger than those in explicit ones, however more computations are required than that of explicit ones as a result of numerical iteration at each time step.

In general-purpose codes for multibody system dynamics, implicit integrators are still the main streams. The better stability property of implicit integrators makes them work well in simulations, since the equations of motions can be severely stiff in practical problems. The backward difference formula (BDF) family of methods was introduced by Gear [1] to solve DAEs, and was later applied to equations of motions of mechanical systems by Orlandea et al. [2]. The BDF methods are of variable orders and variable step-sizes, which give them unbelievable potentials for further improvements. The BDF methods were well studied by

A. Kecskeméthy and F. Geu Flores (Eds.): ECCOMAS 2019, COMPUTMETHODS 53, pp. 348–355, 2020.
https://doi.org/10.1007/978-3-030-23132-3_42

Petzold and her colleagues, and organized in their monograph [3], in which BDF methods are successfully applied to DAEs with index-3 as well as their equivalences of index-2 [4]. Generalized-α family of methods, including HHT integrator [5], generalized-α Index-3 integrator [6] and HHT/generalized-α Index-2 integrator [7], works also very well in practice problems especially for systems with a wide range of frequency oscillations; their easy implementations render them very popular in the academic literatures and software implementations. The implicit Runge-Kutta family of integrators [8] can also be implemented efficiently to solve the DAEs in constrained mechanical systems. In implicit integrators, analytical or numerical Jacobians are required in each time step of iterations, such that the constraint equations are perfectly satisfied. However, the thrashing or even non-convergence issues could arise when solving problems with severe discontinuities such as those in contact/impact problems.

Explicit integrators have been developed as a compensation to the applications of implicit integrators. Since no numerical iterations are required, explicit integrators would be more efficient than implicit ones in very large scale problems; they can also handle the discontinuity issues effectively. However, the constraint equations are not strictly satisfied in explicit integrators, and stabilization techniques can be adopted to enforce the constraints. Gear [10] had realized the great potential of the explicit integrators, and suggested to use penalty methods to regularize the DAEs such that they are suitable to be solved by explicit ordinary differential equation (ODE) integrators. New explicit ODE integrators have been developed to solve ODEs with specific stiff types. The Runge-Kutta-Chebyshev (RKC) integrators for ODEs were first proposed by Verwer et al. [11] in 1990. Medovikov developed a code DUMKA3 based on their approach [9]. Abdulle and Medovikov improved this method, and developed the codes for second order [12] and fourth order formulas [13]. New approaches on regularization of the equations of motions of the constrained mechanical systems are also developed by Braun and Goldfarb [16] recently, such that the equations of motions can be simulated by an explicit integrator.

In this work, a computationally efficient explicit integrator is proposed to simulate the dynamics of multibody systems. A constraint violation stabilization technique is introduced to regularize the DAEs of constrained mechanical systems, and the RKC integrator is adopted to simulate the dynamics. In Sect. 2, the details of the explicit integrator are described. In Sect. 3, the proposed method is illustrated with practical example, and compared to several popular methods, i.e., generalized-α, BDF-I3 and BDF-I2, showing the efficiency of the explicit integrator. Furthermore, the pros and cons of the current explicit integrator are discussed in Sect. 4.

2 Description of the Integrator

The standard form of an index-3 differential-algebraic equation (DAE) for multibody system dynamics is

$$\begin{aligned} \mathbf{M}(t,\boldsymbol{q})\ddot{\boldsymbol{q}} + \boldsymbol{\Phi}_{\boldsymbol{q}}^T(t,\boldsymbol{q})\boldsymbol{\lambda} = \boldsymbol{f}(t,\boldsymbol{q},\dot{\boldsymbol{q}}) \\ \boldsymbol{\Phi}(t,\boldsymbol{q}) = \mathbf{0} \end{aligned} \tag{1}$$

and Gear et al. [4] proved that this index-3 DAE is equivalent to an index-2 DAE with much better numerical properties

$$\begin{aligned} \mathbf{M}(t,\boldsymbol{q})\dot{\boldsymbol{v}} + \boldsymbol{\Phi}_{\boldsymbol{q}}^T(t,\boldsymbol{q})\boldsymbol{\lambda} = \boldsymbol{f}(t,\boldsymbol{q},\boldsymbol{v}) \\ \boldsymbol{\Phi}_{\boldsymbol{q}}(t,\boldsymbol{q})\boldsymbol{v} + \boldsymbol{\Phi}_t(t,\boldsymbol{q}) = \mathbf{0} \\ \dot{\boldsymbol{q}} + \boldsymbol{\Phi}_{\boldsymbol{q}}^T(t,\boldsymbol{q})\boldsymbol{\mu} = \boldsymbol{v} \\ \boldsymbol{\Phi}(t,\boldsymbol{q}) = \mathbf{0} \end{aligned} \tag{2}$$

where the velocity constraints are enforced to be satisfied. However, the third equation of Eq. (2) is not affine invariant, such that the results of simulations could be different for different unit systems of generalized coordinates q's. In this work, the third equation of Eq. (2) will be replaced by its affine invariant form [9]

$$\mathbf{M}(t,\boldsymbol{q})\dot{\boldsymbol{q}} + \boldsymbol{\Phi}_{\boldsymbol{q}}^T(t,\boldsymbol{q})\boldsymbol{\mu} = \mathbf{M}(t,\boldsymbol{q})\boldsymbol{v} \tag{3}$$

to achieve better numerical performance in practice. Following exactly the same procedure in [4], one can prove that the resultant equations are still equivalent to Eq. (1) after this replacement.

In order to solve the equations by an explicit integrator, a simple regularization technique is necessary to transform the constraint equations into ODEs. For a constraint equation as follows

$$f(t,\boldsymbol{x}) = 0 \tag{4}$$

the following ODE

$$\dot{f} + \omega f = f_{\boldsymbol{x}}(t,\boldsymbol{x})\dot{\boldsymbol{x}} + f_t(t,\boldsymbol{x}) + \omega f(t,\boldsymbol{x}) = 0 \tag{5}$$

will be a good stabilization ordinary differential equation. If the positive stabilization parameter ω is large enough, any drifting violation of the constraint (4) will be pulled back to the constraint manifold such that no violation will actually happen during the numerical integration. In this approach, the velocity constraint equation, i.e., the second equation in Eq. (2), becomes

$$\boldsymbol{\Phi}_{\boldsymbol{q}}\dot{\boldsymbol{v}} = \boldsymbol{\gamma} - \omega\left(\boldsymbol{\Phi}_{\boldsymbol{q}}\boldsymbol{v} + \boldsymbol{\Phi}_t\right) \tag{6}$$

where

$$\boldsymbol{\gamma}(t,\boldsymbol{q},\boldsymbol{v}) = -\left(\boldsymbol{\Phi}_{\boldsymbol{q}}\boldsymbol{v}\right)_{\boldsymbol{q}}\boldsymbol{v} - 2\boldsymbol{\Phi}_{\boldsymbol{q}t}\boldsymbol{v} - \boldsymbol{\Phi}_{tt} \tag{7}$$

and the position constraint equation, i.e., the fourth equation in Eq. (2), becomes

$$\boldsymbol{\Phi}_{\boldsymbol{q}}\dot{\boldsymbol{q}} = -\boldsymbol{\Phi}_t - \omega\boldsymbol{\Phi} \tag{8}$$

Moreover, in engineering practices, nonholonomic constraint equations, as well as constraint equations in the control equations, can be transformed similarly. As a result, the governing equations in Eq. (2) become

$$\begin{pmatrix} \mathbf{M}, & \boldsymbol{\Phi}_q^T \\ \boldsymbol{\Phi}_q, & \mathbf{0} \end{pmatrix} \begin{pmatrix} \dot{\boldsymbol{q}}, & \dot{\boldsymbol{v}} \\ \boldsymbol{\mu}, & \boldsymbol{\lambda} \end{pmatrix} = \begin{pmatrix} \mathbf{M}\boldsymbol{v}, & \boldsymbol{f} \\ -\boldsymbol{\Phi}_t - \omega\boldsymbol{\Phi}, & \boldsymbol{\gamma} - \omega\left(\boldsymbol{\Phi}_q\boldsymbol{v} + \boldsymbol{\Phi}_t\right) \end{pmatrix} \tag{9}$$

The mass matrix $\mathbf{M}$ and constraint matrix $\boldsymbol{\Phi}_q$ are usually sparse, and it is easier to solve Eq. (9) directly to yield the ODEs

$$\dot{\boldsymbol{v}} = \hat{\boldsymbol{f}}\left(t, \boldsymbol{q}, \boldsymbol{v}, \omega\right), \quad \dot{\boldsymbol{q}} = \hat{\boldsymbol{g}}\left(t, \boldsymbol{q}, \boldsymbol{v}, \omega\right) \tag{10}$$

In practice, Eq. (9) is preferred, because when quaternions are adopted to describe the rotations, the mass matrix could be singular in Lagrange equations. In order to eliminate the numerical drifting errors in the violation of constraints, it is better to pick large values of ω such as $\omega = 10000$ or 100000, but the large stabilization parameter ω will introduce an large-magnitude eigenvalue on the negative real axis, and usually small step-sizes are required in an explicit ODE integrator to keep $-h\omega$ within the stability zone. Fortunately, the recent developed RKC integrators [12, 13] have a good stability feature which can eliminate this small step-size requirement. A schematic of the stability regions of a RKC integrator is shown in Fig. 1, where the stable domain has a long tail on the negative real axis in the complex plane. The step-sizes in an explicit ODE integrator are selected adaptively in such a way that all $h\lambda$ are in its stable domain, where λ is the eigenvalue of system. Moreover, the bound of stability domain can be adjusted by its number of stage s arbitrarily and adaptively in the negative real axis. In this work, the fourth order RKC ODE integrator [13] is suggested to solve Eq. (9), whose stability bound on the negative real axis $-l_s$ is asymptotically to be $-0.3550 \times s^2$. The stability bounds of this RKC integrator for different stage number s are shown in Table 1. The prominent stability feature of this RKC explicit ODE integrator allows to select the most efficient stage of the formulas and the most efficient step-size at each time-step. Furthermore, in order to improve the computational efficiency of the fourth order RKC

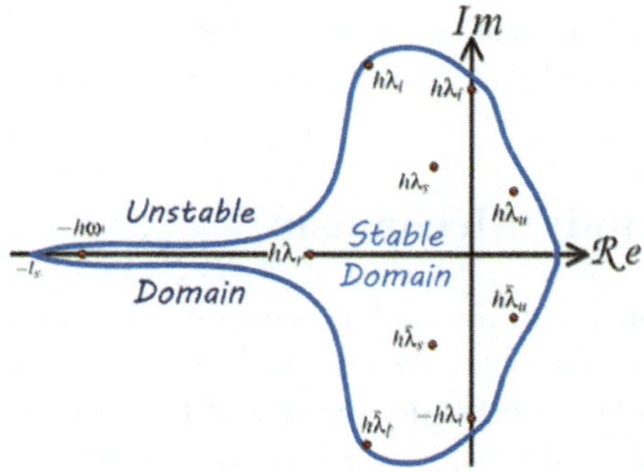

Fig. 1. A schematic of the stability regions of RKC integrators.

Table 1. Stability regions of the 4th order RKC integrator in the negative real axis

s	5	10	20	50	100	250	500	750
l_s	6.00	32.45	138.36	879.89	3538.13	22184.5	88747.0	199685
l_s/s^2	0.2399	0.3245	0.3459	0.3520	0.3538	0.3550	0.3550	0.3550

[a] The datas are calculated in [13]

ODE integrator on the integration of Eq. (9), the following modifications of the algorithms in [13] are applied:

1. In multibody systems, the mass matrix $\mathbf{M}(t, \boldsymbol{q})$ and the constraint matrix $\boldsymbol{\Phi}_q(t, \boldsymbol{q})$ usually vary slow with time, and during an integration step from t_n to $t_{n+1} = t_n + h$, it is accurate enough to replace all the time variant matrices $\mathbf{M}(t, \boldsymbol{q})$ and $\boldsymbol{\Phi}_q(t, \boldsymbol{q})$ by the constant matrices

$$\bar{\mathbf{M}} = \mathbf{M}\left(t_n + \frac{h}{2}, \boldsymbol{q}^{(0)}_{n+\frac{1}{2}}\right), \quad \bar{\boldsymbol{\Phi}}_q = \boldsymbol{\Phi}_q\left(t_n + \frac{h}{2}, \boldsymbol{q}^{(0)}_{n+\frac{1}{2}}\right) \tag{11}$$

   where $\boldsymbol{q}^{(0)}_{n+\frac{1}{2}} = \boldsymbol{q}_n + \frac{h}{2}v_n + \frac{h^2}{8}\dot{v}_n$ is an estimation of $\boldsymbol{q}$ at time $t_n + \frac{h}{2}$. More complicated predictors, such as those in BDF integrators or the generalized-α integrator, can also be adopted here for the prediction of $\boldsymbol{q}^{(0)}_{n+\frac{1}{2}}$;
2. It is safe to set the largest eigenvalue of the system to be $-\omega$. In practice, one needs only to make sure that $-h_{max}\omega$ is in the stable domain, and the stabilization of the simulation will be guaranteed. Hence, although the step size adjustment is adaptive during the simulation, the maximum step size should be prescribed beforehand;
3. The output of the acceleration $\dot{v}$ and Lagrange multiplier $\boldsymbol{\lambda}$ are calculated by

$$\begin{pmatrix} \mathbf{M} & \boldsymbol{\Phi}_q^T \\ \boldsymbol{\Phi}_q & \mathbf{0} \end{pmatrix} \begin{pmatrix} \dot{v} \\ \boldsymbol{\lambda} \end{pmatrix} = \begin{pmatrix} \boldsymbol{f} \\ \gamma \end{pmatrix} \tag{12}$$

   instead of Eq. (9) to get rid of the influences of ω;
4. For each stage s, the coefficients and error estimation parameters are provided in [13], so the local truncation error can be easily estimated. The adaptive step-sizes are adjusted by the error estimations, as described in [18].

3 The Seven Body Mechanism

The seven body mechanism, a standard benchmark example [19] in multibody system dynamics, is verified here by current integrator. The details of geometrical, force parameters and inertia parameters of the model are described in [9].

The equations of motions, as well as constraint equations, are derived by Lagrange equations in [9]. The DAEs are solved by the Matlab versions of three approaches with relative error threshold 10^{-4}: the current method

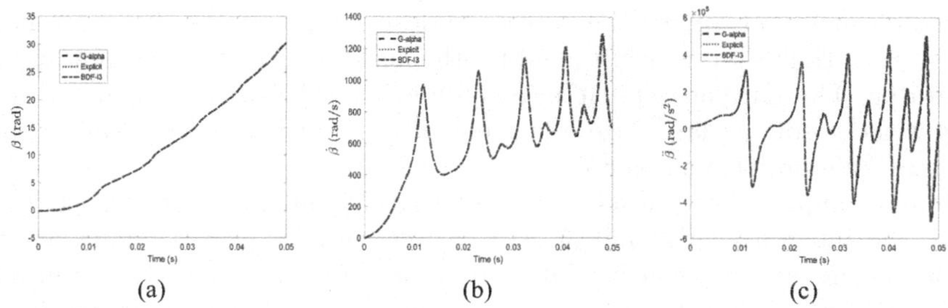

Fig. 2. Comparisons of (a) the rotation angle, (b) angular velocity, and (c) angular acceleration in the Seven Body Mechanism, calculated by current explicit integrator, the generalized-α integrator, and the BDF integrator.

(with $\omega = 10000$), the generalized-α method (with $\rho = 0.5$), and backward difference formulation (BDF) methods. The results agree well as shown in Fig. 2. The steps and times in simulations are compared in Table 2, where the current approach is faster than the generalized-α method. In this example, the analytical Jacobian can be easily derived, which is a sparse matrix, and the LU decomposition of a Jacobian can be reused in multiple steps in the generalized-α method and the BDF methods. It is also necessary to point out, all the constraints in this example are geometric constraints, and the violations of constraints are within the magnitude of 10^{-7} in the current explicit integrator approach, which is accurate enough for applications. In this problem, the computation in each time step in the explicit integrator is more than those in the implicit integrators, because in the explicit integrator approach, 5 evaluations of the equations and 1 LU decomposition is required in each time-step, while in the implicit integrator approaches, an LU decomposition of the Jacobian can be reusable in multi-time-steps, and the Newton iteration usually converges in 2 or 3 iterations. This is not the case in general, because evaluations of Jacobians and their LU decompositions are computational bottle necks for implicit integrators in large scale problems.

Table 2. Steps and Time comparison for the Seven Body Mechanism

Integrator	Generalized-α	Explicit	BDF-I3	BDF-I2
Steps	1240	286	462	267
Time (s)	0.869	0.454	0.317	0.228

4 Discussions

In recursive formulations [20], the mass matrix and the constraint matrix are not provided explicitly as the expressions in Eq. (9), but derived in a recursive procedure such that solving the full-sized linear equation in Eq. (9) is not necessary at each time step, which is good for the performance of the current explicit integrator. The current explicit integrator can resolve the constraint drifting issue without solving large-size linear equations, which will significantly speed up large scale problem simulations.

In most implicit integrators, if the violations of constraints are not extremely small, the integrators will fail [3,9]. However, this integrator failure will not happen in the current explicit ODE integrator, and it is not worth to put a too large ω parameter to enforce the violations of constraints extremely small.

As is well known, Cartesian coordinate formulations with a full-sized sparse equations do not work efficiently in high speed simulations of constrained mechanical systems [17] such as real-time simulation of robots [21]. The yielding equations of motions in recursive formulations are usually nonstiff, and analytical Jacobian matrices can not be easily derived, such that the current explicit integrator could be a good option in real-time simulations of the dynamics in those constrained mechanical systems.

5 Conclusions

The current explicit integrator works faster than the generalized-α integrator in the simulations of nonstiff differential algebraic equations of multibody dynamics systems. Any type of constraint equations can be easily regularized by the current stabilization method, and the yielding system equations can be simulated by the RKC explicit integrators. Violations of constraint equations are small for practical applications, but might not be as small as those in implicit integrators. Although only holonomic constraints are fully discussed in this work, the same technique can be directly applied to nonholonomic constraints, as well as constraints from control or other subjects.

Acknowledgements. This work is supported by the Nature and Science Foundations of China (NSFC) under grant number 11772101.

References

1. Gear, C.W.: Simultaneous numerical solution of differential algebraic equations. IEEE Trans. Circ. Theory **CT–18**(1), 89–95 (1971)
2. Orlandea, N., Chace, M.A., Calahan, D.A.: A sparsity-oriented approach to the design of mechanical systems, Part I and II, Paper No. 76-DET-19 and 76-DET-20. Presented at ASME Mechanical Conferences, Montreal, Quebec, Canada, October 1976
3. Brenan, K.E., Campbell, S.L., Petzold, L.R.: Numerical Solution of Initial Value Problems in Differential Algebraic Equations, 2nd edn. SIAM Classics in Applied Mathematics, vol. 14. SIAM, Philadelphia (1997)
4. Gear, C.W., Gupta, G.A., Leimkuhler, B.: Automatic integration of Euler-Lagrange equations with constraints. J. Comput. Appl. Math. **12–13**, 77–90 (1985)

5. Negrut, D., Rampalli, R., Ottarsson, G., Sajdak, A.: On an implementation of the Hilber-Hughes-Taylor method in the context of index 3 differential-algebraic equations of multibody dynamics. J. Comput. Nonlinear Dyn. **2**(1), 73–85 (2007)
6. Arnold, M., Brüls, O.: Convergence of the generalized-alpha scheme for constrained mechanical systems. Multibody Syst. Dyn. **18**(2), 185–202 (2007)
7. Negrut, D., Jay, L.O., Khude, N.: A discussion of low-order numerical integration formulas for rigid and flexible multibody dynamics. J. Comput. Nonlinear Dyn. **4**(2), 149–160 (2008)
8. Hairer, E., Roche, L., Lubich, C.: The Numerical Solution of Differential-Algebraic Systems by Runge-Kutta Methods. Springer, Heidelberg (1989)
9. Hairer, E., Wanner, G.: Solving Ordinary Differential Equations II: Stiff and Differential-Algebraic Problems, 2nd edn. Springer, Heidelberg (1996)
10. Gear, C.W.: Towards explicit methods for differential algebraic equations. BIT **46**, 505–514 (2005)
11. Verwer, J.H., Hundsdorfer, W.H., Sommeijer, B.P.: Convergence properties of the Runge-Kutta-Chebyshev method. Numer. Math. **57**, 157–178 (1990)
12. Abdulle, A., Medovikov, A.A.: Second order Chebyshev methods based on orthogonal polynomials. Numer. Math. **90**, 1–18 (2001)
13. Abdulle, A.: Fourth order chebyshev methods with recurrence relation. SIAM J. Sci. Comput. **23**(6), 2041–2054 (2002)
14. Ostermeyer, G.P.: On Baumgarte stabilization for differential algebraic equations. In: Haug, E.J., Deyo, R.C. (eds.) Real-Time Integration Methods for Mechanical System Simulation, pp. 193–207 (1990)
15. Bauchau, O.A., Laulusa, A.: Review of contemporary approaches for constraint enforcement in multibody systems. J. Comput. Nonlinear Dyn. **3**(1), 011005 (2008)
16. Braun, D.J., Goldfarb, M.: Simulation of constrained mechanical systems Part I: an equation of motion, and Part II: explicit numerical integration. ASME J. Appl. Mech. **79**(4), 041017–041018 (2012)
17. Haug, E.J., Yen, J.: Generalized coordinate partitioning methods for numerical integration of differential-algebraic equations of dynamics. In: Haug, E.J., Deyo, R.C. (eds.) Real-Time Integration Methods for Mechanical System Simulation, pp. 97–114 (1990)
18. Hairer, E., Nørsett, S.P., Wanner, G.: Solving Ordinary Differential Equations I: Nonstiff Problems, pp. 166–167. Springer, Heidelberg (1987)
19. Schiehlen, W.: Multibody Systems Handbook, pp. 10–15. Springer, Heidelberg (1990)
20. Bae, D.S., Lee, J.K., Cho, H.J., Yae, H.: An explicit integration method for real time simulation of multibody vehicle models. Comput. Methods Appl. Mech. Eng. **187**(1–2), 337–350 (2000). https://doi.org/10.1016/s0045-7825(99)00138-3
21. Featherstone, R., Orin, D.: Robot dynamics: equations and algorithms. In: Proceedings of the 2000 IEEE International Conference on Robotics and Automation, San Francisco, CA, vol. 1, pp. 826–834 (2000). https://doi.org/10.1109/ROBOT.2000.844153

Optimization and Sensitivity Analysis

The Discrete Hamiltonian-Based Adjoint Method for Some Optimization Problems in Multibody Dynamics

Paweł Maciąg(✉), Paweł Malczyk, and Janusz Frączek

Division of Theory of Machines and Robots,
Institute of Aeronautics and Applied Mechanics,
Faculty of Power and Aeronautical Engineering,
Warsaw University of Technology,
Nowowiejska Str. 24, 00-665 Warsaw, Poland
{pmaciag,pmalczyk,jfraczek}@meil.pw.edu.pl

Abstract. The determination of various parameters or control input signals satisfying particular performance criteria is often addressed with optimization techniques where one aims at minimizing certain quantity, which may be implicitly dependent on the dynamic response of a system. Such an approach requires an efficient and reliable method of gradient calculation. The adjoint method is an effective procedure specifically designed for such calculations. This paper presents a discrete Hamiltonian–based adjoint method which allows one to find the gradient of the performance index in multibody systems' optimization. Hamilton's equations of motion are discretized by means of trapezoidal rule and incorporated into a discrete system of adjoint equations. Explicit formula for the gradient of the cost functional is derived and exploited in an exemplary optimal control problem.

1 Introduction

Optimization methods are commonly exploited in the field of multibody dynamics in various aspects, e.g. in the early design stage or in the problems associated with an inverse dynamics task, which is especially important in the robotics field. A broad class of engineering problems involves the determination of input signals which must be supplied to the system [2]. Optimal control methods associated with a direct optimization approach tend to generate a large number of design variables which have to be treated by optimization procedure [1]. The adjoint method allows the computation of the gradient of the performance measure and exhibits a significantly lower computational overhead when compared to other gradient calculation techniques, especially when the number of design variables (or input signals) is large.

In general, the adjoint method requires a solution of a system of differential–algebraic equations (DAEs) backwards in time starting from a prescribed boundary condition in order to obtain a set of adjoint variables which are necessary

A. Kecskeméthy and F. Geu Flores (Eds.): ECCOMAS 2019, COMPUTMETHODS 53, pp. 359–366, 2020.
https://doi.org/10.1007/978-3-030-23132-3_43

to determine the gradient [4,8]. Such an approach is based on continuous cost functional as well as a time integration of equations of motion (EOM). On the other hand, one can derive the discrete system of adjoint equations based on the discretized cost functional and EOM [3,7]. Therefore, the mathematical models of both dynamic and adjoint systems display certain similar features, which can be exploited to improve numerical efficiency.

On the other hand, the performance of the optimization algorithm may strongly depend on robustness of the formulation of EOM. Herein, Hamilton's canonical equations of motion are employed to evaluate the dynamic response of a multibody system (MBS). This method possesses many appealing numerical features [5], such as reduced differential index of the resultant DAEs, that ought to be exploited in the development of an efficient and reliable adjoint-based method.

This paper presents an introductory work that demonstrates an effort to develop an adjoint method for the gradient calculation when a multibody system is modeled by using constrained Hamilton's canonical equations. The primary importance of the paper is to look carefully at the advantages such a connection may provide in the context of efficiency and stability of calculations.

2 Discrete Hamilton's Equations of Motion

The starting point for the analysis of MBS is a set of Hamilton's equations of motion, where the system Lagrangian has been modified in order to impose the constraint equations directly at the velocity level:

$$\mathbf{p}^* = \mathbf{M}(\mathbf{q})\,\dot{\mathbf{q}} + \mathbf{\Phi}_{\mathbf{q}}^T \boldsymbol{\sigma}, \tag{1a}$$

$$\dot{\mathbf{p}}^* = \mathbf{Q}(\mathbf{q}, \dot{\mathbf{q}}, \mathbf{u}) + \dot{\mathbf{\Phi}}_{\mathbf{q}}^T \boldsymbol{\sigma}, \tag{1b}$$

$$\dot{\mathbf{\Phi}}(\mathbf{q}, \dot{\mathbf{q}}) = \mathbf{\Phi}_{\mathbf{q}}\,\dot{\mathbf{q}} = \mathbf{0}. \tag{1c}$$

Here, $\mathbf{q} \in \mathcal{R}^n$ is a vector of generalized coordinates, $\mathbf{\Phi} \in \mathcal{R}^m$ indicates a vector of nonlinear holonomic constraints imposed on the system, $\mathbf{\Phi}_{\mathbf{q}} \in \mathcal{R}^{m \times n}$ denotes a constraint Jacobian matrix, $\mathbf{Q} \in \mathcal{R}^n$ indicates a vector of external forces, $\mathbf{u} \in \mathcal{R}^k$ describes input control signals, $\mathbf{M} \in \mathcal{R}^{n \times n}$ is a mass matrix, and the quantity $\boldsymbol{\sigma} \in \mathcal{R}^m$ denotes constraint impulsive forces. The vector of generalized momenta is defined as $\mathbf{p} = \mathbf{M}\dot{\mathbf{q}}$, whereas Eq. (1) is based on augmented momenta $\mathbf{p}^* \in \mathcal{R}^n$ which incorporate impulsive forces distributed along constrained directions indicated by the Jacobian matrix (see Eq. (1a)).

Equation (1) constitutes a set of $2n + m$ nonlinear DAEs which will be discretized in order to obtain unknown variables, i.e. $\mathbf{p}_i^*, \mathbf{q}_i, \boldsymbol{\sigma}_i$ at all time instances $i = \{1, 2, \cdots N\}$. The trapezoidal method, which is known to be an A-stable integrator, can be employed to tackle a large class of problems arising in MBS dynamics. In the cases of highly stiff problems, this property may be to weak, however, it has been observed to reliably address the solution of DAEs for the systems presented herein. The trapezoidal rule reads as:

$$\begin{aligned} \dot{\mathbf{q}}_i - \frac{2}{\Delta t}\mathbf{q}_i + \hat{\dot{\mathbf{q}}}_{i-1} = \mathbf{0}, \quad \hat{\dot{\mathbf{q}}}_{i-1} = -\frac{2}{\Delta t}\mathbf{q}_{i-1} - \dot{\mathbf{q}}_{i-1}, \\ \dot{\mathbf{p}}_i^* - \frac{2}{\Delta t}\mathbf{p}_i^* + \hat{\dot{\mathbf{p}}}_{i-1} = \mathbf{0}, \quad \hat{\dot{\mathbf{p}}}_{i-1} = -\frac{2}{\Delta t}\mathbf{p}_{i-1}^* - \dot{\mathbf{p}}_{i-1}^*. \end{aligned} \tag{2}$$

This approximation converts the original DAE (1) into a set of nonlinear algebraic equations expressed at i^{th} time instance:

$$\begin{aligned} -\mathbf{p}_i^* + \frac{2}{\Delta t}\mathbf{M}_i\mathbf{q}_i + \mathbf{M}_i\hat{\dot{\mathbf{q}}}_{i-1} + \boldsymbol{\Phi}_{\mathbf{q}_i}^T\boldsymbol{\sigma}_i = \mathbf{0}, \\ -\frac{2}{\Delta t}\mathbf{p}_i^* - \hat{\dot{\mathbf{p}}}_{i-1} + \mathbf{Q}_i + \frac{2}{\Delta t}\left(\boldsymbol{\Phi}_{\mathbf{q}_i}\mathbf{q}_i\right)_{\mathbf{q}_i}^T\boldsymbol{\sigma}_i + \left(\boldsymbol{\Phi}_{\mathbf{q}_i}\hat{\dot{\mathbf{q}}}_{\mathbf{i-1}}\right)_{\mathbf{q}_i}^T\boldsymbol{\sigma}_i = \mathbf{0}, \\ \frac{2}{\Delta t}\boldsymbol{\Phi}_{\mathbf{q}_i}\mathbf{q}_i + \boldsymbol{\Phi}_{\mathbf{q}_i}\hat{\dot{\mathbf{q}}}_{i-1} = \mathbf{0}. \end{aligned} \tag{3}$$

A combined set of equations (2) and (3) may be expressed succinctly in the following form: $\mathbf{f}(\mathbf{y}_i, \mathbf{y}_{i-1}, \mathbf{u}_i) = \mathbf{f}_i = \mathbf{0}$. Given the initial condition $\mathbf{q}(0) = \mathbf{q}, \mathbf{p}^*(0) = \mathbf{p}_0^*$, Eq. (3) can be propagated forward in time with the aid of, e.g., Newton method. This approach requires the definition of a tangent matrix to the Eq. (3), i.e. $\frac{\partial \mathbf{f}_i}{\partial \mathbf{y}_i} = \mathbf{f}_{\mathbf{y}_i}^i$, where $\mathbf{y}_i = [\dot{\mathbf{q}}_i^T, (\dot{\mathbf{p}}^*)_i^T, \mathbf{q}_i^T, (\mathbf{p}^*)_i^T, \boldsymbol{\sigma}_i^T]^T$. An iterative scheme must be established at every time–step, which will yield an update to the next time instance of unknown variables $\mathbf{y}_{i+1}$:

$$[\mathbf{f}_{\mathbf{y}_i}^i]^j \Delta\mathbf{y}_{i+1}^{j+1} = -[\mathbf{f}_i]^j. \tag{4}$$

Thus, $\mathbf{y}_{i+1}$ may be successively updated as $\mathbf{y}_{i+1}^{j+1} = \mathbf{y}_{i+1}^j = \Delta\mathbf{y}_{i+1}^{j+1}$, where j refers to the iterator of Newton–Raphson procedure.

3 Discrete Hamiltonian-Based Adjoint Method

Let us consider an optimal control problem in which the aim is to find a sequence of control inputs $[u_0, u_1 \cdots u_N]$ that minimize the following performance measure:

$$J = \sum_{i=0}^{N-1} h_i(\mathbf{y}_i, \mathbf{u}_i) + S(\mathbf{y}_N), \tag{5}$$

where h_i is a value of discretized cost function at i^{th} time–step, and S denotes a terminal cost used to prescribe particular configuration of the state at the final time. Provided that Eq. (3) is fulfilled by all state and control variables, one can premultiply its LHS by a vector of arbitrary *adjoint variables* $\mathbf{w}_i \in \mathcal{R}^{4n+m}$ and add it to Eq. (5) without introducing any quantitative changes:

$$\hat{J} = h_0 + \mathbf{w}_0\mathbf{f}_0 + \sum_{i=1}^{N-1}\left(h_i(\mathbf{y}_i, \mathbf{u}_i) + \mathbf{w}_i^T\mathbf{f}_i\right) + \mathbf{w}_N^T\mathbf{f}_N + S(\mathbf{y}_N). \tag{6}$$

The variation of Eq. (6) with respect to state variables and control input signals at each time instant reads as:

$$\begin{aligned}\delta\hat{J} &= h^0_{\mathbf{y}_0}\delta\mathbf{y}_0 + h^0_{\mathbf{u}_0}\delta\mathbf{u}_0 + \left(\mathbf{w}_N^T\mathbf{f}_{\mathbf{y}_N}^{\,N} + S_{\mathbf{y}_N}\right)\delta\mathbf{y}_N + \mathbf{w}_N^T\mathbf{f}_{\mathbf{u}_N}^{\,N}\delta\mathbf{u}_N \\ &+ \sum_{i=1}^{N-1}\Big(h^i_{\mathbf{y}_i}\delta\mathbf{y}_i + h^i_{\mathbf{u}_i}\delta\mathbf{u}_i + \mathbf{w}_i^T\mathbf{f}_{\mathbf{y}_i}^{\,i}\delta\mathbf{y}_i + \mathbf{w}_i^T\mathbf{f}_{\mathbf{y}_{i-1}}^{\,i}\delta\mathbf{y}_{i-1} + \mathbf{w}_i^T\mathbf{f}_{\mathbf{u}_i}^{\,i}\delta\mathbf{u}_i\Big),\end{aligned} \tag{7}$$

where the following abbreviation is used: $\mathbf{a}^i_{\mathbf{b}_j} = \frac{\partial \mathbf{a}_i}{\partial \mathbf{b}_j}$ ($\mathbf{a}$, $\mathbf{b}$ denote a pair of arbitrary vectors). By performing a simple index shift, Eq. (7) can be rewritten in a more favorable form:

$$\begin{aligned}\delta\hat{J} &= \left(h^0_{\mathbf{y}_0} + \mathbf{w}_1^T\mathbf{f}_{\mathbf{y}0}^{\,i}\right)\delta\mathbf{y}_0 + h^0_{\mathbf{u}_0}\delta\mathbf{u}_0 + \left(\mathbf{w}_N^T\mathbf{f}_{\mathbf{y}_N}^{\,N} + S_{\mathbf{y}_N}\right)\delta\mathbf{y}_N + \mathbf{w}_N^T\mathbf{f}_{\mathbf{u}_N}^{\,N}\delta\mathbf{u}_N \\ &+ \sum_{i=1}^{N-1}\Big(\left(h^i_{\mathbf{y}_i} + \mathbf{w}_i^T\mathbf{f}_{\mathbf{y}_i}^{\,i} + \mathbf{w}_{i+1}^T\mathbf{f}_{\mathbf{y}_i}^{\,i+1}\right)\delta\mathbf{y}_i + \left(h^i_{\mathbf{u}_i} + \mathbf{w}_i^T\mathbf{f}_{\mathbf{u}_i}^{\,i}\right)\delta\mathbf{u}_i\Big).\end{aligned} \tag{8}$$

Equation (8) presents a relation between the variation of the performance measure and the variations of all variables expressed at each time instance. Since the adjoint variables are at this point arbitrary, one can equate the expressions staying next to the state variations to zero. This condition uniquely defines all adjoint variables and generates the following sets of algebraic equations:

$$\left(\mathbf{f}_{\mathbf{y}_i}^{\,i}\right)^T\mathbf{w}_i = -\left(h^i_{\mathbf{y}_i}\right)^T - \left(\mathbf{f}_{\mathbf{y}_i}^{\,i+1}\right)^T\mathbf{w}_{i+1}, \qquad \left(\mathbf{f}_{\mathbf{y}_N}^{\,N}\right)^T\mathbf{w}_N = -\left(\mathbf{S}_{\mathbf{y}_N}\right)^T. \tag{9}$$

Jacobi matrices that appear in Eq. (9) have the following explicit form:

$$\mathbf{f}_{\mathbf{y}_i}^{\,i} = \begin{bmatrix} \mathbf{I} & \mathbf{0} & -\gamma\mathbf{I} & \mathbf{0} & \mathbf{0} \\ \mathbf{0} & \mathbf{I} & \mathbf{0} & -\gamma\mathbf{I} & \mathbf{0} \\ \mathbf{0} & \mathbf{0} & [\mathbf{f}_3^{\,i}]_{\mathbf{q}_i} & -\mathbf{I} & \mathbf{\Phi}_{\mathbf{q}_i}^T \\ \mathbf{0} & \mathbf{0} & [\mathbf{f}_4^{\,i}]_{\mathbf{q}_i} & -\gamma\mathbf{I} & \dot{\mathbf{\Phi}}_{\mathbf{q}_i}^T \\ \mathbf{0} & \mathbf{0} & \dot{\mathbf{\Phi}}_{\mathbf{q}_i}^{\,i} & \mathbf{0} & \mathbf{0} \end{bmatrix}, \quad \mathbf{f}_{\mathbf{y}_i}^{\,i+1} = \begin{bmatrix} \mathbf{I} & \mathbf{0} & \gamma\mathbf{I} & \mathbf{0} & \mathbf{0} \\ \mathbf{0} & \mathbf{I} & \mathbf{0} & \gamma\mathbf{I} & \mathbf{0} \\ -\mathbf{M}_i & \mathbf{0} & -\gamma\mathbf{M}_i & \mathbf{0} & \mathbf{0} \\ [\mathbf{f}_4^{\,i+1}]_{\dot{\mathbf{q}}_i} & \mathbf{I} & [\mathbf{f}_4^{\,i+1}]_{\mathbf{q}_i} & \gamma\mathbf{I} & \mathbf{0} \\ -\mathbf{\Phi}_{\mathbf{q}_i} & \mathbf{0} & -\gamma\mathbf{\Phi}_{\mathbf{q}_i} & \mathbf{0} & \mathbf{0} \end{bmatrix}, \tag{10}$$

where $\mathbf{I} \in \mathcal{R}^{n\times n}$ is the identity matrix, and the abbreviated entries read as:

$$\begin{aligned} &[\mathbf{f}_3^{\,i}]_{\mathbf{q}_i} = \gamma\mathbf{M}_i + \left(\mathbf{\Phi}_{\mathbf{q}}^T\boldsymbol{\sigma}_i\right)_{\mathbf{q}_i}, && [\mathbf{f}_4^{\,i}]_{\mathbf{q}_i} = \mathbf{Q}^i_{\mathbf{q}_i} + \gamma\mathbf{Q}^i_{\dot{\mathbf{q}}_i} + \left[[\dot{\mathbf{\Phi}}_{\mathbf{q}_i}^{\,i}]^T\boldsymbol{\sigma}_i\right]_{\mathbf{q}_i}, \\ &[\mathbf{f}_4^{\,i+1}]_{\dot{\mathbf{q}}_i} = -\mathbf{S} - \mathbf{Q}^i_{\dot{\mathbf{q}}_i}, && [\mathbf{f}_4^{\,i+1}]_{\mathbf{q}_i} = -\gamma\left(\mathbf{S} + \mathbf{Q}^i_{\dot{\mathbf{q}}_i}\right), \\ &\dot{\mathbf{\Phi}}_{\mathbf{q}_i}^{\,i} = \gamma\left(\mathbf{\Phi}_{\mathbf{q}_i}\mathbf{q}_i\right)_{\mathbf{q}_i} + \left(\mathbf{\Phi}_{\mathbf{q}_i}\hat{\dot{\mathbf{q}}}_{i-1}\right)_{\mathbf{q}_i}, \ \gamma = \frac{2}{\Delta t} = \frac{\partial\dot{\mathbf{q}}_i}{\partial\mathbf{q}_i}, \\ &\mathbf{S} = \left[(\mathbf{\Phi}_{\mathbf{q}_{i+1}}\mathbf{q}_i)^T\boldsymbol{\sigma}_{i+1}\right]_{\mathbf{q}_i} = \left[(\mathbf{\Phi}_{\mathbf{q}_{i+1}}\dot{\mathbf{q}}_i)^T\boldsymbol{\sigma}_{i+1}\right]_{\dot{\mathbf{q}}_i}. \end{aligned}$$

The adjoint variables obtained by solving Eq. (9) significantly simplify the expression (8). The gradient of the performance measure (5) can thus be evaluated as follows:

$$\nabla J(\mathbf{u}_i) = \left(h^i_{\mathbf{u}_i} + \mathbf{w}_i^T\mathbf{f}_{\mathbf{u}_i}^{\,i}\right)\Delta t, \quad \nabla J(\mathbf{u}_N) = \mathbf{w}_N^T\mathbf{f}_{\mathbf{u}_N}^{\,N}\Delta t, \tag{11}$$

where $i = \{0, 1, \cdots N-1\}$. Let us note, that this procedure allows solving problems associated with optimal design as well. If we consider $\mathbf{u}$ as a vector of static design parameters instead of a control sequence explicitly dependent on time, the formula for a gradient becomes:

$$\nabla J(\mathbf{u}) = h^0_{\mathbf{u}_0} + \sum_{i=0}^{N-1} \left(h^i_{\mathbf{u}_i} + \mathbf{w}_i^T \mathbf{f}^{\,i}_{\mathbf{u}_i}\right) \Delta t. \tag{12}$$

4 Numerical Example

An optimal control problem of an underactuated system is presented as an illustrative example to demonstrate the performance of the developed method [9]. The goal is to identify a control input force $u = F(t)$ acting on a cart, which allows a swing–up maneuver of a triple pendulum attached to the cart. The layout of the system is presented in Fig. 1. The mass of a cart is $m_c = 0.8\,\mathrm{kg}$ and its moment of inertia about z axis (perpendicular to the plane of motion) is $J_c = 0.034\,\mathrm{kg\,m^2}$. Three identical bodies are attached sequentially to the cart via revolute joints. The mass of each link is equal to $m_p = 0.4\,\mathrm{kg}$ and moment of inertia about z axis is $J_p = 0.083\,\mathrm{kg\,m^2}$. Additionally, a linear frictional force is introduced into translational and one of revolute joints, where $c_{trans} = 0.5\,\frac{Ns}{m}$ and $c_{rot} = 0.2\,\frac{Nsm}{rad}$. Let us note, that in this simple case, friction is included in the EOM in the form of generalized force. The system is initially at rest in a lower equilibrium configuration, i.e. $\varphi_1 = \varphi_2 = \varphi_3 = -\frac{\pi}{2}$ rad.

The optimization procedure has been supplied with a constant initial sequence of input control variables equal to $F_0(t) = 10\,N$. Since the trajectory of the links is not provided in any form, the cost function can be reduced to a terminal term from Eq. (5), which reads as: $J = 0.5 \cdot \left[(\varphi_1 - \varphi_v)^2 + (\varphi_2 - \varphi_v)^2 + (\varphi_3 - \varphi_v)^2\right]_{t_f}$, where $\varphi_v = \frac{\pi}{2}$ rad can be associated with vertical orientation of each link. The steepest descent method is employed for the optimization, along with a backtracking algorithm as a line–search procedure defining an optimal step size [10]. The gradient obtained with the adjoint method is exploited as a search direction, i.e. the following control inputs are determined as $\mathbf{u}_{i+1} = -\alpha_{i+1} \nabla J(\mathbf{u}_i)$, where α_{i+1} is a step size obtained via the backtracking procedure. Figure 2 presents four snapshots of the system's motion (marked with red color) at consequent time–steps. For better readability, initial and final position of the bodies are added as a reference. The cart in the first frame (i.e. when $t = 1.26\,\mathrm{s}$) is at its right–most position from which it moves left–wards in the following frames.

Figure 3 depicts a history of iterations of the optimization procedure: blue circles show the value of a cost functional with respect to the number of iterations, whereas red x-shaped markers present first–order optimality $||\nabla J||_{\mathrm{inf}}$ on a logarithmic scale. The procedure keeps the cost close to zero after 300 iterations. Figure 4 shows the time–histories of position ($\mathbf{\Phi}(\mathbf{q})$), velocity ($\dot{\mathbf{\Phi}}(\mathbf{q}, \dot{\mathbf{q}})$), and acceleration–level algebraic constraints ($\ddot{\mathbf{\Phi}}(\mathbf{q}, \dot{\mathbf{q}}, \ddot{\mathbf{q}})$). The dynamic response is recorded for the starting iteration. Since velocity–level constraints are explicitly included in the EOM, $\dot{\mathbf{\Phi}}$ is very close to zero, i.e. up to 10^{-14}. Some constraint violation errors arise at the position level, however, their value generally lies between 10^{-5} and 10^{-4}, which does not introduce significant numerical errors. During the optimization process, the shape of the input force acting on the cart is determined. Figure 5 presents the snapshots of the identified force throughout the process: initial, two intermediate ones, and the one obtained at the final iteration.

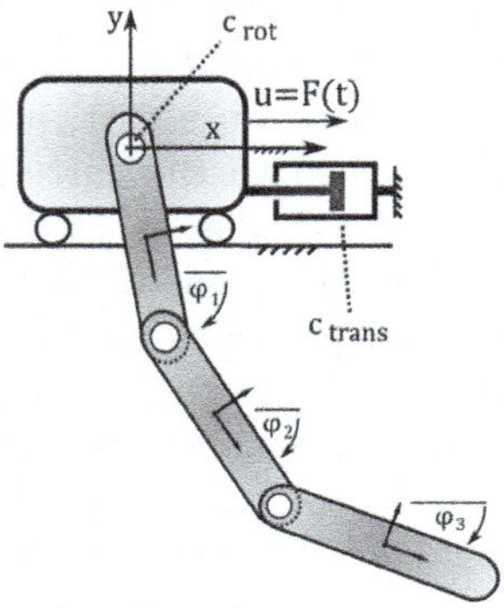

Fig. 1. Layout of the MBS

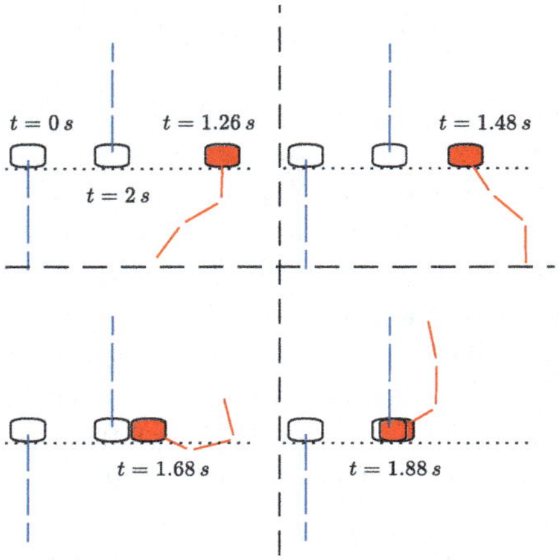

Fig. 2. Consequent snapshots of the system acting under the identified control input

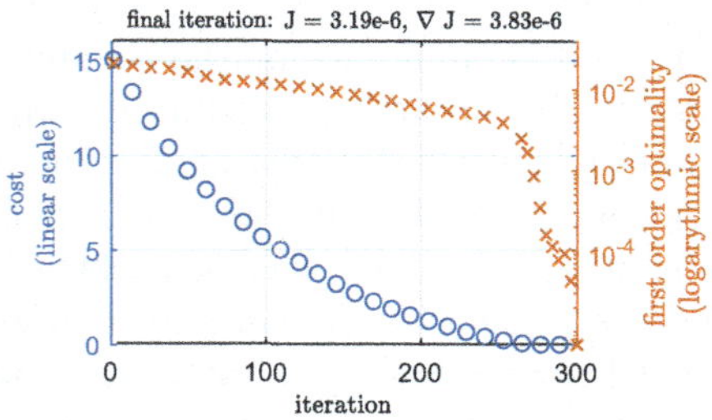

Fig. 3. The progress of the optimization process

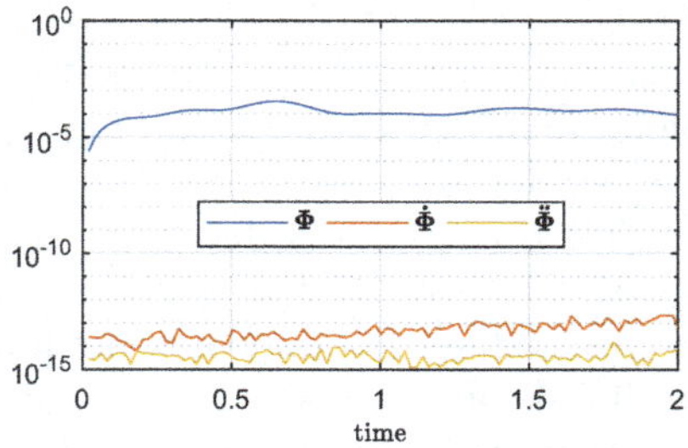

Fig. 4. Holonomic constraints

The accuracy of the adjoint method can be compared with a rather reliable, yet remarkably inefficient method of finite differencing. Figure 6 presents a time variant functions of the gradient calculated with both complex finite differences and adjoint method. Both plots are recorded for the initial point of the optimization procedure. The maximum relative error between the gradients, which occurs

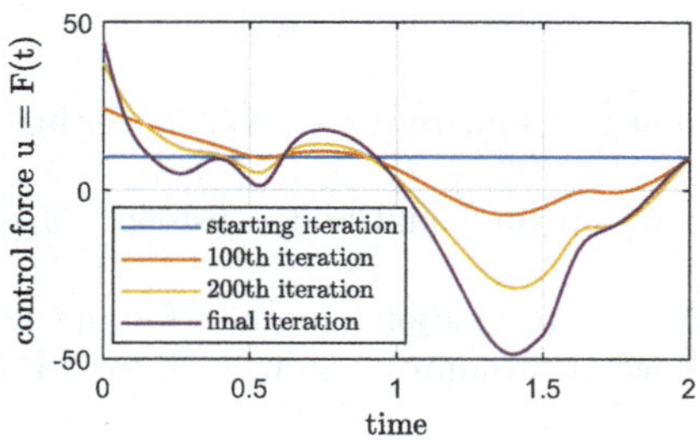

Fig. 5. Identified force acting on the cart recorded at following iterations

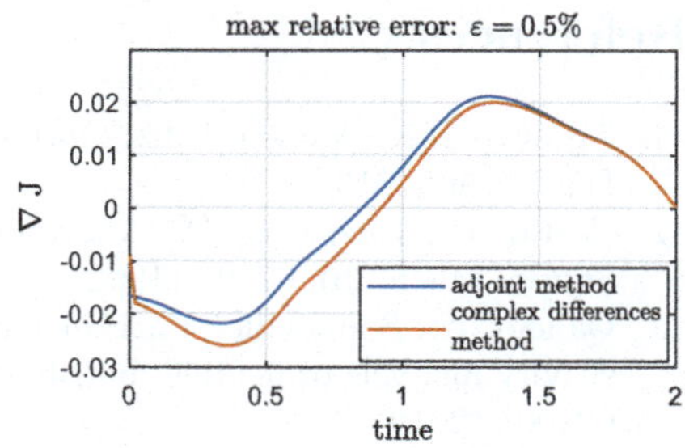

Fig. 6. Time variant functions of the gradient calculated with different methods

at $t = 0.52\,\mathrm{s}$ is equal to 0.5%. Let us compare the time required to calculate the gradient from $t_0 = 0\,\mathrm{s}$ to $t_f = 2\,\mathrm{s}$ with a constant step size equal to $\Delta t = 0.02\,\mathrm{s}$. Here, a standard PC has been used, which yielded the following execution times: $T_{adjoint} = 0.35\,\mathrm{s}$ and $T_{complex} = 150\,\mathrm{s}$. This significant discrepancy comes from the fact, that the adjoint method solves $4n + m$ Eq. (9) for all design variables. On the other hand, finite differences method requires to perform a forward solution of dynamic equations (1) for each design parameter, which is by far more time–consuming task compared to the execution of the adjoint method.

5 Conclusions

This paper presents an introductory effort to elaborate systematic methods of efficient gradient calculation for a multibody system. The forward dynamics problem is addressed with an efficient solution strategy based on Hamilton's canonical equations. A trapezoidal integration rule is incorporated into the equations of motion in order to get its discretized version. Furthermore, the solution of a discrete adjoint system is used to efficiently and reliably compute a gradient vector. A representative optimal control problem is considered in the text, that illustrates an application of the method developed here. The sample test case is an underactuated multi–link pendulum on a cart. Numerical results are compared against the finite differences method. The proposed formulation is amenable to parallelization according to the Hamiltonian based divide and conquer scheme [5,6]. These areas are of ongoing research for the authors.

Acknowledgments. This work has been supported by National Science Center under grant No. 2018/29/B/ST8/00374. The first author would also like to acknowledge the support of the Institute of Aeronautics and Applied Mechanics funds for scientific research.

References

1. Agrawal, S.K., Fabien, B.C.: Optimization of Dynamic Systems, vol. 70. Springer, Dordrecht (2013)
2. Bestle, D., Eberhard, P.: Analyzing and optimizing multibody systems. Mech. Struct. Mach. **20**, 67–92 (1992)
3. Callejo, A., Sonneville, V., Bauchau, O.A.: Discrete adjoint method for the sensitivity analysis of flexible multibody systems. J. Comput. Nonlinear Dyn. **14**(2), 021001 (2019)
4. Cao, Y., Li, S., Petzold, L., Serban, R.: Adjoint sensitivity analysis for differential-algebraic equations: the adjoint DAE system and its numerical solution. SIAM J. Sci. Comput. **24**(3), 1076–1089 (2003)
5. Chadaj, K., Malczyk, P., Frączek, J.: A parallel Hamiltonian formulation for forward dynamics of closed-loop multibody systems. Multibody Syst. Dyn. **39**(1) (2017). https://doi.org/10.1007/s11044-016-9531-x
6. Chadaj, K., Malczyk, P., Frączek, J.: A parallel recursive Hamiltonian algorithm for forward dynamics of serial kinematic chains. IEEE Trans. Robot. **33**(3), 647–660 (2017). https://doi.org/10.1109/TRO.2017.2654507
7. Lauß, T., Oberpeilsteiner, S., Steiner, W., Nachbagauer, K.: The discrete adjoint gradient computation for optimization problems in multibody dynamics. J. Comput. Nonlinear Dyn. **12**(3), 031016 (2017)
8. Maciąg, P., Malczyk, P., Frączek, J.: Optimal design of multibody systems using the adjoint method. In: Dynamical Systems Theory and Applications, pp. 241–253. Springer, Cham (2017). https://doi.org/10.1007/978-3-319-96601-4_22
9. Nachbagauer, K., Oberpeilsteiner, S., Sherif, K., Steiner, W.: The use of the adjoint method for solving typical optimization problems in multibody dynamics. J. Comput. Nonlinear Dyn. **10**(6) (2015). https://doi.org/10.1115/1.4028417
10. Nocedal, J., Wright, S.: Numerical Optimization. Springer, Heidelberg (2006)

Dynamic Parameters Optimization and Identification of a Parallel Robot

Taha Houda(✉), Ali Amouri(✉), Lotfi Beji(✉), and Malik Mallem(✉)

IBISC Laboratory, University of Evry/Paris Saclay,
40 Rue du Pelvoux, 91020 Evry, France
{Taha.Houda,Ali.Amouri,Lotfi.Beji,Malik.Mallem}@univ-evry.fr

Abstract. A common issue of the parallel robot is that has a large number of dynamic parameters, which requires a lot of processing time, whether in dynamic modeling, identification or control. The optimization and estimation of inertial parameters of a large DoF number of robotic system is crucial to tune the model-based control law in order to improve the robot accuracy. In this paper, we present an optimized number of dynamic parameters, called base inertial parameters. As a result, only 90 base inertial parameters affect the evr@ simulator instead of 210 standard one. Torque signals evaluation from experimental parallel platform and the developed analytical form show the effectiveness of the obtained results.

1 Introduction

It is well known that dynamic parameters play an important role in describing the behavior of a multibody system such as robot manipulators [1]. Hence, an optimized analytical form of the dynamic model must be developed, and the number of mathematical operations should be reduced. Thus, a control based on an optimized dynamic model can be build and easily implemented in real time. The Gouph-Stewart platform has been extensively studied in the literature from modeling and control views due to the high stiffness, payload capacity and high accuracy that occur [2,3]. In addition, the Gough-Stewart platform has retained the attention of industries, and a lot of tests and applications have been developed. One cites, for example, simulators for land vehicles in the automotive industry, flight simulators in the fields of aeronautics and submarine simulators as well as they have been used as manipulators. The Ibisc laboratory, through the IRA2 team, is currently working on the dynamic optimization and control in the mixed reality environment of the Gough-Stewart platform. As shown by Fig. 1, the parallel platform (6-DOF) is fixed on two sliding joints, and where the operator should be attached to the upper platform. This platform was conducted and built by our team.

Consequently, we are dealing with an 8-DOF motion system which is attended to be used as a simulator for people with reduced mobility (disabled person) [4]. The system's objectives are to create a real feeling motion when the human is

A. Kecskeméthy and F. Geu Flores (Eds.): ECCOMAS 2019, COMPUTMETHODS 53, pp. 367–374, 2020.
https://doi.org/10.1007/978-3-030-23132-3_44

merged into mixed reality environments. This system will be used in several fields such as rehabilitation for people with motor disabilities, rescue environments and a sport simulator for educational purposes. The combination of the small displacement asserted by the parallel platform and the large displacements of sliders, all of these allow a large operator's workspace for several types of sports, such as ski, wake and jet ski simulations. Consequently, an agreement is necessary such that the upper platform trajectories integrating disabled human objectives are stable (to ensure safety). Hence, we resume our contribution by the following: to improve the controllability, performance and reliability of the parallel simulator, a description of the Evr@ simulator was carried out in Sect. 2, a dynamic optimization of the simulator parameters has been detailed through the energy model in Sect. 3, followed by a detailed dynamic parameters identification in Sect. 4.

2 System Description

The multi-body structure consists of 6-DoF Gough-Stewart platform mounted on a 2-DoF slider, these two sliders carry both the 6-DoF platform and the axes as shown in the Fig. 1.

As it is known, the Gough-Stewart platform consists of six legs and two platforms, where the lower platform fixed on the two sliders called the base platform and the other called the upper platform, each leg is connected on one side, to the base platform by a cardan joint and on the other side by a spherical joint, each of the 8 joints is actuated using a hybrid Nanotec PD4 servomotor.

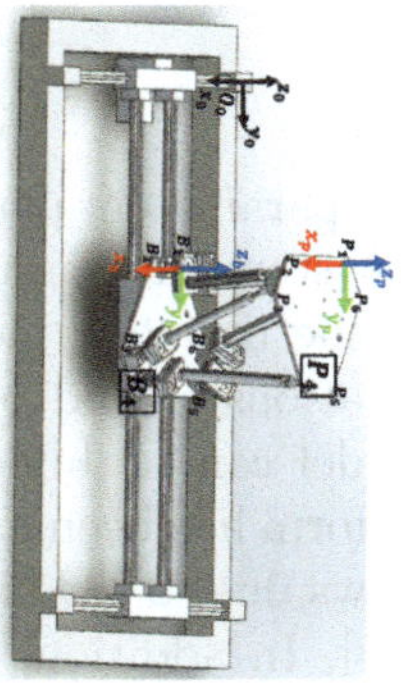

Fig. 1. Evr@ simulator

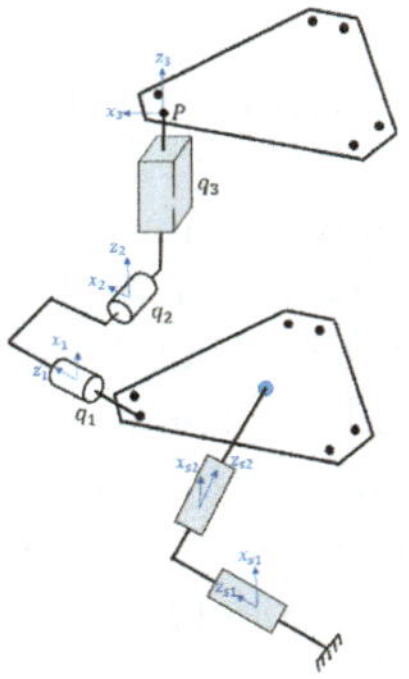

Fig. 2. Leg parameterization

The rotational movement of the motors mounted on the leg is transformed to translational movement due to the high-quality ball screw-up system. The 8 brushless motor are controlled by CAN-Open Protocol. The motor has three embedded sensors for position, velocity and torque. The information is collected with a sampling time of about 0.0196 s, which corresponds to 51 Hz sampling frequency. All the system has been designed, manufactured and programmed within the IRA2 team.

3 Dynamic Parameters Optimization

Such a procedure requires a dynamic based control procedure where all system parameters must to be minimized and identified [5]. With respect to (Denavit-Hartenberg) parametrization and following notations given by Khalil [6], the dynamic of the studied system depends on 210 dynamic parameters knowns as standard inertial parameters (10 parameters per link, 2 sliding links, 3 links per leg and upper platform link). In the following, we detail the methods carried to optimize the number of dynamic parameters.

Proposition 1. *For a given link, in the link energy, if the column of h_k is equal to zero or a constant number, the corresponding inertial parameter can be eliminated.*

Proof. Let us introduce the total energy H_k of the kth link

$$H_k = E_k + U_k = (e_k + u_k)\xi_k^{in} = h_k\xi_k^{in} \tag{1}$$

where E_k is the kinetic energy of the kth link,

$$E_k = \frac{1}{2}[\omega_k^T J_k\omega_k + M_k V_k^T V_k + 2MS_k^T(V_k \times \omega_k)] \tag{2}$$

$$= e_k\xi_k^{in} \tag{3}$$

and U_k is the potential energy of the kth link,

$$U_k = -M_k\ {}^0g^T({}^0P_k + {}^0A_k{}^kS_k) \tag{4}$$

$$= u_k\xi_k^{in} \tag{5}$$

where e_k and u_k are functions of the kinematic variables, given in Appendix. The link k standard inertial parameters and h_k vector are defined by:
$\xi_k^{in} = [XX_k\ XY_k\ XZ_k\ YY_k\ YZ_k\ ZZ_k\ MX_k\ MY_k\ MZ_k\ M_k]^T \in \mathbb{R}^{10\times 1}$.
$h_k = [h_{XX_k}\ h_{XY_k}\ h_{XZ_k}\ h_{YY_k}\ h_{YZ_k}\ h_{ZZ_k}\ h_{MX_k}\ h_{MY_k}\ h_{MZ_k}\ h_{M_k}]^T \in \mathbb{R}^{10\times 1}$.

Therefore, k^{th} link energy H_k is as:

$$H_k = \sum_{j=1}^{r_2} h_k^j\ \xi_k^{j,in} \tag{6}$$

where r_2 is the number of inertial parameters by link, $r_2 = 10$, h_k^j is the jth element of h_k vector corresponding to k^{th} link and $\xi_i^{j,in}$ is the jth element of the inertial vector corresponding to i^{th} link.

Let us take the lth column of link k $h_k^l = const$, we can write H_k as

$$H_k = \sum_{j=1}^{l-1} h_k^j\ \xi_k^{j,in} + \sum_{j=l+1}^{r_2} h_k^j\ \xi_k^{j,in} + const \tag{7}$$

Since the dynamic effect of such link is seen by the differentiation of the total energy ΔH_k, therefore we can set $\xi_k^{l,in} = 0$ without changing the value of ΔH_k. This means that $\xi_k^{l,in}$ has no effect on the dynamic model, then it can be eliminated.

Proposition 2. *For one leg, in the link energy, if the* h_k *column can be expressed as function of the previous link'energy function* h_{k-1}*, then* ξ_k^{in} *can be grouped with* ξ_{k-1}^{in}*, and another grouped inertial parameter* $\xi_{R,k-1}^{in}$ *appears.*

Proof. The robot leg energy H_{leg} is as:

$$H_{leg} = \sum_{i=1}^{r_1} \sum_{j=1}^{r_2} h_i^j \; \xi_i^{j,in} \tag{8}$$

where r_1 is the number of link by leg, in Evr@ simulator, $r_1 = 3$, h_i^j is the jth element of the energy vector corresponding to i^{th} link and $\xi_i^{j,in}$ is the jth element of the inertial vector corresponding to i^{th} link, so $h_i \xi_i^{in} = \sum_{j=1}^{r_2} h_i^j \; \xi_i^{j,in}$.

In one leg, let us consider a possible energy linear dependence between l^{th} column of the link k and link $k-1$, which can take the following form $h_k^l = h_{k-1}\alpha_{k-1}$, where $\alpha_{k-1} \in \mathbb{R}^{10\times 1}$. Then,

$$\begin{aligned} H_{leg} &= \sum_{i=1}^{k-2} \sum_{j=1}^{r_2} h_i^j \; \xi_i^{j,in} + \sum_{i=k+1}^{r_1} \sum_{j=1}^{r_2} h_i^j \; \xi_i^{j,in} + \sum_{j=1}^{r_2} h_{k-1}^j \; \xi_i^{j,in} + \sum_{j=1}^{r_2} h_k^j \; \xi_i^{j,in} \\ &= \sum_{i=1}^{k-2} \sum_{j=1}^{r_2} h_i^j \; \xi_i^{j,in} + \sum_{i=k+1}^{r_1} \sum_{j=1}^{r_2} h_i^j \; \xi_i^{j,in} \\ &\quad + \sum_{j=1}^{l-1} h_k^j \; \xi_i^{j,in} + \sum_{j=l+1}^{r_2} h_k^j \; \xi_i^{j,in} + h_{k-1}\xi_{k-1}^{in} + h_{k-1}\alpha_{k-1}\xi_k^{l,in} \\ &= \sum_{i=1}^{k-2} \sum_{j=1}^{r_2} h_i^j \; \xi_i^{j,in} + \sum_{i=k+1}^{r_1} \sum_{j=1}^{r_2} h_i^j \; \xi_i^{j,in} \\ &\quad + \sum_{j=1}^{l-1} h_k^j \; \xi_i^{j,in} + \sum_{j=l+1}^{r_2} h_k^j \; \xi_i^{j,in} + h_{k-1}(\xi_{k-1}^{in} + \alpha_{k-1}\xi_k^{l,in}) \\ &= \sum_{i=1}^{k-2} \sum_{j=1}^{r_2} h_i^j \; \xi_i^{j,in} + \sum_{i=k+1}^{r_1} \sum_{j=1}^{r_2} h_i^j \; \xi_i^{j,in} + \sum_{j=1}^{l-1} h_k^j \; \xi_i^{j,in} + \sum_{j=l+1}^{r_2} h_k^j \; \xi_i^{j,in} + h_{k-1}\xi_{R,k-1}^{in} \end{aligned} \tag{9}$$

where a grouped parameter can be identified, $\xi_{R,k-1}^{in} = \xi_{k-1}^{in} + \alpha_{k-1}\xi_k^{l,in}$

Proposition 3. *The 8 DoF optimized dynamics parameters are of number 90 parameters instead of 210 standard inertial parameters.*

Proof. At First, let us note that all the link velocities are computed using the recursive Newton-Euler relations such as presented in [6,7]

$$\begin{aligned} {}^k\omega_k &= {}^kR_{k-1} \; {}^{k-1}\omega_{k-1} + \overline{\sigma}_k \dot{q}_k {}^k a_k = {}^k\omega_{k-1} + \sigma_k \dot{q}_k {}^k a_k \\ {}^kV_k &= {}^kR_{k-1}({}^{k-1}V_{k-1} + {}^{k-1}\omega_{k-1} \times {}^{k-1}P_k) + \sigma_k \dot{q}_k {}^k a_k \end{aligned}$$

All symbols definitions are given in appendix. Form the definition of the kinetic and potential energies, the h_k elements can be formulated as given in Table 1.

The columns of h_k which are constant:

1. for sliding joint: h_{XX}, h_{XY}, h_{XZ}, h_{YY}, h_{YZ}, h_{ZZ}, h_{MX}, h_{MY}, h_{MZ} (9 parameters eliminated)
2. for the first rotational link of each leg: h_{XX}, h_{XY}, h_{XZ}, h_{YY}, h_{YZ} (5 parameters eliminated by leg)

Now as a direct application of Proposition 1, a 48 ($9 \times 2 + 5 \times 6$) standard inertial parameters are eliminated. For example, the first element of h_k vector of the sliding joint is equal to $h_s^1 = \frac{1}{2}\omega_{1,s}\omega_{1,s}$, as the sliding joint has no rotational movement which means $\omega_{1,s} = 0$. Consequently, $h_s^1 = 0$ then the XX_s parameter is eliminated.

For one leg, the h_k columns which have linear dependence with the h_{k-1} previous link, are for the first rotational link: h_{MZ}, h_M, for the second rotational link: h_{YY} h_{MZ}, h_M and for the prismatic link: h_{XX}, h_{XY}, h_{XZ}, h_{YY}, h_{YZ}, h_{ZZ}, h_M. After application of Proposition 2, 72 standard inertial parameters are grouped, 12 by leg. Example, for one leg (according to the notation of the Figs. 1 and 2),

$$^2\omega_2 = \begin{bmatrix} \sin\theta_2\ \dot{q}_1 \\ \cos\theta_2\ \dot{q}_1 \\ \dot{q}_2 \end{bmatrix}, {}^3\omega_3 = \begin{bmatrix} \sin\theta_2\ \dot{q}_1 \\ \dot{q}_2 \\ -\cos\theta_2\ \dot{q}_1 \end{bmatrix}; \tag{10}$$

we can find that $h_3^1 = (\cos\theta_2\ \dot{q}_1)^2$, $h_2^4 = (\cos\theta_2\ \dot{q}_1)^2$ which gives us $h_3^1 = h_2\alpha_2$ where $\alpha_2 = [0\ 0\ 0\ 1\ 0\ 0\ 0\ 0\ 0\ 0]^T$. Consequently, The XX_3 inertial parameter is grouped with link-2 inertial parameters $\xi_{R,2}^{in} = \xi_2^{in} + \alpha_2\xi_3^{l,in}$ following the equation (9).

4 Dynamic Parameters Identification

An energy model based on the optimized dynamic parameters has been applied in the following form, so we can write:

$$\begin{aligned} \Delta H &= T^T dq \\ &= T^T \dot{q}\ dt \end{aligned} \tag{11}$$

where,

$$T = \Gamma - diag(sign(\dot{q}))F_s - diag(\dot{q})F_v \tag{12}$$

After integration,

$$\int_{t_1}^{t_2} T^T \dot{q}\ dt = H(t_2) - H(t_1), \tag{13}$$

Hence, the total energy can be expressed linearly as function of the inertial and friction parameters, we can write:

$$\mathbf{Y} = \int_{t_1}^{t_2} \Gamma^T \dot{q}\ dt = \Delta H + \Delta H_{F_v} + \Delta H_{F_s} \tag{14}$$

$$= \Delta h \xi^{in} + \Delta h_{F_v} F_v + \Delta h_{F_s} F_s \tag{15}$$

Therefore, the identification energy model form is as follow (Fig. 3),

$$\mathbf{Y} = \int_{t_1}^{t_2} \Gamma^T \dot{q}\, dt = [\Delta h\ \Delta h_{F_v}\ \Delta h_{F_s}] \begin{bmatrix} \xi^{in} \\ F_v \\ F_s \end{bmatrix} \tag{16}$$

$$\mathbf{Y}(\Gamma, \dot{q}) = W(q, \dot{q}) \begin{bmatrix} \xi^{in} \\ F_v \\ F_s \end{bmatrix} + \rho \tag{17}$$

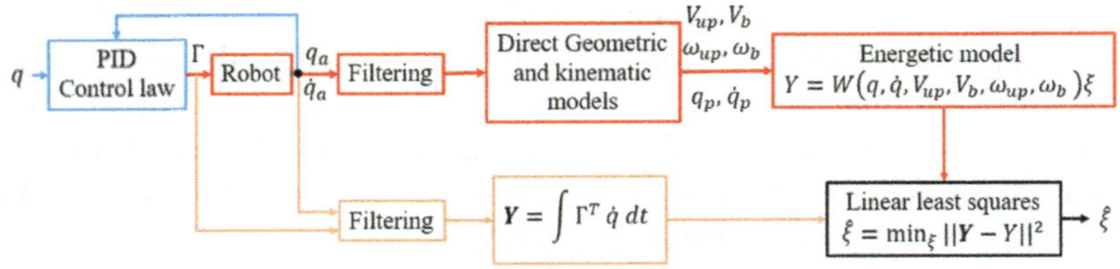

Fig. 3. Schematic diagram of the energy model.

$\Gamma \in \mathbb{R}^8$ is the vector of generalized torques, $q = [q_a, q_p]$, $\dot{q} = [\dot{q}_a, \dot{q}_p]$ where $q_a \in \mathbb{R}^8, q_p \in \mathbb{R}^{12}$ are the active and passive joint positions, $(V_b \times \omega_b)$ and $(V_{up} \times \omega_{up}) \in (\mathbb{R}^3 \times \mathbb{R}^3)$ are the speed vectors of the base and the upper platform respectively, $\xi = [\xi^{in}, \xi^f]$ where ξ^{in},$\xi^f = [F_v, F_s]^T$ are the inertial and friction parameters respectively, F_v, F_s are the viscous and dry friction coefficients.

Several exciting trajectories has been applied with the highest velocity to take around 50 000 equations (>500 × number of parameters) to make sure that the trajectories have enough information, all detected variables such as displacements, velocities and torques were filtered in order to eliminate the high frequency which is harmful for the identification procedure.

A least square method has been used to solve this overdetermined system

$$\hat{\xi} = [W^T W]^{-1}\, W^T\, \mathbf{Y} \tag{18}$$

the precision of the identified parameter $\hat{\xi}$ is depend on the persistence of the observation matrix [8,9].

The identification results of two sliding joints are shown in the Table 1.

Table 1. Two sliding joints inertial parameters

Axes	M (Kg)	$F_{v_1}(N.s/mm)$	$F_{s_1}(N)$	$F_{v_2}(N.s/mm)$	$F_{s_2}(N)$
First	33,7	76,38	9,14	72,38	11.23
Second	20,56	37,30	9,60	42,98	10,44

$F_{v_1}, F_{v_2}, F_{s_1}, F_{s_2}$ Viscous and sec frictions in the first and second directions respectively.

In order to verify the identification results, a torque construction from the model using the Lagrangian formulation and the identified parameters is performed. The Lagrangian is equal to the total kinetic energy differentiated by the total potential energy:

$$L = E - U = e_k \hat{\xi}_k^{in} - u_k \hat{\xi}_k^{in}; \quad \Gamma_{Model} = \Gamma_{\hat{\xi}^{in}} + \Gamma_{\hat{\xi}^f} \tag{19}$$

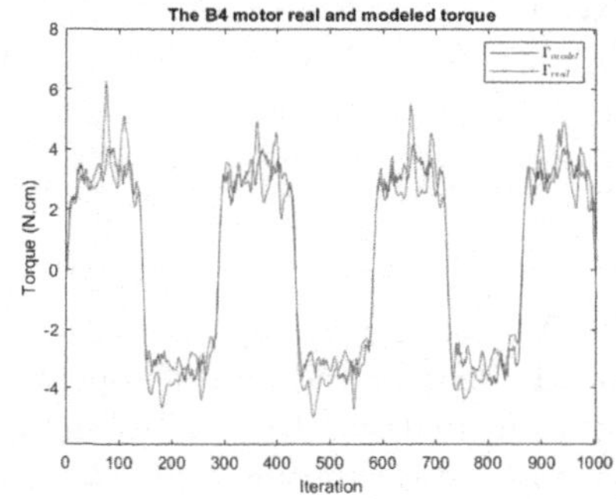

An example of torque detected and constructed of leg B_4P_4 (Fig. 1) is shown beside.

The B_4P_4 axis friction was taken in two directions to take into account the dissymmetry of the friction model, we note 1 and 2 for the first and the second direction, respectively: $F_{s_1} = 1,39\ N, F_{v_1} = 56,98\ N.s/mm, F_{s_2} = 1,33\ N,$ $F_{v_2} = 58,31\ N.s/mm$, Finally, the grouped mass of upper platform and last 6 links is $M_{R_p} = 6.15$ kg.

5 Conclusion and Perspective

The results of this work benefit from an adequate parallel robot's parametrization studied in [6]. Consequently, the 210 standard inertial parameters are reduced to 90 base inertial parameters only. The given procedure has led to an optimized dynamic/energetic model where the implementation in real time is possible. Obtained torques, experimentally sensed and analytically computed, are comparable. The next steps and perspectives will necessarily focus in the multi-body system control using the identified dynamic parameters and study of the hemiplegic skier's reaction and his rehabilitation enhancement in immersion environment.

Appendix

See Table 2.

Table 2. e_k and u_k vector elements

ξ_k^{in}	XX	XY	XZ	YY	YZ
e_k	$\frac{1}{2}\omega_{1,k}\omega_{1,k}$	$\frac{1}{2}\omega_{1,k}\omega_{2,k}$	$\frac{1}{2}\omega_{1,k}\omega_{3,k}$	$\frac{1}{2}\omega_{2,k}\omega_{2,k}$	$\frac{1}{2}\omega_{2,k}\omega_{3,k}$
u_k	0	0	0	0	0
ξ_k^{in}	ZZ	MX	MY	MZ	M
e_k	$\frac{1}{2}\omega_{3,k}\omega_{3,k}$	$\omega_{3,k}V_{2,k} - \omega_{2,k}V_{3,k}$	$\omega_{1,k}V_{3,k} - \omega_{3,k}V_{1,k}$	$\omega_{2,k}V_{1,k} - \omega_{1,k}V_{2,k}$	$\frac{1}{2}{}^kV_k^T{}^kV_k$
u_k	0	$-{}^0g^T{}^0s_k$	$-{}^0g^T{}^0n_k$	$-{}^0g^T{}^0a_k$	$-{}^0g^T{}^0P_k$

Where, ${}^k\omega_k = [\omega_{1,k}\ \omega_{2,k}\ \omega_{3,k}]^T$, ${}^kV_k = [V_{1,k}\ V_{2,k}\ V_{3,k}]^T$ and ${}^0R_k$ and ${}^0P_k$ are the matrix and vector appearing in the transformation matrix ${}^0T_k$ with the rotation matrix ${}^0R_k = [{}^0s_k{}^0n_k{}^0a_k]$, a_k unit vector along Z_k direction, ${}^kJ_k$ - (3×3) inertia tensor of link k with respect to frame F_k. M_k and MS_k are the mass and first moment of inertia of link k respectively.

References

1. Khalil, W., Guegan, S.: Inverse and direct dynamic modeling of Gough-Stewart robots. IEEE Trans. Robot. Autom. **20**, 754–762 (2004)
2. Merlet, J.-P.: Parallel Robots. Kluwer, Dordrecht (2000)
3. Harib, K., Srinivasan, K.: Kinematic and dynamic analysis of Stewart platform-based machine tool structures. Robotica **21**, 541–554 (2003)
4. Amouri, A., Ababsa, F.-E.: Sliding movement platform for mixed reality application. In: 7th IFAC Symposium on Mechatronic Systems, Loughborough, United Kingdom, vol. 49(21), pp. 662–667, September 2016
5. Khalil, W., Gautier, M., Lemoine, P.: Identification of the payload inertial parameters of industrial manipulators. In: IEEE International Conference on Robotics and Automation - ICRA 2007, Rome, Italy, pp. 4943–4948, April 2007
6. Khalil, W., Dombre, E.: Modeling, Identification and Control of Robots. Hermès Penton, London (2002)
7. Luh, J.Y.S., Walker, M.W., Paul, R.C.P.: On-line computational scheme for mechanical manipulators. Trans. ASME J. Dyn. Syst. Meas. Control **102**(2), 69–76 (1980)
8. Kozłowski, K.: Experimental identification of dynamic parameters for a class of geared robots. Robotica **14**, 561–574 (1996)
9. Vandanjon, P.O., Gautier, M., Desbats, P.: Identification of robots inertial parameters by means of spectrum analysis. In: Proceedings of 1995 IEEE International Conference on Robotics and Automation, Nagoya, Japan (1995). https://doi.org/10.1109/ROBOT.1995.525715

Partial Shaking Force Balancing of 3-RRR Parallel Manipulators by Optimal Acceleration Control of the Total Center of Mass

Jing Geng[1,2(✉)] and Vigen Arakelian[1,2]

[1] LS2 N-ECN UMR 6004, 1 rue de la Noë, BP 92101, 44321 Nantes, France
{jing.geng, vigen.arakelyan}@ls2n.fr
[2] INSA-Rennes/Mecaproce, 20 av. des Buttes de Coesmes, CS 70839, 35708 Rennes, France
{jing.geng, vigen.arakelyan}@insa-rennes.fr

Abstract. This paper deals with the problem of shaking forces balancing of 3-RRR planar parallel manipulators. It is known that this problem can be solved by optimal mass redistributions of the moving links, i.e. by adding counterweights or auxiliary structures. In this paper, the reduction of shaking forces of 3-RRR planar parallel manipulators is accomplished by the optimal trajectory planning of the common center of mass of the manipulator, which is carried out by "bang-bang" profile. Such a method allows a considerable reduction in shaking forces without adding any counterweight. Using such a solution, the disadvantages of adding counterweights have been avoided. An increase in the mass of moving links usually leads to an increase in shaking moments and input torques. It has been shown via numerical simulations that the use of the "bang-bang" profile is a more efficient not only for shaking forces minimization but also for minimization of shaking moments and input torques. The efficiency of the suggested balancing approach is illustrated by numerical simulations carried out via ADAMS software.

1 Introduction

Balancing of inertia forces is an important issue in the design of high-speed robot manipulators. The aim of the balancing is the reduction of variable dynamic loads on the manipulator frame, which are sources of vibrations. Different approaches and solutions devoted to the shaking force balancing of robot manipulators have been developed and documented [1]. However, the main methods of balancing are based on the redistribution of moving masses of robot manipulators that is accompanied by an unavoidable increase of the total mass of moving links.

The redistribution of moving masses can be reached by adding counterweights [2–5] or auxiliary structures [6, 7]. The optimization of inertia forces based on the minimization of the total joint actuator efforts has also been studied [8]. In the study [9] has been proposed a balancing approach based on the optimal trajectory planning of the common center of mass of the manipulator by "bang-bang" profile. The aim of the developed balancing method consists in the fact that the manipulator is controlled not by applying end-effector trajectories but by planning the displacements of the total

A. Kecskeméthy and F. Geu Flores (Eds.): ECCOMAS 2019, COMPUTMETHODS 53, pp. 375–382, 2020.
https://doi.org/10.1007/978-3-030-23132-3_45

mass center of moving links. The trajectories of the total mass center of moving links are defined as a straight line between the initial and final positions of the end-effector. Then, the motion between these positions is parameterized with "bang-bang" motion profile. Such a motion generation allows the reduction of the maximal value of the center of mass acceleration and, consequently, leads to the reduction in the shaking force. In [10], has been mentioned that despite the obvious advantages concerning the balancing without adding counterweights, it has also some drawbacks. Therefore, in studies [10–12], has been proposed to combine the balancing through mass redistribution and the balancing via center of mass acceleration control.

In the present study, the authors propose to apply the balancing method described in [9] on 3-RRR planar parallel manipulators. Counterweights are not added on manipulator's links, and the reduction of inertial forces is carried out only by optimal trajectory and acceleration generation of the common center of masses of the manipulator.

2 Shaking Force Balancing of 3-RRR Planar Parallel Manipulator

A kinematic scheme of the 3-RRR planar parallel manipulator is shown in Fig. 1. The output axis *P(x, y)* of the platform is connected to the base by three legs, each of which consists of three revolute joints and two links. The three legs are connected to an equilateral triangle with a uniform redistribution of mass with the revolute joints. In each of the three legs, the revolute joint connected to the base is actuated. Such a manipulator can position the platform freely in a plane and orient it.

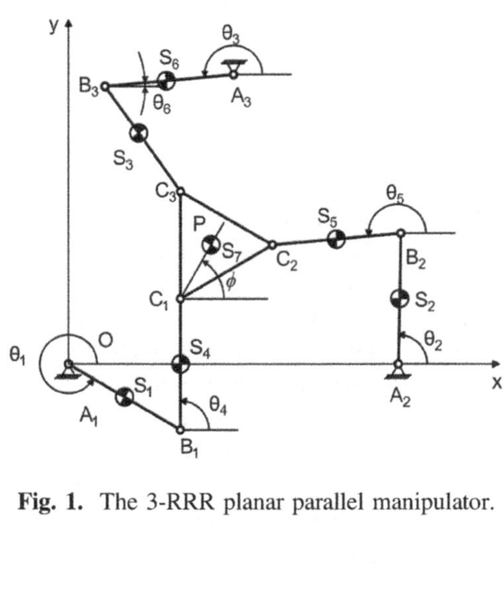

Fig. 1. The 3-RRR planar parallel manipulator.

2.1 Problem Formulation

In the given 3-RRR planar parallel mechanism each actuated joint is denoted as $A_i\,(i = 1,2,3)$, the other end of each actuated link is denoted as $B_i\,(i = 1,2,3)$ and the three vertices $C_i\,(i = 1,2,3)$ of the platform are respectively connected with three legs. A fixed global reference system Oxy is located at A_1 with the y-axis perpendicular to A_1A_2 and the x-axis directed along A_1A_2. The lengths of links are denoted as $l_1 = A_1B_1 = A_2B_2 = A_3B_3$, $l_2 = B_1C_1 = B_2C_2 = B_3C_3$ and $l_3 = C_1C_2 = C_1C_3 = C_2C_3$. The locations of the centers of mass of the linkages $S_i\,(i = 1,2,\ldots6)$ are denoted as $r_1 = l_{A_1S_1} = l_{A_2S_2} = l_{A_3S_3}$, $r_2 = l_{B_1S_4} = l_{B_2S_5} = l_{B_3S_6}$. The center of mass of the platform S_7 is in the geometric center of the equilateral triangle $\Delta C_1C_2C_3$.

The input parameters of such a manipulator are defined by the joint angles $\theta_i(i = 1,2,3)$ and the output parameters by the pose of the moving platform, i.e. its orientation ϕ and position of one point of the moving platform, by example, the centre of mass of the moving platform with parameters x and y.

The coordinates of the common center of mass of the 3-RRR planar parallel manipulator can be expressed as:

$$x_S = \frac{\sum_{i=1}^{7} x_{Si}m_i}{M} \tag{1}$$

$$y_S = \frac{\sum_{i=1}^{7} y_{Si}m_i}{M} \tag{2}$$

with

$$x_{Si} = x_{Ai} + r_1 \cos\theta_i, i = 1,2,3 \tag{3}$$

$$y_{Si} = y_{Ai} + r_1 \sin\theta_i, i = 1,2,3 \tag{4}$$

$$x_{Si} = x_{Ai} + l_1 \cos\theta_{i-3} + r_2 \cos\theta_i, i = 4,5,6 \tag{5}$$

$$y_{Si} = y_{Ai} + l_1 \sin\theta_{i-3} + r_2 \sin\theta_i, i = 4,5,6 \tag{6}$$

$$x_{S7} = x_{A1} + l_1 \cos\theta_1 + l_2 \cos\theta_4 + \left(l_3\big/\sqrt{3}\right)\cos\phi \tag{7}$$

$$y_{S7} = y_{A1} + l_1 \sin\theta_1 + l_2 \sin\theta_4 + \left(l_3\big/\sqrt{3}\right)\sin\phi \tag{8}$$

where, $M = \sum_{i=1}^{7} m_i$ and x_{Si}, y_{Si} are the coordinates of the centers of mass of the corresponding links $(i = 1,\ldots,7)$.

The shaking forces $\mathbf{F}^{sh}$ of the 3-RRR planar parallel manipulator can be written in the form:

$$\mathbf{F}^{sh} = M\ddot{s} \tag{9}$$

where, $\ddot{s}$ is the acceleration of the total mass centre.

The shaking force balancing via mass redistribution consists in adding counterweights in order to keep the total mass centre of moving links stationary [1]. In this case, $\ddot{s} = 0$ for any configuration of the manipulator and, as a result, the shaking force is cancelled. It is obvious that the adding of supplementary masses as counterweights is not desirable because it leads to the increase of the total mass, of the overall size of the manipulator, the efforts in joints, the shaking moment and the input torques. Therefore, in the present study, as in [9], it is proposed to minimize the shaking force via reduction of the total mass centre acceleration:

$$\max|\ddot{s}| \rightarrow \min_{s(t)} \tag{10}$$

i.e. to apply an optimal control of the total mass centre of moving links that allows one to reduce the maximal value of its acceleration.

For this purpose, let's consider the control of the 3-RRR planar parallel manipulator through of its centre of mass. To ensure it, let's assume that the centre of mass moves along a straight line between its initial and final positions. Thus, the motion profile used on this path will define the values of inertial forces. Hence, to minimize the maximum value of the acceleration of the total mass centre and, as a result, inertial forces, the "bang-bang" profile should be used [13]. Consequently, by reducing the acceleration of the centre of mass of the 3-RRR planar parallel manipulator, a decrease in its inertial forces is achieved.

However, it should be remembered that the "bang-bang" law is based on theoretical considerations. In reality, the actuators are unable to achieve discontinuous efforts. Therefore, this law should be modified by a trapezoidal profile in order to take into account the actuators properties in terms of maximal admissible effort variations. However, it was observed that for reasonable actuator parameters, the minimizations obtained in the cases of the "bang-bang" and trapezoidal profiles are very close (less than 1%). The detailed discussion about it can be found in [9].

To achieve the shaking force balancing through the above described approach, it is necessary to find out the relationship between the input angles and the centre of mass positions of the planar 3-RRR planar parallel manipulator.

2.2 Relationship Between the Input Angles and the Total Center of Mass of the 3-RRR Planar Parallel Manipulator

In order to control the manipulator according to the method described above, it is necessary to establish the relationship $S = S(\theta_1, \theta_2, \theta_3)$, i.e. for the given position and the law of motion of the common centre of mass of the manipulator determine its input angles $\theta_1, \theta_2, \theta_3$. Then, by means of the obtained input angles $\theta_1, \theta_2, \theta_3$ via forward kinematics determine the position of the output axis *P(x, y)*. For this purpose, let's establish the relationship between the common center of mass of the manipulator and its input parameters.

Let's start this issue with the initial and final positions $P(x, y)$ of the platform x_i, y_i and x_f, y_f, as well as its orientation ϕ (ϕ_i, ϕ_f) that are known. So, by invers kinematics [14], the input angles corresponding to these positions will be determined: θ_{1i}, θ_{1f}, θ_{2i}, θ_{2f} and θ_{3i}, θ_{3f}. The correspondent values of the common centre of mass of the manipulator can also be found: x_{Si}, y_{Si} and x_{Sf}, y_{Sf}. Subsequently, a straight line connecting the initial and final positions of the comment centre mass of the manipulator can be established and its trajectory planning by "bang-bang" profile can be ensured: $S = S(t)$, i.e. $x = x(t)$ and $y = y(t)$.

Let's now consider the relationship between $S(x, y, \phi)$ and the input angles $\theta_1, \theta_2, \theta_3$.

Taking into account the geometric relationships of the 3-RRR planar manipulator, the following relations can be defined:

$$x_S = \frac{\sum_{i=1}^{7} x_{Si} m_i}{M} \tag{11}$$

$$y_S = \frac{\sum_{i=1}^{7} y_{Si} m_i}{M} \tag{12}$$

$$\tan \phi = \frac{y_{C2} - y_{C1}}{x_{C2} - x_{C1}} + \frac{\pi}{6} \tag{13}$$

where, x_{C1}, y_{C1}, x_{C2}, y_{C2} are the coordinates of the axes C_1 and C_2 of the platform (see Fig. 1).

Some observations showed that it is better to add to these Eqs. 3 loop closure equations and obtain 6 nonlinear equations relatively 6 unknown:

$$f_i(\theta_1, \theta_2, \theta_3, \theta_4, \theta_5, \theta_6) = 0, \quad (i = 1, 2, 3, \ldots 6) \tag{14}$$

Thus, the input angles $\theta_1, \theta_2, \theta_3$ of the 3-RRR planar parallel manipulator can be found via numerical methods by solving nonlinear equations (14).

3 Simulation Results

To create a CAD model and carry out simulations via ADAMS software, the parameters of the 3RRR planar parallel manipulator are the following: $l_1 = l_2 = l_3 = l_4 = l_5 = l_6 = 0.18$ m; $x_{A1} = 0$; $y_{A1} = 0$; $x_{A2} = 0.46$ m; $y_{A2} = 0$; $x_{A3} = 0.23$ m; $y_{A3} = 0.4$ m; $l_3 = l_{C1C2} = l_{C1C3} = l_{C2C3} = 0.15$ m. The masses of links: $m_i = 1\text{kg} (i = 1, \ldots, 6)$ and $m_7 = 3$ kg. The locations of the centers of mass S_i $(i = 1, \ldots, 6)$ of the moving links are the following: $r_{S1} = r_{S2} = r_{S3} = r_{S4} = r_{S5} = r_{S6} = 0.09$ m. The trajectory of the output axis P of the platform is given by its initial position P_i with the coordinates: $xi = 0.230$ m, $yi = 0.133$ m and the final position P_f with the coordinates: $xf = 0.330$ m, $yf = 0.133$ m. For the examined case, the orientation angle ϕ of the platform is zero for whole trajectory. The corresponding input angles are determined via inverse kinematics: $\theta_{1i} = -0.5277$, $\theta_{1f} = -0.3834$, $\theta_{2i} = 1.5678$, $\theta_{2f} = 0.8476$, $\theta_{3i} = 3.6676$, $\theta_{3f} = 4.2572$. The coordinates of the common centre of mass of the manipulator are determined: $x_{Si} = 0.23$ m, $y_{Si} = 0.1329$ m, $x_{Sf} = 0.3029$ m, $y_{Sf} = 0.1227$ m.

Let us now connect the initial and final positions of the common centre of mass of the manipulator by the straight line and generate its trajectory planning by "bang-bang" profile. The variations of the input angles determined from Eq. (14) are shown in Fig. 2. The trajectories of the platform and the common center of mass of the manipulator are given in Fig. 3.

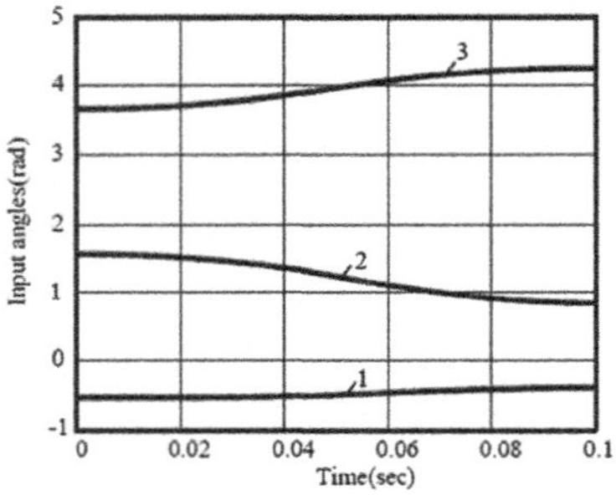

Fig. 2. Variations of the input angles ensuring the displacement of the common center of mass of the manipulator by the straight line with "bang-bang" profile.

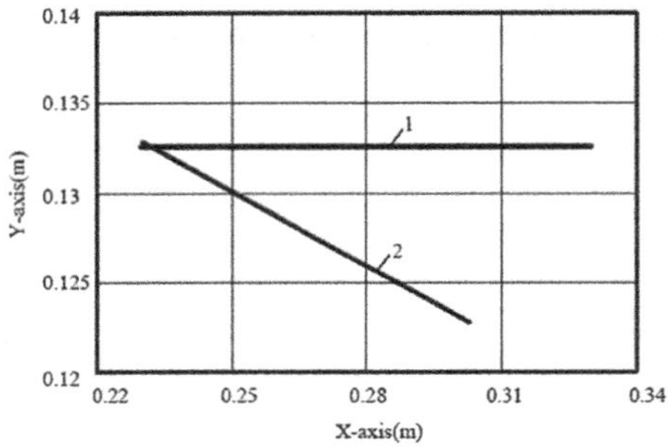

Fig. 3. Trajectories of the platform and the common center of mass.

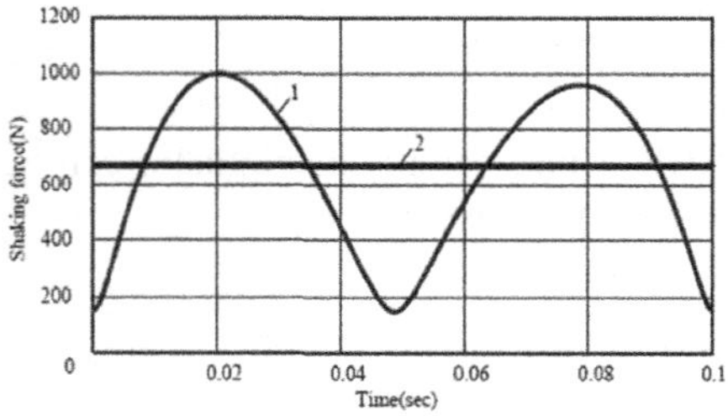

Fig. 4. Variations of shaking forces for two studied cases.

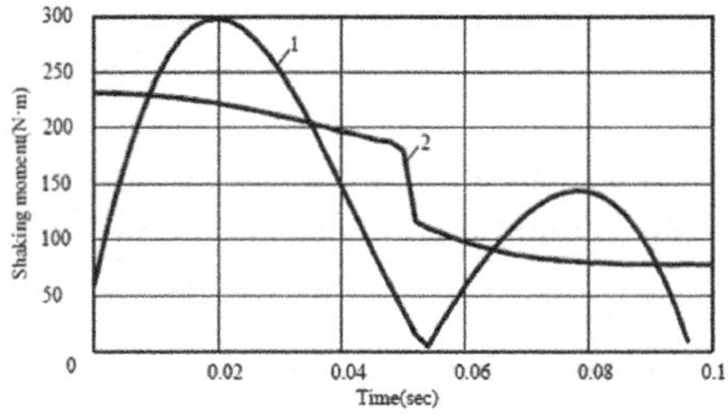

Fig. 5. Variations of shaking moments for two studied cases.

Figures 4 and 5 show the variations of shaking forces and the shaking moments for two studied cases: (1) the displacement of the platform of the unbalanced manipulator by the straight line with fifth order polynomial profile and (2) the generation of the motion via the displacement of the manipulator center mass by "bang-bang" profile.

It should be noted that the purpose of this study is the reduction of shaking forces. However, our observations show that such balancing is very favorable from the point of view of reducing shaking moments and input torques.

The simulation results show that the generation of the motion via the displacement of the manipulator center of mass by "bang-bang" profile allows the reduction of the shaking force of the unbalanced manipulator up to 32.7% what is the main goal. However, the shaking moment has also reduced up to 22.3% and the input torque up to 16.5%.

4 Conclusions

It is known that high-speed manipulators can generate significant fluctuating forces that can be source for vibrations. These variable dynamic forces can be easily cancelled by adding counterweights. However, the main drawback of the shaking force balancing by counterweights is the increase of the inertia of moving links, and consequently, the increase of shaking moments and input torques.

In this paper, the shaking force balancing based on the optimal motion planning of the manipulators' common center of mass is further developed and applied on 3-RRR planar parallel manipulators. This simple and effective balancing method is based on the optimal motion generation of the manipulator center of masses. The aim of the method consists in the fact that the manipulator is controlled not by applying end-effector trajectories but by the generation of the motion via the displacement of the manipulator center mass by "bang-bang" profile. The trajectory planning of the common center of mass is defined as straight line having "bang-bang" acceleration motion profile. Such a motion generation allows the reduction of the maximum value of the center of mass acceleration and, consequently, the reduction in the shaking force. It should be mentioned that such a solution is also very favorable for reduction of the shaking moment because it is carried out without adding counterweights. It is more effective than the generation of motion by fifth order polynomial profile. The efficiency of the considered balancing method has been illustrated via numerical simulations.

References

1. Arakelian, V., Briot, S.: Balancing of Linkages and Robot Manipulators. Advanced Methods with Illustrative Examples. Springer, Switzerland (2015)
2. Agrawal, S., Fattah, A.: Reactionless space and ground robots: novel design and concept studies. Mech. Mach. Theory **39**, 25–40 (2004)
3. Bayer, A., Merk, G.: Industrial robot with a weight balancing system. EP, 2301727 (2011)
4. Wang, J., Gosselin, C.: Static balancing of spatial three-degree-of-freedom parallel mechanisms. Mech. Mach. Theory **34**(3), 437–452 (1999)

5. Gosselin, C.: Smart Devices and Machines for Advanced Manufacturing. Springer, London (2008)
6. Fattah, A., Agrawal, S.: Design arm simulation of a class of spatial reactionless manipulators. Robotica **3**, 75–81 (2005)
7. Van derWijk, V., Herder, J.: Dynamic balancing of Clavel's delta robot. Computational Kinematics, pp. 315–322. Springer (2009)
8. Khatib, O.: Inertial properties in robotic manipulation: an object-level framework. Int. J. Robot. Res. **14**(1), 19–36 (1995)
9. Briot, S., Arakelian, V., Le Baron, J.: Shaking force minimization of high-speed robots via centre of mass acceleration control. Mech. Mach. Theory **57**, 1–12 (2012)
10. Arakelian, V.: Design of partially balanced 5R planar manipulators with reduced center of mass acceleration (RCMA). Robot Design, Dynamics and Control, pp. 113–122. Springer (2016)
11. Arakelian, V., Geng, J., Le Baron, J.-P.: Synthesis of balanced 3-RRR planar parallel manipulators. In: Proceedings of the 19th International Conference on Robotics and Computer Integrated Manufacturing, Prague, Czech Republic, 4–5 September, vol. 4(9), pp. 37–43 (2017)
12. Arakelian, V., Geng, J., Fomin, A.: Inertia forces minimization in planar parallel manipulators via optimal control. J. Mach. Manuf. Reliab. **47**(7), 303–309 (2018)
13. Khalil, W., Dombre, E.: Modeling, identification and control of robots. Hermes Sciences Europe (2002)
14. Williams II, R.L., Shelley, H.B.: Inverse kinematics for planar parallel manipulators. In: Proceedings of the 1997 ASME Design Technical Conferences (DETC'97), Sacramento, California, USA, 14–17 September 1997

Energy Expenditure Minimization for a Delta-2 Robot Through a Mixed Approach

Giovanni Carabin, Ilaria Palomba(✉), Erich Wehrle, and Renato Vidoni

Faculty of Science and Technology, Free University of Bozen-Bolzano, piazza Univesità 5, 39100 Bolzano, Italy
{giovanni.carabin,ilaria.palomba,erich.wehrle,renato.vidoni}@unibz.it

Abstract. Energy efficiency is an important goal of robotic design, especially within the framework of Industry 4.0. In this paper, a methodology is proposed to reduce the energy consumption and is then demonstrated on a two-degree-of-freedom parallel robot (i.e. a Delta-2 robot) performing cyclic pick-and-place operations of a predefined trajectory. We define a *mixed approach* as a methodology that exploits two kinds of energy reducing systems: springs for storing elastic energy and capacitors for recovering braking energy. Using an optimization-based design methodology, two torsional springs with optimum stiffness values are coupled with energy-recuperating drive axles, leading to a reduction of motor torque required. Results show that this approach allows for a significant reduction in energy expenditure.

1 Introduction

The design and control of mechatronic and robotic manufacturing systems is increasingly applied to the development of green production systems, i.e energy efficient and cost-effective systems with less environmental impact. Several solutions have been proposed to reduce energy consumption in industrial robots and mechatronic systems [1].

Systems of higher energy efficiency can be obtained, for example, thanks to the introduction of new technologies for the energy recovery and regeneration [2]. The main idea behind these devices is to harvest the energy during the braking phase, which is typically turned into heat energy and therefore "wasted". This energy can then be stored and provided to the system when required.

Another approach to increase the efficiency of mechatronic systems consists of exploiting the *natural motion* [3]. This requires the modification of the physical system by adding elastic elements, which are properly tuned in order to exploit the free vibration modes of the mechanical system to perform a cyclic task. Energy necessary for carrying out the task is supplied by the elastic elements instead of the actuators. Specifically, the motion of the system is achieved by the transformation of the potential elastic energy stored into the springs into kinetic energy. Although such an approach greatly reduces the energy expenditure, the

A. Kecskeméthy and F. Geu Flores (Eds.): ECCOMAS 2019, COMPUTMETHODS 53, pp. 383–390, 2020.
https://doi.org/10.1007/978-3-030-23132-3_46

exploitation of the natural motion may lead to large deviations from a prescribed trajectory. Indeed, the system is modified so that its free-vibration response matches the required cycle task in terms of initial and final positions of the trajectory and cycle time, but not in terms of the path [4]. Therefore, the system motion may be inadequate for the task execution, as constraints and obstacles cannot be considered.

However, adding springs to a mechanical system has the benefit of reducing motor torque for systems beyond natural motion [5]. The idea of adding springs in robotic systems to reduce the actuator effort was first adopted for gravity-balancing techniques [6].

In this paper, a *mixed approach* is investigated for enhancing the energy performance of a Delta-2 robot operating in cyclical tasks. We define *mixed approach* as the exploitation of both elastic-energy storing systems end energy-recovering devices. Further, a design optimization is used to determine the values of the elastic elements to add to the system.

It should be noted that there are further methodologies to minimize energy expenditure. These include the proper planning of system motion [7–10]. In this work, the trajectory is fixed a priori and has the objective of providing an easily and quickly usable solution for energy minimization. Conversely, the implementation of ad hoc trajectories may require the use of more sophisticated controllers, i.e. traditional robot controllers are limited in terms of trajectory generation (e.g. controllers generally implement linear or circular path, while more complex structures are not so common).

The numerical results of the Delta-2 robot show that the consumed energy can be significantly reduced. Therefore, the proposed approach leads to new effective solutions and results where the system has alternating or cyclic motion, as is the case with of pick-and-place operations.

2 Modeling of Delta-2 Robot

The mechanism considered is a Delta-2 robot, a planar multibody system with two degrees of freedom (DoF) where the end-effector is kinematically constrained

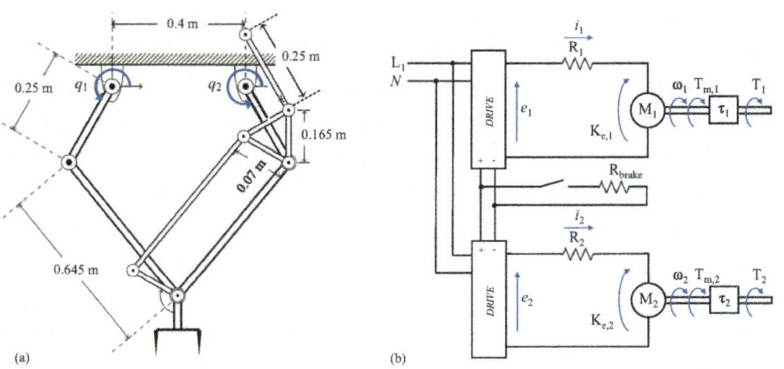

Fig. 1. Delta-2 robot: (a) kinematic schematic, (b) electrical schemamtic.

to always have the same orientation (see Fig. 1(a)). A laboratory prototype of this system is available in the *Smart Mini Factory* laboratory of Free University of Bozen-Bolzano (unibz) [11]. All data in the mechanical (Sect. 2.1) and electrical (Sect. 2.2) models refer to this system.

2.1 Mechanical Model

The kinematic scheme of the robot is provided in Fig. 1(a). The dynamic model, implemented with MATLAB, also considers all gravitational, inertial and payload forces, and is formulated as follows:

$$\mathbf{M}(\mathbf{q})\ddot{\mathbf{q}} + \mathbf{C}(\mathbf{q}, \dot{\mathbf{q}})\dot{\mathbf{q}} + \mathbf{G}(\mathbf{q}) + \mathbf{K}(\mathbf{q} - \mathbf{q}_0) = \mathbf{T}, \tag{1}$$

where $\mathbf{M}(\mathbf{q})$, $\mathbf{C}(\mathbf{q}, \dot{\mathbf{q}})$ and $\mathbf{G}(\mathbf{q})$ are the mass matrix, the centrifugal force matrix and gravity vector, respectively, while $\mathbf{q} = \{q_1\ q_2\}^{\mathrm{T}}$, $\dot{\mathbf{q}}$ and $\ddot{\mathbf{q}}$ are the position, velocity and acceleration vectors of the actuated joint coordinates. Further, $\mathbf{T} = \{T_1\ T_2\}^{\mathrm{T}}$ is the actuation force vector, $\mathbf{K} = \mathrm{diag}(k_1, k_2)$ is the diagonal matrix containing the stiffnesses of the two torsional springs to be determined and $\mathbf{q}_0 = \{q_{0,1}\ q_{0,2}\}^{\mathrm{T}}$ is the vector containing the equilibrium position of such springs, which is also to be determined.

2.2 Electrical Model

The electrical schematic of the Delta-2 robot is shown in Fig. 1(b). The electrical energy of each DC motor drive can be computed as follows:

$$E_j = \int_0^T P_j dt = \int_0^T e_j \cdot i_j dt \qquad j = 1, 2, \tag{2}$$

where e_j is the supply voltage and i_j the current flowing in the windings. Such quantities can be computed by means of the electrical equations of DC motor drivers:

$$e_j = R_j \cdot i_j + K_{e,j} \cdot \omega_j \tag{3}$$
$$i_j = (J_{mj}\alpha_j + \tau_j T_j)/K_{t,j}, \tag{4}$$

where R_j is the resistance of the motor, $K_{t,j}$ is the motor constant and $K_{e,j}$ is the motor velocity constant. In Eq. 3 ω_j and α_j are the motor angular velocity and acceleration, respectively; J_{mj} is the motor inertia, τ_j is the transmission ratio and T_j is the mechanical torque computed by means of the dynamic model in Eq. 1. The electrical energy in Eq. 2 is computed by taking regeneration into account. When P_j is positive, the motor takes energy from the network; conversely when negative, energy is recovered in a capacitor.

3 Design Optimization

The design problem is defined by finding the values of stiffness of the springs k_1 and k_2 as well as the initial positions $q_{0,1}$ and $q_{0,2}$ such that the electrical energy required by the two motors E_1 and E_2 is minimized for a given end-effector trajectory. We formulate the minimum-energy design problem as a bounded yet unconstrained nonlinear optimization problem of the following form:

$$\min_{\mathbf{x}\in\chi} \{f(\mathbf{x})\} \tag{5}$$

$$\begin{aligned} \text{where} \quad & f(\mathbf{x}) := E_1(\mathbf{x}) + E_2(\mathbf{x}) \\ & \mathbf{x} := \left[k_1\ k_2\ q_{0,1}\ q_{0,2}\right]^{\mathrm{T}} \\ & \chi := \left[\mathbf{x}^L, \mathbf{x}^U\right] \\ & \mathbf{x}^U := \left[k^U\ k^U\ q_0^U\ q_0^U\right]^{\mathrm{T}} \\ & \mathbf{x}^L := \left[k^L\ k^L\ q_0^L\ q_0^L\right]^{\mathrm{T}}. \end{aligned} \tag{6}$$

The numerical values for the design domain χ are given by the bounds $\mathbf{x}^U$ and $\mathbf{x}^L$ and are assigned the following numerical values:

$$k^L = 0.1\,\frac{N\cdot m}{rad};\quad k^U = 20.0\,\frac{N\cdot m}{rad};\quad q_0^L = -3\pi\,rad;\quad q_0^U = 3\pi\,rad. \tag{7}$$

It should be noted that a small value for the lower-bound of the spring stiffness is used for numerical stability.

For this optimization, the second-order algorithm NLPQLP [12,13] is used via a link via DesOptPy [14] which is in turn based on pyOpt [15]. To limit the amount of implementation effort, numerical gradients are used, which are calculated via forward finite differences.

4 Results

As a test case to evaluate the effectiveness of the approach, a simple pick-and-place operation is simulated (see Fig. 2). In such a task, the end-effector moves from one location to another and returns back in $T = 1.5$ s. This trajectory follows linear paths (Fig. 2a) and has a payload of 2 kg. A double-S speed profile in the workspace has been choosen for the trajectory, considering acceleration and deceleration times equal to $1/3\,T_i$ and acceleration ramp rise and fall times equal to $0.1\,T_i$, with T_i time period of each trajectory segment (see Fig. 2b and c). In Fig. 3, the corresponding position, velocity and acceleration profiles of actuated joints are shown.

For the numerical optimization, several starting points in the design domain were investigated for the solution of the optimization problem. All lead to the same design and although this does not guarantee convexity, though in terms of practicality, it confirms the validity of gradient-based optimization. The initial design starting from the midpoint of the design domain is reported. From this initial design $\boldsymbol{x}_0\left\{10.5\ 10.5\ 0\ 0\right\}$, the optimization algorithm needs 42 iterations

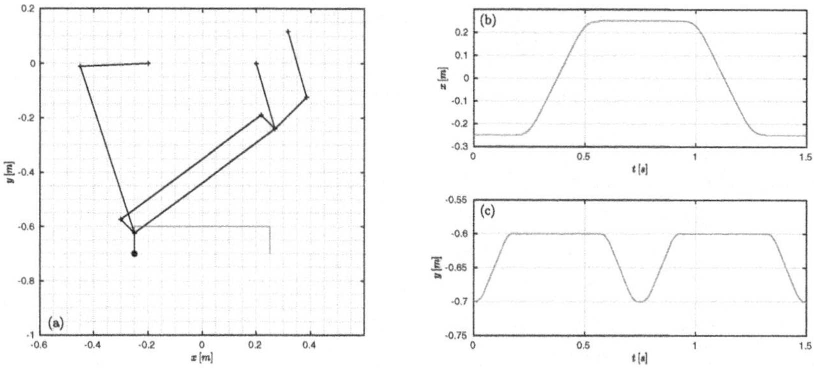

Fig. 2. End-effector trajectory path: (a) Cartesian space, (b) x-coordinate vs time, (c) y-coordinate vs time.

and 228 evaluations to find the optimum: $k_1 = 4.0\,\text{Nm/rad}$, $q_{0,1} = 1.73\,\text{rad}$, $k_2 = 4.4\,\text{Nm/rad}$, and $q_{0,2} = 7.78\,\text{rad}$.

Having imposed a symmetric trajectory within the manipulator workspace, very similar values for the stiffness k_1 and k_2 have been obtained. The slight difference is due to the presence of the end-effector guidance mechanism that loads mainly on the second motor. Even though not immediately visible, the equilibrium angular positions are also symmetrical with respect to the vertical axis. Indeed, the angular work ranges for the first and the second actuated joints are $\left(\frac{3}{4}\pi, \frac{7}{4}\pi\right)$ and $\left(\frac{5}{4}\pi, \frac{9}{4}\pi\right)$, respectively. In this optimal configuration, the springs always apply a torque that pushes the two arms in oppositely oriented outward directions and therefore support the load against gravity.

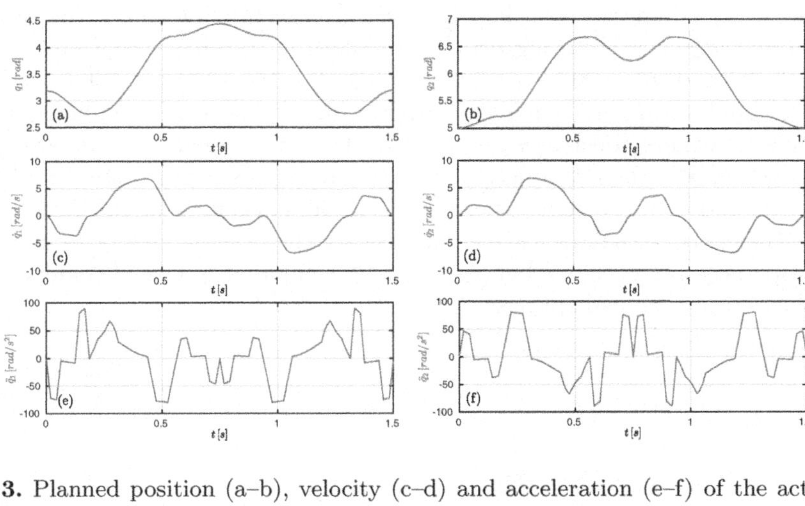

Fig. 3. Planned position (a–b), velocity (c–d) and acceleration (e–f) of the actuated joints.

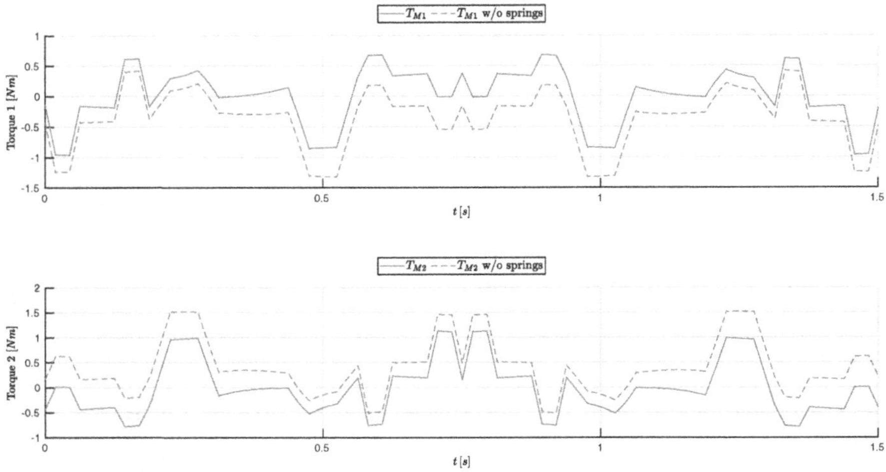

Fig. 4. (a) Torque profile of motor 1, (b) torque profile of motor 2.

The actuation torque profiles generated by the two motors are shown with continuous lines in Fig. 4. This figure also shows motor torques necessary to move the system without torsional springs denoted by dashed lines. It would be noted that two gearboxes with a 1/20 transmission ratio are interposed between each motor and the system. The comparison between the two torque profiles clearly shows that the addition of the torsional springs to the system decreases the mean torque values and hence reduces the energy expenditure.

In order to provide a clear overview of the evidence of the benefit of the co-exploitation of either energy-recovering devices and elastic-energy storing systems, Table 1 shows the values of the consumed energy per cycle for four different cases: (1) no energy reducing devices, (2) only braking recovering devices, (3) only elastic-energy storing devices, (4) both energy-recovering and elastic-energy storing devices. Starting from the baseline case in which there are neither torsional springs nor energy recovered during the motor braking phases, the consumed energy decreases about 70% if both are considered. Moreover, this demonstrates that the introduction of torsional springs to the system leads to a further reduction of consumed energy. This is valid also for systems that already reduce the energy expenditure, e.g via recovery the braking energy as in case 3. Case 4 reduces the consumed energy by 43% compared to case 3 thanks to the introduction of torsional springs.

Table 1. Consumed energy of different cases, where (•) denotes use of recovered braking energy or torsional springs, while (∘) does not. Note that the energy reduction uses case 1 as the baseline for the comparison.

Case	Recovered braking energy	Torsional springs	E [J]	Energy reduction [%]
1	∘	∘	33.4	-
2	•	∘	17.3	48.2
3	∘	•	19.7	41.0
4	•	•	9.8	70.7

5 Conclusion

In this work a mixed approach to reduce the energy consumption of a Delta-2 robot performing cyclic operations has been proposed. In particular, torsional springs coupled in parallel with two energy-recuperator drive axles are exploited to reduce the torques supplied by the motors and, thus, the consumed energy. The dynamic model of a Delta-2 robot has been developed and implemented in MATLAB, while the design optimization is controlled via Python to find both the values of spring stiffness and equilibrium angular positions. As a test case, a pick-and-place operation has been chosen considering a prescribed trajectory.

The results demonstrate how this technique has distinct advantages in terms of reduction of consumed energy. This is also shown on systems, which are outfitted with further energy reduction systems. It was possible to reduce the energy expenditure by 43% for a system with braking energy recuperation, or by 50% for a system with torsional springs with optimized stiffness values.

Though this approach leads to a significant reduction in energy consumption, the torsional springs are designed for a specific operation, i.e. path, cycle time, stopping time, initial and final positions. In this sense, future works will consider techniques including variable stiffness springs in order to reduce energy expenditure for more general cases with the goal of reducing or even removing the dependency on a specific task.

Acknowledgement. This work was partially supported by the Free University of Bozen-Bolzano funds within the project TN2803: "Mech4SME3: Mechatronics for Smart Maintenance and Energy Efficiency Enhancement".

References

1. Carabin, G., Wehrle, E., Vidoni, R.: A review on energy-saving optimization methods for robotic and automatic systems. Robotics **6**(4), 39 (2017)
2. Kapoor, R., Parveen, C.: Comparative study on various KERS. Lect. Notes Eng. Comput. Sci. **3**, 1969–1973 (2013)
3. Babitsky, V.I., Shipilov, A.: Resonant Robotic Systems. Springer, Berlin (2003)

4. Barreto, J.P., Schler, F.J., Corves, B.: The concept of natural motion for pick and place operations. In: New Advances in Mechanisms, Mechanical Transmissions and Robotics, Mechanisms and Machine Science, p. 46 (2017)
5. Boscariol, P., Boschetti, G., Gallina, P., Passarini, C.: Spring design for motor torque reduction in articulated mechanisms. Mech. Mach. Sci. **49**, 557–564 (2018)
6. Wang, J., Gosselin, C.M.: Static balancing of spatial three-degree-of freedom parallel mechanisms. Mech. Mach. Theory **34**(3), 437–452 (1999)
7. Boscariol, P., Carabin, G., Gasparetto, A., Lever, N., Vidoni, R.: Generation, energy-efficient point-to-point trajectory, for industrial robotic machines. In: Proceedings of the ECCOMAS-Multibody Dynamics, Catalonia, Spain, Barcelona, 29 June–2 July (2015)
8. Richiedei, D., Trevisani, A.: Analytical computation of the energy-efficient optimal planning in rest-to-rest motion of constant inertia systems. Mechatronics **39**, 147–159 (2016)
9. Carabin, G., Vidoni, R., Wehrle, E.: Energy saving in mechatronic systems through optimal point-to-point trajectory generation via standard primitives. Mech. Mach. Sci. **68**, 20–28 (2019)
10. Boscariol, P., Richiedei, D.: Energy-efficient design of multipoint trajectories for Cartesian robots. Int. J. Adv. Manuf. Technol. (2019, in press)
11. Gualtieri, L., Rojas, R., Carabin, G., Palomba, I., Rauch, E., Vidoni, R., Matt, D.T.: Advanced automation for SMEs in the I4.0 revolution: engineering education and employees training in the smart mini factory laboratory. In: IEEE International Conference on Industrial Engineering and Engineering Management (IEEM 2018) (2019)
12. Dai, Y.H., Schittkowski, K.: A sequential quadratic programming algorithm with non-monotone line search. Pacific J. Optim. **4**, 335–351 (2008)
13. Schittkowski, K., NLPQLP: a fortran implementation of a sequential quadratic programming algorithm with distributed and non-monotone line search. User's guide, version 4.0 (2013), Department of Computer Science, University of Bayreuth
14. Wehrle, E.J., Xu, Q., Baier, H.: Investigation, optimal design and uncertainty analysis of crash-absorbing extruded aluminium structures. Procedia CIRP **18**, 27–32 (2014)
15. Perez, R., Jansen, P., Martins, J.: PyOpt: a python-based object-oriented framework for nonlinear constrained optimization. Struct. Multi. Optim. **45**, 101–118 (2012)

Training a Four Legged Robot via Deep Reinforcement Learning and Multibody Simulation

Simone Benatti(✉), Alessandro Tasora, and Dario Mangoni

Department of Engineering and Architecture,
Università degli Studi di Parma, Parco Area delle Scienze, 181/A, 43124 Parma, Italy
simone.benatti@studenti.unipr.it,
{alessandro.tasora,dario.mangoni}@unipr.it

Abstract. In this paper we use the *Proximal Policy Optimization* (PPO) deep reinforcement learning algorithm to train a Neural Network to control a four-legged robot in simulation. Reinforcement learning in general can learn complex behavior policies from simple state-reward tuples datasets and PPO in particular has proved its effectiveness in solving complex tasks with continuous states and actions. Moreover, since it is model-free, it is general and can adapt to changes in the environment or in the robot itself.

The virtual environment used to train the agent was modeled using our physics engine *Project Chrono*. Chrono can handle non smooth dynamics simulation allowing us to introduce stiff leg-ground contacts and using its Python interface *Pychrono* it can be interfaced with the Machine Leaning framework *TensorFlow* with ease. We trained the Neural Network until it learned to control the motor torques, then various policy Neural Network input state choices have been compared.

1 Introduction

Neural networks has opened a wide range of new possibilities to robotic control. The last few years have seen a number of significant results accomplished thanks to the application of Deep Reinforcement Learning (DRL) to robotic tasks. Robots controlled by Neural Networks (NN) can solve tasks that would be hard to accomplish via hard coded policies, involving visual inputs from cameras and object manipulation [4]. In addition, they benefits from improved adaptivity, thus being able to work in unstructured environments.

Although the advantages brought by this approach are undeniable, the reinforcement learning process require numerous agent-environment interactions to converge. For this reason DRL has enhanced the interest in multibody simulation, since the use of virtual environments reduces time and costs of the process; that being said, because of the gap between reality and simulation the transfer of the learned policy on the physical robot can become troublesome.

In recent studies [5] the simulation of stiff contacts has been challenged using smooth contacts and domain randomization. While this approach has led to

A. Kecskeméthy and F. Geu Flores (Eds.): ECCOMAS 2019, COMPUTMETHODS 53, pp. 391–398, 2020.
https://doi.org/10.1007/978-3-030-23132-3_47

remarkable results, showing the possibilities brought to robotics by DRL, modeling stiff contacts could further improve the potential of this approach. In our work, we leverage our open-source physical engine Project Chrono [10] which is capable of simulating non-smooth dynamics, a feature that is especially welcome when dealing with simulations of walking robots. To this end we developed a Python wrapper that enables the inter-operation of the multibody software with the TensorFlow AI framework and, as a benchmark, we provide the case of a four-legged robot being trained to walk straight.

2 Problem Description

In this work we wanted a simplified model of a four-legged robot to learn to walk straight as fast as possible, without touching the ground with its abdomen. The robot is composed of a central spherical body and 4 legs and 4 ankles constrained with revolute joints, thus the system number of degrees of freedom is 14. Each hinge is actuated, therefore the agent can control the 8 torques acting on the joints (Fig.1).

Fig. 1. Four-Legged walker

The contact with the ground (a smooth floor) is due to 4 spherical contact shapes (feet) at the and of each ankle. Differently from other widespread Python modules used for benchmarking Deep Learning algorithms in robotic tasks [1] in our work we simulated non-smooth dynamics, therefore the feet-ground contact is stiff.

In addition, we wanted the learning model to be model-free in order to achieve a general solution. Model-based approaches require a model able to estimate the future state given the present state, of course this would set a limit upon the generality and flexibility of the approach.

3 Applied Methods

3.1 Proximal Policy Optimization

Reinforcement Learning (RL) [9] is a machine learning technique in which the software agent can interact with the *environment* taking *actions* to maximize a *reward.* The learning process is based on iterative interactions between the agent and the environment: at each step the agents observes the system state s and gets a reward r. According to the *policy* π the agents takes an action a, influencing the environment, then new states and reward are observed and the process is repeated. The s, a, r tuples collected are used to fit a function approximator (typically a NN). The application of Deep Neural Networks to RL is called Deep Reinforcement Learning (DRL). In Policy Gradient methods (such as the one used in this paper) the NN trained in the process is the Policy itself. In order to fit NN parameters several algorithms have been developed. For its capability of solving complex tasks with continuous states and actions, and robotic tasks in particular, the Proximal Policy Optimization algorithm (PPO) [8] has been chosen. PPO is an actor-critic algorithm. The *actor* is the policy π_θ, which is the NN described by the set of parameters θ prescribing the action depending on the state. The *critic* VF_ς is another NN estimating the Value Function $V^\pi(s_t)$ which is the expected value of the sum of discounted rewards, also depending on the state.

3.1.1 Critic: Estimating the Advantage Function

We maximize the expected sum of γ-discounted rewards by maximizing the advantage function $A^\pi(s_t, a_t)$, which is the advantage from taking the action a_t at timestep t compared to following the policy π. It can be expressed as:

$$A^\pi(s, a) = r_t - V^\pi(s_t) + \gamma V^\pi(s_{t+1}) \tag{1}$$

To estimate the Advantage Function we follow the approach proposed by Schulman [7], in which a NN is used to evaluate the Value Function $V^\pi(s_t)$. It should be noted that since we collect the reward at each step we have the exact Value Function (the sum of discounted rewards) at each timestep, so we can operate standard Supervised Learning on the critic, minimizing the squared error between the observed Value Function and the output of $VF_\varsigma(s_t)$. To get an estimate of $A^\pi(s_t, a_t)$ from $V^\pi(s_t)$ we evaluate the terms $\hat{A}_t^{(k)}$ defined below:

$$\hat{A}_t^{(k)} := \sum_{l=0}^{k-1} \gamma^l \delta_{t+l}^V = -V(s_t) + r_t + \gamma r_{t+1} + .. + \gamma^{k-1} r_{t+k-1} + \gamma^k V(s_{t+k}) \tag{2}$$

where $\delta_{t+l}^V = r_{t+l} - V_{t+l} + \gamma V_{t+l+1}$ is the Temporal Difference (TD) residual of V with discount γ.

Now, evaluating the λ exponentially-weighted average we obtain the generalized advantage estimator GAE(γ, λ):

$$\begin{aligned} \hat{A}_t^{GAE(\gamma,\lambda)} :=& (1-\lambda)\left(\hat{A}_t^{(1)} + \lambda \hat{A}_t^{(2)} + \lambda^2 \hat{A}_t^{(3)} + ...\right) \\ =& \sum_{l=0}^{\infty} (\gamma\lambda)^l \delta_{t+l}^V \end{aligned} \tag{3}$$

3.1.2 Actor: Penalty-Based PPO

We implemented the penalized version [2] of the PPO algorithm, in which the parameter update maximizes the objective function 4:

$$\sum_{t=0}^{\infty}\gamma^t\frac{\pi_\theta(a_t|s_t)}{\pi_{\theta_{old}}(a_t|s_t)}A(s_t,a_t) - \beta_k D_{KL}(\theta_{old}||\theta) - \eta\max(0, D_{KL}(\theta_{old}||\theta) - 2KL_{targ})^2 \tag{4}$$

Where

- θ is the set of weights (or parameters) defining the Neural Network.
- the subscript *old* refers to the weights at the beginning of the policy update step.
- $D_{KL}(\theta_{old}||\theta)$ is the Kullback Leibler divergence between the old and he latest policy. This is a measure of the extent of the policy update.
- β_k is the penalty coefficient, adapted during training as suggested in [8].
- KL_{targ} is the target KL divergence between the old ant the new policy.

To the Trust-Region objective [6] (the first term of the sum) a KL-Penalty and KL_{target} loss are applied to control the policy update, preventing the algorithm from taking too large parameters update. This approach allows to perform multiple update epochs at each optimization step without the risk of diverging too much from the old policy.

3.2 Neural Network Architecture

The NN used for the policy has 3 hidden layers between inputs and outputs. The input is the robot kinematic state $q, \dot{q}$ and 4 more inputs given by feet contact. Considering the total dimension of the input we should have 32 (14*2+4) neurons in the input layer, but since the planar position if the center of gravity is not fed to the network we only have 30. The actions are the 8 motor torques. More specifically the output is a set of 8 means that, together with a set of 8 variances, defines a multivariate Gaussian from which torque values are sampled. Value Function NN has 3 hidden layer as well and the same state inputs but has obviously only 1 output (the Value Function). We used *tanh* as activation function for both NNs and ADAM optimizer to fit their parameters.

3.3 Reward Shaping

Shaping reward bonuses and penalties leads the policy to undertake wanted behaviors and avoid the unwanted. More specifically we introduced a *power* cost to avoid energy-wasting policies and a *joints at limit* cost to avoid local minima. On the other hand we rewarded the *advance speed* of the robot towards the target and the robot *survival* (the agent survives as long as its abdomen does not touch the floor). In RL literature the set of interactions between two successive resets (we reset the simulation whenever the robot falls or reaches the target distance) is usually referred to as *episode*.

3.4 Workflow

The NN training process can be divided into three phases:

- **Policy Rollout:** We launch a set (batch) of simulations in which the robot interacts with the environment using the latest policy π_θ. The whole of state, action and reward tuples collected during an episode is called *Trajectory*.
- **Trainset Building:** The batch of trajectories is gathered and we add $A^\pi(a, s)$ and the discounted sum of future reward at each timestep.
- **Weights Updating:** we feed the dataset to both the actor and the critic optimizers in order to update θ an ζ.

//Training loop pseudocode:
while training **do**
  //Batch:
  1. **For** episode in batch **do**
    //Episode:
    1. **while** not done **do**
      //Timestep:
      1. Perform an action according to π_θ
      2. Get state and reward
  2. Collect state, action reward tuples from the episodes (trajectories)
  3. Evaluate $V^\pi(s)$ according to VF_ζ at each timestep
  4. Evaluate $A^\pi(a,s)$ using GAE at each timestep
  5. Update θ with (s,a,A^π) tuples according to eq. 4
  6. Update ζ minimizing error with the discounted sum of rewards

---

We experienced that the simulation of large number of episodes might require long CPU times. This led us to experiment parallelization techniques. The data collecting process can be parallelized over several CPU cores running multiple parallel episodes and hence effectively speeding up the learning process.

4 Obtained Results and Discussion

After about 10000 episodes the NN learned to control the robot to quickly walk straight without falling and touching the ground (Figs. 2 and 3).

Time Feature
Initially we provided an additional input, feeding the simulation time to the agent. Since the final timesteps of the episode have lower expected sum of future rewards knowing when the episode is about to finish helps the critic estimate the Value Function. On the other hand, this approach adds an input increasing the NN size and consequently adding weights to train, moreover it limits the policy flexibility training it to walk a specific path length. We trained the NN

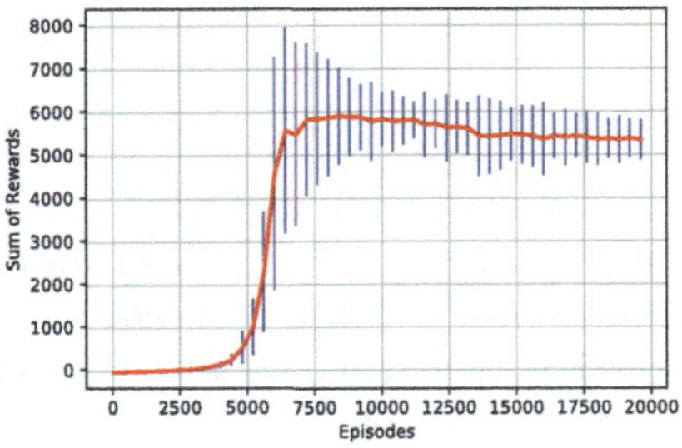

Fig. 2. Episode rew. sum (10 batches mean and var)

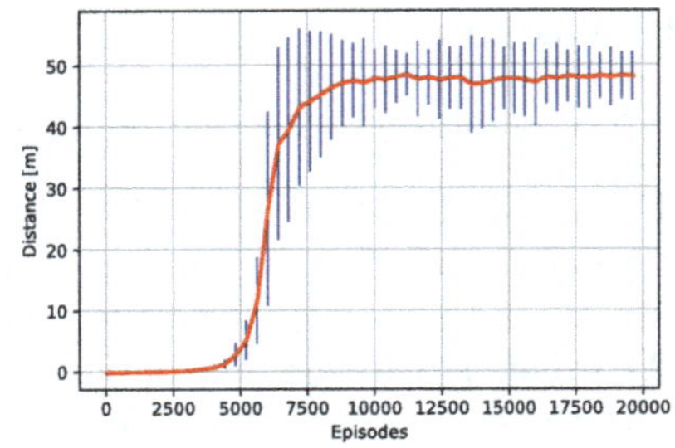

Fig. 3. Episode travelled dist. (10 batches mean and var)

both with and without time feature and comparing the results (Figs. 4 and 5) it can be noted that the latter converges earlier and it does not suffer from the absence of the time input.

Euler Angles vs Quaternions
We investigated how the choice of quaternions instead of Euler Angles affects the policy learning process. Quaternions, unlike Euler angles do not suffer problems of singularity, the drawback is an additional input node and consequently more weights to optimize. As showed in Figs. 6 and 7 the smaller NN taking Euler Angles as inputs learns faster as expected. It must be considered though, that while the advantage of this approach is limited, it can be undertaken only far from EulerAngles singularity conditions. For these reason, this approach is advisable only in particular cases after due consideration.

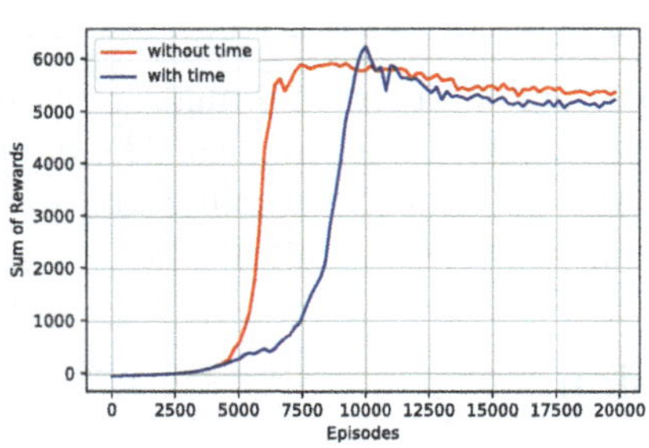

Fig. 4. Episode rew. sum (10 batches mean)

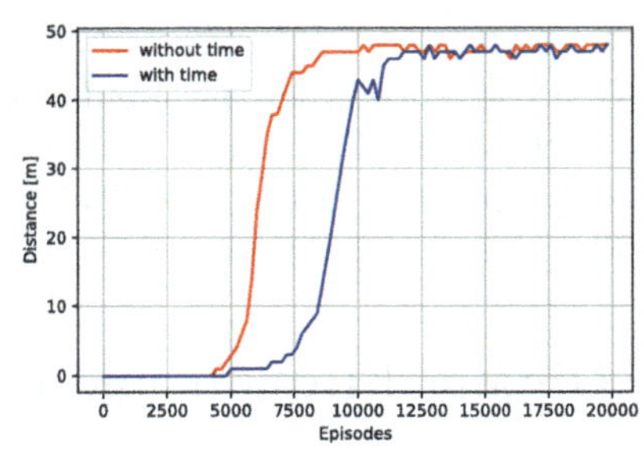

Fig. 5. Episode travelled dist. (10 batches mean)

Input Scaling
As suggested by LeCun [3], backpropagating the error is more efficient when observation are centered and scaled. The case of legged walker makes no exception, as showed in Figs. 8 and 9. The number of episodes needed to converge doubles and the distance reached is shorter in the unscaled case.

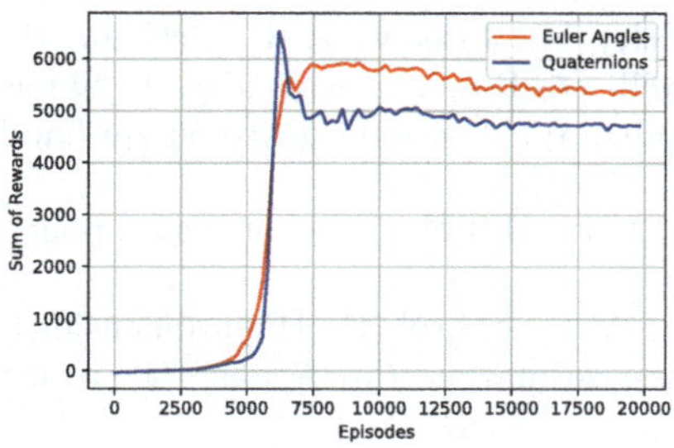

Fig. 6. Episode rew. sum (10 batches mean)

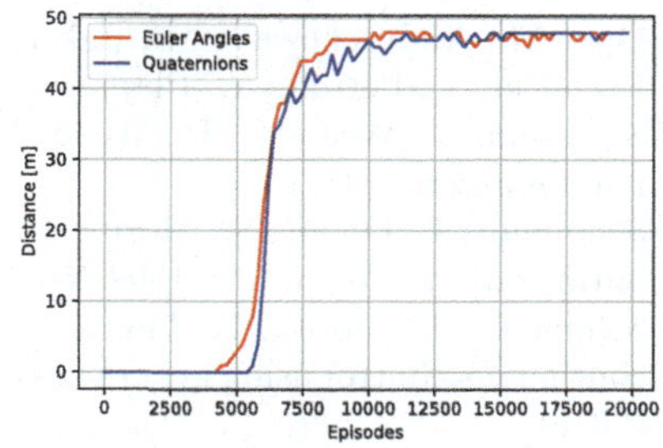

Fig. 7. Episode travelled dist. (10 batches mean)

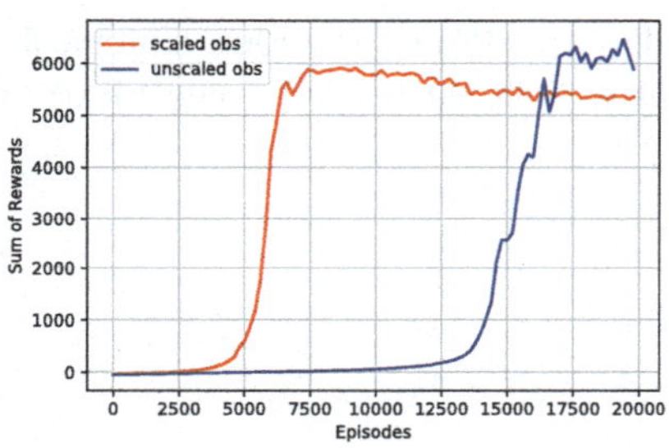

Fig. 8. Episode rew. sum (10 batches mean)

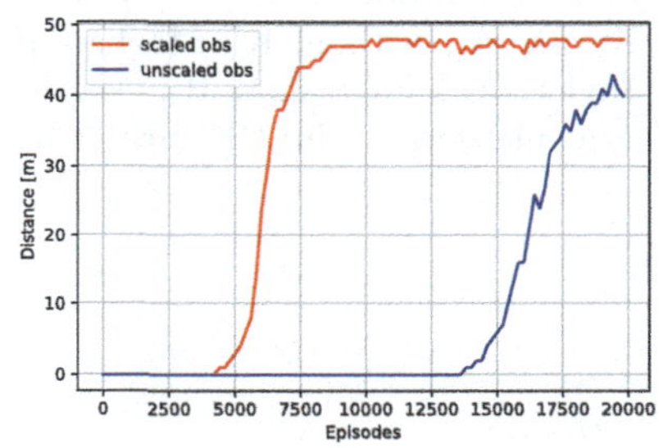

Fig. 9. Episode travelled dist. (10 batches mean)

5 Conclusion

This work highlighted the synergy between Multibody Simulation and Deep Reinforcement Learning, since simulations can provide the large amount of data needed by DRL in order to achieve a good policy. Furthermore, this approach could be further enhanced introducing MPI parallelization and render inputs with Convolutional Neural Networks. Applying these improvements we could train robots to navigate in complex environments, avoiding obstacles and moving on rough terrain.

References

1. Brockman, G., Cheung, V., Pettersson, L., Schneider, J., Schulman, J., Tang, J., Zaremba, W.: Openai gym. *CoRR*, abs/1606.01540 (2016)
2. Heess, N., Dhruva, T.B., Sriram, S., Lemmon, J., Merel, J., Wayne, G., Tassa, Y., Erez, T., Wang, Z., Eslami, S.M.A., Riedmiller, M.A., Silver, D.: Emergence of locomotion behaviours in rich environments. *CoRR*, abs/1707.02286 (2017)
3. LeCun, Y.A., Bottou, L., Orr, G.B., Müller, K.-R.: Efficient BackProp, pp. 9–48. Springer, Heidelberg (2012)
4. Levine, S., Pastor, P., Krizhevsky, A., Quillen, D.: Learning hand-eye coordination for robotic grasping with deep learning and large-scale data collection. *CoRR*, abs/1603.02199 (2016)

5. OpenAI., Andrychowicz, M., Baker, B., Chociej, M., Jozefowicz, R., McGrew, B., Pachocki, J., Petron, A., Plappert, M., Powell, G., Ray, A., Schneider, J., Sidor, S., Tobin, J., Welinder, P., Weng, L., Zaremba, W.: Learning dexterous in-hand manipulation (2018)
6. Schulman, J., Levine, S., Moritz, P., Jordan, M.I., Abbeel, P.: Trust region policy optimization. *CoRR*, abs/1502.05477 (2015)
7. Schulman, J., Moritz, P., Levine, S., Jordan, M.I., Abbeel, P.: High-dimensional continuous control using generalized advantage estimation. *CoRR*, abs/1506.02438 (2015)
8. Schulman, J., Wolski, F., Dhariwal, P., Radford, A., Klimov, O.: Proximal policy optimization algorithms. *CoRR*, abs/1707.06347 (2017)
9. Sutton, R.S., Barto, A.G.: Introduction to Reinforcement Learning, 1st edn. MIT Press, Cambridge (1998)
10. Tasora, A., Serban, R., Mazhar, H., Pazouki, A., Melanz, D., Fleischmann, J., Taylor, M., Sugiyama, H., Negrut, D.: Chrono: an open source multi-physics dynamics engine. In: HPCSE (2015)

Efficient Simulation and Real-Time Applications

Two General Index-3 Semi-Recursive Formulations for the Dynamics of Multibody Systems

Daniel Dopico Dopico(✉), Álvaro López Varela, and Alberto Luaces Fernández

Laboratorio de Ingeniería Mecánica, Universidade da Coruña,
Mendizábal s/n, 15403 Ferrol, Spain
{ddopico,alvaro.lopez1,aluaces}@udc.es

Abstract. A couple of decades ago, the dynamic simulation of complex multibody systems in real-time was an objective difficult to achieve. Nowadays other type of problems, more demanding in terms of computational time, need to be solved like, e.g., the design optimization or the optimal control of multibody systems requiring the fastest dynamics formulations available. MBSLIM (Multibody Systems at Laboratorio de Ingenieria Mecanica) multibody library includes some global formulations for the dynamics of multibody systems. The extension of the library to accommodate topological formulations in relative (joint) coordinates is in progress and the implementation of two of them is described in this work. With this extension in the scope, some topological formulations derived in the past are revisited, generalized and reformulated. The need for generalization of the previously published formulations, was detected because the equations proposed were not general enough to be integrated in an all-purpose multibody library in natural coordinates like MBSLIM, especially because both set of coordinates need to coexist, the definition of the mechanisms has to be the original one and the library has to be automatic and all the existing models have to work with the new approach. Moreover the new solver takes advantage of some problems solved in natural coordinates, like the initial position and velocity problems for closed-loop systems. Finally to test the new equations two benchmark problems are presented: a spatial slider-crank mechanism and the chain and anchor system of a ship.

1 Introduction

In the last decades, the incessant evolution of computers and simulation algorithms made possible to achieve the real time simulation of complex multibody systems on standard PCs. Therefore the challenges in multibody dynamics have changed and the scope is aimed at other types of problems, more demanding in terms of computational time like, e.g., the sensitivity analysis, the design optimization or the optimal control of multibody systems.

A. Kecskeméthy and F. Geu Flores (Eds.): ECCOMAS 2019, COMPUTMETHODS 53, pp. 401–408, 2020.
https://doi.org/10.1007/978-3-030-23132-3_48

Two main families of approaches for the dynamics of multibody systems are possible [1], the global methods define the position of each body independently of the rest of the bodies in the kinematic chain, while the topological methods define the position of each body with respect to the previous bodies in the kinematic chain. Both families of methods have advantages and drawbacks. MBSLIM [2] is a multibody library with forward kinematics, forward and inverse dynamics, static equilibrium and sensitivity analysis of multibody systems, among other capabilities. The library includes some dynamics global formulations on the basis of a natural coordinates formalism plus some angle and distance coordinates (mixed coordinates). The extension of the library to accommodate topological formulations in relative (joint) coordinates is in progress and the implementation of two of them is described in this work based on the works [3–5].

2 Topological ALI3-P Formulations in Relative (Joint) Coordinates

The topological ALI3-P formulations in relative (joint) coordinates presented here, are based on the one derived in [4,5]. The description and expressions provided in previous publications, lack the generality needed for a general implementation in the framework of *MBSLIM* multibody library, because of some of the assumptions and simplifications taken. The topological formulation of this section shares the same equations already presented in [3,6].

2.1 Kinematic Relations for Open Chain Multibody Systems

For a point i in body i and a point $i-1$ in body $i-1$ (Fig. 1), the general expressions of the relative velocities are the following,

$$\dot{\mathbf{r}}_i = \dot{\mathbf{r}}_i^i = \dot{\mathbf{r}}_i^{i-1} + \dot{\mathbf{r}}_i^{i,i-1} \tag{1a}$$

$$\dot{\mathbf{r}}_i^{i-1} = \dot{\mathbf{r}}_{i-1} + \boldsymbol{\omega}_{i-1} \wedge \left(\mathbf{r}_i - \mathbf{r}_{i-1}\right) \tag{1b}$$

$$\boldsymbol{\omega}_i = \boldsymbol{\omega}_{i-1} + \boldsymbol{\omega}_i^{i-1} \tag{1c}$$

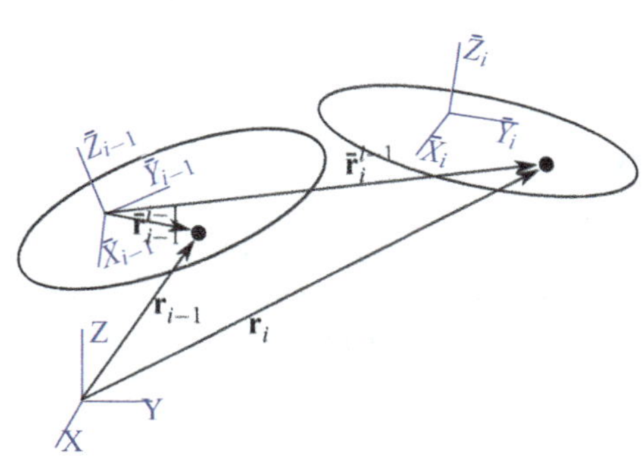

Fig. 1. Kinematics of the relative motion (bodies $i-1$ and i)

where (1b) is the velocity of point i as belonging to body $i-1$, and $\dot{\mathbf{r}}_i^{i,i-1}$ is the relative velocity of point i of body i with respect to body $i-1$, both expressed in global coordinates of the ground.

Taking derivatives in (1), similar relations for accelerations can be obtained,

$$\ddot{\mathbf{r}}_i = \ddot{\mathbf{r}}_i^i = \ddot{\mathbf{r}}_i^{i-1} + \ddot{\mathbf{r}}_i^{i,i-1} + 2\boldsymbol{\omega}_{i-1} \wedge \dot{\mathbf{r}}_i^{i,i-1} \tag{2a}$$

$$\ddot{\mathbf{r}}_i^{i-1} = \ddot{\mathbf{r}}_{i-1} + \dot{\boldsymbol{\omega}}_{i-1} \wedge (\mathbf{r}_i - \mathbf{r}_{i-1}) + \boldsymbol{\omega}_{i-1} \wedge [\boldsymbol{\omega}_{i-1} \wedge (\mathbf{r}_i - \mathbf{r}_{i-1})] \tag{2b}$$

$$\dot{\boldsymbol{\omega}}_i = \dot{\boldsymbol{\omega}}_{i-1} + \boldsymbol{\alpha}_i^{i-1} + \boldsymbol{\omega}_{i-1} \wedge \boldsymbol{\omega}_i^{i-1} \tag{2c}$$

where $\ddot{\mathbf{r}}_i^{i-1}$ is acceleration of point i with body $i-1$ and $\ddot{\mathbf{r}}_i^{i,i-1}$ is the relative acceleration of point i of body i in its relative motion with respect to body $i-1$. The last term $2\boldsymbol{\omega}_{i-1} \wedge \dot{\mathbf{r}}_i^{i,i-1}$ represents the Coriolis or complementary acceleration.

Gathering all the previous equations in matrix form, kinematic recursive relations with the following form can be obtained.

$$\mathbf{V}_i = \mathbf{B}_i^v \mathbf{V}_{i-1} + \mathbf{b}_i^v \dot{\mathbf{z}}_k \tag{3a}$$

$$\dot{\mathbf{V}}_i = \mathbf{B}_i^v \dot{\mathbf{V}}_{i-1} + \mathbf{b}_i^v \ddot{\mathbf{z}}_k + \mathbf{d}_i^v \tag{3b}$$

$$\mathbf{B}_i^v = \begin{bmatrix} \mathbf{I} & \tilde{\mathbf{r}}_{i-1} - \tilde{\mathbf{r}}_i \\ \mathbf{0} & \mathbf{I} \end{bmatrix} \; ; \; \mathbf{b}_i^y \dot{\mathbf{z}}_k = \begin{bmatrix} \dot{\mathbf{r}}_i^{i,i-1} \\ \boldsymbol{\omega}_i^{i-1} \end{bmatrix} \; ; \tag{3c}$$

$$\mathbf{b}_i^y \ddot{\mathbf{z}}_k + \mathbf{d}_i^y = \begin{bmatrix} \boldsymbol{\omega}_{i-1} \wedge [\boldsymbol{\omega}_{i-1} \wedge (\mathbf{r}_i - \mathbf{r}_{i-1})] + \ddot{\mathbf{r}}_i^{i,i-1} + 2\boldsymbol{\omega}_{i-1} \wedge \dot{\mathbf{r}}_i^{i,i-1} \\ \boldsymbol{\alpha}_i^{i-1} + \boldsymbol{\omega}_{i-1} \wedge \boldsymbol{\omega}_i^{i-1} \end{bmatrix} \tag{3d}$$

Terms $\mathbf{b}_i^y \dot{\mathbf{z}}_k$ and $\mathbf{b}_i^y \ddot{\mathbf{z}}_k + \mathbf{d}_i^y$ are particular for each type of kinematic joint because they depend on the relative velocities and accelerations $\dot{\mathbf{r}}_i^{i,i-1}$, $\boldsymbol{\omega}_i^{i-1}$, $\ddot{\mathbf{r}}_i^{i,i-1}$ and $\boldsymbol{\alpha}_i^{i-1}$ which have to be parametrized in terms of the relative (joint) coordinates, $\dot{\mathbf{z}}_k$. The details for different types of joints are omitted here for brevity.

2.2 RTdyn0 Dynamics Recursive Equations

The virtual power principle for a multibody system composed of n_b bodies,

$$\sum_{i=1}^{n_b} \begin{bmatrix} \dot{\mathbf{r}}_G^{i*} \\ \boldsymbol{\omega}_i^* \end{bmatrix} \left(\begin{bmatrix} m_i \mathbf{I}_3 & \mathbf{0} \\ \mathbf{0} & \mathbf{J}_i^G \end{bmatrix} \begin{bmatrix} \ddot{\mathbf{r}}_G^i \\ \dot{\boldsymbol{\omega}}_i \end{bmatrix} - \begin{bmatrix} \mathbf{f}_i \\ \mathbf{n}_i^G - \boldsymbol{\omega}_i \wedge \mathbf{J}_i^G \boldsymbol{\omega}_i \end{bmatrix} \right) = \mathbf{Y}^{*\mathrm{T}} \left[\mathbf{M}\dot{\mathbf{Y}} - \mathbf{Q} \right] = 0 \tag{4}$$

$$\mathbf{M} = \begin{bmatrix} \mathbf{M}_1 & \mathbf{0} & \dots & \mathbf{0} \\ \mathbf{0} & \mathbf{M}_2 & \dots & \mathbf{0} \\ \vdots & \vdots & \ddots & \vdots \\ \mathbf{0} & \mathbf{0} & \mathbf{0} & \mathbf{M}_{n_b} \end{bmatrix} ; \quad \mathbf{Q} = \begin{bmatrix} \mathbf{Q}_1 \\ \mathbf{Q}_2 \\ \vdots \\ \mathbf{Q}_{n_b} \end{bmatrix} ; \quad \mathbf{Y}_i = \begin{bmatrix} \mathbf{Y}_1 \\ \mathbf{Y}_2 \\ \vdots \\ \mathbf{Y}_{n_b} \end{bmatrix} \tag{5}$$

where the star indicates virtual velocities, $\mathbf{J}_i^G = \mathbf{A}_i \bar{\mathbf{J}}_i^G \mathbf{A}_i^{\mathrm{T}}$ is the inertia tensor in the center of mass, $\bar{\mathbf{J}}_i^G$ is the local (constant) inertia tensor, $\mathbf{f}_i$ is the external forces vector over the body and $\mathbf{n}_i^G$ is the external torque vector with respect to the center of mass. All the magnitudes are expressed in the global frame.

It can be proved, by means of the recursive kinematics expressions (3) particularized for the center of mass of each body as reference point, $\mathbf{r}_i = \mathbf{r}_G^i$ and $\mathbf{r}_{i-1} = \mathbf{r}_G^{i-1}$, the following linear relations between Cartesian and relative velocities and accelerations,

$$\mathbf{Y} = \mathbf{R}\dot{\mathbf{z}} \tag{6}$$

$$\mathbf{Y}^* = \mathbf{R}\dot{\mathbf{z}}^* \tag{7}$$

$$\dot{\mathbf{Y}} = \mathbf{R}\ddot{\mathbf{z}} + \dot{\mathbf{R}}\dot{\mathbf{z}} \tag{8}$$

where, $\mathbf{z} = \left[\mathbf{z}_1{}^{\mathrm{T}}\ \mathbf{z}_2{}^{\mathrm{T}} \cdots \mathbf{z}_{n_b}{}^{\mathrm{T}}\right] \in \mathbb{R}^{n_j}$ with n_j the total number of joint coordinates.

Then Eq. (4) becomes,

$$\dot{\mathbf{z}}^{*\mathrm{T}}\left[\left(\mathbf{R}^{\mathrm{T}}\mathbf{M}\mathbf{R}\right)\ddot{\mathbf{z}} - \mathbf{R}^{\mathrm{T}}\left(\mathbf{Q} - \mathbf{M}\dot{\mathbf{R}}\dot{\mathbf{z}}\right)\right] = 0 \tag{9}$$

All the terms involved in Eq. (9) can be calculated recursively. The details about these derivations are too long to be covered here and they are left to the reader.

2.3 RTdyn1 Dynamics Recursive Equations

The RTdyn1 formulation results from choosing the point coincident with the origin of each body as reference point $\mathbf{r}_i = \mathbf{r}_0^i$. The following kinematic relations hold, between the reference point, $\mathbf{r}_0^i$, and the center of mass, $\mathbf{r}_G^i$, and between Cartesian and relative velocities and accelerations:

$$\mathbf{Y}_i^* = \begin{bmatrix}\dot{\mathbf{r}}_G^{i*} \\ \boldsymbol{\omega}_i^*\end{bmatrix} = \begin{bmatrix}\mathbf{I} & -\tilde{\mathbf{r}}_G^i \\ \mathbf{0} & \mathbf{I}\end{bmatrix}\begin{bmatrix}\dot{\mathbf{r}}_0^i \\ \boldsymbol{\omega}_i^*\end{bmatrix} = \mathbf{D}_i\mathbf{Z}_i^* \tag{10}$$

$$\dot{\mathbf{Y}}_i = \begin{bmatrix}\ddot{\mathbf{r}}_G^i \\ \dot{\boldsymbol{\omega}}_i\end{bmatrix} = \begin{bmatrix}\mathbf{I} & -\tilde{\mathbf{r}}_G^i \\ \mathbf{0} & \mathbf{I}\end{bmatrix}\begin{bmatrix}\ddot{\mathbf{r}}_0^i \\ \dot{\boldsymbol{\omega}}_i\end{bmatrix} + \begin{bmatrix}\boldsymbol{\omega}_i \wedge \left(\boldsymbol{\omega}_i \wedge \mathbf{r}_G^i\right) \\ \mathbf{0}\end{bmatrix} = \mathbf{D}_i\dot{\mathbf{Z}}_i + \mathbf{e}_i \tag{11}$$

$$\mathbf{Z} = \mathbf{R}^z\dot{\mathbf{z}} \tag{12}$$

$$\mathbf{Z}^* = \mathbf{R}^z\dot{\mathbf{z}}^* \tag{13}$$

$$\dot{\mathbf{Z}} = \mathbf{R}^z\ddot{\mathbf{z}} + \dot{\mathbf{R}}^z\dot{\mathbf{z}} \tag{14}$$

Applying the previous relations to the virtual power equations (4), a single equation, very similar to (9), is obtained,

$$\dot{\mathbf{z}}^{*\mathrm{T}}\left[\left(\mathbf{R}^{z\mathrm{T}}\mathbf{M}^z\mathbf{R}^z\right)\ddot{\mathbf{z}} - \mathbf{R}^{z\mathrm{T}}\left(\mathbf{Q}^z - \mathbf{M}^z\dot{\mathbf{R}}^z\dot{\mathbf{z}}\right)\right] = 0 \tag{15}$$

2.4 Topological ALI3-P: Equations of Motion

In this section, the extension to non-minimal coordinates of the RTdyn0 equations derived in Sect. 2.2, is accomplished. For the RTdyn1 equations the extension is similar, since both formulations have very similar final expressions. Writing the RTdyn0 equations, (9), in a more compact form:

$$\dot{\mathbf{z}}^{*\mathrm{T}}\left[\mathbf{M}^d\ddot{\mathbf{z}} - \mathbf{Q}^d\right] = 0 \tag{16}$$

$$\mathbf{M}^d = \mathbf{R}^{\mathrm{T}}\mathbf{M}\mathbf{R} \tag{17}$$

$$\mathbf{Q}^d = \mathbf{R}^{\mathrm{T}}\left(\mathbf{Q} - \mathbf{M}\dot{\mathbf{R}}\dot{\mathbf{z}}\right) \tag{18}$$

For closed chain systems, open loop systems in non-minimal coordinates, or system with explicit constraints relating the states, the virtual velocities, $\dot{\mathbf{z}}$, in (9) are not independent, therefore virtual and real states are related by some constraints,

$$\boldsymbol{\Phi}(\mathbf{q}, \mathbf{z}, t) = \mathbf{0} \tag{19}$$

$$\dot{\boldsymbol{\Phi}} = \mathbf{0} \Rightarrow \qquad (\boldsymbol{\Phi}_{\mathbf{q}}\mathbf{q}_{\mathbf{z}} + \boldsymbol{\Phi}_{\mathbf{z}})\,\dot{\mathbf{z}} = -\boldsymbol{\Phi}_t = \mathbf{b} \tag{20}$$

$$\ddot{\boldsymbol{\Phi}} = \mathbf{0} \Rightarrow (\boldsymbol{\Phi}_{\mathbf{q}}\mathbf{q}_{\mathbf{z}} + \boldsymbol{\Phi}_{\mathbf{z}})\ddot{\mathbf{z}} = -\left(\dot{\boldsymbol{\Phi}}_{\mathbf{q}}\mathbf{q}_{\mathbf{z}} + \boldsymbol{\Phi}_{\mathbf{q}}\dot{\mathbf{q}}_{\mathbf{z}} + \dot{\boldsymbol{\Phi}}_{\mathbf{z}}\right)\dot{\mathbf{z}} - \dot{\boldsymbol{\Phi}}_t = \mathbf{c} \tag{21}$$

$$\delta\boldsymbol{\Phi}^*(\delta\mathbf{q}, \delta\mathbf{z}) = \mathbf{0} \Rightarrow \qquad (\boldsymbol{\Phi}_{\mathbf{q}}\mathbf{q}_{\mathbf{z}} + \boldsymbol{\Phi}_{\mathbf{z}})\,\delta\mathbf{z}^* = \mathbf{0} \tag{22}$$

Observe that the previous dependencies cover the general case in which there are some loop closing constraints, typically dependent on the natural coordinates $\mathbf{q} \in \mathbb{R}^{n_c}$, some Euler parameters normalization constraints for the spherical and floating joints which directly depend on $\mathbf{z}$ and even some rheonomic driving constraints or any other user constraint. This approach is very powerful because it means that the topological formulation can take advantage of all the constraint equations (and their derivatives) implemented in MBSLIM for natural coordinates.

Applying the ALI3-P scheme, presented in [7]:

$$\left[\mathbf{M}\ddot{\mathbf{z}} + (\boldsymbol{\Phi}_{\mathbf{q}}\mathbf{q}_{\mathbf{z}} + \boldsymbol{\Phi}_{\mathbf{z}})^{\mathrm{T}}\left(\boldsymbol{\lambda}^{*(i+1)} + \boldsymbol{\alpha}\boldsymbol{\Phi}\right)\right]_{n+1} = \mathbf{Q}_{n+1} \tag{23a}$$

$$\boldsymbol{\lambda}^{*(i+1)}_{n+1} = \boldsymbol{\lambda}^{*(i)}_{n+1} + \boldsymbol{\alpha}\boldsymbol{\Phi}^{(i+1)}_{n+1};\ i > 0 \tag{23b}$$

where $\boldsymbol{\alpha}$ is a diagonal matrix that contains the penalty factors associated with the constraints, n is the time step index, i is the iteration index of the approximate Lagrange multipliers $\boldsymbol{\lambda}^*_{n+1}$.

Proceeding like in the global formulation, the following expressions are valid for the projections of velocities and accelerations:

$$\left(\mathbf{M} + (\boldsymbol{\Phi}_{\mathbf{q}}\mathbf{q}_{\mathbf{z}} + \boldsymbol{\Phi}_{\mathbf{z}})^{\mathrm{T}}\boldsymbol{\alpha}(\boldsymbol{\Phi}_{\mathbf{q}}\mathbf{q}_{\mathbf{z}} + \boldsymbol{\Phi}_{\mathbf{z}})\right)\dot{\mathbf{z}}^{(i+1)} = \mathbf{M}\dot{\mathbf{z}}^* - (\boldsymbol{\Phi}_{\mathbf{q}}\mathbf{q}_{\mathbf{z}} + \boldsymbol{\Phi}_{\mathbf{z}})^{\mathrm{T}}\boldsymbol{\alpha}\boldsymbol{\Phi}_t \tag{24}$$

$$\begin{aligned}&\left(\mathbf{M} + (\boldsymbol{\Phi}_{\mathbf{q}}\mathbf{q}_{\mathbf{z}} + \boldsymbol{\Phi}_{\mathbf{z}})^{\mathrm{T}}\boldsymbol{\alpha}(\boldsymbol{\Phi}_{\mathbf{q}}\mathbf{q}_{\mathbf{z}} + \boldsymbol{\Phi}_{\mathbf{z}})\right)\ddot{\mathbf{z}}^{(i+1)} \\ &= \mathbf{M}\ddot{\mathbf{z}}^* - (\boldsymbol{\Phi}_{\mathbf{q}}\mathbf{q}_{\mathbf{z}} + \boldsymbol{\Phi}_{\mathbf{z}})\,\boldsymbol{\alpha}\left(\left(\dot{\boldsymbol{\Phi}}_{\mathbf{q}}\mathbf{q}_{\mathbf{z}} + \boldsymbol{\Phi}_{\mathbf{q}}\dot{\mathbf{q}}_{\mathbf{z}} + \dot{\boldsymbol{\Phi}}_{\mathbf{z}}\right)\dot{\mathbf{z}} + \dot{\boldsymbol{\Phi}}_t\right)\end{aligned} \tag{25}$$

3 Numerical Experiments

3.1 Spatial Slider-Crank Mechanism

The first numerical experiment presented is the spatial slider-crank mechanism, [8], described in the Library of Computational Benchmark Problems of the IFToMM Technical Committee for Multibody Dynamics based on the example included in [9]. The system is composed of 3 moving bodies: the crank, connected to the ground by revolute joint; the slider, connected to the ground (fixed body) by a prismatic joint; the connecting rod, joined to the crank by a spherical joint and to the slider by a Cardan joint.

A simulation of 5 s was run, under the action of gravity and with the same initial conditions indicated in [8]. The position and velocity of the slider and the crank angle are shown in Fig. 2. The results of the topological formulation perfectly match the global formulation ones and the benchmark reference solution.

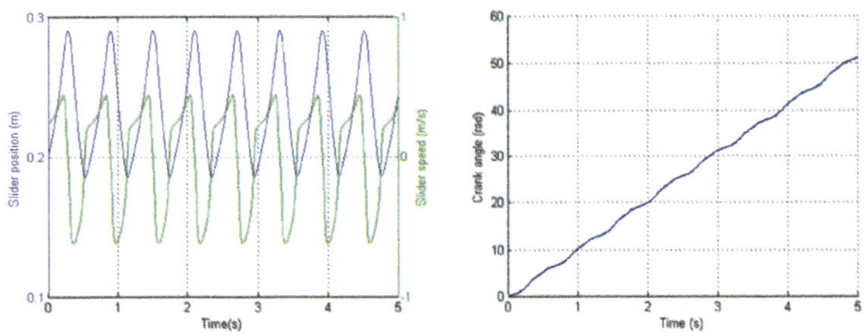

Fig. 2. Slider position and velocity (left); crank angle (right)

3.2 Chain and Anchor System of a Ship

The second example is the simulation of an anchor maneuver of a ship, recently described in [6]. The system is composed of the chain, with a large number of links of different types, and the anchor composed of two bodies linked with a revolute joint. The contacts between the different links of the chain are considered by means of Cardan joints with friction. The system is subjected to the tension force on the chain, the gravity forces, the constraint forces and the contact forces with the hull of the ship. The contacts between anchor and hull or links and hull use the algorithms described in [6].

Figure 3 shows two instants of the simulation of the anchor maneuver with the representation of contact forces between chain, anchor and hull.

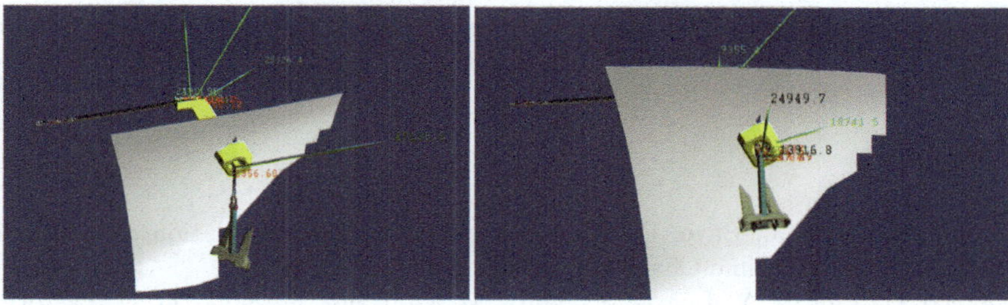

Fig. 3. Simulation of the anchor maenuver

4 Conclusions

The Augmented Lagrangian formulation with projections of velocities and accelerations (ALI3-P formulation) was successfully combined with the recursive calculation of the dynamic terms based on the topology of the system. The result are two semi-recursive ALI3-P topological formulations in joint (relative) coordinates.

The formulations have the generality needed to be implemented in a general multibody dynamics code, in fact the resulting code was implemented in the framework of MBSLIM, a generic library for the dynamics of multibody systems in natural coordinates, whose functionality was extended to accommodate these recursive formulations in joint coordinates. Part of the merit of new implementation is that all the methods, the global one in natural coordinates and the topological ones in relative coordinates are perfectly integrated, compatible and share most of the code to evaluate forces, constraint equations, etc, in fact the new solver can take advantage of some problems solved in natural coordinates, like the multibody model definition, the initial position and initial velocity problems for closed-loop systems or the static equilibrium position problem.

In order to prove the new implementation a batch of numerical experiments have been tested covering different situations: open and closed-loop systems, 2D and 3D systems, all the kinematic joints implemented, additional user constraints (including rheonomic) and different types of forces (spring-damper, contact, tire, etc.). The solutions obtained using the global formulation and the topological formulations have been compared between them (and with reference solutions whenever they exist) in order to guarantee the correctness and accuracy of the new topological implementation. Two representative examples have been chosen to illustrate these tests: the first one is a spatial slider-crank mechanism chosen because it is a 3D example which covers a good variety of kinematic joints; the second one is the simulation of the anchor maneuver of a ship, chosen because it represents a complex, realistic and industrial multibody system with complex phenomena which perfectly illustrates the capabilities of the method and the recent implementation.

Acknowledgements. The support of the Spanish Ministry of Economy and Competitiveness (MINECO) under project DPI2016-81005-P is greatly acknowledged.

References

1. Garcia de Jalon, J., Bayo, E.: Kinematic and Dynamic Simulation of Multibody Systems: The Real-time Challenge. Springer, New York (1994)
2. Dopico, D., Luaces, A., Lugrís, U., Saura, M., González, F., Sanjurjo, E., Pastorino, R.: Mbslim: Multibody systems en laboratorio de ingeniería mecánica, 2009–2016
3. Dopico, D., González, F., Cuadrado, J., Kövecses, J.: Determination of holonomic and nonholonomic constraint reactions in an index-3 augmented Lagrangian formulation with velocity and acceleration projections. J. Comput. Nonlinear Dyn. **9**(4), 041006–041006 (2014). https://doi.org/10.1115/1.4027671
4. Dopico, D.: Formulaciones semi-recursivas y de penalización para la dinámica en tiempo real de sistemas multicuerpo. Ph.D. thesis, Universidade da Coruña (2004)
5. Cuadrado, J., Dopico, D., Gonzalez, M., Naya, M.: A combined penalty and recursive real-time formulation for multibody dynamics. J. Mech. Des. Trans. ASME **126**(4), 602–608 (2004). https://doi.org/10.1115/1.1758257
6. Dopico, D., Luaces, A., Saura, M., Cuadrado, J., Vilela, D.: Simulating the anchor lifting maneuver of ships using contact detection techniques and continuous contact force models. Multibody Syst. Dyn. (2019). https://doi.org/10.1007/s11044-019-09670-8
7. Dopico, D., González, F., Cuadrado, J., Kovecses, J.: Determination of holonomic and nonholonomic constraint reactions in an index-3 augmented Lagrangian formulation with velocity and acceleration projections. J. Comput. Nonlinear Dyn. **9**(4), 041006–041006 (2014). https://doi.org/10.1115/1.4027671
8. I. T. C. for Multibody Dynamics. Library of computational benchmark problems (2014). http://www.iftomm-multibody.org/benchmark
9. Haug, E.J.: Computer Aided Kinematics and Dynamics of Mechanical Systems: Basic Methods. Allyn and Bacon, Prentice Hall College Div, Boston (1989)

Real-Time Capable Calculation of Reaction Forces of Multibody Systems Using Optimized Bushings on the Example of a Vehicle Wheel Suspension

Frédéric Etienne Kracht(✉) and Dieter Schramm

Faculty of Engineering, University of Duisburg-Essen,
Lotharstr.1, 47057 Duisburg, Germany
{frederic.kracht,dieter.schramm}@uni-due.de

Abstract. This paper presents an object-oriented modeling method capable of simulating the dynamics including the reaction forces of multibody systems with kinematic loops in hard real-time, called RTOOM. The modeling method describes the system by explicit equations, which can be solved numerically stable with a standard explicit numerical integrator with fixed step size. By knowing the application and the desired accuracy, the model can be adapted to fit the problem. Algebraic loops are resolved with low-pass filters parameterized for the frequency range of the application. Bushings with optimized spring and damping constants are used to avoid iterative methods for solving kinematics loops. For the optimization, a high accurate, non-manipulated and non-real-time multibody model is used. The optimization targets are stability, computing time and accuracy. The double wishbone suspension of the Formula Student racing car A40-02 of the University of Duisburg-Essen is used as an example. It has been successfully proven that a simulation up to 30 Hz with a required step size of 1 ms can be achieved. The simulation results show a very good accuracy up to 15 Hz with a deviation of the force below 4% and the acceleration below 7%. If the parameterization of the bushings remains the same, the accuracy is still acceptable even at higher frequencies.

1 Introduction

The calculation of reaction forces is required for many applications, such as the dimensioning of joints and bearings, for strength estimation or wear prognoses. Reaction forces must sometimes be taken into account for the calculation of deformations. When applying d'Alembert's principle for the description in minimal coordinates, however, the reaction forces are eliminated when calculating the equations of motion. If, on the other hand, the Newton-Euler equations for multibody systems are used to obtain the reaction forces without applying d'Alembert's principle, the result is no longer a pure differential equation system (ODE). Instead, the result is a coupled differential-algebraic equation system (DAE). DAE causes numerical difficulties and can only be solved directly with specialized numerical methods, see [2] or [3]. This problem solution does not guarantee hard real-time capability, since the use of

A. Kecskeméthy and F. Geu Flores (Eds.): ECCOMAS 2019, COMPUTMETHODS 53, pp. 409–416, 2020.
https://doi.org/10.1007/978-3-030-23132-3_49

differential-algebraic equation solvers is necessary. For some applications, however, real-time capability is mandatory, e.g. when used in a driving simulator or when controlling an active chassis in a real vehicle. In addition, kinematic loops in the topology of the system lead to further problems, which also affect the real-time capability of the simulation. The closing conditions on position level are usually only available as implicit equations and can in this case only be solved iteratively, e.g. using the Newton-Raphson method [4]. Only then, the equations of motion can be set up in minimal coordinates.

This paper presents an object-oriented modeling method (RTOOM) to calculate reaction forces of a multibody system with closed kinematic loops in hard real-time, which is derived from real comfort bearings used in vehicle wheel suspensions. The bodies are modeled object-oriented on the basis of [1]. For cutting kinematic loops, rigid joints are replaced by dynamic bushing elements with spring coefficient and damping. Linear one-, three-dimensional and non-linear bushings are used [5]. This results in a certain loss of accuracy due to translation shift, but standard ODE solver can be used. This, in turn, ensures the hard real-time capability of the system. In order to adjust the accuracy of the ideal solution and at the same time ensure high numerical stability, the bushings are optimized mathematically. A multibody model based on [6] calculates the reaction forces numerically and serves as the basis for the optimization. The application example is the double wishbone wheel suspension at the front right of the Formula Student racing car A40-02 of the University of Duisburg-Essen (see Fig. 1). Bodies in a suspension system typically have two to four link points, where the bodies are connected by rigid joints or bushings. Geometry, center of gravity, inertia tensors, and masses are taken from the CAD data. All joints are ideal and are described by holonomic bounding equations. With the tie rod (part 8-9), the system has two kinematic loops.

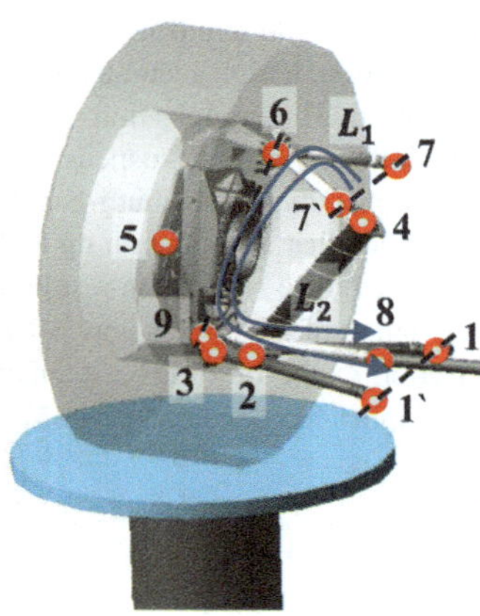

No.	Type of Joint
1	Pivot Joint
1`	Pivot Joint
2	Pivot Joint
3	Ball Joint
4	Pivot Joint
5	Pivot Joint
6	Ball Joint
7	Pivot Joint
7`	Pivot Joint
8	Ball Joint
9	Ball Joint

Fig. 1. Double-wishbone wheel suspension [1]

2 Modeling

In the following, the modeling of the multibody system is presented. The different types of bushings are then introduced. It is also shown how the reaction forces of the comparative model are calculated. All models are created in MATLAB SIMULINK.

2.1 Real-Time Object-Oriented Modeling (RTOOM)

An object-oriented method according to [1] is used as the basis for modeling the rigid bodies. The dynamics of the rigid-body system is described using the Newton-Euler equations. The angular acceleration of the body ${}^{i}\boldsymbol{\alpha}$ can be calculated in the body fixed coordinate system i. The generalized coordinates of the body $\boldsymbol{x}(t) = (\psi \quad \theta \quad \varphi)^T$ which are the cardan angles, are obtained from the kinematic differential equations using the rotatory Jacobian matrix $\boldsymbol{H}_R$ with

$$\ddot{\boldsymbol{x}}(t) = \boldsymbol{H}_R^{-1}\left({}^{i}\boldsymbol{\alpha} - \dot{\boldsymbol{H}}_R\dot{\boldsymbol{x}}(t)\right). \quad (1)$$

$\boldsymbol{H}_R$ is singular if the body hasn't three rotational degrees of freedom. In this case, the equation is explicitly resolved to $\ddot{\boldsymbol{x}}(t)$ using a mathematical manipulation software (e.g. MAPLE).

The system outputs of the body are on the one hand the accelerations ${}^{i}\boldsymbol{a}_j$ and velocities ${}^{i}\boldsymbol{v}_j$ of the contact points $j = 2, \ldots, n$ and on the other hand the reaction force ${}^{i}\boldsymbol{F}_1$

$${}^{i}\boldsymbol{F}_1 = m\boldsymbol{E}_3{}^{i}\boldsymbol{a}_1 - m\left({}^{i}\boldsymbol{r}_{CM} \times {}^{i}\boldsymbol{\alpha}\right) + m\left({}^{i}\boldsymbol{\omega} \times \left({}^{i}\boldsymbol{\omega} \times {}^{i}\boldsymbol{r}_{CM}\right)\right) - \sum_{j=2}^{n} {}^{i}\boldsymbol{F}_j - {}^{i}\boldsymbol{F}_G. \quad (2)$$

where m is the mass of the body and ${}^{i}\boldsymbol{F}_j$ describes the applied forces at point j. ${}^{i}\boldsymbol{F}_G$ represents the weight force. ${}^{i}\boldsymbol{a}_1$ denotes the translational acceleration at point 1.${}^{i}\boldsymbol{\omega}$ is the angular velocity and ${}^{i}\boldsymbol{\alpha}$ is the angular acceleration. The vector ${}^{i}\boldsymbol{r}_{CM}$ is the vector pointing from A to the center of gravity CM in the body-fixed coordinate system. $\boldsymbol{E}_3$ represents the $[3 \times 3]$ unit matrix.

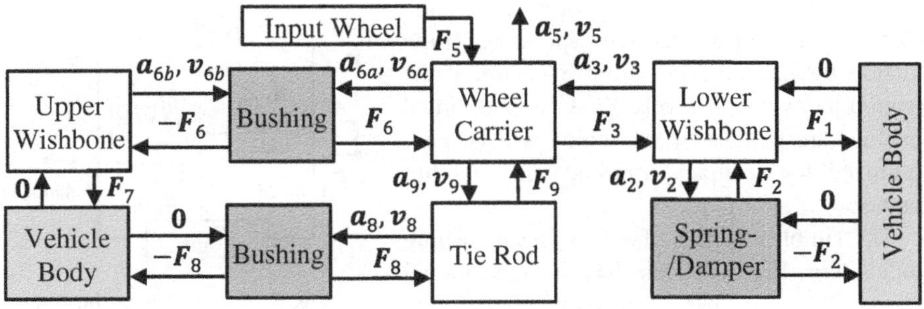

Fig. 2. Block structure of the suspension

The outputs must be transformed in the global coordinate system E, so that the parts can be connected to each other. To create the outputs, the equations are resolved in MAPLE as C-code. This C-code is used in MATLAB/SIMULINK S-functions, which have the advantage of being more efficient and versatile than embedded MATLAB functions, for example. The double wishbone suspension is now assembled using the individual objects. A bushing is used to model each of the two kinematic loops (see Fig. 2). The pivot joints in points 1 and 7 are fixed ($\boldsymbol{a}_1 = \boldsymbol{v}_1 = \boldsymbol{a}_7 = \boldsymbol{v}_7 = 0$). The input of the system is the force $\boldsymbol{F}_5$, which is impressed by the tire.

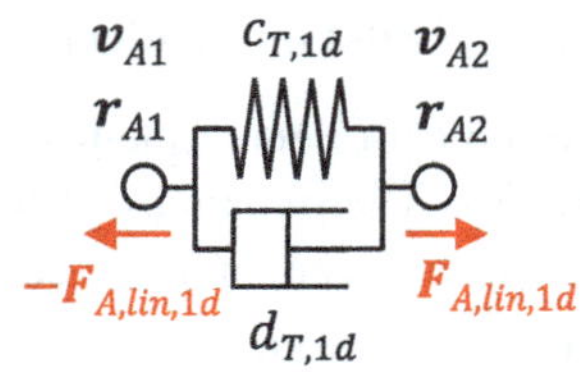

Fig. 3. 1-D bushing

Bushings are rubber bearings for the wheel guidance and are usually used in wheel suspensions to increase comfort and avoid annoying noises. However, in this suspension, these force elements do not occur. Here they are used to cut the kinematic loops and thus avoid iterative procedures to solve the implicit closing conditions at acceleration level. Starting from the Baumgarte stabilization method, the velocity and position are determined by integrating the closing condition at acceleration level. The Baumgarte parameters correspond to the spring and damper constants, so that an associated force can be determined by the deviation of velocity and position. Instead of the parameters of the standard Baumgarte method, more complex approaches are used to determine non-linear spring damper forces.

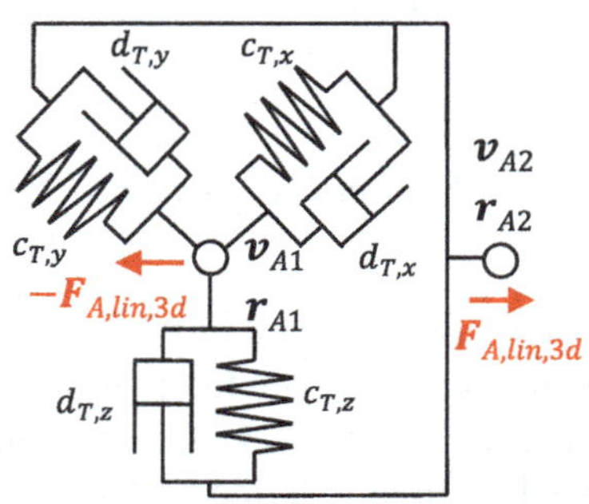

Fig. 4. 3-D linear bushing

In general, each type of bushings is a spring/damper element. A delta of position and velocity results in a force corresponding to the reaction force. However, an error occurs at the position and velocity levels. First, the individual bushings are modeled here, which will be later optimized for accuracy, runtime, and stability manner.

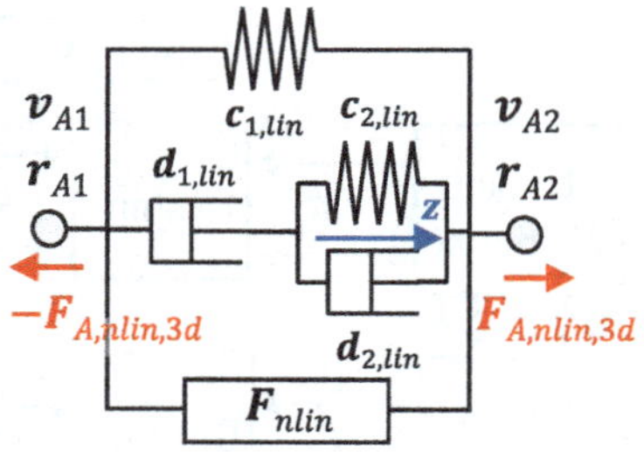

Fig. 5. 3-D nonlinear bushing

The simplest type is the 1-D linear bushing model (see Fig. 3). This has a translational stiffness $c_{T,Bush,1d}$ and a translational damping $d_{T,Bush,1d}$.

The 3-D linear bushing model is modeled as a spring/damper element with a three-dimensional orientation (see Fig. 4). It's also called the Kelvin-Voigt model. This has a translational stiffness $\boldsymbol{c}_{T,3d} = \begin{bmatrix} c_{T,x} & c_{T,y} & c_{T,z} \end{bmatrix}^T$ and a translational damping $\boldsymbol{d}_{T,3d} = \begin{bmatrix} d_{T,x} & d_{T,y} & d_{T,z} \end{bmatrix}^T$.

The nonlinear model in Fig. 5 described here has a modular structure [5]. It consists of a linear module consisting of spring and damper elements and a nonlinear module connected in parallel which describes the amplitude dependence by use of a memory function. The equations are based on [5] and are vectoral here

$$\begin{aligned} \boldsymbol{F}_{A,nlin,3d} = \left(\boldsymbol{c}_{1,lin}^T \boldsymbol{E}_3\right)\left(\Delta \boldsymbol{r}_A + \left(\left(\left(\boldsymbol{c}_{1,lin}^T \boldsymbol{E}_3\right)^{-1} \boldsymbol{d}_{1,lin}\right)^T \boldsymbol{E}_3\right)\left(\Delta \boldsymbol{v}_A - \dot{\boldsymbol{z}}\right)\right) + \\ \int\left(\left(\boldsymbol{E}_3 + \left(\left(\boldsymbol{\rho}_M^T \boldsymbol{E}_3\right)\left|\Delta \boldsymbol{r}_A - \boldsymbol{r}_{Um_i}\right|\right)^T \boldsymbol{E}_3\right)^{-1} \left(\left(\boldsymbol{R}_M \boldsymbol{\rho}_M^T\right)\Delta \boldsymbol{v}_A^T\right)\right) dt, \end{aligned} \tag{3}$$

with the derivative of the internal path coordinate of the linear module $\boldsymbol{z}$

$$\dot{\boldsymbol{z}} = \left(\boldsymbol{E}_3 + \left(\left(\boldsymbol{d}_{1,lin}^T \boldsymbol{E}_3\right)^{-1} \boldsymbol{d}_{2,lin}\right)^T \boldsymbol{E}_3\right)^{-1} \left(\Delta \boldsymbol{v}_A^T - \left(\left(\left(\boldsymbol{d}_{1,lin}^T \boldsymbol{E}_3\right)^{-1} \boldsymbol{c}_{2,lin}\right)^T \boldsymbol{E}_3\right)\boldsymbol{z}\right), \tag{4}$$

where $\boldsymbol{R}_M$ and $\boldsymbol{\rho}_M$ are auxiliary parameters. $\boldsymbol{r}_{Um_i}$ is the path coordinate of the i-th reversal point. Starting from a reversal point, the nonlinear force builds up in the opposite direction so that the sign (+, −) of the movement is always taken into account.

The next step is to increase the accuracy, runtime, and stability with focus on the application. The model should have an application range of 0–30 Hz. Furthermore, a standard explicit integration method with a constant step size h must be used. The Bogacki-Shampine (ode3) integration method has proven to be a robust and fast method, which will also be used here. According to the Nyquist-Shannon sampling theorem, the maximum step size of the numerical solver is obtained first in order to be able to simulate the occurring frequencies of the signal. For $f_{max} = 30\,\text{Hz}$ the largest possible step size is $h_{max} = 1/(2f_{max}) = 1/60\,\text{s} = 0.017\,\text{s}$. As described at the beginning, iterative procedures are not adequate to be used here. Equation (2) is an algebraic equation. If the reaction force is now passed on to another component, an algebraic loop results, which must be solved iteratively. In the system presented here, shown in Fig. 2, these are $\boldsymbol{F}_3$ and $\boldsymbol{F}_9$. In order to guarantee real-time capability, a PT_1 behavior of the build-up of reaction forces is used instead and corresponds to a first order low-pass filter. The physical equivalent is the dynamic friction. The time constant τ is determined via the desired cut-off frequency f_C, so it must be $\tau < 1/(2\pi f_c) = 0.005305\,\text{s}$. The phase shift is $-\pi/4$ and the gain is $1/\sqrt{2} \approx -3\,\text{dB}$.

As already described, the bushing serves to replace the closing conditions. On the one hand the greater the stiffness and damping, the higher the accuracy of the position and velocity. On the other hand, in order to ensure numerical stability, the nonlinear differential equation system must not become too stiff. These stiff differential equations lead to numerical instabilities if the step size of the numerical solver is too large. If the numerical integration method is retained, only the step size can be reduced to stabilize the system. The reduction of stiffness and damping ensures that the eigenvalues of the nonlinear system reach the stability area of the numerical solver largely. However, the smaller step size leads to increased computing time. Therefore, the parameters must be selected in such a way that the accuracy is sufficient and the system is nevertheless hard real-time capable. For this purpose, an optimization process is the obvious choice. The two stiffness and damping parameters for the linear 1-D bushing model, 6 parameters for the linear 3-D bushing model and 18 parameters for the 3-D nonlinear bushing model are the optimization parameters for each of the two bushings. The objective function is the distance between the bushing force $\boldsymbol{F}_6$ simulated with the RTOOM model and the reaction force determined with the model from Sect. 2.2. The nonlinear method of least squares with the Levenberg-Marquardt algorithm is used, which achieved the best results compared to other tested optimization methods. In addition, an attempt is made to increase the step size without making the model unstable.

2.2 Reaction Equations of Ideal Systems

To validate and optimize the RTOOM model, a reference multibody model was created, which calculates the reaction forces and dynamics without using additional force elements and low-pass filters. However, the model includes algebraic loops and is therefore dependent on iterative procedures. The algebraic loop solver uses the trust-region algorithm. The model is therefore not real-time capable but highly accurate. To determine the generalized reaction forces $\boldsymbol{g}(t)$ the principle of virtual work with the corresponding orthogonality relation is applied. Thus, according to [6] the linear algebraic system of equations results as

$$\left(\overline{\boldsymbol{Q}}^T \overline{\overline{\boldsymbol{M}}}^{-1} \overline{\boldsymbol{Q}}\right) \boldsymbol{g}(t) = \widehat{\boldsymbol{k}}(y, \dot{y}, t) - \widehat{\boldsymbol{q}}(y, \dot{y}, t), \tag{5}$$

where $\boldsymbol{y}$ is the position vector. $\overline{\boldsymbol{Q}}$ is the distribution matrix, which can be determined by partial differentiation from the implicit constraints. $\overline{\overline{\boldsymbol{M}}}$ is the mass matrix. $\widehat{\boldsymbol{k}}$ is the vector of the influence of the generalized gyroscopic forces on the reaction forces and $\widehat{\boldsymbol{q}}$ is the vector that characterizes the influence of the generalized impressed forces on the reaction forces. The position vector and its derivation are determined by the equations of motion for ordinary multibody systems using d'Alembert's principle with the equation

$$\left(\overline{\boldsymbol{J}}^T \overline{\overline{\boldsymbol{M}}}\, \overline{\boldsymbol{J}}\right) \ddot{y}(t) = \boldsymbol{q}(y, \dot{y}, t) - \boldsymbol{k}(y, \dot{y}, t), \tag{6}$$

where $\overline{\boldsymbol{J}}$ is the global Jacobi matrix, $\boldsymbol{k}$ the vector of the generalized gyro forces and $\boldsymbol{q}$ the vector of generalized applied forces. The constraint equations are implicitly set at position, speed and acceleration level [4].

The individual matrices and vectors are symbolically evaluated using Maple and then copied as C-code to Matlab/Simulink S-functions. The parameterization is identical to the RTOOM method in Sect. 2.1. The model was successfully validated with an Adams simulation of the same system.

3 Results

Within this section, the results of the RTOOM are presented with different bushing models compared to the reference multibody model. In order to simulate the entire frequency spectrum and also the maximum amplitudes of the input, an input signal of the force $\boldsymbol{F}_5$ is chosen, which simulates the frequencies 0–30 Hz with increasing simulation time. At the same time, the amplitude is continuously increased. So, the time dependent input force is

$$\boldsymbol{F}_5(t) = \boldsymbol{g}_{amp} t \sin\left(t^2 f_{max} \pi / t_{sim}\right), \tag{7}$$

with a simulation time of $t_{sim} = 10\,\mathrm{s}$ and an amplitude gain $\boldsymbol{g}_{amp} = (-10 \quad -10 \quad 100)^T$. The same input applies to the reference model. This simulates all the typical excitation frequencies and a wide range of amplitudes of the 300 kg racing car.

The output quantities considered are the force $\boldsymbol{F}_6$ of the bushing and the acceleration of the wheel center $\boldsymbol{a}_5$. First, the two bushings were optimized and then the simulation comparison was executed. Every bushing model was simulated with and without low-pass filter, whereby some simulations were not stable despite parameter adjustment. The step size h_{max} was increased until a stable simulation was no longer possible. The time constant of the filter τ was chosen as small as possible to minimize the phase shift and gain error of the model. The coefficient of determination R^2 is used to determine the fitting quality of the RTOOM simulation data with the high-precision reference simulation data. Furthermore, the relative error ε_{rel} of the RTOOM simulation is evaluated. The calculation time was also recorded. An attempt was made to avoid possible other processes on the CPU. However, it should be noted that only a relative comparison can be made with this analysis, since it cannot be ruled out that other processes slow down the calculation. Nevertheless, a statement can be made as to which approach (system of equations, integrator and algebraic loop solver) works most efficiently. Furthermore, it is only possible to compile the model as C-code and execute it on a real-time system if the low-pass filter and bushings are used.

The results are summarized in Table 1. The 3-D linear bushing model with low-pass filtering has proven to be the most suitable model in terms of computing time. The 3-D linear bushing model without filtering has the highest accuracy, but requires an algebraic equation solver. The increase in accuracy results from the non-existing phase shift. Although the 3-D nonlinear bushing model has a higher accuracy for calculating the reaction forces, it cannot be simulated stably at acceptable step sizes with filtering and is therefore not real-time capable. The simulation results are still strongly dependent on the frequency range. Especially in the frequency range 0–5 Hz very good results can be achieved for the 3-D linear bushing model with low pass filtering. For 0–5 Hz R^2 is 0.9991 and the relative error is 1.008% for $\boldsymbol{F}_6$. Furthermore, for the acceleration R^2 is 0.9791 and the relative error is 6.514%. The range 5–15 Hz shows a R^2 of 0.9866 and a relative error of 4.187%. R^2 for acceleration results as follows 0.8644 and the relative error is 6.102%. In order to further increase the accuracy of this model, the bushing parameters for the respective frequency range must be adapted so that a different parameterization is available depending on the excitation. If one optimizes the parameters for the different frequency ranges one obtains for the force in the range 0–5 Hz a R^2 of 0.9992. In the range 5–15 Hz R^2 is 0.9974 with an error of 1.973%. For the range 15–30 Hz, a R^2 of 0.9701 is achieved at 8.733% relative error.

Table 1. Simulation results (CPU: Intel Core i7-3635QM 2,40 GHz)

Bushing Typ	Low-Pass τ in s	h_{max} in s	R^2 of $\boldsymbol{F}_6$	ε_{rel} of $\boldsymbol{F}_6$	R^2 of $\boldsymbol{a}_5$	ε_{rel} of $\boldsymbol{a}_5$	t_{COM} in s
1-D linear	0.00125	0.00092	0.8384	19.316%	< 0	20.899%	4.58
3-D linear	0.00125	0.00100	0.9404	10.825%	0.4282	17.245%	4.34
	no	0.00053	0.9832	4.526%	0.5341	15.682%	90.08
3-D non-linear	no	0.00100	0.8610	10.427%	< 0	19.812%	49.57
Reference model	no	0.00100					136.03

4 Summary

Within this paper, it was successfully shown that it is possible to simulate multibody systems with kinematic loops in real-time using bushings. RTOOM is a method that is object-oriented and solves kinematic loops with bushings. The parameters of the bushings have been optimized in terms of runtime, stability, and accuracy, so that hard real-time remains guaranteed with problem-adjusted accuracy. In addition, it has been demonstrated that it is possible to avoid iterative solution methods for algebraic loops and to increase the maximum stable step size by low-pass filtering. The method was exemplified with a double wishbone suspension and a maximum excitation of 30 Hz. The model is hard real-time capable, and a very good accuracy was achieved for frequencies till 15 Hz. For excitation frequencies above 15 Hz the accuracy is acceptable. This makes the RTOOM modeling method suitable for applications that require hard real-time and have a defined application range.

References

1. Kracht, F.E., Saba, M., Schramm, D.: Real-time calculation of reaction forces and elasticities in vehicle wheel suspensions. In: 14th International Symposium on Advanced Vehicle Control (AVEC'18), Beijing, China (2018)
2. Eich-Soellner, E., Führer, C.: Numerical Methods in Multibody Dynamics. Vieweg Teubner (2013)
3. Simeon, B.: Computational Flexible Multibody Dynamics: A Differential-Algebraic Approach. Differential-Algebraic Equations Forum. Springer, Heidelberg (2013)
4. Woernle, C.: Mehrkörpersysteme: Eine Einführung in die Kinematik und Dynamik von Systemen starrer Körper. Springer DE (2011)
5. Pfeffer, P., Hofer, K.: Einfaches nichtlineares Modell für Elastomerund Hydrolager zur Optimierung der Gesamtfahrzeug-Simulation. ATZ-Automobiltechnische Zeitschrift **104**(5), 442–451 (2002)
6. Schramm, D.: Ein effizienter Algorithmus zur numerischen Berechnung der Zwangskräfte mechanischer Systeme. Z. Angew. Math. U. Mecha. 66 (1986)

A Machine Learning Approach for Minimal Coordinate Multibody Simulation

Andrea Angeli[1,2](✉), Frank Naets[1,2], and Wim Desmet[1,2]

[1] Department Mechanical Engineering, KU Leuven, Celestijnenlaan 300, 3001 Leuven, Belgium
{andrea.angeli,frank.naets,wim.desmet}@kuleuven.be
[2] DMMS Lab, Flanders Make, Lommel, Belgium

Abstract. Over the years, a wide range of generalized coordinates have been proposed to describe the motion of rigid and flexible multibody systems. Depending on the type of formulation, a different equation structure is obtained for the model. Most formulations rely on a redundant number of Degrees Of Freedom (DOFs) and some associated constraints, leading to a set of Differential-Algebraic Equations (DAEs) to model the system dynamics. On the other hand, the 'Minimal Coordinate' formulation describes the dynamics through a minimal amount of DOFs and leads to a system of Ordinary Differential Equations (ODEs). For many applications, this ODE structure is an important benefit, as it enables a natural integration for state-estimation and model-based control. The backside of this approach is that it is generally not-straightforward to find a minimal number of parameters to unequivocally describe the system configurations, especially for complex mechanisms. In this work, a machine learning approach based on Auto-Encoders is proposed to find a non-linear transformation that leads to a minimal parameterization of the motion. It is shown that such non-linear transformation can be used to project into minimal coordinates while its inverse permits to perform the simulation in the reduced dimension and re-obtain the original coordinates.

1 Introduction

Over the past decades, several formulations have been proposed to describe the motion of rigid and flexible multibody systems. For rigid mechanisms, such formalisms can be divided in three main categories according to the number and typology of Degrees Of Freedom (DOFs): usage of a minimal amount of coordinates as the 'Minimal Coordinates (MC)' [1] or 'Joint Coordinates (JC)' [2], use of angle and position coordinates as the 'Cartesian Coordinates (CC)' [3] and use of only position coordinates as the 'Natural Coordinates (NC)' [4]. The formulations have been extended to describe flexible mechanisms subject to small deformations as,

A. Kecskeméthy and F. Geu Flores (Eds.): ECCOMAS 2019, COMPUTMETHODS 53, pp. 417–424, 2020.
https://doi.org/10.1007/978-3-030-23132-3_50

for example, the 'Global Modal Parameterization (GMP)' [5,6], the 'Floating Frame of Reference (FFR)' [7], and the 'Flexible Natural Coordinates Formulation (FNCF)' [8].

Depending on the type of formulation used, a different equation structure is obtained for the model. Currently, most of the formulations rely on a redundant number of Degrees Of Freedom (DOFs) and some associated constraints, leading to a set of Differential-Algebraic Equations (DAEs) to model the system dynamics. On the other hand, the MC formulation describes the dynamics through a minimal amount of DOFs and leads to a system of Ordinary Differential Equations (ODEs). For many applications, this ODE structure is an important benefit, as it enables a natural integration for state-estimation and model-based control. The backside of this approach is that it is generally not-straightforward to find a minimal number of parameters to unequivocally describe the system configurations, especially for complex mechanisms.

In this work, starting from a reference multibody simulation expressed in Natural Coordinates, a machine learning approach is proposed to reduce the model to Minimal Coordinates. An Auto-Encoder (AE) neural network is used to find a non-linear transformation that leads to a minimal parameterization of the rigid motion. The inverse function can then be used to perform the simulation of the model dynamics in the reduced dimension and re-obtain the full-order coordinates.

The paper is organised as follows: Sect. 2 describes the proposed approach and Sect. 3 presents an application example; Sect. 4 reports some concluding remarks.

2 Auto-Encoders for Minimal Parameterization of Multibody Systems

This section describes the proposed procedure: in the first part Auto-Encoders are introduced, while in the second part their application to multibody dynamics is presented.

2.1 Auto-Encoders

Recently, novel non-linear dimensionality reduction techniques have been proposed to address the limitations of classic linear methods such as Principal Component Analysis (PCA) [9]. In particular Auto-Encoders, a sub-class of Neural Networks, seem promising to achieve such aim [10]. Here, we provide a brief overview of the key concepts of these methods.

2.1.1 Neural Networks

In general, a Neural Network (NN) is the composition of several basic units as the one reported in Fig. 1.

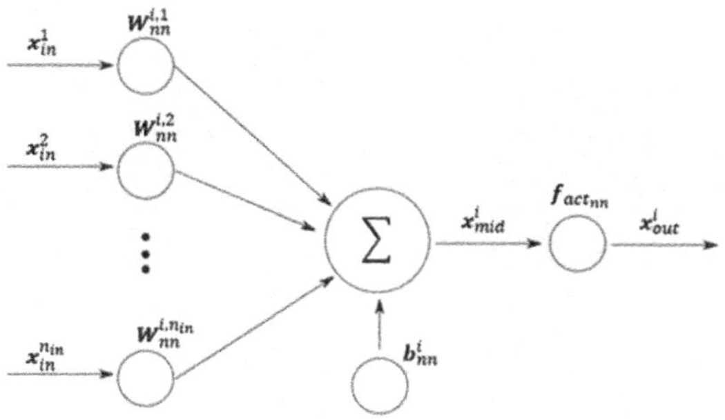

Fig. 1. A Neural Network unit. The terms of the input $\boldsymbol{x}_{in}$ are multiplied by the weights in $\boldsymbol{W}_{nn}$ and added to the bias in $\boldsymbol{b}_{nn}$ before going through the activation function $\boldsymbol{f}_{act_{nn}}$. The full layer will be composed by $i = 1, \ldots, n_{out}$ units to obtain the n_{out} terms of the output $\boldsymbol{x}_{out}$

A vector $\boldsymbol{x}_{in} \in \mathbb{R}^{n_{in}}$ is the input to the NN layer. It is multiplied by the weight matrix $\boldsymbol{W}_{nn} \in \mathbb{R}^{n_{out} \times n_{in}}$ and a bias vector $\boldsymbol{b}_{nn} \in \mathbb{R}^{n_{out}}$ is added to obtain $\boldsymbol{x}_{mid} \in \mathbb{R}^{n_{out}}$. A so-called 'activation function' $\boldsymbol{f}_{act_{nn}}$ is then applied to each element i of $\boldsymbol{x}_{mid}$, to obtain the output vector $\boldsymbol{x}_{out} \in \mathbb{R}^{n_{out}}$:

$$\boldsymbol{x}_{out} = \boldsymbol{f}_{act_{nn}}(\boldsymbol{x}_{mid}) = \boldsymbol{f}_{act_{nn}}(\boldsymbol{W}_{nn}\, \boldsymbol{x}_{in} + \boldsymbol{b}_{nn}) \tag{1}$$

The choice of the activation function typically depends on the application, but the most common are: the linear function (lin), the Rectified Linear Unit (ReLU) and the sigmoid or logistic function (sig). They are, respectively, reported below:

$$\boldsymbol{x}_{out} = \boldsymbol{f}_{act_{lin}}(\boldsymbol{x}_{mid}) = \boldsymbol{x}_{mid} \tag{2}$$

$$\boldsymbol{x}_{out} = \boldsymbol{f}_{act_{ReLU}}(\boldsymbol{x}_{mid}) = \max(0,\ \boldsymbol{x}_{mid}) \tag{3}$$

$$\boldsymbol{x}_{out} = \boldsymbol{f}_{act_{sig}}(\boldsymbol{x}_{mid}) = \frac{1}{1 + e^{-\boldsymbol{x}_{mid}}} \tag{4}$$

The parameters of the NN $\mathscr{P}_{nn} = \{\boldsymbol{W}_{nn},\ \boldsymbol{b}_{nn}\}$ will be trained (or optimized) to minimize a certain cost function as, for example, the difference between $\boldsymbol{x}_{out}$ and a reference $\boldsymbol{x}_{ref}$.

If more layers are stacked so that the output of a layer is the input to following one, the NN is called 'deep' and it typically allows to approximate more complex functions. For a full review of 'Deep Learning' methods, the reader is referred to [11].

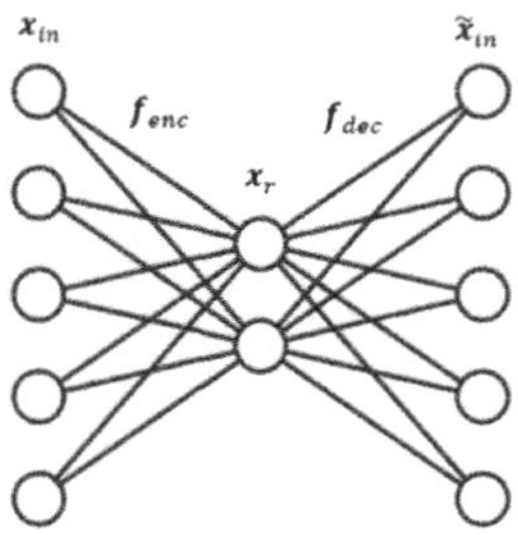

Fig. 2. The structure of a shallow undercomplete Auto-Encoder

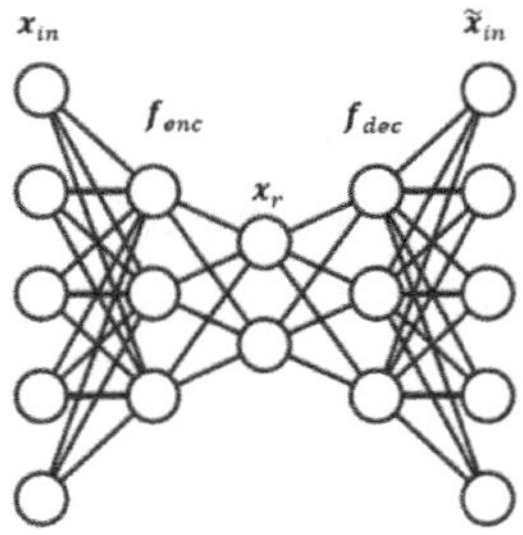

Fig. 3. The structure of a deep undercomplete Auto-Encoder

2.1.2 Auto-Encoder Structure

An Auto-Encoder (AE) is a Neural Network with a particular, symmetric structure. Its 'shallow undercomplete' version is shown in Fig. 2. It is composed by an encoding function $\boldsymbol{f}_{enc}$ where 'shallow' indicates that such function is composed by a single layer of neural units while 'undercomplete' implies that $\boldsymbol{f}_{enc}$ shrinks the input $\boldsymbol{x}_{in} \in \mathbb{R}^{n_{in}}$ to $\boldsymbol{x}_r \in \mathbb{R}^{n_r}$ with $n_r < n_{in}$. Then, its specular decoding function $\boldsymbol{f}_{dec}$ aims to reconstruct the output $\tilde{\boldsymbol{x}}_{in} \in \mathbb{R}^{n_{in}}$ as close as possible to the input $\boldsymbol{x}_{in}$:

$$\boldsymbol{x}_r = \boldsymbol{f}_{act_{enc}}(\boldsymbol{W}_{enc}\, \boldsymbol{x}_{in} + \boldsymbol{b}_{enc}) \tag{5}$$

$$\tilde{\boldsymbol{x}}_{in} = \boldsymbol{f}_{act_{dec}}(\boldsymbol{W}_{dec}\, \boldsymbol{x}_r + \boldsymbol{b}_{dec}) \tag{6}$$

The network parameters $\mathscr{P}_{AE} = \{\boldsymbol{W}_{enc} \in \mathbb{R}^{n_r \times n},\ \boldsymbol{b}_{enc} \in \mathbb{R}^{n_r},\ \boldsymbol{W}_{dec} \in \mathbb{R}^{n \times n_r},\ \boldsymbol{b}_{dec} \in \mathbb{R}^{n}\}$ are trained to minimize the difference between the output and the input:

$$\underset{\mathscr{P}_{AE}}{\operatorname{argmin}}\, ||\tilde{\boldsymbol{x}}_{in}(\mathscr{P}_{AE}) - \boldsymbol{x}_{in}||^2 \tag{7}$$

The bottleneck $n_r < n_{in}$ acts as regularization, preventing the AE from simply copying the inputs and, instead, forcing it to learn a relevant reduced parametrization.

It can be noted that if $\boldsymbol{f}_{act_{enc}}$ and $\boldsymbol{f}_{act_{dec}}$ are set as linear, the Auto-Encoder aims to reproduce PCA. In fact, the obtained AE weights will span the same subspace as the principal components; they would correspond if additional constraints are imposed on the AE such as weights tied $\boldsymbol{W}_{dec} = \boldsymbol{W}_{enc}^{\mathrm{T}}$ and orthogonal $\boldsymbol{W}_{enc}\, \boldsymbol{W}_{enc}^{\mathrm{T}} = \boldsymbol{I}^{n_r}$, where $\boldsymbol{I}^{n_r}$ is the identity matrix of dimension n_r. However, the order of the singular vectors based on singular values is, in general, not guaranteed with Auto-Encoders.

In this work we will use a deep Auto-Encoder as in Fig. 3 with non-linear activation functions and more layers to obtain a non-linear version of PCA to describe the redundant DOFs of a multibody model as a function of the identified minimal coordinates.

2.2 Minimal Coordinate Multibody Simulation

Here, the procedure is presented. It consists of data collection from a reference multibody simulation to train the Auto-Encoder. Then, the AE encoding function is used to obtain the Minimal Coordinate parametrization, while the AE decoding function is used to obtain the reduced-order model, perform the reduced simulation and, finally, reconstruct the full coordinates.

2.2.1 Reference Simulation for AE Training

In order to effectively apply the proposed method, consistent coordinates are necessary, meaning that they should consist of purely displacement coordinates rather than a combination of displacements and rotations. Thus, the starting point is a multibody model expressed in Natural Coordinates, where only position DOFs are used:

$$\boldsymbol{M}\,\ddot{\boldsymbol{x}} + \boldsymbol{f}_{spring} + \boldsymbol{f}_{damper} + \boldsymbol{\Phi}_x^{\mathrm{T}}\,\boldsymbol{\lambda} = \boldsymbol{f}_{ext} \tag{8}$$

$$\boldsymbol{\Phi}(\boldsymbol{x}) = \boldsymbol{0} \tag{9}$$

where, given n DOFs and n_c constraints, $\boldsymbol{x} \in \mathbb{R}^n$ is the vector of Natural Coordinates of the system and the double-dot accent indicates the second time derivative, $\boldsymbol{M} \in \mathbb{R}^{n\times n}$ is a constant mass matrix, $\boldsymbol{f}_{ext} \in \mathbb{R}^n$ is the vector of external forces, $\boldsymbol{f}_{spring}$ and $\boldsymbol{f}_{damper}$ are the forces given respectively by spring and damper elements, $\boldsymbol{\lambda} \in \mathbb{R}^{n_c}$ is the vector of Lagrange multipliers, $\boldsymbol{\Phi} \in \mathbb{R}^{n_c}$ is the constraint vector and $\boldsymbol{\Phi}_x$ is its Jacobian with respect to $\boldsymbol{x}$.

A reference simulation of such model is performed at timesteps $\mathscr{T} = \{0, \ldots, t_{n_t-1}\}$ and given $\boldsymbol{x}^i = \boldsymbol{x}(\mathscr{T} = t_{i-1})$ the data are collected in the form:

$$\mathscr{X} = \{\boldsymbol{x}^1,\ \ldots,\ \boldsymbol{x}^{n_t}\}, \qquad \mathscr{F}_{ext} = \{\boldsymbol{f}^1_{ext},\ \ldots, \boldsymbol{f}^{n_t}_{ext}\}$$

$$\mathscr{F}_{spring} = \{\boldsymbol{f}^1_{spring},\ \ldots, \boldsymbol{f}^{n_t}_{spring}\}, \qquad \mathscr{F}_{damper} = \{\boldsymbol{f}^1_{damper},\ \ldots, \boldsymbol{f}^{n_t}_{damper}\}$$

Each set $\mathscr{X}$, $\mathscr{F}_{ext}$, $\mathscr{F}_{spring}$, $\mathscr{F}_{damper} \in \mathbb{R}^{n\times n_t}$ contains n_t samples drawn from a (set of) reference simulation(s) and is stored together with the matrix $\boldsymbol{M}$.

$\mathscr{X}$ is used as input to train the Auto-Encoder parameters $\mathscr{P}_{AE}$ in order to minimize the mean squared reconstruction error:

$$\underset{\mathscr{P}_{AE}}{\operatorname{argmin}} \frac{1}{n_t} \sum_{i=1}^{n_t} \left(\tilde{\boldsymbol{x}}^i(\mathscr{P}_{AE}) - \boldsymbol{x}^i\right)^2,\ \forall \boldsymbol{x}^i \in \mathscr{X} \tag{10}$$

Sigmoid activation functions as in Eq. 4 are used in order to ensure that the AE functions have consistent derivatives as required by their evaluation in Sect. 2.2.2. The 'Root Mean Square Propagation (RMSprop)' algorithm [12], a variation of 'Stochastic Gradient Descent (SGD)', is used for the AE parameter optimization. The dimension of the AE bottleneck n_r is set equal to the known number of Minimal Coordinates of the system. The procedure returns $\boldsymbol{f}_{enc}$ that gives the MC vector $\boldsymbol{x}_r$ and $\boldsymbol{f}_{dec}$ that backprojects it to the Auto-encoder Coordinates (AC) $\tilde{\boldsymbol{x}}$:

$$\boldsymbol{x}_r = \boldsymbol{f}_{enc}(\boldsymbol{x}) \tag{11}$$

$$\boldsymbol{x} \approx \tilde{\boldsymbol{x}} = \boldsymbol{f}_{dec}(\boldsymbol{x}_r) \tag{12}$$

2.2.2 Minimal Coordinate Model

The kinetic energy $\mathscr{K}$ of the system can be described as a function of the obtained Minimal Coordinates:

$$\dot{\boldsymbol{x}} \approx \frac{\partial \boldsymbol{f}_{dec}}{\partial \boldsymbol{x}_r} \dot{\boldsymbol{x}}_r \tag{13}$$

$$\mathscr{K} = \frac{1}{2} \dot{\boldsymbol{x}}_r^{\mathrm{T}} \left(\frac{\partial \boldsymbol{f}_{dec}}{\partial \boldsymbol{x}_r} \right)^{\mathrm{T}} \boldsymbol{M} \frac{\partial \boldsymbol{f}_{dec}}{\partial \boldsymbol{x}_r} \dot{\boldsymbol{x}}_r \tag{14}$$

Leading to the following inertial forces in the case of a single minimal coordinate $\boldsymbol{x}_r$:

$$\boldsymbol{f}_{m,r} = \left(\frac{\partial \boldsymbol{f}_{dec}}{\partial \boldsymbol{x}_r} \right)^{\mathrm{T}} \boldsymbol{M} \frac{\partial \boldsymbol{f}_{dec}}{\partial \boldsymbol{x}_r} \ddot{\boldsymbol{x}}_r \tag{15}$$

$$\boldsymbol{f}_{g,r} = \left(\frac{\partial \boldsymbol{f}_{dec}}{\partial \boldsymbol{x}_r} \right)^{\mathrm{T}} \boldsymbol{M} \frac{\partial^2 \boldsymbol{f}_{dec}}{\partial {\boldsymbol{x}_r}^2} \dot{\boldsymbol{x}}_r^2 \tag{16}$$

In case of spring-damper elements in the system, their forces are modeled through additional Neural Networks. In particular, a 'NN potential function' $\boldsymbol{f}_{u,r}(\boldsymbol{x}_r)$ is built with parameters $\mathscr{P}_U$ trained to approximate the NC spring forces projected into Minimal Coordinates:

$$\underset{\mathscr{P}_U}{\operatorname{argmin}} \frac{1}{n_t} \sum_{i=1}^{n_t} \left(\frac{\partial \boldsymbol{f}^i_{u,r}(\mathscr{P}_U)}{\partial {\boldsymbol{x}_r}^i} - \left(\frac{\partial \boldsymbol{f}_{dec}}{\partial {\boldsymbol{x}_r}^i} \right)^{\mathrm{T}} \boldsymbol{f}^i_{spring} \right)^2, \forall \boldsymbol{f}^i_{spring} \in \mathscr{F}_{spring} \tag{17}$$

$$\boldsymbol{x}_r^i = \boldsymbol{f}_{enc}(\boldsymbol{x}^i), \ \forall \boldsymbol{x}^i \in \mathscr{X} \tag{18}$$

Similarly, the parameters $\mathscr{P}_D$ of a 'NN damper function' $\boldsymbol{f}_{d,r}(\boldsymbol{x}_r, \dot{\boldsymbol{x}}_r)$ are trained to approximate the NC damping forces projected into Minimal Coordinates:

$$\underset{\mathscr{P}_D}{\operatorname{argmin}} \frac{1}{n_t} \sum_{i=1}^{n_t} \left(\boldsymbol{f}^i_{d,r}(\mathscr{P}_D) - \left(\frac{\partial \boldsymbol{f}_{dec}}{\partial {\boldsymbol{x}_r}^i} \right)^{\mathrm{T}} \boldsymbol{f}^i_{damper} \right)^2, \forall \boldsymbol{f}^i_{damper} \in \mathscr{F}_{damper} \tag{19}$$

Hence, the procedure allows to describe the dynamics as a function of the Minimal Coordinates:

$$\boldsymbol{f}_{m,r}(\boldsymbol{x}_r, \ddot{\boldsymbol{x}}_r) + \boldsymbol{f}_{g,r}(\boldsymbol{x}_r, \dot{\boldsymbol{x}}_r) + \frac{\partial \boldsymbol{f}_{u,r}(\boldsymbol{x}_r)}{\partial \boldsymbol{x}_r} + \boldsymbol{f}_{d,r}(\boldsymbol{x}_r, \dot{\boldsymbol{x}}_r) = \left(\frac{\partial \boldsymbol{f}_{dec}}{\partial \boldsymbol{x}_r} \right)^{\mathrm{T}} \boldsymbol{f}_{ext} \tag{20}$$

At the first timestep $i = 1$, the NC initial position $\boldsymbol{x}^{i=1}$ is reduced to $\boldsymbol{x}_r^{i=1}$ with the encoding function $\boldsymbol{f}_{enc}$ while the initial velocity and acceleration are supposed zero. The derivatives of the decoding function are calculated through automatic differentiation and used to obtain the reduced order model. The dynamics equation is integrated using a central difference scheme, obtaining $\boldsymbol{x}_r^{i=2}$ and $\tilde{\boldsymbol{x}}^{i=2} = \boldsymbol{f}_{dec}(\boldsymbol{x}_r^{i=2})$. The procedure is repeated until $i = n_t - 1$.

3 Application Example

The methodology is demonstrated on the rigid model of a MacPherson suspension shown in Fig. 4. It consists of 6 bodies: a lower control arm is linked to the chassis by two spherical joints and to the steering knuckle by a spherical joint; the knuckle is linked to the tie-rod by a spherical joint and to the strut by a prismatic joint and a spring-damper element. The system is loaded through a time varying vertical force on the knuckle.

The procedure described in Sect. 2 is followed with the AE trained to find the mapping to a single ($n_r = 1$) Minimal Coordinate. The comparison of the dynamics between the original full coordinates and the Auto-Encoder approximation is shown in Fig. 5.

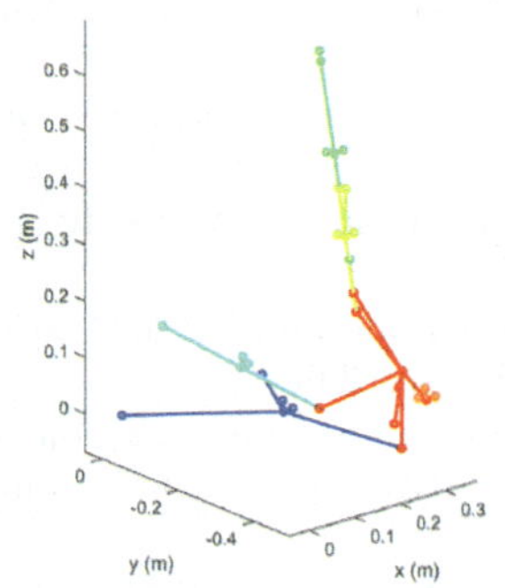

Fig. 4. The suspension model

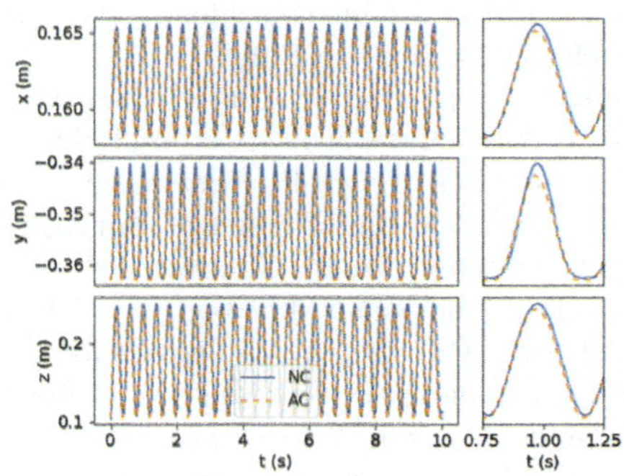

Fig. 5. On the left, comparison of the simulation for the knuckle centre of gravity in 'Natural Coordinates (NC)' $\boldsymbol{x}$ and 'Auto-encoder Coordinates (AC)' $\tilde{\boldsymbol{x}}$. On the right, close-up to show the not perfect match

These results show a relatively close match between the original NC model motion and the AE projected model. However, some differences exist, in particular at maximum displacement levels.

4 Conclusions

With respect to redundant Degree-Of-Freedom approaches for multibody simulations, the Minimal Coordinate formulation has some benefits such as the possibility to express the dynamics as Ordinary Differential Equations.

However, it is often infeasible to set up the required analytic relations between the motion of all bodies as a function of the MC in the case of closed-loop topologies.

In this work, a machine learning approach that reduces a multibody model to Minimal Coordinates is proposed. It is based on a deep Auto-Encoder that trains a non-linear encoding function in order to retrieve the minimal parameters and an inverse decoding function that is used to describe the dynamics in the reduced space and backproject to the full coordinates. An application example is shown.

Acknowledgements. The Research Fund KU Leuven, the Flanders Innovation & Entrepreneurship Agency within the ENDURANCE project and The Research Foundation—Flanders (FWO) are gratefully acknowledged for their support.

References

1. Hiller, M., Kecskemethy, A.: Dynamics of multibody systems with minimal coordinates. In: Computer-Aided Analysis of Rigid and Flexible Mechanical Systems, pp. 61–100. Springer, Dordrecht (1994)
2. Issa, S.M., Arczewski, K.P.: Kinematics and dynamics of multibody system based on natural and joint coordinates using velocity transformations. J. Theor. Appl. Mech. **36**(4), 905–918 (1998)
3. Haug, E.J.: Computer Aided Kinematics and Dynamics of Mechanical Systems, vol. 1. Allyn and Bacon, Boston (1989)
4. de Jalón, J.G., Bayo, E.: Kinematic and Dynamic Simulation of Multibody Systems: The Real-time Challenge. Springer, New York (2012)
5. Brüls, O., Duysinx, P., Golinval, J.-C.: A model reduction method for the control of rigid mechanisms. Multibody Syst. Dyn. **15**(3), 213–227 (2006)
6. Brüls, O., Duysinx, P., Golinval, J.-C.: The global modal parameterization for non-linear model-order reduction in flexible multibody dynamics. Int. J. Numer. Methods Eng. **69**(5), 948–977 (2007)
7. Shabana, A.A.: Dynamics of Multibody Systems. Cambridge University Press, Cambridge (2013)
8. Vermaut, M., Naets, F., Desmet, W.: A flexible natural coordinates formulation (FNCF) for the efficient simulation of small-deformation multibody systems. Int. J. Numer. Methods Eng. **115**(11), 1353–1370 (2018)
9. Van Der Maaten, L., Postma, E., Van den Herik, J.: Dimensionality reduction: a comparative. J. Mach. Learn. Res. **10**, 66–71 (2009)
10. Charte, D., Charte, F., García, S., del Jesus, M.J., Herrera, F.: A practical tutorial on autoencoders for nonlinear feature fusion: taxonomy, models, software and guidelines. Inf. Fusion **44**, 78–96 (2018)
11. Goodfellow, I., Bengio, Y., Courville, A.: Deep Learning. MIT Press, London (2016). http://www.deeplearningbook.org
12. Hinton, G.: Neural networks for machine learning. coursera, [video lectures] (2012)

Efficient Particle Simulation Using a Two-Phase DEM-Lookup Approach

Jonathan Jahnke[1(✉)], Stefan Steidel[1], Michael Burger[1], and Bernd Simeon[2]

[1] Fraunhofer Institute for Industrial Mathematics ITWM, Fraunhofer-Platz 1, 67663 Kaiserslautern, Germany
{jonathan.jahnke,stefan.steidel,michael.burger}@itwm.fraunhofer.de
[2] Felix-Klein-Zentrum für Mathematik, Paul-Ehrlich-Straße, Gebäude 31, 67663 Kaiserslautern, Germany
simeon@mathematik.uni-kl.de

Abstract. We present a fast, data-based approach to predict soil-tool forces. In an expensive offline phase, we perform time-consuming particle simulations based on the Discrete Element Method (DEM). Relevant tool parameters, more specifically cutting depth, angle of incidence and velocity in longitudinal direction are varied to obtain a significant data range of the tool forces and moments. The data is stored in a structure, which we call a Lookup Table (LUT). In an online phase, we use the tool parameters to access the Lookup Table data and to obtain a meaningful approximation of the soil-tool interaction forces.

1 Introduction

In modern vehicle engineering, it is common practice to use numerical models and simulation techniques to efficiently assess a vehicle design under realistic loads and conditions at an early stage in the development process. In particular, within the field of construction machinery, this is a highly challenging task, mainly because of three reasons. First, since the considered systems are very complex; second, the environmental conditions and the resulting loads are substantially varying: an excavator, for example, is digging soil of various kinds with strongly varying properties. Third, the usage is highly variable, because the applications of a construction machine are very versatile and, additionally, the operators differ often drastically in their behavior and their experience. In the past years, the Fraunhofer ITWM has developed the discrete element based software suite GRAPE (GRAnular Physics Engine) that is able to model and simulate soil of different kinds [1–3]. Moreover, the software and the underlying model have been coupled successfully to an MBS wheel loader model within a specifically tailored co-simulation scheme [4,5]. Those steps address the first two mentioned challenges. The next natural step is to additionally reflect the usage variability and the variability caused by different operator types. To achieve this, our goal is to couple GRAPE with an MBS model of a machine and ITWM's driving simulator RODOS® [6]. This realizes efficiently a human-in-the-loop scenario under reproducible conditions. To reach this last step, a soil simulation is

A. Kecskeméthy and F. Geu Flores (Eds.): ECCOMAS 2019, COMPUTMETHODS 53, pp. 425–432, 2020.
https://doi.org/10.1007/978-3-030-23132-3_51

necessary that fulfills the needed accuracy criteria and is real-time capable. In this contribution, we propose a two-phase procedure based on a Lookup Table approach that satisfies those two objectives.

In Sect. 2 we present the particle model, which constitutes a summary of previous work. In Sect. 3, we outline the key idea of the DEM-Lookup approach, forming the main innovation of this contribution. In Sect. 4, we show a numerical example and give an outlook on future work in Sect. 5.

2 Discrete Element Method and Contact Model

In this section, we shortly describe the Discrete Element Method in general. Thereafter, we summarize our model based on previous publication [3,4]. Slight changes in our code lead to an adaption of the model description.

The Discrete Element Method describes particles as rigid bodies which interact according to some contact law. The contact forces, together with gravitational forces and other external forces acting on a single particle are summed up to a force vector $\mathbf{f}_i$, and the equation of motion of each particle is computed according to Newton's second law

$$m_i \ddot{\mathbf{x}}_i = \mathbf{f}_i. \tag{1}$$

Here m_i denotes the mass of the i-th particle, $\mathbf{x}_i$ its position, $\dot{\mathbf{x}}_i$ its velocity and $\ddot{\mathbf{x}}_i$ its acceleration vector. Other models consider also rotational degrees of freedom, thus the Euler Equations have to be considered to incorporate moments on the particle level. Our soil model is based upon spheres with different radii r_i and we neglect rotation, as it is known to have little effect on the draft forces of cohesionless material and increases simulation speed significantly [1,3]. The model is a linearization of Hertzian contact [7], in which two particles are treated as rigid bodies implying a non-penetration penalty term. The error due to the linearization is known to be small when interested in the bulk behaviour of granular matter [8]. If interpenetration occurs, that is, the overlap $\delta_{ij} = r_i + r_j - \|\mathbf{x}_i - \mathbf{x}_j\|$ of the i-th and the j-th particle is larger or equal than zero, a linear, non-negative spring-dashpot penalty term in normal direction is activated. It reads

$$F_{ij}^N = k_{ij}^N \delta_{ij} + d_{ij}^N \dot{\delta_{ij}} \qquad \text{and} \qquad \mathbf{f}_{ij}^N = F_{ij}^N \mathbf{n}_{ij}, \tag{2}$$

where the normal unit vector at the contact point is defined by

$$\mathbf{n}_{ij} = \frac{\mathbf{x}_i - \mathbf{x}_j}{\|\mathbf{x}_i - \mathbf{x}_j\|}.$$

Here, k_{ij}^N and d_{ij}^N describe the stiffness and damping coefficients respectively. The normal force F_{ij}^N is then applied on both particles in the respective normal direction in order to achieve a repulsive effect. The initial contact point $\mathbf{c}_i = \mathbf{c}_j$ is saved in local coordinates for each particle. If no penetration occurs for three consecutive time-steps, the contact points are reset to zero. If on the other hand, the particles translate in tangential direction during non-negative interpenetration δ_{ij}, the contact points $\mathbf{c}_i$ and $\mathbf{c}_j$ differ. Let

$$P_{\perp} : \mathbf{R}^3 \longrightarrow \mathrm{Mat}_{3,3}(\mathbf{R}) \tag{3}$$

$$\mathbf{x} \longmapsto \mathbf{I}_{3,3} - \frac{\mathbf{x}\mathbf{x}^{\top}}{\|\mathbf{x}\|^2} \tag{4}$$

be the projection into the tangent space of $\mathbf{x}$. Then the tangential displacement $\boldsymbol{\xi}_{ij}$ is defined as

$$\boldsymbol{\xi}_{ij} := P_{\perp}(\mathbf{x}_i - \mathbf{x}_j) \cdot (\mathbf{c}_i - \mathbf{c}_j), \tag{5}$$

that is we project the vector $(\mathbf{c}_i - \mathbf{c}_j)$ onto the tangential contact plane. The tangential part of the interaction force then reads

$$\mathbf{f}_{ij}^T = -k_{ij}^T \boldsymbol{\xi}_{ij} - d_{ij}^T \dot{\boldsymbol{\xi}}_{ij} \qquad \text{and} \qquad F_{ij}^T = \|\mathbf{f}_{ij}^T\|. \tag{6}$$

The stiffness constants k_{ij}^N and k_{ij}^T and the damping constants d_{ij}^N and d_{ij}^T in normal and tangential direction depend on the particle radius in order to obtain scale-invariance [9]. We describe here just the constants for the normal force law, in tangential direction the constants are defined in an analogous way. We set

$$k_{ij}^N = \frac{\pi}{4}(r_i + r_j)E_N, \tag{7}$$

where E_N describes the normal elasticity modulus between two particles. Furthermore the damping d_{ij}^N is set to a certain percentage D_N of the critical damping value of a damped oscillator. More specifically, we set

$$d_{ij}^N = D_N \cdot 2 \cdot \sqrt{k_{ij}^N m_{ij}}, \qquad \text{where} \qquad m_{ij} = \frac{m_i \cdot m_j}{m_i + m_j}. \tag{8}$$

The effective mass m_{ij} stems from the consideration of two particles as one damped oscillator. Typically, we relate the tangential stiffness to the normal stiffness, see for example [1]. The tangential force is limited by Coulomb's friction law to $F_{ij}^T \leq \mu F_{ij}^N$, with μ being the local friction coefficient. If the tangential force exceeds this value, slipping friction occurs.

3 DEM-Lookup Approach

The nature of soil forces acting on earthmoving tools is a highly non-linear process. In fact, as one considers measurements of excavation or soil digging with exactly the same input, the output may change drastically due to different drivers and slightly differing maneuvers [4]. It is thus reasonable to approximate this highly non-linear behaviour, at best with similar frequency and magnitude characteristics, using Lookup Tables.

To achieve real-time capable simulation speed, we use a two-phase computation method. In an offline phase, time-consuming DEM simulations are executed, thereby varying a set of tool-parameters. Thereupon, forces and moments are extracted and saved in a Lookup Table. The subsequent online phase consists in finding a meaningful approximation for a given soil-tool state, using the Lookup Table data. More specifically, we perform a set of DEM simulations varying some relevant tool parameters, e.g. the tool position, velocity and rotation. The particle properties representing the parametrization of a specific soil remain unchanged. The DEM simulations are executed in parallel on a high-performance computing cluster. The parameter variation is performed in a loop such that the table data is systematically generated.

Within a single base simulation, i.e. when filling one Lookup Table entry, the respective tool parameters remain constant. The results, i.e the force and moment time-series are gathered and reduced to expectation value and variance per component, containing the main information for a specific tool state. The initial results, where the tool enters into the soil and the surcharge, that is the accumulated soil in front of the tool is increasing, is cut off. Because we require at least 10 particles between tool and boundary to obtain a meaningful force output, we make sure, that within the final simulation state, the tool is fully surrounded by particles and far enough from reaching the boundary. The resulting data structure resembles a multi-dimensional table, where one entry, consisting of mean forces, moments and the respective variances, belongs to a specific set of tool parameters. In the online phase, the tool is moved and the relevant parameters such as position, velocity and rotation are gathered to find a good approximation using the data from the Lookup Table. We achieve this, performing a k-nearest neighbor search and using weighted means [10].

4 Numerical Example

As a first example, we present a rectangular plate moving through gravel, which has been presented in previous work [11], see Fig. 1. Experiments have been performed in cooperation with the soil laboratory at Technical University Kaiserslautern, thus the accuracy of our simulation results can be assured by comparison to the measurement data. In this example, the parameter space is of dimension 3, that is we modify the cutting depth d in vertical z-direction. Furthermore, the angle θ around the global lateral y-axis, as well as the plate's speed v in longitudinal x-direction are varied, compare also Fig. 2. This choice is motivated by earthmoving equations, where the velocity and the cutting depth d have a quadratic influence, see [1,12–14]. The force equation with the terms of interest reads

$$F = (\rho g d^2 N_\rho + \rho g v^2 d N_v + q d N_q) w, \tag{9}$$

depending on the the bulk density ρ, the gravity g, cohesion c and surcharge q. The coefficients N_ρ for the passive earth pressure, N_v for the velocity and N_q for the surcharge depend on different angles via trigonometric identities, see [15]. The Lookup Table can be written as a non-linear mapping

$$\begin{aligned} L_{LUT} : \mathscr{P} &\longrightarrow \mathscr{F}, \\ (d, \theta, v) &\longmapsto (F, M) \end{aligned} \tag{10}$$

with parameter space $\mathscr{P} \subset \mathbf{R} \times [0, 2\pi] \times \mathbf{R}^+$ and $\mathscr{F} \subset \mathbf{R}^6$.

(a) Testbed at soil laboratory TUK (b) DEM-simulation of testbed

Fig. 1. Visualization of numerical example

(a) Variation of the cutting depth (b) Variation of the plate angle (c) Variation of the plate velocity

Fig. 2. Visualization of the variation of different tool parameters

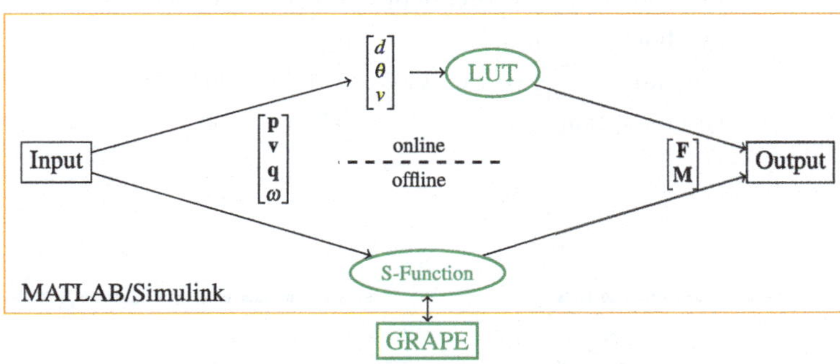

Fig. 3. Program structure of DEM-Lookup approach: in the lower half, the classical DEM-framework used in the offline phase; in the upper half, the online phase using the Lookup Table

4.1 Offline Phase

During a time-consuming offline phase, we perform full DEM-simulations for different tool-parameters within a MATLAB/Simulink environment, see also Fig. 3. The parameter range for our example Lookup Table can be observed in Table 1. The simulation time for one table entry corresponds to 2 s, in total we perform 60 DEM-simulations. The number of basis simulations is problem specific and depends on the chosen parameters and of the desired accuracy. The force sampling rate corresponds to 1 kHz. The acquired time series are then processed and the expectation value and the standard deviation are computed. In Fig. 4, the sensitivity of the mean force in longitudinal x-direction with respect to rake angle θ and cutting depth d can be observed.

4.2 Online Phase

In the online phase, see Fig. 3, we load the Lookup Table into the MATLAB workspace. In each time-step, within our example the tool parameters (d, θ, v), obtained from the current tool position p, the tool velocity v, the current orientation represented by a quaternion q and its angular velocity ω, is used to search for a close approximation from the Lookup Table. In the current scenario, we perform a nearest neighbor search and compute force and moment vectors which are normally distributed, based on the mean and variance of the respective Lookup cell, see Fig. 5 for a trial simulation. Our numerical experiments show that loading the Lookup data-structure takes up most of the time. Once, the simulation scenario is set up, accessing the data is very fast and we aspire for real-time capability.

Table 1. Computing requirements for offline phase

Parameter	Cutting depth	Angle	Velocity
Symbol	d [m]	θ [°]	v [m/s]
Interval	$[-0.25, -0.05]$	$[-45, 75]$	$[0.5, 2.0]$
Discretization	0.1	0.5	30

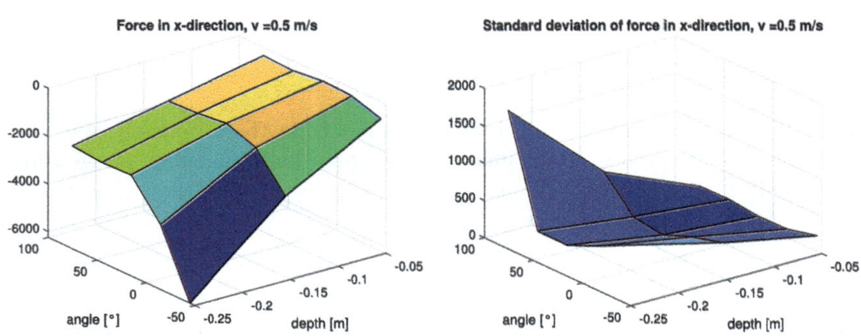

Fig. 4. Expectation value and standard deviation of force magnitude in x-direction for different depth d and angle θ for fixed velocity $v = 0.5$ m/s

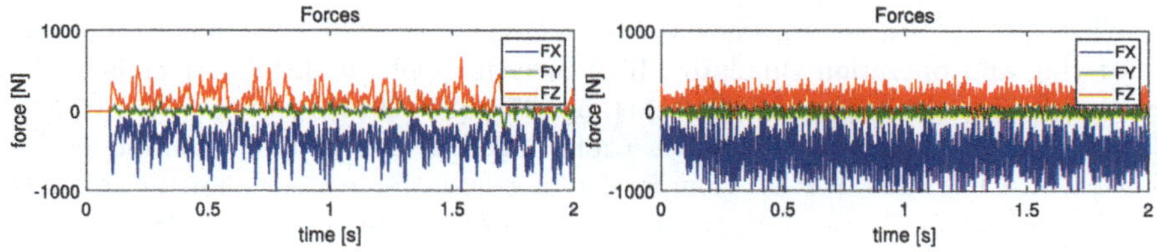

Fig. 5. Force time series for parameters $(-15°, 1.5\ \mathrm{m/s}, -0.05\ \mathrm{m})$ with GRAPE (left) and random Lookup Table approach (right)

5 Conclusion and Outlook

We present an innovative solution to obtain fast, real-time capable draft force predictions, combining classical DEM simulation and a Lookup Table approach. We will further investigate this approach to develop soil simulation within a driving simulator. Therefore, the more complex tool geometry of an excavator bucket has to be used and a different set of parameters has to be chosen to cover also the amount of material filling the bucket. Additionally, more involved approximation techniques within the online phase will be studied. Limitations of the DEM-Lookup approach arise when unprecedented scenarios occur in the online phase with sparse data coverage in the Lookup Table. A difficult endeavor will be the increasing size of the parameter space, when further properties, e.g. acceleration, angular velocity and bucket loading will be incorporated, as the Lookup Table size increases drastically. In the future, we might consider more sophisticated data-based approaches, e.g. supervised machine learning.

An alternative path towards real-time capable draft force prediction is the study of earthmoving equations, e.g. Eq. (9). However, there may arise instabilities and the required accuracy may only be achieved for a limited parameter range.

References

1. Obermayr, M., Dressler, K., Vrettos, C., Eberhard, P.: Prediction of draft forces in cohesionless soil with the discrete element method. J. Terramech. **48**(5), 347–358 (2011)
2. Obermayr, M., Dressler, K., Vrettos, C., Eberhard, P.: A bonded-particle model for cemented sand. Comput. Geotech. **49**, 299–313 (2013)
3. Obermayr, M.: Prediction of load data for construction equipment using the discrete element method. Ph.D thesis, Universität Stuttgart. Shaker, Aachen (2013)
4. Balzer, M., Burger, M., Däuwel, T., Ekevid, T., Steidel, S., Weber, D.: Coupling DEM particles to MBS wheel loader via co-simulation. In: Proceedings of the 4th Commercial Vehicle Technology Symposium, Kaiserslautern, pp. 479–488 (2016)
5. Burger, M., Dressler, K., Ekevid, T., Steidel, S., Weber, D.: Coupling a dem material model to multibody construction equipment. In: Proceedings of the ECCOMAS Thematic Conference on Multibody Dynamics 2017, pp. 417–424 (2017)

6. Kleer, M., Gizatullin, A., Pena Viña, E., Dreßler, K.: The Fraunhofer robot-based driving and operation simulator. In: Proceedings of the 3rd Commercial Vehicle Technology Symposium (CVT 2014), pp. 377–386 (2014)
7. Johnson, K.L.: Contact Mechanics. Cambridge University Press, Cambridge (1985)
8. Renzo, A.D., Maio, F.P.D.: Comparison of contact-force models for the simulation of collisions in dem-based granular flow codes. Chem. Eng. Sci. **59**(3), 525–541 (2004)
9. Feng, Y., Han, K., Owen, D., Loughran, J.: On upscaling of discrete element models: similarity principles. Eng. Comput. **26**(6), 599–609 (2009)
10. Barber, D.: Bayesian Reasoning and Machine Learning. Cambridge University Press, Cambridge (2012)
11. Kleinert, J., Obermayr, M., Balzer, M.: Modeling of large scale granular systems using the discrete element method and the non-smooth contact dynamics method: a comparison. In: Proceedings of the ECCOMAS Thematic Conference on Multibody Dynamics 2013, pp. 191–200 (2013)
12. McKyes, E.: Soil Cutting and Tillage. Developments in Agricultural Engineering. Elsevier (1985)
13. Reece, A.R.: The fundamental equation of earth-moving mechanics. In: Proceedings of the Institution of Mechanical Engineers, vol. 179, pp. 16–22 (1964)
14. Wilkinson, A., DeGennaro, A.: Digging and pushing lunar regolith: classical soil mechanics and the forces needed for excavation and traction. J. Terramech. **44**(2), 133–152 (2007)
15. Cannon, H.N.: Extended Earthmoving with an Autonomous Excavator (1999)

DARTS - Multibody Modeling, Simulation and Analysis Software

Abhinandan Jain(✉)

Jet Propulsion Laboratory, California Institute of Technology,
4800 Oak Grove Drive, Pasadena, CA 91109, USA
jain@jpl.nasa.gov

Abstract. This paper describes the *Dynamics Algorithms for Real-Time Simulation (DARTS)* software for multibody dynamics modeling, analysis and simulation. DARTS is in use for closed-loop simulation for aerospace, ground vehicle, robotics applications and large, multi-scale molecular dynamics applications. DARTS is designed for high-fidelity multibody dynamics, fast computational speed, to handle run-time configuration changes, and to provide a broad family of computational algorithms for analysis and model based control. This paper describes DARTS capabilities, novel aspects of its architecture and design, and application examples.

1 Introduction

This paper describes the current generation of the *Dynamics Algorithms for Real-Time Simulation (DARTS)* software for multibody dynamics modeling, analysis and simulation [1]. DARTS adopts a minimal coordinates approach similar to the SimBody [2] and RBDL [3] tools for solving the equations of motion for time domain simulations. However the goals of DARTS are broader. It is designed to handle rigid/flexible multibody dynamics, arbitrary system topologies, smooth and non-smooth dynamics, run-time configuration changes, and also provide a full complement of computational algorithms needed for dynamics analysis and model based control with fast computational performance. While the DARTS object-oriented implementation is in C++, a rich Python interface is available for all the classes and methods in the system. This allows users full flexibility in defining and configuring the model as desired and to even modify the model topology and properties during run-time. DARTS is in use for the dynamics simulation for aerospace, ground vehicle, robotics and multi-scale molecular dynamics applications [4]. Section 2 describes the dynamics formulation approach, Sect. 3 the software architecture and key features, and Sect. 4 describes application examples. Due to space limitations, this paper limits itself to an overview of key aspects of DARTS, with details to follow in an expanded future publication.

A. Kecskeméthy and F. Geu Flores (Eds.): ECCOMAS 2019, COMPUTMETHODS 53, pp. 433–441, 2020.
https://doi.org/10.1007/978-3-030-23132-3_52

2 Structure-Based Dynamics

DARTS algorithms are based on minimal coordinates dynamics formulations. Minimal coordinate dynamics reduce, if not avoid, the need for explicit constraints, DAE integration methods and constraint stabilization techniques. DARTS is based on the *Spatial Operator Algebra (SOA)* [5] methodology which uses mathematical operator techniques to exploit the structure of minimal coordinate models for analysis and to develop low-cost recursive computational algorithms. Using SOA, analytical operator expressions can be derived for the factorization of the mass matrix and its inverse for arbitrary tree multibody systems. These expressions form the basis for a broad range of low-cost computational algorithms including the well known, $O(\mathcal{N})$ recursive methods for the low-cost solution of the equations of motion for tree-topology systems [5,6]. These algorithms are structure-based and consist of scatter/gather recursions that proceed across the bodies in the system topology. This property allows DARTS to easily handle run-time structural changes in the system topology such as from the attachment/detachment and addition/deletion of bodies. Such structural changes are common in aerospace separation and deployment scenarios, during robotics manipulation and in model coarsening strategies for large-scale molecular dynamics simulations. The algorithms accommodate such changes with recursions simply following the new system topology. DARTS supports general multibody system models including:

Serial/tree rigid/flexible systems: DARTS includes $O(\mathcal{N})$ *articulated-body inertia (ABI)* recursive method for solving the equations of motion of arbitrary tree-topology multibody systems [5,7]. The computational cost of these methods scales linearly with the number of bodies. SOA shows that these structure-based recursive methods for rigid multibody systems directly extend to flexible bodies as well. While some low level computational details change, the overall forward dynamics algorithm structure remains the same. The algorithm implemented within DARTS support rigid and flexible body dynamics. Assumed mode, small deformation models are used for flexible bodies.

Closed-chain topology systems: Recursive methods are not directly usable for system topologies that include loops. One approach is to introduce loop-cuts to obtain a spanning tree. The low-cost $O(\mathcal{N})$ recursive method are used for the spanning tree dynamics, followed by additional steps to correct the solution for the loop-cuts. Within DARTS, this *tree-augmented (TA)* method also uses a SOA-based low-cost method for computing the operational space inertias needed for the loop-cuts during the correction phase [8]. An alternative to the non-minimal coordinate TA approach is the SOA-based *constraint embedding (CE)* technique for closed-loop systems. The CE method transforms the graph-topology systems into a minimal coordinate tree-topology system using compound, variable-geometry bodies. Unlike projection methods, the transformation is structure-preserving so that the $O(\mathcal{N})$ ABI recursive methods and ODE integrators can be used even for closed-loop systems. Both the CE and TA dynamics solution methods are available within DARTS.

Contact/collision dynamics: For non-smooth dynamics involving contact and collisions, DARTS includes support for penalty-based methods. Also available are impulsive complementarity based solution methods that exploit minimal coordinate to reduce the size of the complementarity problem to the number of loop-cut and unilateral contact constraints [9]. Both linear and non-linear complementarity solution algorithms are available in DARTS [10].

3 Architectural Features

Frames layer: The set of *frames* of interest within a simulation can easily number in the hundreds or more. DARTS includes general purpose frame transforms layer that supports the on demand computation of relative pose, velocity and accelerations for any pair of frames in the system. This layer removes the need for special purpose methods for specific frame pair properties. A notable feature is that the values are computed only when requested to avoid the cost of keeping the frames tree fully updated. Transform values are cached and reused, and recomputed only when they are no longer current. The frames layer allows reattachment, as well as the creation and deletion of frames at run-time.

Multibody elements: The key elements of a multibody model are the set of *bodies*, connector *hinges* and *nodes* (physical points of interest on bodies). Within DARTS, each of these entities are distinct C++ objects with associated data and methods. Body and node classes are derived from the frame base class. In addition to supporting a full set of hinge types, a *custom* hinge type allows users to go beyond the pre-defined types and create hinges consisting of an arbitrary sequence of articulation degrees of freedom. Bilateral loop constraints are defined via a hinge that specifies the permissible constrained motion.

Structural changes: DARTS allows run-time structural changes to the system topology such as from the detachment and reattachment of bodies. Also bodies can be deleted and added on the fly. These features are needed for robotics manipulation scenarios, and for events such as heat-shield separation and deployment in aerospace applications. The structure-based SOA algorithms continue to work by simply switching to the new multibody system topology.

Subgraphs: While dynamics computations are usually for the full multibody system, the SOA structure-based algorithms can be limited to connected *subgraphs* of bodies within the system. Important examples of such subgraphs are individual vehicles and robot limbs arms. Within DARTS, the full multibody system is itself a subgraph containing all the bodies. Virtually all of the computational algorithms (including forward dynamics) can be invoked on just the bodies in a subgraph - while ignoring all external bodies. Maintaining the distinction between the multibody model and the computational subgraph has other important benefits. The CE method for closed-loop systems relies on a transformation of the multibody graph into a tree topology system. Within DARTS, the transformed graph is just another subgraph, and using the CE method consists of simply invoking the forward dynamics algorithm on this transformed subgraph.

Python interface: DARTS uses the open-source SWIG tool [11] to auto-generate a comprehensive Python interface for all the C++ classes and their methods. The Python interface allows users to conveniently set up simulations using Python scripts without losing the speed benefits from the compiled C++ layer. The comprehensive Python interface gives users a high degree of flexibility in tailoring the simulation to individual simulation needs and interacting with it during run-time.

Dynamics solvers and integrators: As discussed above, DARTS supports multiple methods for solving the equations of motion. While the $O(\mathcal{N})$ ABI algorithm can be used for tree-topology forward dynamics, the TA algorithm requires a different procedure since it uses non-minimal coordinates. Only integration schemes compatible with the selected solution method can be used. Thus while a minimal coordinates solution method can be used with (any) ODE integrator, a non-minimal coordinates method requires a DAE solver or a Baumgarte like constraint error management scheme. DARTS includes solver C++ classes to support the pairing of a solution method with a compatible numerical integrator.

Visualization: DARTS includes 3D graphics visualization of the multibody model. Even without any graphics data, there is a built in capability to generate a stick figure 3D graphics model that can be animated and articulated. This is useful for debugging and adjusting the model during the early model definition stages. In addition, primitive geometry shapes and CAD parts can be attached to nodes on the bodies to provide a richer graphical representation. Different graphics engines (e.g. Ogre3D, Blender) can be used to render the graphics.

DARTS also includes features such as built in nonlinear solvers for solving loop constraints and state initialization, methods for linearizing the dynamics model for use in control design and analysis, and GUIs for browsing, adjusting and interactively animating the model.

3.1 Computational Algorithms and Features

In addition to the forward dynamics solution methods described above, DARTS includes several other algorithms and features with key ones listed below:

Jacobians: The Jacobian matrix is commonly used in robotics. It maps generalized velocities into the spatial velocity of frames attached to the mechanism. DARTS provides methods to compute the Jacobian matrix map from any subset of degrees of freedom to any set of frames attached to the multibody system.

Hybrid dynamics: The forward dynamics problem solves for the generalized accelerations resulting from specified generalized forces on the system. DARTS supports a more general $O(\mathcal{N})$ recursive hybrid dynamics algorithm where the inputs for any subset of the degrees of freedom can instead be generalized accelerations, and the algorithm will compute the unknown generalized forces for these degrees of freedom. Specifying a degree of freedom to be of *prescribed* type reverses the input/output computation for a degree of freedom. The hybrid algorithm reduces to the standard $O(\mathcal{N})$ ABI forward dynamics algorithm when none of the degrees of freedom are prescribed, and to the $O(\mathcal{N})$ Newton-Euler inverse dynamics algorithm when all degrees of freedom are prescribed. The prescribed property of a degree of freedom can be changed on the fly.

Operational space inertia: For a body node, the *operational space inertia (OSC)* matrix represents the mass matrix of the system reflected to the node. This is an important quantity for task-space and whole-body motion control in robotics, as well for the TA closed-chain dynamics solution method. The SOA algorithms provide a low-cost recursive algorithm for computing OSC matrices [5,12], and DARTS uses it for computing the OSC for any set of nodes specified by the user.

Composite body inertias, kinetic energy, momentum: DARTS includes methods for computing the configuration dependent *composite body*, i.e. combined spatial inertia, momentum and kinetic energy properties for any sub-set of bodies in the system. These methods are useful for computing the properties of individual vehicles, robotic arms etc.

CM frame: In certain applications, it is important to keep track of the center of mass (CM) location of a subgraph. DARTS automatically creates a CM frame for each subgraph that tracks the CM location of the bodies as the configuration evolves with time. This allows users to use the standard frame level transform methods to easily track the location of the CM for any subgraph.

Interbody forces: The recursive $O(\mathcal{N})$ ABI forward dynamics algorithms do not require, nor compute, interbody spatial forces. However there are times when these force values are needed. The SOA algorithm provides an inexpensive expression for computing this value for any pair of connected bodies [13] that is implemented within DARTS.

Pruning bodies: The bodies within a complex multibody model typically match the structure of the mass property and kinematics input deck for the system. This can result in more bodies than are essential to the dynamics computations, and add to the computational cost. DARTS allows users to prune the bodies in a multibody model by freezing and coalescing rigidly connected pairs of bodies.

DCA algorithm: The divide-and-conquer algorithm (DCA) is an alternative forward dynamics methods that is amenable to parallelization [14]. DARTS includes an implementation of this method for use with large multibody models (e.g. molecular dynamics).

Linearization: Since multibody dynamics models are inherently nonlinear, linearized dynamics models are often required for control system design and analysis. DARTS provides methods for automatically computing and exporting linearized models for multibody models.

Statistical dynamics: Statistical properties play a key role in molecular dynamics simulations. The use of multibody methods for such simulations requires generalizations of the classical *equipartition principle* to systems with constraints in order to distribute the thermal energy evenly across the coupled system degrees of freedom. DARTS implements SOA techniques for such a generalized equipartition principle [15]. Also, it is known that the use of multibody methods introduces biases in the statistical properties which can be overcome with the additional use of a *Fixman potential*. DARTS implements SOA methods for incorporating this potential within molecular dynamics simulations [16].

4 Applications

Engineering simulations include vehicle dynamics together with models for sensor/actuator devices, the environment and closing the loop with control software. For such simulation applications, DARTS serves as the vehicle dynamics module within the *DARTS Shell (Dshell)* component-based simulation framework [17]. Key application areas for DARTS and Dshell are described below.

Aerospace: The *DSENDS* tool is an adaptation of the DARTS/Dshell toolkit for spacecraft flight dynamics which uses component models for aerodynamics, engines, thrusters, gravity, fuel consumption, ephemerides etc. DSENDS is in use for closed-loop guidance and control for orbiters, landers and launch vehicle simulations for NASA missions such as Cassini, Mars Pathfinder, Mars Science Laboratory, Phoenix, InSight, Mars 2020, Space Launch System [18]. A recent application has been for rotorcraft simulation for the upcoming NASA Mars Helicopter technology demonstration mission [19].

Ground Vehicles: The *ROAMS* tool is an adaptation of the DARTS/Dshell toolkit for the simulation of autonomous ground vehicles such as NASA planetary rovers [20]. ROAMS includes models for vehicle suspensions, wheel/soil interaction, autonomy sensors such as cameras and lidars, and motion control software. ROAMS has also been used for simulating terrestrial vehicles such as the HMMWV, MRZR4 and other robot mobility platforms [21]. The constraint embedding methods have been especially useful for modeling the dynamics of the double wishbone and trailing arm suspensions in these vehicles.

Robotics and Embedded Use: DARTS has been used for simulating robotics systems such as manipulators and legged/humanoid mobile robots. Some of these applications use the contact/collision dynamics capabilities within DARTS. Another application area within robotics is its use as a modeling layer embedded within the robot control software. The large variety of algorithms supported by the SOA framework makes it especially suitable for addressing a broad and diverse set of model-based computations using low-cost recursive methods. This adaptation of DARTS, called *RoboDarts*, serves as a fast, cross-cutting versatile layer that computes model-based data for the control, execution, perception, estimation and planning modules within robot control software [22,23]. The queries can range from basic frame to frame pose transforms, gravity compensation computations, load balancing for multi-arm manipulation, trajectory and grasp planning, and feed forward control inputs for gait management. The structure-based algorithms are also able to easily accommodate the time-varying configuration and constraints that are common in robotic tasks.

Computational Workbench: *PyCraft* is a multibody dynamics analysis computational workbench based on DARTS [24]. The goal of PyCraft is to provide an interactive environment for the numerical evaluation of operators and operator expressions from SOA theory. With DARTS providing the modeling layer, PyCraft implements C++ operator classes for all the key operators in SOA. Overloaded arithmetic operations for these operators are available for evaluating mathematical operator expressions from the command line. Thus for instance, operator expressions for the mass matrix and mass matrix inverse from SOA can be evaluated directly for any multibody model loaded into DARTS. The PyCraft computational workbench enables the easy evaluation and testing of results from SOA mathematical analysis. PyCraft supports the full range of computations provided by SOA including advanced techniques for computing sensitivities of operators and the mass matrix.

Molecular Dynamics: Another research application area for DARTS has been that of molecular dynamics simulations of bio-molecular systems. Conventional atom-level simulation methods suffer from small time steps dictated by the stiff bond stretching degrees of freedom. By eliminating these degrees of freedom, DARTS enables the use of larger simulation time steps. The *GNEIMO* [25] methods and software make use of DARTS for these applications and include multi-scale methods that significantly increase the simulation duration and sampling for these very large dynamical systems. Segments of the molecular model can be frozen and thawed on the fly to manage the coarseness of the model dynamics. The emphasis of molecular dynamics simulation is usually on statistical properties. Statistically correct initialization of the system state is based on an extension of the equipartition principle [15]. Furthermore, GNEIMO also supports the use of the Fixman potential for correcting statistical biases introduced by the internal constraints [16].

Acknowledgement. The research described in this paper was performed at the Jet Propulsion Laboratory (JPL), California Institute of Technology, under a contract with the National Aeronautics and Space Administration[1].

References

1. Jain, A., Man, G.K.: Real–time simulation of the Cassini spacecraft using DARTS: functional capabilities and the spatial algebra algorithm. In: 5th Annual Conference on Aerospace Computational Control. Jet Propulsion Laboratory, Pasadena, CA, August 1992
2. Sherman, M.A., Seth, A., Delp, S.L.: Simbody: multibody dynamics for biomedical research. Procedia IUTAM **2**, 241–261 (2011)
3. Felis, M.L.: RBDL: an efficient rigid-body dynamics library using recursive algorithms. Autonomous Robots, 1–17 (2016)

4. Dynamics and Real-Time Simulation (DARTS) Lab (2019). http://dartslab.jpl.nasa.gov/
5. Jain, A.: Robot and Multibody Dynamics: Analysis and Algorithms. Springer, Berlin (2011)
6. Featherstone, R.: Rigid Body Dynamics Algorithms. Springer, New York (2008)
7. Jain, A.: Unified formulation of dynamics for serial rigid multibody systems. J. Guidance Control Dyn. **14**(3), 531–542 (1991)
8. Jain, A., Crean, C., Kuo, C., Quadrelli, M.B.: Efficient constraint modeling for closed-chain dynamics. In: The 2nd Joint International Conference on Multibody System Dynamics, Stuttgart, Germany (2012)
9. Jain, A.: Contact dynamics formulation using minimal coordinates. In: Terze, Z. (ed.) Multibody Dynamics, pp. 93–121. Springer (2014)
10. Mylapilli, H., Jain, A.: Evaluation of complementarity techniques for minimal coordinate contact dynamics. In: Proceedings of the ASME Design Engineering Technical Conference, vol. 6 (2014)
11. Simplified Wrapper and Interface Generator (SWIG) (2019). http://swig.org/
12. Rodriguez, G., Jain, A., Kreutz-Delgado, K.: Spatial operator algebra for multibody system dynamics. J. Astronaut. Sci. **40**(1), 27–50 (1992)
13. Jain, A.: Computing inter-body constraint forces in recursive multibody dynamics. In: The 5th Joint International Conference on Multibody System Dynamics, Lisboa, Portugal (2018)
14. Featherstone, R.: A divide-and-conquer articulated-body algorithm for parallel O(log(n)) calculation of rigid-body dynamics. Part 1: basic algorithm. Int. J. Robot. Res. **18**(9), 867–875 (1999)
15. Jain, A., Park, I.-H., Vaidehi, N.: Equipartition principle for internal coordinate molecular dynamics. J. Chem. Theory Comput. **8**(8), 2581–2587 (2012)
16. Jain, A., Kandel, S., Wagner, J., Larsen, A.B., Vaidehi, N.: Fixman compensating potential for general branched molecules. J. Chem. Phys. **139**(24), 244103 (2013)
17. Cameron, J.M., Balaram, J., Jain, A., Kuo, C., Lim, C., Myint, S.: Next generation simulation framework for robotic and human space missions. In: AIAA SPACE Conference and Exposition 2012 (2012)
18. Cameron, J.M., et al.: DSENDS: multi-mission flight dynamics simulator for NASA missions. In: AIAA SPACE 2016, Long Beach, CA (2016)
19. Grip, H., et al.: Flight control system for NASA's mars Helicopter. In: Proceedings of the AIAA Science and Technology Forum and Exposition (2018)
20. Jain, A., Balaram, J., Cameron, J.M., Guineau, J., Lim, C., Pomerantz, M., Sohl, G.: Recent developments in the ROAMS planetary rover simulation environment. In: IEEE 2004 Aerospace Conference, Big Sky, Montana (2004)
21. Jain, A., Guineau, J., Lim, C., Lincoln, W., Pomerantz, M., Sohl, G., Steele, R.: Roams: planetary surface rover simulation environment. In: International Symposium on Artificial Intelligence, Robotics and Automation in Space (i-SAIRAS 2003), Nara, Japan, May 2003
22. Jain, A.: Structure based modeling and computational architecture for robotic systems. In: 2013 IEEE International Conference on Robotics and Automation (2013)

23. Hudson, N., Myint, S., Kuo, C., Matthies, L., Backes, P., Hebert, P., Fuchs, T., Burdick, J.: End-to-end dexterous manipulation with deliberate interactive estimation. In: Proceedings - IEEE International Conference on Robotics and Automation (2012)
24. Jain, A.: PyCraft: an analytical and computational workbench for system level multibody dynamics. In: The 4th Joint International Conference on Multibody System Dynamics, Montreal, Canada (2016)
25. Larsen, A.B., et al.: GneimoSim: a modular internal coordinates molecular dynamics simulation package. J. Comput. Chem. **35**(31), 2245–2255 (2014)

Applications in Vehicle Dynamics and Aerospace Devices

Optimization of Geometric Parameters and Stiffness of Multi-Universal-Joint Drive Shaft Considering the Dynamics of Driveline

Xingyang Lu, Tongli Lu(✉), and Jianwu Zhang

School of Mechanical Engineering, Shanghai Jiao Tong University, Shanghai 200240, People's Republic of China
{xyl996, tllu}@sjtu.edu.cn

Abstract. The design of multi-universal-joint drive shaft is essential for alleviating vehicle vibration. This paper proposes a novel and practical optimization method to reduce the vibration caused by multi-universal-joint drive shaft. Specifically, a dynamic model of a powertrain system including engine, shafts and universal joints is built in MATLAB which is an inalienable part of the fitness function when optimizing. Further, based on the model above, Genetic Algorithm (G.A.) is applied to optimize the drive shaft's parameters: spatial positions and phase angles of universal joints and thickness of shafts' shells. In each iteration of the G.A., the model is simulated under critical condition which is found by a field test of an experimental vehicle. What's more, to verify the effectiveness of the optimization method, results from G.A. is adopted in a full-vehicle model built in ADAMS, indicating that the peak of torsional vibration is reduced greatly.

1 Introduction

Multi-universal-joint drive shafts are widely applied in front-engine, rear-drive (FR) and 4-wheel-drive (4WD) vehicles. Nonetheless, due to the inherent property of a universal joint, neither torque nor rotational speed can be transmitted smoothly. With the increasing demand of comfort and quietness when driving, NVH (noise, vibration and harshness) issues attract more and more attentions. However, most studies about NVH performance of the drive shaft with universal joints are focused on quasi-static or partially dynamic condition.

Asokanthan [1] investigated the nonlinear vibration of a single universal joint, considering the stiffness and inertias of the input and output shafts via an asymptotic method as well as numerical simulation. Chang [2] studied influence of parameters to the instability of a universal joint. To be specific, an amplitude equation produced from higher order stability map is used to determine the region of stable solutions. Li [3] optimized the phase angles of a two-universal-joints drive shaft based on the integrated transit function of the velocity, ignoring shaft flexibility and other parameters, like spatial position. Wu [4] researched on attenuating the floor vibration by arranging phase angles with a simplified model. What's more, the optimization scheme was

A. Kecskeméthy and F. Geu Flores (Eds.): ECCOMAS 2019, COMPUTMETHODS 53, pp. 445–453, 2020.
https://doi.org/10.1007/978-3-030-23132-3_53

validated by an experiment, but the shaft stiffness and other optimizable parameters are left out of consideration.

Based on these studies, the model in this paper considers the comprehensive dynamics of the driveline, which considers not only the multi universal joints, but also the disturbance from engine, the deformation of shafts and the equivalent inertia of the vehicle body. According to the model, three categories of parameters are chosen as optimization variables: spatial positions, phase angles of joints and thickness of shafts' shell.

This paper is organized in the following manner. First, a field experiment on a FR vehicle is conducted to determine the critical condition. Next, a mathematic model of the driveline is built in MATLAB, including engines, clutches, wheels and most importantly, a multi-universal-joint drive shaft. Each single shaft of the driveline system is treated as a torsional spring. Then, a proposed optimization problem is solved by Genetic Algorithm based on the model built above. Last, the optimization results are applied in a full vehicle model built in ADAMS to verify the validity of the presented optimization method.

2 An Experiment of a FR Vehicle Related to NVH

As discussed, the optimization process would focus on the critical condition, namely the condition when the NVH performance is worst. Thus, an experiment is conducted in this section aiming to find the critical condition. The experimental vehicle is a FR vehicle equipped with a three-universal-joint drive shaft as shown in Fig. 1(a). Three universal joints of the driveline are shown in Fig. 1(b)–(d).

(a) Vehicle (b) First joint (c) Second joint (d) Third joint

Fig. 1. Pictures of the experimental vehicle and three universal joints

To evaluate the NVH performance of the vehicle, three acceleration sensors are installed on the front, middle and rear floor of the vehicle, respectively. A series of tests are conducted including acceleration conditions with different gears ($3\sim5$th gear, $900\sim3000$ rpm) and cruising conditions at different velocities (20–80 km/h). Signals from sensors are collected by a BBM PAK device and processed in MATLAB. It is affirmed that the acceleration condition under 4th gear is the critical condition in which the acceleration is significantly greater than others. The experimental results of vertical acceleration of front, middle, and rear floor at critical condition are shown in Fig. 2.

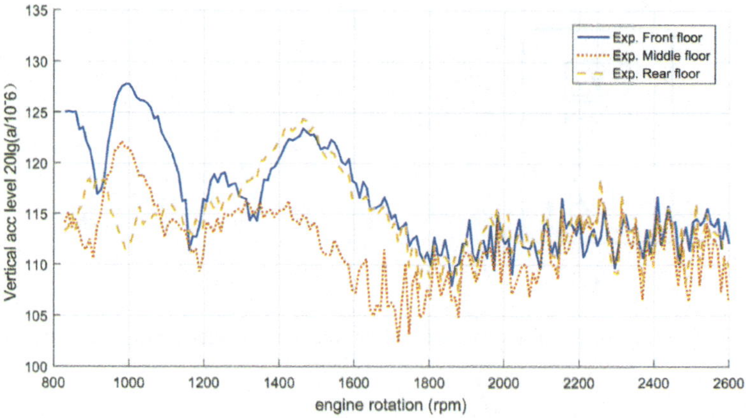

Fig. 2. Vertical acceleration level of front, middle and rear floor of the vehicle in an experiment of acceleration condition at 4th gear

3 Driveline Model Considering Universal Joints

In this section, a mathematic driveline model is established in MATLAB. Note that the model is an essential part of the fitness function of G.A. since the output of the model could be an evaluation of the driveline vibration.

3.1 Engine Modelling

Instead of setting the engine input as an ideal angular velocity or ideal torque, this paper takes the disturbance of the engine torque into consideration. A single piston model could be simplified as a slider-crank mechanism as shown schematically in Fig. 3(a).

The ratio of the piston force and crank torque could be obtained according to the dynamics of the slider-crank mechanism [5]:

$$\frac{T_c}{F_p} = -c\left(\sin(\theta) + \frac{\sin(2\theta)}{2\sqrt{\frac{r^2}{c^2} - \sin^2(\theta)}}\right) \tag{1}$$

where F_p is the piston force, T_c is the crank torque, c is the crank length, θ is the crank angle and r is the connecting rod length.

Meanwhile, cylinder pressure is a periodical function of θ, and could be obtained by circulating cylinder pressure test. Then, F_p in the engine model could be accurately expressed by an interpolation curve with respect to θ. The circulation cylinder pressure curve of the experimental vehicle is presented in Fig. 3(b).

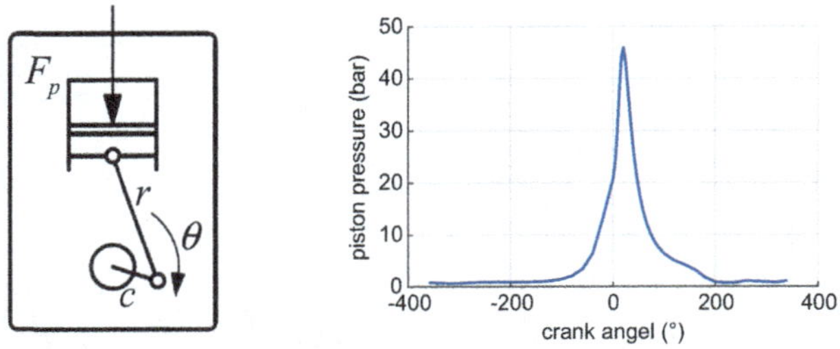

(a) (b)

Fig. 3. Single piston model and circulation cylinder pressure curve

3.2 Driveline Modelling

The driveline mainly contains a clutch, a transmission, a differential assembly, multiple universal joints and shafts as shown schematically in Fig. 4.

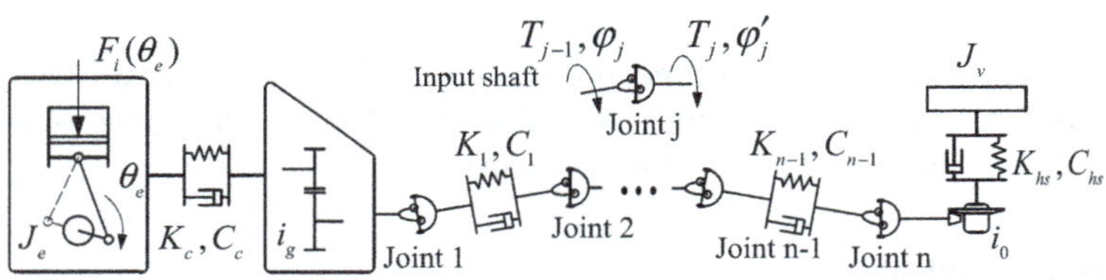

Fig. 4. General dynamic driveline model with n-universal-joint drive shaft

To simplify the model, the torsion damp of the clutch is considered as a torsional spring with damper. Based on the engine model built in Subsect. 3.1, the motion equation of the power assembly is obtained as:

$$T_1 = i_g\left(T_e - J_e\ddot{\theta}_e\right) = K_{hs}(\theta_e - \varphi_1 i_g) + C_{hs}\left(\dot{\theta}_e - \dot{\varphi}_1 i_g\right) \tag{2}$$

where T_1, φ_1 is the torque and rotation angle of the input shaft of the first universal joint, respectively; i_g is the gear ratio of the transmission; J_e is the equivalent inertia of the engine including the inertias of flywheel, crank axle and valve mechanism; K_{hs} and C_{hs} are the stiffness and damping of the clutch.

T_e is the output torque of the engine which is the sum of four crank torques with different offset angles. Thus, T_e could be expressed as:

$$T_e = T_c(\theta_e) + T_c(\theta_e + \pi) + T_c(\theta_e + 2\pi) + T_c(\theta_e + 3\pi) \tag{3}$$

where θ_e is the crank axle angle;

Torque and velocity transit functions of universal joints are [3]:

$$\frac{T_{j+1}}{T_j} = 1/\frac{\dot{\varphi}'_j}{\dot{\varphi}_j} = \frac{1-\sin^2(\alpha_j)\cos^2(\varphi_j)}{\cos(\varphi_j)}, \; j = 1, 2, \cdots, n \tag{4}$$

where T_j, φ_j are the torque and rotation angle of the input shaft of the jth universal joint, respectively, except T_n represents the input torque of the differential, $\varphi_j^{'}$ is the angle of the output shaft of the jth joint and α_j is the bend angle of the jth joint.

Each shaft in the driveline system is considered as a torsional spring with damper, which may excite resonance in the driveline. According to mechanics of materials, the stiffness of each shaft is related to the thickness of the shaft's shell and could be expressed as:

$$K_j = \frac{\pi G D^4}{32 l_j}\left(1 - \left(\frac{D-2h}{D}\right)^4\right) \tag{5}$$

where G is the shear modulus of the shaft's material, D is the outer diameter of the shaft, h is the thickness of the shaft and l_j is the length of the shaft.

Based on the principle of torque balance, the motion equations for shafts and wheels are:

$$\begin{cases} T_j = K_j\left[\left(\varphi'_{j-1} + \delta_{j-1,j} - \gamma_{j-1,j}\right) - \varphi_j\right] + C_j(\dot{\varphi}'_{j-1} - \dot{\varphi}_j), \; j = 2, 3, \cdots, n-1 \\ T_{hs} = J_v \ddot{\varphi}_v = K_{hs}(\varphi'_n/i_0 - \varphi_v) + C_{hs}(\dot{\varphi}'/i_0 - \dot{\varphi}_v) \end{cases} \tag{6}$$

where K_j, C_j are the stiffness and damping of the jth shaft, K_{hs}, C_{hs} are the stiffness and damping of the half axles, $\delta_{j-1,j}$ is the phase angle defined as the torsion angel difference between the output shaft of the (j−1)[th] joint and the input shaft of the jth joint, $\gamma_{j-1,j}$ is the angle difference between principle planes of the (j−1)[th] and j[th] universal joint where principle plane is defined as the plane determined by two axes of input and output shaft.

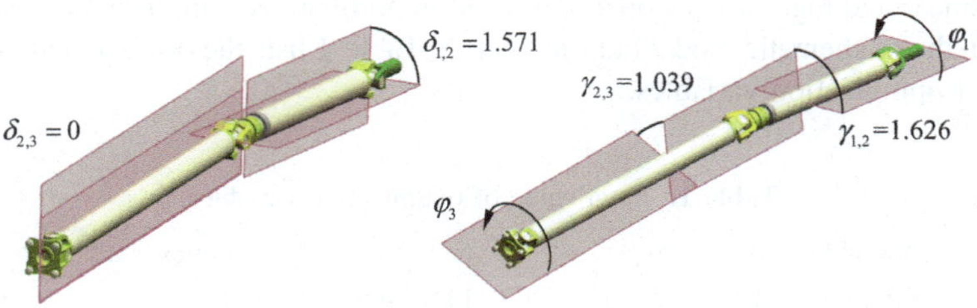

Fig. 5. $\delta_{j-1,j}$ and $\gamma_{j-1,j}$ in the CAD model

$\delta_{j-1,j}$ is one of the parameters to be optimized whereas $\gamma_{j-1,j}$ could be obtained from the spatial positions of joints. Figure 5 shows the concepts and values of $\delta_{j-1,j}$ and $\gamma_{j-1,j}$ in the original drive shaft.

4 Optimization with Genetic Algorithm

As the first and third universal joints are determined by the locations of transmission and rear axle, of which the positions are relatively fixed. Thus the optimization variables are set as:

$$X = [\delta_{1,2}, \delta_{2,3}, x_2, y_2, z_2, h] \tag{7}$$

where x_2, y_2, z_2 are coordinates of the second joint, h is the thickness of the drive shaft shell.

According to the periodic property of the universal joint, the bounds of $\delta_{1,2}, \delta_{2,3}$ are $0 \leq \delta_{1,2}, \delta_{2,3} \leq \pi$. Considering the spatial constrains of the chassis, the bounds of $\Delta x_2, \Delta y_2, \Delta z_2$ are set as $-50 \leq \Delta x_2 \leq 50, -30 \leq \Delta y_2 \leq 30, 0 \leq \Delta z_2 \leq 30$. The initial value of h is designed to ensure the strength of the shaft, thus the lower bound of h is same with its initial value, namely 1.8 mm. The upper bound of h is set as 3 mm for the reason that too much weight of shaft may lead to other problems such as whirling motion.

To assess the vibration level of the driveline, a simulation of acceleration condition with 4th gear is carried out every time the optimization variables change. Acceleration indicates the vibration level that human intuitively feels, while torque is the direct cause of internal parts vibration. Thereby, the fitness function of G.A. combines standard deviations (σ) of wheel acceleration and half shaft torque recorded from the simulation:

$$F = \kappa_1 \sigma(\ddot{\varphi}_a) + \kappa_2 \sigma(T_a) \tag{8}$$

where $\kappa_1 = 1, \kappa_2 = 0.01$ are weight coefficients.

Genetic Algorithm which avoids plunging into local optima is employed to solve the nonlinear and high dimensional optimization problem, accomplished by MATLAB based on the mathematic model built in Sect. 3. Table 1 lists the original and optimized values of optimization variables.

Table 1. Final value of optimization variables

Variable	$\delta_{1,2}$	$\delta_{2,3}$	x_2	y_3	z_2	h	Fitness value
Original	1.571	0	2371.7	13.8	469.4	1.80	1.478
Optimized	1.265	0.547	2424.8	41.6	497.3	2.64	1.335

The fitness value with respect to the number of generation is displayed in Fig. 6. The best fitness value converges to 1.335 around the 65th generation and is reduced about 10% compared to the mean value of the 1st generation.

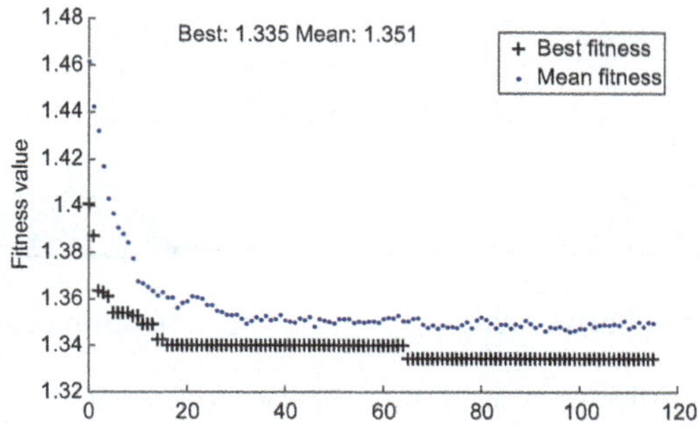

Fig. 6. Convergence of fitness value via G.A.

5 Verification Based on a Full Vehicle Model

The optimization method is based on the driveline model. However, the goal of the optimization is to reduce the vibration caused by multi-universal-joint drive shaft on a real vehicle. Further, a real vehicle model considering the rigid-flexible coupling dynamics containing finite element model of driveline, universal joints and other subsystems is built in ADAMS (Fig. 7) to verify the validity of the optimization.

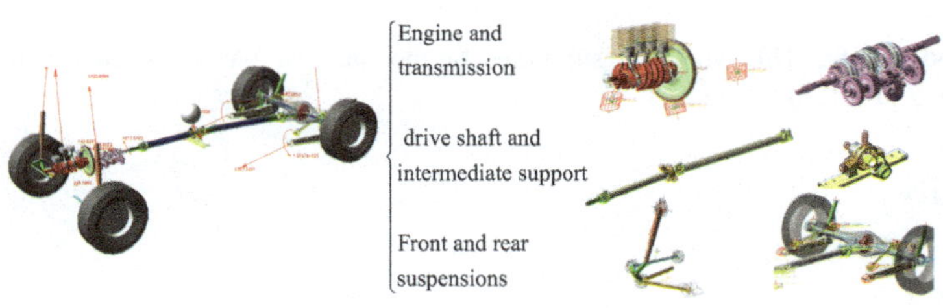

Fig. 7. ADAMS full vehicle model and subsystems

First, the original drive shaft is employed in the full vehicle model. Then, a simulation would be conducted in ADAMS under acceleration condition at 4th gear known as critical condition. Next, the optimized drive shaft is tested on the proposed vehicle model under critical condition as well. It could be seen that the main peak of torsional vibration is 2.3×10^4 °/s^2 with the original drive shaft and is 2.0×10^4 °/s^2 with the optimized one, which means the main peak is reduced by 13%.

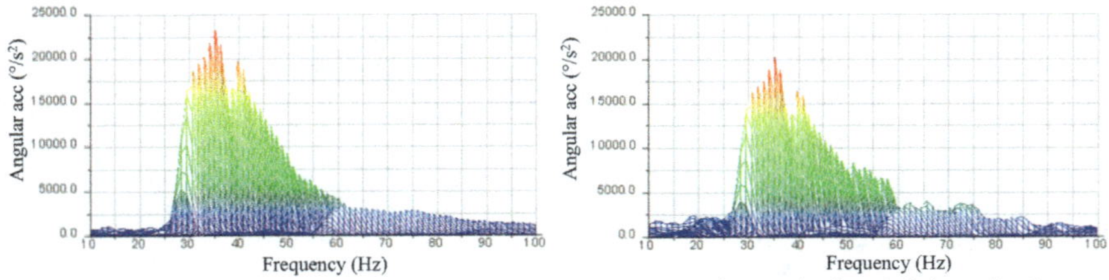

Fig. 8. Time-frequency analysis of angular acceleration at the third joint

6 Conclusion

- In this paper, an optimization method of multi-universal-joint drive shaft considering dynamics of the driveline and critical condition is proposed. Geometric parameters and thickness of the shaft's shell are optimized. Besides, the validity of this method is demonstrated by a full vehicle model.
- Instead of using a so called equivalent angle, the mathematic dynamic model built in Sect. 3 accurately describe the motion of a drive shaft equipped with universal joints. Thus it is valuable for similar research and analysis of multi-universal-joint drive shaft.

Acknowledgments. This work was supported by the National Natural Science Fund of China (Grant No. 51675326).

Appendix

See Table 2

Table 2. Values of parameters

Parameter	Symbol	Value
Rod length (m)	r	0.127
Crank length (m)	c	0.042
Equivalent engine inertia (kg · m^2)	J_e	0.095
Equivalent vehicle inertia (kg · m^2)	J_v	135
Reduction ratios of transmission and differential	i_g, i_0	1, 4.3
Stiffness of clutch (Nm/rad)	K_c	739
Initial stiffness drive shafts (Nm/rad)	K_1, K_2	3.87e4, 2.72e4
Stiffness of half shafts (Nm/rad)	K_{hs}	2.1e4
Damping of clutch (Nm/(rad/s))	C_c	5.6
Damping of drive shafts (Nm/(rad/s))	C_1, C_s	0.07, 0.07
Damping of half shafts (Nm/(rad/s))	C_{hs}	0.05
Initial phase angles	$\delta_{1,2}$, $\delta_{2,3}$	1.571, 0
Initial angel differences between principle planes	$\gamma_{1,2}$, $\gamma_{2,3}$	1.626,1.039

References

1. Asokanthan, S.F., Meehan, P.A.: Non-linear vibration of a torsional system driven by a Hooke's joint. J. Sound Vib. **233**(2), 297–310 (2000)
2. Chang, S.I.: Torsional instabilities and non-linear oscillation of a system incorporating a Hooke's joint. J. Sound Vib. **229**(4), 993–1002 (2000)
3. Li, Y., Ding, X, Li, B.: Optimization design of phase angle of double cross shaft universal joint shaft. Automob. Tech. (2012)
4. Wu, G., Shi, W., Chen, Z.: The effect of multi-universal coupling phase on torsional vibration of drive shaft and vibration of vehicle. SAE International, United States (2013)
5. Christoloukas, D., Savaidis, A.: Theoretical dynamic simulation software for slider crank mechanism of V8 engines. In: Materialwissenschaft und Werkstofftechnik (2016)

Application of Multibody Dynamics in the Modelling of a Limited-Slip Differential

Michal Hajžman(✉), Radek Bulín, and Štěpán Dyk

NTIS - New Technologies for the Information Society, University of West Bohemia, Univerzitni 8, 306 14 Pilsen, Czech Republic
mhajzman@kme.zcu.cz, {rbulin,sdyk}@ntis.zcu.cz

Abstract. The paper is aimed at suitable approaches to the modelling and dynamical analysis of a special class of automotive differentials called limited-slip differentials. The design and function of the differential are briefly introduced and suitable multibody approaches to its modelling are proposed and discussed. The modelling formalism is based on Cartesian coordinates, while rigid bodies and their mutual interaction characterized by contact and friction forces and torques should be considered. Chosen numerical results documenting the effects of various differential model parameters are shown.

1 Introduction

Approaches of multibody dynamics can be advantageously used in problems of dynamics of various machines in order to improve their performance and analyse important properties. This paper is aimed at suitable approaches to the modelling and dynamical analysis of a special class of automotive differentials called limited-slip differentials.

Each modern car is equiped with a differential [2], which has a function to split the driving torque and to allow different angular speeds for driven wheels. This is important when the car is taking a turn and the inner wheel is rolling on smaller radius than the outer wheel [4]. The biggest drawback of standard (so called open) differentials is that the torque is always equally transmitted to both wheels and it could lead to the situation that the first wheel loose the traction on a slippery surface while the second one has proper traction but doesn't have enough power to move a vehicle. This drawback is overcome by limited-slip differentials, where the torque is not split equally but according to the differential type and its setting.

There exist several main types of limited-slip differentials [5]. The design of so called clutch-type limited-slip differential is explained in the next section. Then the possible approaches to the multibody modelling of such differential are introduced and described in more detail. The fourth chapter deals with the results of numerical simulations and with the documentation of the effects of various differential model parameters. This work is motivated by the cooperation

A. Kecskeméthy and F. Geu Flores (Eds.): ECCOMAS 2019, COMPUTMETHODS 53, pp. 454–462, 2020.
https://doi.org/10.1007/978-3-030-23132-3_54

with a motorsport company designing their own rally car and by the development of a proper drive system for formula student applications.

2 Design of a Limited Slip Differential

For sake of the explanation of limited-slip differential function, the schematical cut through a typical clutch-type limited-slip differential is shown in Fig. 1. The input torque from the engine goes to the differential housing which has inner splines and gearing couplings with two pressure rings and half of the clutch plates on both sides. The second half of the clutch plates is coupled with planets, again by means of spline gearing. The satellite carrier is in the contact with the pressure rings and there are two kinds of forces—one group of forces acts in tangential direction and it causes the transmission of torque, the second group acts in axial direction and this leads to the axial pre-stressing of the clutch packs. Increasing axial forces in clutch packs means more friction and more locking, which is expected favourable phenomena.

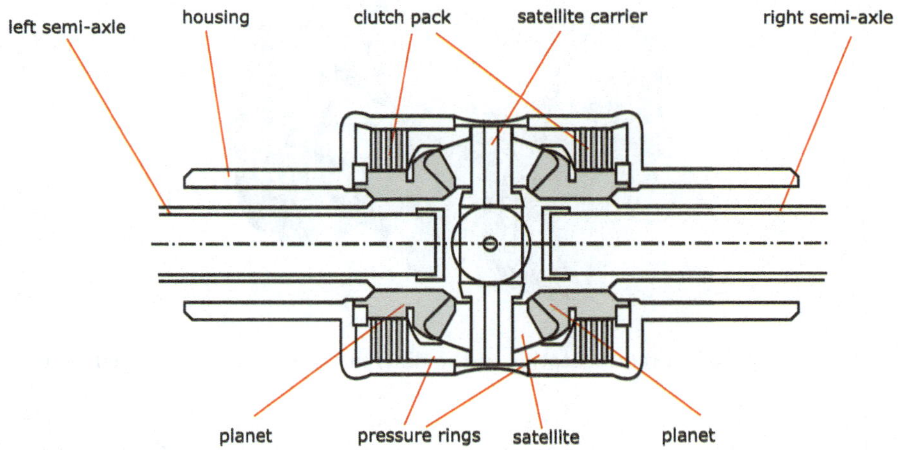

Fig. 1. Scheme of a limited-slip differential of a clutch-type

3 Multibody Modelling of a Differential

The approach used in this work is based on generalized (Cartesian) coordinates, which lead to the mathematical model in the form of a set of differential-algebraic equations (DAEs). Holonomic rheonomic constraints between the coordinates described by vector $\mathbf{q}$ can be written using the vector notation $\boldsymbol{\Phi}(\mathbf{q}, t) = \mathbf{0}$ and for further use in the formulation it must be differentiated to obtain the

Jacobian matrix $\boldsymbol{\Phi}_q$. Common mathematical model can be expressed as the set of differential-algebraic equations of index one in the form

$$\begin{bmatrix} \mathbf{M} & \boldsymbol{\Phi}_q^T \\ \boldsymbol{\Phi}_q & \mathbf{0} \end{bmatrix} \begin{bmatrix} \ddot{\mathbf{q}} \\ -\boldsymbol{\lambda} \end{bmatrix} = \begin{bmatrix} \mathbf{g}(\mathbf{q}, \dot{\mathbf{q}}, t) \\ \boldsymbol{\gamma}(\mathbf{q}, \dot{\mathbf{q}}, t) \end{bmatrix} \tag{1}$$

by the double differentiation of the constraint equations with respect to time. Vector $\boldsymbol{\gamma}(\mathbf{q}, \dot{\mathbf{q}}, t)$ represents the remaining terms after the constraints differentiation. Solution of equations of motion (1) can be based e.g. on elimination of Lagrange multipliers [3] and further direct integration of the underlying ordinary differential equation.

Vector of Lagrange multipliers $\boldsymbol{\lambda}$ is introduced in Eq. (1). Matrix $\mathbf{M}$ is the global mass matrix of the multibody system and vector $\mathbf{g}(\mathbf{q}, \dot{\mathbf{q}}, t)$ contains centrifugal and Coriolis inertia forces, elastic and damping forces and other externally applied forces including the gravity.

Fig. 2. Geometrical representation of the differential model with internal gears

Effects of contacts and impacts is included in vector $\mathbf{g}(\mathbf{q}, \dot{\mathbf{q}}, t)$. In case of interaction of bodies in the limited-slip differential system, it is necessary to consider elastic and viscous components of contact forces in normal direction and also friction forces in tangential direction. Gear couplings were modelled by means of force approach, where meshing forces generate torques between coupled gear wheels. Various models of contact forces were presented in [1]. Alternative treatment of contact problems was shown by authors of [6]. The normal contact forces in the presented model were considered as viscous elastic and the friction was modelled using Coulomb approach. The visualization of the differential internal parts is in Fig. 2 and the model kinematical scheme is shown in Fig. 3. Rigid bodies are depicted by rectangles, while constraints and other interactions are depicted by circles with letters for particular constraints.

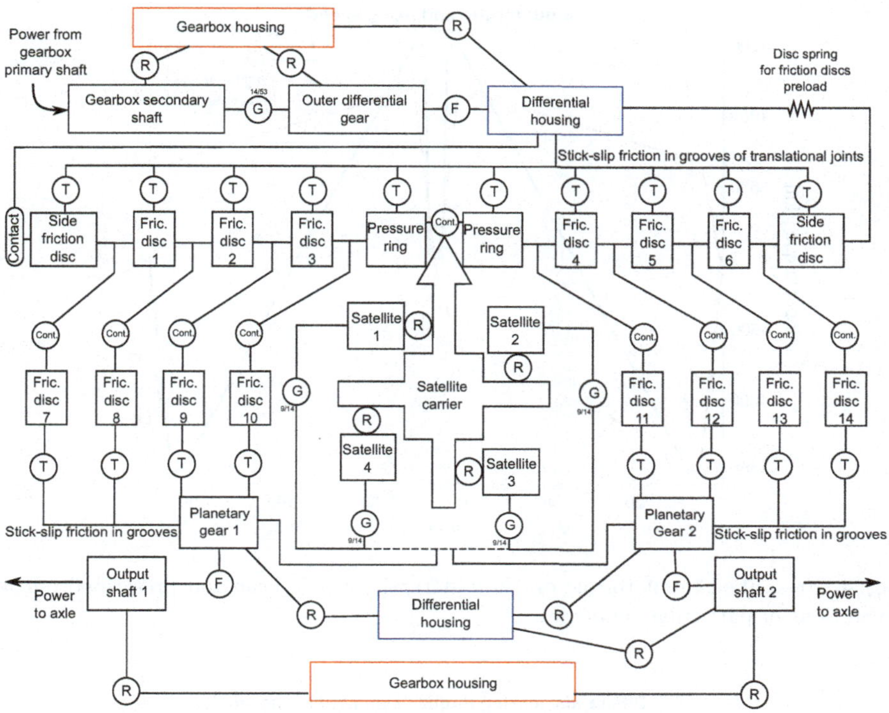

Fig. 3. Kinematical scheme of the limited-slip differential multibody model

4 Results of Numerical Simulations and Parametric Studies

The developed multibody differential model can be utilized in wide range of simulation tasks as a single part or as a part of the whole vehicle model including tires. In order to generate typical locking characteristics the model was employed in nonlinear dynamical analyses with prescribed input torque and output revolutions (see Fig. 4), which are motivated by typical experimental tests performed to evaluate differential behaviour.

Several figures (Figs. 5, 6, 7, 8, 9 and 10) with locking characteristics and locking ratios for various parameters are further shown. The locking characteristics are composed of total torque $T_R + T_L$ transmitted by both semi-axles, which is shown on the horizontal axis, and the difference $|T_R - T_L|$ of theses torques, which is shown on the vertical axis. The figures with locking ratios have the same horizontal axis while the vertical axis means ratio $|T_R - T_L|/(T_R + T_L)$. All figures contain two curves for two combinations of higher and lower speeds (see Fig. 4) prescribed for semi-axles—L > P means that the curve with higher values is prescribed for the left semi-axle and the curve with lower values is prescribed for the right semi-axle, L < P means the opposite case.

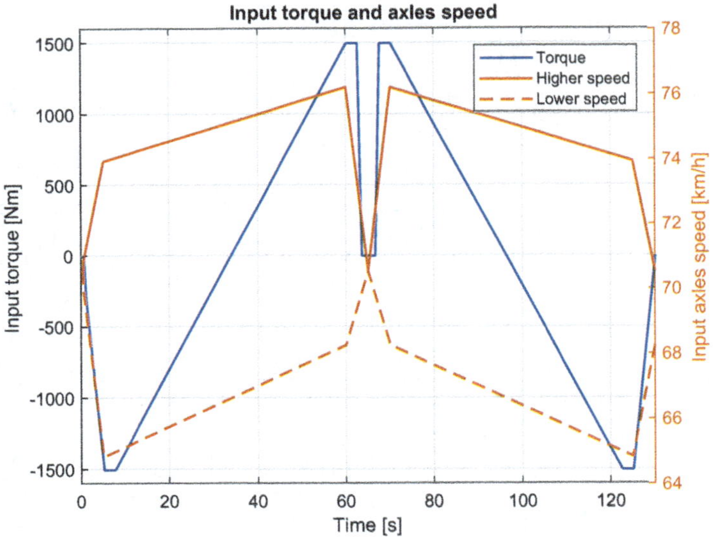

Fig. 4. Prescribed input torque on the differential crown gear and prescribed output revolutions of particular semiaxles

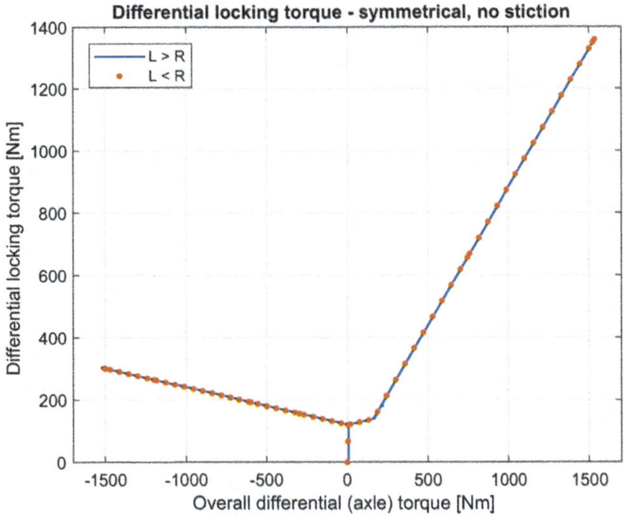

Fig. 5. Locking characteristics for symmetrical friction

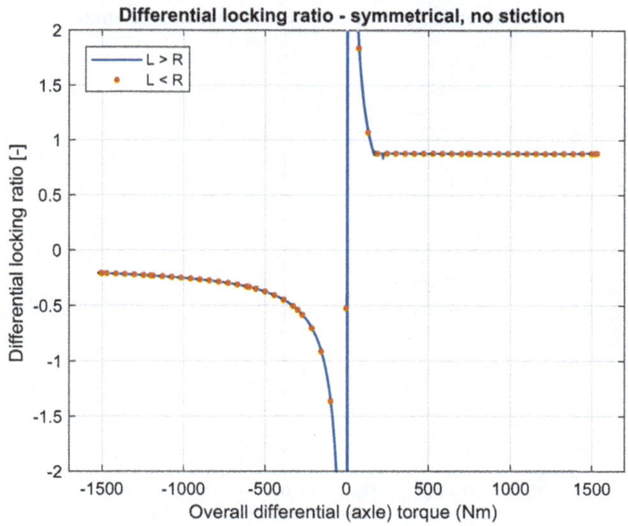

Fig. 6. Calculated locking ratio for symmetrical friction

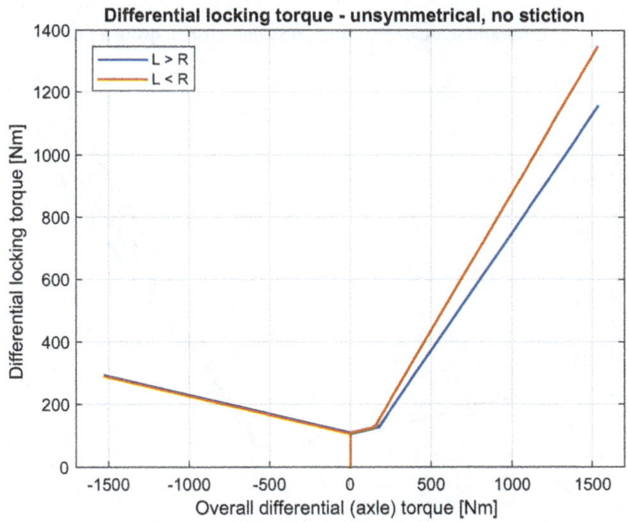

Fig. 7. Locking characteristics for unsymmetrical friction without stiction on grooves

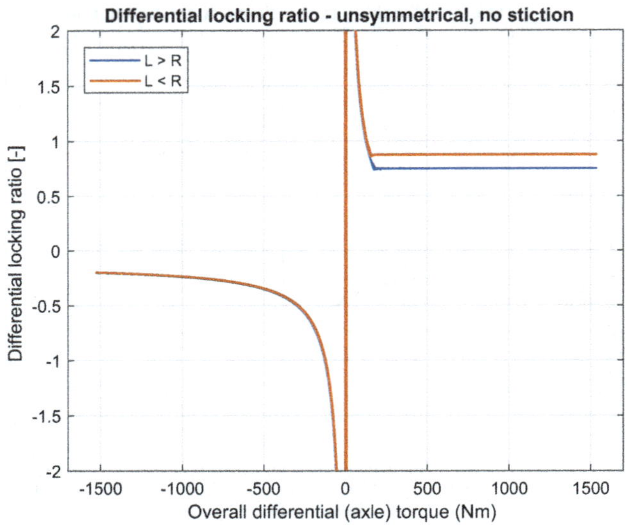

Fig. 8. Calculated locking ratio for unsymmetrical friction without stiction on grooves

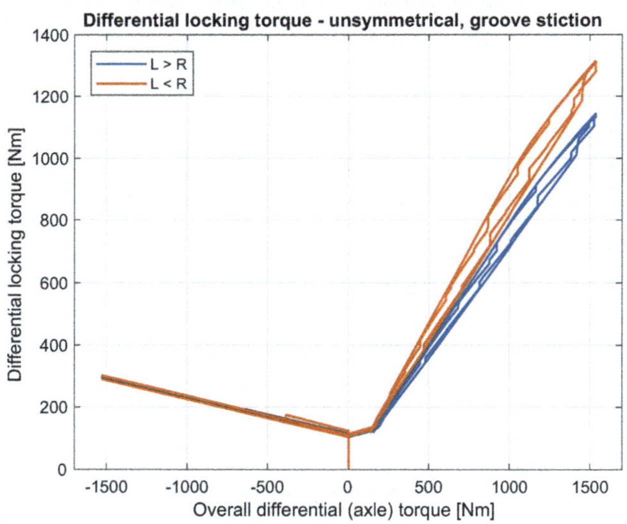

Fig. 9. Locking characteristics for unsymmetrical friction with stiction on grooves

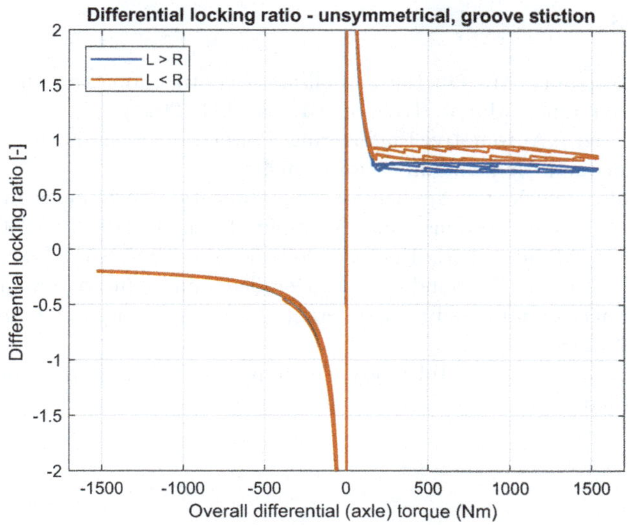

Fig. 10. Calculated locking ratio for unsymmetrical friction with stiction on grooves

Figures 5 and 6 were obtained for the model with same values of friction diameters on both sides of the differential and without consideration of stiction in grooves, axial friction is based on sliding only. The results shown in Figs. 7 and 8 were calculated with different values of friction diameters on both sides (the friction diameter on one side is linearly dependent on axial force) and again without stiction. The values of friction diameters are related to the differential design and setting. The last two Figs. 9 and 10 were obtained for the most complex case.

5 Conclusions

This paper is aimed at the multibody modelling and numerical analysis of a special component of car driving systems called limited-slip differential. Such device has a function to ensure a traction of the vehicle in various limit cases for which a standard differential loose traction. The employed multibody modelling formalism is based on Cartesian coordinates. Design parts are modelled as rigid bodies and their mutual interaction characterized by contact and friction forces and torques was implemented. Obtained characteristics and locking ratios serve as a suitable criterion to evaluate differential properties. Multibody models contribute to the explanation of various effects influenced by design parameters.

Acknowledgements. The research was supported by the project LO1506 of the Czech Ministry of Education, Youth and Sports under the program NPU I.

References

1. Bulín, R., Hajžman, M.: On the modelling of contact forces in the framework of rigid body dynamics. Manuf. Technol. **14**, 136–141 (2014)
2. Crolla, D.A. (ed.): Automotive Engineering: Powertrain, Chassis System and Vehicle Body, Butterworth-Heinemann, Oxford (2009)
3. Hajžman, M., Polach, P.: Application of stabilization techniques in the dynamic analysis of multibody systems. Appl. Comput. Mech. **1**, 479–488 (2007)
4. Milliken, W.F., Milliken, D.L.: Race Car Vehicle Dynamics. SAE, Warrendale (1995)
5. Morselli, R., Zanasi, R., Sandoni, G.: Detailed and reduced dynamic models of passive and active limited-slip car differentials. Math. Comput. Modell. Dyn. Syst. **12**, 347–362 (2006)
6. Pfeiffer, F., Glocker, C.: Multibody Dynamics with Unilateral Contacts. Wiley, Weinheim (2004)

Lateral Dynamics of Vehicles on a “Steerable” Roller Test Stand

Thomas Tentrup[1], Burkhard Corves[2], Jörg Neumann[1], Werner Krass[1], Jan-Lukas Archut[2(✉)], and Jascha Norman Paris[2]

[1] Dürr Assembly Products GmbH, Köllner Straße 122-128, 66346 Püttlingen, Germany
Thomas.Tentrup@durr.com

[2] Institute of Mechanism Theory, Machine Dynamics and Robotics at RWTH Aachen University, Steinbachstraße 53B, 52074 Aachen, Germany
archut@igmr.rwth-aachen.de

Abstract. Semi-autonomous and autonomous driving vehicles will pose new challenges for functional tests in the End of Line area, requiring novel roller test stands. In comparison to conventional vehicles, a higher amount of functionalities has to be validated. Furthermore, functional tests will be performed driverless. Therefore, a novel roller test stand with steerable roller sets has been developed that allows to control the position of the vehicle in the test stand fully automatically, even if there is a steering angle. The concept of the roller test stand is shown. The lateral dynamics of a vehicle on the “steerable” roller test stand is analyzed by means of a simple multibody model. The design of the control structure via virtual commissioning is illustrated. A concept for the application for functional tests is presented.

1 Introduction

In the End of Line area, i.e. at the end of the production of a vehicle, functional tests are necessary to validate the correct function of all mechatronic systems. Due to possible production problems or hardware and software failures, faultless operation of sensors and electronic devices cannot be guaranteed. This becomes even more important for semi-autonomous or autonomous driving vehicles as their systems are safety-critical in a high manner. Thus, the development of such vehicles will affect these test procedures.

Conventionally, functional tests are performed on classical roller test stands, which allow for evaluating systems regarding longitudinal dynamics of vehicles. Figure 1 shows an example for such a conventional roll/brake/ABS test stand. The test stand consists of four roller sets, separately for each wheel and equipped with motors to simulate the driving resistance. Since the roller sets are fixed in their orientation, the vehicle can only drive straight ahead. Thus, a driver is needed who applies small steering inputs at the steering wheel to prevent the vehicle from drifting sideways and being ejected from the test stand. The driver

A. Kecskeméthy and F. Geu Flores (Eds.): ECCOMAS 2019, COMPUTMETHODS 53, pp. 463–470, 2020.
https://doi.org/10.1007/978-3-030-23132-3_55

has furthermore to accelerate, brake and change gears accordingly to the test procedure. Additionally, he moves the car in the End of Line area and drives the vehicle into and out of the test stand.

Fig. 1. Conventional roller test stand *x-road*

Semi-autonomous or autonomous driving vehicles will pose new challenges for the functional testing in the End of Line area, which can be categorized as follows. Firstly, vehicles will move autonomously in the End of Line area. Secondly, functional tests will be performed fully automatically without any driver input. Thirdly, a huge amount of sensors in comparison to conventional vehicles has to be calibrated with high accuracy. Fourthly, the systems of semi-autonomous or autonomous driving vehicles do not only intervene in the longitudinal dynamics of a vehicle but also in its lateral dynamic. Thus, new test procedures regarding these systems have to be performed.

In the past, conventional roller test stands have already been used for testing advanced driver assistance systems (ADAS). In [4] and [2], vehicles to be tested were placed onto a roller test stand and embedded into a virtual environment. This environment was created by means of robot vehicles that move in front of the vehicle to represent actually surrounding vehicles. A more advanced multi-sensor simulation software including LCD screen and CAN bus interface was used in [1] to create the virtual environment for a vehicle on a roller test stand. These methods focus on testing and evaluating systems during the development process of advanced driver assistance systems. Furthermore, only systems regarding longitudinal dynamics can be tested.

Thus, novel roller test stands are needed to address the new challenges for functional testing in the End of Line area. They have to allow both for driverless testing and for testing of autonomous steering systems. The control principle of maintaining a central position of the roller test stand has therefore to be reversed. This task cannot longer be executed by a driver, but has to be performed by the roller test stand, even if there is a steering angle. Consequently, a novel roller test stand with "steerable" roller sets has been developed by the company Dürr Assembly Products GmbH.

This contribution introduces the novel roller test stand and gives insight in the collaborative development process between Dürr Assembly Products GmbH and the Institute of Mechanism Theory, Machine Dynamics and Robotics at RWTH Aachen University. It particularly focuses on the lateral dynamic behavior of the vehicle and the design process of the control structure. In Sect. 2, the lateral dynamic of vehicles on roller test stands is analysed. This analysis is then used for developing the novel roller test stand, whose concept is described in Sect. 3. Virtual commissioning was used during the development process to design the control of the test stand, illustrated in Sect. 4. Section 5 shows then how the novel test stand can be used for the dynamic validation of semi-autonomous and autonomous driving vehicles in the End of Line area.

2 Lateral Dynamic Behaviour of Vehicles on Roller Sets

For designing a "steerable" roller test stand an understanding of the lateral dynamic behavior of a vehicle on roller sets is necessary. A simple multibody model of a vehicle on a roller test stand as shown in Fig. 2 is therefore modelled. It consists of a steerable axle, commonly the front axle of a vehicle, that is placed with both wheels on two separate roller sets. The axle is modelled as one single rigid body with reduced mass M that equals the axle load. It is furthermore constrained to move in lateral direction. At both ends of the axle, the front wheels are added as coupler elements between the front axle and the roller set. The steering angle α_s about the vertical axis can be separately prescribed for both wheels. The camber angle is set to 0°, thus corresponding effects are not considered in the model. Both roller sets are fixed and rotate with a constant angular velocity, corresponding to a pre-defined driving velocity v_x.

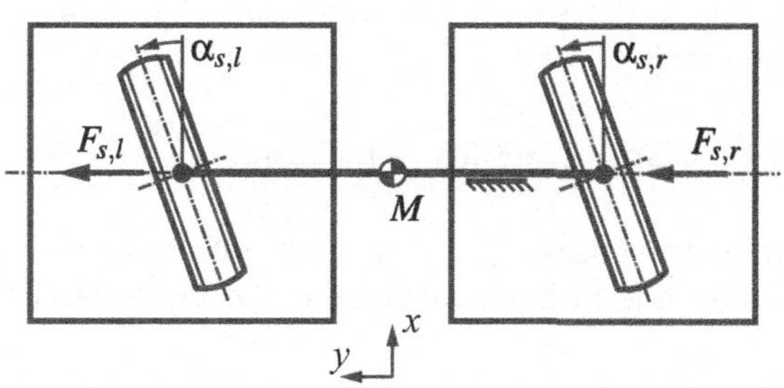

Fig. 2. Model of a steerable axle on roller sets

The lateral tyre forces $F_{s,l}$ and $F_{s,r}$ acting on the axle are described by a stationary tyre model, illustrated in Fig. 3(a) [3]. It describes the relation between the lateral tyre force and the slip angle α in a linear way. The slip angle

α is defined as the angle between the actual velocity vector v of the wheel center relative to the ground and the circumferential speed v_c of the wheel. Due to their directional derivation, shear stresses occur in the wheel contact surface, resulting in a lateral tyre force F_α. For sufficiently small slip angles, this force can be linearized with respect to the slip angle:

$$F_\alpha = c_\alpha \cdot \alpha, \tag{1}$$

where c_α is the lateral tyre stiffness. Considering small-angle approximation, the force F_s perpendicular to the velocity v equals approximately the force F_α.

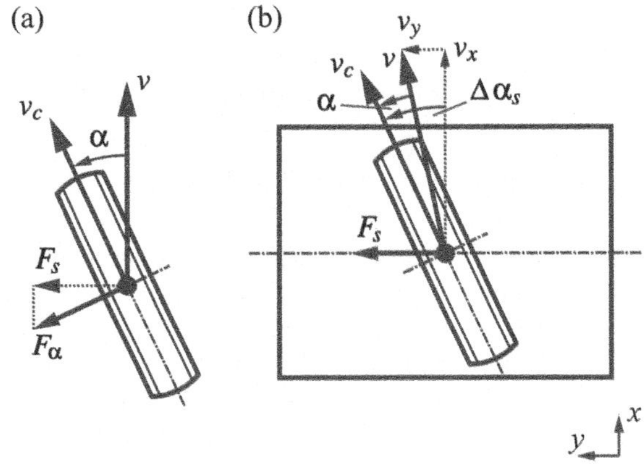

Fig. 3. (a) Lateral tyre force due to a slip angle, (b) situation on a roller set

Figure 3(b) shows now the situation of the tyre on a roller set. The velocity v of the wheel center relative to the roller can be given by both components v_x in longitudinal direction and v_y in lateral direction. Furthermore, the wheel is twisted relative to the roller set by the angle $\Delta\alpha_s$. Consequently, the resulting slip angle α is given by

$$\alpha = \Delta\alpha_s - \arctan\left(\frac{v_y}{v_x}\right) \approx \Delta\alpha_s - \frac{180°}{\pi} \cdot \frac{v_y}{v_x} \tag{2}$$

for sufficiently small velocities v_y. Taking this into consideration, the newton equation for the front axle in lateral direction can be formulated as

$$\frac{dv_y}{dt} = 2 \cdot C_\alpha \cdot \left(\frac{\alpha_{s,l} + \alpha_{s,r}}{2} - \frac{180°}{\pi} \cdot \frac{v_y}{v_x}\right), \tag{3}$$

where $\alpha_{s,l}$ and $\alpha_{s,r}$ are the slip angles at the left and right tyre, respectively. C_α is the related lateral tyre stiffness $C_\alpha = c_\alpha / M$.

Figure 4 shows the resulting lateral displacement s_y from a simultaneous step input of the angles $\alpha_{s,l}$ and $\alpha_{s,r}$ from 0° to 1°. The lateral tyre stiffness C_α is chosen to 2.5 m/s^2·°). The slip angle increases suddenly causing an acceleration in lateral direction. With increasing lateral velocity, slip angle and subsequently lateral acceleration decrease until a steady state with constant velocity is reached. A lateral displacement of 200 mm is achieved within 0.88 s respectively 0.51 s.

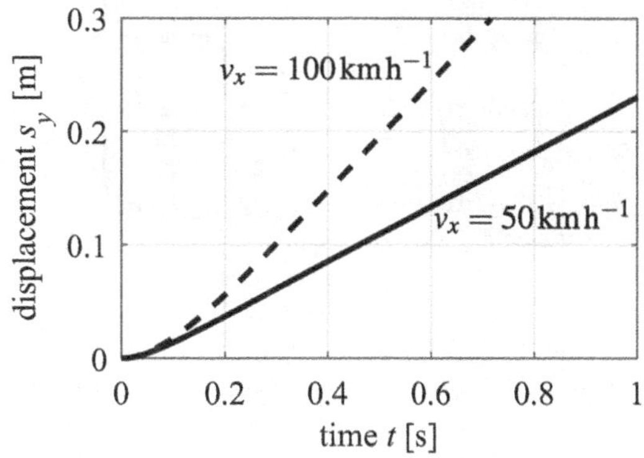

Fig. 4. Resulting displacement s_y from a simultaneous step input of the angles $\alpha_{s,l}$ and $\alpha_{s,r}$

3 Concept of the "Steerable" Roller Test Stand

As analyzed with the vehicle model described before, lateral tyre forces cause a lateral motion of a vehicle in the roller test stand. They are, among other effects, induced by a slip angle between the wheels and the roller sets. In conventional roller test stands, the driver applies small steering inputs to adjust the relative angle between the wheels and the roller sets and thus the slip angle. By doing so, he maintains a central position of the vehicle in the roller test stand. Since systems of semi-autonomous and autonomous driving vehicles perform steering interventions autonomously, e.g. to make a lane change, the relative angle between the wheels and the roller sets has now to be adjusted by the roller test stand. Thus, the novel roller test stand has to feature steerable roller sets. Figure 5 shows the structure of this "steerable" roller test stand, called *x-road curve*.

It is an extension of an existing conventional roller test stand, where the front axle of the test stand is replaced by two separate roller sets. They are individually mounted on floating plates, which are rotatable relative to the frame. A linear motor with internal force and distance measuring system actuates each roller set. The kinematic is designed for a maximum wheel load of 1000 kg and allows a maximum steering angle of ±10°.

Both roller sets at the front axle are equipped with a line laser. They measure the rotation and position of the corresponding wheel relative to the roller set by tracking the rim. The structure of the rear axle is not changed in comparison to the conventional roller test stand, but it is extended by two point lasers. They are directed at the vehicle body and detect the lateral position of the whole vehicle within the roller test stand.

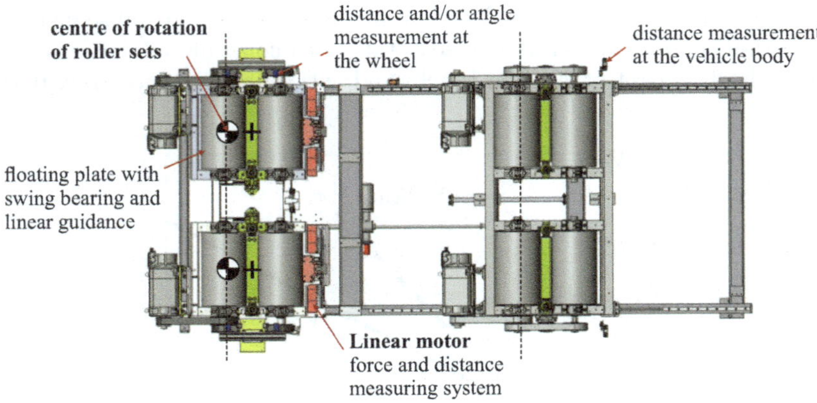

Fig. 5. Novel roller test stand *x-road curve*

4 Design of the Control Structure

A closed-loop control is designed for actuating the linear motors according to the position of the vehicle in the roller test stand. It is specified to restrict the maximum lateral motion to ± 200 mm for a maximum steering angle of $\pm 10°$ at a maximum driving speed of $130\,\mathrm{kmh}^{-1}$. The control structure takes the actual speed of the rollers, the actual position and orientation of the front wheels and the vehicle body as well as the actual position of the linear motors as input. A nominal position of the linear motors is then calculated, so that the relative angle between tyre and rollers is minimized and the vehicle will be centered in the roller test stand. The nominal position is then given to the converter. The control is implemented on the Dürr AP automation system *x-line* and operates at a cycle time of roughly 0.1 s.

Since the development of the roller test stand is carried out under great time and budget pressure, virtual commissioning is used to proof its overall feasibility and to develop the control structure. To this end, a simulation model of the whole system is built as illustrated in Fig. 6. It consists of several models of the motor including converter and encoder, the laser measurement system, data transmission via Ethernet and the kinematics of the roller sets. The vehicle model as described in Sect. 2 replaces the actual vehicle. Properties of each component, e.g. transfer functions, latencies, measure frequencies, measure accuracies, limits of the actuators and cycle time are considered in the model. The behavior of the individual components is then determined via a defined stimulation of the system, i.e. a steering angle α_s with defined amplitude and frequency. Driving speed v_x is held constant during a simulation. The system reaction, that means the resulting lateral motion of the vehicle s_y, is then evaluated for different control structures. In this way, a suitable control structure is designed and controller coefficients are determined to meet the defined requirements.

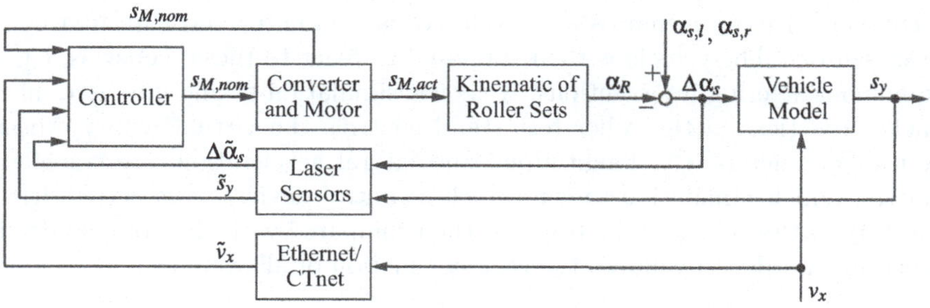

Fig. 6. Simulation model of the whole test stand

This control structure is then used for controlling the actual roller test stand. Physical tests are performed taking the pre-determined controller coefficients as starting values. Comparison of simulation results with these tests proves that the model is well suited to evaluate the system performance for this application. Thereby, time-consuming and risky test situations are avoided as far as possible. Only a fine tuning with different vehicles and different driving maneuvers is necessary, resulting in a faster commissioning of the actual roller test stand.

5 Concept for the Application of the "Steerable" Roller Test Stand in the End of Line Area

As outlined in Sect. 1, new test procedures are necessary for validating the mechatronic systems of semi-autonomous and autonomous driving vehicles. The novel roller test stand can be integrated into a validation concept as follows. The vehicle is placed onto the roller test stand and embedded into a virtual reality as illustrated in Fig. 7.

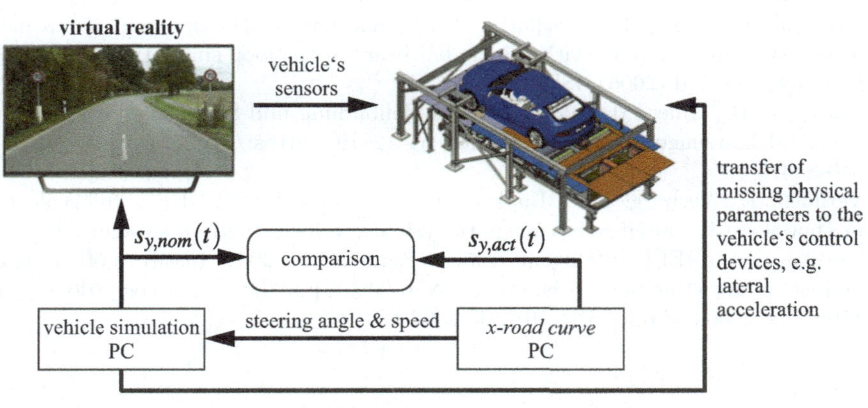

Fig. 7. Concept for the validation of semi-autonomous and autonomous driving vehicles in the End of Line area

Different driving scenarios are displaced on a monitor and captured by the ADAS sensors. The vehicle is then supposed to react to these scenarios, e.g. by autonomous steering interventions. A vehicle simulation taking into account the vehicle's reaction on the roller test stand controls the virtual reality. Missing physical parameters, e.g. longitudinal and lateral acceleration, are transferred from the vehicle simulation to the vehicle's control devices. The actual driving trajectory of the vehicle to be tested is then compared with the nominal driving trajectory in order to validate the correct function of all systems.

6 Conclusion

This contribution presents the novel roller test stand *x-road curve* developed at Dürr Assembly Products GmbH. It features steerable roller sets at the front axle to meet new challenges for End of Line testing, arising from semi-autonomus and autonomous vehicles. The lateral dynamic of vehicles on roller test stands is analyzed. The vehicle model takes into account lateral tyre forces due to a slip angle between tyre and roller set. These forces are actively controlled by the roller test stand to center the vehicle in the test stand. The development of the control structure using virtual commissioning is illustrated. Control structure and controller coefficients determined through simulation show good suitability for controlling the actual roller test stand. Only few adaptions during tests with different vehicles and driving maneuvers are necessary. A concept for functional tests of advanced driver assistance systems in a virtual environment is shown.

References

1. Galko, C., Rossi, R., Savatier, X.: Vehicle-Hardware-In-The-Loop System for ADAS Prototyping and Validation in International Conference on Embedded Computer Systems: Architectures, Modeling and Simulation (SAMOS XIV), 2014 Galuzzi, C. (ed.). IEEE, Piscataway, NJ, 2014), pp. 329–334. ISBN: 978-1-4799-3770-7
2. Gietelink, O., Ploeg, J., de Schutter, B., Verhaegen, M.: Development of advanced driver assistance systems with vehicle hardware-in-the-loop simulations. Veh. Syst. Dyn. **44**, 569–590 (2006). ISSN: 0042-3114
3. Schramm, D., Hiller, M., Bardini, R.: Modellbildung und Simulation der Dynamik von Kraftfahrzeugen. Springer, Heidelberg (2010). https://doi.org/10.1007/978-3-540-89315-8
4. Verhoeff, L., Verburg, D.J., Lupker, H.A., Kusters, L.: VEHIL: a full-scale test methodology for intelligent transport systems, vehicles and subsystems. In: Proceedings of the IEEE Intelligent Vehicles Symposium, 2000 (Institute of Electrical and Electronics Engineers, Piscataway, N.J, 2000), pp. 369–375. https://doi.org/10.1109/IVS.2000.898371, ISBN: 0-7803-6363-9

Dynamic Interaction of Heavy Duty Vehicles and Expansion Joints

Daniel Rill[1(✉)], Christiane Butz[1], and Georg Rill[2]

[1] Maurer Engineering GmbH, Frankfurter Ring 193, 80807 München, Germany
{d.rill,c.butz}@maurer.eu
[2] OTH Regensburg, Galgenbergstr. 30, 93053 Regensburg, Germany
georg.rill@oth-regensburg.de

Abstract. The "Smart Bridge (Intelligente Brücke)" project cluster, initiated by the German Federal Highway Research Institute (Bundesanstalt für Straßenwesen, BASt) and the Federal Ministry of Transport and Digital Infrastructure (BMVI), focuses on "smart" monitoring devices that allow an efficient and economic maintenance management of bridge infrastructures. Among the participating projects, the one presented herein focuses on the development of a smart expansion joint, to assess the traffic parameters on site. This is achieved by measuring velocity and weight of crossing vehicles. In reference measurements, performed with a three-axle truck and a typical tractor semi-trailer combination with five axles in total, it was shown that the interaction between the vehicle and the expansion joint is highly dynamic and depends on several factors. To get more insight into this dynamic problem, a virtual test rig was set up. Although nearly all vehicle parameters had to be estimated, the simulation results conform very well with the measurements and are robust to vehicle parameter variations. In addition, they indicate a significant influence of the expansion joint dynamic to the peak values of the measured wheel loads, in particular on higher driving velocities. By compensating the relevant dynamic effects in the measurements, a "smart" data processing algorithm makes it possible to determine the actual vehicle weights in random traffic with reliability and appropriate accuracy.

1 Introduction

The bridge equipped with important aspects of the "Smart Bridge" is part of the greater program "Digital Test Bed Motorway", which encompasses numerous projects under the common motto "Mobility 4.0" and stretches along the A9 between Nuremberg and Munich.

The structure (BW402e), built in 2015 at the motorway interchange Nuremberg, includes various sensor-equipped components, which are developed and investigated in several projects by a consortium. One aim is to better understand the correlation between bridge condition and traffic effects, in order to develop more accurate models for condition prognosis.

A. Kecskeméthy and F. Geu Flores (Eds.): ECCOMAS 2019, COMPUTMETHODS 53, pp. 471–478, 2020.
https://doi.org/10.1007/978-3-030-23132-3_56

Among these components, one of the expansion joints of the bridge (located between the bridge deck and eastern abutment) was outfitted with sensors.

To gather calibration data, reference runs with two different trucks with defined weight and several velocity settings were conducted. Based on these data sets, an evaluation software was created, which continuously gathers and assesses the data to determine the traffic characteristics.

2 Measurement Setup

The smart expansion joint (installed and maintained by Maurer Group) was specially developed, based on a standard modular Swivel-Joist-Joint. While maintaining basic requirements, such as watertightness and conformity to building codes (by singular assessment according to TL/TP FÜ(2005)), the "smart" joint has some modifications to facilitate the intended measurements [1]. Mainly, the lamellae are physically separated between the lanes, so traffic can be evaluated for both lanes independently. The joint is also equipped with diamond shaped plates on top of the lamellae, which in this case serve a dual purpose: First, as in standard devices, they provide a reduction of the tire contact noise and second, they allow for a smoother force progression during each crossing event. The left image in Fig. 1 shows the working principle of the "smart" joint. Integrated into the control and support mechanism are sliding bearings with integrated load sensors. These measure the vertical loads on both ends of each support bar. Also, the first and third lamellae of each segment are connected to draw-wire sensors, which measure the current total gap width. This setup allows not only for a measurement of the total force which acts on the joint, but by comparing the load peaks on the front and rear side with the joint opening, the velocity of the vehicles can be calculated.

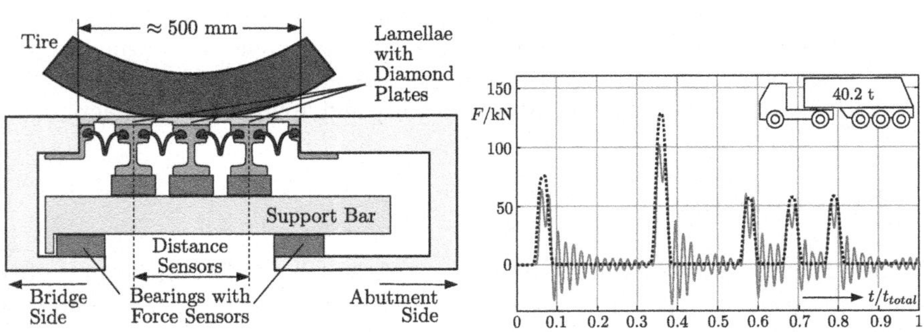

Fig. 1. Schematic longitudinal section of smart expansion joint with sensor locations and deviations between measurement at 90 km/h (solid line) and 5 km/h (dotted line)

From the data obtained by reference runs, it became apparent that the relation between static and (dynamic) measured load is dependent on the vehicle's velocity, cf. right plot in Fig. 1.

3 Virtual Test Rig

Appropriate models are required to describe the dynamic interaction between heavy duty vehicles and expansion joints.

For decades the multibody system approach has been successfully applied to all kind of vehicles, [2]. Complex vehicle models which may be set up by commercial multibody software packages or by appropriate modeling approaches will produce accurate simulation results only if a reliable set of parameters is available. In this particular case, however, just the wheel bases and the total mass of the test vehicles measured by the static axle loads were known. That is why rather simple two-dimensional models, like the planar model presented in [3] that describes the vertical dynamics of an agricultural tractor, were used to simulate the three-axle truck and the tractor semi-trailer combination, respectively. It turned out that the compliance of the expansion gap plays an important rule in the interaction between the vehicle and the expansion joint. A lumped mass approach to the expansion joint that takes relevant body deformations into account results in a comparatively lean three-dimensional model matching the accuracy of the vehicle models and providing an excellent run time performance.

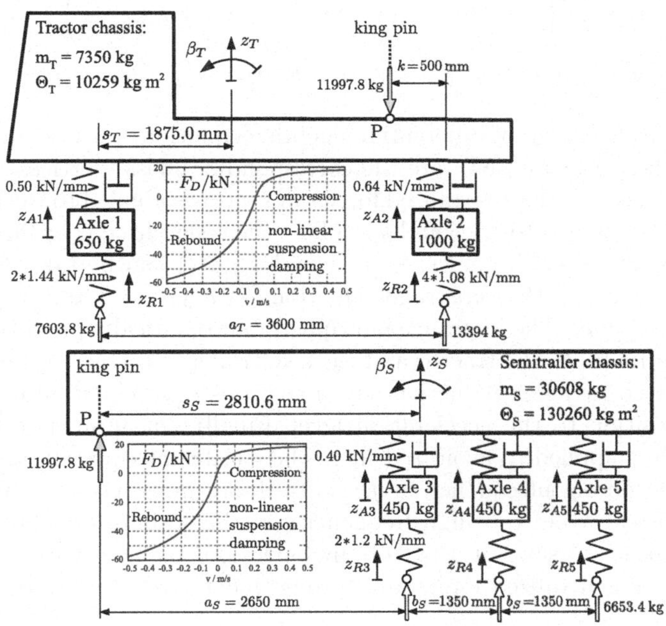

Fig. 2. Two-dimensional model of a tractor/semitrailer combination with model parameters

The two-dimensional model for the tractor/semi-trailer combination is separated into two parts, cf. Fig. 2. The first model has $f_T = 4$ degrees of freedom that are represented by the tractor chassis hub motion z_T, the tractor chassis pitch motion β_T, and the vertical displacements of the front z_{A1} and the rear z_{A2} axles. A three-axled truck is then modeled simply by adding a second rear axle with the corresponding vertical axle displacement z_{A3} as an additional degree of freedom. The second model represents the semitrailer that has no front axle but three rear axles. It has $f_S = 5$ degrees of freedom represented by the semitrailer chassis hub motion z_S, the semitrailer chassis pitch motion β_S, and the vertical displacements z_{A3}, z_{A4}, z_{A5} of the three rear axles. The two vehicle parts are coupled by a constraint equation provided by the revolute joint located at point P on the king pin line that approximates the real 5th-wheel coupling in this two-dimensional approach. Only the wheel bases and the static axle loads are known. The derived and estimated parameters of the tractor and the semi-trailer are provided in Fig. 2. Here, also the mass of the unloaded truck had to be estimated in order to derive the mass of the trailer and the load at the kingpin. The pneumatic coupling between the three rear axles of the trailer is neglected because it will act rather slowly and hence will not influence a rather fast expansion joint crossing maneuver. The equations of motion for the tractor and the semi-trailer result in two coupled sets of second order differential equations

$$\mathbf{M}_T\,\mathbf{y}_T = \mathbf{q}_T + \lambda\left(\frac{dg}{d\mathbf{y}_T}\right)^T \quad \text{and} \quad \mathbf{M}_S\,\mathbf{y}_S = \mathbf{q}_S + \lambda\left(\frac{dg}{d\mathbf{y}_S}\right)^T \quad \text{as well as}$$
$$g = g(\mathbf{y}_T, \mathbf{y}_S) = z_T + (a_T - s_T - k)\,\beta_T - (z_S - s_S\,\beta_S) = 0 \tag{1}$$

that represent a system of differential algebraic equations (DAEs) in an index 3 notation, where $\mathbf{M}_T$ and $\mathbf{M}_S$ are the corresponding mass matrices, the vectors $\mathbf{q}_T$ and $\mathbf{q}_S$ represent the generalized forces and torques applied to the tractor and the semitrailer, the vectors $\mathbf{y}_T$ and $\mathbf{y}_S$ collect the generalized coordinates of each vehicle part. Small pitch angles $\beta_T \ll 1$ and $\beta_S \ll 1$ can be taken for granted here which simplifies the constraint equation $g(\mathbf{y}_T, \mathbf{y}_S) = 0$ and its derivatives dg/dy_T and dg/dy_S. The Lagrange multiplier λ automatically provides appropriate coupling forces to the tractor and the semitrailer, [4]. DAE systems of index 3 can be solved by index-reduction where the constraint equation is transferred from the position to the acceleration level. Finally, an approriate Baumgarte stabilization is applied to avoid a drift in the constraint equation, [5]. The chosen stabilization eigenfrequency of $f_C = 10$ Hertz and a viscous damping rate of $\zeta_C = 0.5$ will avoid too high frequencies in the stabilization but still grant a sufficient fast and smooth decay of any deviations in the constraint equation. The vectors of generalized forces and torques incorporate the suspension forces

$$F_{Si} = c_{Si}\,u_i + F_D\,(v_i) \quad \text{with} \quad F_D\,(v_i) = \begin{cases} d_0\,\frac{v_i}{1+p_C v_i} & \text{if } v_i \geq 0 \\ d_0\,\frac{v_i}{1-p_R v_i} & \text{if } v_i < 0 \end{cases} \tag{2}$$

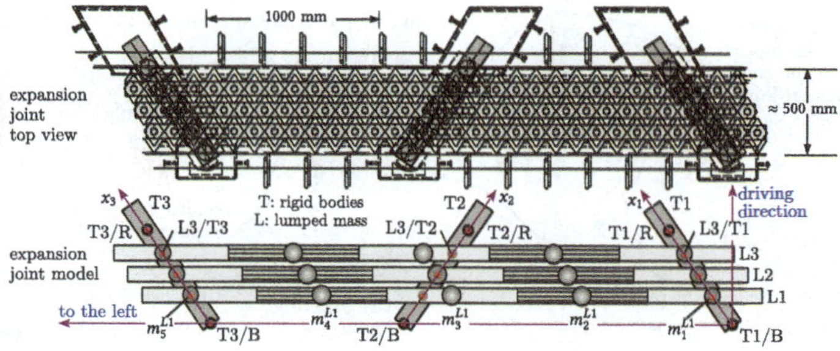

Fig. 3. Top view of one part of the expansion joint and corresponding lumped mass model

where the parameter c_{Si} denotes the stiffness of the corresponding suspension springs and d_0 the overall suspension damping constant at $v_i = 0$. The parameters $p_C > 0$ and $p_R > p_C$ generate a degressive damping characteristic that differs in compression and rebound mode. Although the suspension stiffness is modeled differently at the front and the rear axles, the same damping parameters were applied at each axle as typical for the layout of heavy duty trucks. The static and dynamic axle loads are provided by

$$F_{Ti}^S = \hat{c}_{Ti}\,(z_{Ri} - z_{Ai}) \quad \text{and} \quad T_D\,\frac{dF_{Ti}^D}{dt} = F_{Ti}^S - F_{Ti}^D \quad \text{or} \quad \frac{dF_{Ti}^D}{dt} = \frac{1}{T_D}\left(F_{Ti}^S - F_{Ti}^D\right) \tag{3}$$

where $\hat{c}_{Ti}$ denotes the overall tire stiffness at each axle, z_{Ri} is the average height of the road or expansion joint respectively and z_{Ai} describes the vertical displacement of the axle. The dynamic loads F_{Ti}^D generated by the first order differential equations are distributed equally to each tire, hence representing the load that is applied to the road or the expansion joint respectively.

The expansion joint, as depicted in Figs. 1 and 3 is modeled by a lumped mass system where each lamella (L1, L2, L3) was approximated by five modal masses (m_1^{L1} ... m_5^{L3}) in order to model the bending eigenmodes. A 4th-order polynomial defined by the vertical displacements of the 5 point masses provides the deflection of each lamella at arbitrary points. The support-bars or traverses (T1, T2, T3) were described by rigid bodies with hub and pitch about their local lateral axis (y_1, y_2, y_3) as degrees of freedom. Small pitch angles can be taken for granted here. The rubber mounts that connect the lamellae with the support-bars (T1/L1, ... T3/L3) and the traverses with the bridge (T1/B, T2/B, T3/B) or the road (T1/R, T2/R, T3/R), respectively, were modeled by simple linear spring damper combinations. Hence, the multibody model of the expansion joint is linear and has in total $f = 3*5+3*2 = 21$ degrees of freedom that reproduce eigenmodes of the vertical vibrations in the range from 98 to 986 Hz.

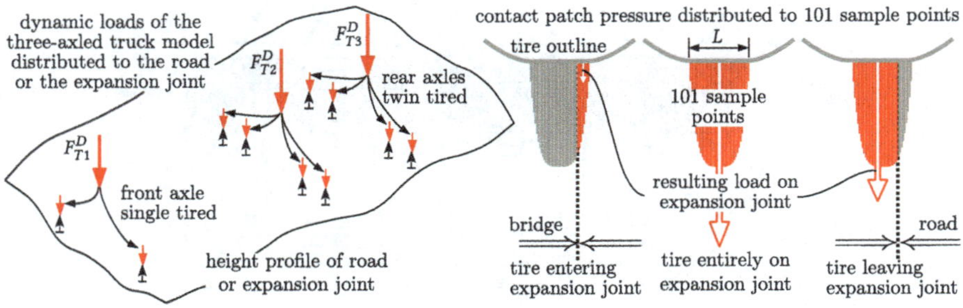

Fig. 4. Interaction between vehicle (three-axled truck) and expansion joint

To couple the two-dimensional vehicle models with the three-dimensional expansion joint model, the dynamic wheel loads, in case of the three-axle truck F_{T1}^D to F_{T3}^D, are equally split into corresponding wheel loads according to the number of tires mounted to each axle, left image in Fig. 4. The wheel or tire loads are distributed over the contact patch. Within the two-dimensional vehicle models, just the load distribution into the longitudinal direction is taken into account. According to [6] the length of the contact patch depends on the wheel load and the unloaded tire radius.

By using the dynamic axle load, this length can be calculated prior to the contact calculation itself.

The contact length L determines the distribution of the load between the bridge deck, expansion joint and the road surface, right image in Fig. 4. To distribute the dynamic tire loads properly, the contact patch was discretized by 101 sample points and the pressure distribution along the contact length was assumed to follow a 4th-order polynominal.

4 Simulation Results

As seen in Fig. 5, the measured forces represented by the peak-values correspond accurately with the steady state axle loads (F_1^0 *to* F_5^0) of the two test vehicles at low velocities ($v \approx 30$ km/h). With increasing driving velocity v, the time history of the resulting force $F = F(t)$ becomes more and more dynamic. On this particular expansion joint the first bending eigen-mode of the lamellae occurs at a frequency of $f \approx 100$ Hz and generates a double peak and a significant transient response that causes fluctuations in the time history of the measured signal even when the axle has passed the expansion joint. The simulation provides the movements of the vehicles too. In all runs, the pitch motions were limited to a few angle seconds which indicates that the most uncertain inertia properties of the truck, tractor, or semitrailer chassis will have no impact on the results. A reduced suspension damping results in quite different chassis and axle movements, but the influence on the forces exposed to the expansion joint is not noticeable.

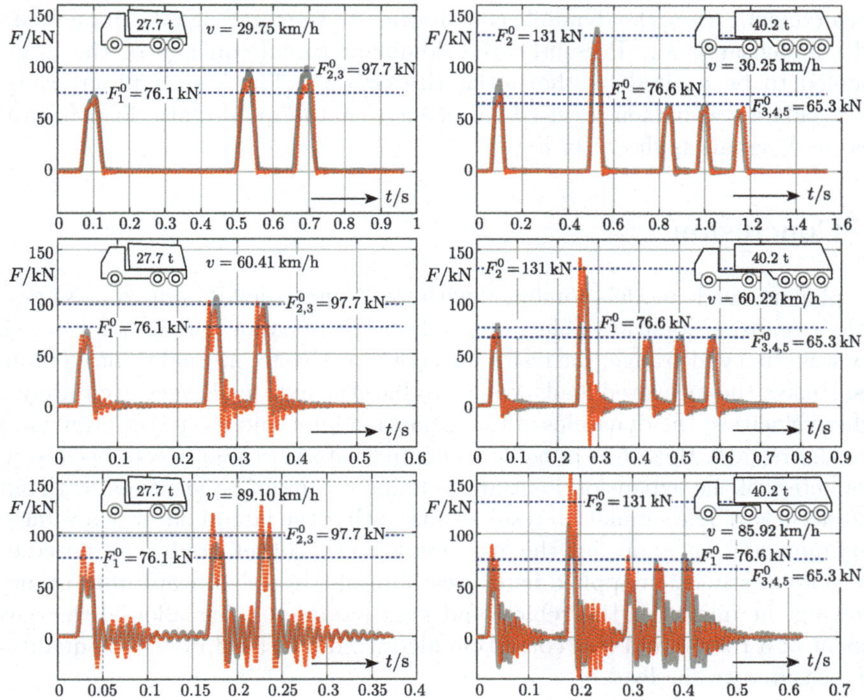

Fig. 5. Comparison between measurements (solid gray) and simulation results (dotted red) at different driving velocities where the dotted blue horizontal lines mark the static axle loads

5 Data Processing

With the insights gained from the virtual experiments, a correction algorithm could be developed, which allows for a compensation of the dynamic influence. As it was found that mainly the vibrations of the joint are responsible for the

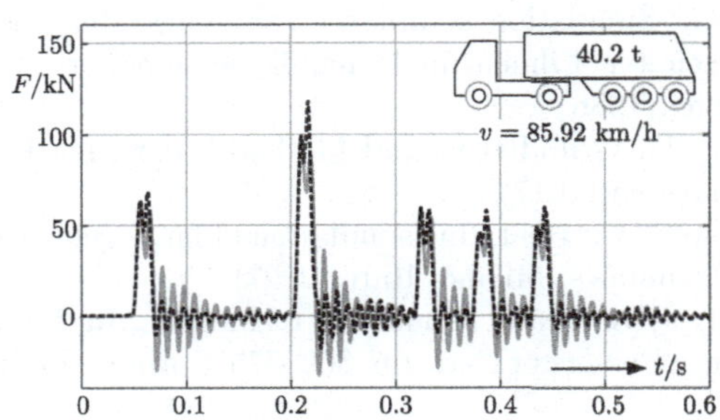

Fig. 6. Smart correction of measured signal. The solid gray line represents the raw measured signal. The dotted black line shows how the compensation algorithm has closely reconstructed the quasi-static values by subtracting the dynamic contribution.

measurement errors, the dynamic contribution to the force signal can be isolated and compensated for. Presently, the dominant eigenfrequency of the joint is modeled to be excited by the initial tire contact. The known progression of this oscillation can then be used to reconstruct the quasi-static load from the measured signal, as shown in Fig. 6.

6 Conclusion

The planar vehicle models combined with the 3-dimensional lumped mass model of the expansion joint are able to reproduce the dynamics of the joint crossing very well. In a wide range, the results are not sensitive to internal vehicle parameters. Hence, the measured loads will reproduce the axle loads very accurately. At higher velocities, the dynamics of the expansion joint and its height profile influence the results. But, it is possible to compensate for these effects because the parameters of the expansion joint are well known or can be identified with sufficient accuracy. The simulation results showed that the lateral offset of a vehicle is also important. By analyzing the measurements this offset can be calculated too. The simple estimation applied here already results in a significant improvement. Knowing the impact of the vehicle and joint parameters has allowed the development of a robust and fast correction algorithm which improves the quality of the measured traffic load.

Acknowledgements. The research is conducted on behalf of Bundesanstalt für Straßenwesen (BASt) and Bundesministerium für Verkehr und digitale Infrastruktur (BMVI). The authors also gratefully acknowledge the support on site by Autobahndirektion Nordbayern.

References

1. Butz, C., Rill, D.: Smart expansion joints and smart bearings embedded in the "Intelligente Brücke im Digitalen Testfeld Autobahn" (Intelligent Bridge of the Digital Motorway Test Bed). In: Proceedings of the 12th - Japanese-German Bridge Symposium, Munich, Germany, 4.9.-7.9 (2018)
2. Gipser, M., Rill, G.: Simulation komplexer Fahrzeugsysteme und Ergebnisdarstellung in rechnererzeugten Filmen, in Dynamik gekoppelter Systeme, VDI-Bericht, vol. 603. VDI-Verlag (1986)
3. Rill, G., Schaeffer, T.: Grundlagen und Methodik der Mehrkörpersimulation, 3rd edn. Vieweg, Wiesbaden (2017)
4. Rill, G., Chucholowski, C.: Real time simulation of large vehicle systems. In: ECCOMAS Multibody Dynamics, Milano, Italy (2007)
5. Rill, G.: Multibody systems and simulation techniques. In: Lugner, P. (ed.) Vehicle Dynamics of Modern Passenger Cars, pp. 309–375. Springer International Publishing (2019)
6. Rill, G.: Road Vehicle Dynamics. Taylor & Francis, Boca Raton (2012)

A Study on the Behaviour of the Rotating Disk with the Damage on the Tread

Yasutaka Maki[1,2(✉)] and Yoshiaki Terumichi[2]

[1] Running Gear Laboratory, Railway Technical Research Institute, 2-8-38 Hikaricho, Kokubunji-shi, Tokyo 185-8540, Japan
maki.yasutaka.37@rtri.or.jp

[2] Department of Mechanical Engineering, Sophia University, 7-1 Kioicho, Chiyoda-ku, Tokyo 102-8554, Japan
y-terumi@sophia.ac.jp

Abstract. As the evaluation method for the extent of the damage of the wheel tread of a railway vehicle, the measurement of its length in the rotational direction has been adopted for the decision of the wheel treatment. It is derived from the assumption that the depth of a wheel flat, which is one of the most serious damages, is proportion to its length in the rotational direction. Although the length of a shallow scrape tends to be long, the profile of the wheel with the long shallow scrape has less amount of irregularities compared with an exact circle. Therefore, in order to clarify the effect of the rotating wheelset with the shallow scrape on a bogie, we focused on the behaviour of the rotating wheel with the shallow scrape and evaluated it by dynamic model simulation and bench tests for an actual bogie. By the numerical simulation, we comprehended that the shallow scrape causes the vertical acceleration of the axlebox and the impact force to increase linearly in accordance with the running speed-up, and the vertical acceleration has a local maximum value around 110 km/h in accordance with a decrease in vertical collision velocity of the wheel with the roller in a higher speed range.

1 Introduction

There are some kinds of damages on the wheel tread of a railway vehicle related to adhesive force between the wheel and the rail. Since a wheel-flat especially causes large impact forces against the rail during the wheel rotating on the rail and damages the railway track and the parts equipped with the bogie in serious cases, the evaluation methods for the wheel-flat were suggested based on the measurements of the bogies and the tracks obtained from running tests. [1, 2] The measurement of its length in the rotational direction has been widely adopted for the decision of the wheel treatment. It was derived from the assumption that the depth of a wheel flat is proportion to its length in the rotational direction. Although the length of a shallow scrape tends to be long, the profile of the wheel with a long shallow scrape has less amount of irregularities compared with an exact circle. The authors built a dynamic model for a real-scale bogie equipped with the wheelset with a wheel flat. [4] In this paper, we focus on the behaviour of the rotating wheelset with a shallow scrape to clarify the effect on the bogie.

A. Kecskeméthy and F. Geu Flores (Eds.): ECCOMAS 2019, COMPUTMETHODS 53, pp. 479–486, 2020.
https://doi.org/10.1007/978-3-030-23132-3_57

2 Numerical Model Simulation

We built a dynamic model for the bench test to grasp and clarify the mechanism of the collision behaviour of the rotating wheelset with the roller rig and the effects on the bogie by the shallow scrape.

2.1 Modelling the Bogie on the Roller Rig

Figure 1 shows the dynamic model composed of rigid bodies which has 22 degrees of freedom. They are shown in Fig. 1 as z_b, φ_b, θ_b for the carbody, x_t, y_t, z_t, φ_t, θ_t, ψ_t for the bogie, z_w, φ_w for the wheelset, z_{box} for the axlebox and z_R,φ_R for a roller. z_{dmp} means the freedom in an oil damper equipped between the bogie frame and the axlebox. Each mass of the rigid body is denoted as m_n (n = 1, 2, 3, 4). The differential equations for the degrees of freedom are integrated numerically by Runge-Kutta-Gill method to obtain displacements and velocities. The reaction force on the wheel by the roller rig is generated with a Hertzian contact spring. Moreover, the model also has equivalent springs in the axlebox which simulate the bending stiffness of the wheel axle. The spring constant and the damping ratio are defined as k_{box} and c_{box} shown in Fig. 1. The k_{box} is gained from the length and the diameter of an L_{beam} long cantilever beam fixed at the joint of the wheel and the axle.

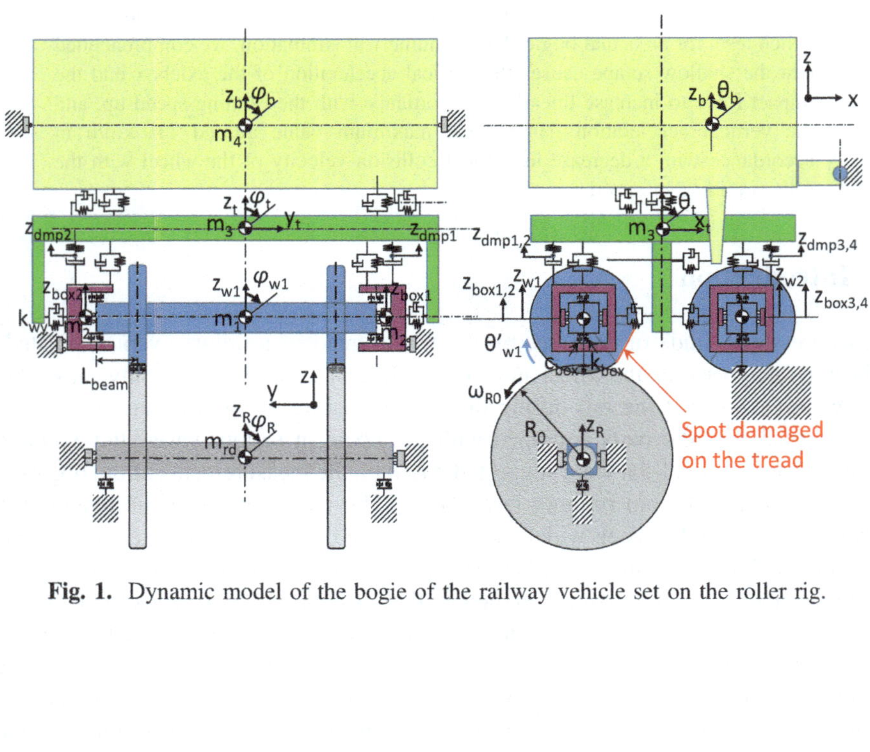

Fig. 1. Dynamic model of the bogie of the railway vehicle set on the roller rig.

2.2 Modelling the Wheel with a Shallow Scrape

The shape of the shallow scrape on the wheel tread is modelled as an arc with a curvature radius of r_1 different from the radius of the wheel of r_0 as shown in Fig. 2. The vertical displacement of the centre of the wheel is derived from the geometrical relationship between the wheel and the roller in the following subsection.

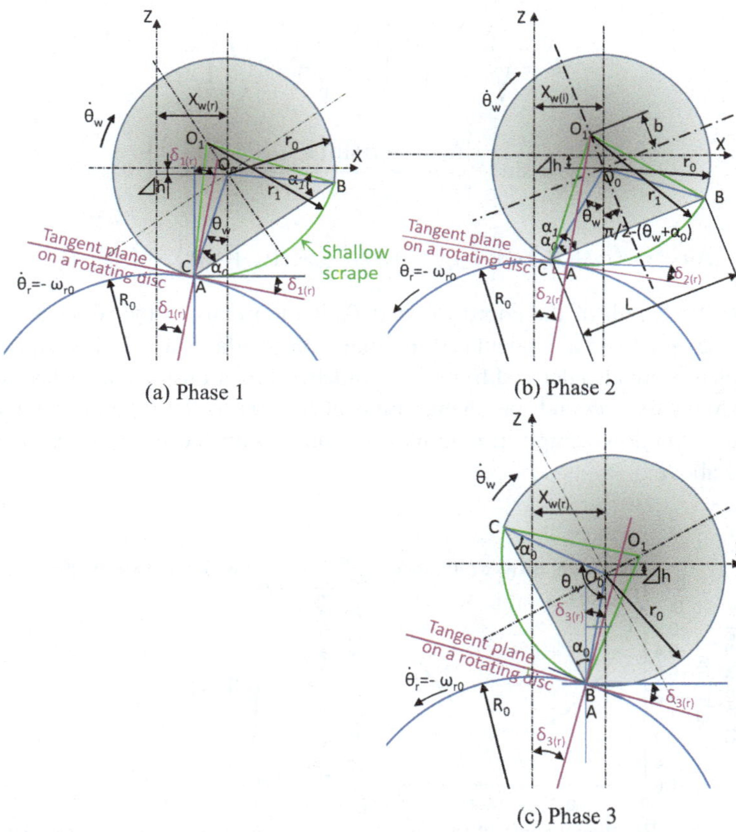

Fig. 2. The vertical displacement (Δh) of the center of the wheel rotating on the roller according to geometrical relationships between the shapes of the roller and the shallow scrape.

2.2.1 Partition of the Geometrical Contact Condition on the Roller

Figure 2 describes three cases of the rotating wheel on the roller considering where the contact point between the shallow scrape and the roller is located. The shallow scrape is formed by the arc BC with the curvature radius r_1 from the centre of the circle O_1. The contact point between the wheel and the roller is given as point A. The angle $\delta_{i(\mathrm{r})}$ (i = 1, 2, 3) is defined as the angle subtended between a perpendicular and a line segment from the centre of the roller to point A. The rotational angle of the wheel θ_w has the origin when the point C lies right below the centre of the wheel O_0. Angle α_0 and angle α_1 are the base angles of the triangles O_0BC and O_1BC.

2.2.2 Effect of the Shallow Scrape on the Vertical Displacement of the Centre of the Wheel Rotating

According to the partition determined by the θ_w and x_w as shown in Fig. 2, the angle $\delta_{i\,(r)}$ and the vertical displacement of the centre of the wheel Δh are derived as follows:

$$b = \sqrt{r_1^2 - \left(\frac{L}{2}\right)^2} - \sqrt{r_0^2 - \left(\frac{L}{2}\right)^2} \tag{1}$$

$$\delta_{2(r)} = \sin^{-1}\left[\frac{X_{w(r)} - b\sin\{\frac{\pi}{2} - (\theta_w + \alpha_0)\}}{R_0 + r_1}\right] \tag{2}$$

$$\Delta h_{(r)} = (r_1 + R_0)\cos\delta_{2(r)} + b\cos\left\{\frac{\pi}{2} - (\theta_w + \alpha_0)\right\} - (r_0 + R_0) \tag{3}$$

where L is the length of the chord BC and R_0 is the radius of the roller as shown in Fig. 2. These equations are calculated to obtain Δh for the phase 2. The equations for the other phases are also derived from the geometrical relationships described in Fig. 2. Figure 3 shows the Δh and the change ratio of the Δh to the rotational angle of the wheel with a shallow scrape in continuous contact with the roller in each phase as mentioned above.

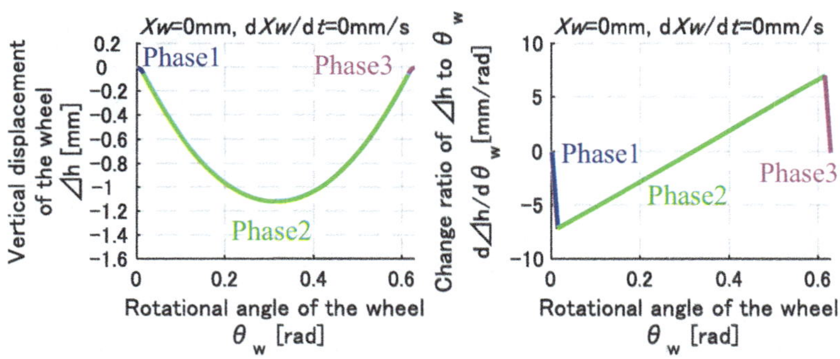

Fig. 3. The vertical displacement Δh and the change ratio of Δh to the rotational angle of the wheel with the shallow scrape having the continuous contact with the roller.

2.2.3 Contact Force Between the Wheel Tread and the Roller

The reaction force on the wheel generated by a Hertzian contact spring [3] and its displacement are given as follows:

$$st_{(i)} = \left\{z_w \mp b_i\varphi_w - (z_R \mp b_i\varphi_R) - \Delta h_{(i)}\right\}\cos\delta_{(i)}\ (i = r, l) \tag{4}$$

$$N_{(i)} = -\text{sign}(st_{(i)})k_{rz}\left|st_{(i)}\right|^{3/2} \tag{5}$$

where the displacement of the Hertzian spring is given by $st_{(i)}$ and normal force on the wheel with respect to the tangent plane on the roller is defined as $N_{(i)}$. Subscript i denotes the right wheel ($i = r$) or left wheel ($i = l$). The symbol of sign($st_{(i)}$) shown in Eq. 5 describes the sign of the displacement of the Hertzian spring. The constant of the Hertzian spring is given by k_{rz}. Longitudinal creep force is also considered to be applied to the wheel in the longitudinal direction.

3 Verification of the Dynamic Model by Experiments

3.1 Experiments on the Roller Rig

The vertical acceleration of an axlebox was measured in an actual bogie in the bench test to verify the dynamic model. A shallow scrape was machined on a part of the wheel tread by a 3D machinery as shown in Fig. 4. The actual bogie was set on the roller rig and the wheelset was driven by the roller up to the speed of 130 km/h.

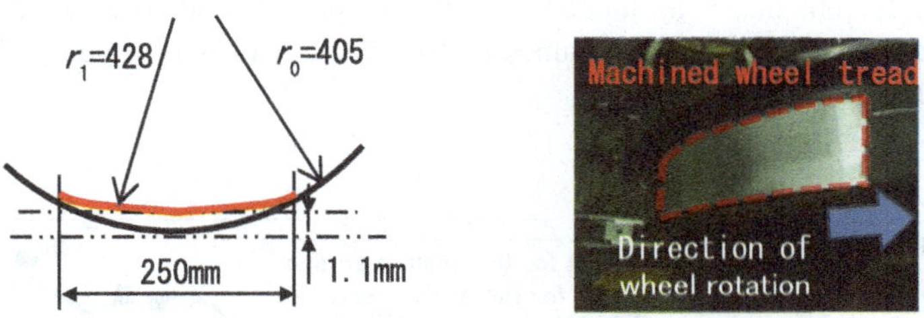

Fig. 4. The shape of the shallow scrape on the wheel tread (left) and the photo of the wheel tread machined and tested in the bench testing (right).

3.2 Numerical Results and Comparison with Experiments

Blue lines in Fig. 5 show the measurements of the vertical acceleration α of the axlebox normalized by the maximum value of α_{max} obtained in the bench testing. Negative acceleration denoted by arrow *A* in Fig. 5 is caused by the first descent of the wheel. Then, the wheel collides with the roller and it causes impulsive α/α_{max} denoted by arrow *B*. The second descent, the same as the first one, causes negative acceleration denoted by arrow *C*, and the second collision occurs at point *D*.

The simulation results are described by black lines in Fig. 5. At the any speeds, the first descents of the axlebox denoted by arrow *A* are reproduced well. Also, the durations from the first descent *A* to the first collision *B* coincide with the measurements. However, there are differences in impulsive α/α_{max} as shown with the arrow *B* and the arrow *D*. The dynamic model estimates impulsive α/α_{max} larger than the measurements. This is because the model overestimates the bending behaviour at the end of the wheelset rather than that obtained in the bench test. In addition, the second

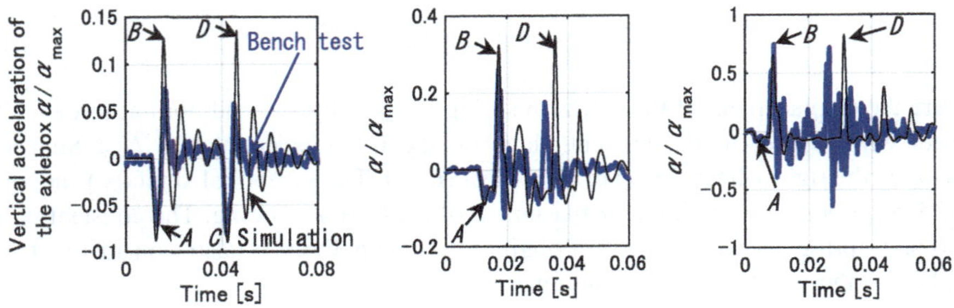

Fig. 5. Comparison of time-domain waveforms of the vertical acceleration of the axlebox with a focus on the effects of the shallow scrape.

collision denoted by arrow *D* appears slightly later than the results obtained in the bench test in a higher speed range than 60 km/h in Figs. 5 (b) and (c). This is caused by the higher bounce of the axlebox due to the overestimation of the bending behaviour mentioned above. Figure 6 shows the maximum value of the α/α_{max} obtained by the bench testing and the simulation for each speed. Although the simulation result is larger than the measurements, it is noticed that the simulation model reproduce trend of a linear increase in the maximum values of the α/α_{max} and it has the local maximum value around 110 km/h.

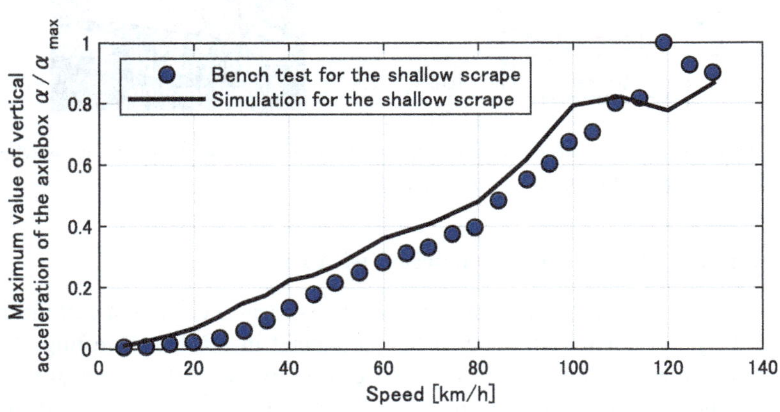

Fig. 6. Comparison of the maximum values of α/α_{max} by the simulation results and the measurements obtained in the bench testing on the roller rig for various speeds.

3.3 The Effects of the Rotating Wheel with the Shallow Scrape on the Contact and the Collision with the Roller

Figure 7 shows the simulation results concerning the contact state of the wheel with the roller. The dashed line coloured in blue means the rotational phase of the wheel in which a contact point coincides with the boundary phase between phase 1 and phase 2.

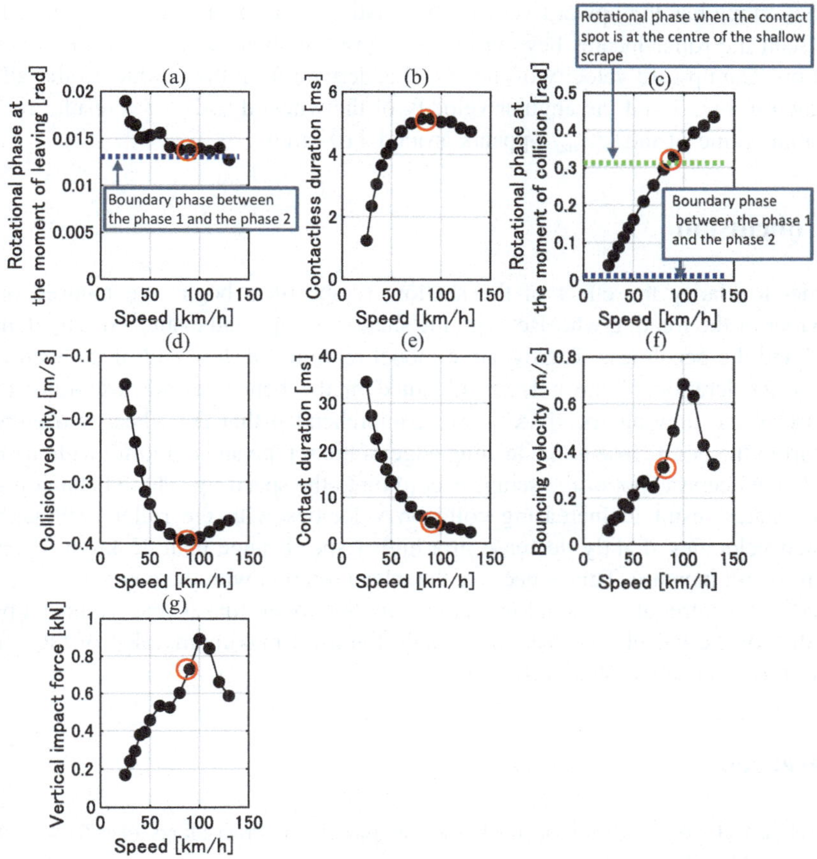

Fig. 7. Simulation results by the dynamic model concerning the contact state of the wheel with the roller at the moment of leaving and colliding with the roller.

The green dashed line shows the rotational phase of the wheel in which the contact spot coincides with the centre of the shallow scrape in the rotational direction. The latter is the rotational angle where the loss of the wheel radius becomes the greatest compared with the exact circle. According as the rotational speed of the wheel increases, the point where the wheel is getting out of touch with the roller shifts ahead to the leading edge of the shallow scrape (Fig. 7a). Moreover, the rotational phase increases at the moment of collision (Fig. 7c), which means the spot of the collision between the wheel and the roller shifts to the tailing edge of the shallow scrape. When the contact spot with the roller coincides with the centre of the shallow scrape mentioned above, the duration of the contactless rotation becomes the longest shown as red circles and the collision velocity gets the highest at a speed of 85 km/h as shown in Fig. 7d. In a higher speed range than 85 km/h, the collision velocity, however, slightly decreases owing to a decrease in the amount of the wheel's descent. Meanwhile, bouncing velocity still increases up to 100 km/h (Fig. 7f). This is because the centre of the wheel with a shallow scrape has the upward velocity by rotating on a roller in case that the contact point with the roller locates beyond the centre of the shallow scrape in the rotational direction. The upward velocity of the wheel is derived from the product of the $\Delta h/\mathrm{d}\theta_w$ as shown in Fig. 3 and the angular velocity of the wheel $\mathrm{d}\theta_w/\mathrm{d}t$. As a result, the local maximum value of the α/α_{max} appears around 110 km/h.

4 Conclusion

In order to clarify the effect of the shallow scrape on a bogie, we focused on the behaviour of the rotating wheelset with the shallow scrape and evaluated it by dynamic model and the bench test. The dynamic model reproduced the trend of the increase in vertical acceleration of the axlebox obtained in the bench testing according to the increase of the running speed. Also, we comprehended that the wheel is leaving the roller from the point close to the leading edge of the scrape and colliding with the roller close to the deepest spot of the scrape according to the speed-up. These behaviours and contact states result in increasing collision velocities with the roller. Although the collision velocities slightly decrease in a higher speed range than 85 km/h, bouncing velocities still increase. This is because the wheel rotation with the contact spot located beyond the centre of the shallow scrape on the roller makes the vertical upward velocities of the wheel increase. As a result, the local maximum value of the vertical impact force appears around 110 km/h.

References

1. AAR: Effect of flat wheels on track and equipment. In: Proceedings of AREA, vol. 53, pp. 423–448 (1953)
2. Betzhold, C.: Erhöhung der Beanspruchung des Eisenbahnoberbaues durch Wechselwirkungen zwuschen Fahrzeug und Oberbau, Glasers Annalen, pp. 108–115, (April 1957) (in German)
3. Stronge, W.: Impact mechanics, The Press Syndicate of the University of Cambridge, pp. 116–119 (2004)
4. Yasutaka, M., Yoshiaki, T.: A study on the contact condition and the collision behavior of a wheelset with a flat set on a roller rig. Transactions of JSME **84**(865) (2018). https://doi.org/10.1299/transjsme.18-00198 (in Japanese)

Multibody Dynamics Analysis of Railway Vehicle with Independently Rotating Wheels Using Negative Tread Conicity

Yu Wang(✉), Shihpin Lin, Hiroshi Tajima, and Yoshihiro Suda

Institute of Industrial Science, The University of Tokyo, 4-6-1 Meguro-ku, Tokyo 153-8505, Japan
{wangyu17, lin, suda}@iis.u-tokyo.ac.jp, cootaji@yahoo.co.jp

Abstract. In order to improve running stability and curving ability of railway vehicle with independently rotating wheels, an innovative structure, independently rotating wheels with negative tread conicity was proposed by one of the authors of this paper. Since the proposal of this structure, dynamics analysis based on simplified model had been implemented by means of simulation and scale model experiment. But precise numerical analysis has not been handled under the framework of multibody dynamics analysis to date. To this end, the multibody dynamics analysis of railway vehicle with independently rotating wheels using negative tread conicity is carried out in this paper. The modeling process is presented in detail, and for dealing with wheel/rail contact problem in railway vehicle, the velocity transformation and differential algebraic equation combination method is exploited. The simulation results indicate that the dynamics performance associate with running stability and curving ability could be improved by virtue of using negative tread conicity wheels.

1 Introduction

The light rail transit (LRT) system is becoming more popular as a kind of public traffic system, especially in urban areas of Japan and Europe, where tight curve section has to be set to adapt street layout. Moreover, considering the accessibility, the platform of LRT needs to be lowered. To meet such specific demands, the wheelset using independently rotating wheels (IRW) was proposed as a substitution of solid axle wheelset. By decoupling left and right wheels and removing the axle, the low-floor design can be achieved, and the curving ability in tight curve section is likely to be considerably improved [1]. However, as the longitudinal creep forces decrease drastically, the self-steering ability of IRW with normal tread shape is insufficient. To regain the self-steering ability like the solid axle wheelset, some creative ideas were put forward such as EDF bogie and EEF bogie [2], which make use of gravity stiffness to generate a restoring moment. But these structures show the feature of complex linkage mechanism, so that were not widely used in practical application. The concept of independently rotating wheel with negative tread conicity (NTCIRW) was proposed by Suda [3] for improving the inherent defect of IRW. This structure changes the direction of

A. Kecskeméthy and F. Geu Flores (Eds.): ECCOMAS 2019, COMPUTMETHODS 53, pp. 487–494, 2020.
https://doi.org/10.1007/978-3-030-23132-3_58

yaw moment generated by gravitational restoring force, without complex linkage mechanism, and the superior performance had been validated by scale model experiment and full model simulation [4]. However, since the proposal of NTCIRW, all the investigations are based on simplified dynamics model. The wheel/rail contact problem using actual wheel and rail shapes hasn't been considered in previous studies, and flange/rail contact, which plays an important role in cornering, was also ignored when establishing dynamics model. To this end, the multibody dynamics analysis of railway vehicle with NTCIRW is implemented in this paper, and the superior running stability and curving ability are confirmed by numerical simulation.

2 Modeling of Wheel Unit with NTCIRW

2.1 Wheel/Rail Contact Problem

Since the most complex process of modeling wheel unit is the part that involves wheel/rail contact, contact problem is firstly dealt with in this section. Generally, wheel/rail contact problem can be solved by constraint approach or elastic approach [5]. In this paper, the constraint contact is adopted to deal with tread/rail contact for determining contact parameters, while the elastic contact is adopted for flange/rail contact to get flange contact force time-efficiently. The wheel used in simulation has conical tread and knife-edge flange and the discrete geometric data of UIC60 rail is created by cubic spline interpolation. As shown in Fig. 1, taking the left contact pair as an example, the contact points on tread and rail are termed as T_3 and P_3, respectively. To describe the position of contact points, commonly four parameters are used, which are the lateral distance p_{T3} from the origin of wheel coordinate system to T_3, the lateral distance p_{P3} from the origin of rail coordinate system to P_3, the angle θ_{T3} from T_3 coordinate system to wheel coordinate system and the traveled arc length s_{P3} of P_3 in trajectory coordinate system.

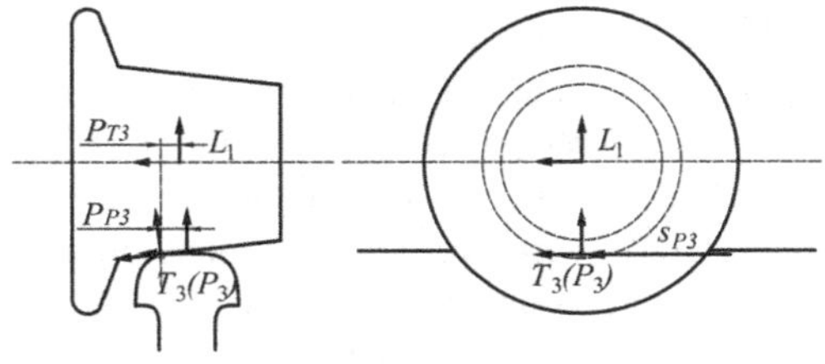

Fig. 1. Contact parameters.

The contact constraints are formulated according to two arguments that vectors from the origin of inertial coordinate system to contact points on wheel and rail are equal and the X and Y axis of wheel and rail contact point coordinate system are parallel respectively. The formulation of left tread/rail contact constraint is given as Eq. (1).

$$\mathbf{\Psi}^{tl} = \begin{bmatrix} \mathbf{D}_X^T \mathbf{C}_{P3T3}(s_{P3}, p_{P3}, \theta_{T3}, p_{T3}, \varphi_{P1D1}) \mathbf{D}_Z \\ \mathbf{D}_Y^T \mathbf{C}_{P3T3}(s_{P3}, p_{P3}, \theta_{T3}, p_{T3}, \varphi_{P1D1}) \mathbf{D}_Z \\ \mathbf{D}_X^T \mathbf{R}_{P3T3}(s_{P3}, p_{P3}, \theta_{T3}, p_{T3}, \varphi_{P1D1}, r_{P1D1Z}) \\ \mathbf{D}_Y^T \mathbf{R}_{P3T3}(s_{P3}, p_{P3}, \theta_{T3}, p_{T3}, \varphi_{P1D1}, r_{P1D1Z}) \\ \mathbf{D}_Z^T \mathbf{R}_{P3T3}(s_{P3}, p_{P3}, \theta_{T3}, p_{T3}, \varphi_{P1D1}, r_{P1D1Z}) \end{bmatrix} = \mathbf{0}_{5\times 1} \tag{1}$$

where $\mathbf{\Psi}^{tl}$ denotes left tread/rail contact constraint, $\mathbf{D}_X = [1,0,0]^T$, $\mathbf{D}_Y = [0,1,0]^T$, $\mathbf{D}_Z = [0,0,1]^T$, $\mathbf{C}_{P3T3}$ is the coordinate transformation matrix from T_3 coordinate system to P_3 coordinate system, and $\mathbf{R}_{P3T3}$ is the vector from P_3 to T_3. It should be noted that $\mathbf{C}_{P3T3}$ and $\mathbf{R}_{P3T3}$ are functions of contact parameters and two nongeneralized coordinates (vertical displacement r_{P1D1Z} and roll angle φ_{P1D1} of the wheel frame). By simultaneously solving the equations of left and right contact constrains using Newton-Rapson algorithm, the contact parameters can be obtained. Then, with the solved contact points and generalized coordinates, the longitudinal, lateral and spin creepage represented in T_3 coordinate system termed as υ_{T3_x}, υ_{T3_y} and φ_{T3} can be calculated according to (2)~(4):

$$\upsilon_{T3\_x} = \frac{\mathbf{D}_X^T \mathbf{C}_{OT3}^T \mathbf{V}_{OT3}}{\mathbf{D}_X^T \mathbf{C}_{OT3}^T \mathbf{V}_{OL1}} = \frac{\mathbf{D}_X^T \mathbf{C}_{L1T3}^T (\mathbf{V}'_{OL1} - \tilde{\mathbf{R}}_{L1T3} \mathbf{\Omega}'_{OL1})}{\mathbf{D}_X^T \mathbf{C}_{L1T3}^T \mathbf{V}'_{OL1}} \tag{2}$$

$$\upsilon_{T3\_y} = \frac{\mathbf{D}_Y^T \mathbf{C}_{OT3}^T \mathbf{V}_{OT3}}{\mathbf{D}_X^T \mathbf{C}_{OT3}^T \mathbf{V}_{OL1}} = \frac{\mathbf{D}_Y^T \mathbf{C}_{L1T3}^T (\mathbf{V}'_{OL1} - \tilde{\mathbf{R}}_{L1T3} \mathbf{\Omega}'_{OL1})}{\mathbf{D}_X^T \mathbf{C}_{L1T3}^T \mathbf{V}'_{OL1}} \tag{3}$$

$$\varphi_{T3} = \frac{\mathbf{D}_Z^T \mathbf{C}_{OT3}^T \mathbf{\Omega}_{OL1}}{\mathbf{D}_X^T \mathbf{C}_{OT3}^T \mathbf{V}_{OL1}} = \frac{\mathbf{D}_Z^T \mathbf{C}_{L1T3}^T \mathbf{\Omega}'_{OL1}}{\mathbf{D}_X^T \mathbf{C}_{L1T3}^T \mathbf{V}'_{OL1}} \tag{4}$$

where ~ is an operator that converses vector $\mathbf{R}_{OA} = \begin{bmatrix} r_{OAX} \\ r_{OAY} \\ r_{OAZ} \end{bmatrix}$ to matrix $\tilde{\mathbf{R}}_{OA} = \begin{bmatrix} 0 & -r_{OAZ} & r_{OAY} \\ r_{OAZ} & 0 & -r_{OAX} \\ -r_{OAY} & r_{OAX} & 0 \end{bmatrix}$. With this operator, the cross product of two vectors can be replaced by dot product. $\mathbf{\Omega}$ denotes angular velocity and $\mathbf{V}$ denotes translational velocity. In this paper, two kinds of nomenclature are used in order to distinguish the representation coordinate system. For example, for $\mathbf{R}_{OA}$, O is the starting point of the vector and the origin of representation coordinate system, while for $\mathbf{R}'_{OA}$, the origin of representation coordinate system is A. These two nomenclatures can be converted to each other by using coordinate transformation matrix $\mathbf{C}_{OA}$. Finally, the creep forces are calculated using saturation model [6], with the creep coefficients as: $\kappa_{11\_t} = 5\times10^5$ N, $\kappa_{22\_t} = 4\times10^5$ N, $\kappa_{23\_t} = 4\times10^2$ N, $\kappa_{11\_f} = 3\times10^4$ N, $\kappa_{22\_f} = 2\times10^4$ N, $\kappa_{23\_f} = 2\times10^2$ N.

For flange/rail contact, the elastic contact is adopted so that the contact constraint is still satisfied but without the last equation of Eq. (1), because the flange is permitted to indent into the rail. Apart from creep forces, which can be calculated followed the same process as that of tread/rail contact, the impact force modeled as a parallel spring-damper element also need to be calculated. The left contact force represented in rail contact point coordinate system is written as:

$$\mathbf{F}'_{OP5Z} = \begin{cases} \mathbf{D}_Z\left[K_f(-\delta)^{3/2} - C_f\dot{\delta}|\delta|\right] & \text{if } \delta < 0 \\ 0 & \text{if } \delta \geq 0 \end{cases} \tag{5}$$

where $\delta = \mathbf{D}_Z^T\mathbf{R}_{P5T5}$, means the distance of nearest points on flange and rail in Z axis of rail contact point coordinate system. Here, T_5 is contact point on left wheel flange, while P_5 is the contact point on left rail. $K_f = 3 \times 10^8$ N/m and $C_f = 5 \times 10^8$ N · s/m are stiffness coefficient and damping coefficient, respectively.

2.2 Dynamics Equation of Wheel Unit

The schematic diagram of wheel unit is shown as left figure of Fig. 2. The wheel unit consists of a wheel frame $\mathbf{D}_1$ and two independently rotating wheels $\mathbf{L}_1$ and $\mathbf{R}_1$. The traveled arc length along the track is used as a generalized coordinate to describe the longitudinal motion of wheel frame. Due to the presence of contact constraints, the vertical displacement and roll angle of the wheel frame are no longer generalized coordinates. In addition, each wheel has one generalized coordinate.

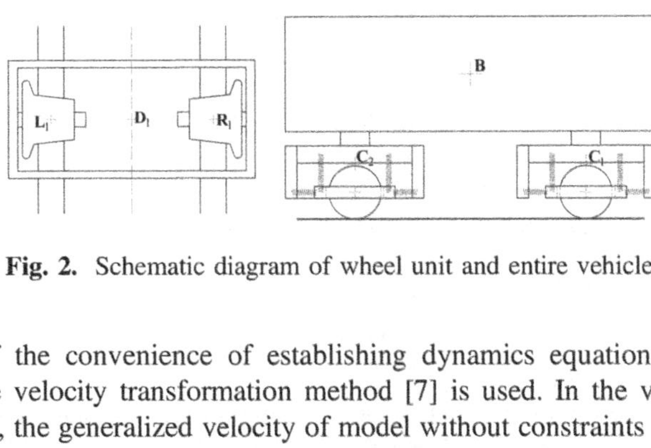

Fig. 2. Schematic diagram of wheel unit and entire vehicle.

In view of the convenience of establishing dynamics equation of model with constraints, the velocity transformation method [7] is used. In the velocity transformation method, the generalized velocity of model without constraints is denoted as **H**, while that with constraints is denoted as **S**. By considering the constraint conditions, **H** can be expressed by **S** as:

$$\mathbf{H} = \mathbf{H}_\mathbf{S}\mathbf{S} + \mathbf{H}_{\bar{\mathbf{S}}} \tag{6}$$

where $\mathbf{H}_\mathbf{S}$ is the Jacobian matrix, and $\mathbf{H}_{\bar{\mathbf{S}}}$ is the remaining part. With these two terms, the dynamics equation can be written as:

$$\mathbf{m}^\mathbf{S}\dot{\mathbf{S}} = \mathbf{f}^\mathbf{S} \tag{7}$$

where $\mathbf{m}^\mathbf{S}$ and $\mathbf{f}^\mathbf{S}$ are calculated according to Eq. (8).

$$\begin{cases} \mathbf{m}^{\mathrm{S}} = \mathbf{H}_{\mathrm{S}}^{T}\mathbf{m}^{\mathrm{H}}\mathbf{H}_{\mathrm{S}} \\ \mathbf{f}^{\mathrm{S}} = \mathbf{H}_{\mathrm{S}}^{T}\left[\mathbf{f}^{\mathrm{H}} - \mathbf{m}^{\mathrm{H}}\left(\frac{d\mathbf{H}_{\mathrm{S}}}{dt}\mathbf{S} + \frac{d\mathbf{H}_{\bar{\mathrm{S}}}}{dt}\right)\right] \end{cases} \tag{8}$$

where $\mathbf{m}^{\mathbf{H}}$ is the mass matrix and consists of the moment of inertia and mass of all bodies. $\mathbf{f}^{\mathbf{H}}$ is the force matrix and consists of equivalent forces and moments represented in each body coordinate system [7].

If removing all the nongeneralized coordinates from $\mathbf{S}$, the $\mathbf{H}_{\mathrm{S}}$ in Eq. (6) will appear, so that its derivative term $\mathbf{H}_{\bar{\mathrm{S}}}$ that contains unsolved generalized acceleration terms will appear. This is the special problem in the modeling of railway vehicle dynamics. To solve this problem, the velocity transformation combined with differential algebraic equation method is proposed in this paper.

Above all, the generalized velocity without consideration of wheel revolute joint constraints has the form as $\mathbf{H} = [\mathbf{\Omega}'_{\mathrm{OD}I}, \mathbf{V}'_{\mathrm{OD}I}, \mathbf{\Omega}'_{\mathrm{OL}I}, \mathbf{V}_{\mathrm{OL}I}, \mathbf{\Omega}'_{\mathrm{OR}I}, \mathbf{V}'_{\mathrm{OR}I}]^{T}$, and the generalized velocity with consideration of wheel revolute joint constraints is $\mathbf{S} = [\mathbf{\Omega}'_{P1D1}, \dot{s}_{P1}, v_{P1D1Y}, v_{P1D1Z}, \dot{\theta}_{D1L1}, \dot{\theta}_{D1R1}]^{T}$. In $\mathbf{S}$, two translational coordinates and three orientation coordinates of wheel frame are defined with respect to trajectory coordinate system. With the arc length s_{P1}, all the necessary trajectory information can be obtained from pre-built spline interpolation function. In the calculation process, the initial values of these generalized positions and generalized velocities are given firstly. Then by applying wheel/rail contact constraint as Eq. (1), the actual values of φ_{P1D1} and r_{P1D1Z} of wheel frame are updated. Hereafter, by differentiating the last equation of Eq. (1) and that of right wheel/rail contact constraint, the real values of roll angular velocity $\dot{\varphi}_{P1D1}$ and vertical velocity v_{P1D1Z} of wheel frame can be solved.

$$\mathbf{\Phi} = \begin{bmatrix} {}^{L}\mathbf{\Phi} \\ {}^{R}\mathbf{\Phi} \end{bmatrix} = \begin{bmatrix} \mathbf{D}_{Z}^{T}\mathbf{V}_{P3T3}(\dot{\varphi}_{P1D1}, v_{P1D1Z}) \\ \mathbf{D}_{Z}^{T}\mathbf{V}_{Q3U3}(\dot{\varphi}_{P1D1}, v_{P1D1Z}) \end{bmatrix} = \mathbf{0}_{2\times 1} \tag{9}$$

where U_3 and Q_3 denote contact points on right tread and rail, respectively. ${}^{L}\mathbf{\Phi}$ and ${}^{R}\mathbf{\Phi}$ are contact constraints on both sides in velocity level, and $\mathbf{\Phi}$ can be further expressed by $\mathbf{S}$ as:

$$\mathbf{\Phi} = \mathbf{\Phi}_{\mathrm{S}}\mathbf{S} + \mathbf{\Phi}_{\bar{\mathrm{S}}} = \mathbf{0}_{2\times 1} \tag{10}$$

where $\mathbf{\Phi}_{S} = \left[{}^{L}\mathbf{\Phi}_{S}{}^{R}\mathbf{\Phi}_{S}\right]^{T}$, $\mathbf{\Phi}_{S} = \left[{}^{L}\mathbf{\Phi}_{S} \quad {}^{R}\mathbf{\Phi}_{S}\right]^{T}$. The Eq. (7) is formulated just considering the revolute joint constraints. With the additional wheel/rail contact constraints, constraint force denoted as $\bar{\mathbf{f}}^{\mathrm{S}}$ is added to this equation, which can also be represented using Lagrange multipliers $\mathbf{\Lambda}$ as $\bar{\mathbf{f}}^{\mathrm{S}} = -\mathbf{\Phi}_{\mathrm{S}}^{T}\mathbf{\Lambda}$. Finally, the dynamics equation of wheel unit established by means of velocity transformation and differential algebraic equation combined method is shown as:

$$\begin{bmatrix} \mathbf{m}^{\mathrm{S}} & \mathbf{\Phi}_{\mathrm{S}}^{T} \\ \mathbf{\Phi}_{\mathrm{S}} & \mathbf{0} \end{bmatrix}\begin{bmatrix} \dot{\mathbf{S}} \\ \mathbf{\Lambda} \end{bmatrix} = \begin{bmatrix} \mathbf{f}^{\mathrm{S}} \\ -\dot{\mathbf{\Phi}}^{R} + \dot{\mathbf{\Phi}}^{\bar{R}} \end{bmatrix} \tag{11}$$

where $\dot{\mathbf{\Phi}}^R = \frac{d\mathbf{\Phi}_S}{\mathrm{dt}} S + \frac{d\mathbf{\Phi}_{\bar{S}}}{\mathrm{dt}}$. $\dot{\mathbf{\Phi}}^{\bar{R}}$ is equal to $\left[\mathbf{D}_Z^T \mathbf{a}_{P3T3}, \mathbf{D}_Z^T \mathbf{a}_{Q3U3}\right]^T$, and the elements are centrifugal acceleration of contact point T_3 and U_3. With velocity transformation and differential algebraic equation combined method, the formulation of dynamics equation of complex multibody system can be carried out under certain procedures.

Numerical simulation of wheel unit with IRW and NTCIRW given the same initial yaw angle and longitudinal velocity of wheel frame is implemented using the fourth-order Runge-Kutta algorithm under the environment of MATLAB. The lateral displacement of wheel unit is plotted in Fig. 3. The black solid lines are output from commercial software SIMPACK for illustrating the correctness of the results. As shown in Fig. 3, the wheel unit with NTCIRW can return to the centerline of track, while the lateral displacement of IRW increases rapidly and finally the wheel unit falls into unstable state. Therefore, the statement that NTCIRW outperforms IRW in terms of running stability can be confirmed.

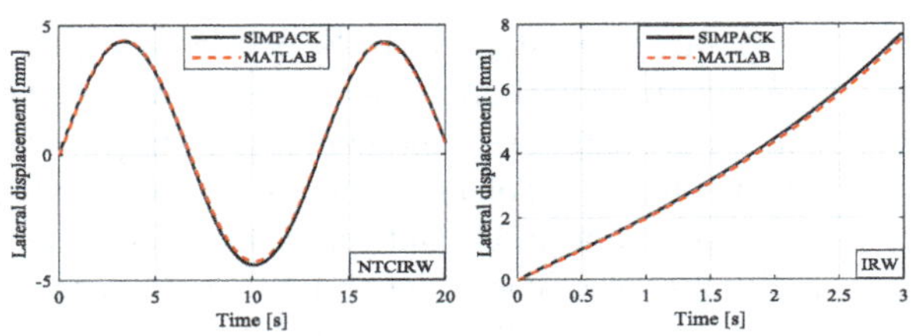

Fig. 3. Lateral displacement of NTCIRW and IRW.

3 Modeling of Entire Railway Vehicle

The schematic diagram of railway vehicle with NTCIRW is shown as the right figure of Fig. 2, consisting of nine body components, which are vehicle body **B**, two bogies $\mathbf{C}_1$ and $\mathbf{C}_2$, two wheel frames with the wheel base as 4 m and four independently rotating wheels. The bogie only has a yaw freedom relative to the vehicle body and connects the wheel frame through parallel spring-damper elements in all three directions. The process of modeling of wheel frame has been presented in Sect. 2, and since vehicle body and bogies simply consist of revolute joint constraints, it is only necessary to use the standard velocity transformation method to establish the dynamics equation by Eqs. (6) ~ (8). The inertia properties of vehicle and parameters of primary suspension system are collected in Tables 1 and 2. The track used in numerical simulation has five segments, that are straight line, entry transition curve, steady curve with radius of 33 m, exit transition curve and ending straight line, in sequential. The track data is discretized by a cubic spline interpolation function.

Table 1. Inertia properties of vehicle.

Body	Mass [kg]	Moment of inertia [$kg \cdot m^2$]		
		Roll inertia	Pitch inertia	Yaw inertia
Wheel frame	300	200	100	300
Wheel	500	24	46.225	24
Bogie	500	560	340	840
Vehicle body	2000	2400	7500	8500

Table 2. Parameters of primary suspension system.

	Longitudinal	Lateral	Vertical	Body-bogie
Stiffness coefficient [kN/m] (NTCIRW)	$k_x = 8000$	$k_y = 6000$	$k_z = 1000$	$k_\psi = 0$
Stiffness coefficient [kN/m] (IRW)				$k_\psi = 150$
Damping coefficient [$kN \cdot s/m$]	$d_x = 1$	$d_y = 1$	$d_z = 1$	$d_\psi = 15$

4 Simulation Results

In the numerical simulation, the vehicle starts moving with the initial longitudinal velocity as 5 m/s. The yaw angle of vehicle body and wheel units are plotted in the first row of Fig. 4. As shown in these two figures, the maximum yaw angle of front wheel unit of vehicle with NTCIRW is about 0.043 rad, and that of rear wheel unit is just 0.021 rad. While the maximum yaw angle of front wheel unit of vehicle with IRW reaches 0.067 rad, and that of rear wheel unit reaches 0.059 rad. Meanwhile, the yaw angle reduces to 0 rad during the steady curve for vehicle with NTCIRW, which is extremely expected from the perspective of reducing wheel wear. In order to evaluate the wheel/rail contact situation, the magnitudes of contact forces acted on NTCIRW and IRW are plotted in the second row of Fig. 4. It can be clearly seen from these

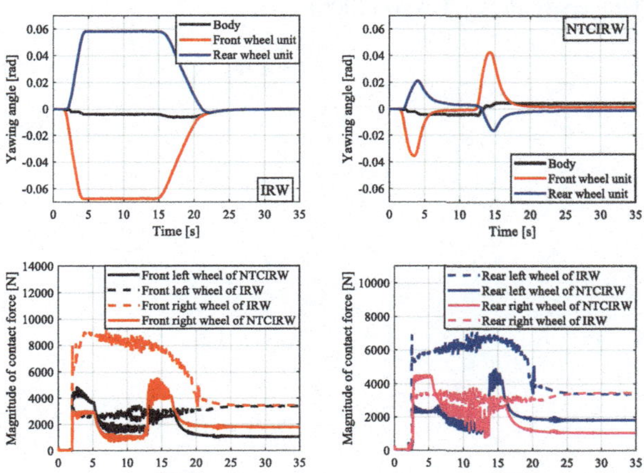

Fig. 4. Simulation results

figures that the contact force of the IRW is much larger than that of the NTCIRW. Therefore, in terms of lower wheel wear and derailment possibility, the curving ability of vehicle with NTCIRW is much better than vehicle with IRW.

5 Conclusion

In this paper, precise dynamics modeling of wheel unit and entire vehicle with independently rotating wheels using negative tread conicity are implemented by multibody dynamics analysis. A feasible multibody dynamics analysis method, which named as velocity transformation and differential algebraic equation combined method is introduced, which is tailored of dealing with wheel/rail contact problem in modeling of railway vehicle. From the simulation results, the statements that the running stability of wheel unit with NTCIRW outperforms wheel unit with IRW and the curving ability of vehicle with NTCIRW outperforms vehicle with IRW are confirmed.

References

Dukkipati, R.V., Narayana Swamy, S., Osman, M.O.M.: Independently rotating wheel systems for railway vehicles-a state of the art review. Veh. Syst. Dyn. 21(1), 297–330 (1992)

Frederich, F.: Dynamics of a bogie with independent wheels. Veh. Syst. Dyn. **19**(Suppl.), 217–232 (1989)

Suda, Y.: The possibility of a new concept of self-steering truck with independently rotating wheels. J-Rail **2008**, 473–476 (2008)

Ejiri, K.: Research on running stability improvement of railway vehicle independently rotating wheel bogie with self-steering ability. Ph.D. thesis, Ibaraki University (2015)

Shabana, A.A., Zaazaa, K.E., Escalona, J.L., Sany, J.R.: Development of elastic force model for wheel/rail contact problems. J. Sound Vib. **269**, 295–325 (2004)

Kalker, J.: A fast algorithm for the simplified theory of rolling contact. Veh. Syst. Dyn. **11**(1), 1–13 (1982)

Tajima, H.: Basics of multibody dynamics: how to set up 3-dimensional motion equations. Tokyo Denki University Press, Tokyo (2006)

A Full-Vehicle Motion Simulator for Railways Applications

Roshan Pradhan[1(✉)], Vishnu Sukumar[2], Subir Kumar Saha[2], and Santosh Kumar Singh[2]

[1] Birla Institute of Technology and Science Pilani, Pilani, India
roshannpradhan@gmail.com
[2] IIT Delhi, New Delhi, India
{Vishnu.Sukumar, saha, singhsp}@mech.iitd.ac.in

Abstract. The present study attempts to develop a 3D multi-body dynamic model of a train. First, a software is developed to simulate the longitudinal dynamics of the train by treating the train body as a point mass moving on a track. It has applications in railway operations like runtime estimation, trajectory calculation, and route planning. Additionally, a 3D dynamic model of the train was built to gain insights into the motion of train components in response to forces arising from wheel-rail interaction and external track inputs. Finally, a method is proposed to realistically capture the motion of an N-compartment train as it traverses along the entire journey, for a typical route used by the trains in India.

1 Introduction

The core of the present work lies in realistically capturing how a train will move as it traverses its route. This involves not only realistic multi-body simulation of 3D train models, but also considerations like the wheel and rail geometric profiles, wheel-rail contact modeling, and incorporation of track inputs like gradient and curvature. Before building a full-train model, the longitudinal dynamics of the train is simulated considering the mass to be focused at a single point. This analysis can be applied directly to practical applications like runtime estimation, route planning, and trajectory calculation. The other considerations mentioned come into play when it is necessary to study the motion of train components like the car-bodies, coupler joints, wheelset axles, suspension springs, etc. In order to design for stability, safety, and ride comfort, an analysis method is needed to gain quantitative information on the motion of these components. Additionally, modeling the dynamics of the entire train helps create a test-bench that can be used for iteratively testing and tuning any new proposed systems that may need to be introduced. Railway agencies all over the world utilize Train Motion Simulator (TMS) packages for route-planning and optimization, runtime and fuel consumption estimation, and system performance testing [1–3]. Further, they usually simulate the performance of those systems that are coupled with the longitudinal motion of the train like the braking pneumatics, power systems, and autonomous train controllers (ATC) [4, 5]. It is clear from the literature that there have been efforts to develop simulation frameworks to test the performance of trains throughout the

A. Kecskeméthy and F. Geu Flores (Eds.): ECCOMAS 2019, COMPUTMETHODS 53, pp. 495–502, 2020.
https://doi.org/10.1007/978-3-030-23132-3_59

journey. Most of the work in this field treats the train body as a point mass upon which tractive, resistive, and braking forces are acting, which poses a limitation when the dynamic behavior of the train as a whole needs to be studied.

On the other hand, the dynamics of trains has been studied extensively in literature both from a contact mechanics and a multi-body dynamics perspective, since it plays a crucial role in ride comfort, ride stability, and safety against derailment [6]. Refined models have been developed by researchers taking into account suspension components such as air springs, Z-links, leaf springs, etc. [7]. Additionally, there exists literature on integrating contact mechanics codes with 3D multibody models to capture the coupled behavior more completely [8]. However, most of the literature does not take into account the external and self-excited forces experienced by the train as it traverses a particular route. In some cases the problem is simplified by modeling the track disturbances as external sinusoidal inputs [9] acting on the axle or the wheels.

This study aims to improve upon the state-of-the-art by better capturing the track inputs faced by the 3D model as it follows a prescribed speed profile over the whole route for a real-world example. This is done by combining the dynamics of the full-train system with the results obtained from the TMS to better represent the full motion of an N-compartment train as it travels along the length of its route.

2 Train Motion Simulator (TMS)

This section describes the development of the TMS package used for computing the train's longitudinal motion. The TMS outputs the estimated runtime and speed profile of the train along its route, for the set of given inputs (a) Speed limit constraints, (b) Sectional track data, (c) Station location and wait time, and (d) Bogie dynamic characteristics.

2.1 Working Principles of the TMS

The equation of motion representing the longitudinal dynamics of a train is given below [6].

$$F = T_e(v) - R_e(v, s) - B_e(v) = M_e\dot{v} \tag{1}$$

where v and s denote the velocity and position of the train respectively; T_e is the tractive force provided by the motor; B_e is the force exerted by the brakes; M_e is the total mass of the train multiplied by a factor α to incorporate the effect of rotational inertia of the wheels.

R_e is the total resistive force acting on the train represented by -

$$R_e(v, s) = R_{start} + R_{roll}(v) + R_{gradient}(s) + R_{curve}(s) \tag{2}$$

where,

$$R_{roll} = \{n_l(Mg)_l(a_l + b_l v + c_l v^2) + n_c(Mg)_c(a_c + b_c v + c_c v^2)\}/1000 \tag{3}$$

$$R_{start} = n_l R_{l_{start}} + n_c R_{c_{start}} \tag{4}$$

$$R_{grad} = \{n_l(Mg)_l + n_c(Mg)_c\}/G \tag{5}$$

$$R_{curve} = (0.0004) \times S\{n_l(Mg)_l + n_c(Mg)_c\} \tag{6}$$

in which subscripts l and c refer to loco and coach respectively; n to the number of compartments; a, b, c represent the rolling resistance coefficients; gradient is represented by 1:G (1 unit rise in vertical level for G units travelled); and curvature S is specified in degrees. R_{start} here denotes the resistance to onset of motion. Based on the length of the section for which the curvature is defined, the radius of curvature can be found by the relation $r = 180 \times length/(\pi S)$.

The longitudinal dynamics can be obtained by solving Eq. (1). Since the net force F acting on the train is not constant, simple kinematic relations cannot be used. Additionally, since F depends on section-wise track layout features like gradient and curvature, it would not be possible to solve Eq. (1) by analytically integrating the equation of motion over velocity. Thus, Finite Difference Method (FDM) was used in order to numerically solve Eq. (1). The FDM scheme can be written in velocity, time, or distance increments. Distance-increment approach was used here since the track layout features are all defined according to position on the track. The algorithm takes the net force acting on the train as constant over a short distance increment to calculate velocity at the next point, and then iterates over the whole route to calculate the train dynamics.

2.2 Simulation Algorithm

The speed profile of the train for a particular route is dependent on route-specific constraints, namely section-wise local speed limits, station locations, and wait times. Based on these constraints, the longitudinal dynamics was calculated by modeling the train as a point mass located at the center of the body.

The input speed limit profile was first modified based on the train length to ensure the entire train obeys the speed limits. Based on this profile, the trajectories were computed using Eq. (1) by considering the dynamic characteristics of the locos and coaches, and the section-wise gradient/curvature data that are required to calculate the net force (Fig. 1).

By taking the 'step-up' corners in the modified speed limit profile as the initial points, forward acceleration trajectories were calculated, whereas by taking the 'step-down' corners as final points backward deceleration trajectories were calculated. At each interval, three speeds are checked at all times: one based on the past speed limits, the current speed limit, and one based on the future speed limits. Among the three groups of speed, the lowest value represents the speed of the train at the location, since it allows the train to travel at the maximum speed for the current location while clearing

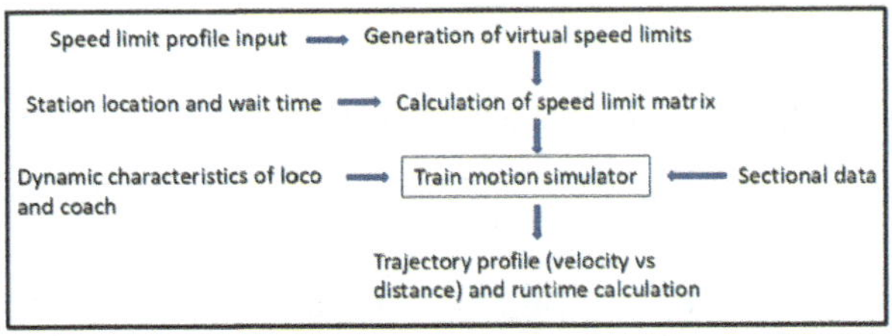

Fig. 1. Schematic illustrating workflow of the TMS.

all the past, current and future speed limits. Refer [3] for a more detailed explanation. While generating the forward trajectories from the 'step-up' corners on the basis of Eq. (1), it was assumed that the train operates at the highest power setting. In reality, the power setting is discretized into typically 8 levels known as notches. While accelerating, the driver ramps up the notches one-by-one for a smoother acceleration. Since this is understood to be a driver prerogative, the simulation assumes that the acceleration takes place at notch 8 throughout. Similar story with backward trajectories being calculated assuming maximum braking. Thus, the final trajectory is typically sharper than what is actually observed from field tests. The results are shown in Fig. 3.

3 Multibody Simulation

This section delves into the development of the 3D multi-body models of the train. The software 'Simscape Multibody' was used as the simulation environment.

3.1 Wheelset-Rail Contact

Contact was not explicitly modeled in this study, due to the computational complexity associated with finite element contact simulation. Constraints are instead imposed on the wheelset axle in such a way as to replicate the effect that actual wheel-rail contact would have on the wheelset center. The simplified wheel-rail model here is that of a rigid conical wheelset undergoing pure rolling on rigid knife-edge rails, as indicated in Fig. 2(b). These simplifications mean that the contact takes place at a single point, and these contact points are separated by a fixed lateral distance $2l_0$, independent of the state of the wheelset.

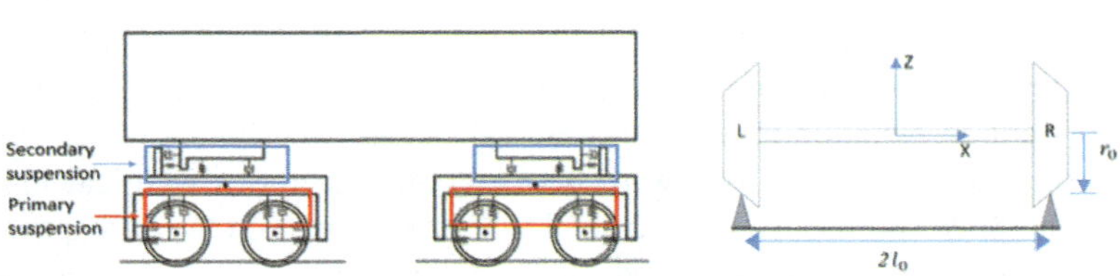

Fig. 2. (a) Mechanical model of the train, from [10]; (b) Wheelset on rails diagram.

In this model, constraints are imposed on the center of the 6-DOF wheelset in order to define its motion kinematically. Three holonomic constraints arise out of the wheelset-rail geometry, and the two nonholonomic constraints represent the contact forces [11]. Since nonholonomic constraints do not reduce the degree of mobility, our model has a total of three DOF. The holonomic constraints are given by –

$$\phi = -\lambda x/l_0 \tag{7}$$

$$z = \lambda x^2/l_0 \tag{8}$$

$$y = r_0\theta \tag{9}$$

The nonholonomic constraints are given by –

$$\dot{x} = -\psi\dot{y} \tag{10}$$

$$\dot{\psi} = \lambda\dot{y}x/(l_0 r_0) \tag{11}$$

where (x, y, z) are the coordinates of the wheelset center according to Fig. 2(b), (θ, ϕ, ψ) represent the spin, roll, and yaw motion respectively, λ is the conicity of the wheelset, l_0 is the half-distance between the knife-edges, and r_0 is the nominal radius at the point of contact.

Equations (7) and (8) arise out of the geometry of the conical wheelset, and Eq. (9) represents the no-slip condition. Equation (10) represents the horizontal component of forward velocity which is proportional to the yaw angle (after applying small angle approximation). Equation (11) represents the rate of change of yaw due to a differential rolling radius between the left and right wheels, caused by the lateral offset x of the wheelset.

Taking forward velocity $\dot{y}$ as a system parameter, we can solve Eqs. (10) and (11) to obtain x and ψ as functions of time. The frequency of oscillation of this dynamical system is termed as the 'Klingel frequency', given by $\omega_{Kl} = \dot{y}\sqrt{\lambda/l_0 r_0}$ [12]. Thus, the hunting oscillation of the train is modeled by kinematically defining the wheelset motion in lateral (x) and yaw (ψ) directions. At a constant velocity $\dot{y}$, the system undergoes SHM with frequency ω_{Kl}, and with the amplitude of oscillation determined by the initial conditions, i.e. the initial offset when hunting motion starts. The time-series outputs of x and ψ are then operated on to determine the rest of the DOFs as functions of time, using Eqs. (7)–(9).

The spin θ of the wheelset is not explicitly modeled, rather the gyroscopic effect of the spin is externally imposed on the wheelset center. The gyroscopic couples about pitch and yaw ($M_{g\phi}$ and $M_{g\psi}$ respectively) are represented by Eqs. (12) and (13), taken from [13], where I_{xx} is the moment of inertia of the wheelset about the spin axis.

$$M_{g\phi} = I_{xx}\dot{y}\dot{\psi}/r_0 \tag{12}$$

$$M_{g\psi} = I_{xx}\lambda\dot{y}\dot{x}/(l_0 r_0) \tag{13}$$

3.2 Upper-Body Model

The upper-body model utilizes the linear spring-damper element between two bodies. This representation is useful since even complicated suspension geometries can be simplified into effective stiffness and damping components in three directions. Both the coaches and the locos were built based on the mechanical model in Fig. 2(a). The loco model consists of six wheelset axles (3 DOF), two bogies (6 DOF), and a car-body (6 DOF). Thus, the total DOFs of the loco model are 36. Similarly, the coach consists of four axles, two bogies, and a car-body, thus totaling 30 DOF. Another crucial difference between the loco and coach models is that in the locos the axles are powered, i.e. a longitudinal force acts on the center of each wheelset. In the coaches, the longitudinal motion of the wheelset is unconstrained.

3.3 Full-Route Model

The track inputs, whether due to gradient, curvature, or lateral disturbances, were modeled using time-varying gravity fields. The field acting upon the workspace was varied according to the values of gradient and curvature. The longitudinal velocity $\dot{y}$ as a function of distance, as obtained from the TMS results, is given as input to the longitudinal DOF of the wheelset axle in the loco model. This is meant to capture the fact that the locos would be driving the motion of the coaches.

4 Results

The results were obtained for the typical route of Marwar Jn.–Abu Rd. (MJ-ABR) in the western part of India, and validated using results from the proprietary software RUNTRAIN used by Indian Railways. In this simulation, four BG-ICF type coaches are hauled by a WDM2 type loco of the Indian Railways.

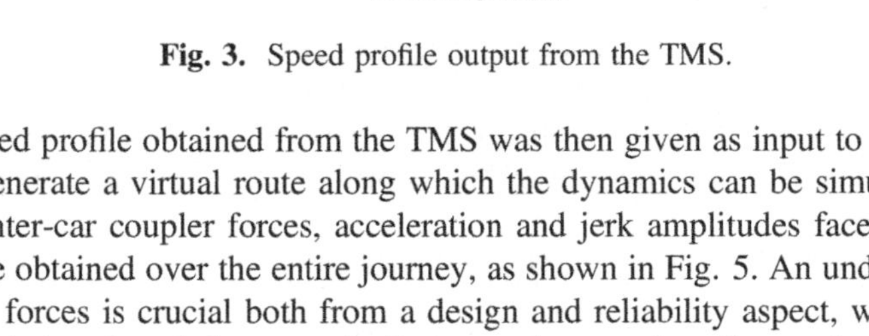

Fig. 3. Speed profile output from the TMS.

The speed profile obtained from the TMS was then given as input to the Simscape model to generate a virtual route along which the dynamics can be simulated. Quantities like inter-car coupler forces, acceleration and jerk amplitudes faced by the car-body can be obtained over the entire journey, as shown in Fig. 5. An understanding of the coupler forces is crucial both from a design and reliability aspect, whereas acceleration and jerk values help ascertain limits on the severity of powering and braking modes of operation from a ride comfort perspective [14].

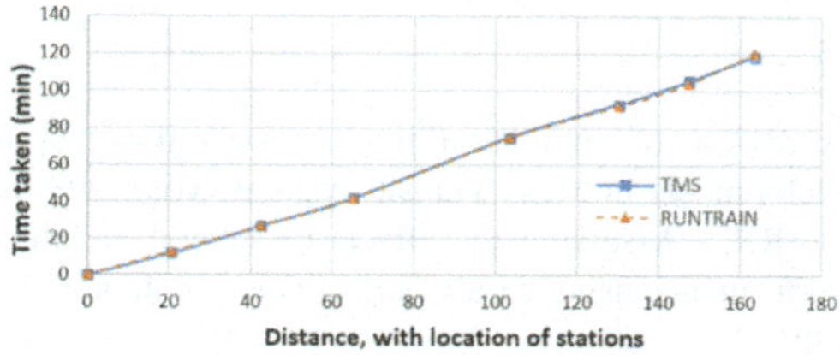

Fig. 4. Comparison of runtime with proprietary software RUNTRAIN.

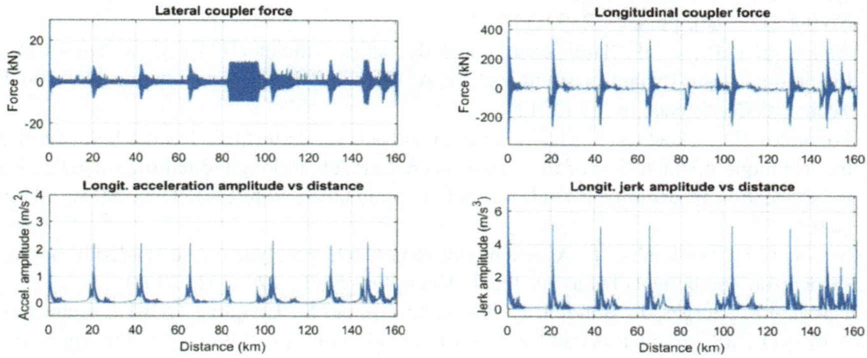

Fig. 5. Results obtained from the full-journey simulation in Simscape.

5 Conclusions

The main focus of this paper has been to gain a better understanding of the dynamical behavior of a train throughout its journey. First, a TMS package was developed to calculate the longitudinal dynamics of the train. Its results were validated using the proprietary software RUNTRAIN. Next, the theory required to model contact conditions was discussed, followed by a description of the upper-body model developed in Simscape. The results from the longitudinal simulator were then integrated into the Simscape model to realistically test the dynamic behavior of a 5-compartment train as it moves along a particular route. The simulator developed in this paper is useful for any future prediction of the full-body dynamics throughout the journey of a train running in a constrained stretch.

Acknowledgements. The authors would like to acknowledge Mr. Vijay Goel and Mr. Sudhakar Kumar of Railway Design and Standards Organisation (RDSO) for their technical inputs.

References

1. Goodman, C.J.: A review of simulation models for railway systems. In: International Conference on Developments in Mass Transit Systems, vol. 1998, pp. 80–85 (1998)
2. Hoyt, E.V., Levary, R.R.: Assessing the effects of several variables on freight train fuel consumption and performance using a train performance simulator. Transp. Res. Part A Gen. **24**(2), 99–112 (1990)
3. Kikuchi, S.: A simulation model of train travel on a rail transit line. J. Adv. Transp. **25**(2), 211–224 (1991)
4. Ku, B.-Y., Jang, J.S.R., Ho, S.-L.: A modulized train performance simulator for rapid transit DC analysis. In: Proceedings of the 2000 ASME/IEEE Joint Railroad Conference (Cat. No.00CH37110), pp. 213–219 (2000)
5. Andersen, D.R., et al.: Train energy and dynamics simulator (TEDS): a state-of-the-art longitudinal train dynamics simulator. In: ASME 2012 Rail Transportation Division Fall Technical Conference, p. 57 (2012)
6. Pogorelov, D., Yazykov, V., Lysikov, N., Oztemel, E., Arar, O.F., Rende, F.S.: Train 3D: the technique for inclusion of three-dimensional models in longitudinal train dynamics and its application in derailment studies and train simulators. Veh. Syst. Dyn. **55**(4), 583–600 (2017)
7. Sayyaadi, H., Shokouhi, N.: A new model in rail–vehicles dynamics considering nonlinear suspension components behavior. Int. J. Mech. Sci. **51**(3), 222–232 (2009)
8. Pascal, J.P., Berger, J., Bondon, F., Clerc, C., Teppe, S.: Coupling OCREC contact code with ADAMS: simulations of one coach at 130mph. In Volume 4: 7th International Conference on Multibody Systems, Nonlinear Dynamics, and Control, Parts A, B and C, pp. 1883–1888 (2009)
9. Dikmen, F.: Vibration analysis of 19 degrees of freedom rail vehicle. Sci. Res. Essays **6**(26), 5600–5608 (2011)
10. Knothe, K., Stichel, S.: Rail Vehicle Dynamics. Springer International Publishing, Cham (2017)
11. Fisette, P.: Railway Vehicle Dynamics (2007). http://hosting.umons.ac.be/html/mecara/grasmech/RailVehicles.pdf
12. Klingel, J.: Über den Lauf von Eisenbahnwagen auf gerader Bahn. In: Organ für die Fortschritte des Eisenbahnwesens, pp. 113–123 (1883)
13. Jawahar, P.M., Gupta, K.N., Raghu, E.: Mathematical modelling for lateral dynamic simulation of a railway vehicle with conventional and unconventional wheelset. Math. Comput. Model. **14**, 989–994 (1990)
14. Powell, J.P., Palacín, R.: Passenger stability within moving railway vehicles: limits on maximum longitudinal acceleration. Urban Rail Transit **1**(2), 95–103 (2015)

Simulation of the Maglev Train Transrapid Traveling on a Flexible Guideway Using the Multibody Systems Approach

Georg Schneider[1](✉), Xin Liang[2], Florian Dignath[3], and Peter Eberhard[1]

[1] Institute of Engineering and Computational Mechanics, University of Stuttgart, Pfaffenwaldring 9, 70569 Stuttgart, Germany
{georg.schneider,peter.eberhard}@itm.uni-stuttgart.de

[2] CRRC Qingdao Sifang Co., Ltd., No. 88 Jinhongdong Road, Chengyang District, Qingdao 266111, People's Republic of China
liangxin.sf@crrcgc.cc

[3] thyssenkrupp Transrapid GmbH, Moosacher Str. 58, 80809 Munich, Germany
florian.dignath@thyssenkrupp.com

Abstract. In this paper, an elastic multibody model is presented describing the vertical dynamics of the maglev train Transrapid. The vehicle is represented by a rigid multibody system with five bodies and ten degrees of freedom, whereas the guideway is modeled as an Euler-Bernoulli beam. The coupling of vehicle and guideway is realized by discretized magnet forces. The control laws for the magnet forces which are responsible for keeping the air gap between vehicle and guideway at a nearly constant value are approximated by PID-T1 control laws. The model is used for simulations of the vehicle crossing the guideway with velocities up to 600 km/h. Based on these simulations, system dynamics like guideway deflections and vehicle accelerations are analyzed.

1 Introduction

"Maglev" is the abbreviation for "magnetic levitation" and describes a method that uses the physical phenomenon of magnetic forces to support an object. One field of application is in transportation systems. The vertical suspension, lateral guidance, and longitudinal acceleration and deceleration of the vehicle are realized by electromagnetic forces. This brings differences compared to wheel-on-rail systems, since there is no contact between vehicle and guideway. Some of the advantages are lack of friction, less wear and tear, less noise emission, and higher velocities.

In China, the Transrapid airport link in Shanghai is running successfully since 2003 until today and the research and development of a maglev train system with a maximum speed of 600 km/h to connect the cities of Jinan and Qingdao in

A. Kecskeméthy and F. Geu Flores (Eds.): ECCOMAS 2019, COMPUTMETHODS 53, pp. 503–510, 2020.
https://doi.org/10.1007/978-3-030-23132-3_60

eastern China is in the scope of the Chinese rolling stock manufacturer CRRC Qingdao Sifang Co., Ltd. at present [1].

Against this background, in this contribution the coupled system of vehicle and guideway is analyzed with respect to its dynamics. A flexible multibody system is set up including a rigid multibody vehicle traveling across an elastic beam. The magnet forces, which ensure the vehicle to levitate above the guideway, are discretized and controlled by a PID-T1 control law as presented in [3]. This model is used to investigate the deflection of the guideway due to the load of several subsequent magnet forces as well as the resulting accelerations transmitted to the car body and thereby to the passengers.

2 State of the Art

In [7], the behavior of a flexible guideway under a passing vehicle is investigated. The guideway is modeled as Euler-Bernoulli beam, whereas the vehicle is represented by a single force or lumped mass, respectively, moving with constant velocity. Two limiting cases are considered: first, a constant force (moving force problem), and second, a constant gap between mass and beam (moving mass problem).

In [2,4], a much more detailed vehicle model with up to 28° of freedom is implemented, consisting of multiple rigid bodies representing the car body, levitation chassis, and levitation magnets, interconnected by spring-damper elements. In turn, the guideway is simplified by means of several rigid bodies, each supported on the foundation by spring-damper elements. Both subsystems, vehicle and guideway, are coupled by the magnetic forces of the levitation magnets distributed along the length of the vehicle. With this rigid multibody system the vertical vibrations are examined caused by a passing Transrapid vehicle transmitted via the guideway into the ground.

In [9], the guideway is represented by an Euler-Bernoulli beam, whereas for the vehicle various models with different levels of detail are implemented, reaching from a simple moving force to a rigid multibody system with 6° of freedom (DOFs), representing the car body with 2 DOFs and four levitation chassis with one DOF each. The latter model is parametrized with data from the Transrapid model TR06. With those models, the influence of the implemented vehicle model on the guideway displacement is analyzed. Furthermore, the effect of guideway irregularities on the guideway displacement and vehicle acceleration is investigated.

3 Simulation Model: Elastic Multibody System

In this contribution, concepts from the approaches described above are combined. On the one hand, the vehicle is modeled as a multibody system of several rigid bodies similar to [2,4,9]. On the other hand, following [7,9], the guideway is modeled as a single-span Euler-Bernoulli beam. The coupled system is shown in Fig. 1.

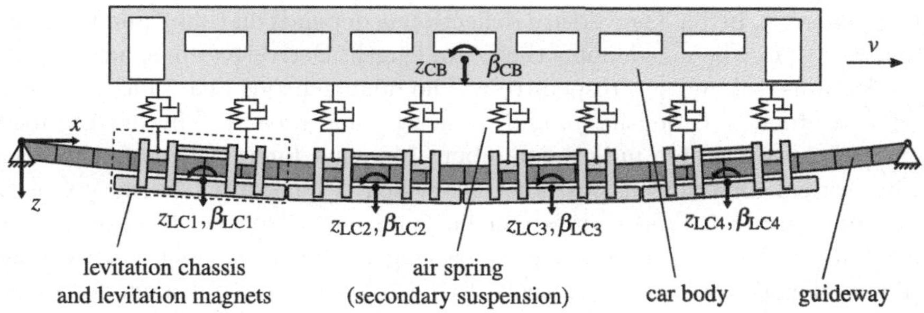

Fig. 1. Elastic multibody system with a vehicle of rigid bodies traveling with constant velocity v across a guideway represented by a flexible beam.

3.1 Vehicle Model: Rigid Multibody System

The vehicle of the Transrapid consists of several sections, e.g. two end sections and one mid section in between. One section is a car body, which is connected to four levitation chassis by four air springs per levitation chassis. Each levitation chassis consists of two levitation frames connected by a longitudinal girder. Along the vehicle, eight levitation magnets as well as seven guidance and one braking magnet are distributed on each side. Each magnet is mounted to two neighboring levitation frames.

For this contribution, just a single mid section is taken into account. The car body is modeled as a rigid body with two degrees of freedom, namely a translational one in vertical direction, z_{CB}, and a rotational one allowing pitching, β_{CB}. Each levitation chassis together with its magnets is modeled as a separate rigid body with a translational and a rotational degree of freedom, z_{LCi} and β_{LCi}, $i = 1(1)4$. The air springs connecting the car body and the levitation chassis are represented by two spring-damper elements per levitation chassis, summarizing two air springs. The mechanical model having ten degrees of freedom is set up with the Matlab-based multibody simulation toolbox Neweul-M^2 [6] using relative kinematics. Based on a system definition file written by the user, the equations of motion are calculated by Neweul-M^2 symbolically in an automated manner. If desired, Neweul-M^2 can then export the equations of motion to C-code and compile this code as an S-function which can then be included in a Simulink model. The vehicle parameters like masses, moments of inertia, and spring and damper constants are chosen to fit the Transrapid model TR08 as it is described in [4]. The vehicle length is 24.768 m.

3.2 Guideway Model: Euler-Bernoulli Beam

The Euler-Bernoulli beam representing the guideway is described by the partial differential equation

$$EI\, w''''(x,t) + \rho A\, \ddot{w}(x,t) = \sum_{\mu} F_{\mu}(t)\, \delta(x - x_{\mu}) \,, \tag{1}$$

to be found, e.g., in [8]. The vertical deflection w depends on time t and position x with $0 \leq x \leq L$, where L denotes the beam length. Derivatives in space and time are written as $(\ )'$ and $(\dot{\ })$, respectively. The homogeneous beam has a constant bending stiffness EI, density ρ, and cross-sectional area A. The moving forces $F_\mu(t)$, $\mu = 1(1)m$, where m is the number of magnet forces acting on the beam, act in vertical direction at positions x_μ, which is considered by means of the Dirac function $\delta(\ )$. Since forces can be transmitted to the discretized beam only at its nodes, a vertical force F_μ, acting on beam element i with length L_i between nodes i and $i+1$, needs to be replaced by equivalent forces and torques acting at the neighboring nodes, cf. Fig. 2. According to [5], the vector of equivalent nodal forces and torques $\mathbf{p}_{\text{equ},i} = [F_{z,i}, M_{y,i}, F_{z,i+1}, M_{y,i+1}]^{\text{T}}$ is obtained by distributing F_μ by means of the vector of Hermite polynomials $\boldsymbol{g}_{\text{H}}$ according to

$$\boldsymbol{p}_{\text{equ},i} = \begin{bmatrix} F_{z,i} \\ M_{y,i} \\ F_{z,i+1} \\ M_{y,i+1} \end{bmatrix} = \begin{bmatrix} F_\mu & 0 & 0 & 0 \\ 0 & -F_\mu L_i & 0 & 0 \\ 0 & 0 & F_\mu & 0 \\ 0 & 0 & 0 & -F_\mu L_i \end{bmatrix} \boldsymbol{g}_{\text{H}} \tag{2}$$

with

$$\boldsymbol{g}_{\text{H}} = \begin{bmatrix} g_1 \\ g_2 \\ g_3 \\ g_4 \end{bmatrix} = \begin{bmatrix} 1 - 3\xi_{\mu,i}^2 + 2\xi_{\mu,i}^3 \\ \xi_{\mu,i} - 2\xi_{\mu,i}^2 + \xi_{\mu,i}^3 \\ 3\xi_{\mu,i}^2 - 2\xi_{\mu,i}^3 \\ -\xi_{\mu,i}^2 + \xi_{\mu,i}^3 \end{bmatrix}, \qquad \xi_{\mu,i} = \frac{x_{\mu,i}}{L_i}, \qquad 0 \leq x_{\mu,i} \leq L_i\,, \tag{3}$$

as plotted in Fig. 3. The position of F_μ on beam element i is described by the normalized local coordinate $\xi_{\mu,i}$.

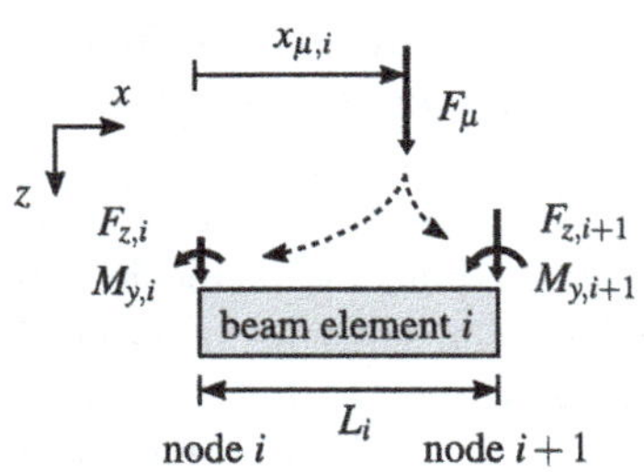

Fig. 2. Force F_μ acting at position $x_{\mu,i}$ on beam element i is replaced by equivalent forces and torques acting on the neighboring nodes i and $i+1$.

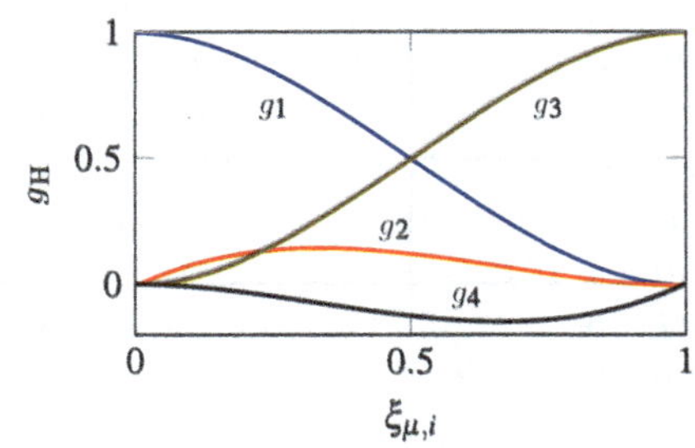

Fig. 3. Plot of Hermite interpolation polynomials $\boldsymbol{g}_{\text{H}} = [g_1, g_2, g_3, g_4]^{\text{T}}$ for $\xi_{\mu,i} \in [0, 1]$.

Since the magnetic force F_μ depends on the air gap, i.e. the distance between the magnet and the guideway, the deflection of the guideway is required right at the position where the magnet force applies. The deflection at an arbitrary position of a beam element can be obtained by interpolation of the deflections and rotations of the neighboring nodes by means of the Hermite polynomials [5]. So, for F_μ acting on beam element i at position $x_{\mu,i}$, the deflection below F_μ writes as

$$w_\mu = [w_i, \varphi_i, w_{i+1}, \varphi_{i+1}]\, \boldsymbol{g}_{\mathrm{H}} \tag{4}$$

with deflections $w_{\{i,i+1\}}$ and rotations $\varphi_{\{i,i+1\}}$ at nodes $\{i, i+1\}$.

Geometry and material parameters of the guideway like beam length, area of cross section, area moment of inertia, Young's modulus, and density are chosen in accordance with [9] and represent the first generation of concrete guideway at the test facility in Emsland (TVE). The beam with a length of 24.768 m is discretized by 24 elements with a element length of approximately 1 m. In contrast to [9], where the beam is reduced to just the first eigenmode, here the first three eigenmodes are taken into account in order to allow asymmetric deflections of the beam. Three eigenmodes are sufficient to describe the behavior of a beam supported at its ends (single span beam), as shown in [7]. According to [8], rotational inertia and shear deformations can be neglected in this case, which justifies using the Euler-Bernoulli beam theory. The deflection at mid span caused by the weight of the beam is 8.6 mm which matches the analytical solution. Its first natural frequency is 6 Hz. The Euler-Bernoulli beam is created and reduced using the Matlab-based model order reduction toolbox MatMorembs, developed at Institute for Engineering and Computational Mechanics, and then included as a flexible body in Neweul-M^2.

3.3 Coupling of Vehicle and Guideway

The Transrapid vehicle is levitating above the guideway due to magnetic forces between the levitation magnets and a reaction rail on the bottom side of the guideway. Therefore, the magnets are attracted to the guideway from below. The air gap between levitation magnets and guideway is kept nearly constant over time by magnet control units varying the voltage applied to the magnets.

In this model, a highly simplified control law for the magnet forces is implemented by means of a PID-T1 controller, as proposed in [3]. The magnet forces, which actually act continuously along the vehicle, are simplified by eight discrete force elements. One force element thereby summarizes the magnet forces of two levitation magnets, one from each side (left and right) of the vehicle. Thus, always two force elements are attached to one levitation chassis in the model.

Both subsystems together, coupled by the magnetic forces of the levitation magnets, thus form a two-dimensional elastic multibody system, cf. Fig. 1. The simulation of the coupled vehicle and guideway is done in Simulink. The Simulink model structure is shown in Fig. 4.

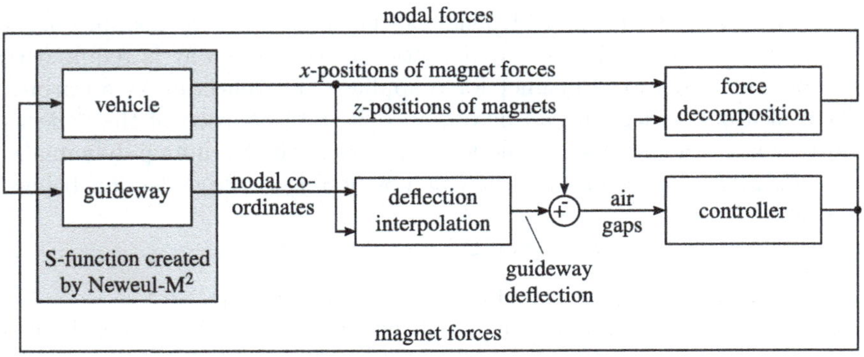

Fig. 4. Model structure of the Simulink model, where the mechanical models of vehicle and guideway are coupled by means of their respective in- and outputs. The transformation between modal and nodal coordinates is done internally by Neweul-M^2 by using the modal projection matrix.

4 Simulation Results

First, the behavior of the guideway under the passing vehicle is investigated. The simulation results for the guideway deflection at mid span versus the position of the first magnet force are shown in Fig. 5 for velocities from quasistatic to 600 km/h. The first magnet force enters the flexible guideway element at $x_{\mathrm{MF,front}} = 0$. Three vertical dashed lines indicate the positions where the last magnet force enters (left line) and leaves (right line) the beam element, and where the first magnet force leaves it (middle line). For the quasi-static case, the maximum deflection occurs for $x_{\mathrm{MF,front}} = 23.22\,\mathrm{m}$. At this position, the beam is loaded with all eight magnet forces and they are distributed symmetrically around the beam mid span. For the considered velocities, the maximum deflection increases with velocity. Especially for $v = 600\,\mathrm{km/h}$, a relatively high overshoot can be observed when the vehicle is about to leave the elastic guideway element, which might be due to near-resonance of the transfer time of each magnet over the elastic beam of 1/6.7 s with its eigenperiod of 1/6 s.

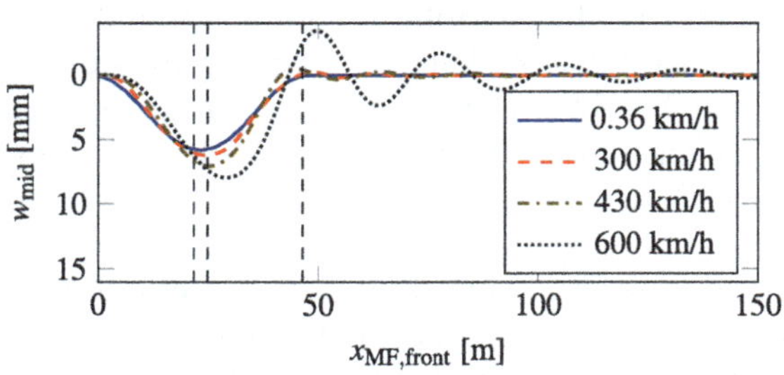

Fig. 5. Deflection of the guideway at midspan versus position of the first (front) magnet force for various velocities.

Second, the behavior of the vehicle is analyzed. In Fig. 6, the displacements and accelerations of the car body and the first (front) levitation chassis are plotted for the same velocities as for the guideway in Fig. 5. In contrast to the guideway, for which the major deflections occur while the vehicle is above it, the car body displacements reach their maxima while the vehicle is leaving or has even left the flexible guideway element already when traveling with 300 km/h or more. Furthermore, the levitation chassis reacts to the guideway deflection much faster and with higher intensity than the car body due to the air springs supporting the car body. The amplitude of car body displacements is even decreasing for higher velocities. This behavior results in smaller vertical accelerations for the car body than for the levitation chassis, which is good with regard to comfort and proves the function of the secondary suspension.

The results in Fig. 6 represent the vehicle's response when traveling over one elastic guideway element, while the guideway before and after this element is considered ideal stiff. This could e.g. be the case, if the elastic element represents a bridge element while the rest of the guideway is constructed at-grade with continuous support.

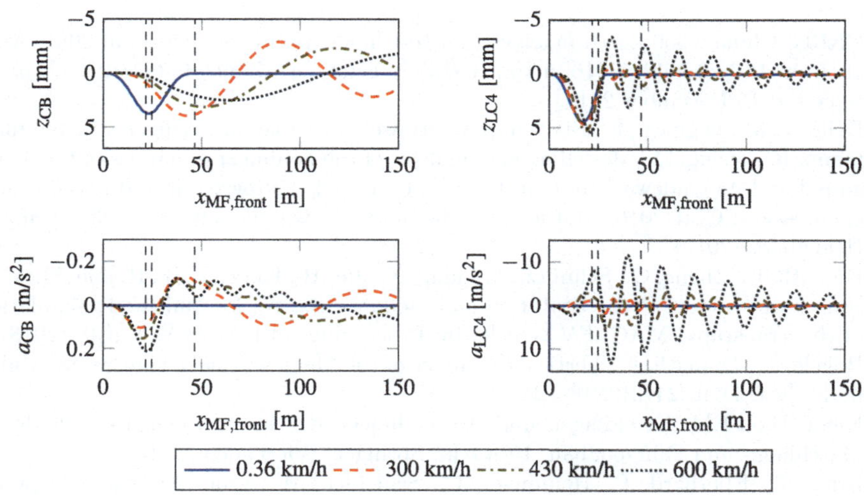

Fig. 6. Displacement and acceleration of the car body z_{CB}, a_{CB} and first (front) levitation chassis z_{LC4}, a_{LC4} versus position of the first (front) magnet force for various velocities.

5 Conclusions

A flexible multibody model is implemented describing the coupled dynamics of a rigid multibody Transrapid maglev vehicle traveling across one elastic guideway element. The distribution of discretized magnet forces to the beam element nodes as well as the interpolation of nodal coordinates to obtain beam deflections at arbitrary positions by means of Hermite polynomials are described. Simulation results show that the car body accelerations are mitigated by the air springs connecting it to the levitation chassis, which means a higher ride comfort for passengers. It can be noted that the vehicle's response to the disturbance caused by the elastic deformation of the guideway element lasts several seconds. For the analysis of a ride with high velocity over a series of such guideway elements, therefore, a guideway length of about 500 m would need to be modeled by the elastic beams presented in this work. This investigation was done for a highly simplified model of an existing maglev train in order to see trends what might happen for very high speeds. In future investigations more detailed models of the real train under development must be utilized in order to make reliable computations and predictions which can serve to design the new high speed guideway.

Acknowledgements. The authors would like to thank Denis Smaitch for his support in the course of his bachelor thesis supervised by GS.

References

1. CRRC: China's 600 km/h maglev train test line expects completion in 2021. Article from 16 November 2016. http://www.crrcgc.cc/en/g7389/s14333/t279572.aspx. Accessed 13 December 2018
2. Dellnitz, M., Dignath, F., Flaßkamp, K., Hessel-von Molo, M., Krüger, M., Timmermann, R., Zheng, Q.: Modelling and analysis of the nonlinear dynamics of the Transrapid and its guideway. In: Günther, M., et al. (eds.) Progress in Industrial Mathematics at ECMI 2010, Mathematics in Industry, vol. 17, pp. 112–123. Springer, Heidelberg (2012)
3. Dignath, F., Zheng, Q., Schmitz, P., Liang, X., Jin, H., Kurzeck, B., Ronde, M.: Fast computable model of the levitation and guidance control for multibody simulation of the Transrapid MAGLEV vehicle. In: Proceedings of RAILWAYS 2018 (2018)
4. Hägele, N., Dignath, F.: Vertical dynamics of the Maglev vehicle Transrapid. Multibody Syst. Dyn. **21**(3), 213–231 (2009)
5. Klein, B.: FEM: Grundlagen und Anwendungen der Finite-Element-Methode im Maschinen- und Fahrzeugbau, 10th edn. Springer, Wiesbaden (2015)
6. Kurz, T., Eberhard, P., Henninger, C., Schiehlen, W.: From Neweul to Neweul-M^2: symbolical equations of motion for multibody system analysis and synthesis. Multibody Syst. Dyn. **24**(1), 25–41 (2010)
7. Meisinger, R.: Simulation of a single and double-span guideway under action of moving MAGLEV vehicles with constant force and constant gap. Series of Publications by Nuremberg Tech Georg Simon Ohm, vol. 14 (2002)
8. Popp, K., Schiehlen, W.: Ground Vehicle Dynamics. Springer, Heidelberg (2010)
9. Ren, S., Romeijn, A., Klap, K.: Dynamic simulation of the maglev vehicle/guideway system. J. Bridg. Eng. **15**(3), 269–278 (2010)

Omni-Vehicle Dynamical Models Mutual Matching for Different Roller–Floor Contact Models

Kirill V. Gerasimov[1], Alexandra A. Zobova[1], and Ivan I. Kosenko[2(✉)]

[1] Lomonosov Moscow State University, Leninskie gory, 1, Moscow 119991, Russia
kirill.gerasimov.msu@gmail.com, azobova@mech.math.msu.su
[2] Moscow Aviation Institute (National Research University), Volokolamskoe shosse, 4, Moscow 125993, Russia
kosenkoii@gmail.com

Abstract. Our goal is to build up an efficient computer model of the omni-vehicle comprising several omni-wheels (each one carries several freely rotating rollers on its periphery); the vehicle rolls without control on a horizontal plane. For simulating contact interaction we use the holonomic non-ideal constraints instead of ideal non-holonomic ones. Although there exist papers on the dynamical models of the vehicle and its control, e. g. [1,2], here we study how the algorithm for computing the contact forces between a particular roller and the plane influences a description of the vehicle dynamics. To simulate dynamics we build up our dynamical models using Modelica language of object-oriented modeling. Earlier we proposed the contact tracking algorithms [3] for the point-contact dry friction regularized by linear saturation function. Using numerical simulations of the models developed we compare the vehicle dynamics for three particular contact models: (a) ideal non-holonomic contact "with impacts", (b) model of viscous friction with the large coefficient of viscosity, and (c) one with the regularized dry friction concentrating mostly on the tangent forces description.

1 Problem Statement

We study the dynamics of an omni-mobile vehicle as a multibody system: it consists of a platform, three similar omni-wheels that are mounted symmetrically on the platform: their axes make the angles of $2\pi/3$ with each other, see Fig. 1. Each omni-wheel consists of a disk and four or five heavy rollers that can freely rotate about their axes tangent to the periphery of the wheel disk. The vehicle moves along a horizontal plane without control. Thus, we separate the dynamical properties of the multibody system from the influence of any control system. The main variables under the study are: the mass center speed value v, the angular velocity of the platform ν_3 and the trajectory projection of the mass center to the floor $(x(t), y(t))$.

A. Kecskeméthy and F. Geu Flores (Eds.): ECCOMAS 2019, COMPUTMETHODS 53, pp. 511–517, 2020.
https://doi.org/10.1007/978-3-030-23132-3_61

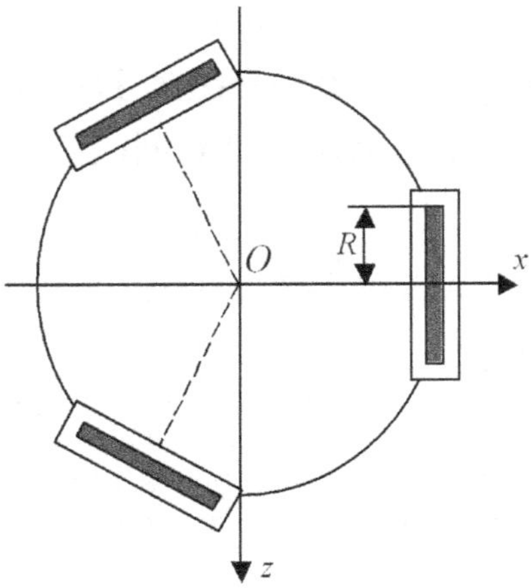

Fig. 1. Top view of the vehicle.

Radii of the vehicle platform and the wheel became 0.15 and 0.05 respectively in dimensionless units. Masses of the platform, the wheel, and roller are 1, 0.15, and 0.05 respectively. The roller moment of inertia with respect to its axis of symmetry was set to $1.6 \cdot 10^{-5}$.

For describing the vehicle dynamics we use a system of the Newton – Euler ODEs

$$\dot{\mathbf{r}} = \mathbf{v}, \quad m\dot{\mathbf{v}} = \mathbf{F} + \mathbf{R}, \quad \dot{\mathbf{q}} = \frac{1}{2}\mathbf{q} \circ (0, \omega_x, \omega_y, \omega_z)^T, \quad I\dot{\boldsymbol{\omega}} + [\boldsymbol{\omega}, I\boldsymbol{\omega}] = \mathbf{M} + \mathbf{L},$$

for each rigid body composing the mechanical system: the vehicle body, three wheels, the set of rollers, four or five rollers per wheel. Here $\mathbf{r}$ is the particular body mass-center with respect to the inertial fixed frame of reference, $\mathbf{q}$ is the quaternion defining orientation of the body, $\circ$ is the quaternion multiplication, $\mathbf{v}$ is the mass-center velocity, $\boldsymbol{\omega}$ is the angular velocity vector being expressed with respect to the body coordinate frame of reference, m is the body mass, I is the matrix of the body central principal tensor of inertia, $\mathbf{F}$ is the total vector of active forces, $\mathbf{R}$ is the total vector of reactions, $\mathbf{M}$ is the total vector of the active forces torques, $\mathbf{L}$ is the total vector of the reaction torques. In addition, one has to define equations of constraints each one of the form $f(\mathbf{r}, \mathbf{v}, \mathbf{q}, \boldsymbol{\omega}) = 0$, and models of the constraints reactions like $g(\mathbf{R}, \mathbf{r}, \mathbf{v}, \mathbf{L}, \mathbf{q}, \boldsymbol{\omega}) = 0$.

Number of rollers per wheel is not principal. We change this number here purposely, simply to demonstrate efficiency of the object-oriented technology for constructing the complicated multilayer dynamical structures, three layers of tree structure in our case, like wheeled vehicles with different rigid body components and certain number of constraints of different kinds. In addition, case of 5 or greater number of rollers per wheel is convenient for the reason of possibility for the contact overlapping when roller changing.

We use the object-oriented technology of Modelica language for unifying the total vehicle model being represented by the system of differential-algebraic equations, DAEs, which is composed usually by differential equations of motion and algebraic equations of constraints. The language compiler reduces the index of the DAE system, and then run-time system simulates the model. One can find in [4] more details of how is possible to arrange the multibody system model architecture representing it as a communication network.

2 Dry Friction

In a current context, the dry friction point-model is the most general and physically motivated one for rigid bodies' contact. At the same time, the difficulties concerning the implementation of switching between modes of sliding and rolling make the dynamical model with dry friction unstable and inefficient. To rid this issue off, one could approximate the dry friction force, for instance, by the arctan trigonometric function:

$$\mathbf{F} = -\mu N \cdot (\mathbf{v}/|\mathbf{v}|) \cdot \arctan(|\mathbf{v}|/\delta)$$

where μ is the friction coefficient, N is the normal reaction, if positive otherwise is zero, $\mathbf{v}$ is the relative slipping velocity, δ is an approximation small parameter. We use here simpler and more efficient approximation

$$\mathbf{F} = -\mu N \cdot (\mathbf{v}/|\mathbf{v}|) \cdot \begin{cases} |\mathbf{v}|/\delta \text{ for } |\mathbf{v}| < \delta, \\ 1 \quad \text{ for } |\mathbf{v}| \geq \delta \end{cases}$$

by the linear saturation function. According to the theorem of Novozhilov [5], the approximation takes place as $\delta \longrightarrow 0$. This latter theorem is based in turn on the known Tikhonov theorem about singularly perturbed systems of ODEs. In addition, regularization by the linear saturation function demonstrates high quality of approximation to the exact dry friction case.

To implement the verification process, we use so-called mass-less (with the rollers of zero inertia and zero radius) non-holonomic model of the vehicle [6] as a limit case. To verify mutual correctness of the mass-less model and approximate dry friction model of ours from above the numeric testings were performed. The only difference for these models is the value of ratio of the roller mass to the whole wheel mass. It turned out that as ratio parameter approaches zero then motion variables of the vehicle and omni-wheels approach the corresponding solutions of Cauchy problem for ODEs of mass-less model being used in [6] where dynamics of rollers is neglected.

Number of rollers per each wheel is equal to 4. Three types of initial conditions were implemented: (a) $\nu_3(0) = \omega_0$ corresponds to the left case of Fig. 2; (b) $\mathbf{v}(0) = (v_0, 0, 0)^T$ corresponds to the central picture of Fig. 2; (b) combination of two previous cases. Describe these types in more details:

(a) Platform of the vehicle has non-zero angular velocity $\omega_0 = 1$ about vertical axis which passes through the vehicle mass-center. The mass-center velocity v_0 is equal to zero which is expected because the vehicle rotates about its axis of symmetry, and mass-center is in the rest.

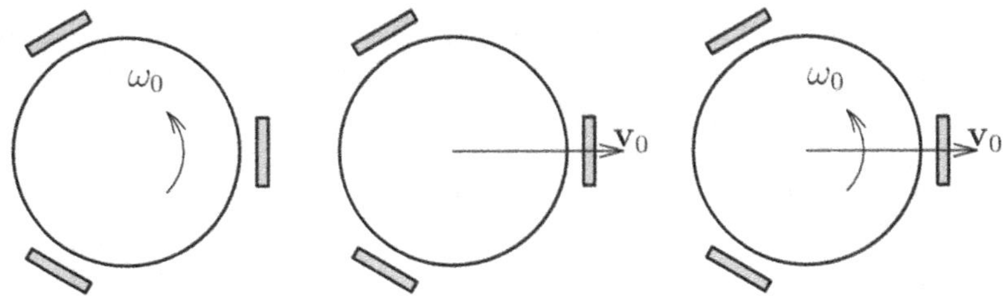

Fig. 2. Types of initial conditions: (left) pure rotation with angular velocity ω_0 of the vehicle body; (central) pure translation with the vehicle body mass-center velocity v_0; (right) case of combined initial conditions. Left and central cases used to verify the dry friction model. Right combined case corresponds to verification the model with viscous friction

(b) The vehicle platform mass-center has initial translatory velocity $v_0 = 1$ in direction of one of wheels. At the same time the platform angular velocity ω_0 is assumed to be zero which is also expected that the mass-center moves along the axis Ox and the vehicle does not rotate.

(c) The vehicle platform has non-zero angular velocity $\omega_0 = 1$ about central vertical axis of symmetry. And also the platform mass center has non-zero linear velocity $v_0 = 1$ in direction from the vehicle center to the center of the first wheel.

3 Non-holonomic Model with Impacts

Another way to model a contact is to impose the constraints of non-sliding between the rollers and the plane. In this case, the number of degree of freedom of the dynamical system decreases and, thus, we get a faster and more stable model. However, the wheels rotation causes switching the contact from one roller to another. The velocity of relative slipping between the roller and the floor can have sufficiently large value when a contacting process starts. Then the slipping process continues up to the instant when rolling begins with zero slipping. So, in the model of ideal permanent rolling, we proposed to replace the process of relative slipping by the instant inelastic impact at contact which happens due to imposing of new contact constraint. This latter case we discuss in details in [7].

4 Viscous Friction

According to known result [8], the dynamics of the system with non-slipping conditions can be obtained as a limit case for the models with viscous friction model

$$\mathbf{F} = -\gamma \mathbf{v}$$

as the viscous coefficient γ tends to infinity. We implemented the omni-vehicle model with viscous friction on OpenModelica software, and it shows proper matching with non-holonomic one.

For the case of viscous friction computations were performed with initial conditions of the third type from the Fig. 2. This time each wheel has five rollers. Dynamical behavior of the model is similar to the non-holonomic model: spiral shape of the mass-center trajectory, increasing of the vehicle platform angular velocity combined with decreasing of its mass-center linear velocity after which angular velocity starts to decrease with subsequent decreasing of the kinetic energy.

Coefficient of viscosity was sufficiently large, namely of value 10^5. Then non-holonomic behavior became more explicit: impact-like mode of the rollers changing and also similarity with non-holonomic model according to [8] while moving along the regular segments of motion.

5 Comparison of the Models and Simulation Results

Some results of computational experiments are shown in Fig. 3. Mass center's initial velocity is aligned with x-axis and the vehicle platform has non-zero angular velocity. Solutions with the same initial conditions for all three models qualitatively match and demonstrate the following common properties: the velocity of translatory motion $v(t)$ of the vehicle decreases fast enough while the angular velocity of its rotation ν_3 firstly increases and then keeps its value much longer, see Fig. 3; trajectory of the vehicle mass-center $(x(t), y(t))$ has a shape of spiral, see Fig. 4. Although the rectilinear motion with zero angular velocity is possible for all three cases of contact model, it turned out that it is unstable, while the rotation about the vertical axis is stable and has a big attraction domain.

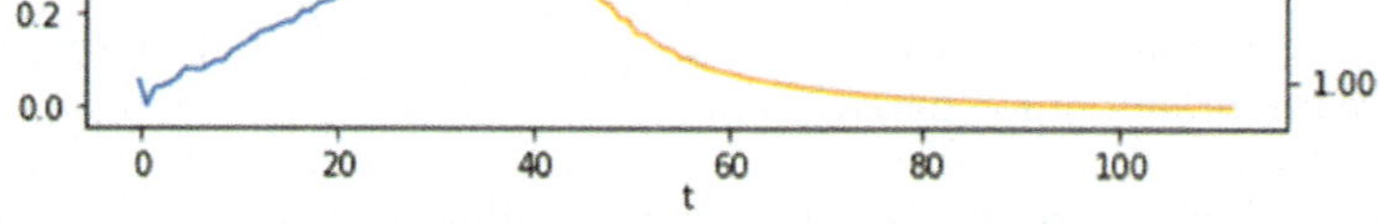

Fig. 3. Absolute value v of the mass center velocity and the platform angular velocity ν_3.

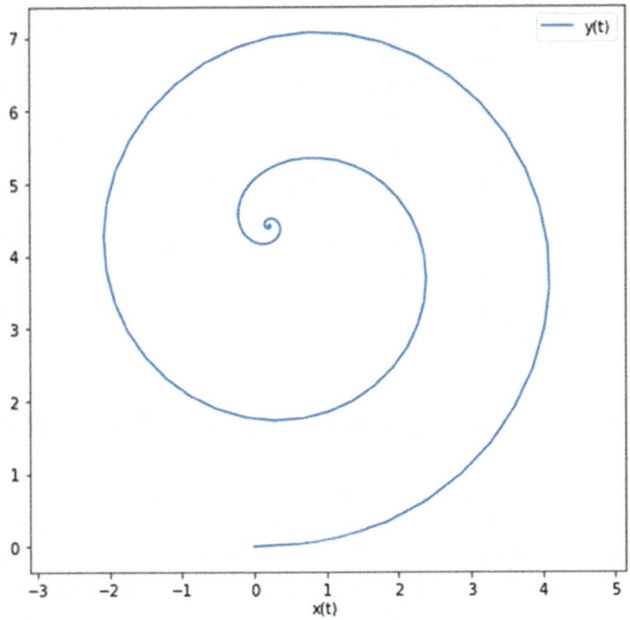

Fig. 4. The projection of the vehicle mass center trajectory on the horizontal plane.

6 Conclusions

One has to note that dry friction model, rather its saturated approximation, seems to be the most natural for the point-contact model between the rigid roller and supporting plane. At the same time, the non-holonomic rolling model itself is a limit case for the viscous friction while its coefficient tends to infinity. The natural question arises: is the dynamics of the considered systems similar while using three models mentioned above:

(a) ideal non-holonomic model of rolling “with impacts”;
(b) model of viscous friction with a large but finite coefficient of viscosity;
(c) approximate model for dry friction.

Computational experiments show that such similarity exists.

Acknowledgements. The work was performed at MSU and supported by RFBR, project 19-01-00140, for the first author and at MAI for the third author.

References

1. Kálmán, V.: Controlled braking for omnidirectional wheels. Int. J. Control Sci. Eng. **3**, 48–57 (2013)
2. Tobolár, J., Herrmann, F., Bünte, T.: Object-oriented modelling and control of vehicles with omni-directional wheels. Comput. Mech. **2009**, 2 p. (2009)
3. Kosenko, I.I., Stepanov, S.Y., Gerasimov, K.V.: Contact tracking algorithms in case of the omni-directional wheel rolling on the horizontal surface. Multibody Syst. Dyn. **45**, 273–292 (2019)
4. Kosenko, I.I.: Physically oriented approach to construct multibody system dynamics models using modelica language. In: Proceedings of Multibody 2007, Multibody Dynamics 2007. An ECCOMAS Thematic Conference, Politecnico di Milano, 25–28 June 2007, Milano, Italy, 20 p. (2007)
5. Novozhilov, I.V.: Fractional Analysis: Methods of Motion Decomposition. Birkhäuser, Boston (1997)
6. Borisov, A.V., Kilin, A.A., Mamaev, I.S.: Dynamics and control of an omniwheel vehicle. Regul. Chaotic Dyn. **20**, 153–172 (2015)
7. Gerasimov, K.V., Zobova, A.A., Kosenko, I.I.: Adjustment of non-holonomic constraints by absolutely inelastic tangent impact in the dynamics of an omni-vehicle. In: ECCOMAS Thematic Conference on Multibody Dynamics, Book of Abstracts, 15th–18th July 2019, Duisburg, Germany (2019)
8. Karapetyan, A.V.: On realizing nonholonomic constraints by viscous friction forces and celtic stone stability. J. Appl. Math. Mech. **1**, 30–36 (1981)

Adjustment of Non-Holonomic Constraints by Absolutely Inelastic Tangent Impact in the Dynamics of an Omni-Vehicle

Alexandra A. Zobova[1(✉)], Kirill V. Gerasimov[1], and Ivan I. Kosenko[2]

[1] Department of Mechanics and Mathematics, Lomonosov Moscow State University, Leninskie gory, 1, Moscow 119991, Russia
azobova@mech.math.msu.su, azobova@gmail.com, kirill.gerasimov.msu@gmail.com

[2] Department of Information Technologies and Applied Mathematics, Moscow Aviation Institute (National Research University), Volokolamskoe shosse, 4, Moscow 125993, Russia
kosenkoii@gmail.com

Abstract. We consider the motion of a vehicle with omni-wheels that moves along a horizontal plane without slipping. The mass of the rollers is non-zero and influences the dynamics of the system. The system is a multibody system of rigid bodies with non-holonomic constraints. The angular speed of the roller that enters in contact with the plane does not obligatorily satisfy the conditions of non-slipping. We propose to consider the vanishing of this slippage as a tangent inelastic impact. We get the analytic formula for the velocities after impact. It allows simulating the motion as a piecewise smooth system with switches.

1 Introduction

We study the dynamics of a vehicle with omni-wheels moving along a horizontal plane. An omni-wheel carries the rollers on its periphery. Their axes are tangent to the circumference of the wheel. Since the rollers rotate freely about their axes, the wheel can move in two principal directions: as an ordinary wheel going from one roller to the next one and in the perpendicular direction when the wheel touches the plane by one particular rotating roller. This property increases the maneuverability of the vehicle. There exist two main approaches to model the omni-wheels – non-holonomic one started from [1]: the mass of the rollers is neglected, and the non-slipping condition in the direction of wheel's plane is assumed (see also [2–4]). The second approach considers the rollers as rigid or deformable bodies and to include their dynamics into the model (e.g., [5]). Here we try to combine these two approaches: the conditions of non-slipping are imposed on massive rollers. In this work, we consider the dynamics of the rollers and suggest the model for switching of contact from one roller to another using impact theory. This work continues the research started in [4, 6, 7].

A. Kecskeméthy and F. Geu Flores (Eds.): ECCOMAS 2019, COMPUTMETHODS 53, pp. 518–525, 2020.
https://doi.org/10.1007/978-3-030-23132-3_62

2 The Statement of the Problem

Here we give a full description of the multibody system under consideration and introduce necessary notations.

2.1 The Structure of Multibody System and Coordinates

The symmetrical vehicle of mass M with N omni-wheels of mass m and radius l, each carrying n rollers of mass μ, moves along a fixed horizontal absolutely rough plane OXY. The centers of omni-wheel belong to a horizontal plane $S\xi\eta$ having the origin S in the center of mass and attached rigidly to the vehicle's platform. The wheel's axes are parallel to the vectors SP_i going to the centers of wheels P_i (Fig. 1) and the distance SP_i equals R. We denote by α_i the angles between $\mathbf{SP}_1 = R\mathbf{e}_\xi$ and $\mathbf{SP}_i$ (we have $\alpha_1 = 0$). So the platform mass center S is the mass center of the whole system, consisting of the platform, the wheels, and the rollers.

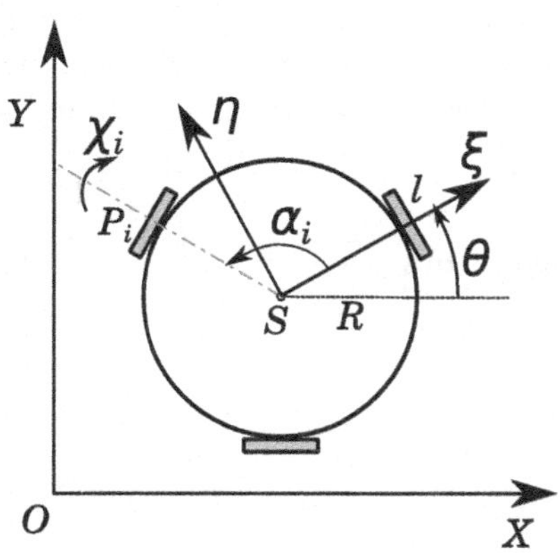

Fig. 1. Vehicle

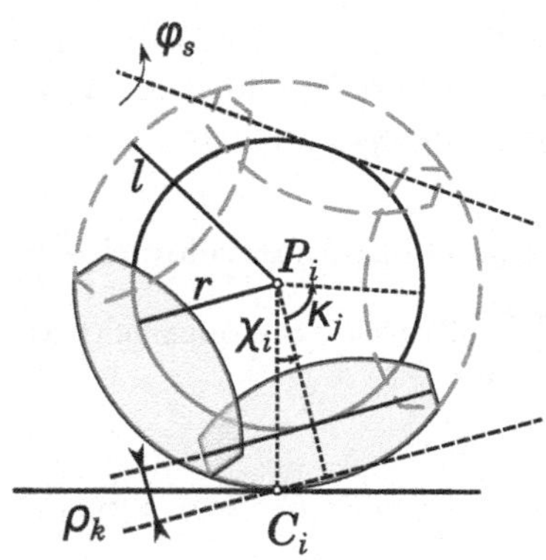

Fig. 2. Rollers on the omni-wheel

Each roller is bounded by a surface of revolution of a circular arc having radius l and subtending angle $\frac{2\pi}{n}$ about the axis remoted from the circle's center on a distance $r < l\cos\frac{\pi}{n}$ (Fig. 2). The same roller's surface but for $r = l\cos\frac{\pi}{n}$ was proposed in [6]. The centers of the rollers belong to the circle of radius r. The angles between the radius-vectors of the centers of the first roller and the center of j-th roller we denote by $\kappa_j = \frac{2\pi(j-1)}{n}$. Consequently, the each wheel's contact point always lies on the circle of radius l and passes from one roller to another each $\frac{2\pi}{n}$-angle of the wheel rotation. Due to overlapping of the rollers (Fig. 2), this geometry of the roller is not possible in real systems where either a gap exists between the rollers (then the contact can be in two points on the edges of the rollers), or the rollers are mounted in two or more rows in parallel planes. However, we propose to use the described roller's geometry that allows considering massive rollers without excessive complications. The rollers' axes are tangent to the circumference of the wheels (Fig. 2).

The multibody system under consideration consists of $N(n+1)+1$ rigid bodies and has a tree structure.

Let x, y be absolute coordinates of mass center S, θ be the course angle between the inertial axis OX and the axis $S\xi$. Let χ_i be the i-th wheel rotation angle counted counterclockwise looking from the outer of the vehicle from the vertical axis to the radius-vector of the center of the first roller, and ϕ_j are the rollers' angles. Thus, we have generalized coordinates

$$q = (x, y, \theta, \{\chi_i\}|_{i=1}^{N}, \{\phi_k\}|_{k=1}^{N}, \{\phi_s\}|_{s=1}^{N(n-1)})^T \in \mathbb{R}^{3+N(n+1)}$$

where firstly go the angles ϕ_k of rollers that are in contact at the moment, and after them the angles ϕ_s of other ("exempted" of constraints) rollers. The index s stands for throughout numbering of the rollers. For j-th roller on i-th wheel we have

$$s(i,j) = (n-1)(i-1) + j - 1 \tag{1}$$

We assume that there is no friction in the rollers' axes; the plane and rollers are absolutely rigid, so the contact between a supporting roller and the plane occurs in one point, and there is no slippage between each roller in contact and the supporting plane.

2.2 The Quasivelocities and the Constraints

We introduce the quasivelocities

$$\nu = (\nu_1, \nu_2, \nu_3, \nu_s)$$

$$v_S = R\nu_1 e_\xi + R\nu_2 e_\eta,\ \nu_3 = \Lambda\dot{\theta}, \quad \nu_s = \dot{\phi}_s,\ s = 1, \ldots, N(n-1), \tag{2}$$

where ν_1, ν_2 are the projections of the velocity of point S on the axes $S\xi$ and $S\eta$, ν_3 is the angular velocity of the platform (up to a constant multiplier Λ which meaning is given in Eq. 9), ν_s are the angular velocities (with respect to the wheel) of the $n(N-1)$ "exempted" rollers. From Eq. (2) follow

$$\dot{x} = R\nu_1 \cos\theta - R\nu_2 \sin\theta, \qquad \dot{y} = R\nu_1 \sin\theta + R\nu_2 \cos\theta$$

The condition of nonslippage means vanishing of the velocities of contact points C_i belonging to the rollers in contact. In quasivelocities we get

$$\dot{\phi}_k = \frac{R}{\rho_k}(\nu_1 \cos\alpha_k + \nu_2 \sin\alpha_k); \quad \rho_k = l\cos\chi_k - r \tag{3}$$

$$\dot{\chi}_i = \frac{R}{l}(\nu_1 \sin\alpha_i - \nu_2 \cos\alpha_i - \frac{\nu_3}{\Lambda}) \tag{4}$$

Here ρ_k is the distance from the roller's axis to C_i (see Fig. 2).

So the following differential constraints are imposed on the system

$$\dot{q} = \mathbf{V}\nu, \quad \mathbf{V} = \mathbf{V}(\theta, \chi_i) = \begin{bmatrix} \mathbf{V}_1 & \mathbf{O} \\ \mathbf{V}_2 & \mathbf{O} \\ \mathbf{V}_3 & \mathbf{O} \\ \mathbf{O} & \mathbf{E} \end{bmatrix}, \tag{5}$$

$$\mathbf{V}_1 = \begin{bmatrix} R\cos\theta & -R\sin\theta & 0 \\ R\sin\theta & R\cos\theta & 0 \\ 0 & 0 & \frac{1}{\Lambda} \end{bmatrix}, \quad \mathbf{V}_2 = \left[\frac{R}{l}\sin\alpha_i \ -\frac{R}{l}\cos\alpha_i \ -\frac{R}{\Lambda l}\right]_{i=1}^{N} \tag{6}$$

$$\mathbf{V}_3 = \left[\frac{R}{\rho_k}\cos\alpha_k \ \frac{R}{\rho_k}\sin\alpha_k \ 0\right]_{k=1}^{N} \tag{7}$$

where $\mathbf{O}$ are zero-matrices of appropriate dimensions, $\mathbf{V}_1$ is 3×3 matrix, $\mathbf{V}_2$ and $\mathbf{V}_3$ are $N \times 3$ matrix. They map $(\nu_1,\ \nu_2,\ \nu_3)$-vector into $(\dot{x},\ \dot{y},\ \dot{\theta})$, $(\dot{\chi}_i)$, $(\dot{\phi}_k)$, respectively.

2.3 Equations of Motion

While each wheel contacts the plane with the same roller the dynamics of motion is governed by the equations in pseudovelocities ν. To obtain them, we use Tatarinov laconic form of equations for the systems with differential constraints. The detailed description of the equations and their properties are given in [7].

The kinetic energy of the system T has the form

$$2T = Mv_S^2 + I_S\dot{\theta}^2 + J\sum_i \dot{\chi}_i^2 + B\sum_{i,j}(\dot{\phi}_{ij}^2 + 2\dot{\theta}\sin(\kappa_j + \chi_i)\dot{\phi}_{ij}) = \dot{q}^{\mathrm{T}}\mathscr{M}\dot{q} \tag{8}$$

Here M and I_S are the mass of the system and its moment of inertia about the vertical (including wheels and the rollers), J is the moment of inertia of each wheel about its axis, B is the rollers' moment of inertia. The matrix $\mathscr{M}$ has the following structure:

$$\mathscr{M} = \begin{bmatrix} M_{11} & \mathbf{O} & M_{131} & M_{132} \\ \mathbf{O} & JE_N & \mathbf{O} & \mathbf{O} \\ M_{131}^T & \mathbf{O} & BE_N & \mathbf{O} \\ M_{132}^T & \mathbf{O} & \mathbf{O} & BE_{N(n-1)} \end{bmatrix}$$

$$M_{11} = \mathrm{diag}(M, M, I_S), \quad M_{13} = [M_{131} \quad M_{132}] = \begin{bmatrix} 0 & \cdots & 0 \\ 0 & \cdots & 0 \\ B\sin\chi_{11} & \cdots & B\sin\chi_{Nn} \end{bmatrix}$$

where E_S is $S \times S$-unity matrix and M_{13} is $Nn \times Nn$ matrix consisting of two submatrices M_{131} (its size is $3 \times N$) and M_{132} ($N(n-1) \times N$). The matrix M_{131} corresponds to the rollers in contact, and M_{132} corresponds to other rollers.

Substituting constraints into kinetic energy gives

$$2T = \nu^{\mathrm{T}}\mathbf{V}^{\mathrm{T}}\mathscr{M}\mathbf{V}\nu = \nu^{\mathrm{T}}\mathscr{M}^*(\chi_i)\nu, \quad \mathscr{M}^*(\chi_i) = \begin{bmatrix} M_{11}^* & \mathbf{V}_1^{\mathrm{T}}M_{132} \\ M_{132}^T\mathbf{V}_1 & BE_{N(n-1)} \end{bmatrix}$$

$$M_{11}^* = (\mathbf{V}_3^{\mathrm{T}}M_{131}^{\mathrm{T}}\mathbf{V}_1 + \mathbf{V}_1^{\mathrm{T}}M_{131}\mathbf{V}_3) + MR^2E_3 + J\mathbf{V}_2^{\mathrm{T}}\mathbf{V}_2 + B\mathbf{V}_3^{\mathrm{T}}\mathbf{V}_3$$
$$(\mathbf{V}_1^{\mathrm{T}}M_{11}\mathbf{V}_1 = MR^2E_3 \Leftrightarrow J_S = MR^2\Lambda^2) \tag{9}$$

The equations of motion by inertia (i.e. without control torque in wheels' axes) are derived in [7] and has the from:

$$\mathscr{M}^*(\chi_i)\dot{\nu} = \Phi(\nu, \chi_i) \tag{10}$$

where $\Phi(\nu, \chi_i)$ is a quadratic form on ν with the coefficients, depending on χ_i. To close this system, one need to add the constraints

$$\dot{\chi} = \mathbf{V}_2\nu, \quad \dot{\chi} = (\dot{\chi}_1, \ldots, \dot{\chi}_N)^{\mathrm{T}}$$

Equation (10) conserve the energy of the system

$$T = h = \text{const}$$

Equation (10) govern the dynamics of the vehicle only during the period when each wheel contacts the ground by one particular roller. However, when a wheel approaches the edge of the first roller $\chi_i = \pm\frac{\pi}{n}$ the contact passes to the next roller. Next section is dedicated to a procedure of recalculating of quasi-velocities after a change of the roller in contact.

3 Change of the Roller in Contact

During the motion of the vehicle, the wheels rotate and the contact points go from a roller to the next one. In general, the condition of non-slipping for the next roller is not fulfilled. Formally, to proceed the calculations, we need to introduce a friction force that stops slipping of this particular roller. However, simulation of the systems with friction is always a non-trivial task as they are governed by stiff ODE and the numerical schemes can be unstable. To avoid the integrating of the system with friction, we propose to consider the short period of constraint imposing as a tangent inelastic impact that lasts a time-interval that is comparatively small to the vehicle dynamics' time-scale.

3.1 Vanishing of Slipping as a Tangent Impact

We assume that the friction force is large enough and slipping of a new roller in contact vanishes instantly. The previous roller (that left the contact zone) starts to rotate freely about its axis. We make the re-numeration of indexes of the rollers so that the roller that is entering the contact gets the number k equals the number of its carrying wheel. The constraints Eq. (3) is broken. We propose a plausible algorithm for calculating of the quasi-velocities ν^+ so that after the impact all constraints are fulfilled. We assume that

1. the impact lasts a period of time $\Delta t \ll \sqrt{R/g}$, so that the changes in generalized coordinates can be neglected $\Delta \mathbf{q} \sim \dot{\mathbf{q}}\Delta t \ll R$, the changes in generalized velocities are finite $\Delta\dot{\mathbf{q}} < \infty$; the slippage ends in time Δt;
2. the interaction of the vehicle with the supporting plane during the impact is reduced to the normal and tangential forces at all contact points of the wheels $\mathbf{R}_i = \mathbf{N}_i + \mathbf{F}_i$, $i = 1, \ldots, N$ (in other words, rolling and spinning torques vanish);

3.2 The Basic Equation of the Impact and Its Solution

The impact problem is formulated in the manner of classical mechanics

$$\mathscr{M}(\dot{\mathbf{q}}^+ - \dot{\mathbf{q}}^-) = \mathbf{Q}$$

with the generalized impact impulses $\mathbf{Q}$ are linear combinations of the tangent reactions $\mathbf{F}$ applied at the points of contact $\mathbf{Q} = \mathbf{KF}$. Matrix $\mathbf{K}$ is a known $[3 + N(n+1)] \times 2N$ matrix depending on the geometry of the vehicle:

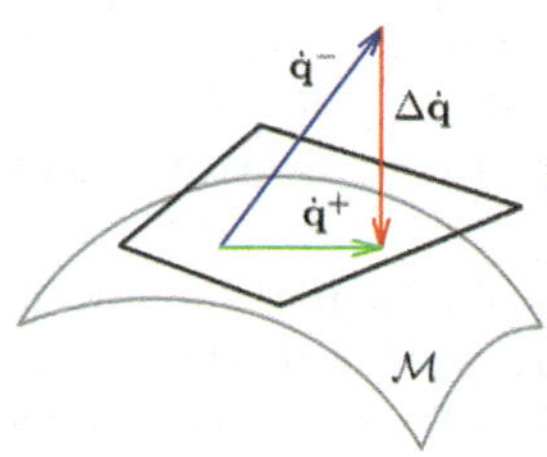

Fig. 3. Geometric interpretation of an absolutely inelastic impact

After the impact, the constraints Eq. 5 are fulfilled, so $\dot{\mathbf{q}}^+ = \mathbf{V}\nu^+$. Summarizing, we get

$$\mathscr{M}(\dot{\mathbf{q}}^+ - \dot{\mathbf{q}}^-) = \mathbf{Q} \quad \Rightarrow \quad \mathscr{M}\mathbf{V}\nu^+ - \mathbf{KF} = \mathscr{M}\dot{\mathbf{q}}^- \tag{11}$$

That is the system of linear equations on vector $(\nu^+, \mathbf{F})$ Besides, assumption 3.1 assures that the friction forces $\mathbf{F}_i$ do not work on the motions without slipping. It gives $\mathbf{V}^{\mathrm{T}}KF = 0$. Multiplying Eq. 11 by $\mathbf{V}^{\mathrm{T}}$ we get the solution:

$$\nu^+ = (\mathscr{M}^*)^{-1}\mathbf{V}^{\mathrm{T}}\mathscr{M}\dot{\mathbf{q}}^- \tag{12}$$

Geometrycally speaking, the first assumption means that the impact is inelastic; it is equivalent to the projection of $\dot{\mathbf{q}}^-$ onto the plane defined by $\mathbf{V}$ in the space of virtual displacements (Fig. 3). The second assumption means that this projection is orthogonal in the kinetic metric.

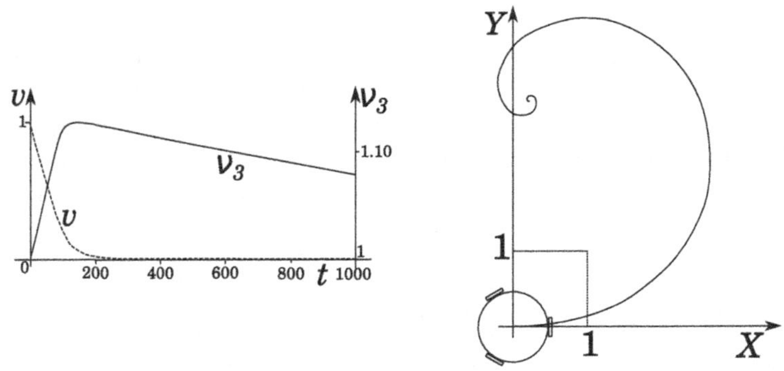

Fig. 4. Results of numerical simulation.

4 Dynamics Simulation and Conclusions

Combining both smooth parts of motion and impacts, we obtain the solutions of motion equations numerically for a vehicle with three wheels $N = 3$, placed symmetrically on a platform, each carrying five rollers $n = 5$ (Fig. 1). The parameters are $R = 0.15$, $r = 0.05$, $M = 1$, $m = 0.15$ and $\mu = 0.05$. The problem has no natural time-scale, so it was taken arbitrary. All the quantities on the graphs are undimensional.

Numerical simulation on smooth parts of motions – Cauchy's problem for Eq. 10 – is performed by Runge-Kutta method. The algorithm proceeds till the change of any roller in contact. Then we renumerate the rollers (see Sect. 3.1), and solve linear algebraic system Eq. 12. Obtained ν^+ are used as initial conditions for the next smooth part of motion.

The results of the numerical simulations are shown in Fig. 4. The most interesting motion is the motion that combines two types of motion: translation of the center of mass and rotation about the vertical axis. We take the following initial conditions $\nu_3(0) = 1$, $\nu_1(0) = 1$, others $\nu_i = 0$, $\mathbf{q} = 0$. On left panel of Fig. 4 we show the dependence of velocity $v(t) = \sqrt{\nu_1^2 + \nu_2^2}$ (its values are shown on the left axis) and $\nu_3(t)$ (right axis). Velocity decreases quite fast till almost zero value, while angular velocity firstly increases and then slowly decreases till full stop of the vehicle. Kinetic energy decreases in accordance to Carnot's theorem. The trajectory of the point S on the plane is a spiral curve shown on the right panel of the figure.

To conclude, we propose a model of an omni-vehicle with massive rollers moving along an absolutely rough horizontal plane. Using the theory of impact, the calculation of the change of velocities when a new roller enters the contact is reduced to solving of a system of linear algebraic equations that has a unique solution including the case of multiple impacts.

Acknowledgements. The work is supported by the Russian Foundation for Basic Research (project 19-01-00140).

References

1. Campion, G., Bastin, G., d'Andréa Novel, B.: Structural properties and classification of kinematic and dynamic models of wheeled mobile robots. IEEE Trans. Robot. Autom. **12**, 47–62 (1996)
2. Balkcom, D.J., Kavathekar, P.A., Mason, M.T.: Time-optimal trajectories for an omni-directional vehicle. Int. J. Robot. Res. **25**(10), 985–999 (2006)
3. Martynenko, Y.G., Formalskii, A.M.: On the motion of a mobile robot with roller-carrying wheels. J. Comput. Syst. Sci. Int. **46**(6), 976–983 (2007)
4. Zobova, A.A., Tatarinov, Y.V.: The dynamics of an omni-mobile vehicle. J. Appl. Math. Mech. **73**(1), 8–15 (2009)
5. Kalman, V.: Controlled braking for omnidirectional wheels. Int. J. Control Sci. Eng. **3**(2), 48–57 (2013)
6. Kosenko, I.I., Gerasimov, K.V.: Object-oriented approach to the construction of an omni vehicle dynamical model. J. Mech. Sci. Technol. **29**(7), 2593–2599 (2015)
7. Gerasimov, K., Zobova, A.A.: On the motion of a symmetrical vehicle with omni-wheels with massive rollers. Mech. Solids **53**(Suppl. 2), S32–S42 (2018)

Multibody Models and Simulations to Assess the Stability of Counterbalance Forklift Trucks

Michele Gardella[1] and Alberto Martini[2(✉)]

[1] School of Engineering, University of Bologna, V. Fontanelle 40, 47121 Forlì, FC, Italy
michele.gardella@studio.unibo.it

[2] Department of Industrial Engineering (DIN), University of Bologna, V.le del Risorgimento 2, 40136 Bologna, BO, Italy
alberto.martini6@unibo.it

Abstract. Assessing the stability of counterbalance forklift trucks is a problem of primary interest, since critical safety issues related to their operation are involved. Indeed, forklifts are the most common vehicles for material handling in industry. The stability limit is typically verified through experimental tests, which are costly and time consuming. This study aims at developing numerical tools to reliably predict the stability limit through simulations. Such tools should permit, on one hand, to partially reduce the amount of experimental tests necessarily required and, on the other hand, to support the design phase of new products since the early stages. A multibody model taking into account the compliance of the tires and of the mast assembly is developed. An experimental campaign to validate the model is ongoing. The preliminary numerical results confirm the model as a promising tool to estimate the stability limit of the forklift with satisfactory accuracy.

1 Introduction

Counterbalance forklift trucks constitute worldwide an essential equipment for material handling in manufacturing and warehousing. One of the most critical tasks to ensure the safety of forklift operation is assessing properly the stability conditions of the vehicle. Indeed, forklift tip over and roll over caused by, respectively, braking and cornering maneuvers, represent the most common accidents for this kind of vehicles [3].

The stability analysis of forklifts is typically performed in accordance with the ISO 22915-2:2018 Standard, which defines four different test conditions, referred to as *Test 1-4* [1]: the vehicle is placed on a tilt-table that is driven in quasi-static conditions to a prescribed slope. The testing conditions of major interest for this study are the ones referred to as *Test 1* and *Test 3* [2]: the former simulates a breaking maneuver with the mast at maximum height carrying the rated load (Fig. 1), for evaluating the longitudinal stability; the latter simulates a turning maneuver at constant speed with the rated load lifted at maximum height (Fig. 2), hence assessing the lateral stability.

The company promoting this research activity, namely Toyota Material Handling Manufacturing Italy SpA (Bologna, Italy), conducts experimental tests to assess the stability of each new model of forklift. Moreover, many vehicles customized with

A. Kecskeméthy and F. Geu Flores (Eds.): ECCOMAS 2019, COMPUTMETHODS 53, pp. 526–533, 2020.
https://doi.org/10.1007/978-3-030-23132-3_63

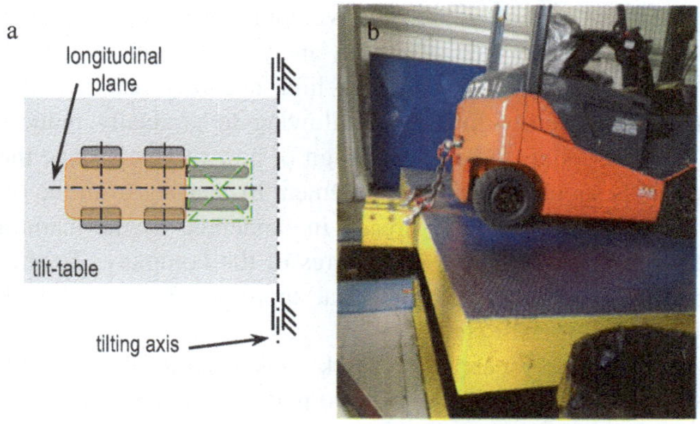

Fig. 1. *Test 1*, (a) schematics, (b) view of the forklift during test.

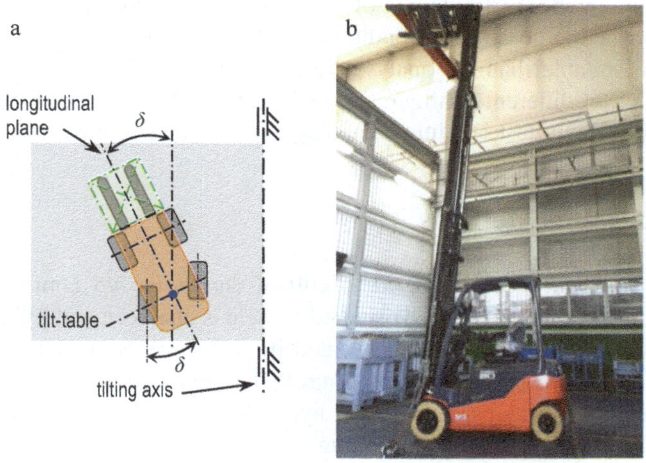

Fig. 2. *Test 3*, (a) schematics, (b) view of the forklift during test.

masts bigger than the standard model and/or special purpose forks (e.g. featuring additional actuators for handling the payload) are also manufactured, each one requiring additional tests for the stability analysis. This results in a large amount of experimental tests to be performed.

Moreover, the test are used by the company to gain further insights into the stability limits, to be exploited for guiding the design of new forklift models. Indeed, the analytical models adopted in the early design phases do not provide accurate enough results, due to the complex deformations affecting the forklift loaded with the nominal load. However, due to safety reasons, the actual tip over/roll over conditions are not generally reached in tests, particularly in *Test 3* (although the vehicle is secured to the tilt-table and/or to a fixed frame by means of chains and/or belts); hence, the stability limit must be extrapolated from the experimental results. This causes higher safety factors to be required in the design process.

The main objective of the research is developing numerical simulation tools based on multibody models to carry out the stability analysis of counterbalance forklift trucks. Firstly, they should be able to predict with high accuracy and reliability the forklift behavior in terms of stability limit, thus allowing to gradually reduce the need for experimental tests as well as to help the design of new products since the early stages. Secondly, such tools should be easy to implement by starting from the other numerical models (e.g. CAD and FE models) already implemented by the manufacturer, hence being well integrated in the design procedures of the company. To the authors' best knowledge, there is no established numerical approach already available in the literature to reliably assess forklift stability.

As a first stage of the research, this work aims at replicating the stability tests, in particular *Test 1* and *Test 3*, for achieving two main goals: the first one is to carry out a deeper investigation of the discrepancies between the experimental results and the analytical models; the second goal is to define the most convenient modelling strategy (i.e. satisfactorily accurate and cost effective) based on a multibody approach, which is reckoned as the main novel outcome of the presented activity. Indeed, although such tests are performed in quasi-static conditions, multibody modelling and simulations are deemed more suitable than FE analyses due to both a higher flexibility for rapidly switching between different configurations of the forklift (e.g. modifying the steering angle) and/or different test conditions, and potentially lower computational time.

2 Materials and Methods

The study focuses on a four-wheeled forklift of the *Traigo 48* family, with electric propulsion (Fig. 3). In particular, the studied vehicle is characterized by a rated load of 1800 kg and a lift height of 5.5 m provided by a 3-stage mast. It is equipped with super-elastic solid tires. The mast tilt is controlled by two hydraulic cylinders; its lift is driven by a system of chains, pulleys and hydraulic actuators. The mast also includes an actuated attachment for fork positioning.

Fig. 3. Side view of the studied forklift.

On the basis of the company know-how, the stability of the loaded forklift with fully extended mast is expected to be primarily affected by two contributions of deflection: (i) nonlinear deformations of the front tires; (ii) flexural deformations of the forks and the mast, with nonlinearities introduced by the translational guides (and additional torsional components in *Test 3*). The former effect should be preponderant in *Test 1*, where the mast deflection is compensated by imposing a proper backward tilt angle on the mast, in order to keep unaltered the longitudinal distance between the front axle and the Center of Mass (CoM) of the load at the beginning of the test. The latter is expected to be predominant in *Test 3*, where the vertical longitudinal plane of the forklift is inclined with respect to the rotational axis of the tilt-table, and maximum backward tilt angle is imposed on the mast (the corresponding angles being 16° and 6° for the studied forklift).

Approximate analytical models have been developed to reckon the stability limit. The model for *Test 1* takes into account the forklift pitch angle associated with the tire compliance (estimated on the basis of the datasheet provided by the tire manufacture) and the load CoM displacement caused by the deflection of the mast (approximated as a cantilever beam). The error on the stability limit predicted by such model is not deemed satisfactory, although not exceeding 10%. As for *Test 3*, only the tire compliance is considered, since reliable flexural-torsional deflections cannot be analytically estimated, due to the complexity of the mast assembly. With this rough approximation, the stability limit is overestimated by about 50%.

2.1 Numerical Multibody Model

A numerical model of the complete forklift is implemented by using the FunctionBay RecurDyn multibody software package. All the components are modelled as rigid bodies derived from the CAD model, with the exception of the mast assembly. The steering system is also included, to set the longitudinal plane of the rear axle wheels parallel to the tilting axis in *Test 3* (Fig. 2). The mass distribution is adjusted to match the actual forklift global CoM, determined by measuring the tire normal forces through load cells.

The tire-ground interactions are modelled by using *solid-to-solid* contacts. In particular, the *Extended Surface* feature is adopted. Indeed, its formulation (namely a function of the maximum penetration depth) allows to define the contact stiffness straightforwardly by setting a spline that fits the static force-displacement curves provided by the tire manufacturer. Hence, it appears convenient in terms of efforts required to implement the model. By contrast, longer computational time may be experienced with respect to visco-elastic forces defined through analytical functions. Moreover, such contacts do not replicate stiction effects, which are actually experienced in tests, since the wheels are braked). Nonetheless, both drawbacks are deemed acceptable. In particular, relative tangential motion between the tire and the ground is expected to be negligible, the tests to be simulated generally not exceeding 30 s.

The mast assembly is modelled by means of a Finite Element (FE) mesh and a *full-flex* formulation (i.e. not reduced). In particular, the three mast sections (i.e. inner, central and outer mast), the attachment (i.e. the fork carrier) and the forks are meshed with shell elements (4-node structural shells, size of about 15 mm). The rollers, composed of a steel outer ring and a PTFE thrust bearing, are modelled as rigid parts. Each roller is connected to its (flexible) shaft through a rigid spider allowing the rotational DOF. The interactions between each roller and the corresponding mast guide are modelled by using two distinct *solid-to-flex* contacts: a *cylinder-to-surface* is implemented to take into account the radial loads; an additional contact models the axial loads. Contact patches with a finer mesh size (about 2 mm) are added to make the contacts work properly. The connections between the mast and the front axle assembly are modelled by means of ideal joints. At this stage, the relative motion between the mast sections and the forks is prescribed by using velocity functions that neglect the compliance of the hydraulic circuit and the transmission chains. The *full-flex* formulation permits to exploit the FE models already available to the company (developed in Ansys Workbench). The increment in the computational cost with respect to a Craig-Bampton reduced model is regarded as acceptable.

The load is a rigid body supported by contact forces exerted by the forks.

2.2 Experimental Tests

Experimental measurements are performed for model updating and validation. In particular, *Test 1* and *Test 3* are carried out with a special setup featuring four load cells installed between the forklift tires and the tilt-table, in order to monitor the normal force acting on each wheel (Fig. 4). It is worth noting that the normal force is expected to become null, for the wheel(s) at the highest level on the tilt-table, when reaching the stability limit.

Fig. 4. Close up of the load cells in *Test 3*.

A further ad hoc experiment is arranged and conducted on the mast assembly to better characterize its flexural-torsional compliance. The test aims at measuring the actual displacements of the application point of the load (a dummy rigid pallet) and of other reference points belonging to the three mast sections, the fork carrier and the forks, with respect to the connections between the mast and the front axle assembly, for different lift heights. The mast assembly is separated from the vehicle and installed on a rigid support connected to the tilt-table, which replicates the orientation of the mast in *Test 3*. The tilt-table is rotated up to a slope of 7% (about 4°), in two loading conditions, namely loaded and unloaded mast. The 3D positions of the points of interest are detected through stereophotogrammetric measurements, performed by using a

8-camera VICON optoelectronic system. To this purpose, proper reflective markers are attached to the locations to be monitored (Fig. 5a).

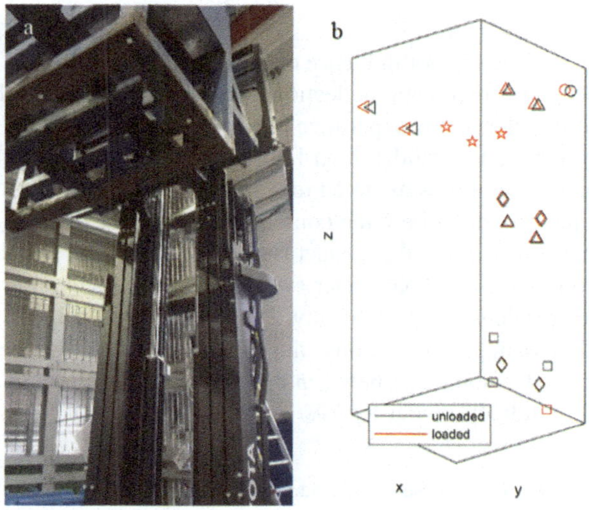

Fig. 5. (a) Close up of the markers in the test setup; (b) comparison between unloaded and loaded mast, for test at 5.5 m lift height, of points of the outer (square), central (diamond), inner (upward triangle), carrier (circle), forks (leftward triangle) and load (star).

The comparison between the displacements of several reference points measured in the tested loading cases, for a lift height of 5.5 m, is reported as example in Fig. 5b (the actual values can not be shown, due to NDA). Although the experimental campaign is still ongoing, the first results confirms that remarkable combined flexural-torsional deformations can be experienced on the top of the extended mast, hence significantly affecting the stability limit.

3 Preliminary Results and Model Verification

The reliability of the implemented multibody model is assessed by evaluating the percentage error ($\varepsilon\%$) affecting the stability limit predicted by the model with respect to the measured data, computed as:

$$\varepsilon\% = \frac{stability_{sim} - stability_{exp}}{stability_{exp}} \cdot 100 \tag{1}$$

where the stability limit is defined as the tilt-table slope corresponding to a condition of null normal force acting on the highest wheel of the forklift, and the subscripts *sim* and *exp* indicate numerical and measured results, respectively. The models are expected to overestimate the stability limit, hence $\varepsilon\%$ being typically positive.

The model has been verified in two steps, on the basis of the preliminary results currently available.

3.1 Test 1

As mentioned in Sect. 2, a minor influence of the mast deflection on the results of *Test 1* is expected, since the initial static deflection is compensated before the test starts by tilting the mast itself. Hence, the experimental results obtained in *Test 1* are exploited to partially validate the contact model describing the tire-ground interaction, whereas, at this stage, the mast assembly is assumed to behave like a rigid body (thus temporarily suppressing the features describing the compliance of the mast).

The spline curve describing the contact stiffness for a normal load between 0 and 5 kN is adjusted to minimize the parameter ε%, through numerical optimization. Indeed, the tire manufacturer does not provide any data for this load range.

After model updating, the stability limit error provided by the simulations is reduced below 5%. A comparison between the results of the approximated model (§2) and the updated multibody model for *Test 1* is reported in Table 1.

Table 1. Results provided by the numerical models.

Test #	Stability limit error (ε%)	
	Approximate model	Multibody model
Test 1	8.3%	3.5%
Test 3	47.0%	18.7%

3.2 Test 3

Since the experimental campaign on the mast assembly is still ongoing, only a partial verification of the model for *Test 3* is performed. In particular, the position of the load CoM is updated on the basis of the first experimental results, whereas modifications in the mass distribution of the mast assembly caused by its deflection are neglected (i.e., the mast assembly is again considered as rigid).

The simulation results obtained from this model exhibit a significant improvement in the stability limit prediction, ε% being lower than 20%. A comparison with the results of the approximated model (§2) is shown in Table 1.

4 Conclusion

A numerical multibody model for the stability analysis of a forklift truck has been developed and its reliability has been partially verified by using the preliminary results of an experimental campaign.

The results provided by the current model are promising and reasonably, after completing its validation, it will permit to predict the stability limit of the studied forklift with high accuracy.

On the basis of the results of this study, a standardized modelling procedure will be defined. Then, a larger set of numerical models will be generated for comparing the results with the company database of the stability test measurements, in order to confirm the reliability of the developed approach.

Acknowledgements. This activity is performed in collaboration with Toyota Material Handling Manufacturing Italy SpA (Bologna, Italy), and in particular with Tommaso Piazza, which are gratefully acknowledged for operative cooperation and use of facilities. EnginSoft SpA (Trento, Italy) is also kindly acknowledged for providing the RecurDyn software package and support for the simulations.

References

1. International Organization for Standardization: Industrial trucks – Verification of stability – Part 2: Counterbalanced trucks with mast. Standard ISO 22915-2 (2018)
2. Larsson, T.J., Lambert, J., Wilde, M., Tully, G., Askew, H., Skinner, W., Carter, B., Martin, T., Kenningham, L.: Industrial forklift trucks - dynamic stability and the design of safe logistics. Saf. Sci. Monit. **7**(1), IV-3 (2003)
3. Lemerle, P., Höppner, O., Rebelle, J.: Dynamic stability of forklift trucks in cornering situations: parametrical analysis using a driving simulator. Veh. Syst. Dyn. **49**(10), 1673–1693 (2011)

Automatic Differentiation in Multibody Helicopter Simulation

Max Kontak[1(✉)], Melven Röhrig-Zöllner[1], Johannes Hofmann[2], and Felix Weiß[2]

[1] Simulation and Software Technology, German Aerospace Center (DLR), Linder Höhe, 51147 Cologne, Germany
{Max.Kontak,Melven.Roehrig-Zoellner}@dlr.de
[2] Institute of Flight Systems, German Aerospace Center (DLR), Brunswick, Germany
{Johannes.Hofmann,Felix.Weiss}@dlr.de

Abstract. In a first approximation, helicopters can be modeled by open-loop multibody systems (MBS). For this type of MBS the joints' degrees of freedom provide a globally valid set of minimal states. We derive the equations of motion in these minimal coordinates and observe that one has to compute Jacobian matrices of the bodies' kinematics with respect to the minimal states. Classically, these Jacobians are derived analytically from a complicated composition of coordinate transformations. In this paper, we will present an alternative approach, where the arising Jacobians are computed by automatic differentiation (AD). This makes the implementation of a simulation code for open-loop MBS more efficient, less error-prone, and easier to extend. To emphasize the applicability of our approach, we provide simulation results for rigid MBS helicopter models.

1 Introduction

Helicopters are fascinating and complex machines especially considering their structural dynamics. They are prone to a number of instabilities, where the dominant ones stem from their means to create lift and propulsion—the rotor. The kinematics of a helicopter rotor are rather complex compared to a propeller. The blades are hinged both in the rotor plane as well as orthogonal to it and they have a variable pitch angle. These hinges and bearings are at different places and can be equipped with damping and/or coupling mechanisms. The rotating mass of a helicopter compared to its takeoff weight is much higher than for a propeller aircraft, while the rotor's dominant eigenfrequencies are lower. This makes the accurate modeling of helicopter dynamics a challenging task.

The large relative motion of a helicopter's components makes a multibody formulation the ideal model for describing it. The dominant load paths of a helicopter are, in its simplest form, a tree connecting the root (fuselage) to the leaves (rotor blades). We call such a system an *open-loop MBS*. Such a system possesses minimal coordinates, which eliminate the constraint equations. For a graphical representation of an open-loop MBS, see Fig. 1.

A. Kecskeméthy and F. Geu Flores (Eds.): ECCOMAS 2019, COMPUTMETHODS 53, pp. 534–542, 2020.
https://doi.org/10.1007/978-3-030-23132-3_64

Using minimal coordinates in helicopter simulation has an important advantage: usually, one is not just interested in the evolution of the system for some initial condition but one needs to find stable flight conditions (trim problem, see, e.g., [9]). This is easier when the system is modeled with a low number of unconstrained states.

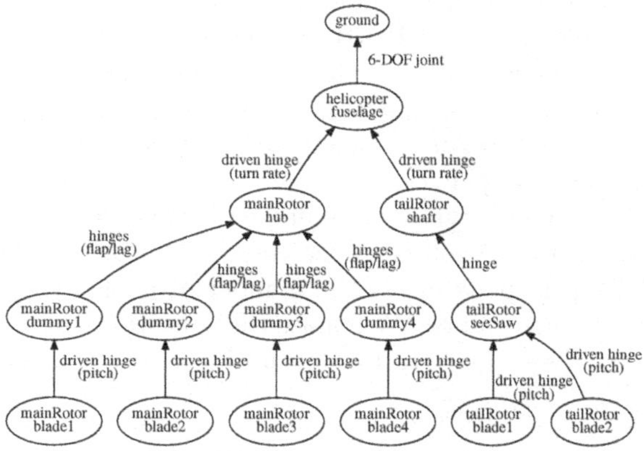

Fig. 1. Simple open-loop MBS graph for a helicopter with a main and a tail rotor. The nodes represent bodies and the edges the different joint types that connect them. A virtual *6-DOF* joint is used to model the rigid body motion of the fuselage. A *Hinge* connects two bodies allowing only motion around one rotational axis (1 DOF). The *flap/lag hinges* connect the rotor hub to the blades and are consisting of two single hinges, whose rotation axes are orthogonal two each other (in total 2 DOFs).

For the formulation of the equations of motion in minimal coordinates, one has to compute the Jacobian matrices of the body motion with respect to the minimal coordinates. Deriving the required terms by hand is quite technical and their implementation is very error-prone. In this paper, we present an approach that uses automatic differentiation (AD) to compute these Jacobians.

There are only a few references, which deal with the application of AD to multibody dynamics. There, automatic differentiation has been used for the following purposes: first, to obtain Jacobians of the multibody kinematics with respect to (e.g., design) parameters. This can be used to optimize the parameters of a multibody system with respect to a certain objective (see, e.g., [2]) or for sensitivity analysis (see, e.g., [1,3]). Secondly, automatic differentiation has been used to compute the derivatives needed for the algebraic part of the time integration of the resulting differential-algebraic equations (see, e.g., [4,5,12]). There has been little effort to automatically generate equations of motion by using AD. One example is [6], where these equations are derived by applying AD in a Lagrangian mechanics setting. To our knowledge, there has not yet been any approach to compute the equations of motion of an open-loop multibody system in minimal coordinates with automatic differentiation.

2 Dynamics of Open-Loop Multibody Systems

General multibody systems consist of multiple bodies, which are connected by an arbitrary number of joints. In this paper, we consider only rigid bodies. We define reference frames at the center of mass of the bodies (*body frames*). We denote the position and velocity of all body frames by $\boldsymbol{r}(t)$ and $\boldsymbol{v}(t)$, respectively. Note that we include the orientation (as a quaternion) and the angular velocity of the body frames in these two vectors. For these kinematic variables, we use a *floating frame of reference* formulation (cf. [11, p. 7]), where we have to consider that $\dot{\boldsymbol{r}} \neq \boldsymbol{v}$. Therefore, we introduce an abstract function $\boldsymbol{f}$, which maps the position, orientation, velocity, and angular velocity to the derivative of the position and orientation, such that $\dot{\boldsymbol{r}} = \boldsymbol{f}(\boldsymbol{r}, \boldsymbol{v})$.

There can be different types of joints with different degrees of freedom (DOFs). For ease of implementation, the free motion of a body is modeled by a joint with 6 DOFs. Depending on the number of DOFs, the joints introduce different numbers of constraints. We denote the collection of all constraint equations by $\boldsymbol{g}(\boldsymbol{r}) = \boldsymbol{0}$. Then, the equations of motion of the constrained multibody system are given by

$$\dot{\boldsymbol{r}} = \boldsymbol{f}(\boldsymbol{r}, \boldsymbol{v}), \qquad \boldsymbol{M}\dot{\boldsymbol{v}} = \boldsymbol{h}(\boldsymbol{r}, \boldsymbol{v}) + \boldsymbol{G}(\boldsymbol{r})^{\mathsf{T}}\boldsymbol{\lambda}, \qquad \boldsymbol{g}(\boldsymbol{r}) = \boldsymbol{0}, \tag{1}$$

where $\boldsymbol{M}$ is the mass matrix, $\boldsymbol{G} := \frac{\partial \boldsymbol{g}}{\partial \boldsymbol{r}}$ is the constraint Jacobian, $\boldsymbol{\lambda}$ is the vector of Lagrangian multipliers, and $\boldsymbol{h}$ is the sum of all forces, including pseudo-forces, which are introduced by the choice of a moving reference frame (cf. [13, p. 16]).

2.1 Equations of Motion in Minimal Coordinates

Assuming sufficient regularity of the constraint functions, there is always a set of locally valid minimal coordinates $\boldsymbol{s}$, $\boldsymbol{u}$ on position and velocity level (cf. [13, pp. 17–18]). That is, there exists functions $\boldsymbol{R}$, $\boldsymbol{V}$ mapping these minimal states to the body states as $\boldsymbol{r} = \boldsymbol{R}(\boldsymbol{s}, \boldsymbol{u})$, $\boldsymbol{v} = \boldsymbol{V}(\boldsymbol{s}, \boldsymbol{u})$, and $\boldsymbol{g}(\boldsymbol{R}(\boldsymbol{s}, \boldsymbol{u})) \equiv \boldsymbol{0}$ is fulfilled for all $\boldsymbol{s}$, $\boldsymbol{u}$. We can employ the chain rule to obtain

$$\dot{\boldsymbol{v}} = \frac{\partial \boldsymbol{V}(\boldsymbol{s}, \boldsymbol{u})}{\partial \boldsymbol{s}}\dot{\boldsymbol{s}} + \frac{\partial \boldsymbol{V}(\boldsymbol{s}, \boldsymbol{u})}{\partial \boldsymbol{u}}\dot{\boldsymbol{u}} = \frac{\partial \boldsymbol{V}(\boldsymbol{s}, \boldsymbol{u})}{\partial \boldsymbol{s}}\boldsymbol{F}(\boldsymbol{s}, \boldsymbol{u}) + \frac{\partial \boldsymbol{V}(\boldsymbol{s}, \boldsymbol{u})}{\partial \boldsymbol{u}}\dot{\boldsymbol{u}},$$

where the function $\boldsymbol{F}$ maps the minimal position and velocity states to the derivative of the minimal position states. Inserting this into the equations of motion (1), multiplying from the left by the transpose of the Jacobian $\boldsymbol{J}_{\boldsymbol{u}}(\boldsymbol{s}, \boldsymbol{u}) := \frac{\partial \boldsymbol{V}(\boldsymbol{s}, \boldsymbol{u})}{\partial \boldsymbol{u}}$, and using $\boldsymbol{J}_{\boldsymbol{u}}^{\mathsf{T}}\boldsymbol{G}^{\mathsf{T}} = \boldsymbol{0}$, which is true due to the chain rule applied to $\boldsymbol{g} \equiv \boldsymbol{0}$, we obtain the equations of motion in minimal coordinates, which reads

$$\dot{\boldsymbol{s}} = \boldsymbol{F}(\boldsymbol{r}, \boldsymbol{u}), \qquad \tilde{\boldsymbol{M}}(\boldsymbol{s}, \boldsymbol{u})\dot{\boldsymbol{u}} = \tilde{\boldsymbol{h}}(\boldsymbol{s}, \boldsymbol{u}), \tag{2}$$

where $\tilde{\boldsymbol{M}} = \boldsymbol{J}_{\boldsymbol{u}}^{\mathsf{T}}\boldsymbol{M}\boldsymbol{J}_{\boldsymbol{u}}$ is the mass matrix in minimal coordinates. The right-hand side $\tilde{\boldsymbol{h}} := \boldsymbol{J}_{\boldsymbol{u}}^{\mathsf{T}}(\boldsymbol{h} - \boldsymbol{M}\boldsymbol{H})$ includes all pseudo accelerations that are induced by the coordinate transformation, which are, by the considerations above, given as $\boldsymbol{H}(\boldsymbol{s},\boldsymbol{u}) := \boldsymbol{J}_s(\boldsymbol{s},\boldsymbol{u})\boldsymbol{F}(\boldsymbol{s},\boldsymbol{u})$, where $\boldsymbol{J}_s(\boldsymbol{s},\boldsymbol{u}) := \frac{\partial \boldsymbol{V}(\boldsymbol{s},\boldsymbol{u})}{\partial \boldsymbol{s}}$.

For open-loop multibody systems, where the bodies can be arranged in a tree (cf. Fig. 1), the degrees of freedom of the joints can be chosen as minimal coordinates, which are valid *globally.*

To implement the equations of motion in minimal coordinates (2), we need the following ingredients: the mappings of minimal joint states to the body states $\boldsymbol{R}$ and $\boldsymbol{V}$ and the Jacobians $\boldsymbol{J}_u(\boldsymbol{s},\boldsymbol{u})$ and $\boldsymbol{J}_s(\boldsymbol{s},\boldsymbol{u})$. The functions $\boldsymbol{R}$ and $\boldsymbol{V}$ can be implemented easily by going through the open-loop tree from top to bottom and applying the coordinate transformations induced by the joints depending on their states. Instead of computing the Jacobians "by hand", these will be computed by automatic differentiation in our implementation, which we discuss below.

3 Automatic Differentiation for Open-Loop Multibody Systems

The Jacobians $\boldsymbol{J}_u(\boldsymbol{s},\boldsymbol{u})$ and $\boldsymbol{J}_s(\boldsymbol{s},\boldsymbol{u})$ consist of derivatives of the function $\boldsymbol{V}$ with respect to components of either $\boldsymbol{s}$ or $\boldsymbol{u}$. Due to the way, in which this function $\boldsymbol{V}$ is calculated, it is a long composition of simple transformations. When computing these Jacobians by hand, it turns out that one has to employ the chain rule many times. The idea behind *automatic differentiation (AD)* is to automate this process (see, e. g., [10, Chapter 2]).

There are different variants of AD. Here, we employ the so-called forward-mode AD which is suitable for cases where the number of inputs is smaller than the number of outputs. The following example illustrates the idea behind AD: detached from our original problem, for two functions $\boldsymbol{f}\colon \mathbb{R}^m \to \mathbb{R}^n$ and $g\colon \mathbb{R}^n \to \mathbb{R}$, we can compute the function value and the derivative of the composition $g \circ \boldsymbol{f}\colon \mathbb{R}^m \to \mathbb{R}$, $\boldsymbol{x} = (x_1, \ldots, x_m)^\mathsf{T} \mapsto g(\boldsymbol{f}(\boldsymbol{x}))$ with respect to a variable x_i at a specific point $\hat{\boldsymbol{x}}$ simultaneously in two steps by introducing temporary variables:

	function value	derivative
step 1:	$\boldsymbol{y} = \boldsymbol{f}(\hat{\boldsymbol{x}})$	$\boldsymbol{v} = \left.\dfrac{\partial \boldsymbol{f}(\boldsymbol{x})}{\partial x_i}\right\vert_{\boldsymbol{x}=\hat{\boldsymbol{x}}}$
step 2:	$z = g(\boldsymbol{y})$	$w = \displaystyle\sum_{j=1}^{n} \left.\dfrac{\partial g(\boldsymbol{y})}{\partial y_j}\right\vert_{\boldsymbol{y}=\boldsymbol{v}} v_j$

after which $z = (g \circ \boldsymbol{f})(\hat{\boldsymbol{x}})$ and $w = \left.\frac{\partial (g\circ \boldsymbol{f})(\boldsymbol{x})}{\partial x_i}\right\vert_{\boldsymbol{x}=\hat{\boldsymbol{x}}}$. Using a library, which supports automatic differentiation (in our case, the Eigen C++ library [7]), one simply implements the left equations for the values and the library automatically keeps track of derivatives via the right equations by overloading the respective arithmetic operators. This makes the development of a multibody dynamics software much easier.

To illustrate this, we show pseudo code for the transformation of minimal states to three-dimensional body states. First, we consider a hinge joint:

```
//! compute the joints' relative position and velocity from minimal states
Kinematics Hinges::relativeKinematics(angle, angleDerivative) {
  // a hinge does not imply any translational relative movement:
  position = Zero(3);
  velocity = Zero(3);
  // a hinge does imply a specific rotational relative movement:
  // create quaternion from angle and rotation axis
  orientation = AngleAxis(angle, axis));
  angularVelocity = angleDerivative * axis;

  return position, orientation, velocity, angularVelocity;
}
```

The actual implementation looks almost like the code above, except for data types and C++ template arguments. Note that this is the *only* part of the software that is specific to hinge-type joints. Other joints can be implemented in a similar manner.

To calculate the absolute motion of all bodies, we need to traverse the open-loop graph and combine the relative joint motion with the motion of the previous body:

```
//! compute absolute positions and velocities from relative states
//!
//! All nodes are sorted by their distance to the root node,
//! so parent nodes are evaluated before child nodes.
Kinematics MultiBody::absoluteKinematics(Kinematics relative) {
  // the first node is the origin:
  position[0] = Zero(3); orientation[0] = Identity();
  velocity[0] = Zero(3); angularVelocity[0] = Zero(3);
  // traverse the open loop graph, ignoring the origin:
  for( i = 1; i < numNodes; i++) {
    // get the index of the previous node
    prevNode = parentNode[i];
    // combine movement of prev. node with relative movement:
    orientation[i] = orientation[prevNode] * relative.orientation[i];
    position[i] = position[prevNode] + orientation[i] * relative.position[i];
    angularVelocity[i] = ...; velocity[i] = ...;
  }
  return position, orientation, velocity, angularVelocity;
}
```

The is just the formula for the relative motion at the position and velocity level. We can iterate through the graph in a linear way as all nodes are sorted by the level (distance to the root node) in the tree. By combining the functions above, one can calculate the three-dimensional body motion from the minimal joint states. When we employ forward-mode AD, we obtain the required Jacobian matrices $\boldsymbol{J}_u$ and $\boldsymbol{J}_s$.

In contrast, if one derives the matrix entries of $\boldsymbol{J}_u$ and $\boldsymbol{J}_s$ analytically (see, e.g., [11]) and puts the resulting equations in code, one obtains a much longer and more complicated implementation: in our preceding implementation the same functionality consists of about 500 lines of code that assemble the Jacobian matrices. This code is difficult to write and to maintain due to possible index errors. For each joint type, several functions are needed (the `relativeKinematics` functionality, but also functions for its derivatives) that need to be consistent with each other.

We also want to point out that our approach makes it much easier to extend the software (e.g., with new joint types) because only the transformations on the position and velocity level from the minimal coordinates to three-dimensional relative motion have to be implemented. This can also be applied to *flexible* multi-body systems where the coordinate transformations depend on the flexible DOFs.

4 Simulation Results

Results are shown for time simulations of two different test cases, both representing the configuration of the light multi-purpose helicopter Bo105. The geometry and mass-related data have been adopted from both [14] and DLR-internal sources.

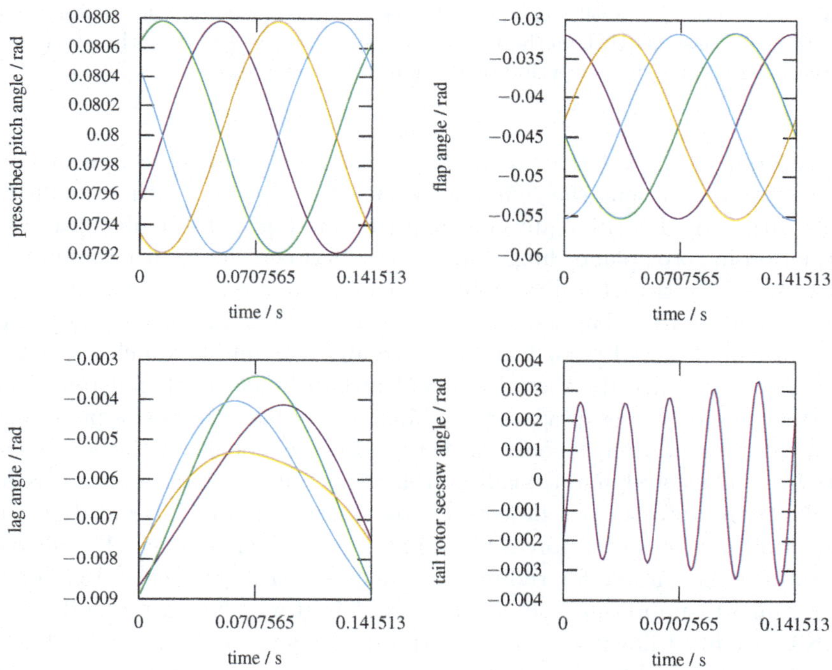

Fig. 2. Hinge angles for a simulation of a free-flying helicopter in about 10 m/s forward flight over one rotor revolution. The flap movement shows periodic behavior with a frequency of 1/rev. The lead-lag movement has a lower eigen-frequency and shows transient behavior. The tail rotor has a higher turn rate (about ∼5.2 times faster than the main rotor). This is reflected in the seesaw movement, which shows a varying amplitude due to the superposition of the different frequencies.

The first test case is an aeromechanic simulation: the MBS consists of a freely moving fuselage, a main rotor, and a tail rotor, the latter two rotating

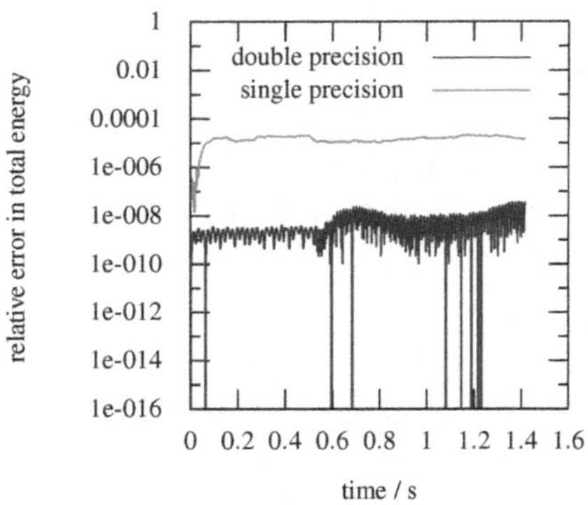

Fig. 3. Relative error in total energy (potential energy, translational and rotational kinetic energy). The graphs show the ratio of the total energy error in relation to the initial total energy for both single and double precision. The system is integrated in time using a classical RK4 method (not energy-conserving) with a time-step size of 100/rev. The results show a period of 10 main rotor revolutions.

with constant speed relative to the fuselage (cf. the kinematic tree in Fig. 1). To accurately represent the dynamic behavior of the four flexible main rotor blades with a rigid MBS, equivalent flap (motion out of rotor plane) and lead-lag (motion in rotor plane) hinges are placed between the rotating hub and the blades (see [15]). Structural damping of lead-lag motion is imprinted via force elements, the lead-lag dampers. Main rotor blade pitch angles are prescribed by driven hinges. The tail rotor features a seesaw, a mounting which is connected to the rotating tail rotor shaft by a central flap hinge. Both tail rotor blades are attached to the seesaw via driven hinges to control the blade pitch angles. For ease of presentation, we only use very simplistic models for the aerodynamic forces here: The aerodynamic loads on the rotors are calculated from the relative velocity of the rotor blades based on a parametrization of lift and drag coefficients. Similarly, we define forces for the fuselage and empennage. The described configuration is suitable for rudimentary analyses of flight mechanics. For time integration we simply use the classical explicit Runge-Kutta method of order 4 (RK4). Figure 2 illustrates the resulting movement of the rotor blades for a helicopter in forward flight with about 10 m/s. One observes that for multi-rotor systems the movement is composed of transient behavior and oscillations of different frequencies. This is due to driven motion with different frequencies (e.g., for main and tail rotor) and due to the different eigenfrequencies of the components (e.g., flap and lead-lag movement). Note that in the movement of the seesaw, we observe a varying amplitude due to the superposition of different frequencies.

The second case is a pure structural analysis. It serves as a demonstration of total energy conservation in the MBS model. No energy sources or sinks (force elements, driven joints) are applied. Consequently, the aerodynamic models as well as the lead-lag dampers are omitted and the driven motion joints are replaced by freely moving joints. Apart from these changes, the configuration equals that of the first test case. The rotor turn rates are defined through an appropriate initial condition. In Fig. 3 we find the relative error in total energy. The error is of the size of the square root of the employed floating point precision and does not seem to grow over the whole simulation time (10 rotor revolutions). So we note that the total energy is approximately conserved very well, although we applied an explicit time integration scheme which does not account for the conservation of energy (like, e.g., a symplectic method would, cf. [8]).

5 Conclusions and Outlook

In this paper, we have discussed the use of automatic differentiation (AD) for the simulation of open-loop multibody systems. Since open-loop systems possess globally valid minimal states, one can formulate the equations of motion in minimal coordinates to eliminate the constraint equations. This formulation incorporates Jacobian matrices of the body kinematics with respect to the joint states. We have shown that these Jacobians can be computed easily by automatic differentiation and illustrated how this improves the design of a multibody simulation software. In particular the resulting code is much shorter, more generic, and can easily be extended (with just a few lines of code) for different kinds of joints and flexible bodies.

Finally, we have presented examples from a helicopter simulation based on a simple rigid MBS helicopter model. The results demonstrate the numerical accuracy of algorithm for the application at hand and underline that open-loop MBS are useful modeling tools for helicopter simulation.

Of course, the results presented in this paper are only a starting point for further research activities. First, we will extend our approach to include different models for flexible bodies. The basic idea here is to equip the bodies with states (representing the flexible motion) and let the automatic differentiation take care of the calculation of all additional required Jacobian matrices. Secondly, we only considered open-loop multibody systems, which is, as we have pointed out, a very good first approximation for helicopter simulations. A better model for helicopters would include certain closed-loop parts, which arise inside the tree structure of an open-loop MBS, for example, the main rotor hub with its pitch control rods. As one still wants to maintain a small set of states (almost minimal coordinates) we will investigate how such "local" closed-loop parts can be integrated in a "global" open-loop MBS.

References

1. Bhalerao, K.D., Poursina, M., Anderson, K.S.: An efficient direct differentiation approach for sensitivity analysis of flexible multibody systems. Multibody Syst. Dyn. **23**, 121–140 (2010)
2. Bischof, C.H.: On the automatic differentiation of computer programs and an application to multibody systems. In: Bestle, D., Schielen, W. (eds.) IUTAM Symposium on Optimization of Mechanical Systems, pp. 41–48. Springer (1996)

3. Callejo, A., Dopico, D.: Direct sensitivity analysis of multibody systems: a vehicle dynamics benchmark. J. Comput. Nonlinear Dynam. **14**, 021004 (2019)
4. Callejo, A., Narayanan, S.H.K., Garcia de Jalon, J., Norris, B.: Performance of automatic differentiation tools in the dynamic simulation of multibody systems. Adv. Eng. Software **73**, 35–44 (2014)
5. Eberhard, P., Bischof, C.: Automatic differentiation of numerical integration algorithms. Math. Comp. **68**, 717–731 (1999)
6. Griffith, D.T., Turner, J.D., Junkins, J.L.: Some applications of automatic differentiation to rigid, flexible, and constrained multibody dynamics. In: Proceedings of IDETC/CIE 2005. ASME (2005)
7. Guennebaud, G., Jacob, B., et al.: Eigen v3 (2010). http://eigen.tuxfamily.org
8. Hairer, E., Lubich, C., Wanner, G.: Geometric Numerical Integration, 2nd edn. Springer, Heidelberg (2006)
9. Johnson, W.: Rotorcraft Aeromechanics. Cambridge Aerospace Series. Cambridge University Press, Cambridge (2013)
10. Naumann, U.: The Art of Differentiating Computer Programs. SIAM, Philadelphia (2012)
11. Schwertassek, R., Wallrapp, O.: Dynamik flexibler Mehrkörpersysteme. Vieweg (1999)
12. Serban, R., Haug, E.J.: Kinematic and kinetic derivatives in multibody systems. Mech. Struct. Mach. **26**, 145–173 (1998)
13. Simeon, B.: Computational Flexible Multibody Dynamics. Springer, Heidelberg (2013)
14. van der Wall, B.G.: Grundlagen der Hubschrauber-Aerodynamik. Springer Vieweg (2015)
15. van der Wall, B.G.: Grundlagen der Dynamik von Hubschrauber-Rotoren. Springer Vieweg (2018)

Author Index

A. Kecskeméthy and F. Geu Flores (Eds.): ECCOMAS 2019, COMPUTMETHODS 53, pp. 543–545, 2020.
https://doi.org/10.1007/978-3-030-23132-3

The manufacturer's authorised representative in the EU is Springer Nature Customer Service Centre GmbH, Europaplatz 3, 69115 Heidelberg, Germany. If you have any concerns regarding our products, please contact ProductSafety@springernature.com

Printed and bound by CPI Group (UK) Ltd, Croydon, CR0 4YY
30/09/2026
02225495-0014